## Graphs of Common Functions and Models

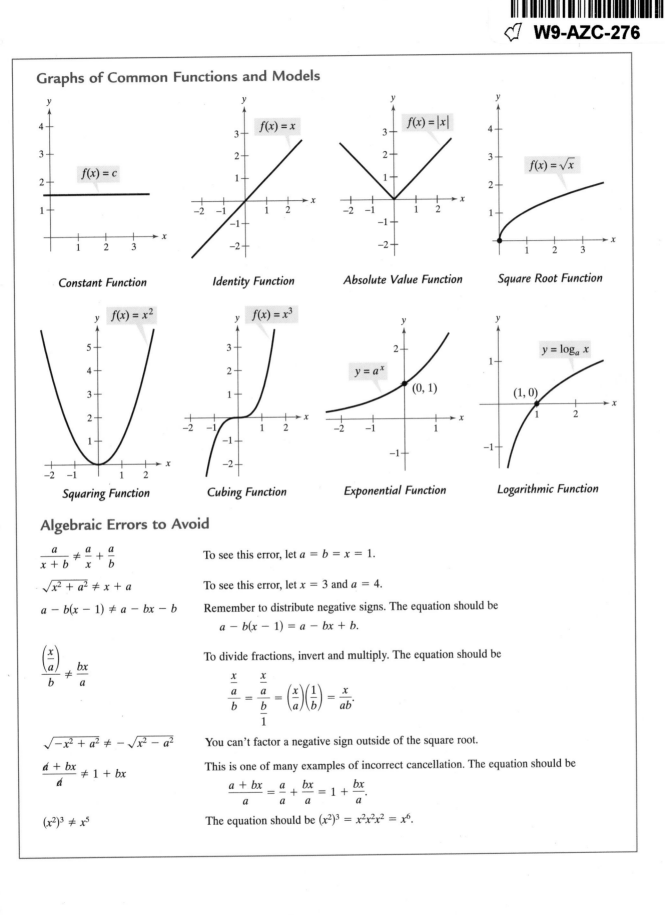

$f(x) = c$

Constant Function

$f(x) = x$

Identity Function

$f(x) = |x|$

Absolute Value Function

$f(x) = \sqrt{x}$

Square Root Function

$f(x) = x^2$

Squaring Function

$f(x) = x^3$

Cubing Function

$y = a^x$ $(0, 1)$

Exponential Function

$y = \log_a x$ $(1, 0)$

Logarithmic Function

## Algebraic Errors to Avoid

$\dfrac{a}{x + b} \neq \dfrac{a}{x} + \dfrac{a}{b}$

To see this error, let $a = b = x = 1$.

$\sqrt{x^2 + a^2} \neq x + a$

To see this error, let $x = 3$ and $a = 4$.

$a - b(x - 1) \neq a - bx - b$

Remember to distribute negative signs. The equation should be
$$a - b(x - 1) = a - bx + b.$$

$\dfrac{\left(\dfrac{x}{a}\right)}{b} \neq \dfrac{bx}{a}$

To divide fractions, invert and multiply. The equation should be
$$\frac{\dfrac{x}{a}}{b} = \frac{\dfrac{x}{a}}{\dfrac{b}{1}} = \left(\frac{x}{a}\right)\left(\frac{1}{b}\right) = \frac{x}{ab}.$$

$\sqrt{-x^2 + a^2} \neq -\sqrt{x^2 - a^2}$

You can't factor a negative sign outside of the square root.

$\dfrac{a + bx}{a} \neq 1 + bx$

This is one of many examples of incorrect cancellation. The equation should be
$$\frac{a + bx}{a} = \frac{a}{a} + \frac{bx}{a} = 1 + \frac{bx}{a}.$$

$(x^2)^3 \neq x^5$

The equation should be $(x^2)^3 = x^2 x^2 x^2 = x^6$.

# College Algebra

## Concepts and Models

**Fourth Edition**

**Ron Larson**
*The Pennsylvania State University*
*The Behrend College*

**Robert P. Hostetler**
*The Pennsylvania State University*
*The Behrend College*

**Anne V. Hodgkins**
*Phoenix College*

**Houghton Mifflin Company**
Boston   New York

Publisher: Jack Shira
Managing Editor: Cathy Cantin
Development Manager: Maureen Ross
Development Editor: Laura Wheel
Assistant Editor: Rosalind Martin
Supervising Editor: Karen Carter
Senior Project Editor: Patty Bergin
Editorial Assistant: Meghan Lydon
Production Technology Supervisor: Gary Crespo
Executive Marketing Manager: Michael Busnach
Senior Marketing Manager: Danielle Potvin
Marketing Associate: Nicole Mollica
Manufacturing Coordinator: Jane Spelman

We have included examples and exercises that use real-life data as well as technology output from a variety of software. This would not have been possible without the help of many people and organizations. Our wholehearted thanks goes to all for their time and effort.

Trademark acknowledgments: TI and CBR are registered trademarks of Texas Instruments, Inc.
Cover image: © Cosmo & Action/Photonica

Printed in the U.S.A.

Library of Congress Catalog Card Number: 2001133293

ISBN: 0-618-22026-7

123456789–VH–05 04 03 02

# Contents

**A Word from the Authors (Preface)**   vii

**Highlights of Features**   ix

**An Introduction to Graphing Utilities**   xx

Chapter 1   **Equations and Inequalities**   1

1.1   Linear Equations   2
1.2   Mathematical Modeling   11
1.3   Quadratic Equations   25
1.4   The Quadratic Formula   36
      **Mid-Chapter Quiz**   46
1.5   Other Types of Equations   47
1.6   Linear Inequalities   58
1.7   Other Types of Inequalities   70
      *Chapter Project:* Salaries in Professional Baseball   80
      **Summary**   81   **Review Exercises**   82   **Test**   86

Chapter 2   **The Cartesian Plane and Graphs**   87

2.1   The Cartesian Plane   88
2.2   Graphs of Equations   99
2.3   Graphing Utilities   109
      **Mid-Chapter Quiz**   118
2.4   Lines in the Plane   119
2.5   Linear Modeling   131
      *Chapter Project:* Demographics   143
      **Summary**   144   **Review Exercises**   145   **Test**   149
      **Cumulative Test: Chapters 1–2**   150

**Chapter 3    Functions and Graphs    151**

3.1    Functions    152

3.2    Graphs of Functions    165

3.3    Transformations of Functions    176

**Mid-Chapter Quiz**    185

3.4    The Algebra of Functions    186

3.5    Inverse Functions    196

*Chapter Project:*  Carbon Monoxide Emissions    206

**Summary**    207    **Review Exercises**    208    **Test**    212


**Chapter 4    Polynomial and Rational Functions    213**

4.1    Quadratic Functions and Models    214

4.2    Polynomial Functions of Higher Degree    226

4.3    Polynomial Division    236

4.4    Real Zeros of Polynomial Functions    246

**Mid-Chapter Quiz**    259

4.5    Complex Numbers    260

4.6    The Fundamental Theorem of Algebra    270

4.7    Rational Functions    278

*Chapter Project:*  Analyzing a Bouncing Ball    290

**Summary**    291    **Review Exercises**    292    **Test**    296


**Chapter 5    Exponential and Logarithmic Functions    297**

5.1    Exponential Functions    298

5.2    Logarithmic Functions    310

5.3    Properties of Logarithms    320

**Mid-Chapter Quiz**    328

5.4    Solving Exponential and Logarithmic Equations    329

5.5    Exponential and Logarithmic Models    339

*Chapter Project:*  Newton's Law of Cooling    351

**Summary**    352    **Review Exercises**    353    **Test**    357

**Cumulative Test: Chapters 3–5**    358

**Chapter 6**   **Systems of Equations and Inequalities**   359

6.1   Systems of Equations   360
6.2   Linear Systems in Two Variables   370
6.3   Linear Systems in Three or More Variables   382
      **Mid-Chapter Quiz**   395
6.4   Systems of Inequalities   396
6.5   Linear Programming   406
      *Chapter Project:* Consumer Credit Outstanding   417
      **Summary**   418   **Review Exercises**   419   **Test**   424


**Chapter 7**   **Matrices and Determinants**   425

7.1   Matrices and Linear Systems   426
7.2   Operations with Matrices   440
7.3   The Inverse of a Square Matrix   454
      **Mid-Chapter Quiz**   464
7.4   The Determinant of a Square Matrix   465
7.5   Applications of Matrices and Determinants   475
      *Chapter Project:* Market Share   484
      **Summary**   485   **Review Exercises**   486   **Test**   490


**Chapter 8**   **Sequences, Series, and Probability**   491

8.1   Sequences and Summation Notation   492
8.2   Arithmetic Sequences and Series   503
8.3   Geometric Sequences and Series   512
      **Mid-Chapter Quiz**   522
8.4   The Binomial Theorem   523
8.5   Counting Principles   530
8.6   Probability   540
8.7   Mathematical Induction   553
      *Chapter Project:* The Multiplier Effect   564
      **Summary**   565   **Review Exercises**   566   **Test**   570
      **Cumulative Test: Chapters 6–8**   571

Appendices   **Appendix A**

Review of Fundamental Concepts of Algebra

A.1   Real Numbers: Order and Absolute Value   A1

A.2   The Basic Rules of Algebra   A8

A.3   Integer Exponents   A18

A.4   Radicals and Rational Exponents   A26

A.5   Polynomials and Special Products   A34

A.6   Factoring   A42

A.7   Fractional Expressions and Probability   A47

**Appendix B**

Conic Sections

B.1   Conic Sections   A56

B.2   Conic Sections and Translations   A68

**Appendix C**

Further Concepts in Statistics

C.1   Data and Linear Modeling   A77

C.2   Measures of Central Tendency and Dispersion   A86

**Answers to Warm Ups, Odd-Numbered Exercises, Quizzes, and Tests**   A95

**Index of Applications**   A163

**Index**   A169

# A Word from the Authors

Welcome to *College Algebra: Concepts and Models*, Fourth Edition. In this revision, we continue to focus on developing students' conceptual understanding of college algebra while offering opportunities for them to hone their problem-solving skills.

We have found that many college algebra students grasp theoretical concepts more easily when they work with them in the context of a real-life situation. Throughout the Fourth Edition, students now have many more opportunities to collect, analyze, and model real data. We updated all real-data application examples or replaced them with new applications that use current data.

## Changes in Exercises

In the Fourth Edition, we placed a special emphasis on revising the exercise sets.

- We added several exercises that use the regression feature of a graphing calculator to model real-life data. For instance, see Exercise 68 in Section 3.1.

- We updated all exercises involving real-life data. For instance, see Exercise 60 in Section 4.2. We also added several new exercises involving real-life data. For instance, see Exercise 91 in Section 5.3.

- We added more exercises that use piecewise-defined functions to model real-life data. For instance, see Exercise 71 in Section 3.2.

- We added more exercises that require the use of inverse functions to answer questions about real-life situations. For instance, see Exercise 60 in Section 3.5.

- We added more exercises that use exponential functions to model real-life data and predict future events. For instance, see Exercise 45 in Section 5.5.

- We added more exercises that use logistics functions to model real-life data and predict future events. For instance, see Exercise 44 in Section 5.5.

- We added exercises that use a linear regression program to model an arithmetic sequence. For instance, see Exercise 69 in Section 8.2.

There are many opportunities in the Fourth Edition for students to use a graphing utility in the problem-solving process—to visualize and explore theoretical concepts, to analyze real data, and to verify alternative solution methods. However, to accommodate a variety of teaching and learning styles, the use of graphing technology is always optional. At the suggestion of our users, the Technology feature uses only generic references, omitting keystrokes for and references to specific calculator models. On pages xx–xxv, we have included "An Introduction to Graphing Utilities." This section addresses the basic functions of graphing calculators that are used in college algebra, including the equation editor and the table, zoom, and trace features.

PREFACE

To encourage mastery and understanding, we have included opportunities for student self-assessment at strategic points throughout each chapter. Learning Objectives at the beginning of each section provide students with a conceptual outline for reference. In the Chapter Summary, these Objectives are linked to the Review Exercises for students who require additional guided practice. Finally, Mid-Chapter Quizzes, Chapter Tests, and Cumulative Tests allow students to test understanding and learning retention at regular intervals.

We hope you will enjoy using the Fourth Edition in your college algebra class. We think its straightforward, readable style, engaging applications, and study tools will appeal to your students and contribute to their success in this course.

Ron Larson

Robert P. Hostetler

Anne Hodgkins

## Highlights of Features

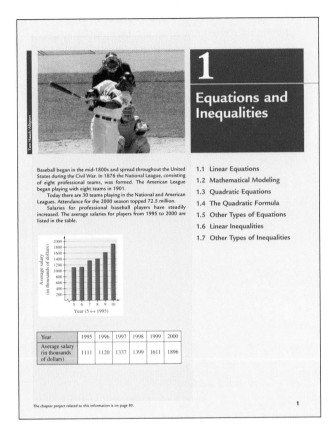

### Chapter Opener

Each chapter begins with a real-life application with current data. At the end of the chapter, students revisit this application as a *Chapter Project* after they have been exposed to the algebra techniques required for solving the problem. A list of section titles outlines the topics covered in the chapter.

### Section Opener

Each section opens with a list of learning *Objectives*, the algebra skills presented in the section. This conceptual outline functions as a useful tool for reference and review for the students and class planning for the instructor. In the *Chapter Summary*, these objectives are linked to *Review Exercises* for additional student practice.

### Additional Features

In addition to features highlighted here, many other carefully crafted learning and assessment tools designed to create a rich learning environment can be found throughout the text. These learning and assessment tools include *Study Tips*, *Historical Notes*, *Math Matters*, *Discussing the Concept*, *Mid-Chapter Quizzes*, *Chapter Tests*, *Cumulative Tests*, and an extensive art program.

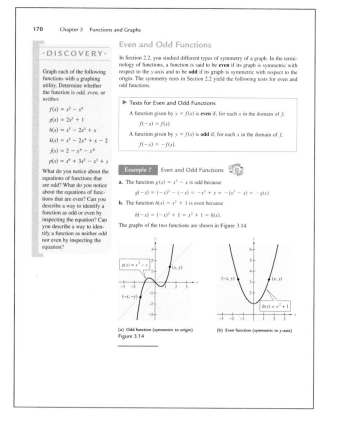

·DISCOVERY·

Graph each of the following functions with a graphing utility. Determine whether the function is *odd, even,* or *neither.*

$f(x) = x^2 - x^4$
$g(x) = 2x^3 + 1$
$h(x) = x^5 - 2x^3 + x$
$k(x) = x^5 - 2x^4 + x - 2$
$j(x) = 2 - x^6 - x^8$
$p(x) = x^9 + 3x^5 - x^3 + x$

What do you notice about the equations of functions that are odd? What do you notice about the equations of functions that are even? Can you describe a way to identify a function as odd or even by inspecting the equation? Can you describe a way to identify a function as neither odd nor even by inspecting the equation?

### Even and Odd Functions

In Section 2.2, you studied different types of symmetry of a graph. In the terminology of functions, a function is said to be **even** if its graph is symmetric with respect to the *y*-axis and to be **odd** if its graph is symmetric with respect to the origin. The symmetry tests in Section 2.2 yield the following tests for even and odd functions.

▶ Tests for Even and Odd Functions

A function given by $y = f(x)$ is **even** if, for each $x$ in the domain of $f$,
$$f(-x) = f(x).$$
A function given by $y = f(x)$ is **odd** if, for each $x$ in the domain of $f$,
$$f(-x) = -f(x).$$

**Example 7** Even and Odd Functions

a. The function $g(x) = x^3 - x$ is odd because
$$g(-x) = (-x)^3 - (-x) = -x^3 + x = -(x^3 - x) = -g(x).$$
b. The function $h(x) = x^2 + 1$ is even because
$$h(-x) = (-x)^2 + 1 = x^2 + 1 = h(x).$$
The graphs of the two functions are shown in Figure 3.14.

(a) Odd function (symmetric to origin)  (b) Even function (symmetric to *y*-axis)
Figure 3.14

## Discovery

*Discovery* activities offer opportunities for the exploration of selected mathematical concepts. Students are encouraged to use techniques such as visualization and modeling to develop their intuitive understanding of theoretical topics. These optional activities can be omitted at the instructor's discretion without affecting the flow of the material.

## Technology

Students are encouraged to use a graphing utility as a problem-solving tool. The text offers many opportunities to visualize concepts, to discover alternative approaches, and to verify the results of other solution methods using technology. However, students are not required to have access to a graphing utility to use this text effectively. The text appropriately addresses both the benefits of using technology and its possible misuse or misinterpretation.

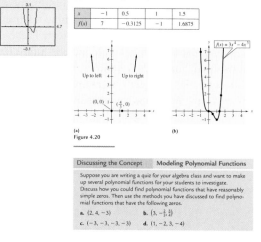

·TECHNOLOGY·

Example 7 uses an "algebraic approach" to describe the graph of the function. A graphing utility is a valuable complement to this approach. Remember that the most important part of using a graphing utility is to find a viewing window that shows all important parts of the graph. For instance, the graph below shows the important parts of the graph of the function in Example 7.

Example 7 shows how the Leading Coefficient Test and zeros of polynomial functions can be used as sketching aids.

**Example 7** Sketching the Graph of a Polynomial Function

Sketch the graph of $f(x) = 3x^4 - 4x^3$.

Solution

Because the leading coefficient is positive and the degree is even, you know that the graph eventually rises to the left and to the right, as shown in Figure 4.20(a). By factoring

$$f(x) = 3x^4 - 4x^3 = x^3(3x - 4)$$

you can see that the zeros of $f$ are $x = 0$ and $x = \frac{4}{3}$ (both of odd multiplicity). So the *x*-intercepts occur at $(0, 0)$ and $(\frac{4}{3}, 0)$. To sketch the graph by hand, find a few additional points, as shown in the table. Then plot the points and complete the graph, as shown in Figure 4.20(b).

| $x$ | $-1$ | 0.5 | 1 | 1.5 |
|---|---|---|---|---|
| $f(x)$ | 7 | $-0.3125$ | $-1$ | 1.6875 |

(a)
(b)
Figure 4.20

| Discussing the Concept | Modeling Polynomial Functions |
|---|---|

Suppose you are writing a quiz for your algebra class and want to make up several polynomial functions for your students to investigate. Discuss how you could find polynomial functions that have reasonably simple zeros. Then use the methods you have discussed to find polynomial functions that have the following zeros.

a. $(2, 4, -3)$  b. $(3, -\frac{4}{3}, \frac{3}{4})$
c. $(-3, -3, -3, -3)$  d. $(1, -2, 3, -4)$

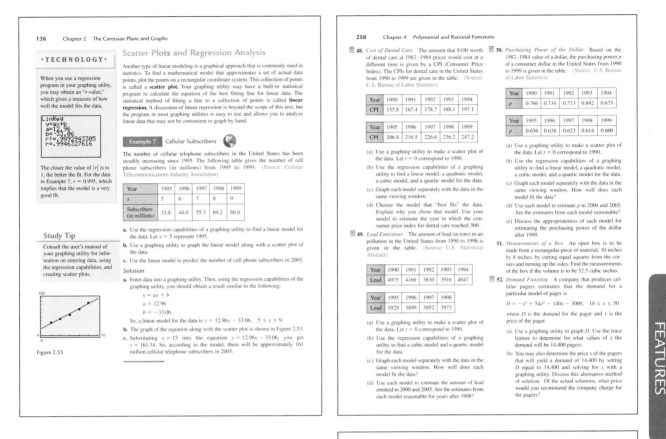

136  Chapter 2  The Cartesian Plane and Graphs

**·TECHNOLOGY·**

When you use a regression program in your graphing utility, you may obtain an "*r*-value," which gives a measure of how well the model fits the data.

```
LinReg
y=ax+b
a=12.96
b=-33.06
r²=.9892943305
r=.9946327616
```

The closer the value of |*r*| is to 1, the better the fit. For the data in Example 7, *r* ≈ 0.995, which implies that the model is a very good fit.

**Study Tip**

Consult the user's manual of your graphing utility for information on entering data, using the regression capabilities, and creating scatter plots.

Figure 2.53

### Scatter Plots and Regression Analysis

Another type of linear modeling is a graphical approach that is commonly used in statistics. To find a mathematical model that approximates a set of actual data points, plot the points on a rectangular coordinate system. This collection of points is called a **scatter plot.** Your graphing utility may have a built-in statistical program to calculate the equation of the best fitting line for linear data. The statistical method of fitting a line to a collection of points is called **linear regression.** A discussion of linear regression is beyond the scope of this text, but the program in most graphing utilities is easy to use and allows you to analyze linear data that may not be convenient to graph by hand.

**Example 7**  Cellular Subscribers

The number of cellular telephone subscribers in the United States has been steadily increasing since 1995. The following table gives the number of cell phone subscribers (in millions) from 1995 to 1999. (Source: Cellular Telecommunications Industry Association)

| Year | 1995 | 1996 | 1997 | 1998 | 1999 |
|------|------|------|------|------|------|
| *x* | 5 | 6 | 7 | 8 | 9 |
| Subscribers (in millions) | 33.8 | 44.0 | 55.3 | 69.2 | 86.0 |

**a.** Use the regression capabilities of a graphing utility to find a linear model for the data. Let *x* = 5 represent 1995.
**b.** Use a graphing utility to graph the linear model along with a scatter plot of the data.
**c.** Use the linear model to predict the number of cell phone subscribers in 2005.

**Solution**

**a.** Enter data into a graphing utility. Then, using the regression capabilities of the graphing utility, you should obtain a result similar to the following.

$$y = ax + b$$
$$a = 12.96$$
$$b = -33.06$$

So, a linear model for the data is $y = 12.96x - 33.06$, $5 \le x \le 9$.

**b.** The graph of the equation along with the scatter plot is shown in Figure 2.53.

**c.** Substituting $x = 15$ into the equation $y = 12.96x - 33.06$, you get $y = 161.34$. So, according to the model, there will be approximately 161 million cellular telephone subscribers in 2005.

---

258  Chapter 4  Polynomial and Rational Functions

**48.** *Cost of Dental Care*  The amount that $100 worth of dental care at 1982–1984 prices would cost at a different time is given by a CPI (Consumer Price Index). The CPIs for dental care in the United States from 1990 to 1999 are given in the table. (Source: U.S. Bureau of Labor Statistics)

| Year | 1990 | 1991 | 1992 | 1993 | 1994 |
|------|------|------|------|------|------|
| CPI | 155.8 | 167.4 | 178.7 | 188.1 | 197.1 |

| Year | 1995 | 1996 | 1997 | 1998 | 1999 |
|------|------|------|------|------|------|
| CPI | 206.8 | 216.5 | 226.6 | 236.2 | 247.2 |

(a) Use a graphing utility to make a scatter plot of the data. Let *t* = 0 correspond to 1990.
(b) Use the regression capabilities of a graphing utility to find a linear model, a quadratic model, a cubic model, and a quartic model for the data.
(c) Graph each model separately with the data in the same viewing window.
(d) Choose the model that "best fits" the data. Explain why you chose that model. Use your model to estimate the year in which the consumer price index for dental care reached 300.

**49.** *Lead Emissions*  The amount of lead (in tons) in air pollution in the United States from 1990 to 1998 is given in the table. (Source: U.S. Statistical Abstract)

| Year | 1990 | 1991 | 1992 | 1993 | 1994 |
|------|------|------|------|------|------|
| Lead | 4975 | 4169 | 3810 | 3916 | 4047 |

| Year | 1995 | 1996 | 1997 | 1998 |
|------|------|------|------|------|
| Lead | 3929 | 3899 | 3952 | 3973 |

(a) Use a graphing utility to make a scatter plot of the data. Let *t* = 0 correspond to 1990.
(b) Use the regression capabilities of a graphing utility to find a cubic model and a quartic model for the data.
(c) Graph each model separately with the data in the same viewing window. How well does each model fit the data?
(d) Use each model to estimate the amount of lead emitted in 2000 and 2005. Are the estimates from each model reasonable for years after 1998?

**50.** *Purchasing Power of the Dollar*  Based on the 1982–1984 value of a dollar, the purchasing power *p* of a consumer dollar in the United States from 1990 to 1999 is given in the table. (Source: U.S. Bureau of Labor Statistics)

| Year | 1990 | 1991 | 1992 | 1993 | 1994 |
|------|------|------|------|------|------|
| *p* | 0.766 | 0.734 | 0.713 | 0.692 | 0.675 |

| Year | 1995 | 1996 | 1997 | 1998 | 1999 |
|------|------|------|------|------|------|
| *p* | 0.656 | 0.638 | 0.623 | 0.614 | 0.600 |

(a) Use a graphing utility to make a scatter plot of the data. Let *t* = 0 correspond to 1990.
(b) Use the regression capabilities of a graphing utility to find a linear model, a quadratic model, a cubic model, and a quartic model for the data.
(c) Graph each model separately with the data in the same viewing window. How well does each model fit the data?
(d) Use each model to estimate *p* in 2000 and 2005. Are the estimates from each model reasonable?
(e) Discuss the appropriateness of each model for estimating the purchasing power of the dollar after 1999.

**51.** *Measurements of a Box*  An open box is to be made from a rectangular piece of material, 10 inches by 8 inches, by cutting equal squares from the corners and turning up the sides. Find the measurements of the box if the volume is to be 52.5 cubic inches.

**52.** *Demand Function*  A company that produces cellular pagers estimates that the demand for a particular model of pager is

$$D = -x^3 + 54x^2 - 140x - 3000, \quad 10 \le x \le 50$$

where *D* is the demand for the pager and *x* is the price of the pager.

(a) Use a graphing utility to graph *D*. Use the trace feature to determine for what values of *x* the demand will be 14,400 pagers.
(b) You may also determine the price *x* of the pagers that will yield a demand of 14,400 by setting *D* equal to 14,400 and solving for *x* with a graphing utility. Discuss this alternative method of solution. Of the actual solutions, what price would you recommend the company charge for the pagers?

---

## Real-Life Applications

A wide variety of real-life applications, many using current, real data are integrated throughout examples and exercises. The icon found in examples indicates a real-life application.

## Regression Exercises

New regression exercises have been included in the Fourth Edition. Mainly utilizing the power of technology, these new regression exercises require students to analyze data, to create regression models, and to determine models of best fit. In addition, students are asked to make predictions and decisions based on the models they create.

---

Review Exercises  295

In Exercises 91–94, sketch the graph of the rational function. As sketching aids, check for intercepts, symmetry, vertical asymptotes, and horizontal asymptotes.

**91.** $P(x) = \dfrac{3 - x}{x + 2}$

**92.** $f(x) = \dfrac{4}{(x - 1)^2}$

**93.** $g(x) = \dfrac{1}{x + 3} + 2$

**94.** $h(x) = \dfrac{-3x}{2x^2 + 3x - 5}$

**95.** *Average Cost*  The cost in dollars of producing *x* units is $C = 100,000 + 0.9x$, and so the average cost per unit is

$$\overline{C} = \frac{C}{x} = \frac{100,000 + 0.9x}{x}, \quad x > 0.$$

Sketch the graph of this average cost function and find the average cost of producing *x* = 1000, *x* = 10,000, and *x* = 100,000 units.

**96.** *Population of Fish*  The Parks and Wildlife Commission introduces 50,000 game fish into a large lake. The population of the fish (in thousands) is

$$N = \frac{20(4 + 3t)}{1 + 0.05t}, \quad t \ge 0$$

where *t* is time in years.

(a) Find the population when *t* = 5, 10, and 25.
(b) What is the limiting number of this population of fish in the lake as time progresses?

**97.** *Human Memory Model*  Consider the learning curve

$$P = \frac{0.5 + 0.9(n - 1)}{1 + 0.9(n - 1)}, \quad n > 0$$

where *P* is the percent of correct responses after *n* trials. Complete the table for this model.

| *n* | 1 | 2 | 3 | 4 | 5 | 6 | 7 | 8 | 9 | 10 |
|-----|---|---|---|---|---|---|---|---|---|----|
| *P* |   |   |   |   |   |   |   |   |   |    |

**98.** *Smokestack Emission*  The cost in dollars of removing *p* percent of the air pollutants in the stack emission of a utility company that burns coal to generate electricity is

$$C = \frac{95,000p}{100 - p}, \quad 0 \le p < 100.$$

(a) Find the cost of removing 25%.
(b) Find the cost of removing 60%.
(c) Find the cost of removing 99%.
(d) According to this model, would it be possible to remove 100% of the pollutants? Explain.

**99.** *Average Recycling Cost*  The cost in dollars of recycling a waste product is

$$C = 450,000 + 6x, \quad x > 0$$

where *x* is the number of pounds of waste. The average recycling cost is

$$\overline{C} = \frac{C}{x} = \frac{450,000 + 6x}{x}, \quad x > 0.$$

(a) Sketch the graph of $\overline{C}$.
(b) Find the average cost of recycling for *x* = 1000 pounds, *x* = 10,000 pounds, and *x* = 100,000 pounds. What can you conclude?

**100.** *Sound Recordings*  The total dollar values *V* (in millions) for sound recordings shipped at manufacturers' suggested list prices in the United States from 1991 to 2000 are given in the table. (Source: The Recording Industry Association of America)

| Year | 1991 | 1992 | 1993 | 1994 | 1995 |
|------|------|------|------|------|------|
| V | 7834 | 9024 | 10,047 | 12,068 | 12,320 |

| Year | 1996 | 1997 | 1998 | 1999 | 2000 |
|------|------|------|------|------|------|
| V | 12,534 | 12,237 | 13,724 | 14,585 | 14,323 |

(a) Use a graphing utility to make a scatter plot of the data. Let *t* = 1 correspond to 1991.
(b) Use the regression capabilities of a graphing utility to find a quadratic model, a cubic model, and a quartic model for the data.
(c) Graph each model separately with the data in the same viewing window. How well does each model fit the data?
(d) Use all three models to estimate the total values of recordings shipped in 2001 and 2002. Does each estimate seem reasonable?

FEATURES

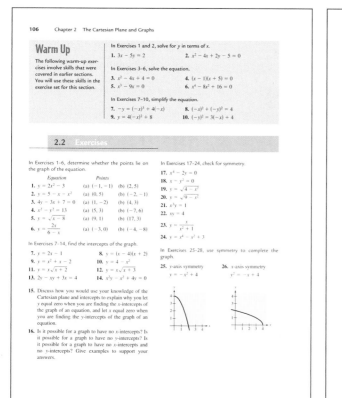

106    Chapter 2   The Cartesian Plane and Graphs

## Warm Up

The following warm-up exercises involve skills that were covered in earlier sections. You will use these skills in the exercise set for this section.

In Exercises 1 and 2, solve for $y$ in terms of $x$.

1. $3x - 5y = 2$
2. $x^2 - 4x + 2y - 5 = 0$

In Exercises 3–6, solve the equation.

3. $x^2 - 4x + 4 = 0$
4. $(x - 1)(x + 5) = 0$
5. $x^3 - 9x = 0$
6. $x^4 - 8x^2 + 16 = 0$

In Exercises 7–10, simplify the equation.

7. $-y = (-x)^3 + 4(-x)$
8. $(-x)^2 + (-y)^2 = 4$
9. $y = 4(-x)^2 + 8$
10. $(-y)^2 = 3(-x) + 4$

### 2.2   Exercises

In Exercises 1–6, determine whether the points lie on the graph of the equation.

|  | Equation | | Points | |
| 1. | $y = 2x^2 - 3$ | (a) $(-1, -1)$ | (b) $(2, 5)$ |
| 2. | $y = 5 - x - x^2$ | (a) $(0, 5)$ | (b) $(-2, -1)$ |
| 3. | $4y - 3x + 7 = 0$ | (a) $(1, -2)$ | (b) $(4, 3)$ |
| 4. | $x^2 - y^2 = 13$ | (a) $(5, 3)$ | (b) $(-7, 6)$ |
| 5. | $y = \sqrt{x - 8}$ | (a) $(9, 1)$ | (b) $(17, 3)$ |
| 6. | $y = \dfrac{2x}{6 - x}$ | (a) $(-3, 0)$ | (b) $(-4, -8)$ |

In Exercises 7–14, find the intercepts of the graph.

7. $y = 2x - 1$
8. $y = (x - 4)(x + 2)$
9. $y = x^2 + x - 2$
10. $y = 4 - x^2$
11. $y = x\sqrt{x + 2}$
12. $y = x\sqrt{x + 3}$
13. $2y - xy + 3x = 4$
14. $x^2 y - x^2 + 4y = 0$

15. Discuss how you would use your knowledge of the Cartesian plane and intercepts to explain why you let $y$ equal zero when you are finding the $x$-intercepts of the graph of an equation, and let $x$ equal zero when you are finding the $y$-intercepts of the graph of an equation.

16. Is it possible for a graph to have no $x$-intercepts? Is it possible for a graph to have no $y$-intercepts? Is it possible for a graph to have no $x$-intercepts and no $y$-intercepts? Give examples to support your answers.

In Exercises 17–24, check for symmetry.

17. $x^4 - 2y = 0$
18. $x - y^2 = 0$
19. $y = \sqrt{4 - x^2}$
20. $y = \sqrt{9 - x^2}$
21. $x^3 y = 1$
22. $xy = 4$
23. $y = \dfrac{x}{x^2 + 1}$
24. $y = x^4 - x^2 + 3$

In Exercises 25–28, use symmetry to complete the graph.

25. $y$-axis symmetry
$y = -x^2 + 4$

26. $x$-axis symmetry
$y^2 = -x + 4$

---

140    Chapter 2   The Cartesian Plane and Graphs

28. *Sales Commission*  A salesperson receives a monthly salary of $2500 plus a commission of 7% of sales. Write a linear equation for the salesperson's monthly wage $W$ in terms of the person's monthly sales $S$.

29. *Height of a Parachutist*  After opening the parachute, the descent of a parachutist follows a linear model. At 2:08 P.M. the height of the parachutist is 7000 feet. At 2:10 P.M. the height is 4600 feet.
   (a) Write a linear equation that gives the height of the parachutist in terms of the time $t$. (Let $t = 0$ represent 2:08 P.M. and let $t$ be measured in seconds.)
   (b) Use the equation in part (a) to find the time when the parachutist will reach the ground.

30. *Distance Traveled by a Car*  You are driving at a constant speed on a freeway. At 4:30 P.M. you drive by a sign that gives the distance to Montgomery, Alabama as 84 miles. At 4:59 P.M. you drive by another sign that says the distance to Montgomery is 56 miles.
   (a) Write a linear equation that gives your distance from Montgomery in terms of $t$. (Let $t = 0$ represent 4:30 P.M. and let $t$ be measured in minutes.)
   (b) Use the equation in part (a) to find the time when you will reach Montgomery.

In Exercises 31–34, match the description with one of the graphs. Also find the slope and describe how it is interpreted in the real-life situation. [The graphs are labeled (a), (b), (c), and (d).]

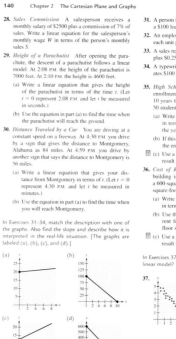

(a)    (b)

(c)    (d)

31. A person is paying $10 per week to a friend to repay a $100 loan.

32. An employee is paid $12.50 per hour plus $1.50 for each unit produced per hour.

33. A sales representative receives $20 per day for food, plus $0.25 for each mile traveled.

34. A typewriter that was purchased for $600 depreciates $100 per year.

35. *High School Enrollment*  A high school had an enrollment of 1200 students in 1990. During the next 10 years the enrollment increased by approximately 50 students per year.
   (a) Write a linear equation giving the enrollment $N$ in terms of the year $t$. (Let $t = 0$ correspond to the year 1990.)
   (b) If this constant rate of growth continues, predict the enrollment in the year 2010.
   (c) Use a graphing utility to confirm graphically the result in part (b).

36. *Cost of Renting*  The monthly rent in an office building is related to the office size. The rent for a 600-square-foot office is $750. The rent for a 900-square-foot office is $1150.
   (a) Write a linear equation giving the monthly rent $r$ in terms of the square footage $x$.
   (b) Use the equation in part (a) to find the monthly rent for an office that has 1300 square feet of floor space.
   (c) Use a graphing utility to confirm graphically the result in part (b).

In Exercises 37–40, can the data be approximated by a linear model?

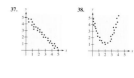

37.    38.

---

142    Chapter 2   The Cartesian Plane and Graphs

48. *Total Sales*  The total sales for Microsoft Corporation from 1995 to 1999 (in millions of dollars) are given by the following ordered pairs. (Source: Microsoft Corp.)

(1995, 5937.0), (1996, 8671.0), (1997, 11,358)
(1998, 14,484), (1999, 19,747)
   (a) Use a graphing utility to make a scatter plot of the data. (Let $y$ represent the total sales in millions of dollars and let $t = 1$ represent 1995.)
   (b) Use two points on the scatter plot to find an equation of the line that approximates the data.
   (c) Use a graphing utility to predict the sales of Microsoft in 2000. Compare the result with the prediction given by the equation in part (b).
   (d) The actual sales for Microsoft Corporation in 2000 were $22,956 million. How accurate were your predictions?

49. *Contracting Purchase*  A contractor purchases a piece of equipment for $36,500. The operating cost is $5.25 per hour for fuel and maintenance, and the operator is paid $11.50 per hour.
   (a) Write an equation giving the cost $C$ of operating the equipment for $t$ hours. (Include the purchase cost.)
   (b) If customers are charged $27 per hour, write an equation for the revenue $R$ derived from $t$ hours of use.
   (c) Write an equation for the profit $P$ derived from $t$ hours of use.
   (d) *Break-Even Point*  Use the result of part (c) to find the number of hours this equipment must be used to yield a profit of 0 dollars.

50. *Contracting Purchase*  A contractor purchases a piece of equipment for $55,000. The operating cost is $5.95 per hour for fuel and maintenance, and the operator is paid $12.25 per hour.
   (a) Write an equation giving the cost $C$ of operating the equipment for $t$ hours. (Include the purchase cost.)
   (b) If customers are charged $35 per hour, write an equation for the revenue $R$ derived from $t$ hours of use.
   (c) Write an equation for the profit $P$ derived from $t$ hours of use.
   (d) *Break-Even Point*  Use the result of part (c) to find the number of hours this equipment must be used to yield a profit of 0 dollars.

51. *Real Estate Purchase*  A real estate office handles an apartment complex with 50 units. When the rent per unit is $380 per month, all 50 units are occupied. However, when the rent is $425 per month, the average number of occupied units drops to 47. Assume that the relationship between the monthly rent $p$ and the demand $x$ is linear.
   (a) Write the equation of the line giving the demand $x$ in terms of the rent $p$.
   (b) Use the equation in part (a) to predict the number of units occupied if the rent is raised to $455.

52. *Sunday Newspapers*  The percents $p$ of United States households purchasing a Sunday newspaper for selected years from 1968 to 1998 are given in the table below. (Source: Television Bureau of Advertising)

| Year | 1968 | 1973 | 1978 | 1983 |
| --- | --- | --- | --- | --- |
| $p$ | 83 | 77 | 72 | 67 |

| Year | 1988 | 1993 | 1998 |
| --- | --- | --- | --- |
| $p$ | 68 | 66 | 61 |

   (a) Use your graphing utility to draw a scatter plot for these data. Let $t = 8$ represent 1968. Do the data appear linear?
   (b) Use the regression capabilities of a graphing utility to develop a linear model for predicting the percent of U.S. households purchasing a Sunday newspaper.
   (c) Use your linear model to predict the percents for 1999 and 2000. Are your predictions reasonable?
   (d) What trend do you see in the data? If you were the manager of a newspaper, what strategies could you use to reverse this trend?

---

## Warm Up

Beginning with Section 1.1, each exercise set is preceded by 10 warm-up exercises. These *Warm Ups* provide students with the opportunity to review and practice previously learned skills necessary to master the new skills presented in the section.

## Exercise Sets

The exercises consist of a variety of computational, conceptual, and applied problems, which are carefully graded in difficulty. The exercise sets in the Fourth Edition include more multi-part, exploratory, modeling, and data analysis, as well as a rich variety of relevant and meaningful applications.

---

**80**    Chapter 1    Equations and Inequalities

## Chapter Project
### Salaries in Professional Baseball

The record-breaking attendance (72.6 million people) for major league baseball was not the only record that was broken in the 2001 season. Barry Bonds (San Francisco) hit 73 home runs, drew 177 walks, and had a slugging percentage of .863 to set records in each category. Rickey Henderson (San Diego) broke the career record for runs with 2248 and extended his career records for walks and stolen bases to 2141 and 1395, respectively. Roger Clemens (New York Yankees) became the first pitcher in baseball history to have a 20–1 start. Albert Pujols (St. Louis) set a new rookie RBI (runs batted in) record with 130. As players reach and exceed previous records, they expect to be compensated accordingly. The average salary for a major league baseball player nearly doubled from 1995 to 2000.

The table below gives the average salaries (in thousands of dollars) for professional baseball players from 1995 to 2000.

| Year | 1995 | 1996 | 1997 | 1998 | 1999 | 2000 |
|---|---|---|---|---|---|---|
| Average salary (in thousands of dollars) | 1111 | 1120 | 1337 | 1399 | 1611 | 1896 |

The average salary $S$ can be modeled by the equation

$$S = 24.29t^2 - 208.3t + 1538, \quad 5 \le t \le 10$$

where $t = 5$ represents 1995. Use this information to investigate the following questions.

1. *Compare the Data*   Make a table that compares the actual average salary for 1995 to 2000 with the average found using the model.
2. *Compare Projected Average Salaries with Actual Average Salaries*   Use the model to predict the average salary in 2001 and in 2002. Use a library or other reference source to find the actual average salary. How well did the model predict the average salary?
3. *Predict Future Average Salaries*   According to this model, when will the average salary reach $4,500,000?
   (a) Answer the question numerically by creating a table of values.
   (b) Answer the question algebraically by solving
   $$4500 = 24.29t^2 - 208.3t + 1538.$$
4. *Find a Model*   Enter the table above in a graphing utility. Enter 5 for 1995, enter 6 for 1996, and so on. Use the quadratic fit program to find a quadratic model for the data. Do you get the same model as the one above?
5. *Research*   Use a reference source to find data that can be closely modeled with a quadratic model. Compare the model with the actual data numerically and graphically.

---

Chapter Summary    **81**

## CHAPTER SUMMARY

After studying this chapter, you should have acquired the following skills. These skills are keyed to the Review Exercises that begin on page 82. Answers to odd-numbered Review Exercises are given in the back of the book.

**1.1** · Classify an equation as an identity or a conditional equation.    Review Exercises 1, 2
· Determine whether a given value is a solution.    Review Exercises 3, 4
· Solve a linear equation in one variable.    Review Exercises 5–12
**1.2** · Use mathematical models to solve word problems.    Review Exercises 13, 23, 24
· Model and solve percent and mixture problems.    Review Exercises 14–16, 21, 22, 25, 26
· Use common formulas to solve geometry and simple interest problems.    Review Exercises 17–20
**1.3** · Solve a quadratic equation by factoring.    Review Exercises 27–30
· Solve a quadratic equation by extracting square roots.    Review Exercises 31–34
· Analyze a quadratic equation.    Review Exercises 35, 36
· Construct and use a quadratic model to solve an application problem.    Review Exercises 37–40
**1.4** · Use the discriminant to determine the number of real solutions of a quadratic equation.    Review Exercises 41, 42
· Solve a quadratic equation using the Quadratic Formula.    Review Exercises 43–51
· Use the Quadratic Formula to solve an application problem.    Review Exercises 52, 53
**1.5** · Solve a polynomial equation by factoring.    Review Exercises 54–57
· Rewrite and solve an equation involving radicals or rational exponents.    Review Exercises 58–61
· Rewrite and solve an equation with fractions or absolute values.    Review Exercises 62–65
· Construct and use a nonquadratic model to solve an application problem.    Review Exercises 66–68
**1.6** · Solve and graph a linear inequality.    Review Exercises 69–74
· Construct and use a linear inequality to solve an application problem.    Review Exercises 75, 76
**1.7** · Solve and graph a polynomial inequality.    Review Exercises 77–79, 83, 84
· Solve and graph a rational inequality.    Review Exercises 80–82, 85, 86
· Determine the domain of an expression involving a square root.    Review Exercises 87–90
· Construct and use a polynomial inequality to solve an application problem.    Review Exercises 91–98

**FEATURES**

---

**208**    Chapter 3    Functions and Graphs

### REVIEW EXERCISES

In Exercises 1–4, decide whether the equation represents $y$ as a function of $x$.

1. $3x - 4y = 12$     2. $y^2 = x^2 - 9$
3. $y = \sqrt{x + 3}$     4. $x^2 + y^2 - 6x + 8y = 0$

In Exercises 5 and 6, decide whether the set represents a function from $A$ to $B$.

$A = \{1, 2, 3\}$    $B = \{-3, -4, -7\}$

Give reasons for your answer.

5. $\{(1, -3), (2, -7), (3, -3)\}$
6. $\{(1, -4), (2, -3), (3, -9)\}$

In Exercises 7–10, evaluate the function and simplify the results.

7. $f(x) = 3x - 5$
   (a) $f(1)$  (b) $f(-2)$  (c) $f(m)$  (d) $f(x + 1)$
8. $f(x) = \sqrt{x + 9} - 3$
   (a) $f(7)$  (b) $f(0)$  (c) $f(-5)$  (d) $f(x + 2)$
9. $f(x) = |x| + 5$
   (a) $f(0)$  (b) $f(-3)$  (c) $f(\frac{1}{2})$  (d) $f(-x^2)$
10. $f(x) = \begin{cases} x^2 + 1, & x \le 2 \\ x + 3, & x > 2 \end{cases}$
   (a) $f(0)$  (b) $f(2)$  (c) $f(3)$  (d) $f(-5)$

In Exercises 11 and 12, find all real values of $x$ such that $f(x) = 0$.

11. $f(x) = \dfrac{2x + 7}{3}$     12. $f(x) = x^3 - 4x$

In Exercises 13–18, find the domain of the function.

13. $f(x) = 3x^2 + 8x + 4$
14. $g(t) = \dfrac{5}{t^2 - 9}$
15. $h(x) = \sqrt{x + 9}$
16. $f(t) = \sqrt[3]{t - 5}$
17. $g(t) = \dfrac{\sqrt{t - 3}}{t - 5}$
18. $h(x) = \sqrt[4]{9 - x^2}$

19. *Reasoning*   A student has difficulty understanding why the domains of
$$h(x) = \frac{x^2 - 4}{x} \quad \text{and} \quad k(x) = \frac{x}{x^2 - 4}$$
are different. How would you explain their respective domains algebraically? How could you use a graphing utility to explain their domains?

20. *Reasoning*   A student has difficulty understanding why the domains of
$$h(x) = \sqrt{x + 1} \quad \text{and} \quad k(x) = \sqrt[3]{x + 1}$$
are different. How would you explain their respective domains algebraically? How could you use a graphing utility to explain their domains?

21. *Volume of a Box*   An open box is to be made from a square piece of material 20 inches on a side by cutting equal squares from the corners and turning up the sides (see figure).
   (a) Write the volume of the box as a function of $x$.
   (b) What is the domain of this function?
   (c) Use a graphing utility to sketch the graph of this function.

22. *Balance in an Account*   A person deposits $5000 in an account that pays 6.25% interest compounded quarterly.
   (a) Write the balance of the account in terms of the time $t$ that the principal is left in the account.
   (b) What is the domain of this function?

23. *Vertical Motion*   The velocity of a ball thrown vertically upward from ground level is given by $v(t) = -32t + 48$, where $t$ is the time in seconds and $v$ is the velocity in feet per second.
   (a) Find the velocity when $t = 1$.
   (b) Find the time when the ball reaches its maximum height. [*Hint*: Find the time when $v(t) = 0$.]
   (c) Find the velocity when $t = 2$.

---

## Chapter Project

*Chapter Projects* are extensions of the applications presented in the Chapter Opener. Real data is previewed at the beginning of the chapter and then analyzed in detail in the *Project* at the end of the chapter. Here the student is guided through a set of multi-part exercises using modeling, graphing, and critical thinking skills to analyze the data.

## Chapter Summary

The *Chapter Summary* is another vehicle for student self-assessment. For the student who needs additional practice and review, each section learning *Objective* is keyed to the appropriate *Review Exercises*.

## Review Exercises

*Review Exercises* at the end of each chapter offer students an additional opportunity for practice and review. Answers to odd-numbered review exercises are given in the back of the text.

## Acknowledgments

We would like to thank our colleagues who have helped us develop this project throughout this and previous editions. Their encouragement, criticisms, and suggestions have been invaluable to us.

## Reviewers

Judith A. Ahrens, Pellisippi State Technical Community College; Sandra Beken, Horry-Georgetown Technical College; Diane Benjamin, University of Wisconsin—Platteville; Kent Craghead, Colby Community College; Carol Edwards, St. Louis Community College at Florissant Valley; Nick Geller, Collin County Community College; Carolyn H. Goldberg, Niagara County Community College; Carl Hughes, Fayetteville State University; Buddy A. Johns, Wichita State University; Annie Jones, John C. Calhoun State Community College; Steven Z. Kahn, Anne Arundel Community College; Claire Krukenberg, Eastern Illinois University; John Kubicek, Southwest Missouri State University; Gael Mericle, Mankato State University; Michael Montano, Riverside Community College; Terrie L. Nichols, Cuyamaca College; Mark Omodt, Anoka-Ramsey Community College; G. Bryan Stewart, Tarrant County Junior College; David Surowski, Kansas State University; Jamie Whitehead, Texarkana College

In addition, we would like to thank the staffs of Larson Texts, Inc. and Meridian Creative Group who assisted with the proofreading of the manuscript, preparing and proofreading the art package, and checking and typesetting the supplements.

A special note of appreciation goes to all the instructors and students who have used previous editions of the text.

On a personal level, we would like to thank our families, especially Deanna Gilbert Larson, Eloise Hostetler, and Jay N. Torok, for their love, patience, and support. Also, special thanks goes to R. Scott O'Neil.

If you have suggestions for improving the text, please feel free to write to us. Over the past 20 years, we have received many useful comments from both instructors and students.

Ron Larson
Robert P. Hostetler
Anne V. Hodgkins

## Supplements

## Resources

*Website (college.hmco.com)*
Additional text-specific study and interactive features for students and instructors can be found at this Houghton Mifflin website.

*BlackBoard Course Cartridge*

*WebCT e-pack*

### For the Student

*Study and Solutions Guide* by Anne V. Hodgkins (Phoenix College) and Dianna L. Zook (Purdue University and Indiana University at Fort Wayne)

*Student Success Organizer* by Emily Keaton
Fill-in study guide keyed to the text by section

*Learning Tools Student CD-ROM*

*SMARTHINKING™.com On-line Tutoring*

*Graphing Technology Guide* by Benjamin N. Levy and Laurel Technical Services

*Instructional Videotapes/DVD* by Dana Mosely

*Instructional Videotapes for Graphing Calculators* by Dana Mosely

### For the Instructor

*Instructor's Annotated Edition*

*Complete Solutions Guide* by Anne V. Hodgkins (Phoenix College) and Dianna L. Zook (Purdue University and Indiana University at Fort Wayne)

*HM Testing* (Windows, Macintosh)
Computerized test generator with algorithmically generated test items

*Test Item File*
Print component of HM Testing, containing one example of each algorithm in the HM Testing program

*HMClassPrep™ Instructor's CD-ROM*

## *College Algebra: Concepts and Models,* Fourth Edition
## Learning Tools Student CD-ROM

The Learning Tools Student CD-ROM that accompanies the text provides students with the set of electronic learning tools described below. These tools provide extra help with examples and help bring mathematics to life with motion and sound. The CD-ROM also provides access to Ace Practice Tests and SMARTHINKING, the online tutoring center.

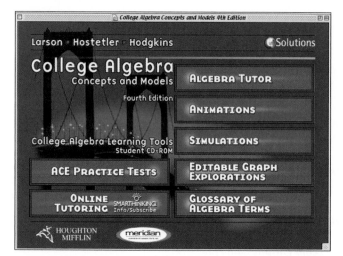

The icon in the text signifies a corresponding Animation, Simulation, or Editable Graph Exploration.

**Animations** use motion and sound to help explain concepts, while allowing students to progress at their own pace.

**Simulations** provide for hands-on experimentation by allowing students to change variables. The program then displays the effects of those changes.

**Editable Graph Explorations** encourage students to explore concepts by using an onscreen graphing calculator. The students can change pre-programmed equations and then graph them to see the effects of the changes.

The **Glossary of Algebra Terms** is a vast and detailed mathematical resource that provides quick access to algebraic terminology.

**Ace Practice Tests** are text-specific tests that provide immediate feedback to students.

**SMARTHINKING** is an online virtual learning assistance center with qualified tutors and independent study resources.

The **Algebra Tutor** provides students with extra step-by-step examples and exercises with diagnostics. In the text, Algebra Tutor help is also signified by the icon beside pertinent examples. The Algebra Tutor gives students three ways to learn the concepts of a section.

- **Understand**—step-by-step example that relates to a specific concept
- **Connect**—step-by-step example that combines concepts within a section
- **Practice**—exercises with diagnostics, providing students with the opportunity to determine potential causes of errors while practicing skills

## *College Algebra: Concepts and Models*, **Fourth Edition**

## Learning Tools Student CD-ROM

Selected examples and concepts throughout the text are identified by the Learning Tools Student CD-ROM icon. The chart on this and the following pages indicates the feature(s) of the CD—Algebra Tutor, Animation, Simulation, and Editable Graph Exploration—that corresponds to the example or concept. (See page xvi for more information about the Learning Tools Student CD-ROM.)

| Chapter | Section | Example/Concept | Algebra Tutor | Animations | Simulations | Editable Graph Explorations |
|---------|---------|-----------------|---------------|------------|-------------|-----------------------------|
| 1 | 1 | Examples 1–7 | ✓ | | | |
| 1 | 2 | Examples 1–3, 6, 8, 9 | ✓ | | | |
| 1 | 3 | Example 1 | ✓ | ✓ | | |
| 1 | 3 | Examples 2–5, 7 | ✓ | | | |
| 1 | 4 | Examples 1, 3–5 | ✓ | | | |
| 1 | 4 | Example 2 | ✓ | ✓ | | |
| 1 | 5 | Examples 2–7, 10 | ✓ | | | |
| 1 | 6 | Graphing Inequalities | | ✓ | | |
| 1 | 6 | Examples 2–6, 8 | ✓ | | | |
| 1 | 7 | Examples 1, 2, 4–6 | ✓ | | | |
| 2 | 1 | Examples 2, 3 | | ✓ | | |
| 2 | 1 | Examples 4–6, 8, 10 | ✓ | | | |
| 2 | 2 | Examples 1, 4 | ✓ | | | ✓ |
| 2 | 2 | Example 3 | | ✓ | | |
| 2 | 2 | Definition of Symmetry | | ✓ | | |
| 2 | 2 | Example 5 | ✓ | ✓ | | ✓ |
| 2 | 2 | Example 6 | ✓ | | | |
| 2 | 2 | Completing the Square | | ✓ | | |
| 2 | 3 | Example 1 | ✓ | | | |
| 2 | 4 | Example 1 | ✓ | | ✓ | |
| 2 | 4 | Examples 2–6 | ✓ | | | |
| 2 | 5 | Example 1 | | | ✓ | |
| 2 | 5 | Example 2 | ✓ | | | |
| 3 | 1 | Examples 2, 3, 5, 6 | ✓ | | | |
| 3 | 2 | Examples 1–3, 5 | ✓ | | | |
| 3 | 2 | Example 7 | ✓ | ✓ | | |
| 3 | 3 | Vertical and Horizontal Shifts | | ✓ | | |
| 3 | 3 | Examples 1, 4, 6 | ✓ | | | |
| 3 | 3 | Example 2 | | ✓ | | |
| 3 | 3 | Example 5 | | ✓ | | |
| 3 | 4 | Examples 2, 4, 6 | ✓ | | | |
| 3 | 4 | Examples 3, 5 | ✓ | ✓ | | |
| 3 | 5 | Examples 2, 3, 5, 7, 8 | ✓ | | | |
| 3 | 5 | Example 6 | | ✓ | | |

## *College Algebra: Concepts and Models,* **Fourth Edition**
## **Learning Tools Student CD-ROM**

| Chapter | Section | Example/Concept | Algebra Tutor | Animations | Simulations | Editable Graph Explorations |
|---|---|---|---|---|---|---|
| 4 | 1 | Examples 1, 3, 4 | ✓ | | | |
| 4 | 1 | Example 2 | ✓ | | | ✓ |
| 4 | 1 | Finding $x$-intercepts | | | | ✓ |
| 4 | 1 | Example 5 | | | ✓ | |
| 4 | 2 | Examples 2, 3, 5, 7 | ✓ | | | |
| 4 | 3 | Examples 1, 5, 6 | ✓ | | | |
| 4 | 3 | Examples 3, 4 | ✓ | ✓ | | |
| 4 | 4 | Examples 2, 4, 5 | ✓ | | | |
| 4 | 4 | Example 3 | ✓ | ✓ | | |
| 4 | 5 | Examples 1–6 | ✓ | | | |
| 4 | 6 | Examples 1–5 | ✓ | | | |
| 4 | 7 | Examples 2–4, 7 | ✓ | | | |
| 4 | 7 | Example 5 | ✓ | ✓ | | |
| 4 | 7 | Slant Asymptotes | | ✓ | | ✓ |
| 5 | 1 | Examples 1, 5, 6, 8 | ✓ | | | |
| 5 | 1 | Example 2 | ✓ | | | ✓ |
| 5 | 1 | Example 3 | | | | ✓ |
| 5 | 1 | Example 4 | ✓ | ✓ | | |
| 5 | 1 | Example 7 | ✓ | | ✓ | |
| 5 | 2 | Examples 1, 2, 7, 8, 10 | ✓ | | | |
| 5 | 2 | Example 4 | | ✓ | | |
| 5 | 2 | Example 6 | ✓ | ✓ | | |
| 5 | 3 | Examples 1, 2, 5–8 | ✓ | | | |
| 5 | 4 | Examples 3–9 | ✓ | | | |
| 5 | 5 | Examples 1, 2, 5, 6 | ✓ | | | |
| 6 | 1 | Examples 1, 3, 6 | ✓ | | | |
| 6 | 1 | Example 5 | ✓ | ✓ | | |
| 6 | 2 | Examples 1–8 | ✓ | | | |
| 6 | 2 | Graphical Interpretation of Solutions | | | ✓ | |
| 6 | 3 | Examples 3–8 | ✓ | | | |
| 6 | 4 | Example 1 | ✓ | | | ✓ |
| 6 | 4 | Examples 2, 3, 5, 7 | ✓ | | | |
| 6 | 4 | Example 4 | ✓ | ✓ | | |
| 6 | 5 | Example 1 | | ✓ | | |
| 6 | 5 | Examples 3, 5–7 | ✓ | | | |

## *College Algebra: Concepts and Models,* Fourth Edition
## Learning Tools Student CD-ROM

| Chapter | Section | Example/Concept | Algebra Tutor | Animations | Simulations | Editable Graph Explorations |
|---|---|---|---|---|---|---|
| 7 | 1 | Examples 1, 2, 4, 6, 8 | ✓ | | | |
| 7 | 1 | Examples 5, 7 | ✓ | ✓ | | |
| 7 | 2 | Examples 1–3, 5, 7, 8 | ✓ | | | |
| 7 | 2 | Example 6 | ✓ | ✓ | | |
| 7 | 3 | Examples 1, 4, 5 | ✓ | | | |
| 7 | 3 | Example 3 | ✓ | ✓ | | |
| 7 | 4 | Examples 1, 3, 4 | ✓ | | | |
| 7 | 5 | Examples 1–3, 5, 6 | ✓ | | | |
| 8 | 1 | Examples 2, 4–7 | ✓ | | | |
| 8 | 2 | Examples 1–3, 5, 7, 8 | ✓ | | | |
| 8 | 3 | Examples 1–3, 5, 6, 8 | ✓ | | | |
| 8 | 3 | Example 7 | ✓ | ✓ | | |
| 8 | 4 | Examples 1, 6, 7 | ✓ | | | |
| 8 | 4 | Example 5 | | ✓ | | |
| 8 | 5 | Examples 4, 5, 7, 9 | ✓ | | | |
| 8 | 6 | Examples 1, 2, 5, 8, 9, 11 | ✓ | | | |
| 8 | 6 | Example 3 | ✓ | | ✓ | |
| 8 | 7 | Examples 1, 4 | ✓ | | | |

LEARNING TOOLS

# An Introduction to Graphing Utilities

Graphing utilities such as graphing calculators and computers with graphing software are very valuable tools for visualizing mathematical principles, verifying solutions to equations, exploring mathematical ideas, and developing mathematical models. Although graphing utilities are extremely helpful in learning mathematics, their use does not mean that learning algebra is any less important. In fact, the combination of knowledge of mathematics and the use of graphing utilities allows you to explore mathematics more easily and to a greater depth. If you are using a graphing utility in this course, it is up to you to learn its capabilities and to practice using this tool to enhance your mathematical learning.

In this text there are many opportunities to use a graphing utility, some of which are described below.

> ▶ **Some Uses of a Graphing Utility**
>
> A graphing utility can be used to
>
> • check or validate answers to problems obtained using algebraic methods.
>
> • discover and explore algebraic properties, rules, and concepts.
>
> • graph functions, and approximate solutions to equations involving functions.
>
> • efficiently perform complicated mathematical procedures such as those found in many real-life applications.
>
> • find mathematical model for sets of data.

In this introduction, the features of graphing utilities are discussed from a generic perspective. To learn how to use the features of a specific graphing utility, consult your user's manual. Additionally, keystroke guides are available for most graphing utilities, and your college library may have a videotape on how to use your graphing utility.

## The Equation Editor

Many graphing utilities are designed to act as "function graphers." In this course, you will study functions and their graphs in detail. You may recall from previous courses that a function can be thought of as a rule that describes the relationship between two variables. These rules are frequently written in terms of $x$ and $y$. For example, the equation $y = 3x + 5$ represents $y$ as a function of $x$.

Figure 1

Many graphing utilities have an equation editor that requires an equation to be written in "$y =$" form in order to be entered, as shown in Figure 1. (You should note that your equation editor screen may not look like the screen shown in Figure 1.) To determine exactly how to enter an equation into your graphing utility, consult your user's manual.

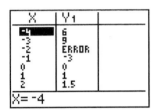

Figure 2

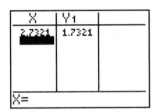

Figure 3

 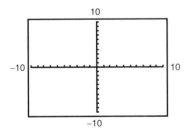

Figure 4

## The Table Feature

Most graphing utilities are capable of displaying a table of values with $x$-values and one or more corresponding $y$-values. These tables can be used to check solutions of an equation and to generate ordered pairs to assist in graphing an equation.

To use the *table* feature, enter an equation into the equation editor in "$y =$" form. The table may have a setup screen, which will allow you to select the starting $x$-value and the table step or $x$-increment. You may then have the option of automatically generating values for $x$ and $y$ or building your own table using the *ask* mode. In the *ask* mode, you enter a value for $x$ and the graphing utility displays the $y$-value.

For example, enter the equation

$$y = \frac{3x}{x + 2}$$

into the equation editor as shown in Figure 2. In the table setup screen, set the table to start at $x = -4$ and set the table step to 1. When you view the table, notice that the first $x$-value is $-4$ and each value after it increases by 1. Also notice that the Y1 column gives the resulting $y$-value for each $x$-value, as shown in Figure 3. The table shows that the $y$-value when $x = -2$ is ERROR. This means that the variable $x$ may not take on the value $-2$ in this equation.

With the same equation in the equation editor, set the table to *ask* mode. In this mode you do not need to set the starting $x$-value or the table step, because you are entering any value you choose for $x$. You may enter any real value for $x$— integers, fractions, decimals, irrational numbers, and so forth. If you enter $x = 1 + \sqrt{3}$, the graphing utility may rewrite the number as a decimal approximation, as shown in Figure 4. You can continue to build your own table by entering additional $x$-values in order to generate $y$-values.

If you have several equations in the equation editor, the table may generate $y$-values for each equation.

## Creating a Viewing Window

A **viewing window** for a graph is a rectangular portion of the coordinate plane. A viewing window is determined by the following six values.

Xmin = the smallest value of $x$
Xmax = the largest value of $x$
Xscl = the number of units per tick mark on the $x$-axis
Ymin = the smallest value of $y$
Ymax = the largest value of $y$
Yscl = the number of units per tick mark on the $y$-axis

When you enter these six values into a graphing utility, you are setting the viewing window. Some graphing utilities have a standard viewing window, as shown in Figure 5.

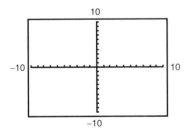

Figure 5

GRAPHING UTILITIES

By choosing different viewing windows for a graph, it is possible to obtain very different impressions of the graph's shape. For instance, Figure 6 shows four different viewing windows for the graph of

$$y = 0.1x^4 - x^3 + 2x^2.$$

Of these, the view shown in part (a) is the most complete.

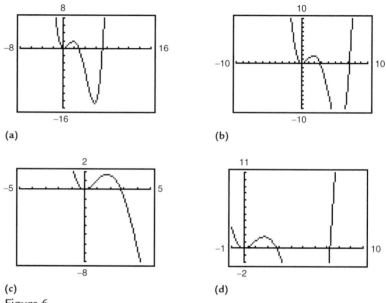

(a)                    (b)

(c)                    (d)

Figure 6

On most graphing utilities, the display screen is two-thirds as high as it is wide. On such screens, you can obtain a graph with a true geometric perspective by using a **square setting**—one in which

$$\frac{\text{Ymax} - \text{Ymin}}{\text{Xmax} - \text{Xmin}} = \frac{2}{3}.$$

One such setting is shown in Figure 7. Notice that the $x$ and $y$ tick marks are equally spaced on a square setting, but not on a standard setting.

Figure 7

To see how the viewing window affects the geometric perspective, graph the semicircles $y_1 = \sqrt{9 - x^2}$ and $y_2 = -\sqrt{9 - x^2}$ in a standard viewing window. Then graph $y_1$ and $y_2$ in a square window. Note the difference in the shapes of the circles.

## Zoom and Trace Features

When you graph an equation, you can move from point to point along its graph using the *trace* feature. As you trace the graph, the coordinates of each point are displayed, as shown in Figure 8. The *trace* feature combined with the *zoom* feature allows you to obtain better and better approximations of desired points on a graph. For instance, you can use the *zoom* feature of a graphing utility to approximate the $x$-intercept(s) of a graph [the point(s) where the graph crosses the $x$-axis]. Suppose you want to approximate the $x$-intercept(s) of the graph of $y = 2x^3 - 3x + 2$.

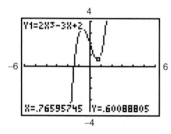

Figure 8

Begin by graphing the equation, as shown in Figure 9(a). From the viewing window shown, the graph appears to have only one $x$-intercept. This intercept lies between $-2$ and $-1$. By zooming in on the intercept, you can improve the approximation, as shown in Figure 9(b). To three decimal places, the solution is $x \approx -1.476$.

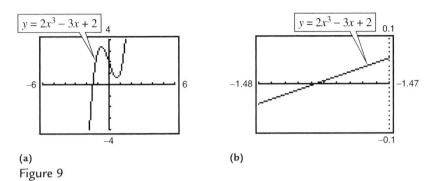

(a)                                                    (b)

Figure 9

Here are some suggestions for using the *zoom* feature.

1. With each successive zoom-in, adjust the $x$-scale so that the viewing window shows at least one tick mark on each side of the $x$-intercept.

2. The error in your approximation will be less than the distance between two scale marks.

3. The trace feature can usually be used to add one more decimal place of accuracy without changing the viewing window.

Figure 10(a) shows the graph of $y = x^2 - 5x + 3$. Figures 10(b) and 10(c) show "zoom-in views" of the two $x$-intercepts. From these views, you can approximate the $x$-intercepts to be $x \approx 0.697$ and $x \approx 4.303$.

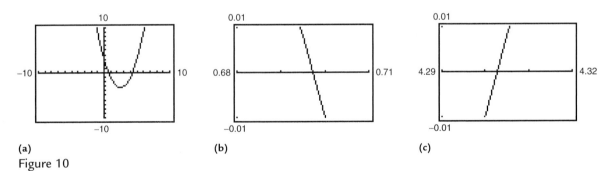

(a)                        (b)                        (c)

Figure 10

## Zero or Root Feature

Using the zero or root feature, you can find the real zeros of functions of the various types studied in this text—polynomial, exponential, and logarithmic functions. To find the zeros of a function such as $f(x) = \frac{3}{2}x - 2$, first enter the function as $y_1 = \frac{3}{4}x - 2$. Then use the zero or root feature, which may require entering lower and upper bound estimates of the root, as shown in Figures 11(a) and 11(b).

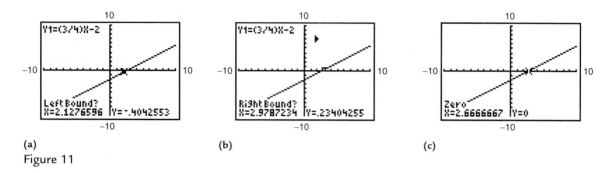

(a)

(b)

(c)

Figure 11

In Figure 11(c), you can see that the zero is $x = 2.6666667 \approx 2\frac{2}{3}$.

## Intersect Feature

To find the points of intersection of two graphs, you can use the *intersect* feature. For instance, to find the points of intersection of the graphs of $y_1 = -x + 2$ and $y_2 = x + 4$, enter these two functions and use the *intersect* feature, as shown in Figure 12.

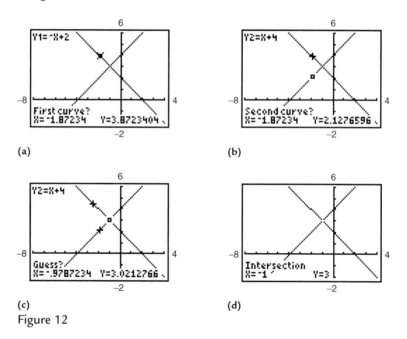

(a)

(b)

(c)

(d)

Figure 12

From Figure 12(d), you can see that the point of intersection is $(-1, 3)$.

# Regression Capabilities

Throughout the text, you will be asked to use the regression capabilities of a graphing utility to find models for sets of data. Most graphing utilities have built-in regression programs for the following.

| *Regression* | *Form of Model* |
|---|---|
| Linear | $y = ax + b$ |
| Quadratic | $y = ax^2 + bx + c$ |
| Cubic | $y = ax^3 + bx^2 + cx + d$ |
| Quartic | $y = ax^4 + bx^3 + cx^2 + dx + e$ |
| Logarithmic | $y = a + b \ln(x)$ |
| Exponential | $y = ab^x$ |
| Power | $y = ax^b$ |
| Logistic | $y = \dfrac{c}{1 + ae^{-bx}}$ |
| Sine | $y = a \sin(bx + c) + d$ |

For instance, you can find the linear regression model for the average hourly wages $y$ (in dollars per hour) of production workers in manufacturing industries from 1987 through 1997 shown in the table.    (Source: U.S. Bureau of Labor Statistics)

| Year | 1987 | 1988 | 1989 | 1990 | 1991 | 1992 |
|---|---|---|---|---|---|---|
| $y$ | 9.91 | 10.19 | 10.48 | 10.83 | 11.18 | 11.46 |

| Year | 1993 | 1994 | 1995 | 1996 | 1997 |
|---|---|---|---|---|---|
| $y$ | 11.74 | 12.06 | 12.37 | 12.78 | 13.17 |

First let $x = 0$ correspond to 1990 and enter the data into the list editor, as shown in Figure 13. Note that the list in the first column contains the years and the list in the second column contains the hourly wages that correspond to the years. Run your graphing utility's built-in linear regression program to obtain the coefficients $a$ and $b$ for the model $y = ax + b$, as shown in Figure 14. So, a linear model for the data is

$$y \approx 0.321x + 10.83.$$

When you run some regression programs, you may obtain an "$r$-value," which gives a measure of how well the model fits the data. The closer the value of $|r|$ is to 1, the better the fit. For the data in the table above, $r \approx 0.999$, which implies that the model is a very good fit.

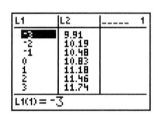

Figure 13

```
LinReg
 y=ax+b
 a=.3213636364
 b=10.82727273
 r²=.9979974124
 r=.9989982044
```

Figure 14

# Equations and Inequalities

Baseball began in the mid-1800s and spread throughout the United States during the Civil War. In 1876 the National League, consisting of eight professional teams, was formed. The American League began playing with eight teams in 1901.

Today there are 30 teams playing in the National and American Leagues. Attendance for the 2000 season topped 72.5 million.

Salaries for professional baseball players have steadily increased. The average salaries for players from 1995 to 2000 are listed in the table.

**1.1  Linear Equations**

**1.2  Mathematical Modeling**

**1.3  Quadratic Equations**

**1.4  The Quadratic Formula**

**1.5  Other Types of Equations**

**1.6  Linear Inequalities**

**1.7  Other Types of Inequalities**

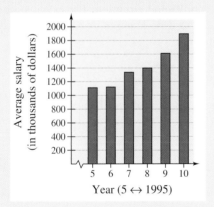

Year (5 ↔ 1995)

| Year | 1995 | 1996 | 1997 | 1998 | 1999 | 2000 |
|------|------|------|------|------|------|------|
| Average salary (in thousands of dollars) | 1111 | 1120 | 1337 | 1399 | 1611 | 1896 |

The chapter project related to this information is on page 80.

## 1.1    Linear Equations

### Objectives

- Classify an equation as an identity or a conditional equation.
- Solve a linear equation in one variable.
- Use a linear model to solve an application problem.

## Equations and Solutions

An **equation** is a statement that two algebraic expressions are equal. Some examples of equations in $x$ are

$$3x - 5 = 7, \qquad x^2 - x - 6 = 0, \qquad \text{and} \qquad \sqrt{2x} = 4.$$

To **solve** an equation in $x$ means that you find all values of $x$ for which the equation is **true.** Such values are called **solutions.** For instance, $x = 4$ is a solution of the equation $3x - 5 = 7$, because $3(4) - 5 = 7$ is a true statement.

An equation that is true for *every* real number in the domain of the variable is an **identity.** Two examples of identities are

$$x^2 - 9 = (x + 3)(x - 3) \qquad \text{and} \qquad \frac{x}{3x^2} = \frac{1}{3x}, \qquad x \neq 0.$$

The first equation is an identity because it is a true statement for all real values of $x$. The second is an identity because it is true for all nonzero real values of $x$.

An equation that is true for just *some* (or even none) of the real numbers in the domain of the variable is called a **conditional equation.** For example, the equation $x^2 - 9 = 0$ is conditional because $x = 3$ and $x = -3$ are the only values in the domain that satisfy the equation.

### Example 1    Classifying Equations

Classify each of the following as an identity or a conditional equation.

**a.** $2(x + 3) = 2x + 6$    **b.** $2(x + 3) = x + 6$    **c.** $2(x + 3) = 2x + 3$

**Solution**

**a.** This equation is an identity because it is true for every real value of $x$.

**b.** This equation is a conditional equation because there are real values of $x$ (such as $x = 1$) for which the equation is not true.

**c.** This equation is a conditional equation because there are no real number values of $x$ for which the equation is true.

Be sure you understand that equations are used in algebra for two distinct purposes: (1) *identities* are usually used to state mathematical properties and (2) *conditional equations* are usually used to model and solve problems that occur in real life.

The icon indicates that you can use the Learning Tools CD that accompanies this text to access an Animation, Simulation, Editable Graph Exploration, or Algebra Tutor help according to the chart on pages xvii–xix.

# Linear Equations in One Variable

The most common type of conditional equation is a **linear equation.**

---

▶ Definition of a Linear Equation

A **linear equation** in one variable $x$ is an equation that can be written in the standard form

$$ax + b = 0$$

where $a$ and $b$ are real numbers with $a \neq 0$.

---

A linear equation has exactly one solution. To see this, consider the following steps. (Remember that $a \neq 0$.)

| | |
|---|---|
| $ax + b = 0$ | Original equation |
| $ax = -b$ | Subtract $b$ from both sides. |
| $x = -\dfrac{b}{a}$ | Divide both sides by $a$. |

So, the equation $ax + b = 0$ has exactly one solution, $x = -b/a$.

To solve a linear equation in $x$, you should isolate $x$ by a sequence of **equivalent** (and usually simpler) equations, each having the same solution as the original equation. The operations that yield equivalent equations come from the basic rules of algebra reviewed in the appendix.

---

▶ Forming Equivalent Equations

A given equation can be transformed into an equivalent equation by one or more of the following steps.

| | *Given Equation* | *Equivalent Equation* |
|---|---|---|
| **1.** Remove symbols of grouping, combine like terms, or simplify one or both sides of the equation. | $2x - x = 4$ <br> $3(x - 2) = 5$ | $x = 4$ <br> $3x - 6 = 5$ |
| **2.** Add (or subtract) the same quantity to (from) *both* sides of the equation. | $x + 1 = 6$ | $x = 5$ |
| **3.** Multiply (or divide) *both* sides of the equation by the same *nonzero* quantity. | $2x = 6$ | $x = 3$ |
| **4.** Interchange sides of the equation. | $2 = x$ | $x = 2$ |

---

**Example 2** Solving a Linear Equation

Solve $3x - 6 = 0$.

**Solution**

| | |
|---|---|
| $3x - 6 = 0$ | Original equation |
| $3x = 6$ | Add 6 to both sides. |
| $x = 2$ | Divide both sides by 3. |

After solving an equation, you should **check each solution** in the *original* equation. For instance, in Example 2, you can check that 2 is a solution by substituting 2 for $x$ in the original equation $3x - 6 = 0$, as follows.

**Check**

| | |
|---|---|
| $3x - 6 = 0$ | Original equation |
| $3(2) - 6 \stackrel{?}{=} 0$ | Substitute 2 for $x$. |
| $6 - 6 = 0$ | Solution checks. ✓ |

## Study Tip

Students sometimes tell us that a solution looks easy when we work it in class, but that they don't see where to begin when trying it alone. Keep in mind that no one—not even great mathematicians—can expect to look at every mathematical problem and immediately know where to begin. Many problems involve some trial and error before a solution is found. To make algebra work for you, you must put in a lot of time, you must expect to try solution methods that end up not working, *and* you must learn from both your successes and your failures.

**Example 3** Solving a Linear Equation

| | |
|---|---|
| $6(x - 1) + 4 = 3(7x + 1)$ | Original equation |
| $6x - 6 + 4 = 21x + 3$ | Distributive Property |
| $6x - 2 = 21x + 3$ | Simplify. |
| $-15x = 5$ | Add 2 and subtract $21x$. |
| $x = -\frac{1}{3}$ | Divide both sides by $-15$. |

The solution is $-\frac{1}{3}$. You can check this as follows.

**Check**

| | |
|---|---|
| $6(x - 1) + 4 = 3(7x + 1)$ | Original equation |
| $6\left(-\frac{1}{3} - 1\right) + 4 \stackrel{?}{=} 3\left[7\left(-\frac{1}{3}\right) + 1\right]$ | Substitute $-\frac{1}{3}$ for $x$. |
| $6\left(-\frac{4}{3}\right) + 4 \stackrel{?}{=} 3\left(-\frac{7}{3} + 1\right)$ | Add fractions. |
| $-\frac{24}{3} + 4 \stackrel{?}{=} -\frac{21}{3} + 3$ | Multiply. |
| $-8 + 4 \stackrel{?}{=} -7 + 3$ | Simplify. |
| $-4 = -4$ | Solution checks. ✓ |

# Equations Involving Fractional Expressions

To solve an equation involving fractional expressions, you can multiply every term in the equation by the least common denominator (LCD) of the terms.

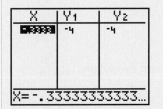
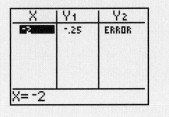
**Example 4**    An Equation Involving Fractional Expressions

$$\frac{x}{3} + \frac{3x}{4} = 2 \qquad \text{Original equation}$$

$$(12)\frac{x}{3} + (12)\frac{3x}{4} = (12)2 \qquad \text{Multiply by least common denominator.}$$

$$4x + 9x = 24 \qquad \text{Reduce and multiply.}$$

$$13x = 24 \qquad \text{Combine like terms.}$$

$$x = \frac{24}{13} \qquad \text{Divide both sides by 13.}$$

The solution is $\frac{24}{13}$. Check this in the original equation.

When an equation is multiplied or divided by a *variable* expression, it is possible to introduce an **extraneous** solution—one that does not satisfy the original equation. In such cases a check is especially important.

**Example 5**    An Equation with an Extraneous Solution

Solve $\dfrac{1}{x - 2} = \dfrac{3}{x + 2} - \dfrac{6x}{x^2 - 4}$.

Solution

The least common denominator is $x^2 - 4 = (x + 2)(x - 2)$. Multiplying each term by this LCD and reducing produces the following.

$$\frac{1}{x - 2} = \frac{3}{x + 2} - \frac{6x}{x^2 - 4}$$

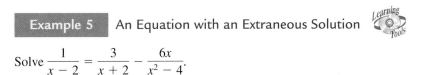

$$x + 2 = 3(x - 2) - 6x, \quad x \neq \pm 2$$

$$x + 2 = 3x - 6 - 6x$$

$$4x = -8$$

$$x = -2$$

After checking $x = -2$, you can see that it yields a denominator of zero. So, $x = -2$ is extraneous, and the equation has *no solution*.

An equation with a *single fraction* on each side can be cleared of denominators by **cross-multiplying,** which is equivalent to multiplying by the least common denominator and then reducing.

### Example 6    Cross-Multiplying to Solve an Equation

Solve $\dfrac{3y - 2}{2y + 1} = \dfrac{6y - 9}{4y + 3}$.

**Solution**

$$\dfrac{3y - 2}{2y + 1} = \dfrac{6y - 9}{4y + 3} \qquad \text{Original equation}$$

$$(3y - 2)(4y + 3) = (6y - 9)(2y + 1) \qquad \text{Cross-multiply.}$$

$$12y^2 + y - 6 = 12y^2 - 12y - 9 \qquad \text{Multiply.}$$

$$13y = -3 \qquad \text{Isolate } y\text{-term on left.}$$

$$y = -\dfrac{3}{13} \qquad \text{Divide both sides by 13.}$$

The solution is $-\frac{3}{13}$. Check this in the original equation.

### Example 7    Using a Calculator to Solve an Equation

Solve $\dfrac{1}{9.38} - \dfrac{3}{x} = \dfrac{5}{0.3714}$.

**Solution**

Roundoff error will be minimized if you solve for $x$ before performing any calculations. The least common denominator is $(9.38)(0.3714)(x)$.

$$\dfrac{1}{9.38} - \dfrac{3}{x} = \dfrac{5}{0.3714}$$

$$(9.38)(0.3714)(x)\left(\dfrac{1}{9.38} - \dfrac{3}{x}\right) = (9.38)(0.3714)(x)\left(\dfrac{5}{0.3714}\right)$$

$$0.3714x - 3(9.38)(0.3714) = (9.38)(5)(x), \quad x \neq 0$$

$$[0.3714 - 5(9.38)]x = 3(9.38)(0.3714)$$

$$x = \dfrac{3(9.38)(0.3714)}{0.3714 - 5(9.38)}$$

$$x \approx -0.225 \qquad \text{Round to three places.}$$

Because of roundoff error, a check of a decimal solution may not yield exactly the same values for both sides of the original equation. The difference, however, should be quite small.

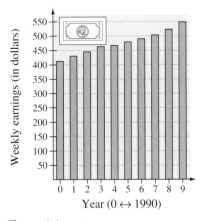

Figure 1.1

# Application

**Example 8**    Weekly Earnings

The median weekly earnings $y$ (in dollars) for workers in the United States from 1990 to 1999 can be modeled by the linear equation $y = 13.7t + 414$ for $0 \le t \le 9$ and where $t = 0$ represents 1990, as shown in Figure 1.1. Use the model to estimate when the median weekly earnings will reach \$660.    (Source: U.S. Bureau of Labor Statistics)

## Solution

$$y = 13.7t + 414 \qquad \text{Given model}$$

$$660 = 13.7t + 414 \qquad \text{Substitute 660 for } y \text{ so you can solve for } t.$$

$$246 = 13.7t \qquad \text{Subtract 414 from both sides.}$$

$$t = \frac{246}{13.7} \approx 18 \qquad \text{Divide both sides by 13.7.}$$

Because $t = 0$ corresponds to 1990, it follows that $t = 18$ corresponds to 2008. So, you can expect the median weekly earnings to reach \$660 by 2008.

---

| Discussing the Concept | Checking a Solution |
|---|---|

The solution $x = 2$ does not check in either of the following original equations. Why? Is there an error? When a solution does not check, what are the possibilities?

**a.**
$$\frac{2}{x-1} = \frac{1}{x-2} - \frac{1}{(x-2)(x-1)} \qquad \text{Original equation}$$

$$\frac{2}{x-1}(x-2)(x-1) = \frac{1}{x-2}(x-2)(x-1) -$$
$$\frac{1}{(x-2)(x-1)}(x-2)(x-1)$$

$$2(x-2) = (x-1) - 1$$

$$2x - 4 = x - 2$$

$$x = 2$$

**b.**
$$\frac{2}{x-1} = \frac{1}{x-6} - \frac{3}{(x-1)(x-6)} \qquad \text{Original equation}$$

$$\frac{2}{x-1}(x-1)(x-6) = \frac{1}{x-6}(x-1)(x-6) -$$
$$\frac{3}{(x-1)(x-6)}(x-1)(x-6)$$

$$2(x-6) = (x-1) - 3$$

$$2x - 6 = x - 4$$

$$x = 2$$

# Warm Up

The following warm-up exercises involve skills that were covered in previous courses. You will use these skills in the exercise set for this section.

In Exercises 1–10, perform the indicated operations and simplify your answer.

**1.** $(2x - 4) - (5x + 6)$  **2.** $(3x - 5) + (2x - 7)$

**3.** $2(x + 1) - (x + 2)$  **4.** $-3(2x - 4) + 7(x + 2)$

**5.** $\dfrac{x}{3} + \dfrac{x}{5}$  **6.** $x - \dfrac{x}{4}$

**7.** $\dfrac{1}{x + 1} - \dfrac{1}{x}$  **8.** $\dfrac{2}{x} + \dfrac{3}{x}$

**9.** $\dfrac{4}{x} + \dfrac{3}{x - 2}$  **10.** $\dfrac{1}{x + 1} - \dfrac{1}{x - 1}$

## 1.1    Exercises

In Exercises 1–6, determine whether the equation is an identity or a conditional equation.

**1.** $2(x - 1) = 2x - 2$

**2.** $3(x + 2) = 3x + 6$

**3.** $2(x - 1) = 3x + 4$

**4.** $3(x + 2) = 2x + 4$

**5.** $2(x + 1) = 2x + 1$

**6.** $3(x + 4) = 3x + 4$

In Exercises 7–14, determine whether the values of $x$ are solutions of the equation.

**7.** $5x - 3 = 3x + 5$

   (a) $x = 0$  (b) $x = -5$  (c) $x = 4$  (d) $x = 10$

**8.** $7 - 3x = 5x - 17$

   (a) $x = -3$  (b) $x = 0$  (c) $x = 8$  (d) $x = 3$

**9.** $3x^2 + 2x - 5 = 2x^2 - 2$

   (a) $x = -3$  (b) $x = 1$  (c) $x = 4$  (d) $x = -5$

**10.** $5x^3 + 2x - 3 = 4x^3 + 2x - 11$

   (a) $x = 2$  (b) $x = -2$  (c) $x = 0$  (d) $x = 10$

**11.** $\dfrac{5}{2x} - \dfrac{4}{x} = 3$

   (a) $x = -\frac{1}{2}$  (b) $x = 4$  (c) $x = 0$  (d) $x = \frac{1}{4}$

**12.** $3 + \dfrac{1}{x + 2} = 4$

   (a) $x = -1$  (b) $x = -2$  (c) $x = 0$  (d) $x = 5$

**13.** $(x + 5)(x - 3) = 20$

   (a) $x = 3$  (b) $x = -2$  (c) $x = 0$  (d) $x = -7$

**14.** $\sqrt[3]{x - 8} = 3$

   (a) $x = 2$  (b) $x = -5$  (c) $x = 35$  (d) $x = 8$

In Exercises 15–52, solve the equation and check your answer. (Some equations have no solution.)

**15.** $x + 10 = 15$

**16.** $7 - x = 18$

**17.** $7 - 2x = 15$

**18.** $7x + 2 = 16$

**19.** $8x - 5 = 3x + 10$

**20.** $7x + 3 = 3x - 13$

**21.** $2(x + 5) - 7 = 3(x - 2)$

**22.** $2(13t - 15) + 3(t - 19) = 0$

**23.** $6[x - (2x + 3)] = 8 - 5x$

**24.** $8(x + 2) - 3(2x + 1) = 2(x + 5)$

**25.** $\dfrac{5x}{4} + \dfrac{1}{2} = x - \dfrac{1}{2}$

**26.** $\dfrac{x}{5} - \dfrac{x}{2} = 3$

**27.** $\frac{3}{2}(z + 5) - \frac{1}{4}(z + 24) = 0$    *4 common denominator*

**28.** $\dfrac{3x}{2} + \dfrac{1}{4}(x - 2) = 10$

**29.** $0.25x + 0.75(10 - x) = 3$

**30.** $0.60x + 0.40(100 - x) = 50$

**31.** $x + 8 = 2(x - 2) - x$

**32.** $3(x + 3) = 5(1 - x) - 1$

**33.** $\dfrac{100 - 4u}{3} = \dfrac{5u + 6}{4} + 6$

**34.** $\dfrac{17 + y}{y} + \dfrac{32 + y}{y} = 100$

**35.** $\dfrac{5x - 4}{5x + 4} = \dfrac{2}{3}$

**36.** $\dfrac{10x + 3}{5x + 6} = \dfrac{1}{2}$

**37.** $10 - \dfrac{13}{x} = 4 + \dfrac{5}{x}$

**38.** $\dfrac{15}{x} - 4 = \dfrac{6}{x} + 3$

**39.** $\dfrac{1}{x - 3} + \dfrac{1}{x + 3} = \dfrac{10}{x^2 - 9}$

**40.** $\dfrac{1}{x - 2} + \dfrac{3}{x + 3} = \dfrac{4}{x^2 + x - 6}$

**41.** $\dfrac{x}{x + 4} + \dfrac{4}{x + 4} + 2 = 0$

**42.** $\dfrac{2}{(x - 4)(x - 2)} = \dfrac{1}{x - 4} + \dfrac{2}{x - 2}$

**43.** $\dfrac{7}{2x + 1} - \dfrac{8x}{2x - 1} = -4$

**44.** $\dfrac{4}{u - 1} + \dfrac{6}{3u + 1} = \dfrac{15}{3u + 1}$

**45.** $\dfrac{3}{x(x - 3)} + \dfrac{4}{x} = \dfrac{1}{x - 3}$

**46.** $3 = 2 + \dfrac{2}{z + 2}$

**47.** $(x + 2)^2 + 5 = (x + 3)^2$

**48.** $(x + 1)^2 + 2(x - 2) = (x + 1)(x - 2)$

**49.** $(x + 2)^2 - x^2 = 4(x + 1)$

**50.** $4(x + 1) - 3x = x + 5$

**51.** $(2x + 1)^2 = 4(x^2 + x + 1)$

**52.** $(2x - 1)^2 = 4(x^2 - x + 6)$

**53.** Explain why a solution of an equation involving fractional expressions may be extraneous.

**54.** What method or methods would you suggest for checking a solution of an equation involving fractional expressions?

In Exercises 55–60, use a calculator to solve the equation. (Round your answer to three decimal places.)

**55.** $0.275x + 0.725(500 - x) = 300$

**56.** $2.763 - 4.5(2.1x - 5.1432) = 6.32x + 5$

**57.** $\dfrac{x}{0.6321} + \dfrac{x}{0.0692} = 1000$

**58.** $\dfrac{2}{7.398} - \dfrac{4.405}{x} = \dfrac{1}{x}$

**59.** $(x + 5.62)^2 + 10.83 = (x + 7)^2$

**60.** $\dfrac{x}{2.625} + \dfrac{x}{4.875} = 1$

**61.** What method or methods would you recommend for checking the solutions to Exercises 55–60 using your graphing utility?

**62.** In Exercises 55–60, your answers are rounded to three decimal places. What effect does rounding have as you check a solution?

In Exercises 63–68, evaluate each expression in two ways. (a) Calculate entirely on your calculator using appropriate parentheses, and then round the answer to two decimal places. (b) Round both the numerator and the denominator to two decimal places before dividing, and then round the final answer to two decimal places. Does the second method introduce an additional roundoff error?

**63.** $\dfrac{1 + 0.73205}{1 - 0.73205}$

**64.** $\dfrac{1 + 0.86603}{1 - 0.86603}$

**65.** $\dfrac{2 - 1.63254}{(2.58)(0.135)}$

**66.** $\dfrac{2 + 0.57735}{1 + 0.57735}$

**67.** $\dfrac{333 + \dfrac{1.98}{0.74}}{4 + \dfrac{6.25}{3.15}}$

**68.** $\dfrac{1.73205 - 1.19195}{3 - (1.73205)(1.19195)}$

**69.** *Television Advertising* The amount of money spent on television advertising from 1990 to 1999 can be approximated by the linear equation

$$y = 2619.2t + 25{,}958, \qquad 0 \le t \le 9$$

where $y$ is the amount in millions of dollars per year and $t$ is the calendar year, with $t = 0$ corresponding to 1990. If this linear pattern continues, when will the amount of television advertising reach \$60,000 million per year? (Source: Universal McCann)

The symbol ▦ indicates an exercise or parts of an exercise in which you are instructed to use a graphing utility.

**70.** *Internet Commerce*   The expected amount of internet commerce in the United States from 2000 to 2004 can be approximated by the linear equation

$$y = 672.37t + 283.3, \qquad 0 \le t \le 4$$

where $y$ is the amount in billions of U.S. dollars and $t = 0$ corresponds to 2000. If this linear pattern continues, in what year will internet commerce in the United States reach 5000 billion U.S. dollars? (Source: Forrester Research)

*Human Height*   In Exercises 71 and 72, use the following information. The relationship between the length of an adult's thigh bone and the height of the adult can be approximated by the linear equations

$$y = 0.432x - 10.44 \qquad \text{Female}$$

$$y = 0.449x - 12.15 \qquad \text{Male}$$

where $y$ is the length of the thigh bone (femur) in inches and $x$ is the height of the adult in inches (see figure).

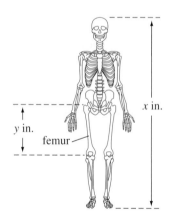

**71.** An anthropologist discovers a thigh bone belonging to an adult human female. The bone is 16 inches long. How tall would you estimate the female to have been?

**72.** From the foot bone of an adult human male, an anthropologist estimates that the male was 69 inches tall. A few feet away from the site where the foot bone was discovered, the anthropologist discovers an adult male thigh bone that is 19 inches long. Is it possible that both bones came from the same person?

*Social Security Benefits*   In Exercises 73 and 74, use the following information. From 1995 to 1999, the average monthly benefit for retired workers can be approximated by

$$y = 621 + 20.3x, \qquad 5 \le x \le 9$$

where $y$ is the average monthly benefit in dollars and $x$ is the year, with $x = 5$ corresponding to 1995. (Source: U.S. Social Security Administration)

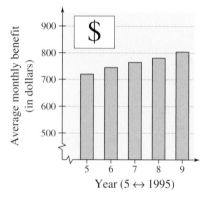

Year (5 ↔ 1995)

**73.** In which year was the average monthly benefit approximately $780?

**74.** If the model continues to be accurate, in what year will the average monthly benefit reach $1000?

*Probability*   In Exercises 75 and 76, use the equation $p + p' = 1$, which states that the sum of the probability that an event *will* occur $p$ and the probability that it *will not* occur $p'$ is 1.

**75.** If the probability that an event will occur is 0.54, what is the probability that the event will not occur?

**76.** If the probability that an event will not occur is 0.87, what is the probability that the event will occur?

## 1.2    Mathematical Modeling

### Objectives

- Construct a mathematical model from a verbal model.
- Model and solve percent and mixture problems.
- Use common formulas to solve geometry and simple interest problems.
- Develop a general problem-solving strategy.

## Introduction to Problem Solving

In this section, you will study ways of using algebra to solve real-life problems. To do this, you will construct one or more equations that represent the real-life problem. This procedure is called **mathematical modeling.**

A good approach to mathematical modeling is to use two stages. In the first stage, the verbal description of the problem is used to form a *verbal model.* Then, after assigning labels to each of the quantities in the verbal model, you form a *mathematical model* or *algebraic equation.*

Verbal description ⟹ Verbal model ⟹ Algebraic equation

When you are trying to construct a verbal model, it is helpful to look for a *hidden equality*—a statement that two algebraic expressions are equal. For instance, in the following example the hidden equality equates your annual income to 24 paychecks and one bonus check.

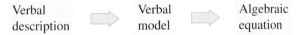

**Example 1**    Using a Verbal Model

You have accepted a job for which your annual salary will be $24,740. This salary includes a year-end bonus of $500. If you are paid twice a month, what will your gross pay be for each paycheck?

### Solution

Because there are 12 months in a year and you will be paid twice a month, it follows that you will receive 24 paychecks during the year.

*Verbal Model:*    Income for year  =  24 paychecks  +  Bonus

*Labels:*    Income for year = 24,740    (dollars)
Amount of each paycheck = $x$    (dollars)
Bonus = 500    (dollars)

*Equation:*    $24,740 = 24x + 500$

Using the techniques discussed in Section 1.1, you will find that the solution is $x = \$1010$. Check that your answer is reasonable. Is $1010 a reasonable answer if you are paid twice a month and earn $24,740 a year?

▶ Translating Key Words and Phrases

| Key Words and Phrases | Verbal Description | Algebraic Statement |
|---|---|---|
| **Consecutive** Next, subsequent | Consecutive integers | $n, n + 1$ |
| **Addition** Sum, plus, greater, increased by, more than, exceeds, total of | The sum of 5 and $x$ Seven more than $y$ | $5 + x$ $y + 7$ |
| **Subtraction** Difference, minus, less than, decreased by, subtracted from, reduced by, the remainder | Four decreased by $b$ Three less than $z$ | $4 - b$ $z - 3$ |
| **Multiplication** Product, multiplied by, twice, times, percent of | Two times $x$ | $2x$ |
| **Division** Quotient, divided by, per | The quotient of $x$ and 8 | $\dfrac{x}{8}$ |

*22,400 - 100*
*x    - 109*

**Example 2**   Constructing Mathematical Models

**a.** A salary of $22,400 is increased by 9%. What is the new salary?

*Verbal Model:*   New salary $=$ 9%(salary) $+$ Salary

*Labels:*   Salary $= 22,400$                                (dollars)
New salary $= S$                               (dollars)
Percent $= 0.09$                   (percent in decimal form)

*Equation:*   $S = 0.09(22,400) + 22,400$

**b.** All items of clothing in a store are reduced by 20%. Find the original price of a suit selling for $140.

*Verbal Model:*   Original price $-$ 20%(original price) $=$ Sale price

*Labels:*   Original price $= p$                                (dollars)
Sale price $= 140$                              (dollars)
Percent $= 0.2$                   (percent in decimal form)

*Equation:*   $p - 0.2p = 140$

## Using Mathematical Models

Study the next several examples carefully. Your goal should be to develop a *general problem-solving strategy.*

**Example 3**    Finding the Percent of a Raise

You accept a job that pays $8 an hour. You are told that after a 2-month probationary period, your hourly wage will be increased to $9 an hour. What percent raise will you receive after the 2-month period?

$8 - 100$

**Solution**

$1 - x$      $x = \frac{1 \cdot 100}{8} = 12.5$

*Verbal Model:*        Raise  =  Percent  ·  Old wage

*Labels:*        Old wage = 8                                        (dollars)
              Raise = 1                                        (dollar)
              Percent = r                        (percent in decimal form)

*Equation:*    1 = r · 8

By solving this equation, you will find that you will receive a raise of $\frac{1}{8}$ = 0.125 or 12.5%.

**Example 4**    Finding the Percent of a Benefit Package

Your annual salary is $22,000. In addition to your salary, your employer also provides the following benefits. The total of this benefit package is equal to what percent of your annual salary?

| | | |
|---|---|---|
| Social security (employer's portion): | 6.20% of salary | $1364 |
| Workmen's compensation: | 0.5% of salary | $110 |
| Unemployment compensation: | 0.75% of salary | $165 |
| Medical insurance: | $2240 per year | $2240 |
| Retirement contribution: | 5% of salary | $1100 |

**Solution**

*Verbal Model:*        Benefit package  =  Percent  ·  Salary

*Labels:*        Salary = 22,000                                        (dollars)
              Benefit package = 4979                                (dollars)
              Percent = r                        (percent in decimal form)

*Equation:*    4979 = r · 22,000

By solving this equation, you will find that your benefit package is equal to r = 4979/22,000, or about 22.6% of your salary.

*In 1998, 16.3% of the population of the United States had no health insurance.* (Source: U.S. Bureau of the Census)

Example 5    Finding the Measurements of a Room

A rectangular family room is twice as long as it is wide, and its perimeter is 84 feet. Find the measurements of the family room.

Solution

For this problem, it helps to sketch a picture (see Figure 1.2).

*Verbal Model:*    $2 \cdot$ Length $+ 2 \cdot$ Width $=$ Perimeter

*Labels:*    Perimeter $= 84$                                      (feet)
Width $= w$                                               (feet)
Length $= l = 2w$                                         (feet)

*Equation:*    $2(2w) + 2w = 84$

$4w + 2w = 84$

$6w = 84$

$w = 14$ feet

$l = 2w = 28$ feet

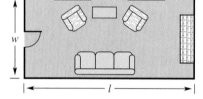

Figure 1.2

The measurements of the room are 14 feet by 28 feet.

Example 6    A Distance Problem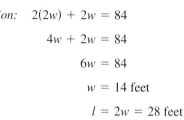

A plane is flying nonstop from New York to San Francisco, a distance of about 2700 miles, as shown in Figure 1.3. After 1.5 hours in the air, the plane flies over Chicago (a distance of 800 miles from New York). How long will it take the plane to fly from New York to San Francisco? (Assume that the plane flies at a constant speed during the entire flight.)

Solution

To solve this problem, use the formula that relates distance, rate, and time. That is, (distance) = (rate)(time). Because it took the plane 1.5 hours to travel a distance of 800 miles, you can conclude that its rate (or speed) must have been

$$\text{Rate} = \frac{\text{distance}}{\text{time}} = \frac{800 \text{ miles}}{1.5 \text{ hours}} \approx 533.33 \text{ miles per hour.}$$

Because the entire trip is 2700 miles, the time for the entire trip is

$$\text{Time} = \frac{\text{distance}}{\text{rate}} = \frac{2700 \text{ miles}}{533.33 \text{ miles per hour}} \approx 5.06 \text{ hours.}$$

Because 0.06 hours represents about 4 minutes, you can conclude that the trip will take about 5 hours and 4 minutes.

Figure 1.3

Another way to solve the distance problem in Example 6 is to use the concept of **ratio and proportion.** To do this, let $x$ represent the time required to fly from New York to San Francisco, and set up the following proportion.

$$\frac{\text{Time to San Francisco}}{\text{Time to Chicago}} = \frac{\text{Distance to San Francisco}}{\text{Distance to Chicago}}$$

$$\frac{x}{1.5} = \frac{2700}{800}$$

$$x = 1.5 \cdot \frac{2700}{800}$$

$$x \approx 5.06$$

Notice how ratio and proportion are used with a result from geometry to solve the problem in the following example.

**Example 7**   An Application Involving Similar Triangles

To determine the height of the Aon Center Building (in Chicago), you measure the shadow cast by the building to be 142 feet long, as shown in Figure 1.4. Then you measure the shadow cast by a 4-foot post and find that its shadow is 6 inches long. How can this information be used to determine the height of the Aon Center Building?

Solution

To find the height of the building, you can use a result from geometry that states that the ratios of corresponding sides of similar triangles are equal.

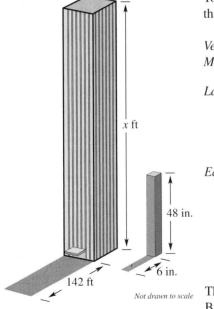

*Verbal Model:*    $\dfrac{\text{Height of building}}{\text{Length of shadow}} = \dfrac{\text{Height of post}}{\text{Length of shadow}}$

*Labels:*

| | |
|---|---|
| Height of building $= x$ | (feet) |
| Length of building's shadow $= 142$ | (feet) |
| Height of post $= 48$ | (inches) |
| Length of post's shadow $= 6$ | (inches) |

*Equation:*    $\dfrac{x}{142} = \dfrac{48}{6}$

$$x = 142 \cdot \frac{48}{6}$$

$$x = 142 \cdot 8$$

$$x = 1136 \text{ feet}$$

The Aon Center Building is about 1136 feet high. (Formerly called the Amoco Oil Building, it is the second highest building in Chicago. The highest is the Sears Tower, which is the tallest building in North America.)

*x* ft

48 in.

6 in.

142 ft

*Not drawn to scale*

Figure 1.4

## Mixture Problems

The next example is called a **mixture problem** because it involves two different unknown quantities that are *mixed* in a specific way. Watch for a *hidden product* in the verbal model.

**Example 8**    A Simple Interest Problem

You have received an inheritance of $10,000 that has been invested in two ways. Part of the money was invested at $9\frac{1}{2}\%$ simple interest and the remainder was invested at 11%. After 1 year, the two investments have returned a combined interest of $1038.50. How much was invested in each type of account?

**Solution**

Simple interest problems are based on the formula $I = Prt$, where $I$ is the interest, $P$ is the principal, $r$ is the annual percentage rate (in decimal form), and $t$ is the time in years.

*Verbal Model:*    $\boxed{\begin{array}{c}\text{Interest}\\\text{from }9\frac{1}{2}\%\end{array}}$ $+$ $\boxed{\begin{array}{c}\text{Interest}\\\text{from }11\%\end{array}}$ $=$ $\boxed{\begin{array}{c}\text{Total}\\\text{interest}\end{array}}$

*Labels:*

| | |
|---|---|
| Amount invested at $9\frac{1}{2}\% = x$ | (dollars) |
| Amount invested at $11\% = 10,000 - x$ | (dollars) |
| Interest from $9\frac{1}{2}\% = Prt = (x)(0.095)(1)$ | (dollars) |
| Interest from $11\% = Prt = (10,000 - x)(0.11)(1)$ | (dollars) |
| Total interest $= 1038.50$ | (dollars) |

*Equation:*

$$0.095x + 0.11(10,000 - x) = 1038.5$$

$$0.095x + 1100 - 0.11x = 1038.5$$

$$-0.015x = -61.5$$

$$x = \$4100$$

So, the amount invested at $9\frac{1}{2}\%$ is $4100 and the amount invested at 11% is

$$10,000 - x = 10,000 - 4100 = \$5900.$$

Check these results in the original statement of the problem.

In Example 8, did you recognize the hidden products in the two terms on the left side of the equation? Both hidden products come from the common formula

$$\boxed{\text{Interest}} = \boxed{\text{Principal}} \cdot \boxed{\text{Rate}} \cdot \boxed{\text{Time}}$$

$$I = Prt.$$

# Common Formulas

Many common types of geometric, scientific, and investment problems use ready-made equations, called **formulas.** Knowing formulas such as those in the following lists will help you translate and solve a wide variety of real-life problems involving perimeter, area, volume, temperature, interest, and principal.

---

▶ **Common Formulas for Area, Perimeter, and Volume**

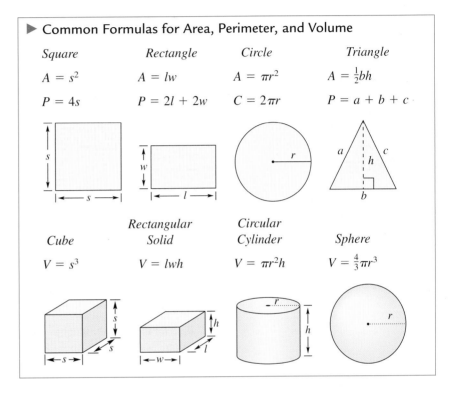

| *Square* | *Rectangle* | *Circle* | *Triangle* |
|---|---|---|---|
| $A = s^2$ | $A = lw$ | $A = \pi r^2$ | $A = \frac{1}{2}bh$ |
| $P = 4s$ | $P = 2l + 2w$ | $C = 2\pi r$ | $P = a + b + c$ |

| *Cube* | *Rectangular Solid* | *Circular Cylinder* | *Sphere* |
|---|---|---|---|
| $V = s^3$ | $V = lwh$ | $V = \pi r^2 h$ | $V = \frac{4}{3}\pi r^3$ |

---

▶ **Miscellaneous Common Formulas**

*Temperature:*    $F$ = degrees Fahrenheit, $C$ = degrees Celsius

$$F = \frac{9}{5}C + 32$$

*Simple interest:*  $I$ = interest, $P$ = principal, $r$ = interest rate, $t$ = time

$$I = Prt$$

*Distance:*    $d$ = distance traveled, $r$ = rate, $t$ = time

$$d = rt$$

When working with applied problems, you often need to rewrite one of the common formulas. For instance, the formula $P = 2l + 2w$ for the perimeter of a rectangle can be rewritten or solved for $w$ to produce $w = \frac{1}{2}(P - 2l)$.

**Example 9**    Using a Formula

A cylindrical can has a volume of 300 cubic centimeters and a radius of 3 centimeters, as shown in Figure 1.5. Find the height of the can.

**Solution**

The formula for the *volume of a cylinder* is $V = \pi r^2 h$. To find the height of the can, solve for $h$. Then, using $V = 300$ centimeters$^3$ and $r = 3$ centimeters, find the height.

$$h = \frac{V}{\pi r^2} = \frac{300 \text{ centimeters}^3}{\pi (3 \text{ centimeters})^2} = \frac{300 \text{ centimeters}^3}{9\pi \text{ centimeters}^2} \approx 10.61 \text{ centimeters}$$

3 cm

*h*

**Figure 1.5**

▶ **Strategy for Solving Word Problems**

1. *Search* for the hidden equality—two expressions said to be equal or known to be equal. A sketch may be helpful.

2. *Write* a verbal model that equates these two expressions.

3. *Assign* numbers to the known quantities and letters (or algebraic expressions) to the variable quantities.

4. *Rewrite* the verbal model as an algebraic equation using the assigned labels.

5. *Solve* the resulting algebraic equation.

6. *Check* to see that the answer satisfies the word problem as stated. (Remember that "solving for *x*" or some other variable may not completely answer the question.)

**Discussing the Concept    Red Herrings**

Applied problems in textbooks usually give precisely the right amount of information that is necessary to solve a given problem. In real life, however, you must often sort through the given information and discard information that is irrelevant to the problem. Such irrelevant information is called a **red herring.** Find the red herrings in the following problem.

From 100 to 200 feet beneath the surface of the ocean, pressure changes at a rate of approximately 4.4 pounds per square inch for every 10-foot change in depth. A diver takes 30 minutes to ascend 25 feet from a depth of 150 feet. What change in pressure does the diver experience?

# Warm Up

The following warm-up exercises involve skills that were covered in previous courses and earlier sections. You will use these skills in the exercise set for this section.

In Exercises 1–10, solve the equation (if possible) and check your answer.

1. $3x - 42 = 0$
2. $64 - 16x = 0$
3. $2 - 3x = 14 + x$
4. $7 + 5x = 7x - 1$
5. $5[1 + 2(x + 3)] = 6 - 3(x - 1)$
6. $2 - 5(x - 1) = 2[x + 10(x - 1)]$
7. $\dfrac{x}{3} + \dfrac{x}{2} = \dfrac{1}{3}$
8. $\dfrac{2}{x} + \dfrac{2}{5} = 1$
9. $1 - \dfrac{2}{z} = \dfrac{z}{z + 3}$
10. $\dfrac{x}{x + 1} - \dfrac{1}{2} = \dfrac{4}{3}$

## 1.2    Exercises

*Creating a Mathematical Model*    In Exercises 1–10, write an algebraic expression for the verbal expression.

1. The sum of two consecutive natural numbers

2. The product of two natural numbers whose sum is 25

3. *Distance Traveled*    The distance traveled in $t$ hours by a car traveling at 50 miles per hour

4. *Travel Time*    The travel time for a plane that is traveling at a rate of $r$ miles per hour for 200 miles

5. *Acid Solution*    The amount of acid in $x$ gallons of a 20% acid solution

6. *Discount*    The sale price for an item that is discounted by 20% of its list price $L$

7. *Geometry*    The perimeter of a rectangle whose width is $x$ and whose length is twice the width

8. *Geometry*    The area of a triangle whose base is 20 inches and whose height is $h$ inches

9. *Total Cost*    The total cost of producing $x$ units for which the fixed costs are $1200 and the cost per unit is $25

10. *Total Revenue*    The total revenue obtained by selling $x$ units at $3.59 per unit

*Using a Mathematical Model*    In Exercises 11–16, write a mathematical model for the number problem, and solve the problem.

11. The sum of two consecutive natural numbers is 525. Find the two numbers.

12. Find three consecutive natural numbers whose sum is 804.

13. One positive number is five times another positive number. The difference between the two numbers is 148. Find the numbers.

14. One number is one-fifth of another number. The difference between the two numbers is 76. Find the numbers.

15. Find two consecutive integers whose product is 5 less than the square of the smaller number.

16. Find two consecutive natural numbers such that the difference of their reciprocals is one-fourth the reciprocal of the smaller number.

**17.** *Weekly Paycheck*    Your weekly paycheck is 15% *more* than your coworker's. Your two paychecks total $645. Find the amount of each paycheck.

**18.** *Monthly Profit*    The total profit for a company in February was 20% *higher* than it was in January. The total profit for the 2 months was $157,498. Find the profit for each month.

**19.** *Weekly Paycheck*    Your weekly paycheck is 15% *less* than your coworker's. Your two paychecks total $645. Find the amount of each paycheck.

**20.** *Monthly Profit*    The total profit for a company in February was 20% *lower* than it was in January. The total profit for the 2 months was $157,498. Find the profit for each month.

*Price Comparison*    In Exercises 21–24, the prices of different items are given for 1917 and 1994. Find the percent increase or decrease for each item.    (Source: *Everything Has Its Price*)

| Item | 1917 | 1994 |
|---|---|---|
| **21.** Three-minute phone call from New York to Los Angeles | $20.70 | $0.75 |
| **22.** Chocolate bar | $0.02 | $0.45 |
| **23.** College tuition | $20.00 | $3880.00 |
| **24.** Movie ticket | $0.15 | $5.45 |

**25.** *Television Viewing*    In 1991 the average time spent watching television per American TV home per day was 7.0 hours. By 1995 the average time had increased by about 4.05%, and by 1999 the average time had increased (over the 1995 time) by an additional 2.06%.    (Source: Nielsen Media Research)

   (a)  Find the average time in 1995.

   (b)  Find the average time in 1999.

**26.** *Highest Football Salaries*    In 1970 the highest paid professional football player (Joe Namath) had a salary of $150,000 per year. By 1980 the highest paid player (Walter Payton) had a salary that was 216.67% higher. By 1990 the highest paid player (Joe Montana) had a salary that was 742.11% higher than the 1980 high. By 2001 the highest paid player (Drew Bledsoe) had a salary that was 427.05% higher than the 1990 high. Find the salaries of Walter Payton in 1980, Joe Montana in 1990, and Drew Bledsoe in 2001.    (Source: *Sport Magazine, The Arizona Republic*)

**27.** *Retirement Nest Eggs*    Based on a survey of 740 companies, the most common types of investments by corporate retirement plans are those shown in the bar graph. How many respondents in the survey used each type of investment? (Most used more than one type of investment.)    (Source: Wyatt Company)

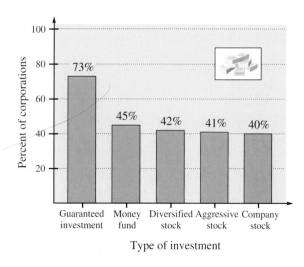

**28.** *School Lunch Snacks*    Based on a survey of 426 mothers, the most common types of snacks put in children's school lunches are those shown in the bar graph. How many respondents put each type of snack in their children's lunches? (Most put in more than one type of snack.)    (Source: Nabisco Brands)

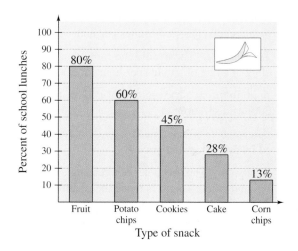

29. *Enjoying Nature*    The Defenders of Wildlife organization conducted a survey in which 300 adults were asked how they enjoyed nature. (Respondents were allowed to answer in more than one category). Of the 300 people, 189 answered "feeding birds," 186 answered "watching wildlife in neighborhood," 129 answered "driving to wildlife site," 108 answered "camping, biking, photography," and 105 answered "hiking." Suppose that one of the 300 people was chosen at random. What is the probability that the person said that feeding birds was a way that he or she enjoyed nature? (Source: Defenders of Wildlife)

30. *Business Headaches*    In a survey, 414 small-business owners were asked what their biggest business headache was. (Respondents were allowed to answer in only one category.) The responses were as follows.

| | |
|---|---|
| Government regulations | 128 |
| Employee relations | 83 |
| Cash flow | 46 |
| Insurance | 37 |
| Paperwork | 25 |
| Taxes | 25 |
| Other | 70 |

Suppose that one of the 414 people was chosen at random. What is the probability that the person said that government regulations was his or her biggest headache? (Source: MasterCard)

31. *Geometry*    A room is 1.5 times as long as it is wide, and its perimeter is 75 feet (see figure). Find the measurements of the room.

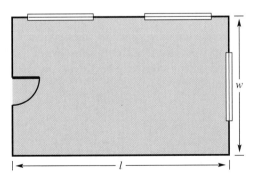

32. *Geometry*    A picture frame has a total perimeter of 3 feet (see figure). The width of the frame is 0.62 times its length. Find the measurements of the frame.

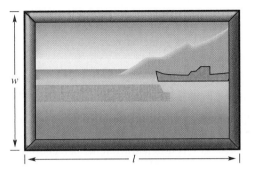

33. *Simple Interest*    You invest $2500 at 7% simple interest. How many years will it take for the investment to earn $1000 in interest?

34. *Simple Interest*    An investment earns $3200 interest over a 7-year period. What is the rate of interest on a $4800 principal investment?

35. *Course Grade*    To get an A in a course you must have an average of at least 90% on four tests that have 100 points each. If your scores on the first three tests were 87, 92, and 84, what must you score on the fourth test to get an A for the course?

36. *Course Grade*    Suppose you are taking a course that has four tests. The first three tests have 100 points each and the fourth test has 200 points. To get an A in the course you must have an average of at least 90% on the four tests. Your scores on the first three tests were 87, 92, and 84. What must you score on the fourth test to get an A for the course?

37. *Loan Payments*    A family has annual loan payments totaling $13,077.75, or 58.6% of its annual income. What is the family's income?

38. *Food Budget*    Suppose your annual budget for food (eating at home and eating at restaurants) is 23.2% of your annual income. During the year, you spent $4832.12 on food. What is your annual income?

39. *List Price*    The price of a swimming pool has been discounted 16.5%. The sale price is $849. Find the original list price of the pool.

40. *List Price*    The price of a compact disc player has been discounted 27.5%. The sale price is $628. Find the original list price of the player.

41. *Discount Rate*    The price of a stereo has been discounted by $150. The sale price is $245. What percent is the discount of the original list price?

**42.** *Discount Rate*   The price of a shirt has been discounted by $20. The sale price is $29.95. What percent is the discount of the original list price?

*Weekly Salary*   In Exercises 43 and 44, use the following information to write a mathematical model and solve. Due to economic factors, your employer has reduced your weekly wage by 15%. Before the reduction, your weekly salary was $425.

**43.** What is your reduced salary?

**44.** What percent raise must you receive to bring your weekly salary back up to $425? Explain why the percent raise is different from the percent reduction.

**45.** *Travel Time*   Suppose you are driving on a freeway to another town that is 150 miles from your home. After 30 minutes, you pass a freeway exit that you know is 25 miles from your home. Assuming that you continue at the same constant speed, how long will it take for the entire trip?

**46.** *Travel Time*   A plane is flying from Orlando to Denver, a distance of about 1950 miles. After 1 hour and 15 minutes, the plane flies over a town that is 600 miles from Orlando. How long will it take the plane to fly from Orlando to Denver? (Assume that the plane flies at a constant speed the entire flight.)

**47.** *Travel Time*   Two cars start at a given point and travel in the same direction at average speeds of 40 miles per hour and 55 miles per hour. How much time must elapse before the cars are 5 miles apart?

**48.** *Catch-Up Time*   Students are traveling in two cars to a football game 135 miles away. The first car leaves on time and travels at an average speed of 45 miles per hour. The second car starts $\frac{1}{2}$ hour later and travels at an average speed of 55 miles per hour. At these speeds, how long will it take the second car of students to catch up to the first car?

**49.** *Travel Time*   Two families meet at a park for a picnic. At the end of the day, one family travels east at an average speed of 42 miles per hour and the other family travels west at an average speed of 50 miles per hour. Both families have approximately 160 miles to travel. Find the time it takes each family to get home.

**50.** *Average Speed*   A truck traveled at an average speed of 55 miles per hour on a 200-mile trip to pick up a load of freight. On the return trip (with the truck fully loaded), the average speed was 40 miles per hour. Find the average speed for the round trip.

**51.** *Radio Waves*   Radio waves travel at the same speed as light, $3.0 \times 10^8$ meters per second. Find the time required for a radio wave to travel from mission control in Houston to NASA astronauts on the surface of the moon $3.86 \times 10^8$ meters away.

**52.** *Distance to a Star*   Find the distance to a star that is 50 light years (distance traveled by light in 1 year) away. (Light travels at 186,000 miles per second.)

**53.** *Height of a Tree*   Suppose that you want to measure the height of a tree that is in your yard. To do this, you measure the tree's shadow and find that it is 25 feet long. You also measure the shadow of a 5-foot lamppost and find its shadow to be 2 feet long (see figure). How tall is the tree?

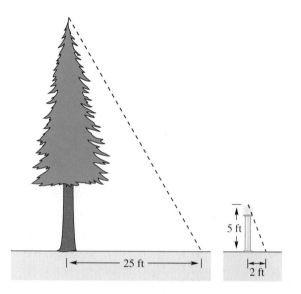

**54.** *Height of a Building*   Suppose that you want to measure the height of a building. To do this, you measure the building's shadow and find that it is 50 feet long. You also measure the shadow of a 4-foot stake and find its shadow to be $3\frac{1}{2}$ feet long. How tall is the building?

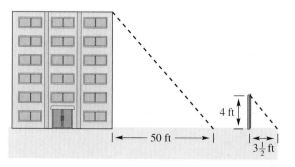

**55.** *Projected Expenses*   From January through May, a company's expenses have totaled $234,980. If the monthly expenses continue at this rate, what will the total expenses for the year be?

**56.** *Projected Revenue*   From January through August, a company's revenues have totaled $345,950. If the monthly revenue continues at this rate, what will the total revenue for the year be?

**57.** *Investment Mix*   Suppose you invest $12,000 in two funds paying $10\frac{1}{2}\%$ and 13% simple interest. The total annual interest is $1447.50. How much is invested in each fund?

**58.** *Investment Mix*   Suppose you invest $25,000 in two funds paying 11% and $12\frac{1}{2}\%$ simple interest. The total annual interest is $2975.00. How much is invested in each fund?

**59.** *Comparing Investment Returns*   You invest $12,000 in a fund paying $9\frac{1}{2}\%$ simple interest and $8000 in a fund for which the interest rate is variable. At the end of the year you receive notification that the total interest for both funds is $2054.40. Find the equivalent simple interest rate on the variable rate fund.

**60.** *Comparing Investment Returns*   You have $10,000 on deposit earning simple interest that is linked to the prime rate. When the prime rate dropped, your rate dropped by $1\frac{1}{2}\%$ for the last quarter of the year. Your total annual interest was $1112.50. What was your interest rate for the first three quarters and for the last quarter?

**61.** *Production Limit*   A company has fixed costs of $10,000 per month and variable costs of $8.50 per unit manufactured. The company has $85,000 available to cover the monthly costs. How many units can the company manufacture? (*Fixed costs* are those that occur regardless of the level of production. *Variable costs* depend on the level of production.)

**62.** *Production Limit*   A company has fixed costs of $10,000 per month and variable costs of $9.30 per unit manufactured. The company has $85,000 available to cover the monthly costs. How many units can the company manufacture? (*Fixed costs* are those that occur regardless of the level of production. *Variable costs* depend on the level of production.)

**63.** *Length of a Tank*   The diameter of a cylindrical propane gas tank is 4 feet (see figure). The total volume of the tank is 603.2 cubic feet. Find the length of the tank.

**64.** *Water Depth*   A trough is 12 feet long, 3 feet deep, and 3 feet wide (see figure). Find the depth of the water when the trough contains 70 gallons. (1 gallon ≈ 0.13368 cubic foot.)

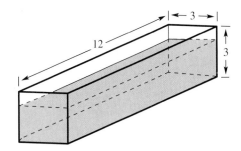

**65.** *Mixture*   A 55-gallon barrel contains a mixture with a concentration of 40%. How much of this mixture must be withdrawn and replaced by 100% concentrate to bring the mixture up to 75% concentration? (See figure.)

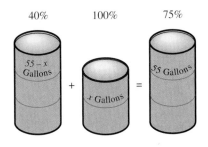

**66.** *Mixture*     A farmer mixed gasoline and oil to make 2 gallons of mixture for his two-cycle chain saw engine. This mixture was 32 parts gasoline and 1 part two-cycle oil. How much gasoline must be added to bring the mixture to 40 parts gasoline and 1 part oil?

**67.** Sometimes, when developing a mathematical model for a problem, the most accurate equation is not used as the model. Explain why and include factors that influence a mathematical model's selection.

*New York City Marathon*     In Exercises 68 and 69, find the average speed of the record-holding runners in the New York City Marathon. The length of the course is 26 miles, 385 yards. (Note that 1 mile = 5280 feet = 1760 yards.)

**68.** Men's record time: 2 hours, 8 minutes

**69.** Women's record time: 2 hours, $24\frac{2}{5}$ minutes

In Exercises 70–83, solve for the variable.

**70.** *Area*     Solve for $h$ in $A = \frac{1}{2}bh$.

**71.** *Perimeter*     Solve for $l$ in $P = 2l + 2w$.

**72.** *Volume*     Solve for $l$ in $V = lwh$.

**73.** *Volume*     Solve for $h$ in $V = \pi r^2 h$.

**74.** *Markup*     Solve for $C$ in $S = C + RC$.

**75.** *Discount*     Solve for $L$ in $S = L - RL$.

**76.** *Interest*     Solve for $r$ in $A = P + Prt$.

**77.** *Interest*     Solve for $P$ in $A = P\left(1 + \dfrac{r}{n}\right)^{nt}$.

**78.** *Area*     Solve for $b$ in $A = \frac{1}{2}(a + b)h$.

**79.** *Area*     Solve for $\theta$ in $A = \dfrac{\pi r^2 \theta}{360}$.

**80.** *Sequence*     Solve for $n$ in $L = a + (n - 1)d$.

**81.** *Series*     Solve for $r$ in $S = \dfrac{rL - a}{r - 1}$.

**82.** *Lateral Surface Area*     Solve for $h$ in $A = 2\pi rh$.

**83.** *Surface Area*     Solve for $r$ in $A = 4\pi r^2$.

## 1.3    Quadratic Equations

### Objectives

- Solve a quadratic equation by factoring.
- Solve a quadratic equation by extracting square roots.
- Construct and use a quadratic model to solve an application problem.

## Solving Quadratic Equations by Factoring

In the first two sections of this chapter, you studied linear equations in one variable. In this and the next section, you will study quadratic equations.

> ▶ **Definition of a Quadratic Equation**
>
> A **quadratic equation** in $x$ is an equation that can be written in the standard form
>
> $$ax^2 + bx + c = 0$$
>
> where $a$, $b$, and $c$ are real numbers with $a \neq 0$. Another name for a quadratic equation in $x$ is a **second-degree polynomial equation in $x$.**

There are three basic techniques for solving quadratic equations: factoring, extracting square roots, and the Quadratic Formula. (The Quadratic Formula is discussed in the next section.) The first technique is based on the Zero-Factor Property given in the appendix.

If $ab = 0$, then $a = 0$ or $b = 0$.          Zero-Factor Property

To use this property, rewrite the left side of the standard form of a quadratic equation as the product of two linear factors. Then find the solutions of the quadratic equation by setting each factor equal to zero.

### Study Tip

The Zero-Factor Property applies *only* to equations written in standard form (in which one side of the equation is zero). So, be sure that all terms are collected on one side *before* factoring. For instance, in the equation

$$(x - 5)(x + 2) = 8$$

it is *incorrect* to set each factor equal to 8. Can you solve this equation correctly?

**Example 1**    Solving a Quadratic Equation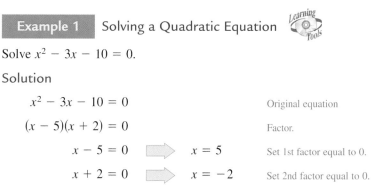

Solve $x^2 - 3x - 10 = 0$.

**Solution**

| | |
|---|---|
| $x^2 - 3x - 10 = 0$ | Original equation |
| $(x - 5)(x + 2) = 0$ | Factor. |
| $x - 5 = 0 \implies x = 5$ | Set 1st factor equal to 0. |
| $x + 2 = 0 \implies x = -2$ | Set 2nd factor equal to 0. |

The solutions are 5 and $-2$. Check these in the original equation.

| Example 2 | Solving a Quadratic Equation by Factoring |

Solve $6x^2 - 3x = 0$.

**Solution**

$$6x^2 - 3x = 0 \qquad \text{Original equation}$$

$$3x(2x - 1) = 0 \qquad \text{Factor.}$$

$$3x = 0 \quad \Longrightarrow \quad x = 0 \qquad \text{Set 1st factor equal to 0.}$$

$$2x - 1 = 0 \quad \Longrightarrow \quad x = \tfrac{1}{2} \qquad \text{Set 2nd factor equal to 0.}$$

The solutions are 0 and $\tfrac{1}{2}$. Check these by substituting in the original equation, as follows.

**Check**

$$6x^2 - 3x = 0 \qquad \text{Original equation}$$

$$6(0)^2 - 3(0) \stackrel{?}{=} 0 \qquad \text{Substitute 0 for } x.$$

$$0 - 0 = 0 \qquad \text{First solution checks. } \checkmark$$

$$6\left(\tfrac{1}{2}\right)^2 - 3\left(\tfrac{1}{2}\right) \stackrel{?}{=} 0 \qquad \text{Substitute } \tfrac{1}{2} \text{ for } x.$$

$$\tfrac{6}{4} - \tfrac{3}{2} = 0 \qquad \text{Second solution checks. } \checkmark$$

If the two factors of a quadratic expression are the same, the corresponding solution is a **double** or **repeated** solution.

| Example 3 | A Quadratic Equation with a Repeated Solution |

Solve $9x^2 - 6x = -1$.

**Solution**

$$9x^2 - 6x = -1 \qquad \text{Original equation}$$

$$9x^2 - 6x + 1 = 0 \qquad \text{Write in standard form.}$$

$$(3x - 1)^2 = 0 \qquad \text{Factor.}$$

$$3x - 1 = 0 \qquad \text{Set repeated factor equal to 0.}$$

$$x = \tfrac{1}{3} \qquad \text{Solution}$$

The only solution is $\tfrac{1}{3}$. Check this in the original equation.

**• TECHNOLOGY •**

To check the solution in Example 3 with your graphing utility, you should first write the equation in standard form.

$$9x^2 - 6x + 1 = 0$$

Then enter the expression $9x^2 - 6x + 1$ into $y_1$ of the equation editor. Now you can use the ASK mode of the table feature of your graphing utility to check the solution.

## Extracting Square Roots

There is a nice shortcut for solving equations of the form $u^2 = d$, where $d > 0$. By factoring, you can see that this equation has two solutions.

$$u^2 = d \qquad \text{Original equation}$$

$$u^2 - d = 0 \qquad \text{Standard form}$$

$$\left(u + \sqrt{d}\right)\left(u - \sqrt{d}\right) = 0 \qquad \text{Factor.}$$

$$u + \sqrt{d} = 0 \quad \Longrightarrow \quad u = -\sqrt{d} \qquad \text{Set 1st factor equal to 0.}$$

$$u - \sqrt{d} = 0 \quad \Longrightarrow \quad u = \sqrt{d} \qquad \text{Set 2nd factor equal to 0.}$$

Solving an equation of the form $u^2 = d$ without going through the steps of factoring is called **extracting square roots.**

---

▶ **Extracting Square Roots**

The equation $u^2 = d$, where $d > 0$, has exactly two solutions:

$$u = \sqrt{d} \qquad \text{and} \qquad u = -\sqrt{d}.$$

These solutions can also be written as $u = \pm\sqrt{d}.$

---

**Example 4**    Extracting Square Roots

Solve $4x^2 = 12$.

**Solution**

$$4x^2 = 12 \qquad \text{Original equation}$$

$$x^2 = 3 \qquad \text{Divide both sides by 4.}$$

$$x = \pm\sqrt{3} \qquad \text{Extract square roots.}$$

The solutions are $\sqrt{3}$ and $-\sqrt{3}$. Check these in the original equation.

**Example 5**    Extracting Square Roots

$$(x - 3)^2 = 7 \qquad \text{Original equation}$$

$$x - 3 = \pm\sqrt{7} \qquad \text{Extract square roots.}$$

$$x = 3 \pm \sqrt{7} \qquad \text{Add 3 to both sides.}$$

The solutions are $3 \pm \sqrt{7}$. Check these in the original equation.

## Ancient Forms of Quadratic Equations

In ancient and medieval times, quadratic equations were classified in three categories.

$$x^2 + px = q$$
$$x^2 = px + q$$
$$x^2 + q = px$$

Solutions of these types of quadratic equations have been found in Babylonian texts over 4000 years old. Quadratic equations of the form $x^2 + px + q = 0$, where $p$ and $q$ are positive, were not solved because such equations have no positive solutions.

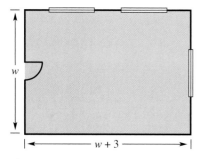

**Figure 1.6**

## Applications

Quadratic equations often occur in problems dealing with area. Here is a simple example.

> A square room has an area of 144 square feet. Find the measurements of the room.

To solve this problem, you can let $x$ represent the length of each side of the room. Then, by solving the equation

$$x^2 = 144$$

you can conclude that each side of the room is 12 feet long. Note that although the equation $x^2 = 144$ has two solutions, $-12$ and $12$, the negative solution makes no sense (for this problem), so you should choose the positive solution.

**Example 6**    Finding the Measurements of a Room

A bedroom is 3 feet longer than it is wide (see Figure 1.6) and has an area of 154 square feet. Find the measurements of the room.

### Solution

You can begin by using the same type of problem-solving strategy that was presented in Section 1.2.

| *Verbal Model:* | Width of room | · | Length of room | = | Area of room |
|---|---|---|---|---|---|

| *Labels:* | Area of room $= 154$ | (square feet) |
|---|---|---|
| | Width of room $= w$ | (feet) |
| | Length of room $= w + 3$ | (feet) |

$$\text{Equation:} \qquad w(w + 3) = 154$$
$$w^2 + 3w - 154 = 0$$
$$(w - 11)(w + 14) = 0$$
$$w - 11 = 0 \implies w = 11$$
$$w + 14 = 0 \implies w = -14$$

Choosing the positive value, you can conclude that the width is 11 feet and the length is $w + 3 = 14$ feet. You can check this in the original statement of the problem as follows.

### Check

The length of 14 feet is 3 feet more than the width of 11 feet. ✓
The area of the room is $(11)(14) = 154$ square feet. ✓

Another application of quadratic equations involves an object that is falling (or vertically projected into the air). The equation that gives the height of such an object is called a **position equation,** and on Earth's surface it has the form

$$s = -16t^2 + v_0t + s_0.$$

In this equation, $s$ represents the height of the object (in feet), $v_0$ represents the initial velocity of the object (in feet per second), $s_0$ represents the initial height of the object (in feet), and $t$ represents the time (in seconds).

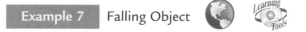

**Example 7**    Falling Object

A construction worker is working on the 24th floor of a building project, as shown in Figure 1.7. The worker accidentally drops a wrench and immediately yells "Look out below!" Could a person at ground level hear this warning in time to get out of the way of the falling wrench?

### Solution

Let's assume that each floor of the building is 10 feet high, so that the wrench was dropped from a height of 240 feet. Because sound travels at about 1100 feet per second, you can assume that a person at ground level hears the warning within 1 second of the time the wrench is dropped. To set up a mathematical model for the height of the wrench, you can use the position equation

$$s = -16t^2 + v_0t + s_0.$$

Because the object is dropped (rather than thrown), you can conclude that the initial velocity is $v_0 = 0$. Moreover, because the initial height is $s_0 = 240$ feet, you have the following model.

$$s = -16t^2 + 240$$

After falling for 1 second, the wrench is at a height of $-16(1^2) + 240 = 224$ feet. After falling for 2 seconds, the wrench is at a height of $-16(2^2) + 240 = 176$ feet. To find the number of seconds it takes the wrench to hit the ground, let the height $s$ be zero and solve the resulting equation for $t$.

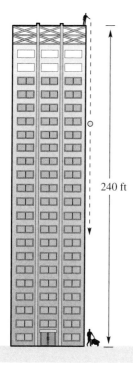

240 ft

$$s = -16t^2 + 240 \qquad \text{Position equation}$$

$$0 = -16t^2 + 240 \qquad \text{Set height equal to 0.}$$

$$16t^2 = 240 \qquad \text{Add } 16t^2 \text{ to both sides.}$$

$$t^2 = 15 \qquad \text{Divide both sides by 16.}$$

$$t = \sqrt{15} \approx 3.87 \qquad \text{Extract positive square root.}$$

The wrench will take about 3.87 seconds to hit the ground. If the person heard the warning 1 second after the wrench was dropped, the person would still have almost 3 seconds to get out of the way.

**Figure 1.7**

A third type of application that often involves a quadratic equation is one dealing with the hypotenuse of a right triangle. Recall from geometry that the sides of a right triangle are related by a formula called the **Pythagorean Theorem.** This theorem states that if $a$ and $b$ are the lengths of the sides of the triangle and $c$ is the length of the hypotenuse (see Figure 1.8),

$$a^2 + b^2 = c^2. \qquad \text{Pythagorean Theorem}$$

Notice how this formula is used in the next example.

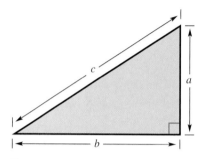

Figure 1.8

**Example 8** Cutting Across the Lawn

Your house is on a large corner lot. Several children in the neighborhood cut across your lawn in such a way that the distance across the lawn is 32 feet, as shown in Figure 1.9. Assuming that Figure 1.9 is drawn to scale, how many feet does a person save by walking across the lawn instead of walking on the sidewalk?

**Solution**

In Figure 1.9, let $x$ represent the length of the shorter part of the sidewalk. Using a ruler, it appears that the length of the longer part of the sidewalk is twice the shorter, so you can represent its length by $2x$.

$$x^2 + (2x)^2 = 32^2 \qquad \text{Pythagorean Theorem}$$
$$5x^2 = 1024 \qquad \text{Add like terms.}$$
$$x^2 = 204.8 \qquad \text{Divide both sides by 5.}$$
$$x = \sqrt{204.8} \qquad \text{Extract positive square root.}$$

The total distance on the sidewalk is

$$x + 2x = 3x = 3\sqrt{204.8} \approx 42.9 \text{ feet.}$$

So, cutting across the lawn saves a person about 10.9 feet of walking.

Figure 1.9

A fourth type of common application of a quadratic equation is one in which a quantity $y$ is changing over time $t$ according to a quadratic model.

**Example 9**    The Cost of a New Car

From 1992 to 1998, the average cost $C$ of a new import car can be modeled by

$$C = 214.3t^2 + 18{,}132, \qquad 2 \le t \le 8$$

where $t = 2$ represents 1992. (See Figure 1.10.) If the average cost of a new import car continued to increase according to this model, in what year would the average cost reach \$55,000?    (Source: U.S. Bureau of Economic Analysis)

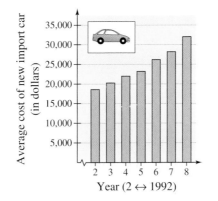

**Figure 1.10**

**Solution**

To solve this problem, let the cost be \$55,000 and solve the equation for $t$.

| | |
|---|---|
| $214.3t^2 + 18{,}132 = 55{,}000$ | Set cost equal to 55,000. |
| $214.3t^2 = 36{,}868$ | Subtract 18,132 from both sides. |
| $t^2 \approx 172.039$ | Divide both sides by 214.3. |
| $t \approx \sqrt{172.039}$ | Extract positive square root. |

The solution is $t \approx 13.1$. However, because $t$ is greater than 13, you can round up to $t = 14$ years. Because $t = 2$ represents 1992, you can conclude that, according to this model, the average cost of a new import car would reach \$55,000 in 2004.

---

**Discussing the Concept    Geometry**

A **Pythagorean triple** is a set of three positive integers $(a, b, c)$ that satisfy the Pythagorean Theorem for right triangles, $a^2 + b^2 = c^2$. One such Pythagorean triple is $(3, 4, 5)$, because $3^2 + 4^2 = 5^2$. How many other Pythagorean triples can you find? (Remember that each member of the triple must be an integer.) Make a list of these triples, and discuss the method you used to find them.

# Warm Up

The following warm-up exercises involve skills that were covered in previous courses and earlier sections. You will use these skills in the exercise set for this section.

In Exercises 1–4, simplify the expression.

**1.** $\sqrt{\frac{7}{50}}$

**2.** $\sqrt{32}$

**3.** $\sqrt{7^2 + 3 \cdot 7^2}$

**4.** $\sqrt{\frac{1}{4} + \frac{3}{8}}$

In Exercises 5–10, factor the expression.

**5.** $3x^2 + 7x$

**6.** $4x^2 - 25$

**7.** $16 - (x - 11)^2$

**8.** $x^2 + 7x - 18$

**9.** $10x^2 + 13x - 3$

**10.** $6x^2 - 73x + 12$

## 1.3 Exercises

In Exercises 1–10, write the quadratic equation in standard form.

**1.** $2x^2 = 3 - 5x$

**2.** $4x^2 - 2x = 9$

**3.** $x^2 = 25x$

**4.** $10x^2 = 90$

**5.** $(x - 3)^2 = 2$

**6.** $12 - 3(x + 7)^2 = 0$

**7.** $x(x + 2) = 3x^2 + 1$

**8.** $x^2 + 1 = \frac{x - 3}{2}$

**9.** $\frac{3x^2 - 10}{5} = 12x$

**10.** $x(x + 5) = 2(x + 5)$

In Exercises 11–22, solve the quadratic equation by factoring.

**11.** $6x^2 + 3x = 0$

**12.** $9x^2 - 1 = 0$

**13.** $x^2 - 2x - 8 = 0$

**14.** $x^2 - 10x + 9 = 0$

**15.** $x^2 + 10x + 25 = 0$

**16.** $16x^2 + 56x + 49 = 0$

**17.** $3 + 5x - 2x^2 = 0$

**18.** $2x^2 = 19x + 33$

**19.** $x^2 + 4x = 12$

**20.** $x^2 + 4x = 21$

**21.** $-x^2 - 7x = 10$

**22.** $-x^2 + 8x = 12$

In Exercises 23–36, solve the equation by extracting square roots. List both the exact answer *and* a decimal answer that has been rounded to two decimal places.

**23.** $x^2 = 16$

**24.** $x^2 = 144$

**25.** $x^2 = 7$

**26.** $x^2 = 27$

**27.** $3x^2 = 36$

**28.** $9x^2 = 25$

**29.** $(x - 12)^2 = 18$

**30.** $(x + 13)^2 = 21$

**31.** $(x + 2)^2 = 12$

**32.** $(x + 5)^2 = 20$

**33.** $12x^2 = 300$

**34.** $6x^2 = 250$

**35.** $3x^2 + 2(x^2 - 4) = 15$

**36.** $x^2 + 3(x^2 - 5) = 10$

In Exercises 37–58, solve the equation by any convenient method.

**37.** $x^2 = 64$

**38.** $7x^2 = 32$

**39.** $x^2 - 2x + 1 = 0$

**40.** $x^2 - 6x + 5 = 0$

**41.** $16x^2 - 9 = 0$

**42.** $11x^2 + 33x = 0$

**43.** $4x^2 - 12x + 9 = 0$

**44.** $x^2 - 14x + 49 = 0$

**45.** $(x + 3)^2 = 81$

**46.** $(x - 5)^2 = 8$

**47.** $4x = 4x^2 - 3$

**48.** $80 + 6x = 9x^2$

**49.** $50 + 5x = 3x^2$

**50.** $144 - 73x + 4x^2 = 0$

**51.** $12x = x^2 + 27$        **52.** $26x = 8x^2 + 15$

**53.** $50x^2 - 60x + 10 = 0$     **54.** $9x^2 + 12x + 3 = 0$

**55.** $(x + 3)^2 - 4 = 0$       **56.** $(x - 2)^2 - 9 = 0$

**57.** $(x + 1)^2 = x^2$         **58.** $(x + 1)^2 = 4x^2$

**59.** Consider the expression $(x + 2)^2$. How would you convince someone in your class that $(x + 2)^2 \neq x^2 + 4$? Give an argument based on the rules of algebra. Give an argument using your graphing utility.

**60.** Consider the expression $\sqrt{a^2 + b^2}$. How would you convince someone in your class that $\sqrt{a^2 + b^2} \neq a + b$? Give an argument based on the rules of algebra or geometry. Give an argument using your graphing utility.

**61.** *Geometry*   A one-story building is 14 feet longer than it is wide (see figure). The building has 1632 square feet of floor space. What are the measurements of the building?

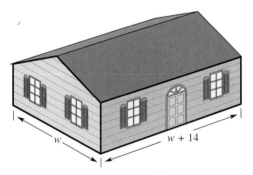

**62.** *Geometry*   A billboard is 10 feet longer than it is high (see figure). The billboard has 336 square feet of advertising space. What are the measurements of the billboard?

**63.** *Geometry*   A triangular sign has a height that is equal to its base. The area of the sign is 4 square feet. Find the base and height of the sign.

**64.** *Geometry*   The building lot shown in the figure has an area of 8000 square feet. What are the measurements of the lot?

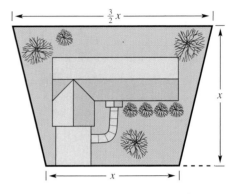

In Exercises 65–68, assume that air resistance is negligible, which implies that the position equation discussed in this section is a reasonable model.

**65.** *Royal Gorge Bridge*   The Royal Gorge Bridge near Canon City, Colorado is the highest suspension bridge in the world. The bridge is 1053 feet above the Arkansas river. If a rock were dropped from the bridge, how long would it take the rock to hit the water?

**66.** *Falling Object*   A rock is dropped from the top of a 200-foot cliff that overlooks the ocean. How long will it take for the rock to hit the water?

**67.** *Olympic Diver*   The high-dive platform in the Olympics is 10 meters above the water. A diver wants to perform an armstand dive, which means she will drop to the water from a handstand position. How long will the diver be in the air? [*Hint:* 1 meter = 3.2808 feet]

**68.** *The Owl and the Mouse*   An owl is circling a field and sees a mouse. The owl folds its wings and begins to dive. If the owl starts its dive from a height of 100 feet, how long does the mouse have to escape?

**69.** *Geometry*   The hypotenuse of an isosceles right triangle is 5 centimeters long. How long are the sides? (An isosceles triangle is one that has two sides of equal length.)

**70.** *Geometry* An equilateral triangle has a height of 1 foot. How long are each of its sides? (*Hint:* Use the height of the triangle to partition the triangle into two right triangles of the same size.)

**71.** *Flying Distance* The cities of Chicago, Atlanta, and Buffalo approximately form the vertices of a right triangle. The distance from Atlanta to Buffalo is about 650 miles and the distance from Atlanta to Chicago is about 565 miles. Use the map in the figure to approximate the flying distance from Atlanta to Buffalo *by way of* Chicago.

**72.** *Depth of a Submarine* The sonar of a navy cruiser detects a submarine that is 3000 feet from the cruiser. The angle formed by the ocean surface and a line from the cruiser to the submarine is 45° (see figure). How deep is the submarine?

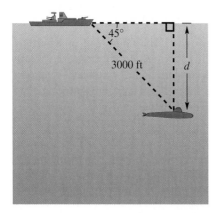

**73.** *Salvage Depth* A marine salvage ship detects a wreck 2000 feet off the bow. If the angle between the ocean surface and a line from the ship to the wreck is 45°, how deep is the wreck?

**74.** *Golf Winnings* The leading golf winnings $W$ (in dollars) from 1992 to 1998 in the United States can be approximated by the model

$$W = 1,218,712 + 19,136.31t^2, \qquad 2 \le t \le 8$$

where $t = 2$ represents 1992 (see figure). (Source: Professional Golfers' Association)

(a) Assuming that the leading winnings continue to follow the pattern given by this quadratic model, in what year would the leading winnings exceed $6,000,000?

(b) In 1999, Tiger Woods had golf winnings that totaled $6,616,585, and in 2000 his golf winnings totaled $9,188,321. Based on this information and your answer in part (a), do you think the model gives a good approximation for 1999 and 2000? Explain your reasoning.

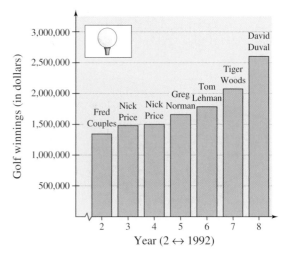

**75.** *Total Revenue* The demand equation for a certain product is $p = 20 - 0.0002x$, where $p$ is the price per unit and $x$ is the number of units sold. The total revenue $R$ for selling $x$ units is given by

$$R = xp = x(20 - 0.0002x).$$

How many units must be sold to produce a revenue of $500,000?

**76.** *Total Revenue* The demand equation for a certain product is $p = 30 - 0.0005x$, where $p$ is the price per unit and $x$ is the number of units sold. The total revenue $R$ for selling $x$ units is given by

$$R = xp = x(30 - 0.0005x).$$

How many units must be sold to produce a revenue of $450,000?

**77.** *Production Cost*   A company determines that the average monthly cost $C$ of raw materials for manufacturing a product line can be modeled by

$$C = 20.75t^2 + 5104, \qquad t \geq 0$$

where $t = 0$ represents 1990. If the average monthly cost of raw materials continues to increase according to this model, in what year will the average monthly cost reach $10,000?

**78.** *Monthly Cost*   A company determines that the average monthly cost $C$ for staffing temporary positions can be modeled by

$$C = 115.35t^2 + 12,072, \qquad t \geq 0$$

where $t = 0$ represents 2000. If the average monthly cost for staffing temporary positions continues to increase according to this model, in what year will the average monthly cost reach $20,000?

**79.** *U.S. Population*   The population of the United States from 1800 to 1890 can be approximated by the model

$$P = 694.2t^2 + 6183.3, \qquad 0 \leq t \leq 9$$

where $P$ is the population (in thousands of people) and $t$ represents the calendar year, with $t = 0$ corresponding to 1800, $t = 1$ corresponding to 1810, and so on (see figure). If this model had continued to be valid up through the present time, in what year would the population of the United States have reached 250,000,000? Judging from the figure, would you say this model was a good representation of the population through 1890? How about through 2000?   (Source: U.S. Bureau of the Census)

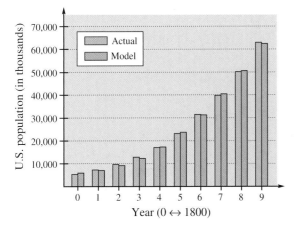

**80.** *U.S. Population*   The population of the United States from 1900 to 2000 can be approximated by the model

$$P = 1948.97t^2 + 97,202, \qquad 0 \leq t \leq 10$$

where $P$ is the population (in thousands of people) and $t$ represents the calendar year, with $t = 0$ corresponding to 1900, $t = 1$ corresponding to 1910, and so on (see figure). If this model continues to be valid, in what year will the population of the United States reach 330,000,000?   (Source: U.S. Bureau of the Census)

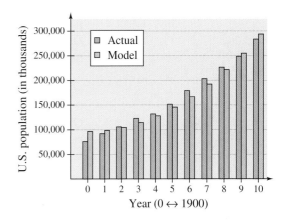

**81.** The U.S. Bureau of the Census predicts that the population in 2050 will be 404,000,000 people. Does the model in Exercise 80 appear to be a valid model for the year 2050?

**82.** *A Short Deck*   A standard deck of cards is missing several cards. One of the cards in the deck, however, is known to be the ace of hearts. A card is drawn from the deck at random, and then replaced, and then a card is drawn again. The probability $p$ that both drawings produce the ace of hearts is given by

$$p^2 \approx 0.000567.$$

How many cards are in the deck? (*Hint:* Write $p$ in the form $1/n$, where $n$ is the number of cards in the deck.)

**83.** *Airline Flight*   Suppose that you are going on a *round-trip* airline flight. You are told that the probability $p$ that both flights will be late is given by

$$p^2 = 0.2016.$$

What is the probability that you will be late on the return trip?

## 1.4    The Quadratic Formula

### Objectives

- Develop the Quadratic Formula by completing the square.
- Use the discriminant to determine the number of real solutions to a quadratic equation.
- Solve a quadratic equation using the Quadratic Formula.
- Use the Quadratic Formula to solve an application problem.

## Development of the Quadratic Formula

In Section 1.3 you studied two methods for solving quadratic equations. These two methods are efficient for special quadratic equations that are factorable or that can be solved by extracting square roots.

There are, however, many quadratic equations that cannot be solved efficiently by either of these two techniques. Fortunately, there is a general formula that can be used to solve *any* quadratic equation. It is called the **Quadratic Formula.** You can develop this important formula by a process called **completing the square,** as follows.

$$ax^2 + bx + c = 0 \qquad \text{Standard form, } a \neq 0$$

$$ax^2 + bx = -c \qquad \text{Subtract } c \text{ from both sides.}$$

$$x^2 + \frac{b}{a}x = -\frac{c}{a} \qquad \text{Divide both sides by } a.$$

$$x^2 + \frac{b}{a}x + \left(\frac{b}{2a}\right)^2 = -\frac{c}{a} + \left(\frac{b}{2a}\right)^2 \qquad \text{Add } \left(\frac{b}{2a}\right)^2 \text{ to both sides.}$$

$$\left(x + \frac{b}{2a}\right)^2 = \frac{b^2 - 4ac}{4a^2} \qquad \text{Simplify.}$$

$$x + \frac{b}{2a} = \pm\sqrt{\frac{b^2 - 4ac}{4a^2}} \qquad \text{Extract square roots.}$$

$$x = \frac{-b \pm \sqrt{b^2 - 4ac}}{2a} \qquad \text{Solutions}$$

This result is summarized below.

### Study Tip

The Quadratic Formula is one of the most important formulas in algebra, and you should memorize it. We have found that it helps to try to memorize a verbal statement of the rule. For instance, you might try to remember the following verbal statement of the Quadratic Formula: "The opposite of $b$, plus or minus the square root of $b$ squared minus $4ac$, all divided by $2a$."

---

▶ **The Quadratic Formula**

The solutions of

$$ax^2 + bx + c = 0, \quad a \neq 0$$

are given by the **Quadratic Formula,**

$$x = \frac{-b \pm \sqrt{b^2 - 4ac}}{2a}.$$

# The Discriminant

In the Quadratic Formula, the quantity under the radical sign, $b^2 - 4ac$, is called the **discriminant** of the quadratic expression $ax^2 + bx + c$.

$b^2 - 4ac$       Discriminant

It can be used to determine the nature of the solutions of a quadratic equation.

---

▶ **Solutions of a Quadratic Equation**

The solutions of a quadratic equation

$$ax^2 + bx + c = 0, \quad a \neq 0$$

can be classified by the discriminant, $b^2 - 4ac$, as follows.

1. If $b^2 - 4ac > 0$, the equation has *two* distinct real solutions.

2. If $b^2 - 4ac = 0$, the equation has *one* repeated real solution.

3. If $b^2 - 4ac < 0$, the equation has *two* distinct imaginary solutions. (In Section 4.5, you will study imaginary numbers.)

---

In case 3 above, the equation has no real solutions.

**Example 1**   **Using the Discriminant**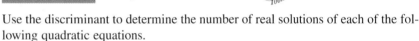

Use the discriminant to determine the number of real solutions of each of the following quadratic equations.

**a.** $4x^2 - 20x + 25 = 0$      **b.** $13x^2 + 7x + 1 = 0$      **c.** $5x^2 = 8x$

**Solution**

**a.** Using $a = 4$, $b = -20$, and $c = 25$, the discriminant is

$$b^2 - 4ac = (-20)^2 - 4(4)(25) = 400 - 400 = 0.$$

So, there is *one* repeated real solution.

**b.** Using $a = 13$, $b = 7$, and $c = 1$, the discriminant is

$$b^2 - 4ac = (7)^2 - 4(13)(1) = 49 - 52 = -3.$$

So, there are *no* real solutions.

**c.** In standard form, this equation is $5x^2 - 8x = 0$, with $a = 5$, $b = -8$, and $c = 0$, which implies that the discriminant is

$$b^2 - 4ac = (-8)^2 - 4(5)(0) = 64.$$

So, there are *two* distinct real solutions.

## Using the Quadratic Formula

When using the Quadratic Formula, remember that *before* the formula can be applied, you must first write the quadratic equation in standard form.

**Example 2**    Two Distinct Solutions

Solve $x^2 + 3x = 9$.

**Solution**

| | |
|---|---|
| $x^2 + 3x = 9$ | Original equation |
| $x^2 + 3x - 9 = 0$ | Write in standard form. |
| $x = \dfrac{-3 \pm \sqrt{(3)^2 - 4(1)(-9)}}{2(1)}$ | Quadratic Formula |
| $x = \dfrac{-3 \pm \sqrt{45}}{2}$ | Simplify. |
| $x = \dfrac{-3 \pm 3\sqrt{5}}{2}$ | Simplify. |

The solutions are

$$x = \frac{-3 + 3\sqrt{5}}{2} \quad \text{and} \quad x = \frac{-3 - 3\sqrt{5}}{2}.$$

Check these in the original equation.

**Example 3**    One Repeated Solution

Solve $8x^2 - 24x + 18 = 0$.

**Solution**

Begin by dividing both sides by the common factor 2.

| | |
|---|---|
| $8x^2 - 24x + 18 = 0$ | Original equation |
| $4x^2 - 12x + 9 = 0$ | Divide both sides by 2. |
| $x = \dfrac{-(-12) \pm \sqrt{(-12)^2 - 4(4)(9)}}{2(4)}$ | Quadratic Formula |
| $x = \dfrac{12 \pm \sqrt{0}}{8}$ | Simplify. |
| $x = \dfrac{3}{2}$ | Repeated solution |

The only solution is $\frac{3}{2}$. Check this in the original equation.

In Exercises 37–46, solve the equation by any convenient method.

**37.** $3x + 4 = 2x - 7$

**38.** $x^2 - 2x + 5 = x^2 - 5$

**39.** $4x^2 - 15 = 25$

**40.** $4x^2 + 2x + 4 = 2x - 8$

**41.** $x^2 + 3x + 1 = 0$

**42.** $x^2 + 3x - 4 = 0$

**43.** $(x - 1)^2 = 9$

**44.** $2x^2 - 4x - 6 = 0$

**45.** $100x^2 - 400 = 0$

**46.** $2x^2 + 4x - 9 = 2(x - 1)^2$

*Writing Real-Life Problems*    In Exercises 47–50, solve the number problem *and* write a real-life problem that could be represented by this verbal model. For instance, an applied problem that could be represented by Exercise 47 is as follows.

*The sum of the length and width of a one-story house is 100 feet. The house has 2500 square feet of floor space. What are the length and width of the house?*

**47.** Find two numbers whose sum is 100 and whose product is 2500.

**48.** One number is 1 more than another number. The product of the two numbers is 72. Find the numbers.

**49.** One number is 1 more than another number. The sum of their squares is 113. Find the numbers.

**50.** One number is 2 more than another number. The product of the two numbers is 440. Find the numbers.

*Cost Equation*    In Exercises 51–54, use the cost equation to find the number of units $x$ that a manufacturer can produce for the cost $C$. (Round your answer to the nearest positive integer.)

**51.** $C = 0.125x^2 + 20x + 5000$     $C = \$14,000$

**52.** $C = 0.5x^2 + 15x + 5000$     $C = \$11,500$

**53.** $C = 800 + 0.04x + 0.002x^2$     $C = \$1680$

**54.** $C = 800 - 10x + \dfrac{x^2}{4}$     $C = \$896$

**55.** *Seating Capacity*    A rectangular classroom seats 72 students. If the seats were rearranged with three more seats in each row, the classroom would have two fewer rows. Find the original number of seats in each row.

**56.** *Measurements of a Corral*    A rancher has 200 feet of fencing to enclose two adjacent rectangular corrals (see figure). Find the measurements such that the enclosed area will be 1400 square feet.

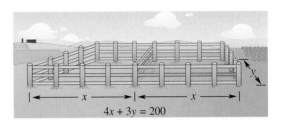

$4x + 3y = 200$

**57.** *Geometry*    An open box is to be made from a square piece of material by cutting 2-inch squares from the corners and turning up the sides (see figure). The volume of the finished box is to be 200 cubic inches. Find the size of the original piece of material.

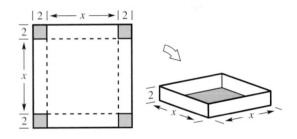

**58.** *Geometry*    An open box (see figure) is to be constructed from 108 square inches of material. Find the measurements of the square base.

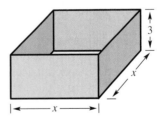

**59.** *On the Moon*    An astronaut on the moon throws a rock straight up into space. The initial velocity is 40 feet per second and the initial height is 5 feet. How long will it take the rock to hit the surface? If the rock had been thrown with the same initial velocity and height on Earth, how long would it remain in the air? (See Example 6.)

**60.** *Hot Air Balloon*    Two people are riding in a hot air balloon that is 200 feet above the ground. One person drops a sack of sand and the balloon starts to rise. One second later, the balloon is 20 feet higher and the other person throws a sack of sand toward the ground (see figure). The position equation for the first sack is $s = -16t^2 + 200$, and (using an initial velocity of $-20$ feet per second) the position equation for the second sack is

$$s = -16t^2 - 20t + 220.$$

Which sack of sand will hit the ground first? (*Hint:* Remember that the first sack was dropped one second before the second sack.)

*Falling Objects*  In Exercises 61 and 62, use the following information. The position equation for falling objects on Earth is of the form

$$s = -16t^2 + v_0 t + s_0$$

where $s$ is the height (in feet), $v_0$ is the initial velocity (in feet per second), $t$ is the time (in seconds), and $s_0$ is the initial height (in feet).

**61.** If a rock were thrown on the surface of Earth with an initial velocity of 27 feet per second from an initial height of 6 feet, would it take a longer or a shorter period of time to reach the ground than it would on the surface of the moon? (See Example 6.)

**62.** Use the Quadratic Formula to find the actual time that the rock would remain in the air.

**63.** *Flying Distance*    A small commuter airline flies to three cities whose locations form the vertices of a right triangle (see figure). The total flight distance (from City A to City B to City C and back to City A) is 1400 miles. It is 600 miles between the two cities that are farthest apart. Find the other two distances between cities.

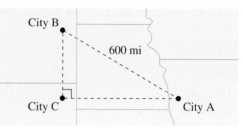

Figure for 63

**64.** *Distance from the Dock*    A windlass is used to tow a boat to a dock. The rope is attached to the boat at a point 15 feet below the level of the windlass (see figure). Find the distance from the boat to the dock when there is 75 feet of rope out.

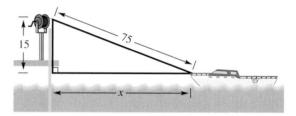

**65.** *Recreational Spending*    The total number of dollars $S$ (in billions) spent on recreation in the United States from 1990 to 1998 can be approximated by the model $S = 1.268t^2 + 17.12t + 278.4$, $0 \le t \le 8$, where $t = 0$ represents 1990. The figure shows the actual spending and the spending represented by the model. (Source: U.S. Bureau of Economic Analysis)

(a) During which year was approximately $400 billion spent on recreation?

(b) According to the model, in what year will the total annual recreational spending reach $820 billion?

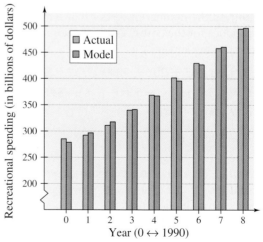

**66.** *Net Sales*    The net sales $S$ (in billions of dollars) of manufacturing corporations in the United States from 1990 to 1999 can be approximated by the model

$$S = 3.88t^2 + 134.3t + 2685, \qquad 0 \le t \le 9$$

where $t = 0$ represents 1990. The figure below shows the actual net sales and the net sales represented by the model. According to this model, in what year should net sales reach $5,000,000,000,000? (Source: U.S. Bureau of the Census)

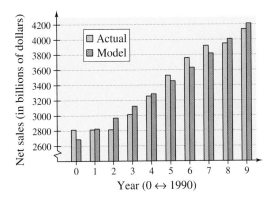

Year ($0 \leftrightarrow 1990$)

**67.** *Internet Access*    The percent $P$ of the public school classrooms with internet access in the United States from 1995 to 1999 can be modeled by

$$P = 1.64t^2 - 8.3t + 7, \qquad 5 \le t \le 9$$

where $t = 5$ represents 1995. According to this model, in what year did the percent of public school classrooms with Internet access reach 88%? Can you use this model to predict the percent of classrooms with Internet access in 2005? Explain. (Source: U.S. National Center for Education Statistics)

**68.** *Physicians' Services*    The amount $A$ of money spent annually for physicians' services in the United States from 1993 to 1998 is approximated by the model

$$A = 0.439t^2 + 3.71t + 171.1, \qquad 3 \le t \le 8$$

where the amount spent is in billions of dollars and $t$ represents the year, with $t = 3$ corresponding to 1993. According to this model, in what year will the amount reach $350 billion? (Source: U.S. Health Care Financing Administration)

**69.** *Flying Speed*    Two planes leave simultaneously from the same airport, one flying due east and the other due south (see figure). The eastbound plane is flying 50 miles per hour faster than the southbound plane. After 3 hours the planes are 2440 miles apart. Find the speed of each plane.

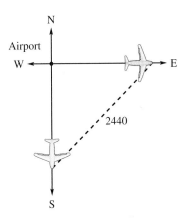

**70.** *Flying Speed*    Two planes leave simultaneously from the same airport, one flying due east and the other due south. The eastbound plane is flying 100 miles per hour faster than the southbound plane. After 2 hours the planes are 1500 miles apart. Find the speed of each plane.

**71.** *Total Revenue*    The demand equation for a certain product is

$$p = 60 - 0.0004x$$

where $p$ is the price per unit and $x$ is the number of units sold. The total revenue $R$ for selling $x$ units is given by $R = xp$. How many units must be sold to produce a revenue of $220,000?

**72.** *Total Revenue*    The demand equation for a certain product is

$$p = 50 - 0.0005x$$

where $p$ is the price per unit and $x$ is the number of units sold. The total revenue $R$ for selling $x$ units is given by $R = xp$. How many units must be sold to produce a revenue of $250,000?

**73.** When the Quadratic Formula is used to solve some problems, such as the problem in Example 5 on page 40, why is only one solution used?

## Mid-Chapter Quiz

Take this quiz as you would take a quiz in class. After you are done, check your work against the answers given in the back of the book.

In Exercises 1–4, solve the equation and check the solution.

**1.** $2(x + 3) - 5(2x - 3) = 5$

**2.** $\dfrac{3x + 3}{5x - 2} = \dfrac{3}{4}$

**3.** $\dfrac{2}{x(x - 1)} + \dfrac{1}{x} = \dfrac{1}{x - 4}$

**4.** $(x + 3)^2 - x^2 = 6(x + 2)$

In Exercises 5 and 6, solve the equation. (Round to three decimal places.)

**5.** $\dfrac{x}{3.057} + \dfrac{x}{4.392} = 100$

**6.** $0.378x + 0.757(500 - x) = 215$

**7.** Describe how you can check your answers to Exercises 1–4 using your graphing utility.

In Exercises 8 and 9, write an algebraic equation for the verbal description. Find the solution if possible and check.

**8.** *Production Limit*   A company has fixed costs of $20,000 per month and variable costs of $7.50 per unit manufactured. The company has $80,000 to cover monthly costs. How many units can the company manufacture?

**9.** *Total Revenue*   The demand equation for a certain product is $p = 60 - 0.0004x$, where $p$ is the price per unit and $x$ is the number of units sold. The total revenue $R$ for selling $x$ units is given by

$$R = xp.$$

How many units must be sold to produce revenue of $250,000?

In Exercises 10–15, solve the equation by the method indicated.

**10.** *Factoring:* $3x^2 + 13x = 10$

**11.** *Extracting roots:* $3x^2 = 15$

**12.** *Extracting roots:* $(x + 3)^2 = 17$

**13.** *Quadratic Formula:* $3x^2 + 7x - 2 = 0$

**14.** *Quadratic Formula:* $2x + x^2 = 5$

**15.** *Quadratic Formula:* $3x^2 - 4.50x - 0.32 = 0$

In Exercises 16 and 17, use the discriminant to determine the number of real solutions of the quadratic equation.

**16.** $x^2 + 3x - 5 = 0$

**17.** $2x^2 - 4x + 9 = 0$

**18.** Describe how you would convince a fellow student that $(x + 3)^2 = x^2 + 6x + 9$.

**19.** A rock is dropped from a height of 250 feet. With no initial velocity, what is the minimum time it will take the rock to hit the ground? Explain.

**20.** An open box 5 inches deep and 180 cubic inches in volume is to be constructed. Find the measurements of the square base.

## 1.5    Other Types of Equations

### Objectives

- Solve a polynomial equation by factoring.
- Rewrite and solve an equation involving radicals or rational exponents.
- Rewrite and solve an equation with fractions or absolute values.
- Construct and use a nonquadratic model to solve an application problem.

## Polynomial Equations

In this section you will extend the techniques for solving equations to nonlinear and nonquadratic equations. At this point in the text, you have only three basic methods for solving nonlinear equations—*factoring, extracting roots,* and the *Quadratic Formula.* So the main goal of this section is to learn to *rewrite* nonlinear equations in a form to which you can apply one of these methods.

**Example 1**    Solving a Polynomial Equation by Factoring

Solve $3x^4 = 48x^2$.

### Solution

The basic approach here is first to write the polynomial equation in standard form with zero on one side, to factor the other side, and then to set each factor equal to zero.

$$3x^4 = 48x^2 \qquad \text{Original equation}$$

$$3x^4 - 48x^2 = 0 \qquad \text{Collect terms on left side.}$$

$$3x^2(x^2 - 16) = 0 \qquad \text{Common monomial factor}$$

$$3x^2(x + 4)(x - 4) = 0 \qquad \text{Difference of two squares}$$

$$3x^2 = 0 \implies x = 0 \qquad \text{Set 1st factor equal to 0.}$$

$$x + 4 = 0 \implies x = -4 \qquad \text{Set 2nd factor equal to 0.}$$

$$x - 4 = 0 \implies x = 4 \qquad \text{Set 3rd factor equal to 0.}$$

You can check these solutions by substituting in the original equation, as follows.

### Check

$$3x^4 = 48x^2 \qquad \text{Original equation}$$

$$3(0)^4 = 48(0)^2 \qquad \text{0 checks. ✔}$$

$$3(-4)^4 = 48(-4)^2 \qquad \text{-4 checks. ✔}$$

$$3(4)^4 = 48(4)^2 \qquad \text{4 checks. ✔}$$

After checking, you can conclude that the solutions are $0$, $-4$, and $4$.

### Study Tip

A common mistake that is made in solving an equation such as that in Example 1 is dividing both sides of the equation by the variable factor $x^2$. This loses the solution $x = 0$. When using factoring to solve an equation, be sure to set each factor equal to zero. Don't divide both sides of an equation by a variable factor in an attempt to simplify the equation.

**Example 2**    Solving a Polynomial Equation by Factoring

Solve $x^3 - 3x^2 - 3x + 9 = 0$.

**Solution**

$$x^3 - 3x^2 - 3x + 9 = 0 \qquad \text{Original equation}$$

$$x^2(x - 3) - 3(x - 3) = 0 \qquad \text{Group terms.}$$

$$(x - 3)(x^2 - 3) = 0 \qquad \text{Factor by grouping.}$$

$$x - 3 = 0 \quad \Longrightarrow \quad x = 3 \qquad \text{Set 1st factor equal to 0.}$$

$$x^2 - 3 = 0 \quad \Longrightarrow \quad x = \pm\sqrt{3} \qquad \text{Set 2nd factor equal to 0.}$$

The solutions are 3, $\sqrt{3}$, and $-\sqrt{3}$. Check these in the original equation.

·DISCOVERY·

What do you observe about the degree of the polynomials in Examples 1, 2, and 3 and the possible number of solutions of the equations? Does your observation apply to the quadratic equations in Sections 1.3 and 1.4?

Occasionally, mathematical models involve equations that are of **quadratic type.** In general, an equation is of quadratic type if it can be written in the form

$$au^2 + bu + c = 0$$

where $a \neq 0$ and $u$ is an algebraic expression.

**Example 3**    Solving an Equation of Quadratic Type

Solve $x^4 - 3x^2 + 2 = 0$.

**Solution**

This equation is of quadratic type with $u = x^2$.

$$(x^2)^2 - 3(x^2) + 2 = 0$$

To solve this equation, you can factor the left side of the equation as the product of two second-degree polynomials.

$$x^4 - 3x^2 + 2 = 0 \qquad \text{Original equation}$$

$$(x^2 - 1)(x^2 - 2) = 0 \qquad \text{Partially factor.}$$

$$(x + 1)(x - 1)(x^2 - 2) = 0 \qquad \text{Completely factor.}$$

$$x + 1 = 0 \quad \Longrightarrow \quad x = -1 \qquad \text{Set 1st factor equal to 0.}$$

$$x - 1 = 0 \quad \Longrightarrow \quad x = 1 \qquad \text{Set 2nd factor equal to 0.}$$

$$x^2 - 2 = 0 \quad \Longrightarrow \quad x = \pm\sqrt{2} \qquad \text{Set 3rd factor equal to 0.}$$

The solutions are $-1$, 1, $\sqrt{2}$, and $-\sqrt{2}$. Check these in the original equation.

## Solving Equations Involving Radicals

The steps involved in solving the remaining equations in this section will often introduce *extraneous solutions,* as discussed in Section 1.1. Operations such as squaring both sides of an equation, raising both sides of an equation to a rational power, or multiplying both sides by a variable quantity all have this potential danger. So, when you use any of these operations, a check is crucial.

| Example 4 | An Equation Involving a Rational Exponent |

Solve $4x^{3/2} - 8 = 0$.

**Solution**

| | |
|---|---|
| $4x^{3/2} - 8 = 0$ | Original equation |
| $4x^{3/2} = 8$ | Add 8 to both sides. |
| $x^{3/2} = 2$ | Isolate $x^{3/2}$. |
| $x = 2^{2/3}$ | Raise both sides to $\frac{2}{3}$ power. |
| $x \approx 1.578$ | Round to three decimal places. |

**Check**

| | |
|---|---|
| $4x^{3/2} - 8 = 0$ | Original equation |
| $4(2^{2/3})^{3/2} \overset{?}{=} 8$ | Substitute $2^{2/3}$ for $x$. |
| $4(2) \overset{?}{=} 8$ | Property of exponents |
| $8 = 8$ | Solution checks. ✓ |

### Study Tip

The basic technique used in Example 4 is to isolate the factor with the rational exponent, and raise both sides to the *reciprocal power.* In Example 5, this is equivalent to isolating the square root and squaring both sides.

| Example 5 | An Equation Involving a Radical |

| | |
|---|---|
| $\sqrt{2x + 7} - x = 2$ | Original equation |
| $\sqrt{2x + 7} = x + 2$ | Isolate the square root. |
| $2x + 7 = x^2 + 4x + 4$ | Square both sides. |
| $0 = x^2 + 2x - 3$ | Standard form |
| $0 = (x + 3)(x - 1)$ | Factor. |
| $x + 3 = 0 \implies x = -3$ | Set 1st factor equal to 0. |
| $x - 1 = 0 \implies x = 1$ | Set 2nd factor equal to 0. |

By checking these values, you can determine that the only solution is 1.

## Equations with Fractions or Absolute Values

To solve an equation involving fractions, multiply both sides of the equation by the least common denominator (LCD) of each term in the equation. This procedure will "clear the equation of fractions." For instance, in the equation

$$\frac{2}{x^2 + 1} + \frac{1}{x} = \frac{2}{x}$$

you can multiply both sides of the equation by $x(x^2 + 1)$. Try doing this and solving the resulting equation. You should obtain one repeated solution: $x = 1$.

---

**Example 6**   An Equation Involving Fractions

Solve $\dfrac{2}{x} = \dfrac{3}{x - 2} - 1$.

**Solution**

For this equation, the LCD of the three terms is $x(x - 2)$, so we begin by multiplying each term in the equation by this expression.

$$\frac{2}{x} = \frac{3}{x - 2} - 1$$

$$x(x - 2)\frac{2}{x} = x(x - 2)\frac{3}{x - 2} - x(x - 2)(1)$$

$$2(x - 2) = 3x - x(x - 2)$$

$$2x - 4 = -x^2 + 5x$$

$$x^2 - 3x - 4 = 0$$

$$(x - 4)(x + 1) = 0$$

$$x - 4 = 0 \quad \Longrightarrow \quad x = 4$$

$$x + 1 = 0 \quad \Longrightarrow \quad x = -1$$

**Check**

$$\frac{2}{x} = \frac{3}{x - 2} - 1 \qquad \text{Original equation}$$

$$\frac{2}{4} \stackrel{?}{=} \frac{3}{4 - 2} - 1 \qquad \text{Substitute 4 for } x.$$

$$\frac{1}{2} = \frac{3}{2} - 1 \qquad \text{4 checks. } \checkmark$$

$$\frac{2}{-1} \stackrel{?}{=} \frac{3}{-1 - 2} - 1 \qquad \text{Substitute } -1 \text{ for } x.$$

$$-2 = -1 - 1 \qquad -1 \text{ checks. } \checkmark$$

The solutions are 4 and $-1$.

---

To solve an equation involving an absolute value, remember that the expression inside the absolute value signs can be positive or negative. This results in *two* separate equations, each of which must be solved. For instance, the equation

$$|x - 2| = 3$$

results in the two equations

$$x - 2 = 3 \quad \text{and} \quad -(x - 2) = 3$$

which implies that the equation has two solutions: 5 and $-1$.

**Example 7**    An Equation Involving Absolute Value

Solve $|x^2 - 3x| = -4x + 6$.

**Solution**

Because the variable expression inside the absolute value signs can be positive or negative, you must solve the following two equations.

*First Equation*

| | |
|---|---|
| $x^2 - 3x = -4x + 6$ | Use positive expression. |
| $x^2 + x - 6 = 0$ | Standard form |
| $(x + 3)(x - 2) = 0$ | Factor. |
| $x + 3 = 0 \implies x = -3$ | Set 1st factor equal to 0. |
| $x - 2 = 0 \implies x = 2$ | Set 2nd factor equal to 0. |

*Second Equation*

| | |
|---|---|
| $-(x^2 - 3x) = -4x + 6$ | Use negative expression. |
| $x^2 - 7x + 6 = 0$ | Standard form |
| $(x - 1)(x - 6) = 0$ | Factor. |
| $x - 1 = 0 \implies x = 1$ | Set 1st factor equal to 0. |
| $x - 6 = 0 \implies x = 6$ | Set 2nd factor equal to 0. |

**Check**

| | |
|---|---|
| $|(-3)^2 - 3(-3)| = -4(-3) + 6$ | $-3$ checks. ✓ |
| $|(2)^2 - 3(2)| \neq -4(2) + 6$ | 2 does not check. ✗ |
| $|(1)^2 - 3(1)| = -4(1) + 6$ | 1 checks. ✓ |
| $|(6)^2 - 3(6)| \neq -4(6) + 6$ | 6 does not check. ✗ |

The solutions are $-3$ and 1.

**Study Tip**

When solving an equation with absolute value, such as in Example 7, write the two equations to be solved. Then solve each equation independently. Be sure to check your solution.

## Applications

It would be impossible to categorize the many different types of applications that involve nonlinear and nonquadratic models. However, from the few examples and exercises that are given below, we hope that you will gain some appreciation for the variety of applications that can occur.

**Example 8**    Reduced Rates

A ski club chartered a bus for a ski trip at a cost of $480. In an attempt to lower the bus fare per skier, the club invited nonmembers to go along. When five non-members joined the trip, the fare per skier decreased by $4.80. How many club members are going on the trip?

### Solution

Begin the solution by creating a verbal model and assigning labels, as follows.

*Verbal Model:*    $\boxed{\text{Cost per skier}} \cdot \boxed{\text{Number of skiers}} = \boxed{\text{Cost of trip}}$

*Labels:*

Cost of trip = 480                                    (dollars)

Number of ski club members = $x$                      (people)

Number of skiers = $x + 5$                            (people)

Original cost per member = $\dfrac{480}{x}$           (dollars per person)

Cost per skier = $\dfrac{480}{x} - 4.80$             (dollars per person)

*Equation:*

$$\left(\frac{480}{x} - 4.80\right)(x + 5) = 480$$

$$\left(\frac{480 - 4.8x}{x}\right)(x + 5) = 480$$

$$(480 - 4.8x)(x + 5) = 480x$$

$$480x - 4.8x^2 - 24x + 2400 = 480x$$

$$-4.8x^2 - 24x + 2400 = 0$$

$$x^2 + 5x - 500 = 0$$

$$(x + 25)(x - 20) = 0$$

$$x + 25 = 0 \implies x = -25$$

$$x - 20 = 0 \implies x = 20$$

Choosing the positive value of $x$, you can conclude that 20 ski club members are going on the trip. Check this in the original statement of the problem.

Interest in a savings account is calculated by one of three basic methods: simple interest, interest compounded $n$ times per year, and interest compounded continuously. The next example uses the formula for interest that is compounded $n$ times per year.

$$A = P\left(1 + \frac{r}{n}\right)^{nt}$$

In this formula, $A$ is the balance in the account, $P$ is the principal (or original deposit), $r$ is the annual percentage rate (in decimal form), $n$ is the number of compoundings per year, and $t$ is the time in years. Later, in Chapter 5, you will study the derivation of this formula for compound interest.

 **Example 9**    Compound Interest

Suppose that when you were born, your grandparents deposited $5000 in a long-term investment in which the interest was compounded quarterly. On your 25th birthday the balance in the account is $25,062.59. What was the annual percentage rate for this investment?

Solution

*Formula:*    $A = P\left(1 + \dfrac{r}{n}\right)^{nt}$

*Labels:*    Amount $= A = 25{,}062.59$  (dollars)
Principal $= P = 5000.00$  (dollars)
Time $= t = 25$  (years)
Compoundings per year $= n = 4$  (compoundings)
Annual rate $= r$  (percent in decimal form)

*Equation:*    $25{,}062.59 = 5000\left(1 + \dfrac{r}{4}\right)^{4(25)}$

$$\frac{25{,}062.59}{5000} = \left(1 + \frac{r}{4}\right)^{100}$$

$$5.0125 \approx \left(1 + \frac{r}{4}\right)^{100}$$

$$(5.0125)^{1/100} \approx 1 + \frac{r}{4}$$

$$1.01625 \approx 1 + \frac{r}{4}$$

$$0.01625 \approx \frac{r}{4}$$

$$0.065 \approx r$$

The annual percentage rate is $0.065 = 6.5\%$. Check this in the original statement of the problem.

**Example 10**   Market Research

The marketing department at a publisher is asked to determine the price of a book. The department determines that the demand for the book depends on the price of the book according to the formula

$$p = 40 - \sqrt{0.0001x + 1}, \qquad x \geq 0$$

where $p$ is the price per book in dollars and $x$ is the number of books sold at the given price. For instance, in Figure 1.13, note that if the price were $39, then (according to the model) no one would be willing to buy the book. On the other hand, if the price were $17.60, 5 million copies could be sold. If the publisher set the price at $12.95, how many copies would be sold?

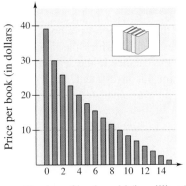

Number of books sold (in millions)

**Figure 1.13**

**Solution**

$$p = 40 - \sqrt{0.0001x + 1} \qquad \text{Given model}$$

$$12.95 = 40 - \sqrt{0.0001x + 1} \qquad \text{Set price at \$12.95.}$$

$$\sqrt{0.0001x + 1} = 27.05 \qquad \text{Isolate radical.}$$

$$0.0001x + 1 = 731.7025 \qquad \text{Square both sides.}$$

$$0.0001x = 730.7025 \qquad \text{Subtract 1 from both sides.}$$

$$x = 7{,}307{,}025 \qquad \text{Divide both sides by 0.0001.}$$

So, by setting the book's price at $12.95, the publisher can expect to sell about 7.3 million copies.

---

**Discussing the Concept**     **Compound Interest**

Suppose that on the day you were born a family member deposited money in an investment that earned interest compounded monthly at the rate of 7% per year. How much money needed to be deposited for the account to be worth $100,000 on your next birthday? How much money needed to be deposited for the account to be worth $1,000,000 on your next birthday?

# Warm Up

The following warm-up exercises involve skills that were covered in previous courses and earlier sections. You will use these skills in the exercise set for this section.

In Exercises 1–10, find the real solutions of the equation.

**1.** $x^2 - 22x + 121 = 0$

**2.** $x(x - 20) + 3(x - 20) = 0$

**3.** $(x + 20)^2 = 625$

**4.** $5x^2 + x = 0$

**5.** $3x^2 + 4x - 4 = 0$

**6.** $12x^2 + 8x - 55 = 0$

**7.** $x^2 + 4x - 5 = 0$

**8.** $4x^2 + 4x - 15 = 0$

**9.** $x^2 - 3x + 1 = 0$

**10.** $x^2 - 4x + 2 = 0$

## 1.5    Exercises

In Exercises 1–58, find the real solutions of the equation. Check your solutions.

**1.** $4x^4 - 18x^2 = 0$

**2.** $20x^3 - 125x = 0$

**3.** $x^3 - 2x^2 - 3x = 0$

**4.** $2x^4 - 15x^3 + 18x^2 = 0$

**5.** $x^4 - 81 = 0$

**6.** $x^6 - 64 = 0$

**7.** $5x^3 + 30x^2 + 45x = 0$

**8.** $9x^4 - 24x^3 + 16x^2 = 0$

**9.** $x^3 - 3x^2 - x + 3 = 0$

**10.** $x^3 + 2x^2 + 3x + 6 = 0$

**11.** $x^4 - x^3 + x - 1 = 0$

**12.** $x^4 + 2x^3 - 8x - 16 = 0$

**13.** $x^4 - 10x^2 + 9 = 0$

**14.** $x^4 - 29x^2 + 100 = 0$

**15.** $x^4 + 5x^2 - 36 = 0$

**16.** $x^4 - 4x^2 + 3 = 0$

**17.** $4x^4 - 65x^2 + 16 = 0$

**18.** $36t^4 + 29t^2 - 7 = 0$

**19.** $x^6 + 7x^3 - 8 = 0$

**20.** $x^6 + 3x^3 + 2 = 0$

**21.** $\sqrt{2x} - 10 = 0$

**22.** $4\sqrt{x} - 3 = 0$

**23.** $\sqrt{x - 10} - 4 = 0$

**24.** $\sqrt{5 - x} - 3 = 0$

**25.** $\sqrt[3]{2x + 5} + 3 = 0$

**26.** $\sqrt[3]{3x + 1} - 5 = 0$

**27.** $2x + 9\sqrt{x} - 5 = 0$

**28.** $6x - 7\sqrt{x} - 3 = 0$

**29.** $x = \sqrt{11x - 30}$

**30.** $2x - \sqrt{15 - 4x} = 0$

**31.** $-\sqrt{26 - 11x} + 4 = x$

**32.** $x + \sqrt{31 - 9x} = 5$

**33.** $\sqrt{x + 1} - 3x = 1$

**34.** $\sqrt{2x + 1} + x = 7$

**35.** $(x - 5)^{2/3} = 16$

**36.** $(x + 3)^{4/3} = 16$

**37.** $(x + 3)^{3/2} = 8$

**38.** $(x^2 + 2)^{2/3} = 9$

**39.** $(x^2 - 5)^{2/3} = 16$

**40.** $(x^2 - x - 22)^{4/3} = 16$

**41.** $\dfrac{1}{x} - \dfrac{1}{x + 1} = 3$

**42.** $\dfrac{x}{x^2 - 4} + \dfrac{1}{x + 2} = 3$

**43.** $\dfrac{20 - x}{x} = x$

**44.** $\dfrac{4}{x} - \dfrac{5}{3} = \dfrac{x}{6}$

**45.** $x = \dfrac{3}{x} + \dfrac{1}{2}$

**46.** $4x + 1 = \dfrac{3}{x}$

**47.** $\dfrac{1}{x} = \dfrac{4}{x - 1} + 1$

**48.** $x + \dfrac{9}{x + 1} = 5$

**49.** $\dfrac{4}{x + 1} - \dfrac{3}{x + 2} = 1$

**50.** $\dfrac{x + 1}{3} - \dfrac{x + 1}{x + 2} = 0$

**51.** $|x + 1| = 2$

**52.** $|x - 2| = 3$

**53.** $|2x - 1| = 5$

**54.** $|3x + 2| = 7$

**55.** $|x| = x^2 + x - 3$

**56.** $|x^2 + 6x| = 3x + 18$

**57.** $|x - 10| = x^2 - 10x$

**58.** $|x + 1| = x^2 - 5$

**59.** *Error Analysis*   Find the error(s) in the following solution.

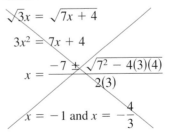

$$\sqrt{3x} = \sqrt{7x + 4}$$

$$3x^2 = 7x + 4$$

$$x = \dfrac{-7 \pm \sqrt{7^2 - 4(3)(4)}}{2(3)}$$

$$x = -1 \text{ and } x = -\dfrac{4}{3}$$

**60.** *Error Analysis*   Find the error(s) in the solution.

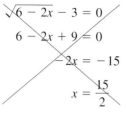

$$\sqrt{6 - 2x} - 3 = 0$$
$$6 - 2x + 9 = 0$$
$$-2x = -15$$
$$x = \frac{15}{2}$$

In Exercises 61–64, use a calculator to find the real solutions of the equation. (Round your answers to three decimal places.)

**61.** $3.2x^4 - 1.5x^2 - 2.1 = 0$

**62.** $7.08x^6 + 4.15x^3 - 9.6 = 0$

**63.** $1.8x - 6\sqrt{x} - 5.6 = 0$

**64.** $4x + 8\sqrt{x} + 3.6 = 0$

**65.** *Sharing the Cost*   A college charters a bus for $1700 to take a group of students to the Fiesta Bowl. When six more students join the trip, the cost per student drops by $7.50. How many students were in the original group?

**66.** *Sharing the Cost*   Three students plan to rent an apartment and share equally on the rent. By adding a fourth person, each person could save $75 a month. How much is the monthly rent?

**67.** *Compound Interest*   A deposit of $2500 reaches a balance of $3544.06 after 5 years. The interest on the account is compounded monthly. What is the annual percentage rate for this investment?

**68.** *Compound Interest*   A sales representative describes a "guaranteed investment fund" that is offered to new investors. You are told that if you deposit $10,000 in the fund you will be guaranteed a return of at least $25,000 after 20 years. (a) If after 20 years you received the minimum guarantee, what annual percentage rate did you receive? (b) If after 20 years you received $35,000, what annual percentage rate did you receive? (Assume that the interest in the fund is compounded quarterly.)

**69.** *Borrowing Money*   Suppose you borrow $100 from a friend and agree to pay the money back, plus $10 in interest, after 6 months. Assuming that the interest is compounded monthly, what annual percentage rate are you paying?

**70.** *Cash Advance*   You take out a cash advance of $500 on a credit card. After 2 months, you owe $515.75. The interest is compounded monthly. What is the annual percentage rate for this loan?

**71.** *Airline Passengers*   An airline offers daily flights between Chicago and Denver. The total monthly cost $C$ of these flights is modeled by

$$C = \sqrt{0.2x + 1}$$

where $C$ is measured in millions of dollars and $x$ is the number of passengers that month in thousands (see figure). The total cost of the flights for a month is 2.5 million dollars. How many passengers flew that month?

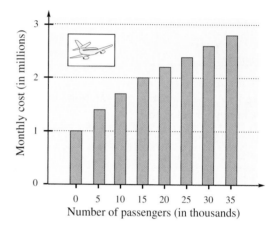

**72.** *Grocery Stores*   The number of grocery stores $G$ (in thousands) in the United States for the period from 1980 to 1998 can be approximated by the model

$$G = \sqrt{31{,}979.8 - 75.09t - 16.318t^2},$$

$$0 \le t \le 18$$

where $t = 0$ represents 1980. (Source: U.S. Department of Agriculture)

(a) Complete the table.

| $t$ | 0 | 5 | 10 | 15 | 18 |
|---|---|---|---|---|---|
| $G$ | | | | | |

(b) According to this model, in what year will there be 140,000 grocery stores in the United States?

**73.** *Life Expectancy*   The life expectancy table (for ages 58–75) used by the U.S. National Center for Health Statistics is modeled by

$$y = \sqrt{0.6632x^2 - 110.55x + 4680.24}$$

where $x$ represents a person's current age and $y$ represents the average number of additional years the person is expected to live (see figure). If a person's life expectancy is estimated to be 20 years, how old is the person? (Source: U.S. National Center for Health Statistics)

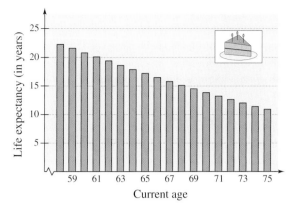

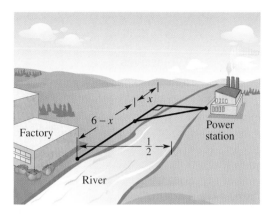

Figure for 76

**74.** *Federal Deficit*   The deficit $D$ (in millions of dollars) for the federal government from 1989 to 1998 can be modeled by

$$D = 251{,}681{,}170 - 1{,}173{,}938t - \frac{12{,}848{,}803{,}000}{t},$$

$$89 \le t \le 98$$

where $t = 89$ represents 1989 (see figure). Predict the year that the deficit will reach $6,000,000 million. (Source: U.S. Office of Management and Budget)

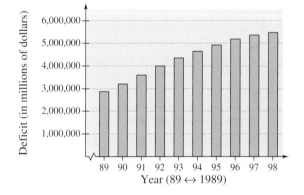

**75.** *Market Research*   The demand equation for a certain product is modeled by

$$p = 30 - \sqrt{0.0001x + 1}$$

where $x$ is the number of units demanded per day and $p$ is the price per unit. Find the demand if the price is set at $13.95.

**76.** *Power Line*   A power station is on one side of a river that is $\frac{1}{2}$ mile wide. A factory is 6 miles downstream on the other side of the river. It costs $18 per foot to run power lines over land and $24 per foot to run them under water. The project's cost is $616,877.27. Find the length $x$ as labeled in the figure.

**77.** *Sailboat Stays*   Two stays for the mast on a sailboat are attached to the boat at two points, as shown in the figure. One point is 10 feet from the mast and the other point is 15 feet from the mast. The total length of the two stays is 35 feet. How high on the mast are the stays attached?

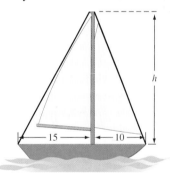

**78.** *Batting Average*   A softball player has been at bat 120 times and has 20 hits. The player's batting average is $20/120 \approx .167$. How many consecutive hits must the player get to obtain a batting average of .200? [*Hint:* (Batting average) = (total hits) ÷ (total times at bat)]

**79.** *Work Rate*   With only the cold water valve open, it takes 8 minutes to fill the tub of a washing machine. With both the hot and cold water valves open, it takes 5 minutes. The time it takes for the tub to fill with only the hot water valve open can be modeled by the equation

$$\frac{1}{8} + \frac{1}{t} = \frac{1}{5}$$

where $t$ is the time (in minutes) that it takes the tub to fill. How long does it take for the tub of the washing machine to fill with only the hot water valve open?

## 1.6    Linear Inequalities

### Objectives

- Write bounded and unbounded intervals using inequalities or interval notation.
- Solve and graph a linear inequality.
- Construct and use a linear inequality to solve an application problem.

## Introduction

Simple inequalities are used to *order* the real numbers. Inequality symbols $<, \leq, >$, and $\geq$ are used to compare two numbers and to denote subsets of real numbers. For instance, the simple inequality

$$x \geq 3$$

denotes all real numbers $x$ that are greater than or equal to 3.

In this section you will expand your work with inequalities to include more involved statements such as

$$5x - 7 < 3x + 9 \quad \text{and} \quad -3 \leq 6x - 1 < 3.$$

As with an equation, you **solve an inequality** in the variable $x$ by finding all values of $x$ for which the inequality is true. Such values are **solutions** and are said to **satisfy** the inequality. The set of all real numbers that are solutions of an inequality is the **solution set** of the inequality.

The set of all points on the real number line that represent the solution set is the **graph** of the inequality. Graphs of many types of inequalities consist of intervals on the real number line. The four different types of **bounded** intervals are summarized below.

*Learning Tools*

---

▶ **Bounded Intervals on the Real Number Line**

Let $a$ and $b$ be real numbers such that $a < b$. The following intervals on the real number line are **bounded.** The numbers $a$ and $b$ are the **endpoints** of each interval.

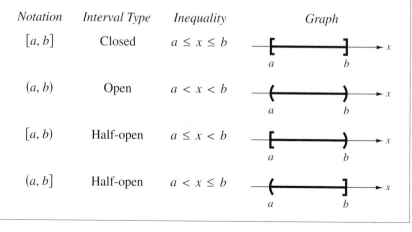

| Notation | Interval Type | Inequality | Graph |
|----------|---------------|------------|-------|
| $[a, b]$ | Closed | $a \leq x \leq b$ | |
| $(a, b)$ | Open | $a < x < b$ | |
| $[a, b)$ | Half-open | $a \leq x < b$ | |
| $(a, b]$ | Half-open | $a < x \leq b$ | |

Note that a closed interval contains both of its endpoints, a half-open interval contains only one of its endpoints, and an open interval does not contain either of its endpoints. Often, the solution of an inequality is an interval on the real line that is **unbounded.** For instance, the interval consisting of all positive numbers is unbounded.

▶ Unbounded Intervals on the Real Number Line

Let $a$ and $b$ be real numbers. The following intervals on the real number line are **unbounded.**

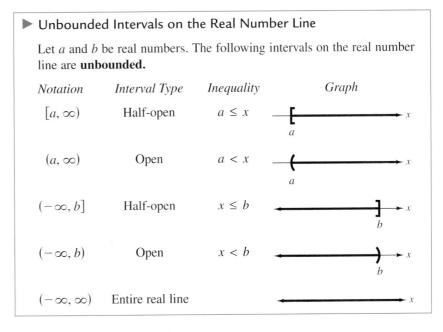

| Notation | Interval Type | Inequality | Graph |
|---|---|---|---|
| $[a, \infty)$ | Half-open | $a \leq x$ | |
| $(a, \infty)$ | Open | $a < x$ | |
| $(-\infty, b]$ | Half-open | $x \leq b$ | |
| $(-\infty, b)$ | Open | $x < b$ | |
| $(-\infty, \infty)$ | Entire real line | | |

The symbols $\infty$ (**positive infinity**) and $-\infty$ (**negative infinity**) do not represent real numbers. They are simply convenient symbols used to describe the unboundedness of an interval such as $(1, \infty)$.

**Example 1**    Intervals and Inequalities

Write an inequality to represent each of the following intervals and state whether the interval is bounded or unbounded.

**a.** $(-3, 5]$      **b.** $(-3, \infty)$      **c.** $[0, 2]$

Solution

**a.** $(-3, 5]$ corresponds to $-3 < x \leq 5$.        Bounded

**b.** $(-3, \infty)$ corresponds to $-3 < x$.        Unbounded

**c.** $[0, 2]$ corresponds to $0 \leq x \leq 2$.        Bounded

## Properties of Inequalities

The procedures for solving linear inequalities in one variable are much like those for solving linear equations. To isolate the variable, you can make use of **properties of inequalities.** These properties are similar to the properties of equality, but there are two important exceptions. When both sides of an inequality are multiplied or divided by a negative number, the direction of the inequality symbol must be reversed. Here is an example.

$$-2 < 5$$ 　　　　　Original inequality

$$(-3)(-2) > (-3)(5)$$ 　　　　　Multiply both sides by $-3$ and reverse the inequality.

$$6 > -15$$ 　　　　　Solution of inequality

Two inequalities that have the same solution set are **equivalent.** For instance, the inequalities

$$x + 2 < 5 \quad \text{and} \quad x < 3$$

are equivalent. The following list describes operations that can be used to create equivalent inequalities.

---

▶ **Properties of Inequalities**

Let $a$, $b$, $c$, and $d$ be real numbers.

1. *Transitive Property*

$$a < b \text{ and } b < c \quad \Longrightarrow \quad a < c$$

2. *Addition of Inequalities*

$$a < b \text{ and } c < d \quad \Longrightarrow \quad a + c < b + d$$

3. *Addition of a Constant*

$$a < b \quad \Longrightarrow \quad a + c < b + c$$

4. *Multiplying by a Constant*

$$\text{For } c > 0, a < b \quad \Longrightarrow \quad ac < bc$$

$$\text{For } c < 0, a < b \quad \Longrightarrow \quad ac > bc$$

---

Each of the properties above is true if the symbol $<$ is replaced by $\leq$. For instance, another form of the multiplication property would be as follows.

$$\text{For } c > 0, a \leq b \quad \Longrightarrow \quad ac \leq bc.$$

$$\text{For } c < 0, a \leq b \quad \Longrightarrow \quad ac \geq bc.$$

## Solving a Linear Inequality

The simplest type of inequality to solve is a **linear inequality** in a single variable. For instance, $2x + 3 > 4$ is a linear inequality in $x$.

As you read through the following examples, pay special attention to the steps in which the inequality symbol is reversed. Remember that when you multiply or divide by a negative number, you must reverse the inequality symbol.

**Example 2**    Solving a Linear Inequality

Solve $5x - 7 > 3x + 9$.

Solution

$$5x - 7 > 3x + 9 \qquad \text{Original inequality}$$
$$5x > 3x + 16 \qquad \text{Add 7 to both sides.}$$
$$5x - 3x > 16 \qquad \text{Subtract } 3x \text{ from both sides.}$$
$$2x > 16 \qquad \text{Combine like terms.}$$
$$x > 8 \qquad \text{Divide both sides by 2.}$$

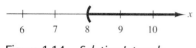

**Figure 1.14**    *Solution Interval:* $(8, \infty)$

The solution set is all real numbers that are greater than 8, which is denoted by $(8, \infty)$. The graph is shown in Figure 1.14.

Checking the solution set of an inequality is not as simple as checking the solutions of an equation. You can, however, get an indication of the validity of a solution set by substituting a few convenient values of $x$ to see whether the original inequality is satisfied.

**Example 3**    Solving a Linear Inequality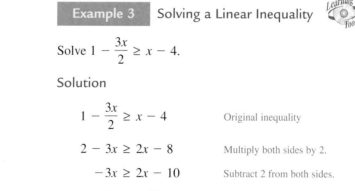

Solve $1 - \dfrac{3x}{2} \geq x - 4$.

Solution

$$1 - \frac{3x}{2} \geq x - 4 \qquad \text{Original inequality}$$
$$2 - 3x \geq 2x - 8 \qquad \text{Multiply both sides by 2.}$$
$$-3x \geq 2x - 10 \qquad \text{Subtract 2 from both sides.}$$
$$-5x \geq -10 \qquad \text{Subtract } 2x \text{ from both sides.}$$
$$x \leq 2 \qquad \text{Divide both sides by } -5 \text{ and reverse inequality.}$$

**Figure 1.15**    *Solution Interval:* $(-\infty, 2]$

The solution set is all real numbers that are less than or equal to 2, which is denoted by $(-\infty, 2]$. The graph is shown in Figure 1.15.

Sometimes it is convenient to write two inequalities as a **double inequality.** For instance, you can write the two inequalities $-4 \le 5x - 2$ and $5x - 2 < 7$ more simply as $-4 \le 5x - 2 < 7$.

**Example 4**    Solving a Double Inequality

To solve the following double inequality, we isolate $x$ as the middle term.

| | |
|---|---|
| $-3 \le 6x - 1 < 3$ | Original inequality |
| $-2 \le 6x < 4$ | Add 1 to all three parts. |
| $-\dfrac{1}{3} \le x < \dfrac{2}{3}$ | Divide by 6 and reduce. |

The solution set is all real numbers that are greater than or equal to $-\frac{1}{3}$ and less than $\frac{2}{3}$. The interval notation for this solution set is

$$\left[-\tfrac{1}{3}, \tfrac{2}{3}\right). \qquad\qquad \text{Solution set}$$

The graph of this solution set is shown in Figure 1.16.

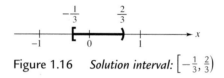

**Figure 1.16**    *Solution interval:* $\left[-\frac{1}{3}, \frac{2}{3}\right)$

The double inequality in Example 4 could have been solved in two parts as follows.

| | |
|---|---|
| $-3 \le 6x - 1$    and | $6x - 1 < 3$ |
| $-2 \le 6x$ | $6x < 4$ |
| $-\dfrac{1}{3} \le x$ | $x < \dfrac{2}{3}$ |

The solution set consists of all real numbers that satisfy *both* inequalities. In other words, the solution set is the set of all values of $x$ for which $-\frac{1}{3} \le x < \frac{2}{3}$.

When combining two inequalities to form a double inequality, be sure that the inequalities satisfy the Transitive Property. For instance, it is *incorrect* to combine the inequalities $3 < x$ and $x \le -1$ as $3 < x \le -1$. This "inequality" is obviously wrong because 3 is not less than $-1$.

## Inequalities Involving Absolute Value

### Study Tip

When an absolute value inequality is solved, the solution may consist of a bounded interval on the real number line (Example 5) or two unbounded intervals on the number line (Example 6).

▶ Solving an Absolute Value Inequality

Let $x$ be a variable or an algebraic expression and let $a$ be a real number such that $a \geq 0$.

**1.** The solutions of $|x| < a$ are all values of $x$ that lie between $-a$ and $a$.

$$|x| < a \qquad \text{if and only if } -a < x < a.$$

**2.** The solutions of $|x| > a$ are all values of $x$ that are less than $-a$ or greater than $a$.

$$|x| > a \qquad \text{if and only if } x < -a \text{ or } x > a.$$

These rules are also valid if $<$ is replaced by $\leq$ and $>$ is replaced by $\geq$.

**Example 5**    Solving an Absolute Value Inequality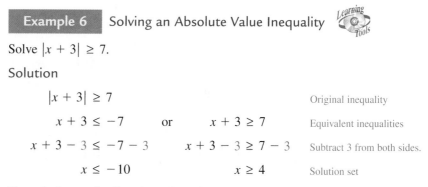

Solve $|x - 5| < 2$.

Solution

$$|x - 5| < 2 \qquad \text{Original inequality}$$
$$-2 < x - 5 < 2 \qquad \text{Equivalent inequalities}$$
$$-2 + 5 < x - 5 + 5 < 2 + 5 \qquad \text{Add 5 to all three parts.}$$
$$3 < x < 7 \qquad \text{Solution set}$$

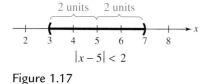

$|x - 5| < 2$

**Figure 1.17**

The solution set consists of all real numbers that are greater than 3 and less than 7, which is denoted by $(3, 7)$. The graph is shown in Figure 1.17.

**Example 6**    Solving an Absolute Value Inequality

Solve $|x + 3| \geq 7$.

Solution

$$|x + 3| \geq 7 \qquad \text{Original inequality}$$
$$x + 3 \leq -7 \qquad \text{or} \qquad x + 3 \geq 7 \qquad \text{Equivalent inequalities}$$
$$x + 3 - 3 \leq -7 - 3 \qquad x + 3 - 3 \geq 7 - 3 \qquad \text{Subtract 3 from both sides.}$$
$$x \leq -10 \qquad\qquad x \geq 4 \qquad \text{Solution set}$$

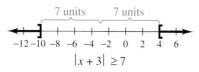

$|x + 3| \geq 7$

**Figure 1.18**

The solution set is all real numbers that are less than or equal to $-10$ *or* greater than or equal to 4, which is denoted by $(-\infty, -10] \cup [4, \infty)$ (see Figure 1.18). The symbol $\cup$ (union) means *or*.

## Applications

**Example 7**    Comparative Shopping

A subcompact car can be rented from Company A for $180 per week with no extra charge for mileage. A similar car can be rented from Company B for $100 per week, plus 20 cents for each mile driven. How many miles must you drive in a week to make the rental fee for Company B more than that for Company A?

**Solution**

*Verbal Model:*    Weekly cost for Company B $>$ Weekly cost for Company A

*Labels:*
Miles driven in one week $= m$                    (miles)
Weekly cost for Company A $= 180$                 (dollars)
Weekly cost for Company B $= 100 + 0.20m$         (dollars)

*Inequality:*
$$100 + 0.2m > 180$$
$$0.2m > 80$$
$$m > 400 \text{ miles}$$

If you drive more than 400 miles in a week, Company A is cheaper.

**Example 8**    Exercise Program

A man begins an exercise and diet program that is designed to reduce his weight by at least 2 pounds per week. At the beginning of the diet the man weighs 225 pounds. Find the maximum number of weeks before the man's weight will reach (or fall below) his goal of 192 pounds.

**Solution**

*Verbal Model:*    Desired weight $\leq$ Current weight $-$ 2 pounds per week $\cdot$ Number of weeks

*Labels:*
Desired weight $= 192$                    (pounds)
Current weight $= 225$                     (pounds)
Number of weeks $= x$                      (weeks)

*Inequality:*
$$192 \leq 225 - 2x$$
$$-33 \leq -2x$$
$$16.5 \geq x \text{ weeks}$$

Losing at least 2 pounds per week, it will take at most $16\frac{1}{2}$ weeks to reach his goal.

**Example 9** Accuracy of a Measurement

You go to a candy store to buy chocolates that cost $9.89 per pound. The scale that is used in the store has a state seal of approval that indicates the scale is accurate to within half an ounce. According to the scale, your purchase weighs one-half pound and costs $4.95. How much might you have been undercharged or overcharged due to an error in the scale?

**Solution**

To solve this problem, let $x$ represent the *true* weight of the candy. Because the state seal indicates that the scale is accurate to within half an ounce (or $\frac{1}{32}$ of a pound), you can conclude that the difference between the exact weight $(x)$ and the scale weight $\left(\frac{1}{2}\right)$ is less than or equal to $\frac{1}{32}$ of a pound. That is,

$$\left| x - \frac{1}{2} \right| \le \frac{1}{32}.$$

You can solve this inequality as follows.

$$-\frac{1}{32} \le x - \frac{1}{2} \le \frac{1}{32}$$

$$\frac{15}{32} \le x \le \frac{17}{32}$$

$$0.46875 \le x \le 0.53125$$

In other words, your "one-half" pound of candy could have weighed as little as 0.46875 pound (which would have cost $4.64) or as much as 0.53125 pound (which would have cost $5.25). So, you could have been undercharged by as much as $0.30 or overcharged by as much as $0.31.

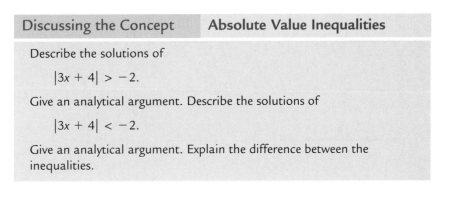

**Discussing the Concept**    **Absolute Value Inequalities**

Describe the solutions of

$$|3x + 4| > -2.$$

Give an analytical argument. Describe the solutions of

$$|3x + 4| < -2.$$

Give an analytical argument. Explain the difference between the inequalities.

# Warm Up

The following warm-up exercises involve skills that were covered in previous courses and earlier sections. You will use these skills in the exercise set for this section.

In Exercises 1–4, determine which of the two numbers is larger.

**1.** $-\frac{1}{2}, -7$

**2.** $-\frac{1}{3}, -\frac{1}{6}$

**3.** $-\pi, -3$

**4.** $-6, -\frac{13}{2}$

In Exercises 5–8, use inequality notation to denote the statement.

**5.** $x$ is nonnegative.

**6.** $z$ is strictly between $-3$ and 10.

**7.** $P$ is no more than 2.

**8.** $W$ is at least 200.

In Exercises 9 and 10, evaluate the expression for the values of $x$.

**9.** $|x - 10|, x = 12, x = 3$

**10.** $|2x - 3|, x = \frac{3}{2}, x = 1$

## 1.6  Exercises

In Exercises 1 and 2, determine whether the values of $x$ are solutions of the inequality.

**1.** $5x - 12 > 0$

(a) $x = 3$   (b) $x = -3$   (c) $x = \frac{5}{2}$   (d) $x = \frac{3}{2}$

**2.** $x + 1 < \frac{2x}{3}$

(a) $x = 0$   (b) $x = 4$   (c) $x = -4$   (d) $x = -3$

In Exercises 3–10, match the inequality with its graph. [The graphs are labeled (a), (b), (c), (d), (e), (f), (g), and (h).]

(a)
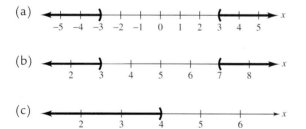

(b)

(c)

(d)

(e)

(f)

(g)

(h)

**3.** $x < 4$

**4.** $x \geq 6$

**5.** $-2 < x \leq 5$

**6.** $0 \leq x \leq \frac{7}{2}$

**7.** $|x| < 4$

**8.** $|x| > 3$

**9.** $|x - 5| > 2$

**10.** $|x + 6| < 3$

In Exercises 11–46, solve the inequality and sketch its graph.

**11.** $3x \geq 9$

**12.** $5x \leq 20$

**13.** $\dfrac{x}{3} < 4$

**14.** $\dfrac{2}{5}x > 7$

**15.** $-10x < 40$

**16.** $-6x > 15$

**17.** $x - 5 \geq 7$

**18.** $x + 7 \leq 12$

**19.** $4(x + 1) < 2x + 3$

**20.** $2x + 7 < 3$

**21.** $2x - 1 \geq 0$

**22.** $3x + 1 \geq 2$

**23.** $3 \leq 2x - 1 < 7$

**24.** $3 > 1 - \dfrac{x}{2} > -3$

**25.** $1 < 2x + 3 < 9$

**26.** $-8 \leq 1 - 3(x - 2) < 13$

**27.** $-4 < \dfrac{2x - 3}{3} < 4$

**28.** $0 \leq \dfrac{x + 3}{2} < 5$

**29.** $\dfrac{3}{4} > x + 1 > \dfrac{1}{4}$

**30.** $-1 < -\dfrac{x}{3} < 1$

**31.** $|x + 3| < 5$

**32.** $\left| \dfrac{2x + 1}{2} \right| < 6$

**33.** $\left| \dfrac{x}{2} \right| > 3$

**34.** $|5x| > 10$

**35.** $|x - 20| \leq 4$

**36.** $|x - 7| < 6$

**37.** $|2x - 5| > 6$

**38.** $2|5 - 3x| + 7 < 21$

**39.** $\left| \dfrac{x - 3}{2} \right| \geq 5$

**40.** $|1 - 2x| < 5$

**41.** $|9 - 2x| - 2 < -1$

**42.** $\left| 1 - \dfrac{2x}{3} \right| < 1$

**43.** $2|x + 10| \geq 9$

**44.** $3|4 - 5x| \leq 9$

**45.** $|x - 5| < 0$

**46.** $|x - 5| \geq 0$

In Exercises 47–54, use absolute value notation to define the solution set.

**47.**

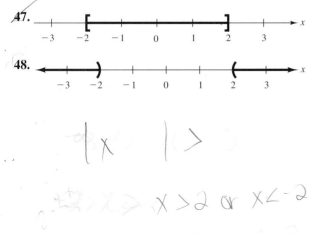

**48.**

**49.**

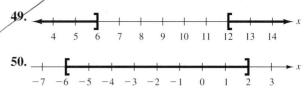

**50.**

**51.** All real numbers at most 10 units from 12

**52.** All real numbers at least 5 units from 8

**53.** All real numbers whose distances from $-3$ are more than 5

**54.** All real numbers whose distances from $-6$ are no more than 7

**55.** *Comparative Shopping*   You can rent a midsize car from Company A for $250 per week with no extra charge for mileage. A similar car can be rented from Company B for $150 per week, plus 25 cents for each mile driven. How many miles must you drive in a week to make the rental fee for Company B greater than that for Company A?

**56.** *Comparative Shopping*   Your department sends its copying to the photocopy center of your company. The photocopy center bills your department $0.10 per page. You are considering buying a departmental copier for $3000. With your own copier the cost per page would be $0.03. The expected life of the copier is 4 years. How many copies must you make in the 4-year period to justify purchasing the copier?

**57.** *Simple Interest*   For $1000 to grow to more than $1250 in 2 years, what must the simple interest rate be?

**58.** *Simple Interest*   For $1000 to grow to more than $1500 in 2 years, what must the simple interest rate be?

**59.** *Weight Loss Program*   A person enrolls in a diet program that guarantees a loss of at least $1\frac{1}{2}$ pounds per week. The person's weight at the beginning of the program is 164 pounds. Find the maximum number of weeks that the person must be in the program before attaining a weight of 128 pounds.

**60.** *Salary Increase*    You accept a new job with a starting salary of $18,500. You are told that you will receive an annual raise of at least $1250. What is the maximum number of years that you must work before your annual salary will be $23,500?

**61.** *Break-Even Analysis*    The revenue $R$ for selling $x$ units of a product is $R = 115.95x$. The cost $C$ of producing $x$ units is $C = 95x + 750$. In order to obtain a profit, the revenue must be greater than the cost.

(a) Complete the table.

| $x$ | 10 | 20 | 30 | 40 | 50 |
|-----|----|----|----|----|----|
| $R$ |    |    |    |    |    |
| $C$ |    |    |    |    |    |

(b) For what values of $x$ will this product return a profit?

**62.** *Break-Even Analysis*    The revenue $R$ for selling $x$ units of a product is $R = 24.55x$. The cost $C$ of producing $x$ units is

$$C = 15.4x + 150,000.$$

In order to obtain a profit, the revenue must be greater than the cost. For what values of $x$ will this product return a profit?

**63.** *Annual Operating Cost*    A utility company has a fleet of vans. The annual operating cost $C$ per van is

$$C = 0.32m + 2300$$

where $m$ is the number of miles traveled by a van in a year. What number of miles will yield an annual operating cost that is less than $10,000?

**64.** *Daily Sales*    A doughnut shop at a shopping mall sells a dozen doughnuts for $2.95. Beyond the fixed costs (rent, utilities, and insurance) of $150 per day, it costs $1.45 for enough materials (flour, sugar, and so on) and labor to produce a dozen doughnuts. If the daily profit varies between $50 and $200, between what levels (in dozens) do the daily sales vary?

**65.** *Cable Television Subscribers*    The number of cable television subscribers $S$ (in thousands) in the United States between 1990 and 2000 can be modeled by

$$S = 1651t + 52,400, \qquad 0 \le t \le 10$$

where $t = 0$ represents 1990. According to this model, in what year will the number of subscribers exceed 75 million (see figure)?    (Source: Nielsen Media Research)

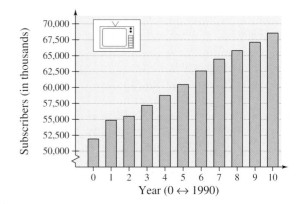

**66.** *Teachers' Salaries*    The average salary $S$ for elementary and secondary school teachers in the United States from 1990 to 1999 can be modeled by

$$S = 960t + 31,900, \qquad 0 \le t \le 9$$

where $t = 0$ represents 1990 (see figure). According to the model, in what year will the average salary exceed $45,000?    (Source: National Education Association)

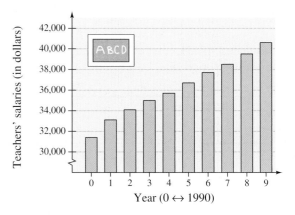

**67.** *Accuracy of Measurement* You buy six T-bone steaks that cost $3.98 per pound. The weight that is listed on the package is 5.72 pounds. If the scale that weighed the package is accurate to within $\frac{1}{2}$ ounce, how much money might you have been under-charged or overcharged?

**68.** *Accuracy of Measurement* You buy a bag of oranges that cost $0.69 per pound. The weight that is listed on the bag is 4.65 pounds. If the scale that weighed the bag is only accurate to within 1 ounce, how much money might you have been under-charged or overcharged?

**69.** *Human Height* The heights $h$ of two-thirds of a population satisfy the inequality

$$|h - 68.5| \leq 2.7$$

where $h$ is measured in inches. Determine the interval on the real number line in which these heights lie.

**70.** *Humidity Control* The specifications for an electronic device state that it is to be operated in a room with relative humidity $h$ defined by

$$|h - 50| \leq 30.$$

What are the minimum and maximum relative humidities for the operation of this device?

## Math Matters • Man and Mouse

A man who falls from a height of 2000 feet (without a parachute) will strike the ground with lethal force. But a mouse can fall from the same height and simply get up and walk away. Why?

The answer is that the speed at which a falling object hits the ground depends partly on the air resistance of the object, which in turn depends on the object's weight and surface area. If the ratio of an object's surface area to its weight is large, then its air resistance will be large. On the other hand, if the ratio of an object's surface area to its weight is small, then its air resistance will be small. This is why a parachute works—a person with a parachute has a much larger surface area (for approximately the same weight) than a person without a parachute. So how does this relate to the falling man and mouse? The falling mouse has a much greater air resistance than the falling man because the ratio of the mouse's surface area to its weight is greater than the ratio of the man's surface area to his weight. To convince yourself that the mouse's ratio is greater than the man's, try the following experiment. Find the ratio of the surface area and weight for the cubes described below.

Notice that as the cube becomes larger, the ratio of its surface area to its weight becomes smaller. (The answers are given in the back of the text.)

| Length of Side | Surface Area | Volume | Density | Weight |
| --- | --- | --- | --- | --- |
| 1 ft | 6 ft$^2$ | 1 ft$^3$ | 1 lb/ft$^3$ | 1 lb |
| 2 ft | 24 ft$^2$ | 8 ft$^3$ | 1 lb/ft$^3$ | 8 lb |
| 3 ft | 54 ft$^2$ | 27 ft$^3$ | 1 lb/ft$^3$ | 27 lb |
| 4 ft | 96 ft$^2$ | 64 ft$^3$ | 1 lb/ft$^3$ | 64 lb |

## 1.7    Other Types of Inequalities

### Objectives

- Use critical numbers to determine test intervals for a polynomial inequality.
- Solve and graph a polynomial inequality.
- Solve and graph a rational inequality.
- Determine the domain of an expression involving a square root.
- Construct and use a polynomial inequality to solve an application problem.

## Polynomial Inequalities

To solve a polynomial inequality such as

$$x^2 - 2x - 3 < 0$$

you can use the fact that a polynomial can change signs only at its zeros (the $x$-values that make the polynomial equal to zero). Between two consecutive zeros a polynomial must be entirely positive or entirely negative. This means that when the real zeros of a polynomial are put in order, they divide the real number line into intervals in which the polynomial has no sign changes. These zeros are the **critical numbers** of the inequality, and the resulting intervals are the **test intervals** for the inequality. For example, the polynomial

$$x^2 - 2x - 3 = (x + 1)(x - 3)$$

has two zeros, $x = -1$ and $x = 3$, and these zeros divide the real number line into three test intervals:

$$(-\infty, -1), \quad (-1, 3), \quad \text{and} \quad (3, \infty).$$

So, to solve the inequality $x^2 - 2x - 3 < 0$, you need only test one value from each of these test intervals. You can use the same basic approach to determine the test intervals for any polynomial.

---

▶ **Finding Test Intervals for a Polynomial**

To determine the intervals on which the values of a polynomial are entirely negative or entirely positive, use the following steps.

1. Find all real zeros of the polynomial, and arrange the zeros in increasing order (from smallest to largest). These zeros are the **critical numbers** of the polynomial.

2. Use the critical numbers of the polynomial to determine its **test intervals.**

3. Choose one representative $x$-value in each test interval and evaluate the polynomial at that value. If the value of the polynomial is negative, the polynomial will have negative values for *every* $x$-value in the interval. If the value of the polynomial is positive, the polynomial will have positive values for *every* $x$-value in the interval.

---

| Example 1 | Finding Test Intervals for a Polynomial  |
|---|---|

Solve $x^2 - x - 6 < 0$.

**Solution**

By factoring the quadratic as

$$x^2 - x - 6 = (x + 2)(x - 3)$$

you can see that the critical numbers are $x = -2$ and $x = 3$. So, the polynomial's test intervals are

$$(-\infty, -2), \quad (-2, 3), \quad \text{and} \quad (3, \infty). \qquad \text{Test intervals}$$

In each test interval, choose a representative $x$-value and evaluate the polynomial.

## Study Tip

The critical numbers act as boundaries between the real numbers that satisfy the inequality and the real numbers that do not satisfy the inequality.

| Interval | $x$-Value | Polynomial Value | Conclusion |
|---|---|---|---|
| $(-\infty, -2)$ | $x = -3$ | $(-3)^2 - (-3) - 6 = 6$ | Positive |
| $(-2, 3)$ | $x = 0$ | $(0)^2 - (0) - 6 = -6$ | Negative |
| $(3, \infty)$ | $x = 4$ | $(4)^2 - (4) - 6 = 6$ | Positive |

From this, you can conclude that the polynomial is positive for all $x$-values in $(-\infty, -2)$ and $(3, \infty)$, and is negative for all $x$-values in $(-2, 3)$. This implies that the solution of the inequality $x^2 - x - 6 < 0$ is the interval $(-2, 3)$, as shown in Figure 1.19.

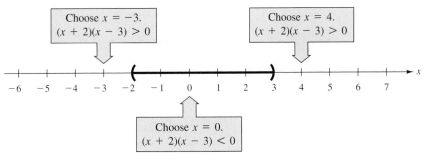

**Figure 1.19**

As with linear inequalities, you can check the reasonableness of a solution by substituting $x$-values into the original inequality. For instance, to check the solution found in Example 1, try substituting several $x$-values from the interval $(-2, 3)$ into the inequality

$$x^2 - x - 6 < 0.$$

Regardless of which $x$-values you choose, the inequality should be satisfied.

In Example 1, the polynomial inequality was given in standard form. Whenever this is not the case, you should begin the solution process by writing the inequality in standard form—with the polynomial on one side and zero on the other.

**Example 2**     Solving a Polynomial Inequality

Solve $x^3 - 3x^2 > 10x$.

**Solution**

Begin by writing the inequality in standard form.

$$x^3 - 3x^2 > 10x \qquad \text{Original inequality}$$

$$x^3 - 3x^2 - 10x > 0 \qquad \text{Write in standard form.}$$

$$x(x - 5)(x + 2) > 0 \qquad \text{Factor.}$$

*Critical numbers:*     $x = -2, x = 0, x = 5$

*Test intervals:*     $(-\infty, -2), (-2, 0), (0, 5), (5, \infty)$

*Test:*     Is $x(x - 5)(x + 2) > 0$?

After testing these intervals, as shown in Figure 1.20, you can see that $x^3 - 3x^2 - 10x$ is positive in the open intervals $(-2, 0)$ and $(5, \infty)$. So, the solution set consists of all real numbers in the intervals $(-2, 0)$ and $(5, \infty)$.

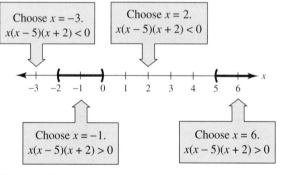

Figure 1.20

When solving a polynomial inequality, be sure you have accounted for the particular type of inequality symbol given in the inequality. For instance, in Example 2, note that the solution consisted of two open intervals because the original inequality contained a "greater than" symbol. If the original inequality had been $x^3 - 3x^2 \geq 10x$, the solution would have consisted of the two intervals $[-2, 0]$ and $[5, \infty)$.

Each of the polynomial inequalities in Examples 1 and 2 has a solution set that consists of a single interval or the union of two intervals. When solving the exercises for this section, you should watch for some unusual solution sets, as illustrated in Example 3.

### Example 3    Unusual Solution Sets

**a.** The solution set of the following inequality consists of the entire set of real numbers, $(-\infty, \infty)$.

$$x^2 + 2x + 4 > 0$$

**b.** The solution set of the following inequality consists of the single real number $\{-1\}$.

$$x^2 + 2x + 1 \leq 0$$

**c.** The solution set of the following inequality is empty.

$$x^2 + 3x + 5 < 0$$

**d.** The solution set of the following inequality consists of all real numbers except the number 2.

$$x^2 - 4x + 4 > 0$$

---

## ·TECHNOLOGY·

**Graphs of Inequalities and Graphing Utilities**   Most graphing utilities can sketch the graph of an inequality. Consult the user's guide to determine the steps for your graphing utility. Once you know how to graph an inequality, you may check solutions by graphing. For example, the solution to

$$x^2 - 5x < 0$$

is the interval $(0, 5)$. When graphed, the solution occurs as an interval above the horizontal axis on the graphing utility, as shown below. The graph does not indicate whether 0 and/or 5 are part of the solution. You must determine whether the endpoints are part of the solution based on the inequality sign and the type of inequality.

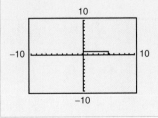

## Rational Inequalities

The concepts of critical numbers and test intervals can be extended to inequalities involving rational expressions. To do this, use the fact that the value of a rational expression can change sign only at its *zeros* (the $x$-values for which its numerator is zero) and its *undefined values* (the $x$-values for which its denominator is zero). These two types of numbers make up the **critical numbers** of a rational inequality.

**Example 4** Solving a Rational Inequality

Solve $\dfrac{2x - 7}{x - 5} \leq 3$.

**Solution**

$$\frac{2x - 7}{x - 5} \leq 3 \qquad \text{Original inequality}$$

$$\frac{2x - 7}{x - 5} - 3 \leq 0 \qquad \text{Write in standard form.}$$

$$\frac{2x - 7 - 3x + 15}{x - 5} \leq 0 \qquad \text{Add fractions.}$$

$$\frac{-x + 8}{x - 5} \leq 0 \qquad \text{Simplify.}$$

*Critical numbers:* $\quad x = 5, x = 8$

*Test intervals:* $\quad (-\infty, 5), (5, 8), (8, \infty)$

*Test:* $\qquad\qquad$ Is $\dfrac{-x + 8}{x - 5} \leq 0$?

After testing these intervals, as shown in Figure 1.21, you can see that the rational expression $(-x + 8)/(x - 5)$ is negative in the open intervals $(-\infty, 5)$ and $(8, \infty)$. Moreover, because $(-x + 8)/(x - 5) = 0$ when $x = 8$, you can conclude that the solution set consists of all real numbers in the intervals

$$(-\infty, 5) \cup [8, \infty). \qquad \text{Solution set}$$

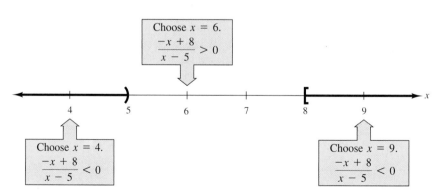

Choose $x = 6$.
$\dfrac{-x + 8}{x - 5} > 0$

Choose $x = 4$.
$\dfrac{-x + 8}{x - 5} < 0$

Choose $x = 9$.
$\dfrac{-x + 8}{x - 5} < 0$

**Figure 1.21**

## Applications

One common application of inequalities comes from business and involves profit, revenue, and cost. The formula that relates these three quantities is

Profit $=$ Revenue $-$ Cost

$$P = R - C.$$

**Example 5**    Increasing the Profit for a Product

The marketing department of a calculator manufacturer has determined that the demand for a new model of calculator is given by

$$p = 100 - 10x, \qquad 0 \le x \le 10 \qquad \text{Demand equation}$$

where the price $p$ per calculator is in dollars and $x$ represents the number of calculators sold, in millions. (If this model is accurate, no one would be willing to pay $100 for the calculator. At the other extreme, the company couldn't *give* away more than 10 million calculators.) The revenue, in millions of dollars, for selling $x$ million calculators is given by

$$R = xp = x(100 - 10x) \qquad \text{Revenue equation}$$

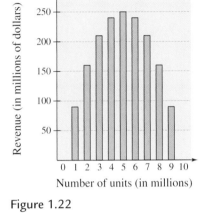

Figure 1.22

as shown in Figure 1.22. The total cost of producing $x$ million calculators is $10 per calculator plus a development cost of $2,500,000. So, the total cost, in millions of dollars, is

$$C = 10x + 2.5. \qquad \text{Cost equation}$$

What price should the company charge per calculator to obtain a profit of at least $190,000,000?

### Solution

*Verbal Model:*    Profit $=$ Revenue $-$ Cost

*Equation:*    $P = R - C$

$$P = 100x - 10x^2 - (10x + 2.5)$$

$$P = -10x^2 + 90x - 2.5$$

To answer the question, you must now solve the inequality

$$-10x^2 + 90x - 2.5 \ge 190$$

Using the techniques described in this section, you can find the solution to be $3.5 \le x \le 5.5$, as shown in Figure 1.23. The prices that correspond to these $x$-values are given by

$$\$45.00 \le p \le \$65.00.$$

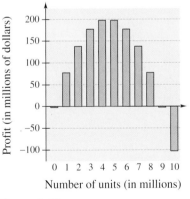

Figure 1.23

Another common application of inequalities is finding the domain of an expression that involves a square root, as shown in Example 6.

| Example 6 | Finding the Domain of an Expression |
|---|---|

Find the domain of the expression $\sqrt{64 - 4x^2}$.

### Solution

Remember that the domain of an expression is the set of all $x$-values for which the expression is defined. Because $\sqrt{64 - 4x^2}$ is defined (has real values) only if $64 - 4x^2$ is nonnegative, the domain is given by $64 - 4x^2 \geq 0$.

$$64 - 4x^2 \geq 0 \qquad \text{Standard form}$$

$$16 - x^2 \geq 0 \qquad \text{Divide both sides by 4.}$$

$$(4 - x)(4 + x) \geq 0 \qquad \text{Factor.}$$

So, the inequality has two critical numbers: $-4$ and $4$. You can use these two numbers to test the inequality as follows.

*Critical numbers:* $\qquad x = -4, x = 4$

*Test intervals:* $\qquad (-\infty, -4), (-4, 4), (4, \infty)$

*Test:* $\qquad$ Is $(4 - x)(4 + x) \geq 0$?

A test shows that $64 - 4x^2$ is greater than or equal to 0 in the *closed interval* $[-4, 4]$. So, the domain of the expression $\sqrt{64 - 4x^2}$ is the interval $[-4, 4]$, as shown in Figure 1.24.

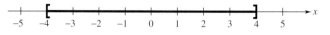

**Figure 1.24**

| Discussing the Concept | Profit Analysis |
|---|---|

Consider the relationship $P = R - C$ described on page 75. Discuss why it might be beneficial to solve $P < 0$ if you owned a business. Use the situation described in Example 5 to illustrate your reasoning.

# Warm Up

The following warm-up exercises involve skills that were covered in previous courses and earlier sections. You will use these skills in the exercise set for this section.

In Exercises 1–10, solve the inequality.

**1.** $-\dfrac{y}{3} > 2$

**2.** $-6z < 27$

**3.** $-3 \le 2x + 3 < 5$

**4.** $-3x + 5 \ge 20$

**5.** $10 > 4 - 3(x + 1)$

**6.** $3 < 1 + 2(x - 4) < 7$

**7.** $2|x| \le 7$

**8.** $|x - 3| > 1$

**9.** $|x + 4| > 2$

**10.** $|2 - x| \le 4$

## 1.7    Exercises

In Exercises 1–28, solve the inequality and sketch its graph.

**1.** $x^2 \le 9$

**2.** $x^2 < 5$

**3.** $x^2 > 4$

**4.** $(x - 3)^2 \ge 1$

**5.** $(x + 2)^2 < 25$

**6.** $(x + 6)^2 \le 8$

**7.** $x^2 + 4x + 4 \ge 9$

**8.** $x^2 - 6x + 9 < 16$

**9.** $x^2 + x < 6$

**10.** $x^2 + 2x > 3$

**11.** $3(x - 1)(x + 1) > 0$

**12.** $6(x + 2)(x - 1) < 0$

**13.** $x^2 + 2x - 3 < 0$

**14.** $x^2 - 4x - 1 > 0$

**15.** $4x^3 - 6x^2 < 0$

**16.** $4x^3 - 12x^2 > 0$

**17.** $x^3 - 4x \ge 0$

**18.** $2x^3 - x^4 \le 0$

**19.** $\dfrac{1}{x} > x$

**20.** $\dfrac{1}{x} < 4$

**21.** $\dfrac{x + 6}{x + 1} < 2$

**22.** $\dfrac{x + 12}{x + 2} \ge 3$

**23.** $\dfrac{3x - 5}{x - 5} > 4$

**24.** $\dfrac{5 + 7x}{1 + 2x} < 4$

**25.** $\dfrac{4}{x + 5} > \dfrac{1}{2x + 3}$

**26.** $\dfrac{5}{x - 6} > \dfrac{3}{x + 2}$

**27.** $\dfrac{1}{x - 3} \le \dfrac{9}{4x + 3}$

**28.** $\dfrac{1}{x} \ge \dfrac{1}{x + 3}$

In Exercises 29–36, find the domain of the expression.

**29.** $\sqrt[4]{4 - x^2}$

**30.** $\sqrt{x^2 - 4}$

**31.** $\sqrt{x^2 - 7x + 12}$

**32.** $\sqrt{144 - 9x^2}$

**33.** $\sqrt{12 - x - x^2}$

**34.** $\sqrt{x^2 + 4}$

**35.** $\sqrt{x^2 - 3x + 3}$

**36.** $\sqrt[4]{-x^2 + 2x - 2}$

In Exercises 37 and 38, consider the domains of the expressions $\sqrt[3]{x^2 - 7x + 12}$ and $\sqrt{x^2 - 7x + 12}$.

**37.** Explain why the domain of $\sqrt[3]{x^2 - 7x + 12}$ consists of all real numbers.

**38.** Explain why the domain of $\sqrt{x^2 - 7x + 12}$ is different from the domain of $\sqrt[3]{x^2 - 7x + 12}$.

In Exercises 39–44, use a calculator to solve the inequality. (Round each number in your answer to two decimal places.)

**39.** $0.4x^2 + 5.26 < 10.2$

**40.** $-1.3x^2 + 3.78 > 2.12$

**41.** $-0.5x^2 + 12.5x + 1.6 > 0$

**42.** $1.2x^2 + 4.8x + 3.1 < 5.3$

**43.** $\dfrac{1}{2.3x - 5.2} > 3.4$

**44.** $\dfrac{2}{3.1x - 3.7} > 5.8$

**45.** *Height of a Projectile*  A projectile is fired straight upward from ground level with an initial velocity of 160 feet per second. During what time period will its height exceed 384 feet?

**46.** *Height of a Projectile*  A projectile is fired straight upward from ground level with an initial velocity of 128 feet per second. During what time period will its height be less than 128 feet?

**47.** *Geometry*    A rectangular playing field with a perimeter of 100 meters is to have an area of at least 500 square meters (see figure). Within what bounds must the length lie?

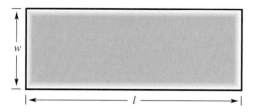

**48.** *Geometry*    A rectangular room with a perimeter of 50 feet is to have an area of at least 120 square feet. Within what bounds must the length lie?

**49.** *Company Profits*    The revenue $R$ and cost $C$ for a product are given by

$$R = x(50 - 0.0002x)$$

and

$$C = 12x + 150,000$$

where $R$ and $C$ are measured in dollars and $x$ represents the number of units sold (see figure).

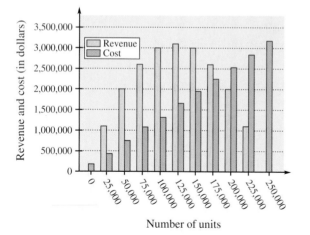

Number of units

(a)  How many units must be sold to obtain a profit of at least $1,650,000?

(b)  The demand equation for the product is

$$p = 50 - 0.0002x$$

where $p$ is the price per unit. What price per unit will produce a profit of at least $1,650,000?

**50.** *Company Profits*    The revenue $R$ and cost $C$ for a product are given by

$$R = x(75 - 0.0005x)$$

and

$$C = 30x + 250,000$$

where $R$ and $C$ are measured in dollars and $x$ represents the number of units sold, in thousands (see figure).

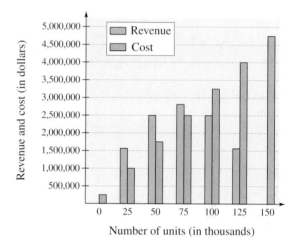

Number of units (in thousands)

(a)  How many units must be sold to obtain a profit of at least $750,000?

(b)  The demand equation for the product is

$$p = 75 - 0.0005x$$

where $p$ is the price per unit. What price per unit will produce a profit of at least $750,000?

**51.** *Compound Interest*    $P$ dollars, invested at interest rate $r$ compounded annually, increases to an amount

$$A = P(1 + r)^2$$

in 2 years. If an investment of $1000 is to increase to an amount greater than $1200 in 2 years, the interest rate must be greater than what percent?

**52.** *Compound Interest*    $P$ dollars, invested at interest rate $r$ compounded annually, increases to an amount

$$A = P(1 + r)^3$$

in 3 years. If an investment of $500 is to increase to an amount greater than $600 in 3 years, the interest rate must be greater than what percent?

**53.** *World Population*   The world population $P$ (in millions) from 1990 to 2000 can be modeled by

$$P = 5285 + 82.9t - 0.34t^2, \quad 0 \le t \le 10$$

where $t = 0$ represents 1990 (see figure). According to this model, in what year will the world population exceed 6,500,000,000?   (Source: U.S. Bureau of the Census)

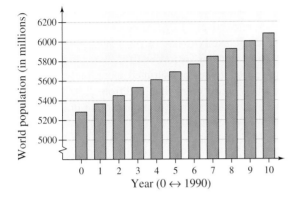

**54.** *Percent of College Graduates*   The percent $p$ of the American population that graduated from college from 1970 to 1999 can be modeled by

$$p = 10.8 + 0.61t - 0.004t^2, \quad 0 \le t \le 29$$

where $t = 0$ represents 1970. According to this model, in what year will the percent of college graduates exceed 30%?   (Source: U.S. Bureau of the Census)

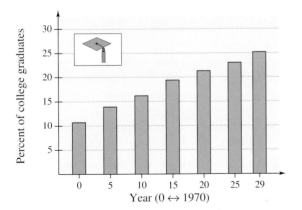

**55.** *Vehicle Registration*   The number of motor vehicle registrations $R$ (in thousands) in the United States from 1990 to 1998 can be modeled by

$$R = 187,122 + 2094.7t + 133.17t^2, \quad 0 \le t \le 8$$

where $t = 0$ represents 1990 (see figure). According to this model, in what year will the number of motor vehicle registrations exceed 275,000,000? (Source: U.S. Federal Highway Administration)

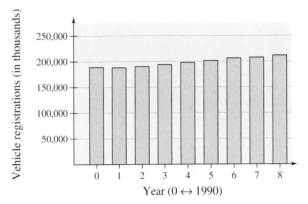

**56.** *New Car Cost*   The average cost $C$ of a new car (in dollars) in the United States from 1992 to 1999 can be modeled by

$$C = 15,305 + 487.4t + 18.95t^2, \quad 2 \le t \le 9$$

where $t = 2$ represents 1992 (see figure). According to this model, in what year will the average cost of a new car exceed $30,000?   (Source: American Automobile Manufacturers Association, Inc.)

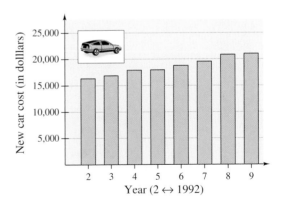

## Chapter Project

## Salaries in Professional Baseball

The record-breaking attendance (72.6 million people) for major league baseball was not the only record that was broken in the 2001 season. Barry Bonds (San Francisco) hit 73 home runs, drew 177 walks, and had a slugging percentage of .863 to set records in each category. Rickey Henderson (San Diego) broke the career record for runs with 2248 and extended his career records for walks and stolen bases to 2141 and 1395, respectively. Roger Clemens (New York Yankees) became the first pitcher in baseball history to have a 20–1 start. Albert Pujols (St. Louis) set a new rookie RBI (runs batted in) record with 130. As players reach and exceed previous records, they expect to be compensated accordingly. The average salary for a major league baseball player nearly doubled from 1995 to 2000.

The table below gives the average salaries (in thousands of dollars) for professional baseball players from 1995 to 2000.

| Year | 1995 | 1996 | 1997 | 1998 | 1999 | 2000 |
|------|------|------|------|------|------|------|
| Average salary (in thousands of dollars) | 1111 | 1120 | 1337 | 1399 | 1611 | 1896 |

The average salary $S$ can be modeled by the equation

$$S = 24.29t^2 - 208.3t + 1538, \qquad 5 \leq t \leq 10$$

where $t = 5$ represents 1995. Use this information to investigate the following questions.

1. *Compare the Data*  Make a table that compares the actual average salary for 1995 to 2000 with the average found using the model.

2. *Compare Projected Average Salaries with Actual Average Salaries*  Use the model to predict the average salary in 2001 and in 2002. Use a library or other reference source to find the actual average salary. How well did the model predict the average salary?

3. *Predict Future Average Salaries*  According to this model, when will the average salary reach $4,500,000?

   (a) Answer the question numerically by creating a table of values.

   (b) Answer the question algebraically by solving

   $$4500 = 24.29t^2 - 208.3t + 1538.$$

4. *Find a Model*  Enter the table above in a graphing utility. Enter 5 for 1995, enter 6 for 1996, and so on. Use the quadratic fit program to find a quadratic model for the data. Do you get the same model as the one above?

5. *Research*  Use a reference source to find data that can be closely modeled with a quadratic model. Compare the model with the actual data numerically and graphically.

## CHAPTER SUMMARY

After studying this chapter, you should have acquired the following skills. These skills are keyed to the Review Exercises that begin on page 82. Answers to odd-numbered Review Exercises are given in the back of the book.

**1.1**
· Classify an equation as an identity or a conditional equation.  Review Exercises 1, 2
· Determine whether a given value is a solution.  Review Exercises 3, 4
· Solve a linear equation in one variable.  Review Exercises 5–12

**1.2**
· Use mathematical models to solve word problems.  Review Exercises 13, 23, 24
· Model and solve percent and mixture problems.  Review Exercises 14–16, 21, 22, 25, 26
· Use common formulas to solve geometry and simple interest problems.  Review Exercises 17–20

**1.3**
· Solve a quadratic equation by factoring.  Review Exercises 27–30
· Solve a quadratic equation by extracting square roots.  Review Exercises 31–34
· Analyze a quadratic equation.  Review Exercises 35, 36
· Construct and use a quadratic model to solve an application problem.  Review Exercises 37–40

**1.4**
· Use the discriminant to determine the number of real solutions of a quadratic equation.  Review Exercises 41, 42
· Solve a quadratic equation using the Quadratic Formula.  Review Exercises 43–51
· Use the Quadratic Formula to solve an application problem.  Review Exercises 52, 53

**1.5**
· Solve a polynomial equation by factoring.  Review Exercises 54–57
· Rewrite and solve an equation involving radicals or rational exponents.  Review Exercises 58–61
· Rewrite and solve an equation with fractions or absolute values.  Review Exercises 62–65
· Construct and use a nonquadratic model to solve an application problem.  Review Exercises 66–68

**1.6**
· Solve and graph a linear inequality.  Review Exercises 69–74
· Construct and use a linear inequality to solve an application problem.  Review Exercises 75, 76

**1.7**
· Solve and graph a polynomial inequality.  Review Exercises 77–79, 83, 84
· Solve and graph a rational inequality.  Review Exercises 80–82, 85, 86
· Determine the domain of an expression involving a square root.  Review Exercises 87–90
· Construct and use a polynomial inequality to solve an application problem.  Review Exercises 91–98

In Exercises 1 and 2, classify the equation as an identity or a conditional equation.

**1.** $5(x - 3) = 2x + 9$      **2.** $3(x + 2) = 3x + 6$

In Exercises 3 and 4, determine whether each given value of $x$ is a solution of the equation.

**3.** $3x^2 + 7x + 5 = x^2 + 9$

(a) $x = 0$  (b) $x = \frac{1}{2}$  (c) $x = -4$  (d) $x = -1$

**4.** $6 + \dfrac{3}{x - 4} = 5$

(a) $x = 5$  (b) $x = 0$  (c) $x = -2$  (d) $x = 7$

In Exercises 5–10, solve the equation (if possible) and check your answer.

**5.** $4(x + 3) - 3 = 2(4 - 3x) - 4$

**6.** $\dfrac{3x - 2}{5x - 1} = \dfrac{3}{4}$

**7.** $(x + 3) + 2(x - 4) = 5(x + 3)$

**8.** $\dfrac{3}{x - 4} + \dfrac{8}{2x + 5} = \dfrac{11}{2x^2 - 3x - 20}$

**9.** $\dfrac{x}{x + 3} - \dfrac{4}{x + 3} + 2 = 0$

**10.** $7 - \dfrac{3}{x} = 8 + \dfrac{5}{x}$

In Exercises 11 and 12, use a calculator to solve the equation for $x$. (Round your answer to three decimal places.)

**11.** $0.375x - 0.75(300 - x) = 200$

**12.** $\dfrac{x}{0.0645} + \dfrac{x}{0.098} = 2$

**13.** Three consecutive even integers have a sum of 42. Find the smallest of these integers.

**14.** *Annual Salary*  Suppose your annual salary is $24,500. You receive a 6% raise. What is your new annual salary?

**15.** *Monthly Profit*  The total profit for a company in June was 15% less than it was in May. The total profit for the 2 months was $225,392. Find the profit for each month.

**16.** *Study Habits*  Based on a survey of 320 math students, the most common study techniques are shown in the graph. How many students in the survey used each type of study technique? (Most students used more than one type of technique.)

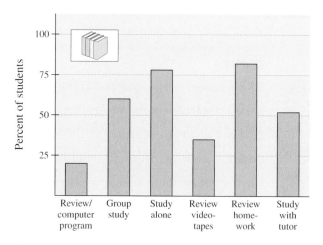

**17.** *Geometry*  A volleyball court is twice as long as it is wide, and its perimeter is 177 feet. Find the measurements of the volleyball court.

**18.** *Geometry*  A room is 1.25 times as long as it is wide, and its perimeter is 90 feet. Find the measurements of the room.

**19.** *Simple Interest*  You deposit $400 in a savings account earning 3% annually. How much interest will you have earned after 1 year?

**20.** *Simple Interest*  You deposit $750 in a money market account. One year later the account balance is $783.75. What was the interest rate for 1 year?

**21.** *List Price*  The price of a microwave oven has been discounted 15%. The sale price is $339.15. Find the original price of the microwave.

**22.** *Discount Rate*  The price of a VCR has been discounted by $85. The sale price is $257. What was the percent discount?

# 2

# The Cartesian Plane and Graphs

Demography is the statistical study of human populations with reference to size, density, distribution, and vital statistics. The United States Bureau of the Census tracks the demographics of the nation and uses these findings to forecast future trends. For example, the Census Bureau predicts that the U.S. population will have increased to about 392 million by the year 2050.

The table below gives the average sizes of U.S. households from 1960 to 2000. These data are also shown in the graph below. (Source: U.S. Bureau of the Census)

Although the overall population of the United States is increasing, from the graph you can see that the average household size is decreasing. Can you think of any possible explanations for this?

**2.1 The Cartesian Plane**

**2.2 Graphs of Equations**

**2.3 Graphing Utilities**

**2.4 Lines in the Plane**

**2.5 Linear Modeling**

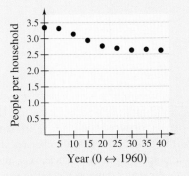

Year (0 ↔ 1960)

| Year | 1960 | 1965 | 1970 | 1975 | 1980 | 1985 | 1990 | 1995 | 2000 |
|------|------|------|------|------|------|------|------|------|------|
| People per household | 3.35 | 3.32 | 3.14 | 2.94 | 2.76 | 2.69 | 2.63 | 2.65 | 2.62 |

The chapter project related to this information is on page 143.

## 2.1    The Cartesian Plane

### Objectives

- Plot points in the Cartesian plane.
- Find the distance between two points in the plane.
- Use the Distance Formula to solve distance and geometry problems.
- Find the midpoint of a line segment.
- Use the Midpoint Formula to solve an application problem.

Corbis-Bettmann

### René Descartes
#### 1596–1650

The Cartesian coordinate plane named after René Descartes was developed independently by another French mathematician, Pierre de Fermat. Fermat's *Introduction to Loci,* written about 1629, was clearer and more systematic than Descartes's *La géométrie.* However, Fermat's work was not published during his lifetime. Consequently, Descartes received the credit for the development of the coordinate plane with the now familiar *x* and *y* axes.

## The Cartesian Plane

Just as you can represent real numbers by points on the real number line, you can represent ordered pairs of real numbers by points in a **rectangular coordinate plane.** This plane is called the **Cartesian plane,** after the French mathematician René Descartes (1596–1650).

The Cartesian plane is formed by two real lines intersecting at right angles, as shown in Figure 2.1. The horizontal number line is the **x-axis** and the vertical number line is the **y-axis.** (The plural of axis is *axes.*) The point of intersection of the two axes is the **origin,** and the axes separate the plane into four **quadrants.**

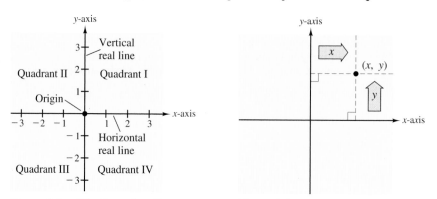

Figure 2.1    *The Cartesian Plane*              Figure 2.2

Each point in the plane corresponds to an **ordered pair** $(x, y)$ of real numbers $x$ and $y$, called the **coordinates** of the point. The first number (or **x-coordinate**) tells how far to the left or right the point is from the vertical axis, and the second number (or **y-coordinate**) tells how far up or down the point is from the horizontal axis, as shown in Figure 2.2.

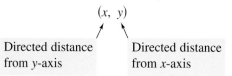

The notation $(a, b)$ to denote an open interval on the real number line and the notation $(x, y)$ to denote a point in the plane are similar but have different meanings. When these notations are used in this text, the nature of the specific problem will show which of the two is being discussed.

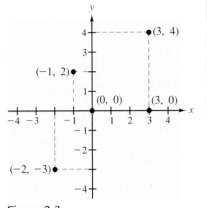

Figure 2.3

### Example 1    Plotting Points in the Cartesian Plane

Plot the points $(-1, 2)$, $(3, 4)$, $(0, 0)$, $(3, 0)$, and $(-2, -3)$ in the Cartesian plane.

### Solution

To plot the point $(-1, 2)$ you can envision a vertical line through $-1$ on the $x$-axis and a horizontal line through $2$ on the $y$-axis. The intersection of these two lines is the point $(-1, 2)$, as shown in Figure 2.3.

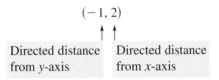

The other four points can be plotted in a similar way.

### Example 2    Shifting Points in the Plane

The triangle shown in Figure 2.4(a) has vertices at the points $(-1, 2)$, $(1, -4)$, and $(2, 3)$. Shift the triangle three units to the right and two units up and find the vertices of the shifted triangle.

### Solution

To shift the vertices three units to the right, add 3 to each $x$-coordinate. To shift the vertices two units up, add 2 to each $y$-coordinate. See Figure 2.4(b).

| *Original Vertices* | *Shifted Vertices* |
|---|---|
| $(-1, 2)$ | $(-1 + 3, 2 + 2) = (2, 4)$ |
| $(1, -4)$ | $(1 + 3, -4 + 2) = (4, -2)$ |
| $(2, 3)$ | $(2 + 3, 3 + 2) = (5, 5)$ |

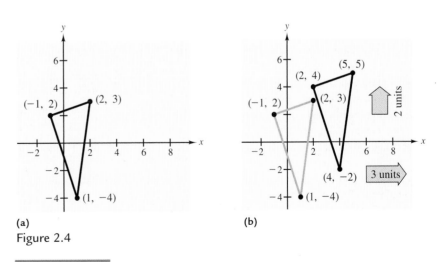

(a)

(b)

Figure 2.4

The value of the rectangular coordinate system is that it allows you to visualize relationships between the variables $x$ and $y$. Today, Descartes's ideas are in common use in virtually every scientific and business-related field.

| Example 3 | Number of Doctorates in Mathematics |

The numbers of doctorates in mathematics granted to United States citizens by universities in the United States in the years 1985 to 2000 are given in the table. Plot these points on a rectangular coordinate system.    (Source: American Mathematical Society)

| Year | 1985 | 1986 | 1987 | 1988 | 1989 | 1990 | 1991 | 1992 |
|---|---|---|---|---|---|---|---|---|
| Degrees | 396 | 386 | 362 | 363 | 411 | 401 | 461 | 430 |

| Year | 1993 | 1994 | 1995 | 1996 | 1997 | 1998 | 1999 | 2000 |
|---|---|---|---|---|---|---|---|---|
| Degrees | 526 | 469 | 567 | 493 | 516 | 586 | 554 | 537 |

*In 1971, nearly 25,000 bachelors degrees in mathematics were earned in the United States. Of these, 38% were earned by women. In 1997, the total number of bachelors degrees in mathematics had fallen to 12,820. Of these, 46.1% were earned by women.* (Source: U.S. National Center for Education Statistics)

**Solution**

The points are shown in Figure 2.5. Note that the break in the $x$-axis indicates that the numbers for the years prior to 1985 have been omitted.

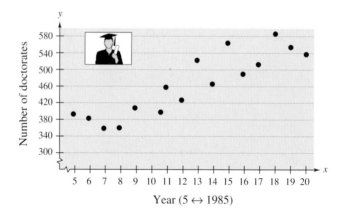

Year (5 ↔ 1985)

Figure 2.5

## The Distance Between Two Points in the Plane

The distance $d$ between two points $a$ and $b$ on the real number line is simply

$$d = |b - a|.$$

The same "absolute value rule" is used to find the distance between two points that lie on the same *vertical* or *horizontal* line in the plane.

**Example 4**   Finding Horizontal and Vertical Distances

**a.** Find the distance between the points $(1, -1)$ and $(1, 4)$.

**b.** Find the distance between the points $(-3, -1)$ and $(1, -1)$.

### Solution

**a.** Because the $x$-coordinates are equal, you can envision a vertical line through the points $(1, -1)$ and $(1, 4)$, as shown in Figure 2.6. The distance between these two points is given by the absolute value of the difference of their $y$-coordinates. That is,

Vertical distance $= |4 - (-1)|$       Subtract $y$-coordinates.

$= 5.$       Simplify.

So, the points are five units apart.

**b.** Because the $y$-coordinates are equal, you can envision a horizontal line through the points $(-3, -1)$ and $(1, -1)$, as shown in Figure 2.6. The distance between these two points is given by the absolute value of the difference of their $x$-coordinates. That is,

Horizontal distance $= |1 - (-3)|$       Subtract $x$-coordinates.

$= 4.$       Simplify.

So, the points are four units apart.

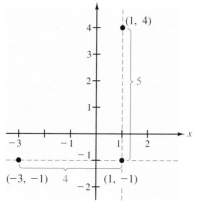

Figure 2.6

The technique used in Example 4 can be used to develop a general formula for finding the distance between two points in the plane. This general formula will work for any two points, even if they do not lie on the same vertical or horizontal line. To develop the formula, you can use the Pythagorean Theorem, which states that for a right triangle, the hypotenuse $c$ and sides $a$ and $b$ are related by the formula

$$a^2 + b^2 = c^2$$       Pythagorean Theorem

as shown in Figure 2.7. (The converse is also true. That is, if $a^2 + b^2 = c^2$, then the triangle is a right triangle.)

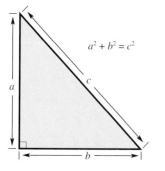

Figure 2.7   *Pythagorean Theorem*

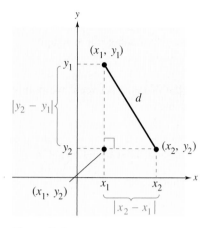

Figure 2.8

To develop a general formula for the distance between two points, let $(x_1, y_1)$ and $(x_2, y_2)$ represent two points in the plane that do not lie on the same horizontal or vertical line. With these two points, a right triangle can be formed, as shown in Figure 2.8. Note that the third vertex of the triangle is $(x_1, y_2)$. Because $(x_1, y_1)$ and $(x_1, y_2)$ lie on the same vertical line, the length of the vertical side of the triangle is $|y_2 - y_1|$. Similarly, the length of the horizontal side is $|x_2 - x_1|$. So, by the Pythagorean Theorem, the distance between $(x_1, y_1)$ and $(x_2, y_2)$ is given by

$$d^2 = |x_2 - x_1|^2 + |y_2 - y_1|^2.$$

Because the distance $d$ must be positive, choose the positive square root and write

$$d = \sqrt{|x_2 - x_1|^2 + |y_2 - y_1|^2}.$$

Finally, replacing $|x_2 - x_1|^2$ and $|y_2 - y_1|^2$ by the equivalent expressions $(x_2 - x_1)^2$ and $(y_2 - y_1)^2$ gives the following formula for the distance between two points in a rectangular coordinate plane.

---

▶ **The Distance Formula**

The distance $d$ between two points $(x_1, y_1)$ and $(x_2, y_2)$ in the coordinate plane is

$$d = \sqrt{(x_2 - x_1)^2 + (y_2 - y_1)^2}.$$

---

| Example 5 | Finding the Distance Between Two Points |  |

Find the distance between the points $(-2, 1)$ and $(3, 4)$.

**Solution**

Let $(x_1, y_1) = (-2, 1)$ and $(x_2, y_2) = (3, 4)$, and apply the Distance Formula.

$$d = \sqrt{[3 - (-2)]^2 + (4 - 1)^2} \qquad \text{Distance Formula}$$

$$= \sqrt{5^2 + 3^2} \qquad \text{Simplify.}$$

$$= \sqrt{25 + 9} \qquad \text{Simplify.}$$

$$= \sqrt{34} \qquad \text{Simplify.}$$

$$\approx 5.83 \qquad \text{Use a calculator.}$$

See Figure 2.9.

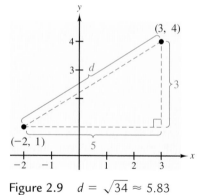

Figure 2.9    $d = \sqrt{34} \approx 5.83$

In Example 5, try letting $(x_1, y_1) = (3, 4)$ and $(x_2, y_2) = (-2, 1)$ to see that the Distance Formula yields the same result.

**Example 6**    Using the Distance Formula in Geometry

Show that the points $(2, 1)$, $(4, 0)$, and $(5, 7)$ are vertices of a right triangle.

**Solution**

The three points are plotted in Figure 2.10. Using the Distance Formula, you can find the lengths of the three sides of the triangle.

$$d_1 = \sqrt{(5 - 2)^2 + (7 - 1)^2} = \sqrt{9 + 36} = \sqrt{45}$$

$$d_2 = \sqrt{(4 - 2)^2 + (0 - 1)^2} = \sqrt{4 + 1} = \sqrt{5}$$

$$d_3 = \sqrt{(5 - 4)^2 + (7 - 0)^2} = \sqrt{1 + 49} = \sqrt{50}$$

Because

$$d_1{}^2 + d_2{}^2 = 45 + 5$$
$$= 50$$
$$= d_3{}^2$$

you can conclude from the Pythagorean Theorem that the triangle is a right triangle.

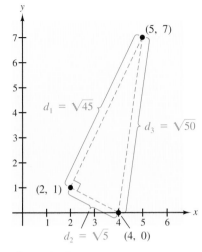

**Figure 2.10**

The next example shows how the Distance Formula can be used to solve a real-life problem.

**Example 7**    The Length of a Football Pass

In a football game, a quarterback throws a pass from the 5-yard line, 20 yards from the sideline. The pass is caught by a wide receiver on the 45-yard line, 50 yards from the same sideline, as shown in Figure 2.11. How long was the pass?

**Solution**

Using the Distance Formula, you can find the distance to be

$$d = \sqrt{(50 - 20)^2 + (45 - 5)^2} \qquad \text{Distance Formula}$$

$$= \sqrt{900 + 1600} \qquad \text{Simplify.}$$

$$= \sqrt{2500} \qquad \text{Simplify.}$$

$$= 50 \text{ yards} \qquad \text{Simplify.}$$

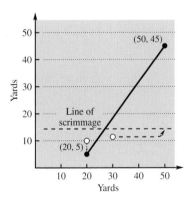

**Figure 2.11**

In Example 7, note that a scale was added to the goal line. When you use coordinate geometry to solve real-life problems, you are free to place the coordinate system in any way that is convenient in the problem.

## Example 8    Using the Distance Formula

Find $x$ such that the distance between $(x, 3)$ and $(2, -1)$ is 5.

### Solution

Using the Distance Formula, you can write the following.

$$\sqrt{(x-2)^2 + (3+1)^2} = 5 \qquad \text{Distance Formula}$$
$$(x^2 - 4x + 4) + 16 = 25 \qquad \text{Square both sides.}$$
$$x^2 - 4x - 5 = 0 \qquad \text{Standard form}$$
$$(x-5)(x+1) = 0 \qquad \text{Factor.}$$
$$x - 5 = 0 \implies x = 5 \qquad \text{Set 1st factor equal to 0.}$$
$$x + 1 = 0 \implies x = -1 \qquad \text{Set 2nd factor equal to 0.}$$

So, you can conclude that each of the points $(5, 3)$ and $(-1, 3)$ lies five units from the point $(2, -1)$, as shown in Figure 2.12.

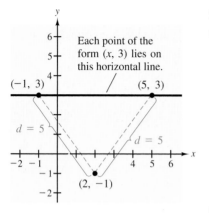

Each point of the form $(x, 3)$ lies on this horizontal line.

**Figure 2.12**

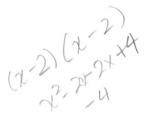

# The Midpoint Formula

The following formula shows how to find the midpoint of the line segment that joins two points.

> ▶ **The Midpoint Formula**
>
> The **midpoint** of the line segment joining the points $(x_1, y_1)$ and $(x_2, y_2)$ in the coordinate plane is
> $$\left( \frac{x_1 + x_2}{2}, \frac{y_1 + y_2}{2} \right).$$

## Example 9    Finding the Midpoint of a Line Segment

Find the midpoint of the line segment joining the points $(-5, -3)$ and $(9, 3)$.

### Solution

Figure 2.13 shows the two given points and their midpoint. By the Midpoint Formula, you have

$$\text{Midpoint} = \left( \frac{-5 + 9}{2}, \frac{-3 + 3}{2} \right) = (2, 0).$$

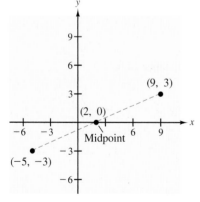

**Figure 2.13**

**Example 10**    Retail Sales

A business had annual retail sales of $240,000 in 1996 and $312,000 in 2002. Find the sales for 1999. (Assume that the annual increase in sales followed a linear pattern. That is, assume that the points representing the sales lie on a straight line, as shown in Figure 2.14.)

**Solution**

To make the computations simpler, let $t = 6$ represent the year 1996 and let $t = 12$ represent the year 2002. Then, if you measure the retail sales in thousands of dollars, the sales for 1996 and 2002 are represented by the points

$$(6, 240)   \text{and}   (12, 312).$$

Because 1999 is midway between 1996 and 2002 and because the growth pattern is linear, you can use the Midpoint Formula to find the 1999 sales.

$$\text{Midpoint} = \left( \frac{6 + 12}{2}, \frac{240 + 312}{2} \right)$$

$$= (9, 276)$$

So, the 1999 sales were approximately $276,000, as indicated in Figure 2.14.

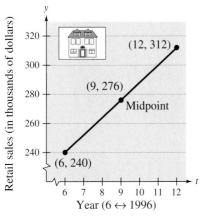

Retail sales (in thousands of dollars)

Year (6 ↔ 1996)

**Figure 2.14**

---

**Discussing the Concept    A Misleading Graph**

Although graphs can help us visualize relationships between two variables, they can also be used to mislead people. The graphs shown below represent the same data points. Which of the two graphs is misleading, and why? Discuss other ways in which graphs can be misleading. Try to find another example of a misleading graph in a newspaper or magazine. Why is it misleading? Why would it be beneficial for someone to use a misleading graph?

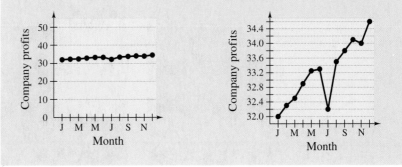

# Warm Up

The following warm-up exercises involve skills that were covered in earlier sections. You will use these skills in the exercise set for this section.

In Exercises 1–6, simplify the expression.

1. $\sqrt{(2-6)^2 + [1-(-2)]^2}$
2. $\sqrt{(1-4)^2 + (-2-1)^2}$
3. $\dfrac{4 + (-2)}{2}$
4. $\dfrac{-1 + (-3)}{2}$
5. $\sqrt{18} + \sqrt{45}$
6. $\sqrt{12} + \sqrt{44}$

In Exercises 7–10, solve the equation.

7. $\sqrt{(4-x)^2 + (5-2)^2} = \sqrt{58}$
8. $\sqrt{(8-6)^2 + (y-5)^2} = 2\sqrt{5}$
9. $\dfrac{x+3}{2} = 7$
10. $\dfrac{-2+y}{2} = 1$

## 2.1    Exercises

In Exercises 1–4, sketch the polygon with the indicated vertices.

1. Triangle: $(-3, 4), (1, -1), (-4, -2)$
2. Triangle: $(0, 3), (-1, -2), (4, 8)$
3. Square: $(2, 4), (5, 1), (2, -2), (-1, 1)$
4. Parallelogram: $(5, 2), (7, 0), (1, -2), (-1, 0)$

*Shifting a Graph*   In Exercises 5 and 6, the figure is shifted in the plane. Find the vertices of the shifted figure.

5.

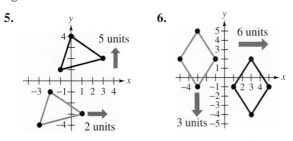

6.

In Exercises 7–10, plot the points and find the distance between them.

7. $(8, -1), (8, 5)$    8. $(-4, 2), (-4, 9)$
9. $(5, -7), (-6, -7)$   10. $(-3, 3), (1, 3)$

11. Find two points in Quadrant I that are two units apart.

12. Find two points in Quadrant II that are three units apart.

In Exercises 13–16, find the length of the hypotenuse in two ways: (a) use the Pythagorean Theorem and (b) use the Distance Formula.

13.

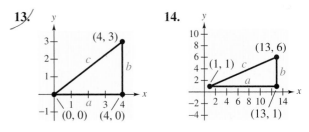

14.

**15.**

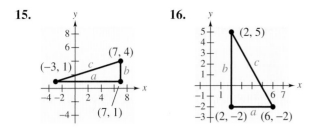

**16.**

In Exercises 17–28, (a) plot the points, (b) find the distance between the points, and (c) find the midpoint of the line segment joining the points.

**17.** $(2, -5), (-6, 1)$

**18.** $(1, 12), (6, 0)$

**19.** $(3, -11), (-12, -3)$

**20.** $(-7, 3), (2, -9)$

**21.** $(-1, 2), (5, 4)$

**22.** $(2, 10), (10, 2)$

**23.** $\left(\frac{1}{2}, 1\right), \left(-\frac{5}{2}, \frac{4}{3}\right)$

**24.** $\left(-\frac{1}{3}, -\frac{1}{3}\right), \left(-\frac{1}{6}, -\frac{1}{2}\right)$

**25.** $(1.7, 8.5), (-3.2, 5.3)$

**26.** $(21.7, 10.2), (7.9, -2.3)$

**27.** $(-36, -18), (48, -72)$

**28.** $(1.451, 3.051), (5.906, 11.360)$

In Exercises 29 and 30, use the Midpoint Formula to estimate the sales for 2000.

**29.**

| Year | 1998 | 2002 |
|------|------|------|
| Sales | $520,000 | $740,000 |

**30.**

| Year | 1998 | 2002 |
|------|------|------|
| Sales | $4,200,000 | $5,650,000 |

In Exercises 31–34, show that the points form the vertices of the indicated polygon.

**31.** Right triangle: $(4, 0), (2, 1), (-1, -5)$

**32.** Isosceles triangle: $(1, -3), (3, 2), (-2, 4)$

**33.** Rhombus: $(0, 0), (1, 2), (2, 1), (3, 3)$

**34.** Parallelogram: $(0, 1), (3, 7), (4, 4), (1, -2)$

In Exercises 35 and 36, find $x$ such that the distance between the points is 15.

**35.** $(3, -4), (x, 5)$          **36.** $(x, 8), (-9, -4)$

In Exercises 37 and 38, find $y$ such that the distance between the points is 20.

**37.** $(-15, y), (-3, -7)$          **38.** $(6, -1), (-10, y)$

In Exercises 39–42, state the quadrant in which $(x, y)$ lies.

**39.** $x > 0$ and $y < 0$          **40.** $x < 0$ and $y < 0$

**41.** $x > 0$ and $y > 0$          **42.** $x < 0$ and $y > 0$

**43.** *Football Pass*   In a football game, the quarterback throws a pass from the 20-yard line, 15 yards from the sideline. The pass is caught on the 50-yard line, 30 yards from the same sideline. How long was the pass?

**44.** *Driving Distance*   Two trucks leave a warehouse at the same time. One truck travels east at a rate of 55 miles per hour and the other travels south at a rate of 65 miles per hour. How far apart are the two trucks after 2 hours? (Assume that both trucks are traveling in a straight line.)

*Average Life Expectancy*   In Exercises 45 and 46, use the figure, which shows the average life expectancy for Americans from 1920 to 1998.   (Source: U.S. National Center for Health Statistics)

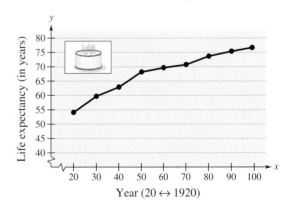

**45.** Estimate the average life expectancy in 1920.

**46.** Estimate the average life expectancy in 1998.

*Gold Prices*    In Exercises 47 and 48, use the figure, which shows the average price of gold from 1975 to 1999.    (Source: U.S. Bureau of Mines)

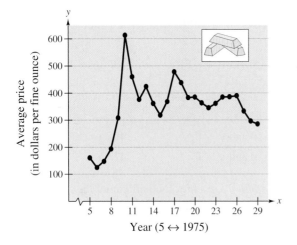

Year (5 ↔ 1975)

**47.** What is the highest price of gold shown in the graph? When did this occur?

**48.** What is the lowest price of gold shown in the graph? When did this occur?

*Fuel Efficiency*    In Exercises 49 and 50, use the figure, which shows the average fuel efficiency for automobiles in the United States from 1980 to 1998.    (Source: U.S. Federal Highway Administration)

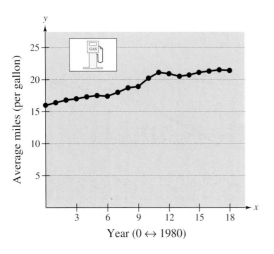

Year (0 ↔ 1980)

**49.** Estimate the percent increase in fuel efficiency from 1980 to 1990.

**50.** Estimate the percent increase in fuel efficiency from 1980 to 1998.

*Football Attendance*    In Exercises 51 and 52, use the figure, which shows the average paid attendance at NFL football games from 1985 to 1998.    (Source: National Football League)

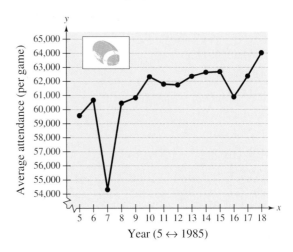

Year (5 ↔ 1985)

**51.** Estimate the increase in attendance from 1985 to 1998.

**52.** Estimate the decrease in attendance from 1986 to 1987.

## 2.2   Graphs of Equations

### Objectives

- Determine whether a point lies on the graph of an equation.
- Sketch the graph of an equation using a table of values.
- Find the $x$- and $y$-intercepts of the graph of an equation.
- Determine the symmetry of a graph.
- Write the equation of a circle in standard form.

## The Graph of an Equation

Frequently, the relationship between two quantities is expressed in the form of an equation. In this section you will study a basic procedure for sketching the graph of an equation. For an equation in variables $x$ and $y$, a point $(a, b)$ is a **solution point** if the substitution $x = a$ and $y = b$ satisfies the equation.

**Example 1**   Solution of an Equation

Determine if $(-1, 0)$ is a solution of the equation $y = 2x^2 - 4x - 6$.

**Solution**

$$y = 2x^2 - 4x - 6 \qquad \text{Original equation}$$

$$0 = 2(-1)^2 - 4(-1) - 6 \qquad \text{Substitute } -1 \text{ for } x \text{ and } 0 \text{ for } y.$$

$$0 = 0 \qquad \text{Simplify.}$$

Both sides of the equation are equivalent, so the point $(-1, 0)$ is a solution.

Most equations have *infinitely* many solution points. The set of all solution points of an equation is its **graph.**

**Example 2**   Sketching the Graph of an Equation

Sketch the graph of $3x + y = 5$ by rewriting the equation as $y = 5 - 3x$ with $y$ isolated on the left. Next, construct a table of values by choosing several values of $x$ and calculating the corresponding values of $y$.

| $x$ | $-1$ | $0$ | $1$ | $2$ | $3$ |
|---|---|---|---|---|---|
| $y = 5 - 3x$ | $8$ | $5$ | $2$ | $-1$ | $-4$ |

Finally, plot the points given in the table and connect them as shown in Figure 2.15. It appears that the graph is a straight line.

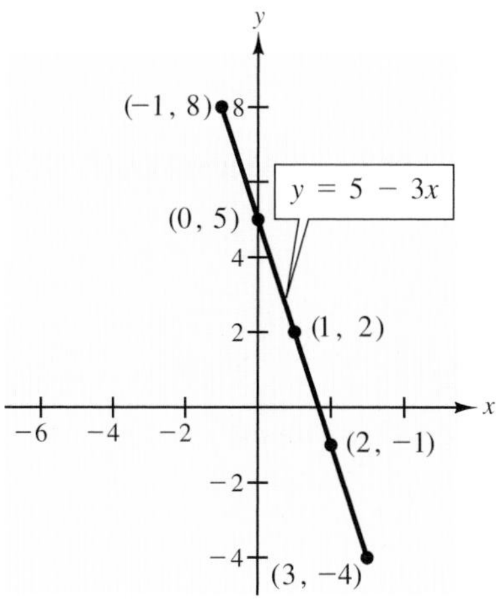

**Figure 2.15**

▶ **The Point-Plotting Method of Graphing**

1. If possible, isolate one of the variables.

2. Construct a table of several solution points.

3. Plot these points in the coordinate plane.

4. Connect the points with a smooth curve.

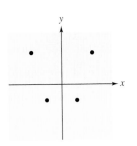

Figure 2.16

Step 4 of the point-plotting method can be difficult. For instance, how would you connect the four points in Figure 2.16? Without further information about the equation, any one of the three graphs in Figure 2.17 would be reasonable. These graphs show that with too few solution points, you can grossly misrepresent the graph of an equation. Throughout this course, you will study many ways to improve your graphing techniques. For now, we suggest that you plot enough points to reveal the essential behavior of the graph.

## Study Tip

When constructing a table, use negative, zero, and positive values for $x$.

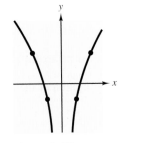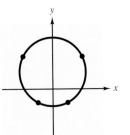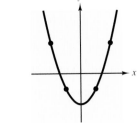

Figure 2.17

**Example 3**    Sketching the Graph of an Equation

Sketch the graph of

$$y = x^2 - 2.$$

### Solution

First, construct a table of values by choosing several convenient values of $x$ and calculating the corresponding values of $y$.

| $x$ | $-2$ | $-1$ | $0$ | $1$ | $2$ | $3$ |
|---|---|---|---|---|---|---|
| $y = x^2 - 2$ | $2$ | $-1$ | $-2$ | $-1$ | $2$ | $7$ |

Next, plot the corresponding solution points. Finally, connect the points with a smooth curve, as shown in Figure 2.18.

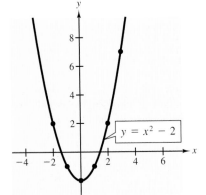

Figure 2.18

## Intercepts of a Graph

When you are sketching a graph, two types of points that are especially useful are those for which either the $y$-coordinate or the $x$-coordinate is zero.

---

▶ Definition of Intercepts

1. The **$x$-intercepts** of a graph are the points at which the graph intersects the $x$-axis. To find the $x$-intercepts, let $y$ be zero and solve for $x$.

2. The **$y$-intercepts** of a graph are the points at which the graph intersects the $y$-axis. To find the $y$-intercepts, let $x$ be zero and solve for $y$.

---

Some graphs have no intercepts and some have several. For instance, consider the three graphs in Figure 2.19.

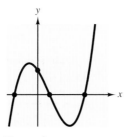
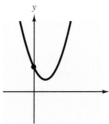
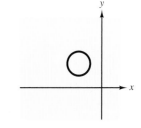

Three $x$-intercepts
One $y$-intercept

No $x$-intercept
One $y$-intercept

No intercepts

Figure 2.19

**Example 4**    Finding $x$- and $y$-Intercepts

Find the $x$- and $y$-intercepts of the graph of

$$y^2 - 3 = x.$$

### Solution

To find the $x$-intercept, let $y = 0$. This produces $-3 = x$, which implies that the graph has one $x$-intercept, which occurs at

$$(-3, 0). \qquad \text{\small $x$-intercept}$$

To find the $y$-intercept, let $x = 0$. This produces $y^2 - 3 = 0$, which has two solutions: $y = \pm\sqrt{3}$. So, the graph has two $y$-intercepts, which occur at

$$\left(0, \sqrt{3}\right) \quad \text{and} \quad \left(0, -\sqrt{3}\right). \qquad \text{\small $y$-intercepts}$$

See Figure 2.20.

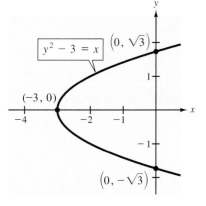

Figure 2.20

## Symmetry

Symmetry with respect to the $x$-axis means that if the Cartesian plane were folded along the $x$-axis, the portion of the graph above the $x$-axis would coincide with the portion below the $x$-axis. Symmetry with respect to the $y$-axis or the origin can be described in a similar manner. Symmetry with respect to the origin means that if the Cartesian plane were rotated $180°$ about the origin, the portion of the graph to the right of the origin would coincide with the portion to the left of the origin. (See Figure 2.21.)

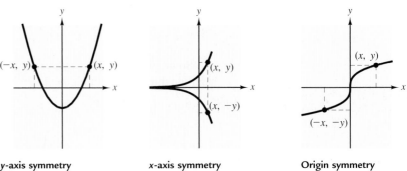

| $y$-axis symmetry | $x$-axis symmetry | Origin symmetry |

Figure 2.21

---

▶ **Definition of Symmetry**   *Learning Tools*

1. A graph is **symmetric with respect to the $y$-axis** if, whenever $(x, y)$ is on the graph, $(-x, y)$ is also on the graph.

2. A graph is **symmetric with respect to the $x$-axis** if, whenever $(x, y)$ is on the graph, $(x, -y)$ is also on the graph.

3. A graph is **symmetric with respect to the origin** if, whenever $(x, y)$ is on the graph, $(-x, -y)$ is also on the graph.

---

Suppose you apply this definition of symmetry to the graph of the equation $y = x^2 - 1$. Replacing $x$ with $-x$ produces the following.

$y = x^2 - 1$   Original equation

$y = (-x)^2 - 1$   Replace $x$ with $-x$.

$y = x^2 - 1$   Replacement yields equivalent equation.

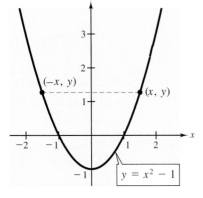

Figure 2.22 *y-Axis Symmetry*

Because the substitution did not change the equation, it follows that if $(x, y)$ is a solution of the equation, then $(-x, y)$ must also be a solution. So, the graph of $y = x^2 - 1$ is symmetric with respect to the $y$-axis, as shown in Figure 2.22.

▶ **Tests for Symmetry**

1. The graph of an equation is symmetric with respect to the *y-axis* if replacing $x$ with $-x$ yields an equivalent equation.

2. The graph of an equation is symmetric with respect to the *x-axis* if replacing $y$ with $-y$ yields an equivalent equation.

3. The graph of an equation is symmetric with respect to the *origin* if replacing $x$ with $-x$ *and* $y$ with $-y$ yields an equivalent equation.

**Example 5**   Using Symmetry as a Sketching Aid

Describe the symmetry of the graph of $x - y^2 = 1$.

**Solution**

Of the three tests for symmetry, the only one that is satisfied by this equation is the test for *x*-axis symmetry.

| | |
|---|---|
| $x - y^2 = 1$ | Original equation |
| $x - (-y)^2 = 1$ | Replace $y$ with $-y$. |
| $x - y^2 = 1$ | Replacement yields equivalent equation. |

So, the graph is symmetric with respect to the *x*-axis. To sketch the graph, plot the points above the *x*-axis and use symmetry to complete the graph, as shown in Figure 2.23.

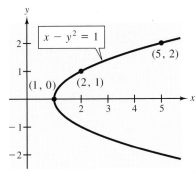

Figure 2.23

## The Equation of a Circle

So far in this section you have studied the point-plotting method and two additional concepts (intercepts and symmetry) that can be used to streamline the graphing procedure. Another graphing aid is *equation recognition,* which is the ability to recognize the general shape of a graph simply by looking at its equation. A circle is one type of graph that is easily recognized.

Figure 2.24 shows a circle of radius $r$ with center at the point $(h, k)$. The point $(x, y)$ is on this circle if and only if its distance from the center $(h, k)$ is $r$. This means that a **circle** in the plane consists of all points $(x, y)$ that are a given positive distance $r$ from a fixed point $(h, k)$. Using the Distance Formula, you can conclude that the point $(x, y)$ lies on the circle if and only if

$$\sqrt{(x - h)^2 + (y - k)^2} = r.$$

By squaring both sides of this equation, you obtain the **standard form of the equation of a circle.**

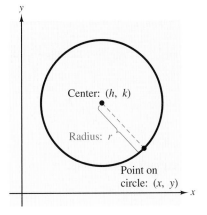

Figure 2.24

> ▶ **Standard Form of the Equation of a Circle**
>
> The **standard form of the equation of a circle** is
>
> $$(x - h)^2 + (y - k)^2 = r^2.$$
>
> The point $(h, k)$ is called the **center** of the circle and the positive number $r$ is called the **radius** of the circle.

The standard form of the equation of a circle whose center is the *origin* is simply

$$x^2 + y^2 = r^2.$$

**Example 6**    Finding an Equation of a Circle

The point $(3, 4)$ lies on a circle whose center is at $(-1, 2)$, as shown in Figure 2.25. Find an equation for the circle.

**Solution**

The radius $r$ of the circle is the distance between $(-1, 2)$ and $(3, 4)$. So, you have

$$r = \sqrt{[3 - (-1)]^2 + (4 - 2)^2}$$

$$= \sqrt{16 + 4} = \sqrt{20}.$$

So, the center of the circle is $(h, k) = (-1, 2)$, the radius is $r = \sqrt{20}$, and you can write the standard form of the equation of the circle as follows.

$$(x - h)^2 + (y - k)^2 = r^2 \qquad \text{Standard form}$$

$$[x - (-1)]^2 + (y - 2)^2 = \left(\sqrt{20}\right)^2 \qquad \text{Let } h = -1, k = 2, \text{ and } r = \sqrt{20}.$$

$$(x + 1)^2 + (y - 2)^2 = 20 \qquad \text{Equation of circle}$$

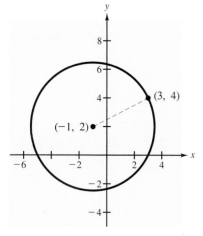

**Figure 2.25**

If you remove the parentheses in the standard equation in Example 6, you obtain the following.

$$(x + 1)^2 + (y - 2)^2 = 20 \qquad \text{Standard form}$$

$$x^2 + 2x + 1 + y^2 - 4y + 4 = 20 \qquad \text{Expand terms.}$$

$$x^2 + y^2 + 2x - 4y - 15 = 0 \qquad \text{General form}$$

The last equation is in the **general form of the equation of a circle.**

$$Ax^2 + Ay^2 + Dx + Ey + F = 0, \qquad A \neq 0$$

The general form of the equation of a circle is less useful than the standard form. For instance, it is not immediately apparent from the general equation shown above that the center is $(-1, 2)$ and the radius is $\sqrt{20}$. To graph the equation of a circle, it is best to write the equation in standard form. You can do this by **completing the square,** as demonstrated in Example 7.

| Example 7 | Completing the Square to Sketch a Circle |

Describe the circle given by

$$4x^2 + 4y^2 + 20x - 16y + 37 = 0.$$

### Study Tip

To complete the square, add the square of half the coefficient of the linear term.

### Solution

Begin by writing the given equation in standard form by completing the square for both the $x$-terms *and* the $y$-terms.

$$4x^2 + 4y^2 + 20x - 16y + 37 = 0 \qquad \text{General form}$$

$$x^2 + y^2 + 5x - 4y + \frac{37}{4} = 0 \qquad \text{Divide by 4.}$$

$$\left(x^2 + 5x + \phantom{xx}\right) + \left(y^2 - 4y + \phantom{xx}\right) = -\frac{37}{4} \qquad \text{Group terms.}$$

$$\left[x^2 + 5x + \left(\frac{5}{2}\right)^2\right] + \left(y^2 - 4y + 2^2\right) = -\frac{37}{4} + \frac{25}{4} + 4 \qquad \text{Complete square.}$$

$$\left(x + \frac{5}{2}\right)^2 + (y - 2)^2 = 1 \qquad \text{Standard form}$$

So, the center of the circle is $\left(-\frac{5}{2}, 2\right)$ and the radius of the circle is 1. Using this information, you can sketch the circle shown in Figure 2.26.

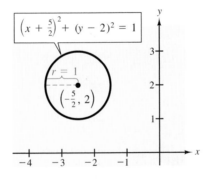

Figure 2.26

| Discussing the Concept | Interpreting an Intercept |

For the graph of each of the following real-life situations, find the indicated intercept and explain its practical significance.

**a.** (*y-intercept*) The trade-in value $y$ of a used car $x$ years after its purchase is $y = 9500 - 2400x$.

**b.** (*x-intercept*) The number of families $y$ demanding child care services each week is related to the price $x$ per hour of care by $y = 7850 - 20x^2$, $x \geq 0$.

Find an example of a real-life situation in which it makes sense for there to be no $x$-intercept.

# Warm Up

The following warm-up exercises involve skills that were covered in earlier sections. You will use these skills in the exercise set for this section.

In Exercises 1 and 2, solve for $y$ in terms of $x$.

**1.** $3x - 5y = 2$

**2.** $x^2 - 4x + 2y - 5 = 0$

In Exercises 3–6, solve the equation.

**3.** $x^2 - 4x + 4 = 0$

**4.** $(x - 1)(x + 5) = 0$

**5.** $x^3 - 9x = 0$

**6.** $x^4 - 8x^2 + 16 = 0$

In Exercises 7–10, simplify the equation.

**7.** $-y = (-x)^3 + 4(-x)$

**8.** $(-x)^2 + (-y)^2 = 4$

**9.** $y = 4(-x)^2 + 8$

**10.** $(-y)^2 = 3(-x) + 4$

## 2.2   Exercises

In Exercises 1–6, determine whether the points lie on the graph of the equation.

| Equation | Points | |
|---|---|---|
| **1.** $y = 2x^2 - 3$ | (a) $(-1, -1)$ | (b) $(2, 5)$ |
| **2.** $y = 5 - x - x^2$ | (a) $(0, 5)$ | (b) $(-2, -1)$ |
| **3.** $4y - 3x + 7 = 0$ | (a) $(1, -2)$ | (b) $(4, 3)$ |
| **4.** $x^2 - y^2 = 13$ | (a) $(5, 3)$ | (b) $(-7, 6)$ |
| **5.** $y = \sqrt{x - 8}$ | (a) $(9, 1)$ | (b) $(17, 3)$ |
| **6.** $y = \dfrac{2x}{6 - x}$ | (a) $(-3, 0)$ | (b) $(-4, -8)$ |

In Exercises 7–14, find the intercepts of the graph.

**7.** $y = 2x - 1$

**8.** $y = (x - 4)(x + 2)$

**9.** $y = x^2 + x - 2$

**10.** $y = 4 - x^2$

**11.** $y = x\sqrt{x + 2}$

**12.** $y = x\sqrt{x + 3}$

**13.** $2y - xy + 3x = 4$

**14.** $x^2y - x^2 + 4y = 0$

**15.** Discuss how you would use your knowledge of the Cartesian plane and intercepts to explain why you let $y$ equal zero when you are finding the $x$-intercepts of the graph of an equation, and let $x$ equal zero when you are finding the $y$-intercepts of the graph of an equation.

**16.** Is it possible for a graph to have no $x$-intercepts? Is it possible for a graph to have no $y$-intercepts? Is it possible for a graph to have no $x$-intercepts and no $y$-intercepts? Give examples to support your answers.

In Exercises 17–24, check for symmetry.

**17.** $x^4 - 2y = 0$

**18.** $x - y^2 = 0$

**19.** $y = \sqrt{4 - x^2}$

**20.** $y = \sqrt{9 - x^2}$

**21.** $x^3y = 1$

**22.** $xy = 4$

**23.** $y = \dfrac{x}{x^2 + 1}$

**24.** $y = x^4 - x^2 + 3$

In Exercises 25–28, use symmetry to complete the graph.

**25.** $y$-axis symmetry
$y = -x^2 + 4$

**26.** $x$-axis symmetry
$y^2 = -x + 4$

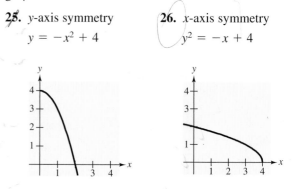

**27.** Origin symmetry

$y = -x^3 + x$

**28.** $y$-axis symmetry

$y = |x| - 2$

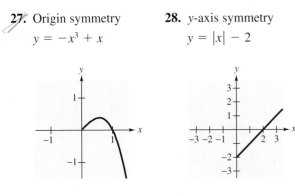

In Exercises 29–34, match the equation with its graph. [The graphs are labeled (a), (b), (c), (d), (e), and (f).]

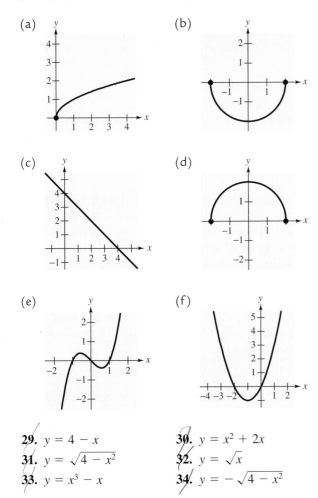

**(a)**  **(b)**  **(c)**  **(d)**  **(e)**  **(f)**

**29.** $y = 4 - x$

**30.** $y = x^2 + 2x$

**31.** $y = \sqrt{4 - x^2}$

**32.** $y = \sqrt{x}$

**33.** $y = x^3 - x$

**34.** $y = -\sqrt{4 - x^2}$

In Exercises 35–54, sketch the graph of the equation. Identify any intercepts and test for symmetry.

**35.** $y = 5 - 3x$

**36.** $y = 2x - 3$

**37.** $y = 1 - x^2$

**38.** $y = x^2 - 1$

**39.** $y = x^2 - 4x + 3$

**40.** $y = -x^2 - 4x$

**41.** $y = x^3 + 2$

**42.** $y = x^3 - 1$

**43.** $y = \dfrac{8}{x^2 + 4}$

**44.** $y = \dfrac{4}{x^2 + 1}$

**45.** $y = \sqrt{x + 1}$

**46.** $y = \sqrt{1 - x}$

**47.** $y = \sqrt[3]{x}$

**48.** $y = \sqrt[3]{x + 1}$

**49.** $y = |x - 4|$

**50.** $y = |x| - 3$

**51.** $x = y^2 - 1$

**52.** $x = y^2 - 4$

**53.** $x^2 + y^2 = 4$

**54.** $x^2 + y^2 = 16$

In Exercises 55–62, find the standard form of the equation of the specified circle.

**55.** Center: $(0, 0)$;  radius: $3$

**56.** Center: $(0, 0)$;  radius: $5$

**57.** Center: $(-4, 1)$;  radius: $\sqrt{2}$

**58.** Center: $\left(0, \frac{1}{3}\right)$;  radius: $\frac{1}{3}$

**59.** Center: $(-1, 2)$;  solution point: $(0, 0)$

**60.** Center: $(3, -2)$;  solution point: $(-1, 1)$

**61.** Endpoints of a diameter: $(-3, 4)$, $(5, -2)$

**62.** Endpoints of a diameter: $(-4, -1)$, $(4, 1)$

In Exercises 63–70, write the equation of the circle in standard form. Then sketch the circle.

**63.** $x^2 + y^2 - 6x + 4y - 3 = 0$

**64.** $x^2 + y^2 - 2x + 6y - 15 = 0$

**65.** $x^2 + y^2 - 2x + 6y + 10 = 0$

**66.** $5x^2 + 5y^2 + 10x + 1 = 0$

**67.** $2x^2 + 2y^2 - 2x - 2y - 3 = 0$

**68.** $4x^2 + 4y^2 - 4x + 2y - 1 = 0$

**69.** $16x^2 + 16y^2 + 16x + 40y - 7 = 0$

**70.** $x^2 + y^2 - 4x + 2y + 3 = 0$

In Exercises 71–74, find the constant $C$ such that the ordered pair is a solution point of the equation.

**71.** $y = C - 4x^2$         $(2, -10)$

**72.** $y = Cx^3$            $(-4, 8)$

**73.** $y = C\sqrt{x + 1}$      $(3, 8)$

**74.** $x + C(y + 2) = 0$    $(4, 3)$

In Exercises 75 and 76, an equation of a circle is written in standard form. Indicate the coordinates of the center of the circle and determine the radius of the circle. Rewrite the equation of the circle in general form.

**75.** $(x - 2)^2 + (y + 3)^2 = 16$

**76.** $\left(x + \frac{1}{2}\right)^2 + (y + 1)^2 = 5$

In Exercises 77 and 78, (a) create a table to compare the data to the values found using the model, (b) sketch a graph comparing the data and the model for that data, and (c) use the model to predict $y$ for the year 2005.

**77.** *Purchasing Power of the Dollar* The table gives the purchasing power of the U.S. dollar for *consumers* from 1989 to 1999. The base year for comparison is 1982, the year in which the purchasing power of the dollar for *producers* was $1.00. (Source: U.S. Bureau of Labor Statistics)

| Year | 1989 | 1990 | 1991 | 1992 |
|------|------|------|------|------|
| Purchasing power | 0.807 | 0.766 | 0.734 | 0.713 |

| Year | 1993 | 1994 | 1995 | 1996 |
|------|------|------|------|------|
| Purchasing power | 0.692 | 0.675 | 0.656 | 0.638 |

| Year | 1997 | 1998 | 1999 |
|------|------|------|------|
| Purchasing power | 0.623 | 0.614 | 0.600 |

A mathematical model that approximates the purchasing power of the dollar is

$$y = \frac{0.769 + 0.032t}{1 + 0.084t}, \quad -1 \le t \le 9$$

where $y$ represents the purchasing power of the dollar and $t = 0$ represents 1990.

**78.** *Life Expectancy* The table gives the life expectancy of a child (at birth) for selected years from 1970 to 2000. (Source: U.S. National Center for Health Statistics)

| Year | 1970 | 1975 | 1980 |
|------|------|------|------|
| Life expectancy | 70.8 | 72.6 | 73.7 |

| Year | 1985 | 1990 | 1995 | 2000 |
|------|------|------|------|------|
| Life expectancy | 74.7 | 75.4 | 75.8 | 77.1 |

A mathematical model for the life expectancy during this period is

$$y = \frac{70.895 + 2.193t}{1 + 0.026t}, \quad 0 \le t \le 30$$

where $y$ represents the life expectancy in years and $t = 0$ represents 1970.

**79.** *Earnings per Share* The earnings per share $y$ (in dollars) for Microsoft Corporation from 1990 to 2000 can be modeled by

$$y = 0.0236t^2 - 0.085t + 0.16, \quad 0 \le t \le 10$$

where $t = 0$ represents 1990. Sketch the graph of this equation. In 2000, Microsoft Corporation projected that the earnings per share in 2001 and 2002 would be $1.80 and $1.95, respectively. Compare Microsoft's projections for 2001 and 2002 with the earnings per share predicted by the model. (Source: Microsoft Corporation)

**80.** *Earnings per Share* The earnings per share $y$ (in dollars) for Wal-Mart Stores from 1990 to 2000 can be modeled by

$$y = 0.105t + 0.19, \quad 0 \le t \le 10$$

where $t = 0$ represents 1990. Sketch the graph of this equation. In 2000, Wal-Mart Stores projected that the earnings per share in 2001 and 2002 would be $1.55 and $1.75, respectively. Compare Wal-Mart's projection for 2001 and 2002 with the earnings per share predicted by the model. (Source: Wal-Mart Stores, Inc.)

## 2.3   Graphing Utilities

### Objectives

- Graph an equation using a graphing utility.
- Determine an appropriate viewing window for the graph of an equation.
- Use a square setting to obtain a true geometric perspective.

The Granger Collection

### Blaise Pascal

1623–1662

Calculators have interested mathematicians for hundreds of years. In 1642, at the age of 18, Blaise Pascal began designing a calculating machine. Over the next few years, Pascal built and sold approximately 50 machines.

## Introduction

In Section 2.2 you studied the point-plotting method for sketching the graph of an equation. One of the disadvantages of the point-plotting method is that to get a good idea about the shape of a graph, you need to plot *many* points. By plotting only a few points, you can badly misrepresent the graph.

For instance, consider the equation $y = \frac{1}{30}x(x^4 - 10x^2 + 39)$. To graph this equation, suppose you calculated only the following five points.

| $x$ | $-3$ | $-1$ | $0$ | $1$ | $3$ |
|---|---|---|---|---|---|
| $y = \frac{1}{30}x(x^4 - 10x^2 + 39)$ | $-3$ | $-1$ | $0$ | $1$ | $3$ |

By plotting these five points, as shown in Figure 2.27(a), you might assume that the graph of the equation is a straight line. This, however, is not correct. By plotting several more points, as shown in Figure 2.27(b), you can see that the actual graph is not straight at all.

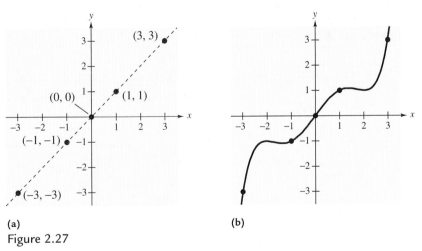

(a)                                                    (b)

Figure 2.27

So, the point-plotting method leaves you with a dilemma. On the one hand, the method can be very inaccurate if only a few points are plotted. But, on the other hand, it is very time consuming to plot a dozen (or more) points. Technology can help you resolve this dilemma. Plotting several points (even hundreds of points) on a rectangular coordinate system is something that a graphing utility can do easily.

## Using a Graphing Utility

There are many different graphing utilities: some are graphing packages for computers and some are hand-held graphing calculators. The steps used to graph an equation with a graphing utility are similar for all types of utilities and are summarized below.

---

▶ **Graphing an Equation with a Graphing Utility**

Before performing the following steps, set your graphing utility so that all the standard defaults are active.

1. Set the viewing window for the graph.

2. Rewrite the equation so that $y$ is isolated on the left side of the equation.

3. Using the equation editor, enter the right side of the equation on the first line of the display. (The first line may be labeled $Y_1 =$.)

4. Press the appropriate key to graph the equation.

---

Graphing utilities are equipped with a *window* feature. This feature allows you to determine the portion of the coordinate plane that you want to view. You can also determine the distance between consecutive tick marks (scale) on each axis. If your calculator is set to the standard graphing defaults, the screen should show the following values.

| | |
|---|---|
| Xmin=-10 | The minimum $x$-value is $-10$. |
| Xmax=10 | The maximum $x$-value is 10. |
| Xscl=1 | The $x$-scale is one unit per tick mark. |
| Ymin=-10 | The minimum $y$-value is $-10$. |
| Ymax=10 | The maximum $y$-value is 10. |
| Yscl=1 | The $y$-scale is one unit per tick mark. |

These settings are summarized visually in Figure 2.28. This viewing window is usually called the **standard viewing window.**

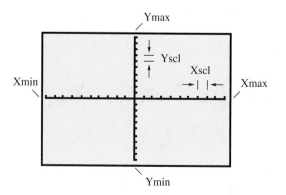

Figure 2.28

The following examples use the standard viewing window. Notice how the window contains all of the intercepts of the graph.

**Example 1**   Graphing an Equation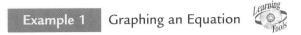

Use a graphing utility to graph $2y + x^3 = 4x$.

**Solution**

To begin, solve the given equation for $y$ in terms of $x$.

$$2y + x^3 = 4x \qquad \text{Original equation}$$

$$2y = -x^3 + 4x \qquad \text{Subtract } x^3 \text{ from both sides.}$$

$$y = -\frac{1}{2}x^3 + 2x \qquad \text{Divide both sides by 2.}$$

Enter the equation in the equation editor of your graphing utility. The equation may require the notation

$$Y_1 = -(1/2)X^{\wedge}3 + 2X.$$

Now graph using the standard viewing window. Your screen should look like the one shown in Figure 2.29.

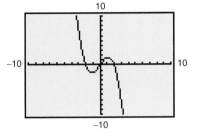

**Figure 2.29**

The next example illustrates how to graph an absolute value equation. Most graphing utilities use the abbreviation "ABS" or "abs" for absolute value. The expression within the absolute value signs should be enclosed in parentheses in the graphing utility. For example, the equation

$$y = \frac{1}{2}|3x + 4|$$

should appear in the equation editor as

$$Y_1 = (1/2)\,\text{abs}\,(3X+4).$$

**Example 2**   Graphing an Equation Involving Absolute Value

Use a graphing utility to graph $y = |x - 3|$.

**Solution**

This equation is already written with $y$ isolated on the left side of the equation. Enter the equation in the equation editor of your graphing utility. The equation may require the notation

$$Y_1 = \text{abs}\,(X-3).$$

Now graph using the standard viewing window. Your screen should look like the one shown in Figure 2.30.

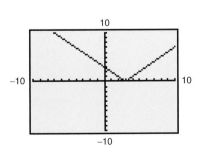

**Figure 2.30**

# Using a Graphing Utility's Special Features

To use your graphing utility to its best advantage, you must learn how to set the viewing window, as illustrated in the next example.

### Example 3    Setting the Viewing Window

Use a graphing utility to graph $y = x^2 + 12$.

### Solution

Enter the equation in the equation editor of your graphing utility and graph. If your graphing utility is set to the standard viewing window, nothing will appear on the screen. The reason for this is that the lowest point on the graph of $y = x^2 + 12$ occurs at the point $(0, 12)$. Using the standard viewing window, you obtain a screen whose largest $y$-value is 10. In other words, none of the graph is visible on a screen whose $y$-values vary between $-10$ and 10, as shown in Figure 2.31(a). To change these settings, enter the following values.

| | |
|---|---|
| Xmin=-10 | The minimum $x$-value is $-10$. |
| Xmax=10 | The maximum $x$-value is 10. |
| Xscl=1 | The $x$-scale is one unit per tick mark. |
| Ymin=-10 | The minimum $y$-value is $-10$. |
| Ymax=30 | The maximum $y$-value is 30. |
| Yscl=5 | The $y$-scale is five units per tick mark. |

Graph the equation with the new viewing window and you will obtain the graph shown in Figure 2.31(b). On this graph, note that each tick mark on the $y$-axis represents five units because you changed the $y$-scale to 5. Also note that the highest point on the $y$-axis is now 30 because you changed the maximum value of $y$ to 30.

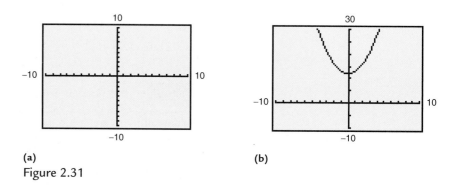

(a)                                    (b)

Figure 2.31

A graphing utility is especially useful for graphing more complicated equations such as

$$y = -0.166 + 0.02x - 0.0953x^2.$$

You must be careful to select a viewing window that shows all of the important features of the graph.

Most graphing utilities have a feature that will change the standard viewing window to a viewing window with a square setting. That is, the tick marks on the x-axis will appear the same distance apart as the tick marks on the y-axis. Look for the square setting in the zoom feature of your graphing utility.

## Study Tip

A square setting is useful anytime you want a graph that has a true geometric perspective. For instance, if you want the graph of $x^2 + y^2 = 25$ to look like a circle, you need to use a square setting. You also need to enter the top and bottom portions of the circle separately, as follows.

$$y = \sqrt{25 - x^2} \qquad \text{Top half}$$
$$y = -\sqrt{25 - x^2} \qquad \text{Bottom half}$$

In Figure 2.33(a) the graph of the circle appears stretched in a standard viewing window, whereas in Figure 2.33(b) the square setting shows the circle in true geometric perspective.

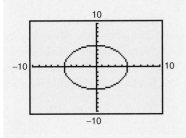

(a)

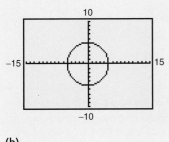

(b)
Figure 2.33

### Example 4   Using a Square Setting

Use a graphing utility to graph $y = x$. The graph of this equation is a straight line that makes a 45° angle with the x-axis and the y-axis. From the graph on your calculator, does the angle appear to be 45°?

### Solution

Graph the equation $y = x$ in a standard viewing window and you will obtain the graph shown in Figure 2.32(a). Note that the angle the line makes with the x-axis doesn't appear to be 45°. The reason for this is that the screen is wider than it is tall. This has the effect of making the tick marks on the x-axis farther apart than the tick marks on the y-axis. To obtain the same distance between tick marks on both axes, you can change the graphing settings from "standard" to "square." The screen should look like that shown in Figure 2.32(b). Note in this figure that the square setting has changed the viewing window so that the x-values vary between −15 and 15.

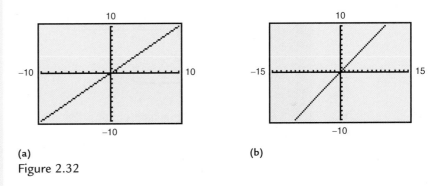

(a)                                         (b)
Figure 2.32

There are many possible square settings on a graphing calculator. To create a square setting, you need the following ratio to be $\frac{2}{3}$.

$$\frac{Ymax - Ymin}{Xmax - Xmin}$$

For instance, the setting in Example 4 is square because $(Ymax - Ymin) = 20$ and $(Xmax - Xmin) = 30$.

**Example 5**    Graphing More than One Equation on the Same Screen

Graph the following equations on the same screen.

$$y = -x + 5, \qquad y = -x, \qquad \text{and} \qquad y = -x - 5$$

**Solution**

Enter all three equations on the first three lines, $Y_1$, $Y_2$, and $Y_3$, of the equation editor. The display may be as follows.

$Y_1 = -X + 5$

$Y_2 = -X$

$Y_3 = -X - 5$

The graphs of the equations are shown in Figure 2.34. Note that the graph of each equation is a straight line, and that the lines are parallel to each other.

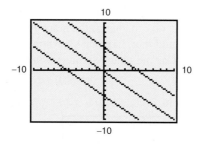

**Figure 2.34**

Most graphing utilities have other convenient graphing aides. You may be able to *box* a portion of the viewing window and enlarge that area of the graph. *Zoom in* and *zoom out* enlarge or shrink the area of the graph centered at the cursor. Additional features may also be available. Check the user's guide for your graphing utility.

| Discussing the Concept | Graphing with Technology |
| --- | --- |

For each of the following equations, describe what you must do before you can graph the equation using a graphing utility.

**a.** $-4y + 32 = 2x$      **b.** $y = \frac{1}{3}(x^3 - 10x)$

**c.** $x^2 + 2y^2 = 16$      **d.** $x^2 = y + 12x$

Now use a graphing utility and a standard viewing window to graph each equation. Enlarge your view of the graph by zooming out or increasing the boundaries of your viewing window several times. For which equations are all of the x-intercepts shown in the standard viewing window?

# Warm Up

The following warm-up exercises involve skills that were covered in earlier sections. You will use these skills in the exercise set for this section.

In Exercises 1–10, solve for $y$ in terms of $x$.

**1.** $3x + y = 4$

**2.** $x - y = 0$

**3.** $2x + 3y = 2$

**4.** $4x - 5y = -2$

**5.** $3x + 4y - 5 = 0$

**6.** $-2x - 3y + 6 = 0$

**7.** $x^2 + y - 4 = 0$

**8.** $-2x^2 + 3y + 2 = 0$

**9.** $x^2 + y^2 = 4$

**10.** $x^2 - y^2 = 9$

## 2.3   Exercises

In Exercises 1–10, use a graphing utility to match the equation with its graph. [The graphs are labeled (a), (b), (c), (d), (e), (f), (g), (h), (i), and (j).]

**1.** $y = x$  d

**2.** $y = -x$  g

**3.** $y = x^2$  a

**4.** $y = -x^2$  f

**5.** $y = x^3$  i

**6.** $y = -x^3$  b

**7.** $y = |x|$  j

**8.** $y = -|x|$  c

**9.** $y = \sqrt{x}$  e

**10.** $y = -\sqrt{x}$  h

(a)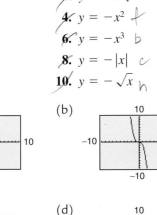

(b)

(c)

(d)

(e)

(f)

(g)

(h)

(i)

(j)

In Exercises 11–30, use a graphing utility to graph the equation. Use a standard setting for each graph.

**11.** $y = x - 5$

**12.** $y = -x + 4$

**13.** $y = -\frac{1}{2}x + 3$

**14.** $y = \frac{2}{3}x + 1$

**15.** $2x - 3y = 4$

**16.** $x + 2y = 3$

**17.** $y = \frac{1}{2}x^2 - 1$

**18.** $y = -x^2 + 6$

**19.** $y = x^2 - 4x - 5$

**20.** $y = x^2 - 3x + 2$

**21.** $y = -x^2 + 2x + 1$

**22.** $y = -x^2 + 4x - 1$

**23.** $2y = x^2 + 2x - 3$

**24.** $3y = -x^2 - 4x + 5$

**25.** $y = |x + 5|$

**26.** $y = \frac{1}{2}|x - 6|$

**27.** $y = \sqrt{x^2 + 1}$

**28.** $y = 2\sqrt{x^2 + 2} - 4$

**29.** $y = \frac{1}{5}(-x^3 + 16x)$

**30.** $y = \frac{1}{8}(x^3 + 8x^2)$

The symbol ▦ indicates an exercise or parts of an exercise in which you are instructed to use a graphing utility.

In Exercises 31–34, use a graphing utility to graph the equation. Does the setting give a good representation of the graph? Explain.

**31.** $y = -2x^2 + 12x + 14$   **32.** $y = -x^2 + 5x + 6$

| | |
|---|---|
| Xmin = -5 | Xmin = -8 |
| Xmax = 10 | Xmax = 4 |
| Xscl = 1 | Xscl = 1 |
| Ymin = -5 | Ymin = -5 |
| Ymax = 35 | Ymax = 15 |
| Yscl = 5 | Yscl = 5 |

**33.** $y = x^3 + 6x^2$   **34.** $y = -x^3 + 16x$

| | |
|---|---|
| Xmin = -10 | Xmin = -6 |
| Xmax = 5 | Xmax = 6 |
| Xscl = 1 | Xscl = 1 |
| Ymin = -4 | Ymin = -25 |
| Ymax = 36 | Ymax = 25 |
| Yscl = 4 | Yscl = 5 |

In Exercises 35 and 36, find a setting on a graphing utility such that the viewing window will be square. In the square viewing window, the following ratio must be $\frac{2}{3}$.

$$\frac{\text{Ymax} - \text{Ymin}}{\text{Xmax} - \text{Xmin}}$$

**35.**
| |
|---|
| Xmin = ? |
| Xmax = 30 |
| Xscl = 2 |
| Ymin = -5 |
| Ymax = 35 |
| Yscl = 5 |

**36.**
| |
|---|
| Xmin = -4 |
| Xmax = 11 |
| Xscl = 1 |
| Ymin = -2 |
| Ymax = ? |
| Yscl = 1 |

In Exercises 37–40, find a setting on a graphing utility such that the graph of the equation agrees with the graph shown.

**37.** $y = -x^2 - 4x + 20$   **38.** $y = x^2 + 12x - 8$

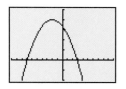   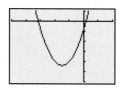

**39.** $y = -x^3 + x^2 + 2x$   **40.** $y = x^3 + 3x^2 - 2x$

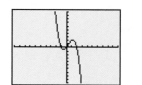   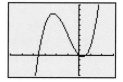

In Exercises 41–44, use a graphing utility to find the number of $x$-intercepts of the equation(s).

**41.** (a) $y = \frac{1}{8}(4x^2 - 32x + 65)$

   (b) $y = \frac{1}{8}(4x^2 + 32x + 63)$

**42.** (a) $y = \frac{1}{4}(-4x^2 + 16x - 15)$

   (b) $y = \frac{1}{4}(-4x^2 + 16x - 17)$

**43.** $y = 4x^3 - 20x^2 - 4x + 61$

**44.** $y = \frac{1}{4}(2x^3 + 6x^2 - 4x + 1)$

**45.** What do you observe about the graphs of the equation in Exercises 41(b) and 42(a)? 41(a) and 42(b)?

**46.** How are the graphs in Exercises 43 and 44 alike? How are the graphs different?

In Exercises 47–50, use a graphing utility to graph the equations on the same screen. Use a square setting that shows a good representation of the graphs. What geometric shape is bounded by the graphs?

**47.** $y = |x| - 4,\ y = -|x| + 4$

**48.** $y = x + |x| - 4,\ y = x - |x| + 4$

**49.** $y = -\sqrt{25 - x^2},\ y = \sqrt{25 - x^2}$

**50.** $y = 6,\ y = -\sqrt{3}x - 4,\ y = \sqrt{3}x - 4$

**51.** Based on 1982 producer prices, the purchasing power of the dollar $y$ for consumers from 1982 to 1999 can be modeled by

$$y = \frac{-0.0643t + 11.195}{0.3771t + 9.829}, \qquad 2 \le t \le 19$$

where $t = 2$ represents 1982.   (Source: U.S. Bureau of Labor Statistics)

(a) Use this model to find the purchasing power of the dollar in 1999.

(b) The purchasing power of the dollar in 1999 was actually 0.600. What does this tell you about the model?

**52.** A mathematical model for a girl's life expectancy at birth from 1970 to 2000 is

$$y = 74.6 + 0.92\sqrt{t}, \qquad 0 \le t \le 30$$

where $y$ represents the life expectancy in years and $t = 0$ represents 1970. (Source: U.S. National Center for Health Statistics)

(a) Using this model, predict the life expectancy in 2010 and in 2020.

(b) Are your answers reasonable? What cautions would you give when using a mathematical model for a time span that is greatly different from the period in which the data for the model were collected?

**53.** Describe how you could use the table feature of your graphing utility to help select an appropriate viewing window for the equation in Exercise 51. If your graphing utility does not have a table feature, is there another way you can obtain ideas for the size of a viewing window?

**54.** A friend uses a graphing utility to graph the equation in Exercise 52. No graph appears on the viewing screen. What might have happened? How would you help your friend find an appropriate viewing window?

*Ever Been Married?*   In Exercises 55–58, use the following models, which relate ages to the percents $y$ of American males and females who have never been married.

$$y = -10.1 + \frac{395.72}{x} + \frac{33{,}601.431}{x^2} \qquad \text{Males}$$

$$y = 7.9 - \frac{1060.17}{x} + \frac{52{,}525.784}{x^2} \qquad \text{Females}$$

In these models, $x$ is the age of the person, and each model is valid for $20 \le x \le 80$. (Source: U.S. Bureau of the Census)

**55.** Use a graphing utility to graph both models on the same screen.

**56.** Write a paragraph describing the relationship between the two graphs that were plotted in Exercise 55.

**57.** Suppose that an American male is chosen at random from the population. If the person is 25 years old, what is the probability that he has never been married?

**58.** Suppose that an American female is chosen at random from the population. If the person is 25 years old, what is the probability that she has never been married?

*Earnings and Dividends*  In Exercises 59–62, use the following model, which approximates the relationship between dividends per share $y$ (in dollars) and earnings per share $x$ (in dollars) for Procter & Gamble Company between 1990 and 2000. (Source: The Procter & Gamble Company)

$$y = 0.0976x^2 + 0.023x + 0.32, \quad 1.03 \le x \le 2.95$$

**59.** Use a graphing utility to graph the model.

**60.** According to the model, what size dividend would Procter & Gamble pay if the earnings per share were $2.00?

**61.** Use the trace feature of your graphing utility to estimate the earnings per share that would produce a dividend per share of $0.80.

**62.** The *payout ratio* for a stock is the ratio of the dividend per share to earnings per share. Use the model to find the payout ratio for an earnings per share of

(a) $1.25      (b) $1.50      (c) $2.00

## Mid-Chapter Quiz

Take this quiz as you would take a quiz in class. After you are done, check your work against the answers given in the back of the book.

In Exercises 1 and 2, (a) plot the points, (b) find the distance between the points, and (c) find the midpoint of the line segment joining the points.

**1.** $(-3, 2)$, $(4, -5)$

**2.** $(1.3, -4.5)$, $(-3.7, 0.7)$

**3.** A business had sales of \$1,330,000 in 1995 and \$1,890,000 in 1999. Estimate the sales in 1997. Explain your reasoning.

**4.** One plane is 200 miles due north of a transmitter and another plane is 150 miles due east of the transmitter. What is the distance between the planes?

In Exercises 5 and 6, describe the figure with the given vertices.

**5.** $(-1, -1)$, $(10, 7)$, $(2, 18)$

**6.** $(-3, -1)$, $(0, 2)$, $(4, 2)$, $(1, -1)$

In Exercises 7–12, sketch the graph of the equation. Identify any intercepts and symmetry.

**7.** $y = 9 - x^2$

**8.** $y = x\sqrt{x + 4}$

**9.** $xy = 9$

**10.** $y = \sqrt{16 - x^2}$

**11.** $x^2 + y^2 = 9$

**12.** $y = |x - 3|$

In Exercises 13 and 14, find the standard form of the equation of the circle.

**13.** Center: $(2, -3)$;    radius: 4

**14.** Center: $\left(0, -\frac{1}{2}\right)$;    radius: 2

**15.** Write the equation $x^2 + y^2 - 2x + 4y - 4 = 0$ in standard form. Then sketch the circle.

In Exercises 16 and 17, use a graphing utility to find the number of $x$-intercepts of the graph of the equation.

**16.** $y = \frac{1}{4}(-4x^2 + 16x - 14)$

**17.** $y = \frac{1}{8}(4x^2 - 32x + 63)$

In Exercises 18–20, use the following model. The earnings per share $y$ (in dollars) for Minnesota Mining and Manufacturing Company from 1990 to 2000 can be modeled by $y = 0.181t + 2.54$, $0 \le t \le 10$, where $t = 0$ represents 1990.    (Source: Minnesota Mining and Manufacturing Co.)

**18.** Determine a convenient graphing utility viewing window.

**19.** Use a graphing utility to graph this model.

**20.** Use the trace feature to predict the earnings per share in 2006.

## 2.4   Lines in the Plane

### Objectives

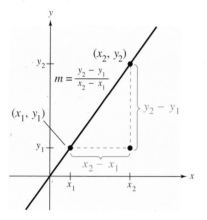

- Determine the slope of a line passing through two points.
- Use the point-slope form to find the equation of a line.
- Use the slope-intercept form to sketch the graph of a line.
- Determine if lines are parallel or perpendicular using slope.
- Write the equation of a line parallel or perpendicular to a given line.

Figure 2.35

## The Slope of a Line

The **slope** of a nonvertical line is a measure of the steepness of the line. Specifically, the slope represents the number of units the line rises or falls vertically for each unit of horizontal change from left to right. For instance, consider the two points $(x_1, y_1)$ and $(x_2, y_2)$ on the line shown in Figure 2.35. As you move from left to right along this line, a change of $y_2 - y_1$ units in the vertical direction corresponds to a change of $x_2 - x_1$ units in the horizontal direction. That is,

$$y_2 - y_1 = \text{the change in } y$$

and

$$x_2 - x_1 = \text{the change in } x.$$

The slope of the line is defined as the quotient of these two changes.

> ▶ Definition of the Slope of a Line
>
> The **slope** $m$ of the nonvertical line passing through the points $(x_1, y_1)$ and $(x_2, y_2)$ is
>
> $$m = \frac{y_2 - y_1}{x_2 - x_1} = \frac{\text{change in } y}{\text{change in } x}$$
>
> where $x_1 \neq x_2$.

The change in $x$ is sometimes called the "run" and the change in $y$ is sometimes called the "rise."

When this formula is used, the *order of subtraction* is important. Given two points on a line, you are free to label either one of them as $(x_1, y_1)$ and the other as $(x_2, y_2)$. However, once this is done, you must form the numerator and denominator using the same order of subtraction.

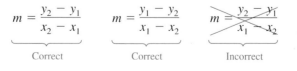

$$m = \frac{y_2 - y_1}{x_2 - x_1} \qquad m = \frac{y_1 - y_2}{x_1 - x_2} \qquad m = \frac{y_2 - y_1}{x_1 - x_2}$$

Correct                    Correct                    Incorrect

| Example 1 | Finding the Slope of a Line Through Two Points |

Find the slopes of the lines passing through the following pairs of points.

**a.** $(-2, 0)$ and $(3, 1)$    **b.** $(-1, 2)$ and $(2, 2)$    **c.** $(0, 4)$ and $(1, -1)$

**Solution**

**a.** Let $(x_1, y_1) = (-2, 0)$ and $(x_2, y_2) = (3, 1)$.

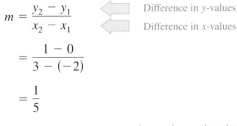

$$m = \frac{y_2 - y_1}{x_2 - x_1}$$    ⟵ Difference in $y$-values
⟵ Difference in $x$-values

$$= \frac{1 - 0}{3 - (-2)}$$

$$= \frac{1}{5}$$

**b.** The slope of the line through $(-1, 2)$ and $(2, 2)$ is

$$m = \frac{2 - 2}{2 - (-1)} = \frac{0}{3} = 0.$$

**c.** The slope of the line through $(0, 4)$ and $(1, -1)$ is

$$m = \frac{-1 - 4}{1 - 0} = \frac{-5}{1} = -5.$$

The graphs of the three lines are shown in Figure 2.36.

## Study Tip

When working with the slope of a line, it helps to remember the following interpretations of slope.

**1.** A line with positive slope *rises* from left to right.

**2.** A line with negative slope *falls* from left to right.

**3.** A line with zero slope is *horizontal*.

**4.** A line with undefined slope is *vertical*.

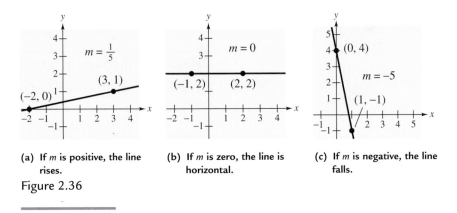

(a) If $m$ is positive, the line rises.

(b) If $m$ is zero, the line is horizontal.

(c) If $m$ is negative, the line falls.

**Figure 2.36**

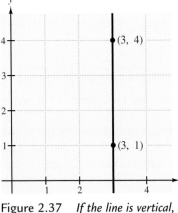

**Figure 2.37**    *If the line is vertical, the slope is undefined.*

The definition of slope does not apply to vertical lines. For instance, consider the points $(3, 4)$ and $(3, 1)$ on the vertical line shown in Figure 2.37. Applying the formula for slope, you have

$$\frac{1 - 4}{3 - 3} = \frac{-3}{0}.$$    Undefined division by zero

Because division by zero is not defined, the slope of a vertical line is not defined.

# The Point-Slope Form

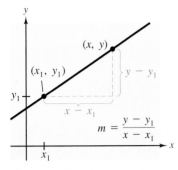

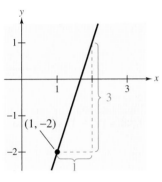

Figure 2.38    *Any two points on a line can be used to determine the slope of the line.*

Figure 2.39

If you know the slope of a line and you also know the coordinates of one point on the line, you can find an equation for the line. For instance, in Figure 2.38, let $(x_1, y_1)$ be a given point on the line whose slope is $m$. If $(x, y)$ is any *other* point on the line, it follows that

$$\frac{y - y_1}{x - x_1} = m.$$

This equation in variables $x$ and $y$ can be rewritten to produce the following **point-slope form** of the equation of a line.

> ▶ **Point-Slope Form of the Equation of a Line**
>
> The **point-slope form** of the equation of the line that passes through the point $(x_1, y_1)$ and has a slope of $m$ is
>
> $$y - y_1 = m(x - x_1).$$

**Example 2**    The Point-Slope Form of the Equation of a Line

Find an equation of the line that passes through $(1, -2)$ and has a slope of 3.

**Solution**

Use the point-slope form with $(x_1, y_1) = (1, -2)$ and $m = 3$.

| | |
|---|---|
| $y - y_1 = m(x - x_1)$ | Point-slope form |
| $y - (-2) = 3(x - 1)$ | Substitute $y_1 = -2$, $x_1 = 1$, and $m = 3$. |
| $y + 2 = 3x - 3$ | Simplify. |
| $y = 3x - 5$ | Equation of line |

The graph of this line is shown in Figure 2.39.

The point-slope form can be used to find the equation of a line passing through two points $(x_1, y_1)$ and $(x_2, y_2)$. First, use the formula for the slope of a line passing through two points. Then, use the point-slope form to obtain the equation

$$y - y_1 = \frac{y_2 - y_1}{x_2 - x_1}(x - x_1).$$

This is sometimes called the **two-point form** of the equation of a line.

| Example 3 | A Linear Model for Sales Prediction |
|---|---|

During the first two quarters of the year, a company had sales of $3.4 million and $3.7 million, respectively.

**a.** Write a linear equation giving the sales $y$ in terms of the quarter $x$.

**b.** Use the equation to predict the sales during the fourth quarter.

Solution

**a.** In Figure 2.40 let $(1, 3.4)$ and $(2, 3.7)$ be two points on the line representing the total sales. The slope of the line passing through these two points is

$$m = \frac{3.7 - 3.4}{2 - 1} = 0.3.$$

By the point-slope form, the equation of the line is as follows.

| | |
|---|---|
| $y - y_1 = m(x - x_1)$ | Point-slope form |
| $y - 3.4 = 0.3(x - 1)$ | Substitute for $y_1$, $m$, and $x_1$. |
| $y = 0.3x - 0.3 + 3.4$ | Simplify. |
| $y = 0.3x + 3.1$ | Equation of line |

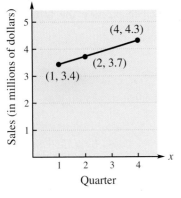

Figure 2.40

**b.** Using the equation from part (a), the fourth quarter sales $(x = 4)$ should be

$$y = 0.3(4) + 3.1 = 4.3 \text{ million dollars.}$$

The estimation method illustrated in Example 3 is called **linear extrapolation.** Note in Figure 2.41(a) that for linear extrapolation, the estimated point lies to the right of the given points. When the estimated point lies *between* two given points, the procedure is called **linear interpolation,** as shown in Figure 2.41(b).

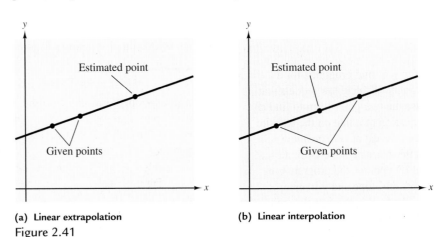

(a) Linear extrapolation

(b) Linear interpolation

Figure 2.41

## Sketching Graphs of Lines

You have seen that to *find the equation of a line* it is convenient to use the point-slope form. This formula, however, is not particularly useful for *sketching the graph of a line*. The form that is better suited to graphing linear equations is the **slope-intercept form** of the equation of a line. To derive the slope-intercept form, write the following.

$$y - y_1 = m(x - x_1) \qquad \text{Point-slope form}$$

$$y = mx - mx_1 + y_1 \qquad \text{Solve for } y.$$

$$y = mx + b \qquad \text{Slope-intercept form}$$

---

▶ **Slope-Intercept Form of the Equation of a Line**

The graph of the equation

$$y = mx + b$$

is a line whose slope is $m$ and whose $y$-intercept is $(0, b)$.

---

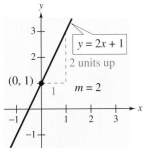

**(a)** When $m$ is positive, the line rises.

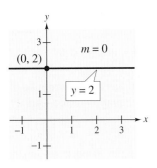

**(b)** When $m$ is zero, the line is horizontal.

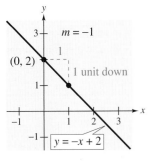

**(c)** When $m$ is negative, the line falls.

**Figure 2.42**

**Example 4**    Sketching the Graph of a Linear Equation

Sketch the graph of each of the following linear equations.

**a.** $y = 2x + 1$    **b.** $y = 2$    **c.** $x + y = 2$

**Solution**

**a.** Because $b = 1$, the $y$-intercept is $(0, 1)$. Moreover, because the slope is $m = 2$, this line *rises* two units for each unit the line moves to the right, as shown in Figure 2.42(a).

**b.** By writing the equation $y = 2$ in the form

$$y = (0)x + 2$$

you can see that the $y$-intercept is $(0, 2)$ and the slope is zero. A zero slope implies that the line is horizontal, as shown in Figure 2.42(b).

**c.** By writing the equation $x + y = 2$ in slope-intercept form,

$$y = -x + 2$$

you can see that the $y$-intercept is $(0, 2)$. Moreover, because the slope is $m = -1$, this line *falls* one unit for each unit the line moves to the right, as shown in Figure 2.42(c).

From the slope-intercept form of the equation of a line, you can see that a horizontal line ($m = 0$) has an equation of the form

$$y = (0)x + b \quad \text{or} \quad y = b. \qquad \text{Horizontal line}$$

This is consistent with the fact that each point on a horizontal line through $(0, b)$ has a $y$-coordinate of $b$, as shown in Figure 2.43.

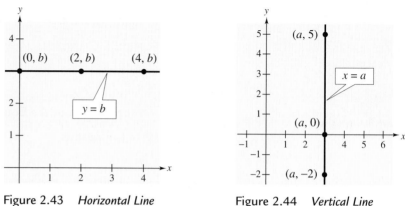

**Figure 2.43** *Horizontal Line*    **Figure 2.44** *Vertical Line*

Similarly, each point on a vertical line through $(a, 0)$ has an $x$-coordinate of $a$, as shown in Figure 2.44. Hence, a vertical line has an equation of the form

$$x = a. \qquad \text{Vertical line}$$

This equation cannot be written in the slope-intercept form because the slope of a vertical line is undefined. However, *every* line has an equation that can be written in the **general form**

$$Ax + By + C = 0 \qquad \text{General form}$$

where $A$ and $B$ are not *both* zero. If $A = 0$ (and $B \neq 0$), the equation can be reduced to the form $y = b$, which represents a horizontal line. If $B = 0$ (and $A \neq 0$), the general equation can be reduced to the form $x = a$, which represents a vertical line.

---

▶ **Summary of Equations of Lines**

  **1.** General form:            $Ax + By + C = 0$

  **2.** Vertical line:             $x = a$

  **3.** Horizontal line:        $y = b$

  **4.** Slope-intercept form:   $y = mx + b$

  **5.** Point-slope form:      $y - y_1 = m(x - x_1)$

## Parallel and Perpendicular Lines

The slope of a line is a convenient tool for determining whether two lines are parallel or perpendicular.

> ▶ Parallel Lines
>
>  Two distinct nonvertical lines are **parallel** if and only if their slopes are equal.

**Example 5**   Equations of Parallel Lines

Find an equation of the line that passes through the point $(2, -1)$ and is parallel to the line $2x - 3y = 5$, as shown in Figure 2.45.

**Solution**

Writing the given equation in slope-intercept form produces the following.

$2x - 3y = 5$        Original equation

$3y = 2x - 5$        Isolate $y$-term.

$y = \dfrac{2}{3}x - \dfrac{5}{3}$        Slope-intercept form

So, the given line has a slope of $m = \tfrac{2}{3}$. Because any line parallel to the given line must also have a slope of $\tfrac{2}{3}$, the required line through $(2, -1)$ has the following equation.

$y - (-1) = \dfrac{2}{3}(x - 2)$        Point-slope form

$y = \dfrac{2}{3}x - \dfrac{4}{3} - 1$        Solve for $y$.

$y = \dfrac{2}{3}x - \dfrac{7}{3}$        Slope-intercept form

Notice the similarity between the slope-intercept form of the original equation and the slope-intercept form of the parallel equation.

Figure 2.45

You have seen that two nonvertical lines are parallel if and only if they have the same slope. Two nonvertical lines are *perpendicular* if and only if their slopes are negative reciprocals of each other. For instance, the lines $y = 2x$ and $y = -\tfrac{1}{2}x$ are perpendicular because one has a slope of 2 and the other has a slope of $-\tfrac{1}{2}$.

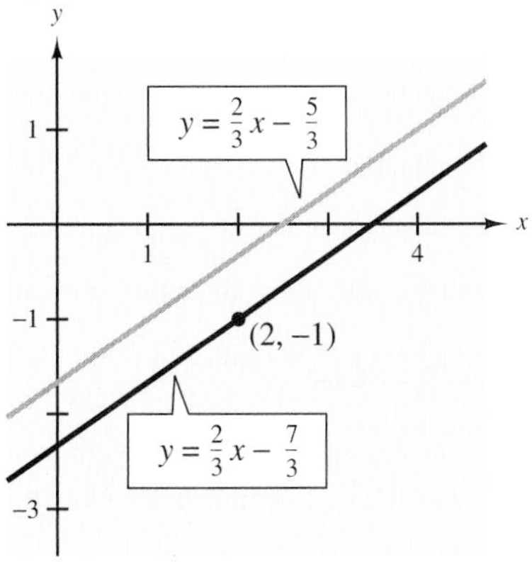

Use a graphing utility to graph $y_1 = \frac{2}{3}x + \frac{5}{2}$ and $y_2 = -\frac{3}{2}x + 2$. When you examine the graphs with a square setting, what do you observe? What do you notice about the slopes of the two lines?

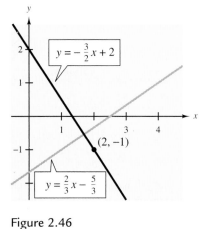

Figure 2.46

▶ **Perpendicular Lines**

Two nonvertical lines are **perpendicular** if and only if their slopes are negative reciprocals of each other. That is,

$$m_1 = -\frac{1}{m_2}.$$

**Example 6**    Equations of Perpendicular Lines

Find an equation of the line that passes through the point $(2, -1)$ and is perpendicular to the line $2x - 3y = 5$.

**Solution**

By writing the given line in the form $y = \frac{2}{3}x - \frac{5}{3}$, you can see that the line has a slope of $\frac{2}{3}$. Hence, any line that is perpendicular to this line must have a slope of $-\frac{3}{2}$ (because $-\frac{3}{2}$ is the negative reciprocal of $\frac{2}{3}$). So, the required line through the point $(2, -1)$ has the following equation.

$$y - (-1) = -\frac{3}{2}(x - 2) \qquad \text{Point-slope form}$$

$$y = -\frac{3}{2}x + 3 - 1 \qquad \text{Solve for } y.$$

$$y = -\frac{3}{2}x + 2 \qquad \text{Slope-intercept form}$$

The graphs of both equations are shown in Figure 2.46.

---

**Discussing the Concept**    **Interpreting Slope**

You think the relationship between daily high temperature $y$ (in degrees Fahrenheit) and time $x$ (in days) is linear. You observe the high temperature on day 0 to be 60°F and the high temperature 15 days later to be 68°F.

**a.** Find the slope of the line passing through these two points.

**b.** Interpret the meaning of the slope in this situation.

**c.** Use linear extrapolation to predict the high temperatures on days 30, 60, 105, and 180. Are all of these predictions reasonable? Why or why not?

# Warm Up

The following warm-up exercises involve skills that were covered in earlier sections. You will use these skills in the exercise set for this section.

In Exercises 1–4, simplify the expression.

**1.** $\dfrac{4 - (-5)}{-3 - (-1)}$

**2.** $\dfrac{-5 - 8}{0 - (-3)}$

**3.** Find $-1/m$ for $m = 4/5$.

**4.** Find $-1/m$ for $m = -2$.

In Exercises 5–10, solve for $y$ in terms of $x$.

**5.** $2x - 3y = 5$

**6.** $4x + 2y = 0$

**7.** $y - (-4) = 3[x - (-1)]$

**8.** $y - 7 = \frac{2}{3}(x - 3)$

**9.** $y - (-1) = \dfrac{3 - (-1)}{2 - 4}(x - 4)$

**10.** $y - 5 = \dfrac{3 - 5}{0 - 2}(x - 2)$

## 2.4   Exercises

In Exercises 1–6, estimate the slope of the line from its graph.

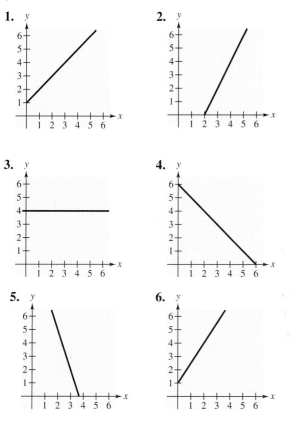

**1.**

**2.**

**3.**

**4.**

**5.**

**6.**

in Exercises 7 and 8, sketch the lines through the given point with the given slopes. Sketch all four lines on the same set of coordinate axes.

| *Point* | *Slopes* |
| --- | --- |

**7.** $(5, -1)$   (a) $-2$   (b) $1$   (c) $0$   (d) $\frac{3}{2}$

**8.** $(-4, -3)$   (a) $-1$   (b) $2$   (c) $\frac{1}{4}$   (d) undefined

In Exercises 9–14, plot the points and find the slope of the line passing through the points.

**9.** $(6, 9), (-4, -1)$

**10.** $(2, 4), (4, -4)$

**11.** $(-6, -1), (-6, 4)$

**12.** $(0, -10), (-4, 0)$

**13.** $\left(-\frac{1}{3}, 1\right), \left(-\frac{2}{3}, \frac{5}{6}\right)$

**14.** $\left(\frac{7}{8}, \frac{3}{4}\right), \left(\frac{5}{4}, -\frac{1}{4}\right)$

In Exercises 15–20, determine if the lines $L_1$ and $L_2$ passing through the pairs of points are parallel, perpendicular, or neither.

**15.** $L_1$: $(0, -1), (5, 9)$;   $L_2$: $(0, 3), (4, 1)$

**16.** $L_1$: $(3, 6), (-6, 0)$;   $L_2$: $(0, -1), \left(5, \frac{7}{3}\right)$

**17.** $L_1$: $(-2, -1), (1, 5)$;   $L_2$: $(1, 3), (5, -5)$

**18.** $L_1$: $(4, 8), (-4, 2)$;   $L_2$: $(3, -5), \left(-1, \frac{1}{3}\right)$

**19.** $L_1$: $(-1, 7), (-6, 4)$;   $L_2$: $(0, 1), (5, 4)$

**20.** $L_1$: $(-1, 3), (2, -5)$;   $L_2$: $(3, 0), (2, -7)$

In Exercises 21–26, use the point on the line and the slope of the line to find three additional points that the line passes through. (Each problem has more than one correct answer.)

| Point | Slope |
|---|---|
| **21.** $(5, -2)$ | $m = 0$ |
| **22.** $(-1, 3)$ | $m$ is undefined |
| **23.** $(5, -6)$ | $m = 1$ |
| **24.** $(10, -6)$ | $m = -1$ |
| **25.** $(-6, -1)$ | $m = \frac{1}{2}$ |
| **26.** $(7, -5)$ | $m = -\frac{2}{3}$ |

In Exercises 27–32, find the slope and $y$-intercept (if possible) of the line specified by the equation.

**27.** $4x - y - 6 = 0$      **28.** $2x + 3y - 9 = 0$

**29.** $8 - 3x = 0$      **30.** $3y + 5 = 0$

**31.** $7x + 6y - 30 = 0$      **32.** $x - y - 10 = 0$

In Exercises 33–40, find an equation of the line passing through the points and use a graphing utility to verify your answer.

**33.** $(7, -4), (-7, 3)$      **34.** $(4, 3), (-4, -4)$

**35.** $(-9, 11), (-9, 14)$      **36.** $(-1, 4), (6, 4)$

**37.** $\left(2, \frac{1}{2}\right), \left(\frac{1}{2}, \frac{5}{4}\right)$      **38.** $(1, 1), \left(6, -\frac{2}{3}\right)$

**39.** $(1, 0.6), (-2, -0.6)$      **40.** $(-8, 0.6), (2, -2.4)$

In Exercises 41–50, find an equation of the line that passes through the point and has the indicated slope. Sketch the line.

| Point | Slope |
|---|---|
| **41.** $(7, 0)$ | $m = 1$ |
| **42.** $(-2, 0)$ | $m = -4$ |
| **43.** $(-3, 6)$ | $m = -2$ |
| **44.** $(-8, 3)$ | $m = -\frac{1}{2}$ |
| **45.** $(4, 0)$ | $m = -\frac{1}{3}$ |
| **46.** $(-2, -5)$ | $m = \frac{3}{4}$ |
| **47.** $(6, -1)$ | $m$ is undefined |
| **48.** $(-10, 4)$ | $m = 0$ |
| **49.** $\left(4, \frac{5}{2}\right)$ | $m = \frac{4}{3}$ |
| **50.** $\left(-\frac{1}{2}, \frac{3}{2}\right)$ | $m = -3$ |

**51.** A fellow student does not understand why the slope of a vertical line is undefined. Describe how you would help this student understand the concept of undefined slope.

**52.** Another student overhears your conversation in Exercise 51 and states, "I do not understand why a horizontal line has zero slope and why that is different from undefined or no slope." Describe how you would explain the concepts of zero slope and undefined slope and why they are different from each other.

In Exercises 53–56, use the *intercept form* of the equation of a line

$$\frac{x}{a} + \frac{y}{b} = 1, \quad a \neq 0, \quad b \neq 0$$

to find the equation of the line with the given intercepts $(a, 0)$ and $(0, b)$.

**53.** $x$-intercept: $(1, 0)$
     $y$-intercept: $(0, -4)$

**54.** $x$-intercept: $(-3, 0)$
     $y$-intercept: $(0, 4)$

**55.** $x$-intercept: $\left(-\frac{1}{6}, 0\right)$
     $y$-intercept: $\left(0, -\frac{2}{3}\right)$

**56.** $x$-intercept: $\left(-\frac{2}{3}, 0\right)$
     $y$-intercept: $\left(0, \frac{1}{2}\right)$

In Exercises 57–62, write an equation of the line through the point (a) parallel to the line and (b) perpendicular to the line. Use a graphing utility to verify your answer.

| Point | Line |
|---|---|
| **57.** $(6, 2)$ | $y - 2x = -1$ |
| **58.** $(-5, 4)$ | $x + y = 8$ |
| **59.** $\left(\frac{1}{4}, -\frac{2}{3}\right)$ | $2x - 3y = 5$ |
| **60.** $\left(\frac{7}{8}, \frac{3}{4}\right)$ | $5x + 3y = 0$ |
| **61.** $(-1, 0)$ | $y = -3$ |
| **62.** $(2, 5)$ | $x = 4$ |

**63.** *Temperature* Find an equation of the line that gives the relationship between the temperature in degrees Celsius, $C$, and degrees Fahrenheit, $F$. Remember that water freezes at $0°$ Celsius ($32°$ Fahrenheit) and boils at $100°$ Celsius ($212°$ Fahrenheit).

**64.** *Temperature*   Use the result of Exercise 63 to complete the table. Is there a temperature for which the Fahrenheit reading is the same as the Celsius reading? If so, what is it?

| $C$ |  | $-10°$ | $10°$ |  |  | $177°$ |
|---|---|---|---|---|---|---|
| $F$ | $0°$ |  |  | $68°$ | $90°$ |  |

**65.** *Simple Interest*   A person deposits $P$ dollars in an account that pays simple interest. After 2 months the balance in the account is $759, and after 3 months the balance in the account is $763.50. Find an equation that gives the relationship between the balance $A$ and the time $t$ in months.

**66.** *Simple Interest*   Use the result of Exercise 65 to complete the table.

| $A$ |  | $759.00 | $763.50 |  |  |  |
|---|---|---|---|---|---|---|
| $t$ | 0 | 1 |  |  | 4 | 5 | 6 |

**67.** *Fourth Quarter Sales*   During the first and second quarters of the year, a business had sales of $145,000 and $152,000, respectively. If the growth of the sales follows a linear pattern, what will the sales be during the fourth quarter?

**68.** *College Enrollment*   A small college had 2546 students in 1997 and 2702 students in 1999. If the enrollment follows a linear growth pattern, how many students will the college have in 2004?

**69.** *Annual Salary*   Suppose that your salary was $28,500 in 1996 and $31,800 in 1999. If your salary follows a linear growth pattern, what salary will you be making in 2003?

**70.** *Endowment Income*   In 1994, the total amount of money endowed to symphony orchestras was $60.4 million. By 1998 the total amount had increased to $111.2 million. If these contributions had continued to increase in a linear pattern, how much money would have been endowed to symphony orchestras in 2001?   (Source: American Symphony Orchestra League, Inc.)

**71.** *American Express Revenue*   In 1997, American Express had a total revenue of $17,760 million. By 1999, the total revenue had reached $21,278 million. If the total revenue had continued to increase in a linear pattern, how much would the revenue have been in 2000? The actual revenue in 2000 was $23,675 million. Do you think the increase in revenue was approximately linear?   (Source: American Express Company)

**72.** *Shopping Centers*   In 1995, the number of shopping centers in the United States was 41,235. By 1998, the number had increased to 43,661. If the number of shopping centers had continued to increase in a linear pattern, what would the number have been in 1999? The actual number of shopping centers in 1999 was 44,426. Do you think the increase in the number of shopping centers was approximately linear?   (Source: National Research Bureau)

**73.** *Computer-Related Expenditures*   In 1994, the total amount of expenditures for personal computers and related products in the United States was $18.0 billion. By 1996, the total expenditures had reached $23.6 billion. If the total expenditures had continued to increase in a linear pattern, how much would the expenditures have been in 1998? The actual total expenditures in 1998 were $30.4 billion. Do you think the increase in expenditures was approximately linear?   (Source: U.S. Bureau of Economic Analysis)

**74.** *Apartment Rent*   The rent for a two-bedroom apartment was $800 in 1998 and $1200 in 2001. If the rent for a two-bedroom apartment follows a linear growth pattern, what is the rate of change per year for the rent? What will the rent be in 2005?

**75.** *Scuba Diving*   Pressure increases at the rate of 1 atmosphere (the weight of air pressure on a body at sea level) for each additional 33 feet of depth under water. At sea level the pressure is 1 atmosphere. At 66 feet below the surface, the total pressure is 3 atmospheres. Write a linear equation to describe the pressure $p$ with respect to depth $d$ below the surface of the sea. What is the rate of change of pressure with depth?   (Source: *PADI Open Water Diver Manual*)

## Math Matters • The Pythagorean Theorem

The Pythagorean Theorem is one of the most famous theorems in mathematics. Its name comes from Pythagoras of Samos (c. 580–500 B.C.), a Greek mathematician who founded a school of mathematics and philosophy at Crotona in what is now southern Italy. The Pythagorean Theorem states a relationship among the three sides of a right triangle. Specifically, the theorem states that the sum of the squares of the two legs of a right triangle is equal to the square of the hypotenuse. This result is usually written as

$$a^2 + b^2 = c^2. \qquad \text{Pythagorean Theorem}$$

One way to prove the Pythagorean Theorem is to use a geometrical argument, as indicated in the series of figures below. In the figure at the left, the areas of the three squares represent the squares of the sides of a right triangle. To prove that $a^2 + b^2 = c^2$, you need to show that the sum of the areas of the two smaller squares is equal to the area of the larger square. One way to do this is to make four copies of the given right triangle, and use the four triangles to form a square whose sides have lengths of $a + b$. By doing this in two different ways, you can conclude that $a^2 + b^2 = c^2$. Do you see why this is true?

According to the *Guinness Book of World Records,* 1995, the Pythagorean Theorem is the "most proven theorem" in mathematics. This publication claims that "*The Pythagorean Proposition* published in 1940 contained 370 different proofs of Pythagoras' theorem, including one by President James Garfield."

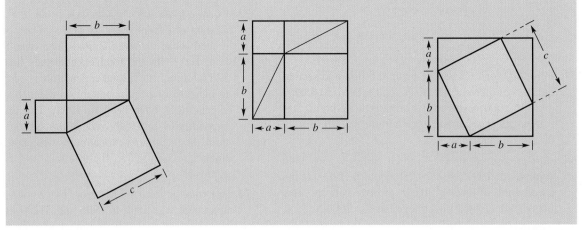

## 2.5    Linear Modeling

### Objectives

- Construct a linear model to relate quantities that vary directly.
- Construct and use a linear model with slope as the rate of change.
- Use a scatter plot to find a linear model that fits a set of data.

## Introduction

The primary objective of applied mathematics is to find equations or **mathematical models** that describe real-world phenomena. In developing a mathematical model to represent actual data, you should strive for two (often conflicting) goals—accuracy and simplicity. That is, you want the model to be simple enough to be workable, yet accurate enough to produce meaningful results.

**Example 1**    A Mathematical Model

The total annual amounts of advertising expenses $y$ (in billions of dollars) in the United States from 1992 to 1998 are given in the table.    (Source: McCann-Erickson, Inc.)

| Year | 2 | 3 | 4 | 5 | 6 | 7 | 8 |
|------|-----|-----|-----|-----|-----|-----|-----|
| $y$ | 132.7 | 139.5 | 151.7 | 162.9 | 175.2 | 187.5 | 201.6 |

A linear model that approximates these data is

$$y = 106.2 + 11.65t, \quad 2 \le t \le 8$$

where $y$ represents the advertising expenses (in billions of dollars) and $t$ represents the year, with $t = 2$ corresponding to 1992. Plot the actual data *and* the model on the same graph. How closely does the model represent the data?

### Solution

The actual data are plotted in Figure 2.47, along with the graph of the linear model. From the figure, it appears that the model is a "good fit" for the actual data. You can see how well the model fits by comparing the actual values of $y$ with the values of $y$ given by the model (these values are labeled $y^*$ in the table below).

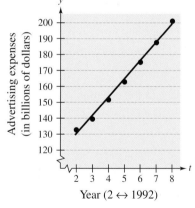

Figure 2.47

| $t$ | 2 | 3 | 4 | 5 | 6 | 7 | 8 |
|------|-----|-----|-----|-----|-----|-----|-----|
| $y$ | 132.7 | 139.5 | 151.7 | 162.9 | 175.2 | 187.5 | 201.6 |
| $y^*$ | 129.5 | 141.2 | 152.8 | 164.5 | 176.1 | 187.8 | 199.4 |

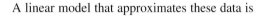

## Direct Variation

There are two basic types of linear models. The more general model has a $y$-intercept that is nonzero: $y = mx + b$, $b \neq 0$. The simpler model, $y = mx$, has a $y$-intercept that is zero. In the simpler model, $y$ is said to **vary directly** as $x$, or be **proportional** to $x$.

---

▶ Direct Variation

The following statements are equivalent.

1. $y$ **varies directly** as $x$.

2. $y$ is **directly proportional** to $x$.

3. $y = mx$ for some nonzero constant $m$.

$m$ is the **constant of variation** or the **constant of proportionality**.

---

| Example 2 | State Income Tax |  |

In Pennsylvania, the state income tax is directly proportional to *gross income*. Suppose you were working in Pennsylvania and your state income tax deduction was $42.00 for a gross monthly income of $1500.00. Find a mathematical model that gives the Pennsylvania state income tax in terms of gross income.

### Solution

*Verbal Model:*    $\dfrac{\text{State}}{\text{income tax}} = m \cdot \dfrac{\text{Gross}}{\text{income}}$

*Labels:*

State income tax = $y$     (dollars)
Gross income = $x$     (dollars)
Income tax rate = $m$     (percent in decimal form)

*Equation:*    $y = mx$

To solve for $m$, substitute the given information into the equation $y = mx$, and then solve for $m$.

$$y = mx \qquad \text{Direct variation model}$$

$$42.00 = m(1500) \qquad \text{Substitute } y = 42.00 \text{ and } x = 1500.$$

$$0.028 = m \qquad \text{Income tax rate}$$

So, the equation (or model) for state income tax in Pennsylvania is $y = 0.028x$. In other words, Pennsylvania has a state income tax rate of 2.8% of gross income. The graph of this equation is shown in Figure 2.48.

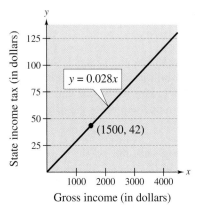

Figure 2.48

Most measurements in the English system and metric system are directly proportional. The next example shows how to use a direct proportion to convert between miles per hour and kilometers per hour.

**Example 3**    The English and Metric Systems

You are traveling at a rate of 64 miles per hour. You switch your speedometer reading to metric units and notice that the speed is 103 kilometers per hour. Use this information to find a mathematical model that relates miles per hour to kilometers per hour.

Solution

If you let $y$ represent the speed in miles per hour and $x$ represent the speed in kilometers per hour, you know that $y$ and $x$ are related by the equation

$$y = mx.$$

You are given that $y = 64$ when $x = 103$. By substituting these values into the equation $y = mx$, you can find the value of $m$.

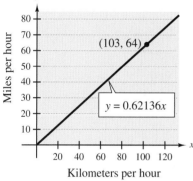

Figure 2.49

| | |
|---|---|
| $y = mx$ | Direct variation model |
| $64 = m(103)$ | Substitute $y = 64$ and $x = 103$. |
| $\dfrac{64}{103} = m$ | Divide both sides by 103. |
| $0.62136 \approx m$ | Use a calculator. |

So, the conversion factor from kilometers per hour to miles per hour is approximately 0.62136, and the model is

$$y = 0.62136x.$$

The graph of this equation is shown in Figure 2.49.

Once you have found a model that converts speeds from kilometers per hour to miles per hour, you can use the model to convert other speeds from the metric system to the English system, as shown in the table.

| Kilometers per hour | 20.0 | 40.0 | 60.0 | 80.0 | 100.0 | 120.0 |
|---|---|---|---|---|---|---|
| Miles per hour | 12.4 | 24.9 | 37.3 | 49.7 | 62.1 | 74.6 |

The conversion equation $y = 0.62136x$ can be approximated by the simpler equation $y = \frac{5}{8}x$. For instance, to convert 40 kilometers per hour, divide by 8 and multiply by 5 to obtain 25 miles per hour.

## Rates of Change

A second common type of linear model is one that involves a known rate of change. In the linear equation

$$y = mx + b$$

you know that $m$ represents the slope of the line. In real-life problems, the slope can often be interpreted as the **rate of change** of $y$ with respect to $x$. Rates of change should always be listed in appropriate units of measure.

**Example 4** Height of a Mountain Climber

A mountain climber is climbing up a 500-foot cliff. By 1 P.M., the mountain climber has climbed 115 feet up the cliff. By 4 P.M., the climber has reached a height of 280 feet, as shown in Figure 2.50. Find the average rate of change of the climber and use this rate of change to find the equation that relates the height of the climber to the time. Use the model to estimate the time when the climber will reach the top of the cliff.

**Solution**

Let $y$ represent the height of the climber and let $t$ represent the time. Then the two points that represent the climber's two positions are

$$(t_1, y_1) = (1, 115) \quad \text{and} \quad (t_2, y_2) = (4, 280).$$

So, the average rate of change of the climber is

$$\text{Average rate of change} = \frac{y_2 - y_1}{t_2 - t_1}$$

$$= \frac{280 - 115}{4 - 1}$$

$$= 55 \text{ feet per hour.}$$

An equation that relates the height of the climber to the time is

$$y - y_1 = m(t - t_1) \qquad \text{Point-slope form}$$

$$y - 115 = 55(t - 1) \qquad \text{Substitute } y_1 = 115, t_1 = 1, \text{ and } m = 55.$$

$$y = 55t + 60. \qquad \text{Linear model}$$

To find the time when the climber reaches the top of the cliff, let $y = 500$ and solve for $t$ to obtain

$$t = 8.$$

So, continuing at the same rate, the climber will reach the top of the cliff at 8 P.M.

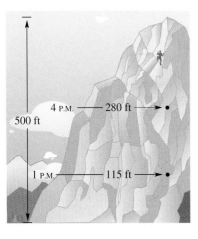

4 P.M. ——— 280 ft

500 ft

1 P.M. ——— 115 ft

**Figure 2.50**

**Example 5**    Population of Omaha, Nebraska

Between 1990 and 2000, the population of the city of Omaha, Nebraska increased at an average rate of approximately 4700 people per year. In 1990, the population was 343,000. Find a mathematical model that gives the population of Omaha in terms of the year, and use the model to estimate the population in 2004. (Source: U.S. Bureau of the Census)

### Solution

Let $y$ represent the population of Omaha, and let $t$ represent the calendar year, with $t = 0$ corresponding to 1990. Letting $t = 0$ correspond to 1990 is convenient because you were given the population in 1990. Now, using the rate of change of 4700 people per year, you have

$$\begin{array}{cc} \text{Rate} & 1990 \\ \text{of change} & \text{population} \\ \downarrow & \downarrow \\ y = \quad mt & + \quad b \end{array}$$

$$y = 4700t + 343{,}000.$$

Using this model, you can predict the 2004 population to be

$$(2004 \text{ population}) = 4700(14) + 343{,}000 = 408{,}800.$$

The graph is shown in Figure 2.51.

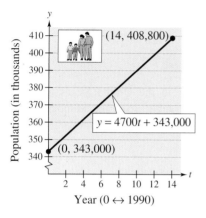

Figure 2.51

**Example 6**    Straight-Line Depreciation

Your company has purchased a \$15,000 machine that has a useful life of 10 years. The salvage value at the end of the 10 years is \$3000. Write a linear equation that describes the book value of the machine each year.

### Solution

Let $V$ represent the value of the machine at the end of the year $t$. You can represent the initial value of the machine by the ordered pair $(0, 15{,}000)$ and the salvage value by the ordered pair $(10, 3000)$. The slope of the line is

$$m = \frac{3000 - 15{,}000}{10 - 0} = -1200$$

which represents the annual depreciation in *dollars per year.* Using the point-slope form, you can write the equation of the line as follows.

$$V - 15{,}000 = -1200(t - 0) \qquad \text{Point-slope form}$$

$$V = -1200t + 15{,}000 \qquad \text{Slope-intercept form}$$

The graph of the equation is shown in Figure 2.52.

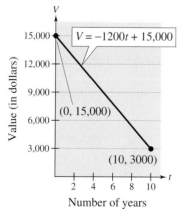

Figure 2.52

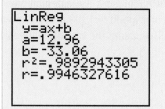
## Study Tip

Consult the user's manual of your graphing utility for information on entering data, using the regression capabilities, and creating scatter plots.

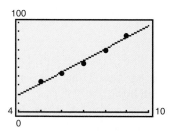

Figure 2.53

# Scatter Plots and Regression Analysis

Another type of linear modeling is a graphical approach that is commonly used in statistics. To find a mathematical model that approximates a set of actual data points, plot the points on a rectangular coordinate system. This collection of points is called a **scatter plot.** Your graphing utility may have a built-in statistical program to calculate the equation of the best fitting line for linear data. The statistical method of fitting a line to a collection of points is called **linear regression.** A discussion of linear regression is beyond the scope of this text, but the program in most graphing utilities is easy to use and allows you to analyze linear data that may not be convenient to graph by hand.

**Example 7** Cellular Subscribers

The number of cellular telephone subscribers in the United States has been steadily increasing since 1995. The following table gives the number of cell phone subscribers (in millions) from 1995 to 1999. (Source: Cellular Telecommunications Industry Association)

| Year | 1995 | 1996 | 1997 | 1998 | 1999 |
|------|------|------|------|------|------|
| $x$ | 5 | 6 | 7 | 8 | 9 |
| Subscribers (in millions) | 33.8 | 44.0 | 55.3 | 69.2 | 86.0 |

**a.** Use the regression capabilities of a graphing utility to find a linear model for the data. Let $x = 5$ represent 1995.

**b.** Use a graphing utility to graph the linear model along with a scatter plot of the data.

**c.** Use the linear model to predict the number of cell phone subscribers in 2005.

### Solution

**a.** Enter data into a graphing utility. Then, using the regression capabilities of the graphing utility, you should obtain a result similar to the following.

$$y = ax + b$$
$$a = 12.96$$
$$b = -33.06$$

So, a linear model for the data is $y = 12.96x - 33.06$, $5 \le x \le 9$.

**b.** The graph of the equation along with the scatter plot is shown in Figure 2.53.

**c.** Substituting $x = 15$ into the equation $y = 12.96x - 33.06$, you get $y = 161.34$. So, according to the model, there will be approximately 161 million cellular telephone subscribers in 2005.

**Example 8**    Prize Money at the Indianapolis 500

The total prize money $p$ (in millions of dollars) awarded at the Indianapolis 500 race from 1993 to 2001 is given in the following table. Construct a scatter plot that represents the data and find a linear model that approximates the data. (Source: Indianapolis Motor Speedway Hall of Fame)

| Year | 1993 | 1994 | 1995 | 1996 | 1997 | 1998 | 1999 | 2000 | 2001 |
|------|------|------|------|------|------|------|------|------|------|
| $p$  | 7.68 | 7.86 | 8.06 | 8.11 | 8.62 | 8.72 | 9.05 | 9.48 | 9.61 |

**Solution**

Let $t = 3$ represent 1993. The scatter plot for the points is shown in Figure 2.54. From the scatter plot, draw a line that approximates the data. Then, to find the equation of the line, approximate two points *on the line*: $(5, 8)$ and $(9, 9)$. The slope of this line is

$$m \approx \frac{p_2 - p_1}{t_2 - t_1} = \frac{9 - 8}{9 - 5} = 0.25.$$

Using the point-slope form, you can determine that the equation of the line is

$$p = 0.25t + 6.75.$$

To check this model, compare the actual $p$-values with the $p$-values given by the model (these values are labeled $p^*$ in the table below).

| $t$ | 3 | 4 | 5 | 6 | 7 | 8 | 9 | 10 | 11 |
|-----|------|------|------|------|------|------|------|------|------|
| $p$ | 7.68 | 7.86 | 8.06 | 8.11 | 8.62 | 8.72 | 9.05 | 9.48 | 9.61 |
| $p^*$ | 7.50 | 7.75 | 8.00 | 8.25 | 8.50 | 8.75 | 9.00 | 9.25 | 9.50 |

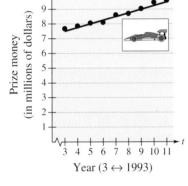

Figure 2.54

## Discussing the Concept    Gathering and Analyzing Data

Measure the height ($h$) and forearm ($f$) of each person in the class with a tape measure. Gather the data in the form ($h, f$) and plot the points on a set of coordinate axes. Do the points appear to follow a linear model? Find an equation for a line that approximately represents the points.

# Warm Up

The following warm-up exercises involve skills that were covered in earlier sections. You will use these skills in the exercise set for this section.

In Exercises 1–4, sketch the graph of the line.

**1.** $y = 2x$

**2.** $y = \frac{1}{2}x$

**3.** $y = 2x + 1$

**4.** $y = \frac{1}{2}x + 1$

In Exercises 5 and 6, find an equation of the line that has the given slope and *y*-intercept.

**5.** Slope: 1; *y*-intercept: $(0, 2)$

**6.** Slope: $\frac{3}{2}$; *y*-intercept: $(0, 3)$

In Exercises 7–10, find an equation of the line that passes through the two points.

**7.** $(1, 3)$ and $(6, 8)$

**8.** $(0, 4)$ and $(7, 10)$

**9.** $(1, 5.2)$ and $(5, 4.7)$

**10.** $(2, 6.5)$ and $(8, 3.6)$

## 2.5     Exercises

**1.** *Employment*   The total numbers of people employed (in thousands) in the United States from 1991 to 1999 are given by the following ordered pairs.

| | |
|---|---|
| (1991, 117,718) | (1997, 129,558) |
| (1992, 118,492) | (1998, 131,463) |
| (1993, 120,259) | (1999, 133,488) |
| (1994, 123,060) | |
| (1995, 124,900) | |
| (1996, 126,708) | |

A linear model that approximates these data is

$$y = 114{,}719 + 2070.7t, \quad 1 \le t \le 9$$

where *y* represents the number of people employed (in thousands) and $t = 1$ represents 1991. Plot the actual data *and* the model on the same graph. How closely does the model represent the data?   (Source: U. S. Bureau of Labor Statistics)

**2.** *Olympic Swimming*   The winning times (in minutes) in the women's 400-meter freestyle swimming event in the Olympics from 1948 to 2000 are given by the following ordered pairs.

(1948, 5.30),  (1952, 5.20)
(1956, 4.91),  (1960, 4.84)
(1964, 4.72),  (1968, 4.53)
(1972, 4.32),  (1976, 4.16)
(1980, 4.15),  (1984, 4.12)
(1988, 4.06),  (1992, 4.12)
(1996, 4.12),  (2000, 4.10)

A linear model that approximates these data is

$$y = 5.30 - 0.024t, \quad 8 \le t \le 60$$

where *y* represents the winning time in minutes and $t = 8$ represents 1948. Plot the actual data *and* the model on the same graph. How closely does the model represent the data?   (Source: *Time Almanac 2001*)

**3.** *Olympic Swimming Revisited*   In Exercise 2, the winning times (in minutes) for the years 1948 to 1972 are relatively high compared with the winning times from 1976 to 2000. Consider the winning times for the years 1976 to 2000 only and use your graphing utility to make a scatter plot. Then use the regression capabilities of your graphing utility to find a best-fitting line for the data. Graph the line on the scatter plot. Use the equation of the line to predict the winning time in this event at the 2004 Summer Olympics. Is your prediction reasonable?

**4.** *Olympic Swimming Revisited*   Use the winning times for the years 1948 to 1972 from Exercise 2 and your graphing utility to make a scatter plot. Then use the regression capabilities of your graphing utility to find a best-fitting line for the data. Graph the line on the scatter plot. Use the equation of the line to predict the winning time in this event at the 2000 Summer Olympics. Is your prediction reasonable? How does this prediction compare with actual winning time in 2000?

*Direct Variation*    In Exercises 5–10, *y* is proportional to *x*. Use the *x*- and *y*-values to find a linear model that relates *y* and *x*.

**5.** $x = 8$, $y = 3$

**6.** $x = 5$, $y = 9$

**7.** $x = 15$, $y = 300$

**8.** $x = 12$, $y = 204$

**9.** $x = 7$, $y = 3.2$

**10.** $x = 11$, $y = 1.5$

**11.** *Simple Interest*    The simple interest that a person receives from an investment is directly proportional to the amount of the investment. By investing $2500 in a bond issue, you obtained an interest payment of $187.50 at the end of 1 year. Find a mathematical model that gives the interest *I* at the end of 1 year in terms of the amount invested *P*.

**12.** *Simple Interest*    The simple interest that a person receives from an investment is directly proportional to the amount of the investment. By investing $5000 in a municipal bond, you obtained interest of $337.50 at the end of 1 year. Find a mathematical model that gives the interest *I* for the municipal bond at the end of 1 year in terms of the amount invested *P*.

**13.** *Property Tax*    Your property tax is based on the assessed value of your property. (The assessed value is often lower than the actual value of the property.) A house that has an assessed value of $50,000 has a property tax of $1840.

(a) Find a mathematical model that gives the amount of property tax *y* in terms of the assessed value *x* of the property.

(b) Use the model to find the property tax on a house that has an assessed value of $85,000.

**14.** *State Sales Tax*    An item that sells for $145.99 has a sales tax of $10.22.

(a) Find a mathematical model that gives the amount of sales tax *y* in terms of the retail price *x*.

(b) Use the model to find the sales tax on a purchase that has a retail price of $540.50.

**15.** *Centimeters and Inches*    On a yardstick, you notice that 13 inches is the same length as 33 centimeters.

(a) Use this information to find a mathematical model that relates centimeters to inches.

(b) Use the model to complete the table.

| Inches | 5 | 10 | 20 | 25 | 30 |
|---|---|---|---|---|---|
| Centimeters | | | | | |

**16.** *Liters and Gallons*    You are buying gasoline and notice that 14 gallons of gasoline is the same as 53 liters.

(a) Use this information to find a linear model that relates gallons to liters.

(b) Use the model to complete the table.

| Gallons | 5 | 10 | 20 | 25 | 30 |
|---|---|---|---|---|---|
| Liters | | | | | |

In Exercises 17–22, you are given the 2002 value of a product *and* the rate at which the value is expected to change during the next 5 years. Use this information to write a linear equation that gives the dollar value of the product in terms of the year. (Let $t = 0$ represent 2002.)

| *2002 Value* | *Rate* |
|---|---|
| **17.** $2540 | $125 increase per year |
| **18.** $156 | $4.50 increase per year |
| **19.** $20,400 | $2000 decrease per year |
| **20.** $45,000 | $2800 decrease per year |
| **21.** $154,000 | $12,500 increase per year |
| **22.** $245,000 | $5600 increase per year |

**23.** *Straight-Line Depreciation*    A business purchases a piece of equipment for $875. After 5 years the equipment will have no value. Write a linear equation giving the value *V* of the equipment during the 5 years.

**24.** *Straight-Line Depreciation*    A business purchases a piece of equipment for $25,000. The equipment will be replaced in 10 years, at which time its salvage value is expected to be $2000. Write a linear equation giving the value *V* of the equipment during the 10 years.

**25.** *Sales Price and List Price*    A store is offering a 15% discount on all items. Write a linear equation giving the sale price *S* for an item with a list price *L*.

**26.** *Sales Price and List Price*    A store is offering a 25% discount on all shirts. Write a linear equation giving the sale price *S* for a shirt with a list price *L*.

**27.** *Hourly Wages*    A manufacturer pays its assembly line workers $11.50 per hour. In addition, workers receive a piecework rate of $0.75 per unit produced. Write a linear equation for the hourly wages *W* in terms of the number of units *x* produced per hour.

**28.** *Sales Commission*  A salesperson receives a monthly salary of $2500 plus a commission of 7% of sales. Write a linear equation for the salesperson's monthly wage $W$ in terms of the person's monthly sales $S$.

**29.** *Height of a Parachutist*  After opening the parachute, the descent of a parachutist follows a linear model. At 2:08 P.M. the height of the parachutist is 7000 feet. At 2:10 P.M. the height is 4600 feet.

    (a) Write a linear equation that gives the height of the parachutist in terms of the time $t$. (Let $t = 0$ represent 2:08 P.M. and let $t$ be measured in seconds.)

    (b) Use the equation in part (a) to find the time when the parachutist will reach the ground.

**30.** *Distance Traveled by a Car*  You are driving at a constant speed on a freeway. At 4:30 P.M. you drive by a sign that gives the distance to Montgomery, Alabama as 84 miles. At 4:59 P.M. you drive by another sign that says the distance to Montgomery is 56 miles.

    (a) Write a linear equation that gives your distance from Montgomery in terms of $t$. (Let $t = 0$ represent 4:30 P.M. and let $t$ be measured in minutes.)

    (b) Use the equation in part (a) to find the time when you will reach Montgomery.

In Exercises 31–34, match the description with one of the graphs. Also find the slope and describe how it is interpreted in the real-life situation. [The graphs are labeled (a), (b), (c), and (d).]

**31.** A person is paying $10 per week to a friend to repay a $100 loan.

**32.** An employee is paid $12.50 per hour plus $1.50 for each unit produced per hour.

**33.** A sales representative receives $20 per day for food, plus $0.25 for each mile traveled.

**34.** A typewriter that was purchased for $600 depreciates $100 per year.

**35.** *High School Enrollment*  A high school had an enrollment of 1200 students in 1990. During the next 10 years the enrollment increased by approximately 50 students per year.

    (a) Write a linear equation giving the enrollment $N$ in terms of the year $t$. (Let $t = 0$ correspond to the year 1990.)

    (b) If this constant rate of growth continues, predict the enrollment in the year 2010.

    (c) Use a graphing utility to confirm graphically the result in part (b).

**36.** *Cost of Renting*  The monthly rent in an office building is related to the office size. The rent for a 600-square-foot office is $750. The rent for a 900-square-foot office is $1150.

    (a) Write a linear equation giving the monthly rent $r$ in terms of the square footage $x$.

    (b) Use the equation in part (a) to find the monthly rent for an office that has 1300 square feet of floor space.

    (c) Use a graphing utility to confirm graphically the result in part (b).

In Exercises 37–40, can the data be approximated by a linear model?

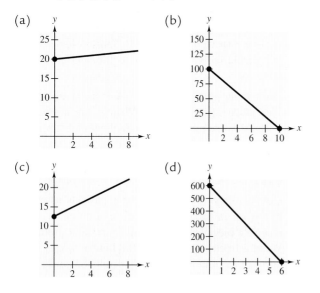

(a) (b) (c) (d)

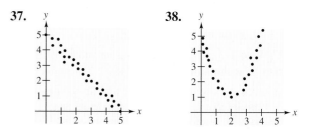

**37.** **38.**

# Chapter Test

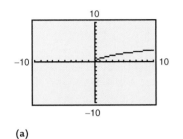

(a)

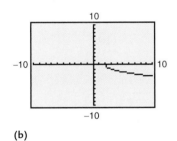

(b)

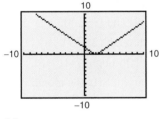

(c)

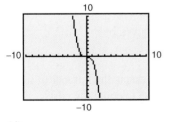

(d)

Figures for 10–13

Take this test as you would take a test in class. After you are done, check your work against the answers given in the back of the book.

In Exercises 1 and 2, find the distance between the points and the midpoint of the line segment connecting the points.

**1.** $(-3, 2), (5, -2)$  **2.** $(3.25, 7.05), (-2.37, 1.62)$

**3.** Find the intercepts of the graph of $y = (x + 5)(x - 3)$.

**4.** Find the intercepts of the graph of $y = x\sqrt{x - 2}$.

**5.** Describe the symmetry of the graph of $y = \dfrac{x}{x^2 - 4}$.

**6.** Find an equation of the line through $(3, 4)$ and $(-2, 4)$.

**7.** Find an equation of the line through $(4, -3)$ with a slope of $\frac{3}{4}$.

**8.** Find an equation of the vertical line through $(2, -7)$.

**9.** Find an equation of the line with intercepts $(3, 0)$ and $(0, -4)$.

In Exercises 10–13, match the equation with its graph. [The graphs are labeled (a), (b), (c), and (d).]

**10.** $y = \sqrt{x}$  **11.** $y = -\sqrt{x - 2}$

**12.** $y = |x - 2|$  **13.** $y = -x^3$

**14.** Write the equation of the circle in standard form and sketch its graph.
$x^2 + y^2 - 4x - 2y - 4 = 0$

**15.** During the first and second quarters of the year, a business had sales of $150,000 and $185,000, respectively. If the sales follow a linear pattern, what will sales be in the third quarter?

**16.** An assembly line worker is paid $8.75 per hour, plus a piecework rate of $0.55 per unit. Write a linear equation for the hourly wages $W$ in terms of the number of units $x$ produced per hour.

**17.** A business purchases a piece of equipment for $25,000. After 5 years, the equipment will be worth only $5000. Write a linear equation that gives the value $V$ of the equipment during the 5 years.

**18.** A store is offering a 30% discount on all summer apparel. Write a linear equation that gives the sale price $S$ of an item in terms of its list price $L$.

**19.** *Retail Trade Sales*  The total retail trade sales $S$ (in billions of dollars) in the United States for the years from 1992 to 1999 are listed in the table. Use a graphing utility to graph a scatter plot of the data and find a linear model for the data.  (Source: Current Business Reports)

| Year | 1992 | 1993 | 1994 | 1995 | 1996 | 1997 | 1998 | 1999 |
|---|---|---|---|---|---|---|---|---|
| $S$ | 1952 | 2082 | 2248 | 2359 | 2502 | 2611 | 2746 | 2995 |

## Cumulative Test: Chapters 1–2

Take this test as you would take a test in class. After you are done, check your work against the answers given in the back of the book.

In Exercises 1–6, solve the equation.

**1.** $4y^2 - 64 = 0$

**2.** $\dfrac{x}{3} + \dfrac{x}{4} = 1$

**3.** $x^4 - 17x^2 + 16 = 0$

**4.** $|2x - 3| = 5$

**5.** $\sqrt{y + 3} + y = 3$

**6.** $2x^2 + x = 5$

**7.** You deposit $5000 in an account that pays 8.5%, compounded monthly. Find the balance after 10 years.

In Exercises 8–10, solve the inequality.

**8.** $\left|\dfrac{2x - 5}{4}\right| \le 3$

**9.** $(x + 4)^2 \le 4$

**10.** $\dfrac{3 + 2x}{4 - x} > 2$

**11.** Find the intercepts of the graph of $y = x\sqrt{x + 4}$.

**12.** Describe the symmetry of the graph of $y = -\sqrt{4 - x^2}$.

**13.** Write the equation of the circle in standard form and sketch its graph.

$$x^2 + y^2 - 6x + 4y - 3 = 0$$

**14.** Find the constant $C$ such that $(-2, 3)$ is a solution point of $y = x^3 + C$.

**15.** Find an equation of the line passing through $(3, -2)$ and $(-1, 5)$.

**16.** Find an equation of the line with slope $\frac{2}{3}$ passing through $(2, -1)$.

**17.** Find an equation of the line with zero slope passing through $(-1, -3)$.

**18.** You have accepted a sales job that pays a base salary of $1000 a month plus a 7% commission on monthly sales $x$. Write a linear model that describes your total monthly salary $y$ in terms of your base salary and commission.

**19.** During the second and third quarters of the year, a business had sales of $210,000 and $230,000, respectively. If the sales growth follows a linear pattern, what will the sales be during the fourth quarter?

**20.** The revenue and cost equations for a product are given by

$$R = x(100 - 0.001x) \quad \text{and} \quad C = 20x + 30,000$$

where $R$ and $C$ are measured in dollars and $x$ represents the number of units sold. How many units must be sold for the revenue to equal the cost?

# 3

# Functions and Graphs

The National Ambient Air Quality Standards (NAAQS) for suspended particulate matter, sulfur dioxide, photo chemical oxidants, carbon monoxide, and nitrogen dioxide were first set by the Environmental Protection Agency (EPA) in 1971. Every five years the NAAQS are reviewed and revised if new health or welfare data indicate that a change is needed. The table below gives the amounts of carbon monoxide emissions, in thousands of tons, in the United States from 1990 to 1998. (Source: *U.S. Statistical Abstract 2000*)

3.1 Functions

3.2 Graphs of Functions

3.3 Transformations of Functions

3.4 The Algebra of Functions

3.5 Inverse Functions

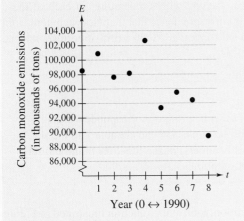

Year (0 ↔ 1990)

| Year | 1990 | 1991 | 1992 | 1993 | 1994 | 1995 | 1996 | 1997 | 1998 |
|---|---|---|---|---|---|---|---|---|---|
| Carbon monoxide emissions (in thousands of tons) | 98,523 | 100,872 | 97,630 | 98,160 | 102,643 | 93,353 | 95,479 | 94,410 | 89,454 |

The chapter project related to this information is on page 206.

## 3.1    Functions

### Objectives

- Determine if an equation or a set of ordered pairs represents a function.
- Use function notation.
- Evaluate a function.
- Find the domain of a function.
- Write a function that relates quantities in an application problem.

## Introduction to Functions

Many everyday phenomena involve two quantities that are related to each other by some rule of correspondence. Here are some examples.

1. The simple interest $I$ earned on $1000 for 1 year is related to the annual percentage rate $r$ by the formula $I = 1000r$.

2. The distance $d$ traveled on a bicycle in 2 hours is related to the speed $s$ of the bicycle by the formula $d = 2s$.

3. The area $A$ of a circle is related to its radius $r$ by the formula $A = \pi r^2$.

   Not all correspondences between two quantities have simple mathematical formulas. For instance, people commonly match up NFL starting quarterbacks with touchdown passes and hours of the day with temperature. In each of these cases, however, there is some rule of correspondence that matches each item from one set with exactly one item from a different set. Such a rule of correspondence is called a **function.**

> ▶ **Definition of a Function**
>
> A **function** $f$ from a set $A$ to a set $B$ is a rule of correspondence that assigns to each element $x$ in the set $A$ exactly one element $y$ in the set $B$. The set $A$ is the **domain** (or set of inputs) of the function $f$, and the set $B$ contains the **range** (or set of outputs).

To get a better idea of this definition, look at the function illustrated in Figure 3.1. This function can be represented by the following set of ordered pairs.

$$\{(1, 9°), (2, 13°), (3, 15°), (4, 15°), (5, 12°), (6, 4°)\}$$

In each ordered pair, the first coordinate is the input and the second coordinate is the output. In this example, note the following characteristics of a function.

1. Each element in $A$ must be matched with an element of $B$.

2. Some elements in $B$ may not be matched with any element in $A$.

3. Two or more elements of $A$ may be matched with the same element of $B$.

The converse of the third statement is not true. That is, an element of $A$ (the domain) cannot be matched with two different elements of $B$.

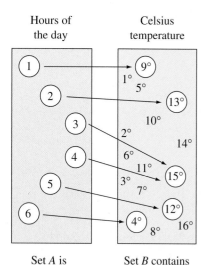

Hours of the day    Celsius temperature

Set $A$ is the domain.
Input: 1, 2, 3, 4, 5, 6

Set $B$ contains the range.
Output: 4°, 9°, 12°, 13°, 15°

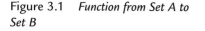

**Figure 3.1    Function from Set A to Set B**

In the following two examples, you are asked to decide whether different correspondences are functions. To do this, you must decide whether each element in the domain *A* is matched with exactly one element in the range *B*. If any element in *A* is matched with two or more elements in *B*, the correspondence is not a function.

### Example 1   Testing for Functions

Let $A = \{a, b, c\}$ and $B = \{1, 2, 3, 4, 5\}$. Which of the following sets of ordered pairs or figures represent functions from set *A* to set *B*?

**a.** $\{(a, 2), (b, 3), (c, 4)\}$

**b.** $\{(a, 4), (b, 5)\}$

**c.**
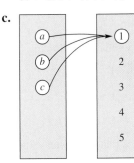

**d.**
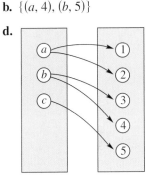

### Solution

**a.** This collection of ordered pairs *does* represent a function from *A* to *B*. Each element of *A* is matched with exactly one element of *B*.

**b.** This collection of ordered pairs *does not* represent a function from *A* to *B*. Not every element of *A* is matched with an element of *B*.

**c.** This figure *does* represent a function from *A* to *B*. It does not matter that each element of *A* is matched with the same element of *B*.

**d.** This figure *does not* represent a function from *A* to *B*. The element *a* in *A* is matched with *two* elements, 1 and 2, of *B*. This is also true of the element *b*.

---

Representing functions by sets of ordered pairs is a common practice in *discrete mathematics*. In algebra, however, it is more common to represent functions by equations or formulas involving two variables. For instance, the equation

$$y = x^2 \qquad \text{\textit{y} is a function of \textit{x}.}$$

represents the variable *y* as a function of the variable *x*. In this equation, *x* is the **independent variable** and *y* is the **dependent variable.** The domain of the function is the set of all values taken on by the independent variable *x*, and the range of the function is the set of all values taken on by the dependent variable *y*.

| Example 2 | Testing for Functions Represented by Equations |

Which of the following equations represent $y$ as a function of $x$?

**a.** $x^2 + y = 1$    **b.** $-x + y^2 = 1$

**Solution**

To determine whether $y$ is a function of $x$, try to solve for $y$ in terms of $x$.

**a.** Solving for $y$ yields the following.

$$x^2 + y = 1 \qquad \text{Original equation}$$
$$y = 1 - x^2 \qquad \text{Solve for } y.$$

To each value of $x$ there corresponds exactly one value of $y$. So, $y$ *is* a function of $x$.

**b.** Solving for $y$ yields the following.

$$-x + y^2 = 1 \qquad \text{Original equation}$$
$$y^2 = 1 + x \qquad \text{Add } x \text{ to both sides.}$$
$$y = \pm \sqrt{1 + x} \qquad \text{Solve for } y.$$

The $\pm$ indicates that to a given value of $x$ there correspond two values of $y$. So, $y$ is *not* a function of $x$.

## Function Notation

When an equation is used to represent a function, it is convenient to name the function so that it can be referenced easily. For example, you know that the equation $y = 1 - x^2$ describes $y$ as a function of $x$. Suppose you give this function the name "$f$." Then you can use the following **function notation.**

| *Input* | *Output* | *Equation* |
|---------|----------|------------|
| $x$ | $f(x)$ | $f(x) = 1 - x^2$ |

The symbol $f(x)$ is read as the **value of $f$ at $x$** or simply **$f$ of $x$.** The symbol $f(x)$ corresponds to the $y$-value for a given $x$. So, you can write $y = f(x)$. Keep in mind that $f$ is the *name* of the function, whereas $f(x)$ is the *value* of the function at $x$. For instance, the function given by

$$f(x) = 3 - 2x$$

has *function values* denoted by $f(-1)$, $f(0)$, $f(2)$, and so on. To find these values, substitute the specified input values into the given equation.

For $x = -1$,  $f(-1) = 3 - 2(-1) = 3 + 2 = 5.$

For $x = 0$,    $f(0) = 3 - 2(0) = 3 - 0 = 3.$

For $x = 2$,    $f(2) = 3 - 2(2) = 3 - 4 = -1.$

Although $f$ is often used as a convenient function name and $x$ is often used as the independent variable, you can use other letters. For instance,

$$f(x) = x^2 - 4x + 7, \quad f(t) = t^2 - 4t + 7, \quad \text{and} \quad g(s) = s^2 - 4s + 7$$

all define the same function. In fact, the role of the independent variable in a function is simply that of a "placeholder." Consequently, the function above could be described by the form

$$f(\quad) = (\quad)^2 - 4(\quad) + 7.$$

### Example 3   Evaluating a Function

Let $g(x) = -x^2 + 4x + 1$ and find the following.

**a.** $g(2)$     **b.** $g(t)$     **c.** $g(x + 2)$

Solution

**a.** Replacing $x$ with 2 in $g(x) = -x^2 + 4x + 1$ yields the following.

$$g(2) = -(2)^2 + 4(2) + 1 = -4 + 8 + 1 = 5$$

**b.** Replacing $x$ with $t$ yields the following.

$$g(t) = -(t)^2 + 4(t) + 1 = -t^2 + 4t + 1$$

**c.** Replacing $x$ with $x + 2$ yields the following.

$$g(x + 2) = -(x + 2)^2 + 4(x + 2) + 1$$
$$= -(x^2 + 4x + 4) + 4x + 8 + 1$$
$$= -x^2 - 4x - 4 + 4x + 8 + 1$$
$$= -x^2 + 5$$

> **Study Tip**
>
> In Example 3(c), note that $g(x + 2)$ is not equal to $g(x) + g(2)$. In general, $g(u + v) \neq g(u) + g(v)$.

A function defined by two or more equations over a specified domain is called a *piecewise-defined* function.

### Example 4   A Piecewise-Defined Function

Evaluate the following function when $x = -1, 0,$ and $1$.

$$f(x) = \begin{cases} x^2 + 1, & x < 0 \\ x - 1, & x \geq 0 \end{cases}$$

Solution

Because $x = -1$ is less than 0, use $f(x) = x^2 + 1$ to obtain

$$f(-1) = (-1)^2 + 1 = 2.$$

For $x = 0$, use $f(x) = x - 1$ to obtain

$$f(0) = (0) - 1 = -1.$$

For $x = 1$, use $f(x) = x - 1$ to obtain $f(1) = (1) - 1 = 0$. The graph of the function is shown in Figure 3.2.

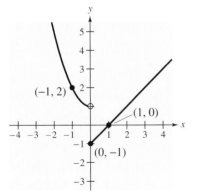

Figure 3.2

# Finding the Domain of a Function

The domain of a function can be described explicitly or it can be *implied* by the expression used to define the function. The **implied domain** is the set of all real numbers for which the expression is defined. For instance, the function given by

$$f(x) = \frac{1}{x^2 - 4}$$

has an implied domain that consists of all real $x$ other than $x = \pm 2$. These two values are excluded from the domain because division by zero is undefined. Another common type of implied domain is that used to avoid even roots of negative numbers. For example, the function given by

$$f(x) = \sqrt{x}$$

is defined only for $x \geq 0$. So, its implied domain is the interval $[0, \infty)$. In general, the domain of a function *excludes* values that would cause division by zero *or* result in the even root of a negative number.

| Example 5 | Finding the Domain of a Function |

*Learning Tools*

Find the domain of each of the following functions.

**a.** $f$: $\{(-3, 0), (-1, 4), (0, 2), (2, 2), (4, -1)\}$     **b.** $g(x) = \dfrac{1}{x + 5}$

**c.** Volume of a sphere: $V = \frac{4}{3}\pi r^3$     **d.** $h(x) = \sqrt{4 - x^2}$

**e.** $r(x) = \sqrt[3]{x + 3}$

### Solution

**a.** The domain of $f$ consists of all first coordinates in the set of ordered pairs.

Domain $= \{-3, -1, 0, 2, 4\}$

**b.** Excluding $x$-values that yield zero in the denominator, the domain of $g$ is the set of all real numbers $x$ such that $x \neq -5$.

**c.** Because this function represents the volume of a sphere, the values of the radius $r$ must be positive. So, the domain is the set of all real numbers $r$ such that $r > 0$.

**d.** This function is defined only for $x$-values for which $4 - x^2 \geq 0$. Using the methods described in Section 1.7, you can conclude that $-2 \leq x \leq 2$. So, the domain is the interval $[-2, 2]$.

**e.** Because the cube root of a number may be positive, zero, or negative, the expression under the radical may be positive, zero, or negative. So, the domain of $r$ is the set of all real numbers, or $(-\infty, \infty)$.

In Example 5(c), note that the domain of a function may be implied by the physical context. For instance, from the equation $V = \frac{4}{3}\pi r^3$, you would have no reason to restrict $r$ to nonnegative values, but the physical context implies that a sphere cannot have a negative radius.

## Applications

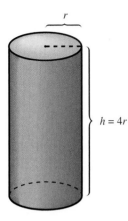

Figure 3.3

| Example 6 | The Measurements of a Container |  |

You work in the marketing department of a soft-drink company and are experimenting with a new soft-drink can that is slightly narrower and taller than a standard can. For your experimental can, the ratio of the height to the radius is 4, as shown in Figure 3.3.

**a.** Express the volume of the can as a function of the radius $r$.

**b.** Express the volume of the can as a function of the height $h$.

### Solution

**a.**  $V = \pi r^2 h = \pi r^2 (4r) = 4\pi r^3$    *V is a function of $r$.*

**b.**  $V = \pi \left(\dfrac{h}{4}\right)^2 h = \dfrac{\pi h^3}{16}$    *V is a function of $h$.*

| Example 7 | The Path of a Baseball |

A baseball is hit 3 feet above ground at a velocity of 100 feet per second and at an angle of 45°. The path of the baseball is given by the function

$$y = -0.0032x^2 + x + 3$$

where $y$ and $x$ are measured in feet, as shown in Figure 3.4. Will the baseball clear a 10-foot fence located 300 feet from home plate?

### Solution

When $x = 300$, the height of the baseball is given by

$$y = -0.0032(300^2) + 300 + 3 = 15 \text{ feet.}$$

So, the ball will clear the fence.

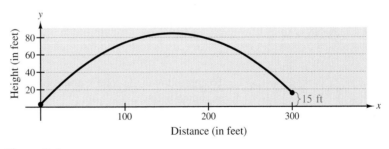

Figure 3.4

*Sales of bicycles and related supplies increased by more than 50 percent in 1997, as bicycle manufacturers broadened their product lines and popularity of the sport soared.*

| Example 8 | Bicycle Sales |
|---|---|

The amount of money spent on bicycles and related supplies in the United States (excluding Hawaii and Alaska) decreased in a linear pattern from 1993 to 1996, as shown in Figure 3.5. Then, in 1997, sales surged and began increasing in a different linear pattern. These two patterns can be approximated by the function

$$S = \begin{cases} 3900 - 112.1t, & 3 \le t \le 6 \\ 4353 + 73.5t, & 7 \le t \le 9 \end{cases}$$

where $S$ represents the amount of money spent (in millions of dollars) and $t$ represents the calendar year, with $t = 3$ corresponding to 1993. Use this function to approximate the total amount of money spent on bicycles and supplies between 1993 and 1999.   (Source: National Sporting Goods Association)

**Solution**

From 1993 to 1996, use the equation $S = 3900 - 112.1t$ to approximate the amount of sales (in millions of dollars).

| Year | 1993 | 1994 | 1995 | 1996 |
|---|---|---|---|---|
| $t$ | 3 | 4 | 5 | 6 |
| $S$ | 3564 | 3452 | 3340 | 3227 |

From 1997 to 1999, use the equation $S = 4353 + 73.5t$ to approximate the amount of sales (in millions of dollars).

| Year | 1997 | 1998 | 1999 |
|---|---|---|---|
| $t$ | 7 | 8 | 9 |
| $S$ | 4868 | 4941 | 5015 |

To approximate the total amount of sales, add the amounts for each of the years as shown below

$$3564 + 3452 + 3340 + 3227 + 4868 + 4941 + 5015 = 28{,}407$$

Because the amount of sales is measured in millions of dollars, you can conclude that the total amount spent on bicycles and related supplies in the United States (excluding Hawaii and Alaska) from 1993 through 1999 was approximately $28,407,000,000.

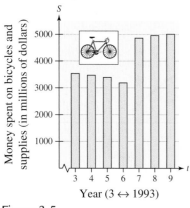

Figure 3.5

▶ **Summary of Function Terminology**

*Function:*  A **function** is a relationship between two variables such that to each value of the independent variable there corresponds exactly one value of the dependent variable.

*Function Notation:*   $y = f(x)$

  $f$ is the **name** of the function.

  $y$ is the **dependent variable.**

  $x$ is the **independent variable.**

  $f(x)$ is the **value of the function at $x$.**

*Domain:*  The **domain** of a function is the set of all values (inputs) of the independent variable for which the function is defined. If $x$ is in the domain of $f$, we say that $f$ is **defined** at $x$. If $x$ is not in the domain of $f$, we say that $f$ is **undefined** at $x$.

*Range:*  The **range** of a function is the set of all values (outputs) assumed by the dependent variable (that is, the set of all function values).

*Implied Domain:*  If $f$ is defined by an algebraic expression and the domain is not specified, the **implied domain** consists of all real numbers for which the expression is defined.

---

| Discussing the Concept | Everyday Functions |
|---|---|

In groups of two or three, identify common real-life functions. Consider everyday activities, events, and expenses, such as long distance telephone calls and car insurance. Here are a couple of examples.

**a.** The statement, "Your happiness is a function of the grade you receive in this course" *is not* a correct mathematical use of the word "function." The word "happiness" is ambiguous.

**b.** The statement, "Your federal income tax is a function of your adjusted gross income" *is* a correct mathematical use of the word "function." Once you have determined your adjusted gross income, your income tax can be determined.

Describe your function in words. Avoid using ambiguous words. Can you find an example of a piecewise-defined function?

# Warm Up

The following warm-up exercises involve skills that were covered in earlier sections. You will use these skills in the exercise set for this section.

In Exercises 1–4, simplify the expression.

**1.** $2(-3)^3 + 4(-3) - 7$

**2.** $4(-1)^2 - 5(-1) + 4$

**3.** $(x + 1)^2 + 3(x + 1) - 4 - (x^2 + 3x - 4)$

**4.** $(x - 2)^2 - 4(x - 2) - (x^2 - 4)$

In Exercises 5 and 6, solve for $y$ in terms of $x$.

**5.** $2x + 5y - 7 = 0$

**6.** $y^2 = x^2$

In Exercises 7–10, solve the inequality.

**7.** $x^2 - 4 \geq 0$

**8.** $9 - x^2 \geq 0$

**9.** $x^2 + 2x + 1 \geq 0$

**10.** $x^2 - 3x + 2 \geq 0$

## 3.1 Exercises

In Exercises 1–4, evaluate the function.

**1.** $f(x) = 6 - 4x$

(a) $f(3) = 6 - 4(\quad)$

(b) $f(-7) = 6 - 4(\quad)$

(c) $f(t) = 6 - 4(\quad)$

(d) $f(c + 1) = 6 - 4(\quad)$

**2.** $f(s) = \dfrac{1}{s + 1}$

(a) $f(4) = \dfrac{1}{(\quad) + 1}$

(b) $f(0) = \dfrac{1}{(\quad) + 1}$

(c) $f(4x) = \dfrac{1}{(\quad) + 1}$

(d) $f(x + 1) = \dfrac{1}{(\quad) + 1}$

**3.** $g(x) = \dfrac{1}{x^2 - 2x}$

(a) $g(1) = \dfrac{1}{(\quad)^2 - 2(\quad)}$

(b) $g(-3) = \dfrac{1}{(\quad)^2 - 2(\quad)}$

(c) $g(t) = \dfrac{1}{(\quad)^2 - 2(\quad)}$

(d) $g(t + 1) = \dfrac{1}{(\quad)^2 - 2(\quad)}$

**4.** $f(t) = \sqrt{25 - t^2}$

(a) $f(3) = \sqrt{25 - (\quad)^2}$

(b) $f(5) = \sqrt{25 - (\quad)^2}$

(c) $f(x + 5) = \sqrt{25 - (\quad)^2}$

(d) $f(2x) = \sqrt{25 - (\quad)^2}$

In Exercises 5–16, evaluate the function and simplify the results.

**5.** $f(x) = 2x - 3$

  (a) $f(1)$            (b) $f(-3)$

  (c) $f(x - 1)$     (d) $f\left(\frac{1}{4}\right)$

**6.** $g(y) = 7 - 3y$

  (a) $g(0)$          (b) $g\left(\frac{7}{3}\right)$

  (c) $g(s)$          (d) $g(s + 2)$

**7.** $h(t) = t^2 - 2t$

  (a) $h(2)$          (b) $h(-1)$

  (c) $h(x + 2)$     (d) $h(1.5)$

**8.** $V(r) = \frac{4}{3}\pi r^3$

  (a) $V(3)$          (b) $V(0)$

  (c) $V\left(\frac{3}{2}\right)$      (d) $V(2r)$

**9.** $f(y) = 3 - \sqrt{y}$

  (a) $f(4)$          (b) $f(100)$

  (c) $f(4x^2)$       (d) $f(0.25)$

**10.** $f(x) = \sqrt{x + 8} + 2$

  (a) $f(-8)$        (b) $f(1)$

  (c) $f(x - 8)$     (d) $f(x + 8)$

**11.** $q(x) = \dfrac{1}{x^2 - 9}$

  (a) $q(4)$          (b) $q(0)$

  (c) $q(3)$          (d) $q(y + 3)$

**12.** $q(t) = \dfrac{2t^2 + 3}{t^2}$

  (a) $q(2)$          (b) $q(0)$

  (c) $q(x)$          (d) $q(-x)$

**13.** $f(x) = \dfrac{|x|}{x}$

  (a) $f(2)$          (b) $f(-2)$

  (c) $f(x^2)$       (d) $f(x - 1)$

**14.** $f(x) = |x| + 4$

  (a) $f(2)$          (b) $f(-2)$

  (c) $f(x^2)$       (d) $f(x + 2)$

**15.** $f(x) = \begin{cases} 2x + 1, & x < 0 \\ 2x + 2, & x \geq 0 \end{cases}$

  (a) $f(-1)$        (b) $f(0)$

  (c) $f(1)$          (d) $f(2)$

**16.** $f(x) = \begin{cases} x^2 + 2, & x \leq 1 \\ 2x^2 + 2, & x > 1 \end{cases}$

  (a) $f(-2)$        (b) $f(0)$

  (c) $f(1)$          (d) $f(2)$

In Exercises 17–22, find all real values of $x$ such that $f(x) = 0$. = find a "y" intercept because $y = f(x)$

**17.** $f(x) = 15 - 3x$      **18.** $f(x) = \dfrac{3x - 4}{5}$

**19.** $f(x) = x^2 - 9$      **20.** $f(x) = x^3 - x$

**21.** $f(x) = \dfrac{3}{x - 1} + \dfrac{4}{x - 2}$

**22.** $f(x) = 2 + \dfrac{3}{x}$

In Exercises 23–34, find the domain of the function.

**23.** $g(x) = 1 - 2x^2$      **24.** $f(x) = 5x^2 + 2x - 1$

**25.** $h(t) = \dfrac{4}{t}$          **26.** $s(y) = \dfrac{3y}{y + 5}$

**27.** $g(y) = \sqrt[3]{y - 10}$    **28.** $f(t) = \sqrt[3]{t + 4}$

**29.** $f(x) = \sqrt[4]{1 - x^2}$     **30.** $g(x) = \sqrt{x + 1}$

**31.** $g(x) = \dfrac{1}{x} - \dfrac{3}{x + 2}$   **32.** $h(x) = \dfrac{10}{x^2 - 2x}$

**33.** $f(x) = \dfrac{\sqrt{x + 1}}{x - 2}$     **34.** $f(s) = \dfrac{\sqrt{s - 1}}{s - 4}$

**35.** Consider $f(x) = \sqrt{x - 2}$ and $g(x) = \sqrt[3]{x - 2}$. Why are the domains of $f$ and $g$ different?

**36.** A student says that the domain of

$$f(x) = \dfrac{\sqrt{x + 1}}{x - 3}$$

is all real numbers except $x = 3$. Is the student correct? Explain.

In Exercises 37–46, decide whether the equation represents $y$ as a function of $x$.

**37.** $x^2 + y^2 = 4$       **38.** $x = y^2$

**39.** $x^2 + y = 4$        **40.** $x + y^2 = 4$

**41.** $2x + 3y = 4$

**42.** $x^2 + y^2 - 2x - 4y + 1 = 0$

**43.** $y^2 = x^2 - 1$        **44.** $y = \sqrt{x + 5}$

**45.** $x^2y - x^2 + 4y = 0$   **46.** $xy - y - x - 2 = 0$

In Exercises 47–50, decide whether the set of ordered pairs represents a function from A to B.

$A = \{0, 1, 2, 3\}$ and $B = \{-2, -1, 0, 1, 2\}$

Give reasons for your answers.

**47.** $\{(0, 1), (1, -2), (2, 0), (3, 2)\}$

**48.** $\{(0, -1), (2, 2), (1, -2), (3, 0), (1, 1)\}$

**49.** $\{(0, 0), (1, 0), (2, 0), (3, 0)\}$

**50.** $\{(0, 2), (3, 0), (1, 1)\}$

In Exercises 51–54, decide whether the set of ordered pairs represents a function from A to B.

$A = \{a, b, c\}$ and $B = \{0, 1, 2, 3\}$

Give reasons for your answers.

**51.** $\{(a, 1), (c, 2), (c, 3), (b, 3)\}$

**52.** $\{(a, 1), (b, 2), (c, 3)\}$

**53.** $\{(1, a), (0, a), (2, c), (3, b)\}$

**54.** $\{(c, 0), (b, 0), (a, 3)\}$

In Exercises 55–58, decide whether the set of figures represents a function from A to B.

$A = \{a, b, c\}$ and $B = \{1, 2, 3, 4\}$

Give reasons for your answers.

**55.**
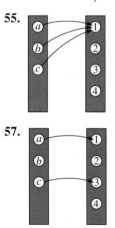
**56.**
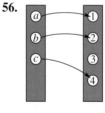

**57.**
**58.**
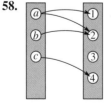

In Exercises 59–62, the domain of f is the set

$A = \{-2, -1, 0, 1, 2\}$.

Write the function as a set of ordered pairs.

**59.** $f(x) = x^2$

**60.** $f(x) = \dfrac{2x}{x^2 + 1}$

**61.** $f(x) = \sqrt{x + 2}$

**62.** $f(x) = |x + 1|$

**63.** *Volume of a Box*   An open box is to be made from a square piece of material 12 inches on a side by cutting equal squares from the corners and turning up the sides (see figure).

(a) Write the volume V as a function of x.

(b) What is the domain of this function?

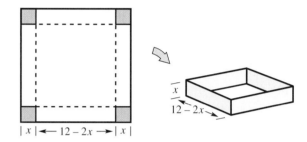

**64.** *Volume of a Package*   A rectangular package to be sent by a postal service can have a maximum combined length and girth (perimeter of a cross section) of 108 inches (see figure).

(a) Write the volume V of a package with maximum length and girth as a function of x.

(b) What is the domain of this function?

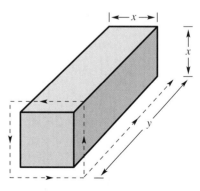

**65.** *Height of a Balloon*   A balloon carrying a transmitter ascends vertically from a point 2000 feet from the receiving station (see figure). Let *d* be the distance between the balloon and the receiving station. Express the height of the balloon as a function of *d*. What is the domain of this function?

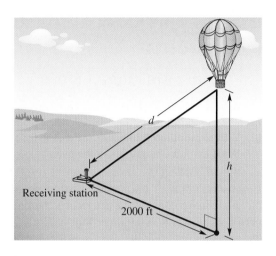

**66.** *Mobile Homes*   The number of mobile homes manufactured for residential use in the United States from 1990 to 1999 can be approximated by the model

$$n = \begin{cases} 2.437t^2 + 13.71t + 179.8, & 0 \le t \le 6 \\ -50.950t^2 + 795.15t - 2731.4, & 7 \le t \le 9 \end{cases}$$

where *n* is the number of mobile homes in thousands and *t* is the year, with *t* = 0 corresponding to 1990 (see figure). Use the model to find the numbers of mobile homes manufactured in 1994 and in 1998. (Source: U.S. Bureau of the Census)

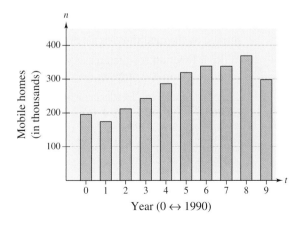

**67.** *Health Care Cost*   The average annual health care cost per consumer unit in the United States from 1990 to 1998 can be approximated by the model

$$C = \begin{cases} -14.083t^3 + 75.36t^2 - 7.3t + 1484, & 0 \le t \le 4 \\ 58.4t + 1432, & 5 \le t \le 8 \end{cases}$$

where *C* is the cost in dollars and *t* is the year, with *t* = 0 corresponding to 1990 (see figure). Use the model to find the average health care costs in 1994 and 1998.   (Source: U.S. Bureau of Labor Statistics)

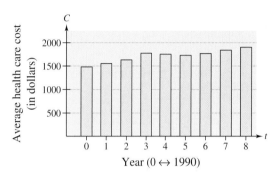

**68.** *Computer Storage Devices*   The total values *V* (in millions of dollars) of shipments of computer storage devices and equipment in the United States for the years 1994 through 1998 are given in the table. (Source: U.S. Bureau of the Census)

| Year | 1994 | 1995 | 1996 | 1997 | 1998 |
|------|------|------|------|------|------|
| V | 5556 | 7903 | 8909 | 8837 | 7248 |

(a) Use a graphing utility to make a scatter plot of the data. Let *t* = 4 correspond to 1994.

(b) Use the regression capabilities of your graphing utility to find a linear model and a quadratic model for the data. Round decimals to two places.

(c) Use each model to approximate the values of shipments for the years 1994 through 1998. Compare the values generated by each model with the actual values given in the table. Which model is a better fit? Justify your answer.

---

The symbol ▦ indicates an exercise or parts of an exercise in which you are instructed to use a graphing utility.

**69.** *Cost of Housing*  The average annual expenditures *E* (in dollars) per consumer unit for shelter in the United States from 1993 to 1998 are given in the following table.  (Source: U.S. Bureau of Labor Statistics)

| Year | 1993 | 1994 | 1995 |
|---|---|---|---|
| Average expenditure (in dollars) | 5415 | 5686 | 5932 |

| Year | 1996 | 1997 | 1998 |
|---|---|---|---|
| Average expenditure (in dollars) | 6064 | 6344 | 6680 |

(a) Use a graphing utility to make a scatter plot of the data. (Suggestion: Let $t = 3$ correspond to 1993.)

(b) Use the regression capabilities of your graphing utility to find a linear model and a quadratic model for the data. Round decimals to two places.

(c) Graph the linear model with the scatter plot. Then graph the quadratic model with the scatter plot. Which model appears to fit the data better? Is one model easier to use than the other? Which model would you choose to use? Explain your answer.

**70.** *Average Cost*  The inventor of a new game believes that the variable cost for producing the game is $0.95 per unit and the fixed costs are $6000. The inventor sells each game for $1.69. Let *x* be the number of games sold.

(a) Write the total cost *C* as a function of the number of games sold.

(b) Write the average cost per unit $\overline{C} = C/x$ as a function of *x*.

(c) Complete the table.

| x | 100 | 1000 | 10,000 | 100,000 |
|---|---|---|---|---|
| $\overline{C}$ | | | | |

(d) Write a paragraph analyzing the data in the table. What do you observe about the average cost per unit as *x* gets larger?

**71.** *Average Cost*  A manufacturer determines that the variable cost for a new product is $1.15 per unit and the fixed costs are $35,000. The product is to be sold for $2.19. Let *x* be the number of units sold.

(a) Write the total cost *C* as a function of the number of units sold.

(b) Write the average cost per unit $\overline{C} = C/x$ as a function of *x*.

(c) Complete the table.

| x | 100 | 1000 | 10,000 | 100,000 |
|---|---|---|---|---|
| $\overline{C}$ | | | | |

(d) Write a paragraph analyzing the data in the table. What do you observe about the average cost per unit as *x* gets larger?

**72.** *Charter Bus Fares*  For groups of 80 or more people, a charter bus company determines the rate per person as rate $= 8 - 0.05(n - 80)$, $n \geq 80$, where the rate is given in dollars and *n* is the number of people.

(a) Express the total revenue *R* for the bus company as a function of *n*.

(b) Complete the table.

| n | 90 | 100 | 110 | 120 | 130 | 140 | 150 |
|---|---|---|---|---|---|---|---|
| R | | | | | | | |

(c) Write a paragraph analyzing the data in the table.

**73.** *Ripples in a Pond*  A stone is thrown into the middle of a calm pond, causing ripples to form in concentric circles. The radius *r* of the outermost ripple increases at the rate of 0.8 foot per second.

(a) Write a function for the radius of the circle formed by the outermost ripple in terms of time *t*.

(b) Write a function for the area *A* enclosed by the outermost ripple. Complete the table.

| Time t | 1 | 2 | 3 | 4 | 5 |
|---|---|---|---|---|---|
| Radius r | | | | | |
| Area A | | | | | |

(c) Compare the ratios $A(2)/A(1)$ and $A(4)/A(2)$. What do you observe? Based on your observation, predict the area when $t = 8$. Verify by checking $t = 8$ in the area function.

## 3.2    Graphs of Functions

### Objectives

- Find the domain and range using the graph of a function.
- Identify the graph of a function using the Vertical Line Test.
- Describe the increasing and decreasing behavior of a function.
- Classify a function as even or odd.
- Identify six common graphs and use them to sketch the graph of a function.

## The Graph of a Function

In Section 3.1 you studied functions from an algebraic point of view. In this section, you will study functions from a geometric perspective. The **graph of a function** $f$ is the collection of ordered pairs $(x, f(x))$ such that $x$ is in the domain of $f$. As you study this section, remember that

$$x = \text{the directed distance from the } y\text{-axis}$$

$$f(x) = \text{the directed distance from the } x\text{-axis}$$

as shown in Figure 3.6. If the graph of a function has an $x$-intercept at $(a, 0)$, then $a$ is a **zero** of the function. In other words, the zeros of a function are the values of $x$ for which $f(x) = 0$. For instance, the function given by $f(x) = x^2 - 4$ has two zeros: $-2$ and $2$.

The **range** of a function (the set of values assumed by the dependent variable) is often more easily determined graphically than algebraically. This technique is illustrated in Example 1.

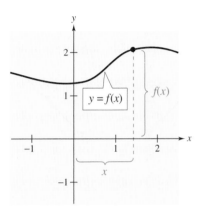

Figure 3.6

**Example 1**    Finding the Domain and Range of a Function

Use the graph of the function $f$, shown in Figure 3.7, to find the following.

**a.** The domain of $f$.

**b.** The function values $f(-1)$ and $f(2)$.

**c.** The range of $f$.

### Solution

**a.** Because the graph does not extend beyond $x = -1$ (on the left) and $x = 4$ (on the right), the domain of $f$ is all $x$ in the interval $[-1, 4]$.

**b.** Because $(-1, -5)$ is a point on the graph of $f$, it follows that

$$f(-1) = -5.$$

Similarly, because $(2, 4)$ is a point on the graph of $f$, it follows that

$$f(2) = 4.$$

**c.** Because the graph does not extend below $f(-1) = -5$ nor above $f(2) = 4$, the range of $f$ is the interval $[-5, 4]$.

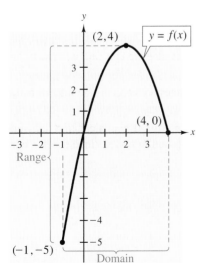

Figure 3.7

By the definition of a function, at most one $y$-value corresponds to a given $x$-value. It follows, then, that a vertical line can intersect the graph of a function at most once. This observation provides a convenient visual test for functions.

---

▶ Vertical Line Test for Functions

A set of points in a coordinate plane is the graph of $y$ as a function of $x$ if and only if no vertical line intersects the graph at more than one point.

---

**Example 2** Vertical Line Test for Functions

Use the Vertical Line Test to decide whether the graphs in Figure 3.8 represent $y$ as a function of $x$.

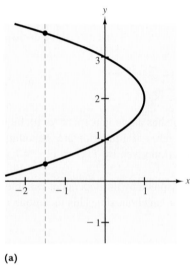

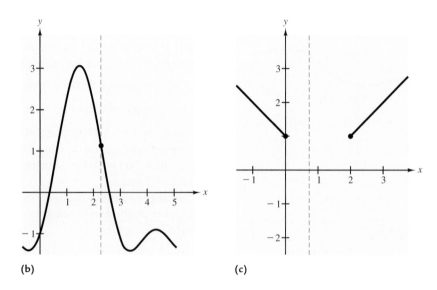

(a)

(b)

(c)

Figure 3.8

### Solution

**a.** This *is not* a graph of $y$ as a function of $x$ because we can find a vertical line that intersects the graph twice.

**b.** This *is* a graph of $y$ as a function of $x$ because every vertical line intersects the graph at most once.

**c.** This *is* a graph of $y$ as a function of $x$. (Note that if a vertical line does not intersect the graph, it simply means that the function is undefined for that particular value of $x$.)

## Increasing and Decreasing Functions

The more you know about the graph of a function, the more you know about the function itself. Consider the graph shown in Figure 3.9. As you move from *left to right*, this graph decreases, then is constant, and then increases.

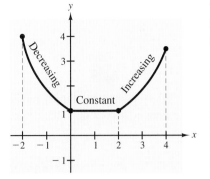

Figure 3.9

> ▶ **Increasing, Decreasing, and Constant Functions**
>
> A function $f$ is **increasing** on an interval if, for any $x_1$ and $x_2$ in the interval, $x_1 < x_2$ implies $f(x_1) < f(x_2)$.
>
> A function $f$ is **decreasing** on an interval if, for any $x_1$ and $x_2$ in the interval, $x_1 < x_2$ implies $f(x_1) > f(x_2)$.
>
> A function $f$ is **constant** on an interval if, for any $x_1$ and $x_2$ in the interval, $f(x_1) = f(x_2)$.

**Example 3**   Increasing and Decreasing Functions

Describe the increasing or decreasing behavior of each function in Figure 3.10.

**Solution**

**a.** This function is increasing over the entire real line.

**b.** This function is increasing on the interval $(-\infty, -1)$, decreasing on the interval $(-1, 1)$, and increasing on the interval $(1, \infty)$.

**c.** This function is increasing on the interval $(-\infty, 0)$, constant on the interval $(0, 2)$, and decreasing on the interval $(2, \infty)$.

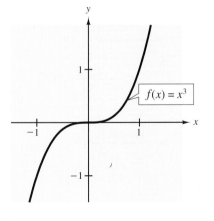

(a)
Figure 3.10

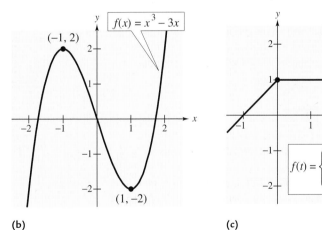

(b)

(c)

To be able to sketch an accurate graph of a function, it is important that you know the points at which the function changes its increasing, decreasing, or constant behavior. These points often identify *maximum* or *minimum* values of the function.

| Example 4 | The Price of One-Family Houses |

During the 1990s, the average price of a new, privately-owned one-family house in the western United States decreased and then increased according to the model

$$C = -164.549t^3 + 3176.34t^2 - 12,554.0t + 147,978, \quad 0 \le t \le 9$$

where $C$ is the average price in dollars and $t$ represents the calendar year, with $t = 0$ corresponding to 1990. According to this model, during which years was the price of one-family houses decreasing? During which years was the price of one-family houses increasing? Approximate the minimum price of a one-family house between 1990 and 1999. (Source: U.S. Bureau of the Census, U.S. Department of Housing and Urban Development)

### Solution

To solve this problem, sketch an accurate graph of the function, as shown in Figure 3.11. From the graph, you can see that the price of a one-family house decreased from 1990 to mid-1992. Then, from mid-1992 to 1999, the price increased. The minimum price during the 10-year period was approximately $135,000.

Figure 3.11

Year (0 ↔ 1990)

Average price (in dollars)

·**T E C H N O L O G Y**·

There are several ways you can use a graphing utility to find the minimum and maximum values of a function. Begin by graphing the function from Example 4

$$C = -164.549t^3 + 3176.34t^2 -$$

$$12,554.0t + 147,978, \quad 0 \le t \le 9$$

as shown at the right. One way to find the minimum value is to use the trace feature. A second way to find the minimum is to use the minimum feature. Using either method, you can determine that the minimum value occurs when $x \approx 2.4$. Check your user's manual for information on the minimum and maximum features of your graphing utility.

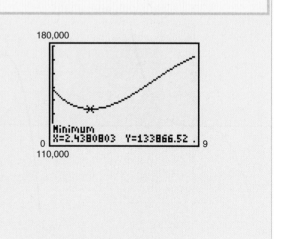

## Step Functions

**Example 5**    The Greatest Integer Function

The **greatest integer function** is denoted by $[\![x]\!]$ and is defined by

$$[\![x]\!] = \text{the greatest integer less than or equal to } x.$$

The graph of this function is shown in Figure 3.12. Note that the graph of the greatest integer function jumps vertically one unit at each integer and is constant (a horizontal line segment) between each pair of consecutive integers. Because of the jumps in its graph, the greatest integer function is an example of a category of functions called **step functions.** Some values of the greatest integer function are as follows.

$$[\![-1]\!] = -1 \qquad [\![-0.5]\!] = -1$$

$$[\![0]\!] = 0 \qquad [\![0.5]\!] = 0$$

$$[\![1]\!] = 1 \qquad [\![1.5]\!] = 1$$

The range of the greatest integer function is the set of all integers.

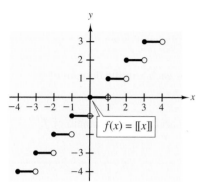

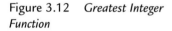

**Figure 3.12    *Greatest Integer Function***

If you use a graphing utility to sketch a step function, you should set the utility to *Dot* mode rather than *Connected* mode.

**Example 6**    The Price of a Telephone Call

The cost of a telephone call between Los Angeles and San Francisco is $0.50 for up to, but not including, the first minute and $0.36 for each additional minute (or portion of a minute). The greatest integer function can be used to create a model for the cost of this call, as follows.

$$C = 0.50 + 0.36[\![t]\!], \quad t > 0$$

where $C$ is the total cost of the call in dollars and $t$ is the length of the call in minutes. Sketch the graph of this function.

### Solution

For calls up to, but not including, 1 minute, the cost is $0.50. For calls between 1 and 2 minutes, the cost is $0.86, and so on.

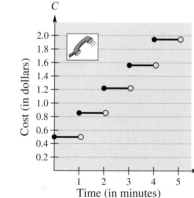

**Figure 3.13**

| *Length of call* | $0 \le t < 1$ | $1 \le t < 2$ | $2 \le t < 3$ | $3 \le t < 4$ | $4 \le t < 5$ |
|---|---|---|---|---|---|
| *Cost of call* | $0.50 | $0.86 | $1.22 | $1.58 | $1.94 |

Using these values, you can sketch the graph shown in Figure 3.13.

·DISCOVERY·

Graph each of the following functions with a graphing utility. Determine whether the function is *odd, even,* or *neither.*

$$f(x) = x^2 - x^4$$

$$g(x) = 2x^3 + 1$$

$$h(x) = x^5 - 2x^3 + x$$

$$k(x) = x^5 - 2x^4 + x - 2$$

$$j(x) = 2 - x^6 - x^8$$

$$p(x) = x^9 + 3x^5 - x^3 + x$$

What do you notice about the equations of functions that are odd? What do you notice about the equations of functions that are even? Can you describe a way to identify a function as odd or even by inspecting the equation? Can you describe a way to identify a function as neither odd nor even by inspecting the equation?

## Even and Odd Functions

In Section 2.2, you studied different types of symmetry of a graph. In the terminology of functions, a function is said to be **even** if its graph is symmetric with respect to the *y*-axis and to be **odd** if its graph is symmetric with respect to the origin. The symmetry tests in Section 2.2 yield the following tests for even and odd functions.

▶ **Tests for Even and Odd Functions**

A function given by $y = f(x)$ is **even** if, for each $x$ in the domain of $f$,

$$f(-x) = f(x).$$

A function given by $y = f(x)$ is **odd** if, for each $x$ in the domain of $f$,

$$f(-x) = -f(x).$$

**Example 7** Even and Odd Functions

**a.** The function $g(x) = x^3 - x$ is odd because

$$g(-x) = (-x)^3 - (-x) = -x^3 + x = -(x^3 - x) = -g(x).$$

**b.** The function $h(x) = x^2 + 1$ is even because

$$h(-x) = (-x)^2 + 1 = x^2 + 1 = h(x).$$

The graphs of the two functions are shown in Figure 3.14.

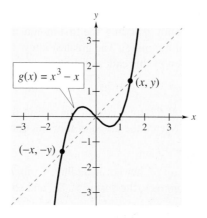

(a) Odd function (symmetric to origin)

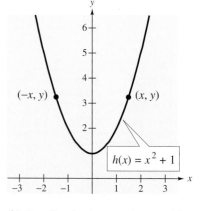

(b) Even function (symmetric to *y*-axis)

Figure 3.14

# Common Graphs

Figure 3.15 shows the graphs of six common functions. You need to be familiar with these graphs. They can be used as an aid to sketching other graphs. For instance, the graph of $f(x) = |x - 2|$, an absolute value function, will be $\vee$ -shaped.

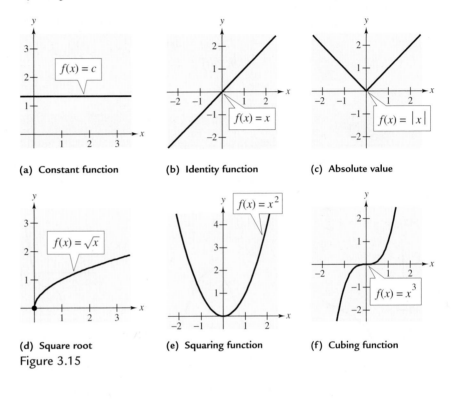

(a) Constant function    (b) Identity function    (c) Absolute value

(d) Square root    (e) Squaring function    (f) Cubing function

Figure 3.15

| Discussing the Concept | Increasing and Decreasing Functions |
|---|---|

Use your school's library or some other reference source to find examples of three different functions that represent the behaviors of quantities between 1991 and 2001. Find one quantity that decreased during the decade, one that increased, and one that was constant. For instance, the value of the dollar decreased, the population of the United States increased, and the land size of the United States remained constant. Can you find three other examples? If the examples you find appear to represent *linear growth or decline,* use the methods described in Section 2.5 to find a linear function $f(x) = a + bx$ that approximates the data.

# Warm Up

The following warm-up exercises involve skills that were covered in earlier sections. You will use these skills in the exercise set for this section.

**1.** Find $f(2)$ for $f(x) = -x^3 + 5x$.  **2.** Find $f(6)$ for $f(x) = x^2 - 6x$.

**3.** Find $f(-x)$ for $f(x) = 3/x$.  **4.** Find $f(-x)$ for $f(x) = x^2 + 3$.

In Exercises 5 and 6, solve the equation.

**5.** $x^3 - 16x = 0$  **6.** $2x^2 - 3x + 1 = 0$

In Exercises 7–10, find the domain of the function.

**7.** $g(x) = 4(x - 4)^{-1}$  **8.** $f(x) = 2x/(x^2 - 9x + 20)$

**9.** $h(t) = \sqrt[4]{5 - 3t}$  **10.** $f(t) = t^3 + 3t - 5$

## 3.2   Exercises

In Exercises 1–8, find the domain and range of the function.

**1.** $f(x) = \sqrt{x - 1}$

**2.** $f(x) = 4 - x^2$

**3.** $f(x) = \sqrt{x^2 - 4}$

**4.** $f(x) = |x - 2|$

**5.** $f(x) = \sqrt{25 - x^2}$

**6.** $f(x) = \dfrac{|x|}{x}$

**7.** $f(x) = x^3 - 1$

**8.** $f(x) = \sqrt{x^2 - 9}$

In Exercises 9–14, use the Vertical Line Test to decide whether $y$ is a function of $x$.

**9.** $y = x^2$

**10.** $y = x^3 - 1$

**11.** $x - y^2 = 0$

**12.** $x^2 + y^2 = 9$

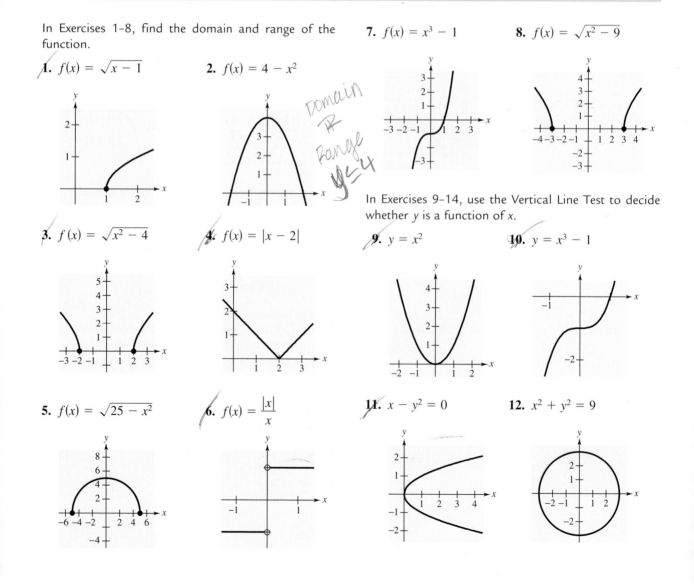

**13.** $x^2 = xy - 1$          **14.** $x = |y|$

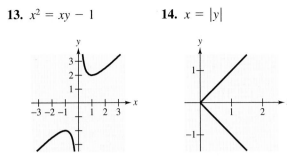

**21.** $y = x\sqrt{x + 3}$          **22.** $y = |x + 1| + |x - 1|$

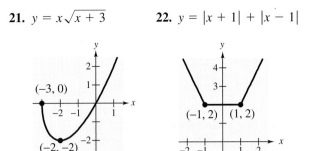

In Exercises 15–22, describe the increasing and decreasing behavior of the function. Find the point or points where the behavior of the function changes.

**15.** $f(x) = 2x$          **16.** $f(x) = x^2 - 2x$

**17.** $f(x) = x^3 - 3x^2$          **18.** $f(x) = \sqrt{x^2 - 4}$

**19.** $f(x) = 3x^4 - 6x^2$          **20.** $f(x) = x^{2/3}$

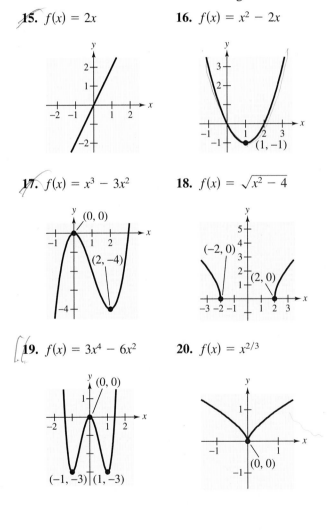

In Exercises 23–28, decide whether the function is even, odd, or neither.

**23.** $f(x) = x^6 - 2x^2 + 3$     **24.** $h(x) = x^3 - 5$

**25.** $g(x) = x^3 - 5x$          **26.** $f(x) = x\sqrt{1 - x^2}$

**27.** $f(t) = t^2 + 2t - 3$      **28.** $g(s) = 4s^{2/3}$

In Exercises 29–40, sketch the graph of the function and determine whether the function is even, odd, or neither.

**29.** $f(x) = 3$          **30.** $g(x) = x$

**31.** $f(x) = 5 - 3x$          **32.** $h(x) = x^2 - 4$

**33.** $g(s) = \dfrac{s^3}{4}$          **34.** $f(t) = -t^4$

**35.** $f(x) = \sqrt{1 - x}$          **36.** $f(x) = x^{3/2}$

**37.** $g(t) = \sqrt[3]{t - 1}$          **38.** $f(x) = |x + 2|$

**39.** $f(x) = \begin{cases} x + 3, & x \le 0 \\ 3, & 0 < x \le 2 \\ 2x - 1, & x > 2 \end{cases}$

**40.** $f(x) = \begin{cases} 2x + 1, & x \le -1 \\ x^2 - 2, & x > -1 \end{cases}$

In Exercises 41–54, sketch the graph of the function.

**41.** $f(x) = 4 - x$          **42.** $f(x) = 4x + 2$

**43.** $f(x) = x^2 - 9$          **44.** $f(x) = x^2 - 4x$

**45.** $f(x) = 1 - x^4$          **46.** $f(x) = \sqrt{x + 2}$

**47.** $f(x) = x^2 + 1$          **48.** $f(x) = -1(1 + |x|)$

**49.** $f(x) = -5$          **50.** $f(x) = \frac{1}{2}(2 + |x|)$

**51.** $f(x) = -\llbracket x \rrbracket$          **52.** $f(x) = 2\llbracket x \rrbracket$

**53.** $f(x) = \llbracket x - 1 \rrbracket$          **54.** $f(x) = \llbracket x + 1 \rrbracket$

In Exercises 55–58, use a graphing utility to graph the function and then estimate the open intervals on which the function is increasing or decreasing.

**55.** $f(x) = x^2 - 4x + 1$     **56.** $f(x) = -x^2 + 6x + 3$

**57.** $f(x) = x^3 - 3x^2$          **58.** $f(x) = -x^3 + 3x + 1$

**59.** *Population of Pennsylvania* The population of the state of Pennsylvania increased in the early 1990s and then began to decrease. A model that approximates the population of Pennsylvania is

$$P = -5.22t^2 + 56.9t + 11,889, \quad 0 \le t \le 9$$

where $P$ represents the population in thousands and $t = 0$ represents 1990. Use the figure to estimate the years that the population was increasing and the years that the population was decreasing. (Source: U.S. Bureau of the Census)

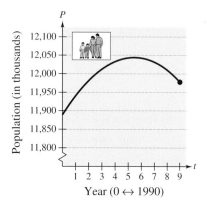

Year (0 ↔ 1990)

**60.** *Population of Rhode Island* The population of the state of Rhode Island decreased during the early 1990s and then began to increase. A model that approximates the population of Rhode Island is

$$P = 0.082t^3 - 0.74t^2 - 1.6t + 1007, \quad 1 \le t \le 9$$

where $P$ represents the population in thousands and $t = 1$ represents 1991. Use the figure to estimate the years that the population was decreasing and the years that the population was increasing. (Source: U.S. Bureau of the Census)

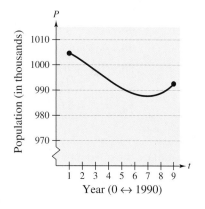

Year (0 ↔ 1990)

**61.** *Company Revenue* A company determines that the total revenue for the years from 1990 to 2000 can be approximated by the function

$$R = -0.035t^3 - 0.03t^2 + 3.25t + 14.41,$$
$$0 \le t \le 10$$

where $R$ is the total revenue in hundreds of thousands of dollars and $t = 0$ represents 1990. Graph the revenue function with a graphing utility and use the trace feature to estimate the years in which the revenue was increasing and the years in which the revenue was decreasing. Find the maximum revenue for the years 1990 to 2000. Find the minimum revenue for the years 1990 to 2000.

**62.** *Radio Stations* From 1990 to 2000, the total number of radio stations in the United States that operated with a country format can be approximated by the function

$$R = 2416 + 108.9t - 16.49t^2 + 0.375t^3,$$
$$0 \le t \le 10$$

where $R$ is the number of radio stations and $t = 0$ represents 1990. Use the figure to estimate the years in which the number of country stations was increasing and the years in which the number of country stations was decreasing. Estimate the maximum number of country stations from 1990 to 2000. Estimate the minimum number of country stations from 1990 to 2000. (Source: M Street Corporation)

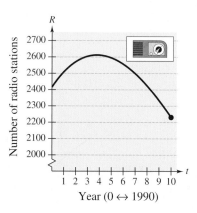

Year (0 ↔ 1990)

**63.** *Reasoning* When finding a maximum or minimum value for Exercises 61 and 62, why should you also check the endpoints of the function?

**64.** *Maximum and Minimum* Use the maximum and minimum features of your graphing utility to verify your answers to Exercises 61 and 62.

**65.** *Maximum Profit*    The marketing department of a company estimates that the demand for a product is given by $p = 100 - 0.0001x$, where $p$ is the price per unit and $x$ is the number of units. The cost of producing $x$ units is given by $C = 350,000 + 30x$, and the profit for producing $x$ units is given by

$$P = R - C = xp - C.$$

Sketch the graph of the profit function and estimate the number of units that would produce a maximum profit.

**66.** *Maximum Profit*    The marketing department of a company estimates that the demand for a product is given by $p = 120 - 0.0001x$, where $p$ is the price per unit and $x$ is the number of units. The cost of producing $x$ units is given by $C = 450,000 + 50x$, and the profit for producing $x$ units is given by

$$P = R - C = xp - C.$$

Sketch the graph of the profit function and estimate the number of units that would produce a maximum profit.

**67.** *Cost of Overnight Delivery*    Suppose that the cost of sending an overnight package from New York to Atlanta is $9.80 for up to, but not including, the first pound and $2.50 for each additional pound (or portion of a pound). A model for the total cost of sending the package is $C = 9.8 + 2.5[\![x]\!]$, $x > 0$, where $C$ is the total cost in dollars and $x$ is the weight of the package in pounds. Sketch the graph of this function.

**68.** *Cost of Overnight Delivery*    Suppose that the cost of sending an overnight package from Los Angeles to Miami is $10.75 for up to, but not including, the first pound and $3.95 for each additional pound (or portion of a pound). A model for the total cost of sending the package is $C = 10.75 + 3.95[\![x]\!]$, $x > 0$, where $C$ is the total cost in dollars and $x$ is the weight in pounds. Sketch the graph of this function.

**69.** *Research and Development*    The total amount spent on basic research and development in the United States from 1960 to 1999 can be approximated by the model

$$C = \begin{cases} 0.09t^2 + 0.15t + 15.8, & 0 \le t \le 19 \\ 8.6t - 107.6, & 20 \le t \le 39 \end{cases}$$

where $C$ is the amount in billions of dollars and $t$ is the year, with $t = 0$ corresponding to 1960. Sketch the graph of this function. (Source: National Science Foundation)

**70.** *Grade Level Salaries*    The 2001 salary for federal employees at the Step 1 level can be approximated by the model

$$S = \begin{cases} 2487.6x + 10,411, & x = 1, 2, \ldots, 10 \\ 9849.0x - 69,381, & x = 11, \ldots, 15 \end{cases}$$

where $S$ is the salary in dollars and $x$ represents the "GS" grade. Sketch a *bar graph* that represents this function. (Source: U.S. Office of Personnel Management)

**71.** *Cable TV Systems*    The numbers of cable television systems in the United States from 1990 to 2000 are given by the following ordered pairs. (Source: *Television and Cable Factbook*)

(1990, 9575), (1991, 10,704), (1992, 11,073),

(1993, 11,108), (1994, 11,214), (1995, 11,215),

(1996, 11,220), (1997, 10,943), (1998, 10,845),

(1999, 10,700), (2000, 10,500)

(a) Use the regression capabilities of a graphing utility to find a cubic model for the data from 1990 to 1996. Let $t = 0$ represent 1990.

(b) Use the regression capabilities of a graphing utility to find a quadratic model for the data from 1997 to 2000. Let $t = 7$ represent 1997.

(c) Use your results from parts (a) and (b) to construct a piecewise model for all of the data.

**72.** *Average Miles per Gallon*    The average numbers of miles per gallon for passenger cars in the United States from 1989 to 1998 are given by the following ordered pairs. (Source: U.S. Federal Highway Administration)

(1989, 18.9), (1990, 20.2), (1991, 21.1),

(1992, 20.9), (1993, 20.5), (1994, 20.7),

(1995, 21.1), (1996, 21.3), (1997, 21.5),

(1998, 21.4)

(a) Use the regression capabilities of a graphing utility to find a quadratic model for the data from 1989 to 1993. Let $t = 9$ represent 1989.

(b) Use the regression capabilities of a graphing utility to find a quadratic model for the data from 1994 to 1998. Let $t = 14$ represent 1994.

(c) Use your results from parts (a) and (b) to construct a piecewise model for all of the data.

## 3.3    Transformations of Functions

### Objectives

- Identify rigid and nonrigid transformations.
- Sketch the graph of a function using transformations and common graphs.
- Write the equation of a function using transformations and common graphs.

## Vertical and Horizontal Shifts

Many functions have graphs that are simple transformations of the common graphs that are summarized on page 171. For example, you can obtain the graph of $h(x) = x^2 + 2$ by shifting the graph of $f(x) = x^2$ *up* two units, as shown in Figure 3.16. In function notation, $h$ and $f$ are related as follows.

$$h(x) = x^2 + 2 = f(x) + 2 \qquad \text{Upward shift of 2}$$

Similarly, you can obtain the graph of $g(x) = (x - 2)^2$ by shifting the graph of $f(x) = x^2$ to the *right* two units, as shown in Figure 3.17. In this case, the functions $g$ and $f$ have the following relationship.

$$g(x) = (x - 2)^2 = f(x - 2) \qquad \text{Right shift of 2}$$

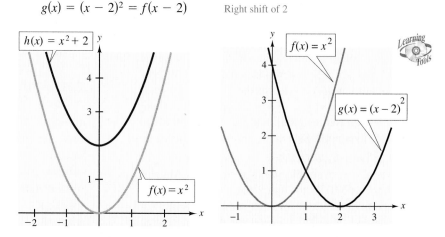

Figure 3.16    *Vertical Shift Upward*    Figure 3.17    *Horizontal Shift to the Right*

---

▶ **Vertical and Horizontal Shifts**

Let $c$ be a positive real number. **Vertical** and **horizontal shifts** in the graph of $y = f(x)$ are represented as follows.

1. Vertical shift $c$ units **upward:**    $h(x) = f(x) + c$

2. Vertical shift $c$ units **downward:**    $h(x) = f(x) - c$

3. Horizontal shift $c$ units to the **right:**    $h(x) = f(x - c)$

4. Horizontal shift $c$ units to the **left:**    $h(x) = f(x + c)$

---

In items 3 and 4, be sure you see that $h(x) = f(x - c)$ corresponds to a *right* shift and $h(x) = f(x + c)$ corresponds to a *left* shift.

Some graphs can be obtained from a combination of vertical and horizontal shifts. This is demonstrated in Example 1(b).

## Study Tip

Vertical and horizontal shifts generate a *family of functions,* each with the same shape but at different locations in the plane. The functions $f$, $g$, and $h$ in Example 1 belong to the family of cubic functions.

**Example 1**    Shifts in the Graph of a Function

Use the graph of $f(x) = x^3$ to sketch the graphs of the following functions.

**a.** $g(x) = x^3 + 1$

**b.** $h(x) = (x + 2)^3 + 1$

### Solution

Relative to the graph of $f(x) = x^3$, the graph of $g(x) = x^3 + 1$ is an upward shift of one unit, and the graph of $h(x) = (x + 2)^3 + 1$ involves a left shift of two units *and* an upward shift of one unit. The graphs of both functions are compared with the graph of $f(x) = x^3$ in Figure 3.18.

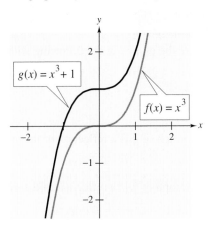

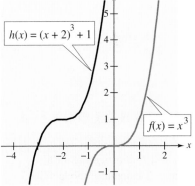

(a)  **Vertical shift: 1 up**

(b)  **Horizontal shift: 2 left; vertical shift: 1 up**

Figure 3.18

## ·DISCOVERY·

The point $(2, 4)$ is on the graph of $f(x) = x^2$. Predict the location of that point if the following transformations are performed.

**a.** $f(x - 4)$

**b.** $f(x) + 1$

**c.** $f(x + 1) - 2$

Use a graphing utility to verify your prediction. Can you find a general description to represent an ordered pair that has been shifted horizontally? vertically?

## ·TECHNOLOGY·

Graphing utilities are ideal tools for exploring translations of functions. Try to predict how the graphs of $g$ and $h$ relate to the graph of $f$. Graph $f$, $g$, and $h$ on the same screen to check your prediction.

**a.** $f(x) = x^2$,   $g(x) = (x - 4)^2$,   $h(x) = (x - 4)^2 + 3$

**b.** $f(x) = x^2$,   $g(x) = (x + 1)^2$,   $h(x) = (x + 1)^2 - 2$

**c.** $f(x) = x^2$,   $g(x) = (x + 4)^2$,   $h(x) = (x + 4)^2 + 2$

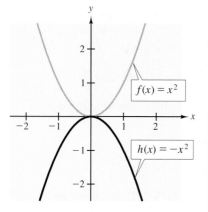

**Figure 3.19    Reflection**

# Reflections

The second common type of transformation is a **reflection.** For instance, if you consider the $x$-axis to be a mirror, the graph of $h(x) = -x^2$ is the mirror image (or reflection) of the graph of $f(x) = x^2$, as shown in Figure 3.19.

▶ **Reflections in the Coordinate Axes**

Reflections in the coordinate axes of the graph of $y = f(x)$ are represented as follows.

1. **Reflection in the $x$-axis:**  $h(x) = -f(x)$

2. **Reflection in the $y$-axis:**  $h(x) = f(-x)$

**Example 2**    Reflections of the Graph of a Function

Compare the graphs of the following with the graph of $f(x) = \sqrt{x}.$

**a.** $g(x) = -\sqrt{x}$

**b.** $h(x) = \sqrt{-x}$

**Solution**

**a.** The graph of $g$ is a reflection of the graph of $f$ in the $x$-axis because

$$g(x) = -\sqrt{x} = -f(x).$$

**b.** The graph of $h$ is a reflection of the graph of $f$ in the $y$-axis because

$$h(x) = \sqrt{-x} = f(-x).$$

The graphs of both functions are shown in Figure 3.20.

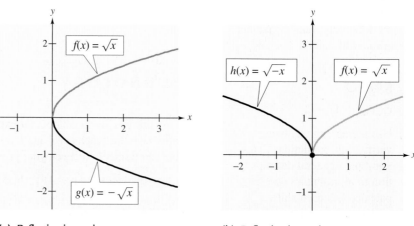

**(a)  Reflection in $x$-axis**

**(b)  Reflection in $y$-axis**

**Figure 3.20**

When sketching the graph of a function involving square roots, remember that the domain must be restricted to exclude negative numbers inside the radical. For instance, here are the domains of the functions in Example 2.

Domain of $g(x) = -\sqrt{x}:$    $x \geq 0$

Domain of $h(x) = \sqrt{-x}:$    $x \leq 0$

**Example 3**    Reflections and Shifts

Use the graph of $f(x) = x^2$ to sketch the graph of each function.

**a.** $g(x) = -(x - 3)^2$

**b.** $h(x) = -x^2 + 2$

**Solution**

**a.** To sketch the graph of $g(x) = -(x - 3)^2$, first shift the graph of $f(x) = x^2$ to the right three units. Then reflect the result in the $x$-axis.

**b.** To sketch the graph of $h(x) = -x^2 + 2$, first reflect the graph of $f(x) = x^2$ in the $x$-axis. Then shift the result upward two units.

The graphs of both functions are shown in Figure 3.21.

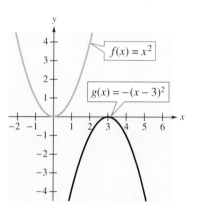

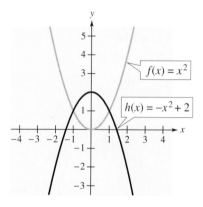

(a)  **Shift and then reflect in the $x$-axis**    (b)  **Reflect in the $x$-axis and then shift**

Figure 3.21

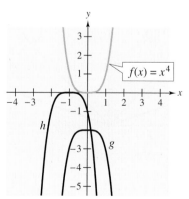

Figure 3.22

**Example 4**    Finding Equations from Graphs

The graphs labeled $g$ and $h$ in Figure 3.22 are transformations of the graph of $f(x) = x^4$. Find an equation for each function.

**Solution**

The graph of $g$ is a reflection in the $x$-axis *followed* by a downward shift of two units of the graph of $f(x) = x^4$. Thus, the equation for $g$ is $g(x) = -x^4 - 2$. The graph of $h$ is a horizontal shift of one unit to the left *followed* by a reflection in the $x$-axis of the graph of $f(x) = x^4$. Thus, the equation for $h$ is $h(x) = -(x + 1)^4$.

**·DISCOVERY·**

Use a graphing utility to graph $f(x) = 2x^2$. Compare this graph with the graph of $h(x) = x^2$. Describe the effect of multiplying $x^2$ by a number greater than 1. Then graph $g(x) = \frac{1}{2}x^2$. Compare this with the graph of $h(x) = x^2$. Describe the effect of multiplying $x^2$ by a number less than 1. Can you think of an easy way to remember this generalization? Use the table feature of your graphing utility to compare the values for $f(x)$, $g(x)$, and $h(x)$. What do you notice? How does this relate to the vertical stretch or vertical shrink of the graph of a function?

## Nonrigid Transformations

Horizontal shifts, vertical shifts, and reflections are **rigid** transformations because the basic shape of the graph is unchanged. These transformations change only the *position* of the graph in the *xy*-plane. A **nonrigid** transformation is one that causes a *distortion*—a change in the shape of the original graph. For instance, a nonrigid transformation of the graph of $y = f(x)$ is represented by $g(x) = cf(x)$, where the transformation is a **vertical stretch** if $c > 1$ and a **vertical shrink** if $0 < c < 1$.

**Example 5**    Nonrigid Transformations

Compare the graphs of the following with the graph of $f(x) = |x|$.

**a.** $h(x) = 3|x|$

**b.** $g(x) = \frac{1}{3}|x|$

Solution

**a.** Relative to the graph of $f(x) = |x|$, the graph of

$$h(x) = 3|x| = 3f(x)$$

is a vertical stretch (each *y*-value is multiplied by 3) of the graph of *f*.

**b.** Similarly, the equation

$$g(x) = \frac{1}{3}|x| = \frac{1}{3}f(x)$$

indicates that the graph of *g* is a vertical shrink $\left(\text{each } y\text{-value is multiplied by } \frac{1}{3}\right)$ of the graph of *f*.

The graphs of both functions are shown in Figure 3.23.

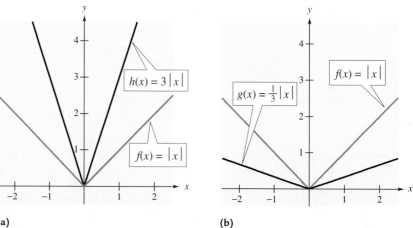

(a)                                      (b)

Figure 3.23

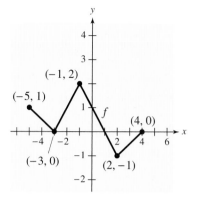

**Figure 3.24**

| Example 6 | Rigid and Nonrigid Transformations |  |

Use the graph of $f$ in Figure 3.24 to sketch the following graphs.

**a.** $g(x) = f(x - 2) + 1$

**b.** $h(x) = \frac{1}{2}f(x)$

**Solution**

**a.** The graph of $g(x)$ is a horizontal shift to the right two units and a vertical shift upward one unit of the graph of $f(x)$. The graph of $g(x)$ is shown in Figure 3.25(a).

**b.** The graph of $h(x)$ is a vertical shrink of the graph of $f(x)$.

For $x = -5$, $h(-5) = \frac{1}{2}f(-5) = \frac{1}{2}(1) = \frac{1}{2}$.

For $x = -3$, $h(-3) = \frac{1}{2}f(-3) = \frac{1}{2}(0) = 0$.

For $x = -1$, $h(-1) = \frac{1}{2}f(-1) = \frac{1}{2}(2) = 1$.

For $x = 2$, $h(2) = \frac{1}{2}f(2) = \frac{1}{2}(-1) = -\frac{1}{2}$.

For $x = 4$, $h(4) = \frac{1}{2}f(4) = \frac{1}{2}(0) = 0$.

The graph of $h(x)$ is shown in Figure 3.25(b).

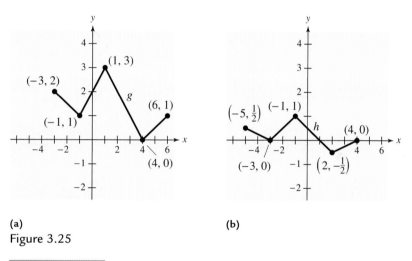

**(a)**                                              **(b)**

**Figure 3.25**

---

| Discussing the Concept | Constructing Transformations |

Use a graphing utility to graph $f(x) = 3 - x^3$. Describe how to alter this function to produce each of the following transformation descriptions. Graph each transformation on the same screen with $f$; confirm that the transformation moves $f$ as described.

**a.** The graph of $f$ shifted upward four units

**b.** The graph of $f$ shifted to the left two units

**c.** The graph of $f$ shifted downward two units and to the right one unit

**d.** The graph of $f$ reflected in the $x$-axis

# Warm Up

The following warm-up exercises involve skills that were covered in earlier sections. You will use these skills in the exercise set for this section.

**1.** Find $f(3)$ for $f(x) = x^2 - 4x + 15$.

**2.** Find $f(-x)$ for $f(x) = 2x/(x - 3)$.

In Exercises 3 and 4, solve the equation.

**3.** $-x^3 + 10x = 0$

**4.** $3x^2 + 2x - 8 = 0$

In Exercises 5–10, sketch the graph of the function.

**5.** $f(x) = -2$

**6.** $f(x) = -x$

**7.** $f(x) = x + 5$

**8.** $f(x) = 2 - x$

**9.** $f(x) = 3x - 4$

**10.** $f(x) = 9x + 10$

## 3.3    Exercises

In Exercises 1–8, use the graph of $f(x) = x^2$ to sketch the graph of the function by hand. Verify with a graphing utility.

**1.** $g(x) = x^2 + 3$

**2.** $g(x) = x^2 - 2$

**3.** $g(x) = (x + 3)^2$

**4.** $g(x) = (x - 4)^2$

**5.** $g(x) = (x - 2)^2 + 2$

**6.** $g(x) = (x + 1)^2 - 3$

**7.** $g(x) = -x^2 + 1$

**8.** $g(x) = -(x - 2)^2$

In Exercises 9–14, use the graph of $f(x) = |x|$ to sketch the graph of the function by hand. Verify with a graphing utility.

**9.** $g(x) = |x| + 2$

**10.** $g(x) = |x - 1|$

**11.** $g(x) = -|x| + 3$

**12.** $g(x) = |x + 2| - 3$

**13.** $g(x) = 4 - |x - 2|$

**14.** $g(x) = |x - 2| + 2$

In Exercises 15–22, use the graph of $f(x) = \sqrt{x}$ to sketch the graph of the function by hand. Verify with a graphing utility.

**15.** $y = \sqrt{x - 2}$

**16.** $y = \sqrt{x + 3}$

**17.** $y = \sqrt{x - 3} + 1$

**18.** $y = \sqrt{x + 5} - 2$

**19.** $y = 2 - \sqrt{x - 4}$

**20.** $y = \sqrt{2x}$

**21.** $y = \sqrt{-x} + 1$

**22.** $y = \sqrt{2x} - 5$

In Exercises 23–30, use the graph of $f(x) = \sqrt[3]{x}$ to sketch the graph of the function by hand. Verify with a graphing utility.

**23.** $y = \sqrt[3]{x} - 1$

**24.** $y = \sqrt[3]{x + 1}$

**25.** $y = 2 - \sqrt[3]{x + 1}$

**26.** $y = -\sqrt[3]{x - 1} - 4$

**27.** $y = \sqrt[3]{x + 1} - 1$

**28.** $y = \frac{1}{2}\sqrt[3]{x}$

**29.** $y = \frac{1}{2}\sqrt[3]{x} - 3$

**30.** $y = 2\sqrt[3]{x - 2} + 1$

In Exercises 31–36, identify the common function and the transformation shown in the graph. Write the equation for the graphed function.

**31.**

**32.**

**33.**

**34.**

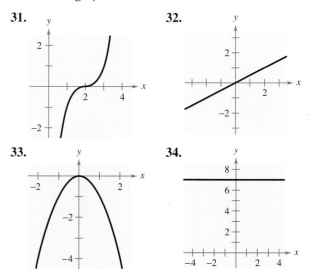

| Example 2 | Finding the Difference of Two Functions |

Given $f(x) = 2x + 1$ and $g(x) = x^2 + 2x - 1$, find $(f - g)(x)$. Then evaluate the difference when $x = 2$.

**Solution**

The difference of the functions $f$ and $g$ is given by

$$(f - g)(x) = f(x) - g(x)$$
$$= (2x + 1) - (x^2 + 2x - 1)$$
$$= -x^2 + 2.$$

When $x = 2$, the value of this difference is

$$(f - g)(2) = -(2)^2 + 2 = -2.$$

In Examples 1 and 2, both $f$ and $g$ have domains that consist of all real numbers. Thus, the domains of $(f + g)$ and $(f - g)$ are also the set of all real numbers. Remember that any restrictions on the domains of $f$ and $g$ must be considered when forming the sum, difference, product, or quotient of $f$ and $g$.

| Example 3 | The Quotient of Two Functions |

Find the domains of $(f/g)(x)$ and $(g/f)(x)$ for the functions

$$f(x) = \sqrt{x} \qquad \text{and} \qquad g(x) = \sqrt{4 - x^2}.$$

**Solution**

The quotient of $f$ and $g$ is given by

$$\left(\frac{f}{g}\right)(x) = \frac{f(x)}{g(x)} = \frac{\sqrt{x}}{\sqrt{4 - x^2}}$$

and the quotient of $g$ and $f$ is given by

$$\left(\frac{g}{f}\right)(x) = \frac{g(x)}{f(x)} = \frac{\sqrt{4 - x^2}}{\sqrt{x}}.$$

The domain of $f$ is $[0, \infty)$ and the domain of $g$ is $[-2, 2]$. The intersection of these two domains is $[0, 2]$, which implies that the domains of $f/g$ and $g/f$ are as follows.

Domain of $\dfrac{f}{g}$: $[0, 2)$

Domain of $\dfrac{g}{f}$: $(0, 2]$

Can you see why these two domains differ slightly?

## Composition of Functions

Another way of combining two functions is to form the **composition** of one with the other. For instance, if $f(x) = x^2$ and $g(x) = x + 1$, the composition of $f$ with $g$ is given by

$$f(g(x)) = f(x + 1) = (x + 1)^2.$$

This composition is denoted as $f \circ g$.

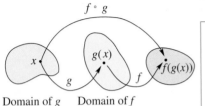

$f \circ g$

$x$    $g$    $g(x)$    $f$    $f(g(x))$

Domain of $g$    Domain of $f$

**Figure 3.26**

---

▶ **Definition of Composition of Two Functions**

The **composition** of the functions $f$ and $g$ is given by

$$(f \circ g)(x) = f(g(x)).$$

The domain of $(f \circ g)$ is the set of all $x$ in the domain of $g$ such that $g(x)$ is in the domain of $f$. (See Figure 3.26.)

---

**Example 4** Composition of Functions

Given $f(x) = x + 2$ and $g(x) = 4 - x^2$, find the following.

**a.** $(f \circ g)(x)$

**b.** $(g \circ f)(x)$

**Solution**

**a.** The composition of $f$ with $g$ is as follows.

$$(f \circ g)(x) = f(g(x)) \qquad \text{Definition of } f \circ g$$
$$= f(4 - x^2) \qquad \text{Definition of } g(x)$$
$$= (4 - x^2) + 2 \qquad \text{Definition of } f(x)$$
$$= -x^2 + 6 \qquad \text{Simplify.}$$

**b.** The composition of $g$ with $f$ is as follows.

$$(g \circ f)(x) = g(f(x)) \qquad \text{Definition of } g \circ f$$
$$= g(x + 2) \qquad \text{Definition of } f(x)$$
$$= 4 - (x + 2)^2 \qquad \text{Definition of } g(x)$$
$$= 4 - (x^2 + 4x + 4) \qquad \text{Expand.}$$
$$= -x^2 - 4x \qquad \text{Simplify.}$$

Note that, in this case, $(f \circ g)(x) \neq (g \circ f)(x)$.

**Example 5**   Finding the Domain of a Composite Function

Find the composition $(f \circ g)(x)$ for the functions

$$f(x) = x^2 - 9 \qquad \text{and} \qquad g(x) = \sqrt{9 - x^2}.$$

Then find the domain of $(f \circ g)$.

**Solution**

The composition of the functions is as follows.

$$(f \circ g)(x) = f(g(x))$$
$$= f\left(\sqrt{9 - x^2}\right)$$
$$= \left(\sqrt{9 - x^2}\right)^2 - 9$$
$$= 9 - x^2 - 9$$
$$= -x^2$$

From this, it might appear that the domain of the composition is the set of all real numbers. This, however, is not true because the domain of $g$ is $[-3, 3]$. So, the domain of $f \circ g$ is $[-3, 3]$.

---

In Examples 4 and 5, you formed the composition of two functions. To "decompose" a composite function, look for an "inner" function and an "outer" function. For instance, the function $h$ given by

$$h(x) = (3x - 5)^3$$

is the composition of $f$ with $g$, where $f(x) = x^3$ and $g(x) = 3x - 5$. That is,

$$h(x) = (3x - 5)^3 = [g(x)]^3 = f(g(x)).$$

In the function $h$, $g(x) = 3x - 5$ is the inner function and $f(x) = x^3$ is the outer function.

**Example 6**   Identifying a Composite Function

Express the function

$$h(x) = \frac{1}{(x - 2)^2}$$

as a composition of two functions.

**Solution**

One way to write $h$ as a composition of two functions is to take the inner function to be $g(x) = x - 2$ and the outer function to be

$$f(x) = \frac{1}{x^2} = x^{-2}.$$

Then you can write

$$h(x) = \frac{1}{(x - 2)^2} = (x - 2)^{-2} = f(x - 2) = f(g(x)).$$

---

**·TECHNOLOGY·**

In Example 5, the domain of the composite function is $-3 \leq x \leq 3$. To convince yourself of this, use a graphing utility to graph

$$y = \left(\sqrt{9 - x^2}\right)^2 - 9$$

as shown in the figure below. Notice that the graphing utility does not extend the graph to the left of $x = -3$ or to the right of $x = 3$.

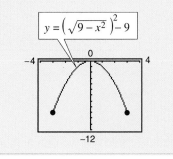

## Applications

**Example 7**    Political Makeup of the U.S. Senate

Consider three functions $R$, $D$, and $I$ that represent the numbers of Republicans, Democrats, and Independents in the U.S. Senate from 1965 to 1999. Sketch the graphs of $R$, $D$, and $I$, and the sum of $R$, $D$, and $I$, in the same coordinate plane. The numbers of Republicans and Democrats in the Senate are shown below.

| Year | Republicans | Democrats | | Year | Republicans | Democrats |
|------|-------------|-----------|--|------|-------------|-----------|
| 1965 | 32 | 68 | | 1983 | 54 | 46 |
| 1967 | 36 | 64 | | 1985 | 53 | 47 |
| 1969 | 43 | 57 | | 1987 | 45 | 55 |
| 1971 | 44 | 54 | | 1989 | 45 | 55 |
| 1973 | 42 | 56 | | 1991 | 44 | 56 |
| 1975 | 37 | 60 | | 1993 | 43 | 57 |
| 1977 | 38 | 61 | | 1995 | 52 | 48 |
| 1979 | 41 | 58 | | 1997 | 55 | 45 |
| 1981 | 53 | 46 | | 1999 | 54 | 45 |

### Solution

The graphs of $R$, $D$, and $I$ are shown in Figure 3.27. Note that the sum of $R$, $D$, and $I$ is the constant function $R + D + I = 100$. This follows from the fact that the number of senators in the United States is 100 (two from each state).

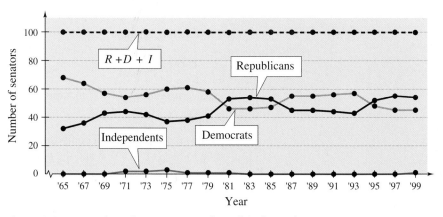

Figure 3.27    *Number of U.S. Senators (by political party)*

**Example 8**    Bacteria Count

The number of bacteria in a refrigerated food is given by

$$N(T) = 20T^2 - 80T + 500, \qquad 2 \le T \le 14$$

where $T$ is the temperature of the food. When the food is removed from refrigeration, the temperature is given by

$$T(t) = 4t + 2, \qquad 0 \le t \le 3$$

where $t$ is the time in hours. Find (a) the composite $N(T(t))$, (b) the number of bacteria in the food when $t = 2$ hours, and (c) the time when the bacteria count reaches 2000.

**Solution**

**a.**  $N(T(t)) = 20(4t + 2)^2 - 80(4t + 2) + 500$

$\qquad\qquad = 20(16t^2 + 16t + 4) - 320t - 160 + 500$

$\qquad\qquad = 320t^2 + 320t + 80 - 320t - 160 + 500$

$\qquad\qquad = 320t^2 + 420$

**b.**  When $t = 2$, the number of bacteria is

$$N = 320(2)^2 + 420 = 1280 + 420 = 1700.$$

**c.**  The bacteria count will reach $N = 2000$ when $320t^2 + 420 = 2000$. By solving this equation, you can determine that the bacteria count will reach 2000 when

$$t \approx 2.2 \text{ hours.}$$

---

## Discussing the Concept    Composition of Functions

The suggested retail price of a new car is $p$ dollars. The dealer has advertised a factory rebate of $1500 and a 7% discount.

**a.**  Write a function $R$, in terms of $p$, giving the cost of the car after receiving the rebate.

**b.**  Write a function $D$, in terms of $p$, giving the cost of the car after receiving the discount.

**c.**  Form the composite functions $(R \circ D)(p)$ and $(D \circ R)(p)$. Explain what each composite function represents.

**d.**  Find $(R \circ D)(21,000)$ and $(D \circ R)(21,000)$. Which function yields the lower price for the car? Explain.

**e.**  Research the rebates and discounts offered by several car dealers in your area. How do they compare with one another?

# Warm Up

The following warm-up exercises involve skills that were covered in earlier sections. You will use these skills in the exercise set for this section.

In Exercises 1–10, perform the indicated operations and simplify the result.

**1.** $\dfrac{1}{x} + \dfrac{1}{1-x}$

**2.** $\dfrac{2}{x+3} - \dfrac{2}{x-3}$

**3.** $\dfrac{3}{x-2} - \dfrac{2}{x(x-2)}$

**4.** $\dfrac{x}{x-5} + \dfrac{1}{3}$

**5.** $(x-1)\left(\dfrac{1}{\sqrt{x^2-1}}\right)$

**6.** $\left(\dfrac{x}{x^2-4}\right)\left(\dfrac{x^2-x-2}{x^2}\right)$

**7.** $(x^2-4) \div \left(\dfrac{x+2}{5}\right)$

**8.** $\left(\dfrac{x}{x^2+3x-10}\right) \div \left(\dfrac{x^2+3x}{x^2+6x+5}\right)$

**9.** $\dfrac{(1/x)+5}{3-(1/x)}$

**10.** $\dfrac{(x/4)-(4/x)}{x-4}$

## 3.4 Exercises

In Exercises 1–4, use the graphs of $f$ and $g$ to graph $h(x) = (f+g)(x)$.

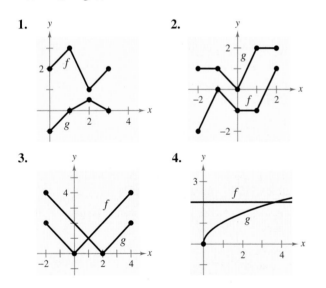

**1.**

**2.**

**3.**

**4.**

In Exercises 5–12, find (a) $(f+g)(x)$, (b) $(f-g)(x)$, (c) $(fg)(x)$, and (d) $(f/g)(x)$. What is the domain of $f/g$?

**5.** $f(x) = x+1$,  $g(x) = x-1$

**6.** $f(x) = 2x-5$,  $g(x) = 1-x$

**7.** $f(x) = x^2$,  $g(x) = 1-x$

**8.** $f(x) = 2x-5$,  $g(x) = 5$

**9.** $f(x) = x^2+5$,  $g(x) = \sqrt{1-x}$

**10.** $f(x) = \sqrt{x^2-4}$,  $g(x) = \dfrac{x^2}{x^2+1}$

**11.** $f(x) = \dfrac{1}{x}$,  $g(x) = \dfrac{1}{x^2}$

**12.** $f(x) = \dfrac{x}{x+1}$,  $g(x) = x^3$

In Exercises 13–24, evaluate the function for $f(x) = x^2 + 1$ and $g(x) = x - 4$.

**13.** $(f+g)(3)$

**14.** $(f-g)(-2)$

**15.** $(f-g)(2t)$

**16.** $(f+g)(t-1)$

**17.** $(fg)(4)$

**18.** $(fg)(-6)$

**19.** $\left(\dfrac{f}{g}\right)(5)$

**20.** $\left(\dfrac{f}{g}\right)(0)$

**21.** $(f-g)(0)$

**22.** $(f+g)(1)$

**23.** $\left(\dfrac{f}{g}\right)(-1) - g(3)$

**24.** $(2f)(5)$

In Exercises 25–28, find (a) $f \circ g$ and (b) $f \circ f$.

**25.** $f(x) = x^2$,  $g(x) = x-1$

**26.** $f(x) = 3x$,  $g(x) = 2x+1$

**27.** $f(x) = 3x+5$,  $g(x) = 5-x$

**28.** $f(x) = x^3$,  $g(x) = \dfrac{1}{x}$

In Exercises 29–36, find (a) $f \circ g$ and (b) $g \circ f$.

**29.** $f(x) = \sqrt{x + 4}$,     $g(x) = x^2$

**30.** $f(x) = \sqrt[3]{x - 1}$,     $g(x) = x^3 + 1$

**31.** $f(x) = \frac{1}{3}x - 3$,     $g(x) = 3x + 1$

**32.** $f(x) = x^4$,     $g(x) = x^4$

**33.** $f(x) = \sqrt{x}$,     $g(x) = \sqrt{x}$

**34.** $f(x) = 2x - 3$,     $g(x) = 2x - 3$

**35.** $f(x) = |x|$,     $g(x) = x + 6$

**36.** $f(x) = x^{2/3}$,     $g(x) = x^6$

In Exercises 37–40, use the graphs of $f$ and $g$ to evaluate the functions.

(a)
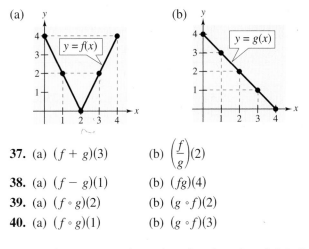
(b)

**37.** (a) $(f + g)(3)$     (b) $\left(\dfrac{f}{g}\right)(2)$

**38.** (a) $(f - g)(1)$     (b) $(fg)(4)$

**39.** (a) $(f \circ g)(2)$     (b) $(g \circ f)(2)$

**40.** (a) $(f \circ g)(1)$     (b) $(g \circ f)(3)$

In Exercises 41–44, determine the domain of (a) $f$, (b) $g$, and (c) $f \circ g$.

**41.** $f(x) = \sqrt{x}$,     $g(x) = x^2 + 1$

**42.** $f(x) = \dfrac{1}{x}$,     $g(x) = x + 3$

**43.** $f(x) = \dfrac{3}{x^2 - 1}$,     $g(x) = x + 1$

**44.** $f(x) = 2x + 3$,     $g(x) = \dfrac{x}{2}$

In Exercises 45–52, find two functions $f$ and $g$ such that $(f \circ g)(x) = h(x)$. (There are many correct answers.)

**45.** $h(x) = (2x + 1)^2$

**46.** $h(x) = (1 - x)^3$

**47.** $h(x) = \sqrt[3]{x^2 - 4}$

**48.** $h(x) = \sqrt{9 - x}$

**49.** $h(x) = \dfrac{1}{x + 2}$

**50.** $h(x) = \dfrac{4}{(5x + 2)^2}$

**51.** $h(x) = (x + 4)^2 + 2(x + 4)$

**52.** $h(x) = (x + 3)^{3/2}$

**53.** *Stopping Distance*   While traveling in a car at $x$ miles per hour, you are required to stop quickly to avoid an accident. The distance the car travels (in feet) during your reaction time is given by $R(x) = \frac{3}{4}x$. The distance traveled while you are braking (in feet) is given by

$$B(x) = \frac{1}{15}x^2.$$

Find the function giving total stopping distance $T$. (*Hint*: $T = R + B$.) Graph the functions $R$, $B$, and $T$ on the same set of coordinate axes for $0 \le x \le 60$.

**54.** *Comparing Profits*   A company has two manufacturing plants, one in New Jersey and the other in California. From 1997 to 2002, the profits for the manufacturing plant in New Jersey have been decreasing according to the function

$$P_1 = 16.97 - 0.43t, \qquad t = 7, 8, 9, 10, 11, 12$$

where $P_1$ represents the profit in millions of dollars and $t = 7$ represents 1997. On the other hand, the profits for the manufacturing plant in California have been increasing according to the function

$$P_2 = 13.36 + 0.56t, \qquad t = 7, 8, 9, 10, 11, 12.$$

Write a function that represents the overall company profits during the 6-year period. Use the *stacked bar graph* in the figure, which represents the total profit for the company during this 6-year period, to determine whether the overall company profits have been increasing or decreasing.

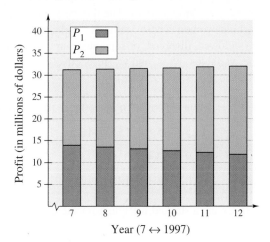

**55.** *Comparing Sales*   You own two fast-food restaurants. During the years 1997 to 2002, the sales for the first restaurant have been decreasing according to the function

$$R_1 = 500 - 0.8t^2, \qquad t = 7, 8, 9, 10, 11, 12$$

where $R_1$ represents the sales for the first restaurant (in thousands of dollars) and $t = 7$ represents 1997. During the same 6-year period, the sales for the second restaurant have been increasing according to the function

$$R_2 = 253.9 + 0.78t, \qquad t = 7, 8, 9, 10, 11, 12.$$

Write a function that represents the total sales for the two restaurants. Use the *stacked bar graph* in the figure, which represents the total sales during this 6-year period, to determine whether the total sales have been increasing or decreasing.

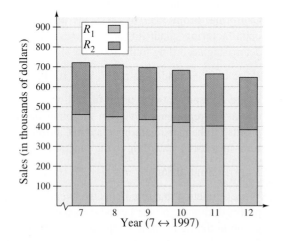

**56.** *Female Labor Force*   Use the table, which gives the marital status of women in the civilian labor force for the years 1994 through 1999. The numbers (in millions) of working women whose status is single, married, or *other* (widowed, divorced, or separated), are represented by the variables $y_1$, $y_2$, and $y_3$, respectively. (Source: U.S Bureau of Labor Statistics)

| Year | 1994 | 1995 | 1996 | 1997 | 1998 | 1999 |
|------|------|------|------|------|------|------|
| $y_1$ | 15.3 | 15.5 | 15.8 | 16.5 | 17.1 | 17.6 |
| $y_2$ | 32.9 | 33.4 | 33.6 | 33.8 | 33.9 | 34.4 |
| $y_3$ | 12.0 | 12.1 | 12.4 | 12.7 | 12.8 | 12.9 |

(a) Create a stacked bar graph for the data.

(b) Use the regression capabilities of a graphing utility to develop linear models for $y_1$, $y_2$, and $y_3$. Let $t = 4$ represent 1994.

(c) Graph the models for $y_1$, $y_2$, $y_3$, and $y_4 = y_1 + y_2 + y_3$ in the same viewing window. Use $y_4$ to predict the total number of women in the work force in the years 2006 and 2009.

**57.** *Energy Consumption*   Use the table, which gives the energy consumption (in quadrillion Btu) in the United States for the years 1990 to 1999. The variables $y_1$, $y_2$, and $y_3$ represent energy use in the residential and commercial, industrial and miscellaneous, and transportation sectors, respectively. (Source: U.S Energy Information Administration)

| Year | 1990 | 1991 | 1992 | 1993 | 1994 |
|------|------|------|------|------|------|
| $y_1$ | 29.48 | 30.14 | 30.03 | 31.12 | 31.37 |
| $y_2$ | 32.15 | 31.80 | 33.01 | 33.30 | 34.35 |
| $y_3$ | 22.54 | 22.13 | 22.47 | 22.89 | 23.52 |

| Year | 1995 | 1996 | 1997 | 1998 | 1999 |
|------|------|------|------|------|------|
| $y_1$ | 32.26 | 33.67 | 33.64 | 33.68 | 34.17 |
| $y_2$ | 34.70 | 35.71 | 35.85 | 35.54 | 36.50 |
| $y_3$ | 23.97 | 24.52 | 24.82 | 25.36 | 25.92 |

(a) Create a stacked bar graph for the data.

(b) Use the regression capabilities of a graphing utility to develop a linear model for $y_1$, $y_2$, and $y_3$, respectively, where $t = 0$ represents 1990.

(c) Graph the models for $y_1$, $y_2$, $y_3$, and $y_4 = y_1 + y_2 + y_3$ in the same viewing window. Use $y_4$ to estimate the total energy consumption in the years 2005 and 2009.

**58.** *Bacteria Count*   The number of bacteria in a refrigerated food product is given by

$$N(T) = 10T^2 - 20T + 600, \qquad 1 \le T \le 20$$

where $T$ is the temperature of the food. When the food is removed from the refrigerator, the temperature is given by

$$T(t) = 3t + 1$$

where $t$ is the time in hours. Find (a) the composite $N(T(t))$ and (b) the time when the bacteria count reaches 1500.

**59.** *Bacteria Count*    The number of bacteria in a refrigerated food product is given by

$$N(T) = 25T^2 - 50T + 300, \qquad 2 \le T \le 20$$

where $T$ is the temperature of the food. When the food is removed from the refrigerator, the temperature is given by

$$T(t) = 2t + 1$$

where $t$ is the time in hours. Find (a) the composite $N(T(t))$ and (b) the time when the bacteria count reaches 750.

**60.** *Troubled Waters*    A pebble is dropped into a calm pond, causing ripples in the form of concentric circles (see figure). The radius (in feet) of the outer ripple is given by $r(t) = 0.6t$, where $t$ is time in seconds after the pebble strikes the water. The area of the circle is given by the function $A(r) = \pi r^2$. Find and interpret $(A \circ r)(t)$.

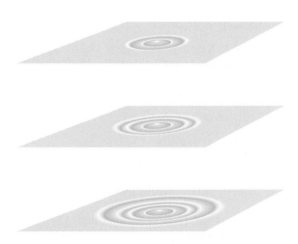

*Price-Earnings Ratio*    In Exercises 61 and 62, the average annual price-earnings ratio for a corporation's stock is defined as the average price of the stock divided by the earnings per share. The average price of a corporation's stock is given as the function $P$ and the earnings per share is given as the function $E$. Find the price-earnings ratio, $P/E$, for the years from 1992 to 2000.

**61.** *McDonald's Corporation*

| Year | P | E |
|------|------|------|
| 1992 | $11.05 | $0.65 |
| 1993 | $13.07 | $0.73 |
| 1994 | $14.36 | $0.84 |
| 1995 | $18.81 | $0.99 |
| 1996 | $23.87 | $1.11 |
| 1997 | $24.15 | $1.15 |
| 1998 | $31.12 | $1.26 |
| 1999 | $42.26 | $1.39 |
| 2000 | $33.43 | $1.46 |

(Source: McDonald's Corp.)

**62.** *Walt Disney Company*

| Year | P | E |
|------|------|------|
| 1992 | $11.42 | $0.51 |
| 1993 | $13.66 | $0.54 |
| 1994 | $14.35 | $0.68 |
| 1995 | $17.14 | $0.84 |
| 1996 | $20.13 | $0.74 |
| 1997 | $25.21 | $0.92 |
| 1998 | $33.84 | $0.90 |
| 1999 | $30.36 | $0.66 |
| 2000 | $35.55 | $0.90 |

(Source: The Walt Disney Company)

**63.** *Cost*    The weekly cost of producing $x$ units in a manufacturing process is given by the function

$$C(x) = 60x + 750.$$

The number of units produced in $t$ hours is given by $x(t) = 50t$. Find and interpret $(C \circ x)(t)$.

**64.** *Cost*    The weekly cost of producing $x$ units in a manufacturing process is given by the function

$$C(x) = 70x + 375.$$

The number of units produced in $t$ hours is given by $x(t) = 40t$. Find and interpret $(C \circ x)(t)$.

**65.** Find the domain of $(f/g)(x)$ and $(g/f)(x)$ for the functions

$$f(x) = \sqrt{x} \quad \text{and} \quad g(x) = \sqrt{9 - x^2}.$$

Why do the two domains differ?

## 3.5    Inverse Functions

### Objectives

- Determine if a function has an inverse.
- Find the inverse of a function.
- Graph a function and its inverse.

## The Inverse of a Function

Recall from Section 3.1 that a function can be represented by a set of ordered pairs. For instance, the function $f(x) = x + 4$ from the set $A = \{1, 2, 3, 4\}$ to the set $B = \{5, 6, 7, 8\}$ can be written as follows.

$$f(x) = x + 4: \ \{(1, 5), (2, 6), (3, 7), (4, 8)\}$$

By interchanging the first and second coordinates of each of these ordered pairs, you can form the **inverse function** of $f$, which is denoted by $f^{-1}$. It is a function from the set $B$ to the set $A$, and can be written as follows.

$$f^{-1}(x) = x - 4: \ \{(5, 1), (6, 2), (7, 3), (8, 4)\}$$

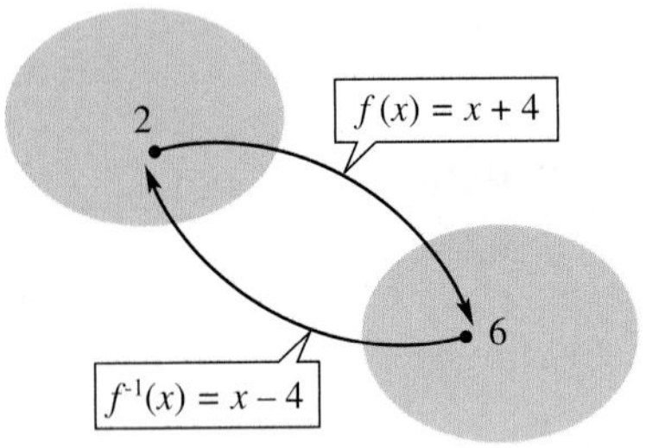

Figure 3.28

Note that the domain of $f$ is equal to the range of $f^{-1}$, and vice versa, as shown in Figure 3.28. Also note that the functions $f$ and $f^{-1}$ have the effect of "undoing" each other. In other words, when you form the composition of $f$ with $f^{-1}$ or the composition of $f^{-1}$ with $f$, you obtain identity functions, as follows.

$$f(f^{-1}(x)) = f(x - 4) = (x - 4) + 4 = x$$
$$f^{-1}(f(x)) = f^{-1}(x + 4) = (x + 4) - 4 = x$$

### Example 1    Finding Inverse Functions Informally

Find the inverse function of $f(x) = 4x$. Then verify that both $f(f^{-1}(x))$ and $f^{-1}(f(x))$ are equal to the identity function.

#### Solution

The given function *multiplies* each input by 4. To "undo" this function, you need to *divide* each input by 4. Thus, the inverse function of $f(x) = 4x$ is

$$f^{-1}(x) = \frac{x}{4}.$$

You can verify that both $f(f^{-1}(x))$ and $f^{-1}(f(x))$ are equal to the identity function as follows.

$$f(f^{-1}(x)) = f\left(\frac{x}{4}\right) = 4\left(\frac{x}{4}\right) = x$$

$$f^{-1}(f(x)) = f^{-1}(4x) = \frac{4x}{4} = x$$

| Example 2 | Finding Inverse Functions Informally  |

Find the inverse function of $f(x) = x - 6$. Then verify that both $f(f^{-1}(x))$ and $f^{-1}(f(x))$ are equal to the identity function.

### Solution

The given function *subtracts* 6 from each input. To "undo" this function, you need to *add* 6 to each input. Thus, the inverse function of $f(x) = x - 6$ is

$$f^{-1}(x) = x + 6.$$

You can verify that both $f(f^{-1}(x))$ and $f^{-1}(f(x))$ are equal to the identity function as follows.

$$f(f^{-1}(x)) = f(x + 6) = (x + 6) - 6 = x$$
$$f^{-1}(f(x)) = f^{-1}(x - 6) = (x - 6) + 6 = x$$

---

The formal definition of the inverse of a function is as follows.

---

▶ **Definition of the Inverse of a Function**

Let $f$ and $g$ be two functions such that

$$f(g(x)) = x \qquad \text{for every } x \text{ in the domain of } g$$

and

$$g(f(x)) = x \qquad \text{for every } x \text{ in the domain of } f.$$

Under these conditions, the function $g$ is the **inverse** of the function $f$. The function $g$ is denoted by $f^{-1}$ (read "$f$-inverse"). Thus,

$$f(f^{-1}(x)) = x \qquad \text{and} \qquad f^{-1}(f(x)) = x.$$

The domain of $f$ must be equal to the range of $f^{-1}$, and the range of $f$ must be equal to the domain of $f^{-1}$.

---

Don't be confused by the use of $-1$ to denote the inverse function $f^{-1}$. In this text, whenever we write $f^{-1}$, we will *always* be referring to the inverse of the function $f$ and *not* to the reciprocal of $f(x)$. That is,

$$f^{-1}(x) \neq \frac{1}{f(x)}.$$

If the function $g$ is the inverse of the function $f$, it must also be true that the function $f$ is the inverse of the function $g$. For this reason, you can say that the functions $f$ and $g$ are *inverses of each other*.

## Example 3   Verifying Inverse Functions

Show that the following functions are inverses of each other.

$$f(x) = 2x^3 - 1 \quad \text{and} \quad g(x) = \sqrt[3]{\frac{x+1}{2}}$$

### Solution

$$f(g(x)) = f\left(\sqrt[3]{\frac{x+1}{2}}\right) = 2\left(\sqrt[3]{\frac{x+1}{2}}\right)^3 - 1$$

$$= 2\left(\frac{x+1}{2}\right) - 1$$

$$= x + 1 - 1$$

$$= x$$

$$g(f(x)) = g(2x^3 - 1) = \sqrt[3]{\frac{(2x^3 - 1) + 1}{2}}$$

$$= \sqrt[3]{\frac{2x^3}{2}}$$

$$= \sqrt[3]{x^3}$$

$$= x$$

·DISCOVERY·

Graph the equations from Example 3 and the equation $y = x$ on your graphing utility using a square viewing window.

$$y_1 = 2x^3 - 1$$

$$y_2 = \sqrt[3]{\frac{x+1}{2}}$$

$$y_3 = x$$

What do you observe about the graphs of $y_1$ and $y_2$?

## Example 4   Verifying Inverse Functions

Which of the functions

$$g(x) = \frac{x-2}{5} \quad \text{and} \quad h(x) = \frac{5}{x} + 2$$

is the inverse of the function $f(x) = \dfrac{5}{x-2}$?

### Solution

By forming the composition of $f$ with $g$, you can see that

$$f(g(x)) = f\left(\frac{x-2}{5}\right) = \frac{5}{[(x-2)/5] - 2} = \frac{25}{x - 12} \neq x.$$

Because this composition is not equal to the identity function $x$, it follows that $g$ *is not* the inverse of $f$. By forming the composition of $f$ with $h$, you have

$$f(h(x)) = f\left(\frac{5}{x} + 2\right) = \frac{5}{(5/x) + 2 - 2} = \frac{5}{5/x} = x.$$

So, it appears that $h$ is the inverse of $f$. You can confirm this by showing that the composition of $h$ with $f$ is also equal to the identity function. (Try doing this.)

# Finding the Inverse of a Function

For simple functions (such as the ones in Examples 1 and 2), you can find inverse functions by inspection. For more complicated functions, however, it is best to use the following guidelines. The key step in these guidelines is switching the roles of $x$ and $y$. This step corresponds to the fact that inverse functions have ordered pairs with the coordinates reversed.

## Study Tip

Note in Step 3 of the guidelines for finding the inverse of a function that it is possible that a function has no inverse. For instance, the function $f(x) = x^2$ has no inverse function.

---

▶ **Finding the Inverse of a Function**

1. In the equation for $f(x)$, replace $f(x)$ by $y$.
2. Interchange the roles of $x$ and $y$.
3. If the new equation does not represent $y$ as a function of $x$, the function $f$ does not have an inverse function. If the new equation does represent $y$ as a function of $x$, solve the new equation for $y$.
4. Replace $y$ by $f^{-1}(x)$.
5. Verify that $f$ and $f^{-1}$ are inverses of each other by showing that the domain of $f$ is equal to the range of $f^{-1}$, the range of $f$ is equal to the domain of $f^{-1}$, and $f(f^{-1}(x)) = x = f^{-1}(f(x))$.

---

**Example 5**    Finding the Inverse of a Function    *Learning Tools*

Find the inverse function of $f(x) = \dfrac{5 - 3x}{2}$.

**Solution**

$$f(x) = \frac{5 - 3x}{2} \qquad \text{Given function}$$

$$y = \frac{5 - 3x}{2} \qquad \text{Replace } f(x) \text{ by } y.$$

$$x = \frac{5 - 3y}{2} \qquad \text{Interchange } x \text{ and } y.$$

$$2x = 5 - 3y \qquad \text{Multiply both sides by 2.}$$

$$3y = 5 - 2x \qquad \text{Isolate the } y\text{-term.}$$

$$y = \frac{5 - 2x}{3} \qquad \text{Solve for } y.$$

$$f^{-1}(x) = \frac{5 - 2x}{3} \qquad \text{Replace } y \text{ by } f^{-1}(x).$$

Note that both $f$ and $f^{-1}$ have domains and ranges that consist of the entire set of real numbers. Check that $f(f^{-1}(x)) = x$ and $f^{-1}(f(x)) = x$.

# The Graph of the Inverse of a Function

The graphs of a function $f$ and its inverse $f^{-1}$ are related to each other in the following way. If the point $(a, b)$ lies on the graph of $f$, then the point $(b, a)$ must lie on the graph of $f^{-1}$, and vice versa. This means that the graph of $f^{-1}$ is a *reflection* of the graph of $f$ in the line $y = x$, as shown in Figure 3.29.

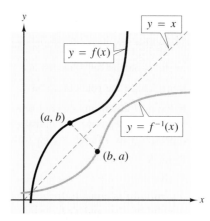

Figure 3.29 *The graph of $f^{-1}$ is a reflection of the graph of $f$ in the line $y = x$.*

**Example 6**    The Graphs of $f$ and $f^{-1}$

Sketch the graphs of the inverse functions $f(x) = 2x - 3$ and $f^{-1}(x) = \frac{1}{2}(x + 3)$ on the same rectangular coordinate system and show that the graphs are reflections of each other in the line $y = x$.

## Solution

The graphs of $f$ and $f^{-1}$ are shown in Figure 3.30. Visually, it appears that the graphs are reflections of each other in the line $y = x$. You can further verify this reflective property by testing a few points on each graph. Note in the following list that if the point $(a, b)$ is on the graph of $f$, then the point $(b, a)$ is on the graph of $f^{-1}$.

| $f(x) = 2x - 3$ | $f^{-1}(x) = \dfrac{1}{2}(x + 3)$ |
|---|---|
| $(-1, -5)$ | $(-5, -1)$ |
| $(0, -3)$ | $(-3, 0)$ |
| $(1, -1)$ | $(-1, 1)$ |
| $(2, 1)$ | $(1, 2)$ |
| $(3, 3)$ | $(3, 3)$ |

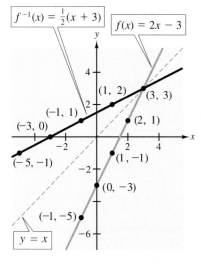

Figure 3.30

In the study tip on page 199, we mentioned that the function

$$f(x) = x^2$$

has no inverse function. What we really meant is that *assuming the domain of f is the entire real line*, the function $f(x) = x^2$ has no inverse function. If, however, we restrict the domain of $f$ to the nonnegative real numbers, then $f$ does have an inverse function, as demonstrated in Example 7.

| **Example 7** | The Graphs of $f$ and $f^{-1}$ |

Sketch the graphs of the inverse functions

$$f(x) = x^2, \quad x \geq 0, \quad \text{and} \quad f^{-1}(x) = \sqrt{x}$$

on the same rectangular coordinate system and show that the graphs are reflections of each other in the line $y = x$.

### Solution

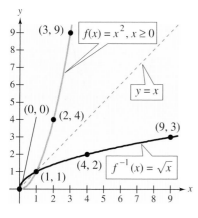

Figure 3.31

The graphs of $f$ and $f^{-1}$ are shown in Figure 3.31. Visually, it appears that the graphs are reflections of each other in the line $y = x$. You can further verify this reflective property by testing a few points on each graph. Note in the following list that if the point $(a, b)$ is on the graph of $f$, then the point $(b, a)$ is on the graph of $f^{-1}$.

| $f(x) = x^2, \quad x \geq 0$ | $f^{-1}(x) = \sqrt{x}$ |
|---|---|
| $(0, 0)$ | $(0, 0)$ |
| $(1, 1)$ | $(1, 1)$ |
| $(2, 4)$ | $(4, 2)$ |
| $(3, 9)$ | $(9, 3)$ |

Try showing that $f(f^{-1}(x)) = x$ and $f^{-1}(f(x)) = x$.

The guidelines for finding the inverse of a function include an *algebraic* test for determining whether a function has an inverse. The reflective property of the graphs of inverse functions gives you a nice *geometric* test for determining whether a function has an inverse. This test is called the **Horizontal Line Test** for inverse functions.

> ▶ Horizontal Line Test for Inverse Functions
>
> A function $f$ has an inverse function if and only if no *horizontal* line intersects the graph of $f$ at more than one point.

**Example 8** Applying the Horizontal Line Test

**a.** The graph of the function $f(x) = x^3 - 1$ is shown in Figure 3.32(a). Because no horizontal line intersects the graph of $f$ at more than one point, you can conclude that $f$ *does* have an inverse function.

**b.** The graph of the function $f(x) = x^2 - 1$ is shown in Figure 3.32(b). Because it is possible to find a horizontal line that intersects the graph of $f$ at more than one point, you can conclude that $f$ *does not* have an inverse function.

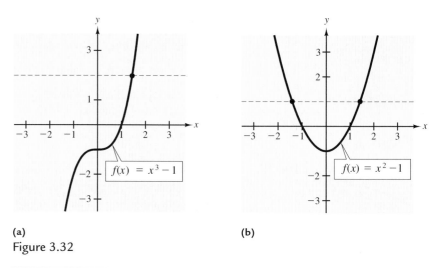

(a)                                      (b)

Figure 3.32

---

**Discussing the Concept**     **Visualizing Inverses**

Sketch the graphs of $f(x) = x^2$ and $g(x) = x$ on graph paper. Fold the graph paper on the line $g(x) = x$ and sketch the reflection of $f(x) = x^2$ about $g(x) = x$. What do you notice about the resulting graph?

Does $f(x)$ pass the Vertical Line Test for functions? That is, is $f(x)$ a function? Does $f(x)$ pass the Horizontal Line Test for inverse functions? That is, is $f^{-1}(x)$ a function? Does $f^{-1}(x)$ pass the Vertical Line Test for functions?

Write a short paragraph discussing what information can be derived from the Vertical Line Test for functions and the Horizontal Line Test for inverse functions.

# Warm Up

The following warm-up exercises involve skills that were covered in earlier sections. You will use these skills in the exercise set for this section.

In Exercises 1–4, find the domain of the function.

**1.** $f(x) = \sqrt[3]{x + 1}$

**2.** $f(x) = \sqrt{x + 1}$

**3.** $g(x) = \dfrac{2}{x^2 - 2x}$

**4.** $h(x) = \dfrac{x}{3x + 5}$

In Exercises 5–8, simplify the expression.

**5.** $2\left(\dfrac{x + 5}{2}\right) - 5$

**6.** $7 - 10\left(\dfrac{7 - x}{10}\right)$

**7.** $\sqrt[3]{2\left(\dfrac{x^3}{2} - 2\right) + 4}$

**8.** $\sqrt[5]{(x + 2)^5} - 2$

In Exercises 9 and 10, solve for $x$ in terms of $y$.

**9.** $y = \dfrac{2x - 6}{3}$

**10.** $y = \sqrt[3]{2x - 4}$

## 3.5    Exercises

In Exercises 1–10, show that $f$ and $g$ are inverse functions algebraically *and* using a graphing utility.

**1.** $f(x) = 2x, \qquad g(x) = \dfrac{x}{2}$

**2.** $f(x) = x - 5, \qquad g(x) = x + 5$

**3.** $f(x) = 5x + 1, \qquad g(x) = \dfrac{x - 1}{5}$

**4.** $f(x) = 3 - 4x, \qquad g(x) = \dfrac{3 - x}{4}$

**5.** $f(x) = x^3, \qquad g(x) = \sqrt[3]{x}$

**6.** $f(x) = \dfrac{1}{x}, \qquad g(x) = \dfrac{1}{x}$

**7.** $f(x) = \sqrt{x - 4}, \qquad g(x) = x^2 + 4, \quad x \geq 0$

**8.** $f(x) = 9 - x^2, \quad x \geq 0$
$g(x) = \sqrt{9 - x}, \quad x \leq 9$

**9.** $f(x) = 1 - x^3, \qquad g(x) = \sqrt[3]{1 - x}$

**10.** $f(x) = \dfrac{1}{1 + x}, \quad x \geq 0$

$g(x) = \dfrac{1 - x}{x}, \quad 0 < x \leq 1$

In Exercises 11–20, use the graph of $y = f(x)$, $y = g(x)$, or $y = h(x)$ to determine whether the function has an inverse function.

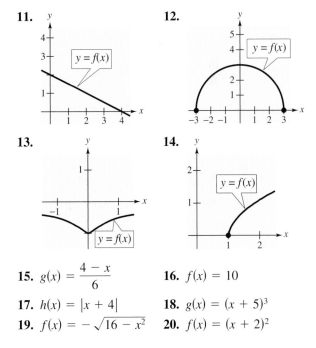

**11.**

**12.**

**13.**

**14.**

**15.** $g(x) = \dfrac{4 - x}{6}$

**16.** $f(x) = 10$

**17.** $h(x) = |x + 4|$

**18.** $g(x) = (x + 5)^3$

**19.** $f(x) = -\sqrt{16 - x^2}$

**20.** $f(x) = (x + 2)^2$

*Error Analysis*    In Exercises 21 and 22, a student has handed in the answer to a problem on a quiz. Find the error(s) in each solution and discuss how to explain each error to the student.

**21.** Find the inverse function $f^{-1}$ of $f(x) = \sqrt{2x - 5}$.

$$f(x) = \sqrt{2x - 5}, \text{ so}$$

$$f^{-1}(x) = \frac{1}{\sqrt{2x - 5}}$$

**22.** Find the inverse function $f^{-1}$ of $f(x) = \frac{3}{5}x + \frac{1}{3}$.

$$f(x) = \frac{3}{5}x + \frac{1}{3}, \text{ so}$$

$$f^{-1}(x) = \frac{5}{3}x - 3$$

In Exercises 23–32, find the inverse function $f^{-1}$ of the function $f$. Then, using a graphing utility, graph both $f$ and $f^{-1}$ on the same screen.

**23.** $f(x) = 2x - 3$          **24.** $f(x) = 3x$

**25.** $f(x) = x^5$             **26.** $f(x) = x^3 + 1$

**27.** $f(x) = \sqrt{x}$        **28.** $f(x) = x^2, \ x \geq 0$

**29.** $f(x) = \sqrt{4 - x^2}, 0 \leq x \leq 2$

**30.** $f(x) = \dfrac{4}{x}$

**31.** $f(x) = \sqrt[3]{x - 1}$     **32.** $f(x) = x^{3/5}$

In Exercises 33–48, determine whether the function has an inverse function. If it does, find its inverse function.

**33.** $f(x) = x^4$             **34.** $f(x) = \dfrac{1}{x^2}$

**35.** $g(x) = \dfrac{x}{8}$        **36.** $f(x) = 3x + 5$

**37.** $p(x) = -4$             **38.** $f(x) = \dfrac{3x + 4}{5}$

**39.** $f(x) = (x + 3)^2, \ x \geq -3$

**40.** $q(x) = (x - 5)^2$          **41.** $h(x) = \dfrac{1}{x}$

**42.** $f(x) = |x - 2|, \ x \leq 2$

**43.** $f(x) = \sqrt{2x + 3}$       **44.** $f(x) = \sqrt{x - 2}$

**45.** $g(x) = x^2 - x^4$          **46.** $f(x) = \dfrac{x^2}{x^2 + 1}$

**47.** $f(x) = 25 - x^2, \ x \leq 0$

**48.** $f(x) = 36 + x^2, \ x \leq 0$

In Exercises 49 and 50, use the graph of $f$ to complete the table and to sketch the graph of $f^{-1}$.

**49.**

| $x$ | 0 | 1 | 2 | 3 | 4 |
|-----|---|---|---|---|---|
| $f^{-1}(x)$ | | | | | |

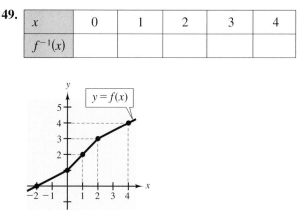

**50.**

| $x$ | 0 | 2 | 4 | 6 |
|-----|---|---|---|---|
| $f^{-1}(x)$ | | | | |

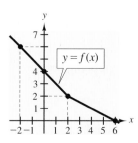

In Exercises 51–54, use the functions

$$f(x) = \frac{1}{8}x - 3 \text{ and } g(x) = x^3$$

to find the value.

**51.** $(f^{-1} \circ g^{-1})(1)$        **52.** $(g^{-1} \circ f^{-1})(-3)$

**53.** $(f^{-1} \circ f^{-1})(6)$         **54.** $(g^{-1} \circ g^{-1})(-4)$

In Exercises 55–58, use the functions

$$f(x) = x + 4 \text{ and } g(x) = 2x - 5$$

to find the composition of functions.

**55.** $g^{-1} \circ f^{-1}$           **56.** $f^{-1} \circ g^{-1}$

**57.** $(f \circ g)^{-1}$            **58.** $(g \circ f)^{-1}$

**59. *Cellular Phones*** The average monthly cellular phone bill $y$ (in dollars) from 1989 to 1998 is given in the table. (Source: Cellular Telecommunications Industry Association)

| Year | 1989 | 1990 | 1991 | 1992 | 1993 |
|------|------|------|------|------|------|
| $y$  | 89.30 | 80.90 | 72.74 | 68.68 | 61.48 |

| Year | 1994 | 1995 | 1996 | 1997 | 1998 |
|------|------|------|------|------|------|
| $y$  | 56.21 | 51.00 | 47.70 | 42.78 | 39.43 |

(a) Use a graphing utility to make a scatter plot for the data. (Let $t = 9$ represent 1989.)

(b) Use the regression capabilities of a graphing utility to find a linear model for the data. Round decimals to two places.

(c) Algebraically find the inverse function of the model in part (b).

(d) Use the inverse function you found in part (c) to estimate the year when the average cellular phone bill was about $25.

**60. *Automated Teller Machines*** The numbers of ATM transactions $n$ (in millions) in the United States from 1990 to 1999 are given in the table. (Source: Bank Network News)

| Year | 1990 | 1991 | 1992 | 1993 | 1994 |
|------|------|------|------|------|------|
| $n$  | 5942 | 6642 | 7537 | 8135 | 9078 |

| Year | 1995 | 1996 | 1997 | 1998 | 1999 |
|------|------|------|------|------|------|
| $n$  | 10,464 | 11,780 | 12,580 | 13,160 | 13,316 |

(a) Use a graphing utility to draw a scatter plot for the data. Let $t = 0$ represent 1990.

(b) Use the regression capabilities of a graphing utility to find a linear model for the data. Round decimals to two places.

(c) Algebraically find the inverse function of the model in part (b).

(d) Use the inverse function you found in part (c) to estimate the year that there will be about 23,000 million ATM transactions.

**61. *Reasoning*** You are teaching another student how to find the inverse function of a one-to-one function. The student you are helping states that interchanging the roles of $x$ and $y$ is "cheating." Discuss how you would use the graphs of $f(x) = x^2$, $x \geq 0$ and $f^{-1}(x) = \sqrt{x}$ to justify that particular step in the process of finding an inverse function.

**62. *Median Family Income*** The median family income in the United States from 1990 to 1998 can be modeled by the function

$$y = 142.02t^2 + 30{,}077, \qquad 0 \leq t \leq 8$$

where $y$ is the median income in dollars and $t = 0$ represents 1990. Solve the equation for $t$ in terms of $y$ and use the result to find the year in which the median family income was $31,355. (Source: U.S. Bureau of the Census)

**63. *Earnings-Dividend Ratio*** From 1990 to 2000, the earnings per share for Wal-Mart Stores were approximately related to the dividends per share by the function

$$f(x) = \sqrt{-0.019 + 0.047x}, \qquad 0.44 \leq x \leq 1.40$$

where $f(x)$ represents the dividends per share (in dollars) and $x$ represents the earnings per share (in dollars). In 1996 Wal-Mart paid dividends of $0.11 per share. Find the inverse function of $f$ and use the inverse function to approximate the earnings per share in 1996. (Source: Wal-Mart Stores, Inc.)

# Chapter Project

## Carbon Monoxide Emissions

To keep air quality within the levels of the NAAQS, the Environmental Protection Agency (EPA) lists specific categories of industry that pose a threat to air quality. All factories in a specified industrial category are required to keep emissions of each specific pollutant type under a specified limit. The table below gives the carbon monoxide emissions in thousands of tons in the United States from 1990 to 1998.

| Year | 1990 | 1991 | 1992 | 1993 | 1994 |
|---|---|---|---|---|---|
| Carbon monoxide emissions (in thousands of tons) | 98,523 | 100,872 | 97,630 | 98,160 | 102,643 |

| Year | 1995 | 1996 | 1997 | 1998 |
|---|---|---|---|---|
| Carbon monoxide emissions (in thousands of tons) | 93,353 | 95,479 | 94,410 | 89,454 |

1. *Graphing the Data*    Use a graphing utility to make a scatter plot for the data.

2. *Selecting a Model*    Use the regression capabilities of a graphing utility to find a quadratic model, a cubic model, and a quartic model for the data. Graph each model along with the data. Do any of the models give a "good" fit?

3. *Reconsidering the Data*    Review the scatter plot of the data. Is there a noticeable break or change in the data? Between what years does the change occur? Divide the data into two data sets, letting the data for 1990 to 1994 be one set and the data for 1995 to 1998 be the other.

4. *Developing a Piecewise-Defined Model*    Make a scatter plot for the data from 1990 to 1994. Based on the new scatter plot, select a regression program and find a model for the data from 1990 to 1994. Repeat this process for the data from 1995 to 1998.

5. *Graphing the Data and the Model*    Use a graphing utility to graph all the data and a piecewise-defined function that describes the data. Consult your user's guide to learn how to enter a piecewise-defined function. How good is the fit now?

6. *Exploring Further*    Use the Internet or a library's reference source to find emission data for ozone, sulfur dioxide, particulate matter, or nitrogen dioxide. Develop a model for the data and determine if a piecewise-defined function gives a good fit.

Allsport Australia/Getty Images

# 4

# Polynomial and Rational Functions

When a ball is dropped and bounces on a surface, the bounces become smaller and smaller until the ball comes to a state of rest. The height of each bounce can be modeled by a quadratic function.

The table below shows data that were collected when a soccer ball was dropped from a height of 5 feet. The table gives the actual heights of the ball at several times during one of its bounces. This data can be modeled by the function

$$f(t) = -13.44t^2 + 51.97t - 47.44,$$

where $t$ is the time in seconds after the ball has been dropped and $f(t)$ is the height of the ball in feet. The graph below shows the model with the scatter plot of the data.

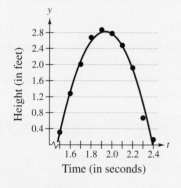

Time (in seconds)

**4.1 Quadratic Functions and Models**

**4.2 Polynomial Functions of Higher Degree**

**4.3 Polynomial Division**

**4.4 Real Zeros of Polynomial Functions**

**4.5 Complex Numbers**

**4.6 The Fundamental Theorem of Algebra**

**4.7 Rational Functions**

| Time (sec) | 1.5 | 1.6 | 1.7 | 1.8 | 1.9 | 2.0 | 2.1 | 2.2 | 2.3 | 2.4 |
|---|---|---|---|---|---|---|---|---|---|---|
| Height (ft) | 0.31 | 1.28 | 2.01 | 2.68 | 2.87 | 2.78 | 2.48 | 1.92 | 0.67 | 0.13 |

The chapter project related to this information is on page 290.

## 4.1 Quadratic Functions and Models

### Objectives

- Sketch the graph of a quadratic function and identify its vertex and intercepts.
- Find a quadratic function given its vertex and a point on the graph.
- Construct and use a quadratic model to solve an application problem.

## The Graph of a Quadratic Function

In this and the next section, you will study the graphs of polynomial functions.

> ▶ **Definition of a Polynomial Function**
>
> Let $n$ be a nonnegative integer and let $a_n, a_{n-1}, \ldots, a_2, a_1, a_0$ be real numbers with $a_n \neq 0$. The function given by
>
> $$f(x) = a_n x^n + a_{n-1} x^{n-1} + \cdots + a_2 x^2 + a_1 x + a_0$$
>
> is called a **polynomial function of $x$ with degree $n$.**

Polynomial functions are classified by degree. For instance, the polynomial function

$$f(x) = a, \quad a \neq 0 \qquad \text{Constant function}$$

has degree 0 and is called a **constant function.** In Chapter 2, you learned that the graph of this type of function is a horizontal line. The polynomial function

$$f(x) = ax + b, \quad a \neq 0 \qquad \text{Linear function}$$

has degree 1 and is called a **linear function.** In Chapter 2, you learned that the graph of the linear function $f(x) = ax + b$ is a line whose slope is $a$ and whose $y$-intercept is $(0, b)$. In this section you will study second-degree polynomial functions, which are called **quadratic functions.**

> ▶ **Definition of a Quadratic Function**
>
> Let $a$, $b$, and $c$ be real numbers with $a \neq 0$. The function of $x$ given by
>
> $$f(x) = ax^2 + bx + c \qquad \text{Quadratic function}$$
>
> is called a **quadratic function.**

The graph of a quadratic function is called a **parabola.** It is "$\cup$"-shaped and can open upward or downward.

## Applications

Many applications involve finding the maximum or minimum value of a quadratic function. By writing the quadratic function $f(x) = ax^2 + bx + c$ in standard form, you can determine that the vertex occurs when $x = -b/2a$.

| Example 5 | The Maximum Height of a Baseball |  |

A baseball is hit 3 feet above ground at a velocity of 100 feet per second and at an angle of $45°$ with respect to the ground. The path of the baseball is given by the function

$$f(x) = -0.0032x^2 + x + 3$$

where $f(x)$ is the height of the baseball (in feet) and $x$ is the distance from home plate (in feet). What is the maximum height reached by the baseball?

### Solution

For this quadratic function, you have

$$f(x) = ax^2 + bx + c = -0.0032x^2 + x + 3.$$

So, $a = -0.0032$ and $b = 1$. Because the function has a maximum when $x = -b/2a$, you can conclude that the baseball reaches its maximum height when it is

$$x = -\frac{b}{2a} = -\frac{1}{2(-0.0032)} = 156.25 \text{ feet}$$

from home plate. At this distance, the maximum height is

$$f(156.25) = -0.0032(156.25)^2 + 156.25 + 3 = 81.125 \text{ feet}.$$

The path of the baseball is shown in Figure 4.8.

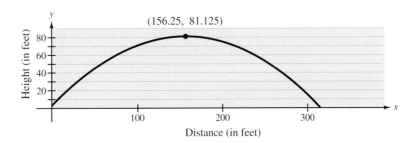

Figure 4.8

---

In Section 2.5 you plotted data points in the coordinate plane and estimated the best fit line. Fitting a quadratic model by this same process would be difficult. Most graphing utilities have a built-in statistical program that easily calculates the best-fitting quadratic for a set of data points. Refer to the user's guide of your graphing utility for the required steps.

**Example 6**    Fitting a Quadratic Function to Data

A study was done to compare the speed $x$ (in miles per hour) with the mileage $y$ (in miles per gallon) of an automobile. The results are shown in the table. Use a graphing utility to plot the data and find the quadratic model that best fits the data. Estimate the speed for which the miles per gallon is greatest.    (Source: Federal Highway Administration)

| Speed $x$ | 15 | 20 | 25 | 30 | 35 | 40 | 45 |
|---|---|---|---|---|---|---|---|
| Mileage $y$ | 22.3 | 25.5 | 27.5 | 29.0 | 28.8 | 30.0 | 29.9 |

| Speed $x$ | 50 | 55 | 60 | 65 | 70 | 75 |
|---|---|---|---|---|---|---|
| Mileage $y$ | 30.2 | 30.4 | 28.8 | 27.4 | 25.3 | 23.3 |

**Solution**

Begin by entering the data into your graphing utility and displaying the scatter plot. From the scatter plot in Figure 4.9(a) you can see that the points have a parabolic trend. Use the quadratic regression capabilities to find the quadratic function that best fits the data. The quadratic equation that best fits the data is

$$y = -0.008x^2 + 0.75x + 13.5.$$

Graph the data and the equation in the same viewing window as shown in Figure 4.9(b). From the graph you can see that the vertex of the graph is approximately (47, 31). So the speed at which the mileage is greatest is 47 miles per hour.

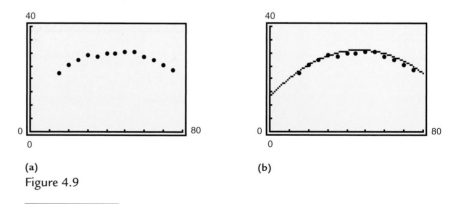

(a)                                                  (b)

Figure 4.9

Example 7    Charitable Contributions

According to a survey conducted by *Independent Sector*, the percent of income that Americans give to charities is related to the amount of income. For families with annual incomes of $100,000 or less, the percent can be modeled by

$$P(x) = 0.0014x^2 - 0.1529x + 5.855, \quad 5 \le x \le 100$$

where $x$ is the annual income in thousands of dollars. According to this model, what income level corresponds to the minimum percent of charitable contributions?

### Solution

There are two ways to answer this question. One is to sketch the graph of the quadratic function, as shown in Figure 4.10. From this graph, it appears that the minimum percent corresponds to an income level of about $55,000. The other way to answer the question is to use the fact that the minimum point of the parabola occurs when $x = -b/2a$.

$$x = -\frac{b}{2a} = -\frac{-0.1529}{2(0.0014)} \approx 54.6.$$

From this $x$-value, you can conclude that the minimum percent corresponds to an income level of about $54,600.

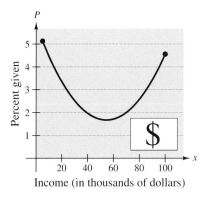

Figure 4.10

| Discussing the Concept | Approximating Intercepts and Extreme Points |
|---|---|

A graphing utility with zoom and trace features can be a useful tool in approximating the intercepts and vertex coordinates of quadratic functions to acceptable degrees of accuracy. Use a graphing utility to locate and approximate the intercepts and vertices of the following quadratic functions to two decimal places. Confirm the coordinates of the vertices algebraically. How do the estimates and the algebraic results compare?

a. $y = x^2 - 3x + 4$     b. $y = x^2 - 17x + 64$

c. $y = 7.52x^2 + 4.16x$     d. $y = -\frac{1}{2}x^2 + 5x - \frac{13}{2}$

e. $y = \frac{x^2}{19} + \frac{13x}{3} + \frac{4}{21}$     f. $y = 14x^2 - 5x + 2$

# Warm Up

The following warm-up exercises involve skills that were covered in earlier sections. You will use these skills in the exercise set for this section.

In Exercises 1–4, solve the equation by factoring.

**1.** $2x^2 + 11x - 6 = 0$      **2.** $5x^2 - 12x - 9 = 0$

**3.** $3 + x - 2x^2 = 0$       **4.** $x^2 + 20x + 100 = 0$

In Exercises 5–10, use the Quadratic Formula to solve the equation.

**5.** $x^2 - 6x + 4 = 0$       **6.** $x^2 + 4x + 1 = 0$

**7.** $2x^2 - 16x + 25 = 0$     **8.** $3x^2 + 30x + 74 = 0$

**9.** $x^2 + 3x + 1 = 0$       **10.** $x^2 + 3x - 3 = 0$

## 4.1 Exercises

In Exercises 1–8, match the quadratic function with its graph. [The graphs are labeled (a), (b), (c), (d), (e), (f), (g), and (h).]

**1.** $f(x) = -(x - 3)^2$       **2.** $f(x) = (x + 5)^2$

**3.** $f(x) = x^2 - 4$        **4.** $f(x) = 5 - x^2$

**5.** $f(x) = (x + 3)^2 - 2$     **6.** $f(x) = (x - 1)^2 - 5$

**7.** $f(x) = -(x + 1)^2 + 3$    **8.** $f(x) = (x - 2)^2 - 4$

In Exercises 9–14, find an equation of the parabola.

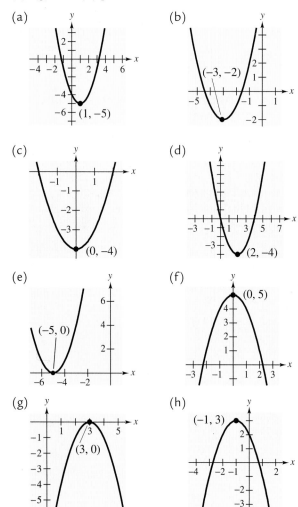

(a)   (b)   (c)   (d)   (e)   (f)   (g)   (h)

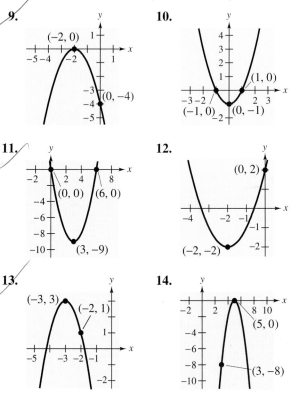

**9.**   **10.**   **11.**   **12.**   **13.**   **14.**

In Exercises 15–32, sketch the graph of the quadratic function. Identify the vertex and intercepts.

**15.** $f(x) = x^2 + 2$

**16.** $f(x) = x^2 - 4$

**17.** $f(x) = 16 - x^2$

**18.** $h(x) = 25 - x^2$

**19.** $f(x) = (x + 5)^2 - 6$

**20.** $f(x) = (x - 6)^2 + 3$

**21.** $h(x) = x^2 - 4x + 2$

**22.** $g(x) = x^2 + 2x + 1$

**23.** $f(x) = -(x^2 + 2x - 3)$

**24.** $f(x) = -(x^2 + 6x - 3)$

**25.** $f(x) = x^2 - x + \frac{5}{4}$

**26.** $f(x) = x^2 + 3x + \frac{1}{4}$

**27.** $f(x) = -x^2 + 2x + 5$

**28.** $f(x) = -x^2 - 4x + 1$

**29.** $h(x) = 4x^2 - 4x + 21$

**30.** $f(x) = 2x^2 - x + 1$

**31.** $f(x) = \frac{1}{4}(x^2 - 16x + 32)$

**32.** $g(x) = \frac{1}{2}(x^2 + 4x - 2)$

In Exercises 33–36, find the quadratic function that has the indicated vertex and whose graph passes through the given point.

**33.** Vertex: $(2, -1)$; point: $(4, -3)$

**34.** Vertex: $(-3, 5)$; point: $(-6, -1)$

**35.** Vertex: $(5, 12)$; point: $(7, 15)$

**36.** Vertex: $(-2, -2)$; point: $(-1, 0)$

In Exercises 37–42, find two quadratic functions whose graphs have the given x-intercepts. Find one function that has a graph that opens upward and another that has a graph that opens downward. (Each exercise has many correct answers.)

**37.** $(2, 0), (-1, 0)$

**38.** $(-4, 0), (0, 0)$

**39.** $(0, 0), (10, 0)$

**40.** $(4, 0), (8, 0)$

**41.** $(-3, 0), \left(-\frac{1}{2}, 0\right)$

**42.** $\left(-\frac{5}{2}, 0\right), (2, 0)$

**43.** *Maximum Area*   The perimeter of a rectangle is 100 feet. Let $x$ represent the width of the rectangle and write a quadratic function that expresses the area of the rectangle in terms of its width. Of all possible rectangles with perimeters of 100 feet, what are the measurements of the one that has the greatest area?

**44.** *Maximum Area*   The perimeter of a rectangle is 400 feet. Let $x$ represent the width of the rectangle and write a quadratic function that expresses the area of the rectangle in terms of its width. Of all possible rectangles with perimeters of 400 feet, what are the measurements of the one that has the greatest area?

**45.** *Maximum Area*   A rancher has 200 feet of fencing with which to enclose two adjacent rectangular corrals (see figure). What measurements will produce a maximum enclosed area?

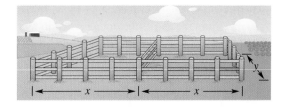

**46.** *Maximum Area*   An indoor physical-fitness room consists of a rectangular region with a semicircle on each end (see figure). The perimeter of the room is to be a 200-meter running track. What measurements will produce a maximum area of the rectangle?

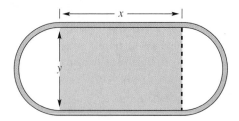

*Maximum Revenue*   In Exercises 47 and 48, find the number of units that produces a maximum revenue. The revenue $R$ is measured in dollars and $x$ is the number of units produced.

**47.** $R = 900x - 0.01x^2$

**48.** $R = 100x - 0.0002x^2$

**49.** *Minimum Cost*   A manufacturer of lighting fixtures has daily production costs of

$$C(x) = 800 - 10x + 0.25x^2$$

where $C$ is the total cost in dollars and $x$ is the number of units produced. How many fixtures should be produced each day to yield a minimum cost?

**50.** *Maximum Profit* The profit for a company is given by

$$P(x) = -0.0002x^2 + 140x - 250,000$$

where $x$ is the number of units produced. What production level will yield a maximum profit?

**51.** *Maximum Height of a Diver* The path of a diver is given by

$$y = -\frac{4}{9}x^2 + \frac{24}{9}x + 10$$

where $y$ is the height in feet and $x$ is the horizontal distance from the end of the diving board in feet (see figure). Use a graphing utility and the trace feature to find the maximum height of the diver.

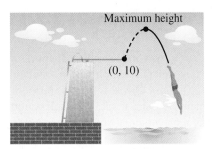

Maximum height

(0, 10)

**52.** *Maximum Height* The winning women's shot put in the 2000 Summer Olympics was recorded by Yanina Korolchik of Belarus. The path of her winning toss was approximately given by

$$y = -0.01707x^2 + 1.0775x + 5$$

where $y$ is the height of the shot in feet and $x$ is the horizontal distance in feet. Use a graphing utility and the trace feature to find how long the winning toss was and the maximum height of the shot.

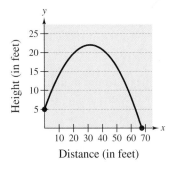

Distance (in feet)

**53.** *Regression Problem* Let $x$ be the number of units (in tens of thousands) that a company produces and let $p(x)$ be the profit (in hundreds of thousands of dollars). The following table gives the profit for different levels of production.

| $x$ | 2 | 4 | 6 | 8 | 10 |
|---|---|---|---|---|---|
| $p(x)$ | 270.5 | 307.8 | 320.1 | 329.2 | 325 |

| $x$ | 12 | 14 | 16 | 18 | 20 |
|---|---|---|---|---|---|
| $p(x)$ | 311.2 | 287.8 | 254.8 | 212.2 | 160 |

(a) Use a graphing utility to make a scatter plot using the data in the table.

(b) Use the regression capabilities of a graphing utility to find a quadratic model for $p(x)$.

(c) Use your graphing utility to graph your model for $p(x)$ with the scatter plot for the data.

(d) Use the model to approximate the level of production that will yield a maximum profit.

(e) With these data and this model, the profit begins to decrease. Discuss how it is possible for production to increase and profit to decrease.

**54.** *Regression Problem* Let $x$ be the angle (in degrees) at which a baseball is hit with no spin at an initial speed of 40 meters per second and let $d(x)$ be the distance (in meters) the ball travels. The following table gives the distances for the different angles the ball is hit. (Source: *The Physics of Sports*)

| $x$ | 10° | 15° | 30° | 36° |
|---|---|---|---|---|
| $d(x)$ | 58.3 | 79.7 | 126.9 | 136.6 |

| $x$ | 42° | 43° | 44° | 45° |
|---|---|---|---|---|
| $d(x)$ | 140.6 | 140.8 | 140.9 | 140.9 |

| $x$ | 48° | 54° | 60° |
|---|---|---|---|
| $d(x)$ | 139.3 | 132.5 | 120.5 |

(a) Use a graphing utility to make a scatter plot using the data in the table.

(b) Use the regression capabilities of a graphing utility to find a quadratic model for $d(x)$.

(c) Use your graphing utility to graph your model for $d(x)$ with the scatter plot for the data.

(d) Use the model to approximate the angle that will result in a maximum distance.

The symbol ▦ indicates an exercise or parts of an exercise in which you are instructed to use a graphing utility.

**55.** *Cigarette Production*    From 1993 to 1999, the number of cigarettes produced in the United States rose and then declined according to the model

$$N(t) = -11.35t^2 + 129.1t + 383, \quad 3 \le t \le 9$$

where $N$ is the number of cigarettes (in billions) and $t$ represents the calendar year, with $t = 3$ corresponding to 1993. Use a graphing utility to determine the year in which the number of cigarettes produced was at its highest.    (Source: U.S. Department of Agriculture)

**56.** *Hydroelectric Power*    The percent $P$ of electrical power produced by hydroelectric generation in the United States from 1994 to 1998 can be approximated by the model

$$P(t) = -0.471t^2 + 5.98t - 8.0, \quad 4 \le t \le 8$$

where $t = 4$ corresponds to 1994. Use a graphing utility to determine the year in which the percent was at its highest.    (Source: U.S. Energy Information Administration)

## Math Matters • Circles and Pi

Pi is a special number that represents a relationship that is found in *all circles*. The relationship is this: The ratio of the circumference of *any* circle to the diameter of the circle always produces the same number—the number denoted by the Greek symbol $\pi$.

Calculation of the decimal representation of the number $\pi$ has consumed the time of many mathematicians (and computers). Because the number $\pi$ is irrational, its decimal representation does not repeat or terminate. Here is a decimal representation of $\pi$ that is accurate to 27 decimal places.

$$\pi \approx 3.141592653589793238462643383.$$

In October 1995, a team of mathematicians in British Columbia discovered a formula for calculating the $n$th digit of $\pi$ without calculating the preceding digits. Run on a computer, the algorithm for the $n$th digit requires that numbers be written in base 16 (hexadecimal representation) instead of base 10 (decimal representation).

The number $\pi$ is used in most formulas that deal with circles. For instance, the area of a circle is given by

$$A = \pi r^2$$

where $r$ is the radius of the circle. Use this formula to determine which of the colored regions inside the circle below has the greater area. Is the area of the red circle larger than that of the blue ring, or do both regions have the same area? (The answer is given in the back of the book.)

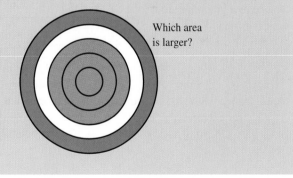

Which area is larger?

## 4.2    Polynomial Functions of Higher Degree

### Objectives

- Identify the characteristics of polynomial functions.
- Apply the Leading Coefficient Test to determine right and left behavior of the graph of a polynomial function.
- Find the real zeros of a polynomial function.
- Sketch the graph of a polynomial function.
- Use a polynomial model to solve an application problem.

## Graphs of Polynomial Functions

### Study Tip

The graphs of polynomial functions of degree greater than 2 are more difficult to analyze than graphs of polynomials of degree 0, 1, or 2. However, using the characteristics presented in this section, together with point plotting, intercepts, and symmetry, you should be able to make reasonably accurate sketches *by hand*. Of course, if you have a graphing utility, the task is easier.

In this section, you will study basic characteristics of the graphs of polynomial functions. The first characteristic is that the graph of a polynomial function is **continuous.** Essentially, this means that the graph of a polynomial function has no breaks, as shown in Figure 4.11(a). Functions with graphs that are not continuous are not polynomial functions, as shown in Figure 4.11(b).

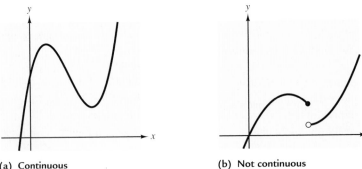

(a)  Continuous          (b)  Not continuous

Figure 4.11

The second characteristic is that the graph of a polynomial function has only smooth, rounded turns, as shown in Figure 4.12(a). A polynomial function cannot have a sharp turn. For instance, the function $f(x) = |x|$, which has a sharp turn at the point $(0, 0)$, as shown in Figure 4.12(b), is not a polynomial function.

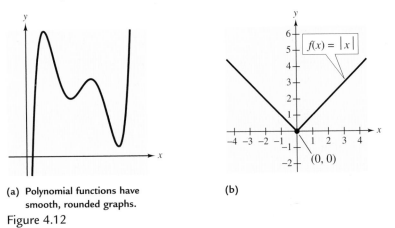

(a)  Polynomial functions have smooth, rounded graphs.          (b)

Figure 4.12

The polynomial functions that have the simplest graphs are monomials of the form $f(x) = x^n$, where $n$ is an integer greater than zero. From Figure 4.13, you can see that when $n$ is *even* the graph is similar to the graph of $f(x) = x^2$ and when $n$ is *odd* the graph is similar to the graph of $f(x) = x^3$. Moreover, the greater the value of $n$, the flatter the graph near the origin.

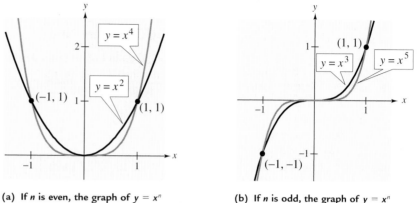

(a) If $n$ is even, the graph of $y = x^n$ touches the axis at the *x*-intercept.

(b) If $n$ is odd, the graph of $y = x^n$ *crosses* the axis at the *x*-intercept.

Figure 4.13

---

**Example 1**    Sketching Transformations of Monomial Functions

**a.** Because the degree of $f(x) = -x^5$ is odd, its graph is similar to the graph of $y = x^3$. In Figure 4.14(a), note that the negative coefficient has the effect of reflecting the graph about the *x*-axis.

**b.** The graph of $h(x) = (x + 1)^4$ is a left shift, by one unit, of the graph of $y = x^4$, as shown in Figure 4.14(b).

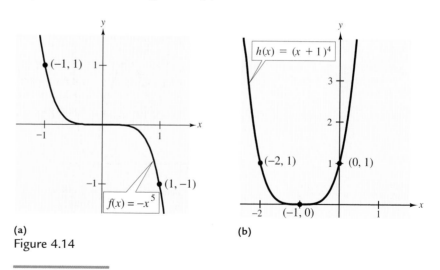

(a)

(b)

Figure 4.14

# The Leading Coefficient Test

For each function, identify the degree of the function and whether it is even or odd. Identify the leading coefficient, and whether the leading coefficient is greater than 0 or less than 0. Use a graphing utility to graph each function. Describe the relationship between the degree and the sign of the leading coefficient of the function and the left- and right-hand behavior of the graph of the function.

(a) $y = x^3 - 2x^2 - x + 1$
(b) $y = 2x^5 + 2x^2 - 5x + 1$
(c) $y = -2x^5 - x^2 + 5x + 3$
(d) $y = -x^3 + 5x - 2$
(e) $y = 2x^2 + 3x - 4$
(f) $y = x^4 - 3x^2 + 2x - 1$
(g) $y = x^2 + 3x + 2$
(h) $y = -x^6 - x^2 - 5x + 4$

In Example 1, note that both graphs eventually rise or fall without bound as $x$ moves to the right. Whether the graph of a polynomial eventually rises or falls can be determined by the function's degree (even or odd) and by its leading coefficient, as indicated in the **Leading Coefficient Test.**

▶ Leading Coefficient Test

As $x$ moves without bound to the left or to the right, the graph of the polynomial function $f(x) = a_n x^n + \cdots + a_1 x + a_0$ eventually rises or falls in the following manner.

**1.** When $n$ is *odd*:

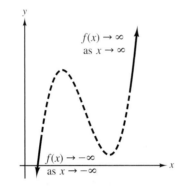

If the leading coefficient is positive $(a_n > 0)$, the graph falls to the left and rises to the right.

If the leading coefficient is negative $(a_n < 0)$, the graph rises to the left and falls to the right.

**2.** When $n$ is *even*:

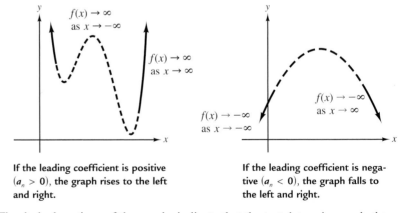

If the leading coefficient is positive $(a_n > 0)$, the graph rises to the left and right.

If the leading coefficient is negative $(a_n < 0)$, the graph falls to the left and right.

The dashed portions of the graphs indicate that the test determines *only* the right and left behavior of the graph.

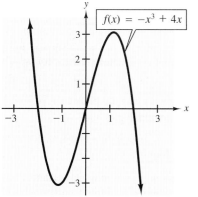

Figure 4.15

**Example 2**    Applying the Leading Coefficient Test

Describe the right- and left-hand behavior of the graph of

$$f(x) = -x^3 + 4x.$$

**Solution**

Because the degree is odd and the leading coefficient is negative, the graph rises to the left and falls to the right, as shown in Figure 4.15.

In Example 2, note that the Leading Coefficient Test tells you only whether the graph *eventually* rises or falls to the right or left. Other characteristics of the graph, such as intercepts and minimum and maximum points, must be determined by means of other tests.

**Example 3**    Applying the Leading Coefficient Test

Describe the right- and left-hand behavior of the graphs of the following functions.

**a.** $f(x) = x^4 - 5x^2 + 4$    **b.** $f(x) = x^5 - x$

**Solution**

**a.** Because the degree is even and the leading coefficient is positive, the graph rises to the left and right, as shown in Figure 4.16(a).

**b.** Because the degree is odd and the leading coefficient is positive, the graph falls to the left and rises to the right, as shown in Figure 4.16(b).

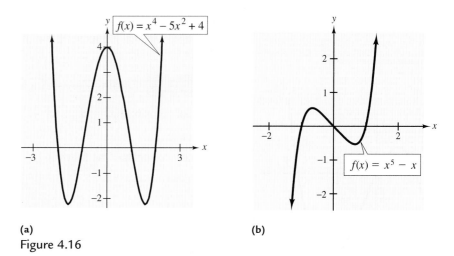

(a)                                                  (b)

Figure 4.16

# Real Zeros of Polynomial Functions

It can be shown that for a polynomial function $f$ of degree $n$, the following statements are true. (Remember that the **zeros** of a function are the $x$-values for which the function is zero.)

1. The graph of $f$ has, at most, $n - 1$ turning points. (Turning points are points at which the graph changes from increasing to decreasing or vice versa.)

2. The function $f$ has, at most, $n$ real zeros. (You will study this result in detail in Section 4.6 on the Fundamental Theorem of Algebra.)

Finding the zeros of polynomial functions is one of the most important problems in algebra. There is a strong interplay between graphical and algebraic approaches to this problem. Sometimes you can use information about the graph of a function to help find its zeros, and in other cases you can use information about the zeros of a function to help sketch its graph.

---

▶ **Real Zeros of Polynomial Functions**

If $f$ is a polynomial function and $a$ is a real number, the following statements are equivalent.

1. $x = a$ is a *zero* of the function $f$.

2. $x = a$ is a *solution* of the polynomial equation $f(x) = 0$.

3. $(x - a)$ is a *factor* of the polynomial $f(x)$.

4. $(a, 0)$ is an *x-intercept* of the graph of $f$.

---

In the equivalent statements above, notice that finding zeros of polynomial functions is closely related to factoring and finding $x$-intercepts.

**Example 4**   Finding Zeros of a Polynomial Function

Find all real zeros of $f(x) = x^3 - x^2 - 2x$.

**Solution**

By factoring, you obtain the following.

$$f(x) = x^3 - x^2 - 2x \qquad \text{Original function}$$

$$= x(x^2 - x - 2) \qquad \text{Remove common monomial factor.}$$

$$= x(x - 2)(x + 1) \qquad \text{Factor completely.}$$

So, the real zeros are $x = 0$, $x = 2$, and $x = -1$, and the corresponding $x$-intercepts are $(0, 0)$, $(2, 0)$, and $(-1, 0)$, as shown in Figure 4.17. In the figure, note that the graph has two turning points. This is consistent with the fact that a third-degree polynomial can have *at most* two turning points.

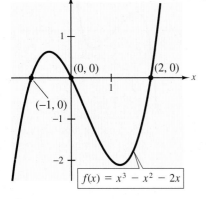

Figure 4.17

**Example 5** Finding Zeros of a Polynomial Function

Find all real zeros of $f(x) = -2x^4 + 2x^2$.

**Solution**

In this case, the polynomial factors as follows.

$$f(x) = -2x^4 + 2x^2 \qquad \text{Original function}$$

$$= -2x^2(x^2 - 1) \qquad \text{Remove common monomial factor.}$$

$$= -2x^2(x - 1)(x + 1) \qquad \text{Factor completely.}$$

So, the real zeros are $x = 0$, $x = 1$, and $x = -1$, and the corresponding $x$-intercepts are $(0, 0)$, $(1, 0)$, and $(-1, 0)$, as shown in Figure 4.18. Note in the figure that the graph has three turning points, which is consistent with the fact that a fourth-degree polynomial can have at most three turning points.

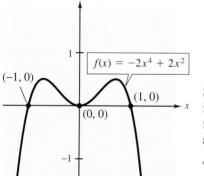

**Figure 4.18**

In Example 5, the real zero arising from $-2x^2 = 0$ is called a **repeated zero.** In general, a factor $(x - a)^k$ yields a repeated zero $x = a$ of **multiplicity** $k$. If $k$ is odd, the graph *crosses* the $x$-axis at $x = a$. If $k$ is even, the graph *touches* the $x$-axis (but does not cross the $x$-axis) at $x = a$. Note how this occurs in Figure 4.18.

**Example 6** Charitable Contributions Revisited

Example 7 in Section 4.1 discussed the model

$$P(x) = 0.0014x^2 - 0.1529x + 5.855, \qquad 5 \le x \le 100$$

where $P$ is the percent of annual income given and $x$ is the annual income in thousands of dollars. Note that this model gives the charitable contributions as a *percent* of annual income. To find the average *amount* that a family gives to charity, you can multiply the given model by the income $1000x$ (and divide by 100 to change from percent to decimal form) to obtain

$$A(x) = 0.014x^3 - 1.529x^2 + 58.55x, \qquad 5 \le x \le 100$$

where $A$ represents the amount of charitable contributions in dollars. Sketch the graph of this function and use the graph to estimate the annual salary for a family that gives $1000 a year to charities.

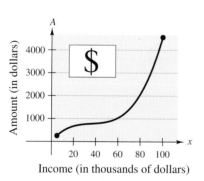

Income (in thousands of dollars)

**Figure 4.19**

**Solution**

The graph of this function is shown in Figure 4.19. From the graph you see that an annual contribution of $1000 corresponds to an annual income of about $59,000.

Example 7 uses an "algebraic approach" to describe the graph of the function. A graphing utility is a valuable complement to this approach. Remember that the most important part of using a graphing utility is to find a viewing window that shows all important parts of the graph. For instance, the graph below shows the important parts of the graph of the function in Example 7.

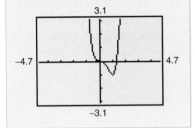

Example 7 shows how the Leading Coefficient Test and zeros of polynomial functions can be used as sketching aids.

**Example 7**    Sketching the Graph of a Polynomial Function

Sketch the graph of $f(x) = 3x^4 - 4x^3$.

Solution

Because the leading coefficient is positive and the degree is even, you know that the graph eventually rises to the left and to the right, as shown in Figure 4.20(a). By factoring

$$f(x) = 3x^4 - 4x^3 = x^3(3x - 4)$$

you can see that the zeros of $f$ are $x = 0$ and $x = \frac{4}{3}$ (both of odd multiplicity). So the x-intercepts occur at $(0, 0)$ and $(\frac{4}{3}, 0)$. To sketch the graph by hand, find a few additional points, as shown in the table. Then plot the points and complete the graph, as shown in Figure 4.20(b).

| $x$    | $-1$ | 0.5     | 1   | 1.5    |
|--------|------|---------|-----|--------|
| $f(x)$ | 7    | $-0.3125$ | $-1$ | 1.6875 |

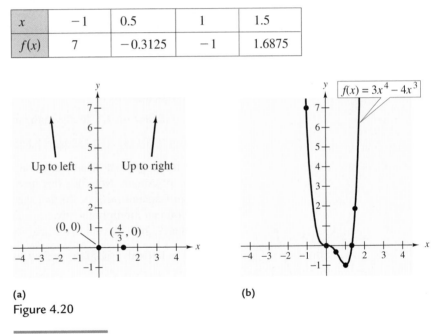

(a)                            (b)

Figure 4.20

**Discussing the Concept        Modeling Polynomial Functions**

Suppose you are writing a quiz for your algebra class and want to make up several polynomial functions for your students to investigate. Discuss how you could find polynomial functions that have reasonably simple zeros. Then use the methods you have discussed to find polynomial functions that have the following zeros.

**a.** $(2, 4, -3)$        **b.** $\left(3, -\frac{2}{3}, \frac{3}{4}\right)$

**c.** $(-3, -3, -3, -3)$        **d.** $(1, -2, 3, -4)$

# Warm Up

The following warm-up exercises involve skills that were covered in earlier sections. You will use these skills in the exercise set for this section.

In Exercises 1–6, factor the expression completely.

1. $12x^2 + 7x - 10$
2. $25x^3 - 60x^2 + 36x$
3. $12z^4 + 17z^3 + 5z^2$
4. $y^3 + 125$
5. $x^3 + 3x^2 - 4x - 12$
6. $x^3 + 2x^2 + 3x + 6$

In Exercises 7–10, find all real solutions of the equation.

7. $5x^2 + 8 = 0$
8. $x^2 - 6x + 4 = 0$
9. $4x^2 + 4x - 11 = 0$
10. $x^4 - 18x^2 + 81 = 0$

## 4.2    Exercises

In Exercises 1–8, match the polynomial function with its graph. [The graphs are labeled (a), (b), (c), (d), (e), (f), (g), and (h).]

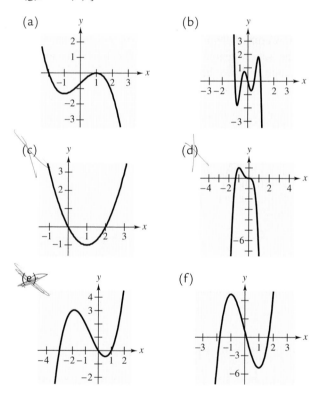

(a)    (b)

(c)    (d)

(e)    (f)

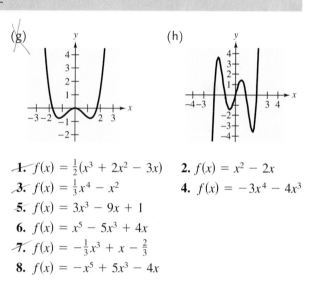

(g)    (h)

1. $f(x) = \frac{1}{2}(x^3 + 2x^2 - 3x)$   2. $f(x) = x^2 - 2x$
3. $f(x) = \frac{1}{3}x^4 - x^2$   4. $f(x) = -3x^4 - 4x^3$
5. $f(x) = 3x^3 - 9x + 1$
6. $f(x) = x^5 - 5x^3 + 4x$
7. $f(x) = -\frac{1}{3}x^3 + x - \frac{2}{3}$
8. $f(x) = -x^5 + 5x^3 - 4x$

In Exercises 9–18, determine the right- and left-hand behavior of the graph of the function.

9. $f(x) = -x^3 + 1$
10. $f(x) = \frac{1}{3}x^3 + 5x$
11. $g(x) = 6 - 4x^2 + x - 3x^5$
12. $f(x) = 2x^5 - 5x + 7.5$
13. $f(x) = 4x^8 - 2$
14. $h(x) = 1 - x^6$
15. $f(x) = 2 + 5x - x^2 - x^3 + 2x^4$
16. $f(x) = \dfrac{3x^4 - 2x + 5}{4}$
17. $h(t) = -\frac{2}{3}(t^2 - 5t + 3)$
18. $f(s) = -\frac{7}{8}(s^3 + 5s^2 - 7s + 1)$

*Algebraic and Graphical Approaches* In Exercises 19–34, find all real zeros of the function algebraically. Then use a graphing utility to confirm your results.

**19.** $f(x) = x^2 - 16$

**20.** $f(x) = 64 - x^2$

**21.** $h(t) = t^2 + 8t + 16$

**22.** $f(x) = x^2 - 12x + 36$

**23.** $f(x) = \frac{1}{3}x^2 + \frac{1}{3}x - \frac{2}{3}$

**24.** $f(x) = \frac{1}{2}x^2 + \frac{5}{2}x - \frac{3}{2}$

**25.** $f(x) = 2x^2 + 4x + 6$

**26.** $g(x) = -5(x^2 + 2x - 4)$

**27.** $f(t) = t^3 - 4t^2 + 4t$

**28.** $f(x) = x^4 - x^3 - 20x^2$

**29.** $g(t) = \frac{1}{2}t^4 - \frac{1}{2}$

**30.** $f(x) = x^5 + x^3 - 6x$

**31.** $f(x) = 2x^4 - 2x^2 - 40$

**32.** $g(t) = t^5 - 6t^3 + 9t$

**33.** $f(x) = x^3 - 3x^2 + 2x - 6$

**34.** $f(x) = x^3 - 4x^2 - 25x + 100$

In Exercises 35–38, use the graph of $y = x^3$ to sketch the graph of the function.

**35.** $f(x) = x^3 + 4$

**36.** $f(x) = (x - 3)^3$

**37.** $f(x) = (x - 2)^3 - 2$

**38.** $f(x) = -(x + 1)^3 + 3$

In Exercises 39–42, use the graph of $y = x^4$ to sketch the graph of the function.

**39.** $f(x) = (x + 3)^4$

**40.** $f(x) = x^4 - 3$

**41.** $f(x) = 4 - x^4$

**42.** $f(x) = \frac{1}{2}(x - 1)^4$

*Analyzing a Graph* In Exercises 43–54, analyze the graph of the function algebraically and use the results to sketch the graph *by hand*. Then use a graphing utility to confirm your sketch.

**43.** $f(x) = -\frac{3}{2}$

**44.** $h(x) = \frac{1}{3}x - 3$

**45.** $f(t) = \frac{1}{2}(t^2 - 4t - 1)$

**46.** $g(x) = -x^2 + 10x - 16$

**47.** $f(x) = 4x^2 - x^3$

**48.** $f(x) = 1 - x^3$

**49.** $f(x) = x^3 - 9x$

**50.** $f(x) = \frac{1}{4}x^4 - 2x^2$

**51.** $g(t) = -\frac{1}{4}(t - 2)^2(t + 2)^2$

**52.** $f(x) = x(x - 2)^2(x + 1)$

**53.** $f(x) = 1 - x^6$

**54.** $g(x) = 1 - (x + 1)^6$

**55.** *Modeling Polynomials* Sketch the graph of a polynomial function that is fourth degree, has a root of multiplicity 2, and has a negative leading coefficient. Sketch another graph under the same conditions but with a positive leading coefficient.

**56.** *Modeling Polynomials* Sketch the graph of a polynomial function that is fifth degree, has a root of multiplicity 2, and has a negative leading coefficient. Sketch another graph under the same conditions but with a positive leading coefficient.

**57.** *Modeling Polynomials* Determine the equation of the fourth-degree polynomial function $f(x)$ whose graph is shown below.

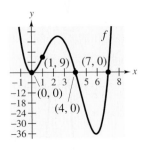

**58.** *Modeling Polynomials* Determine the equation of the third-degree polynomial function $g(x)$ whose graph is shown below.

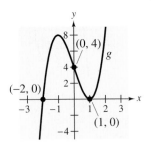

**59.** *Ice Cream*    The numbers of gallons $G$ (in millions) of regular ice cream produced in the United States from 1993 to 1999 are given in the table.    (Source: U.S. Department of Agriculture)

| Year | 1993 | 1994 | 1995 | 1996 |
|------|------|------|------|------|
| $G$ | 866 | 876 | 862 | 879 |

| Year | 1997 | 1998 | 1999 |
|------|------|------|------|
| $G$ | 914 | 935 | 954 |

(a) Use a graphing utility to make a scatter plot of the data. Let $t = 3$ represent 1993.

(b) Use the regression capabilities of a graphing utility to find a quadratic model and a quartic model for $G(t)$.

(c) Graph each model with the data. Make a table to compare the actual values to the values given by each model.

(d) Based on your table, which model do you think represents the data more accurately? Explain your answer.

**60.** *Total Investment Capital*    From 1991 to 2000, the total investment capital for Northwest Natural Gas Company increased by about 390 million dollars. A model that approximates the total investment capital for this corporation is

$$C = -0.1970t^3 + 5.113t^2 + 8.79t + 499.1,$$
$$1 \leq t \leq 10$$

where $C$ is the total investment capital (in millions of dollars) and $t = 1$ represents 1991.    (Source: Northwest Natural Gas Co.)

(a) Use a graphing utility to graph the model.

(b) Estimate the year (between 1991 and 2000) in which the total investment capital was the least.

**61.** *Advertising Expenses*    The total revenue for a soft-drink company is related to its advertising expense by the function

$$R = \frac{1}{50,000}(-x^3 + 600x^2), \qquad 0 \leq x \leq 400$$

where $R$ is the total revenue in millions of dollars and $x$ is the amount spent on advertising (in tens of thousands of dollars). Use the graph of this function to estimate the point on the graph at which the function is increasing most rapidly. This point is called the *point of diminishing returns* because any expenses above this amount will yield less return per dollar invested in advertising.

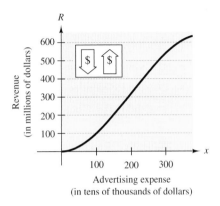

Figure for 61

**62.** *Advertising Expenses*    The total revenue for a hotel corporation is related to its advertising expense by the function

$$R = -0.148x^3 + 4.889x^2 - 17.778x + 125.185,$$
$$0 \leq x \leq 20$$

where $R$ is the total revenue in millions of dollars and $x$ is the amount spent on advertising (in millions of dollars). Use the accompanying graph of this function to estimate the point on the graph at which the function is increasing most rapidly. This point is called the *point of diminishing returns* because any expenses above this amount will yield less return per dollar invested in advertising.

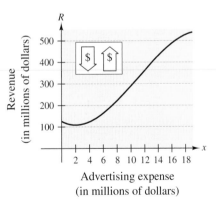

**63.** *Think About It*    Use a graphing utility to graph the following functions: $f(x) = x^2$, $g(x) = x^4$, $h(x) = x^6$. Do the three functions have a common shape? Are their graphs identical? Why or why not?

## 4.3    Polynomial Division

### Objectives

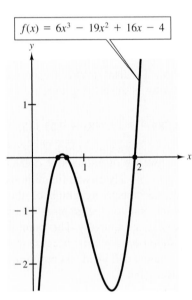

- Divide one polynomial by a second polynomial using long division.
- Simplify a rational expression using long division.
- Use synthetic division to divide two polynomials.
- Use the Remainder Theorem and synthetic division to evaluate a polynomial.
- Use the Factor Theorem to factor a polynomial function.

## Long Division of Polynomials

In this section, you will study two procedures for *dividing* polynomials. These procedures are especially valuable in factoring and finding the zeros of polynomial functions. To begin, suppose you are given the graph of

$$f(x) = 6x^3 - 19x^2 + 16x - 4.$$

Notice that a zero of $f$ occurs at $x = 2$, as shown in Figure 4.21. Because $x = 2$ is a zero of the polynomial function $f$, you know that $(x - 2)$ is a factor of $f(x)$. This means that there exists a second-degree polynomial $q(x)$ such that

$$f(x) = (x - 2) \cdot q(x).$$

To find $q(x)$, you can use **long division,** as illustrated in Example 1.

$f(x) = 6x^3 - 19x^2 + 16x - 4$

**Figure 4.21**

**Example 1**    Long Division of Polynomials

Divide the polynomial $6x^3 - 19x^2 + 16x - 4$ by $x - 2$, and use the result to factor the polynomial completely.

**Solution**

$$
\begin{array}{r}
6x^2 - 7x + 2 \\
x - 2 \overline{\smash{)}\ 6x^3 - 19x^2 + 16x - 4} \\
\underline{6x^3 - 12x^2} \qquad\qquad\qquad\quad \text{Multiply: } 6x^2(x-2). \\
-7x^2 + 16x \qquad\qquad \text{Subtract.} \\
\underline{-7x^2 + 14x} \qquad\qquad \text{Multiply: } -7x(x-2). \\
2x - 4 \qquad \text{Subtract.} \\
\underline{2x - 4} \qquad \text{Multiply: } 2(x-2). \\
0 \qquad \text{Subtract.}
\end{array}
$$

From this division, you can conclude that

$$6x^3 - 19x^2 + 16x - 4 = (x - 2)(6x^2 - 7x + 2)$$

and by factoring the quadratic $6x^2 - 7x + 2$, you have

$$6x^3 - 19x^2 + 16x - 4 = (x - 2)(2x - 1)(3x - 2).$$

Note that the factorization shown in Example 1 agrees with the graph shown in Figure 4.21 in that the three $x$-intercepts occur at $x = 2$, $x = \frac{1}{2}$, and $x = \frac{2}{3}$.

In Example 1, $x - 2$ is a factor of the polynomial $6x^3 - 19x^2 + 16x - 4$, and the long division process produces a remainder of zero. Often, long division will produce a nonzero remainder. For instance, if you divide $x^2 + 3x + 5$ by $x + 1$, you obtain the following.

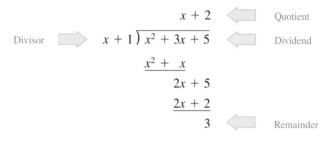

In fractional form, you can write this result as follows.

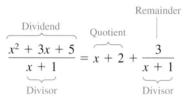

This example illustrates the following well-known theorem called the **Division Algorithm.**

---

▶ **The Division Algorithm**

If $f(x)$ and $d(x)$ are polynomials such that $d(x) \neq 0$, and the degree of $d(x)$ is less than or equal to the degree of $f(x)$, there exist unique polynomials $q(x)$ and $r(x)$ such that

$$f(x) = d(x)q(x) + r(x)$$

Dividend    Quotient
    Divisor        Remainder

where $r(x) = 0$ *or* the degree of $r(x)$ is less than the degree of $d(x)$. If the remainder $r(x)$ is zero, $d(x)$ **divides evenly** into $f(x)$.

---

The Division Algorithm can also be written as

$$\frac{f(x)}{d(x)} = q(x) + \frac{r(x)}{d(x)}.$$

In the Division Algorithm, the rational expression $f(x)/d(x)$ is **improper** because the degree of $f(x)$ is greater than or equal to the degree of $d(x)$. On the other hand, the rational expression $r(x)/d(x)$ is **proper** because the degree of $r(x)$ is less than the degree of $d(x)$.

**Example 2**   Long Division of Polynomials

Divide $x^3 - 1$ by $x - 1$.

**Solution**

Because there is no $x^2$-term or $x$-term in the dividend, you need to line up the subtraction by using zero coefficients (or leaving spaces) for the missing terms.

$$
\begin{array}{r}
x^2 + \phantom{1}x + 1 \\
x - 1 \overline{\smash{)}\, x^3 + 0x^2 + 0x - 1} \\
\underline{x^3 - \phantom{1}x^2} \phantom{+0x-1} \\
x^2 \phantom{+0x-1} \\
\underline{x^2 - \phantom{1}x} \phantom{-1} \\
x - 1 \\
\underline{x - 1} \\
0
\end{array}
$$

So, $x - 1$ divides evenly into $x^3 - 1$ and you can write

$$\frac{x^3 - 1}{x - 1} = x^2 + x + 1.$$

You can check the result of a division problem by multiplying. For instance, in Example 2, try checking that $(x - 1)(x^2 + x + 1) = x^3 - 1$.

**Example 3**   Long Division of Polynomials

Divide $2x^4 + 4x^3 - 5x^2 + 3x - 2$ by $x^2 + 2x - 3$.

**Solution**

$$
\begin{array}{r}
2x^2 \phantom{xxxx} + 1 \\
x^2 + 2x - 3 \overline{\smash{)}\, 2x^4 + 4x^3 - 5x^2 + 3x - 2} \\
\underline{2x^4 + 4x^3 - 6x^2} \phantom{+3x-2} \\
x^2 + 3x - 2 \\
\underline{x^2 + 2x - 3} \\
x + 1
\end{array}
$$

Note that the first subtraction eliminated two terms from the dividend. When this happens, the quotient skips a term. So, you can write

$$\frac{2x^4 + 4x^3 - 5x^2 + 3x - 2}{x^2 + 2x - 3} = 2x^2 + 1 + \frac{x + 1}{x^2 + 2x - 3}.$$

# Synthetic Division

There is a nice shortcut for long division by polynomials of the form $x - k$. The shortcut is called **synthetic division.** We summarize the pattern for synthetic division of a cubic polynomial as follows. (The pattern for higher-degree polynomials is similar.)

---

▶ **Synthetic Division (for a Cubic Polynomial)**

To divide $ax^3 + bx^2 + cx + d$ by $x - k$, use the following pattern.

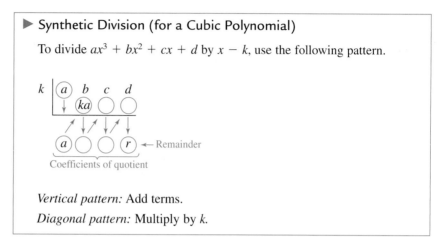

*Vertical pattern:* Add terms.

*Diagonal pattern:* Multiply by $k$.

---

Synthetic division works *only* for divisors of the form $x - k$. You cannot use synthetic division to divide a polynomial by a quadratic, a cubic, or any other higher-degree polynomial.

**Example 4**    Using Synthetic Division

Use synthetic division to divide $x^4 - 10x^2 - 2x + 4$ by $x + 3$.

### Solution

You should set up the array as follows. Note that a zero is included for each missing term in the dividend.

Divisor: $x + 3$     Dividend: $x^4 - 10x^2 - 2x + 4$

$$
\begin{array}{r|rrrrr}
-3 & 1 & 0 & -10 & -2 & 4 \\
   &   & -3 & 9 & 3 & -3 \\
\hline
   & 1 & -3 & -1 & 1 & 1 \\
\end{array}
$$

← Remainder: 1

Quotient: $x^3 - 3x^2 - x + 1$

So, you have

$$
\frac{x^4 - 10x^2 - 2x + 4}{x + 3} = x^3 - 3x^2 - x + 1 + \frac{1}{x + 3}.
$$

## Remainder and Factor Theorems

The remainder obtained in the synthetic division process has an important interpretation found in the Remainder Theorem.

> ▶ **The Remainder Theorem**
>
> If a polynomial $f(x)$ is divided by $x - k$, the remainder is
>
> $$r = f(k).$$

The Remainder Theorem tells you that synthetic division can be used to evaluate a polynomial function. That is, to evaluate a polynomial function $f(x)$ when $x = k$, divide $f(x)$ by $x - k$. The remainder will be $f(k)$, as illustrated in Example 5.

**Example 5**    Using the Remainder Theorem

Use the Remainder Theorem to evaluate the following function at $x = -2$.

$$f(x) = 3x^3 + 8x^2 + 5x - 7$$

**Solution**

Using synthetic division, you obtain the following.

$$
\begin{array}{r|rrrr}
-2 & 3 & 8 & 5 & -7 \\
   &   & -6 & -4 & -2 \\
\hline
   & 3 & 2 & 1 & -9
\end{array}
$$

Because the remainder is $r = -9$, you can conclude that

$$f(-2) = -9.$$

This means that $(-2, -9)$ is a point on the graph of $f$. Try checking this by substituting $x = -2$ in the original function.

Another important theorem is the Factor Theorem, which is stated below.

> ▶ **Factor Theorem**
>
> A polynomial $f(x)$ has a factor $(x - k)$ if and only if $f(k) = 0$.

Example 6    Factoring a Polynomial

Show that $(x - 2)$ and $(x + 3)$ are factors of the polynomial

$$f(x) = 2x^4 + 7x^3 - 4x^2 - 27x - 18.$$

Then find the remaining factors of $f(x)$.

### Solution

Using synthetic division with 2 and $-3$ *successively*, you obtain the following.

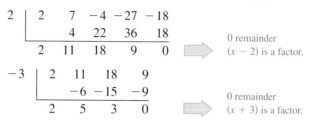

0 remainder
$(x - 2)$ is a factor.

0 remainder
$(x + 3)$ is a factor.

Because the resulting quadratic expression factors as

$$2x^2 + 5x + 3 = (2x + 3)(x + 1)$$

the complete factorization of $f(x)$ is

$$f(x) = (x - 2)(x + 3)(2x + 3)(x + 1).$$

Note that this factorization implies that $f$ has four real zeros:

$$2, -3, -\tfrac{3}{2}, \text{ and } -1.$$

This is confirmed by the graph of $f$, which is shown in Figure 4.22.

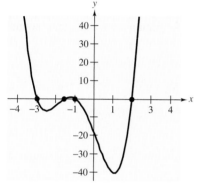

Figure 4.22

---

▶ **Results of Synthetic Division**

The remainder $r$, obtained in the synthetic division of $f(x)$ by $x - k$, provides the following information.

**1.** The remainder $r$ gives the value of $f$ at $x = k$. That is, $r = f(k)$.

**2.** If $r = 0$, $(x - k)$ is a factor of $f(x)$.

**3.** If $r = 0$, $(k, 0)$ is an $x$-intercept of the graph of $f$.

---

Throughout this text, we have emphasized the importance of developing several problem-solving strategies. In the exercises for this section, try using more than one strategy to solve several of the exercises. For instance, if you find that $x - k$ divides evenly into $f(x)$, try sketching the graph of $f$. You should find that $(k, 0)$ is an $x$-intercept of the graph.

## Application

Example 7 Tax Liability

The 1999 yearly federal income tax liability for an employee who was single and claimed no dependents is given by the function

$$y = 0.0000012x^2 + 0.124x - 1036, \quad 10,000 \le x \le 75,000$$

where $y$ represents the tax liability in dollars and $x$ represents the yearly salary. Find a function that gives the tax liability as a *percent* of the yearly salary. (Source: U.S. Department of the Treasury)

### Solution

Because the yearly salary is given by $x$ and the tax liability is given by $y$, the percent (in decimal form) of yearly salary that the person owes in federal income tax is

$$P = \frac{y}{x}$$

$$= \frac{0.0000012x^2 + 0.124x - 1036}{x}$$

$$= 0.0000012x + 0.124 - \frac{1036}{x}.$$

The graphs of these functions are shown in Figure 4.23(a) and (b). Note in Figure 4.23(b) that as a person's yearly salary increases, the percent that he or she must pay in federal income tax also increases.

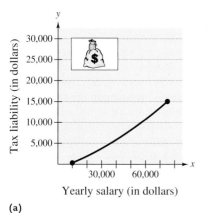

**(a)**

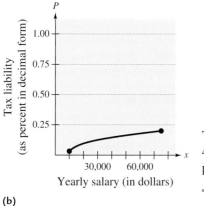

**(b)**

Figure 4.23

| Discussing the Concept | Understanding the Remainder Theorem |
| --- | --- |

The Remainder Theorem states: "If a polynomial $f(x)$ is divided by $x - k$, the remainder is $r = f(k)$." Given $f(x) = x^3 + 6x^2 + 5x - 12$ with the real zeros $-4$, $-3$, and $1$, explain algebraically and graphically what it means if $r = 0$. Explain algebraically and graphically what it means if $r \ne 0$. What is the relationship between the Remainder Theorem and the $x$-intercepts and factors of $f(x)$?

So, you can conclude that the given polynomial has *no* rational zeros. Note from the graph of $f$ in Figure 4.24 that $f$ does have one real zero (between $-1$ and $0$). By the Rational Zero Test, you know that this real zero is *not* a rational number.

| **Example 2** | Rational Zero Test with Leading Coefficient of 1 |

Find the rational zeros of

$$f(x) = x^4 - x^3 + x^2 - 3x - 6.$$

**Solution**

Because the leading coefficient is 1, the possible rational zeros are the factors of the constant term.

   *Possible rational zeros:* $\pm 1, \pm 2, \pm 3, \pm 6$

A test of these possible zeros shows that $x = -1$ and $x = 2$ are the only two that work. Check the others to be sure.

If the leading coefficient of a polynomial is not 1, the list of possible rational zeros can increase dramatically. In such cases the search can be shortened in several ways: (1) a programmable calculator can be used to speed up the calculations; (2) a graph, either by hand or with a graphing utility, can give a good estimate of the locations of the zeros; and (3) synthetic division can be used to test the possible rational zeros.

To see how to use synthetic division to test the possible rational zeros, let's take another look at the function

$$f(x) = x^4 - x^3 + x^2 - 3x - 6$$

given in Example 2. To test that $x = -1$ and $x = 2$ are zeros of $f$, you can apply synthetic division, as follows.

$$
\begin{array}{r|rrrrr}
-1 & 1 & -1 & 1 & -3 & -6 \\
   &   & -1 & 2 & -3 & 6 \\
\hline
   & 1 & -2 & 3 & -6 & 0 \\
\end{array}
$$

$$
\begin{array}{r|rrrr}
2 & 1 & -2 & 3 & -6 \\
  &   & 2 & 0 & 6 \\
\hline
  & 1 & 0 & 3 & 0 \\
\end{array}
$$

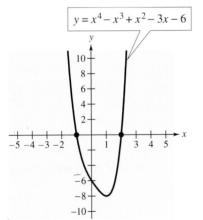

$y = x^4 - x^3 + x^2 - 3x - 6$

**Figure 4.25**

So, you have

$$f(x) = (x + 1)(x - 2)(x^2 + 3).$$

Because the factor $(x^2 + 3)$ produces no real zeros, you can conclude that $x = -1$ and $x = 2$ are the only *real* zeros of $f$. This is verified in the graph of $f$ shown in Figure 4.25.

Finding the first zero is often the hardest part. After that, the search is simplified by using the lower-degree polynomial obtained in synthetic division. Once the lower-degree polynomial is quadratic, either factoring or the Quadratic Formula can be used to find the remaining zeros.

### Example 3   Using the Rational Zero Test

Find the rational zeros of $f(x) = 2x^3 + 3x^2 - 8x + 3$.

Solution

The leading coefficient is 2 and the constant term is 3.

$$\text{Possible rational zeros: } \frac{\text{Factors of 3}}{\text{Factors of 2}} = \frac{\pm 1, \pm 3}{\pm 1, \pm 2} = \pm 1, \pm 3, \pm \frac{1}{2}, \pm \frac{3}{2}$$

By synthetic division, you can determine that $x = 1$ is a zero.

$$
\begin{array}{r|rrrr}
1 & 2 & 3 & -8 & 3 \\
  &   & 2 & 5 & -3 \\
\hline
  & 2 & 5 & -3 & 0
\end{array}
$$

So, $f(x)$ factors as

$$f(x) = (x - 1)(2x^2 + 5x - 3) = (x - 1)(2x - 1)(x + 3)$$

and you can conclude that the zeros of $f$ are $x = 1$, $x = \frac{1}{2}$, and $x = -3$.

### Example 4   Using the Rational Zero Test

Find all the real zeros of $f(x) = 10x^3 - 15x^2 - 16x + 12$.

Solution

The leading coefficient is 10 and the constant term is 12.

$$\text{Possible rational zeros: } \frac{\text{Factors of 12}}{\text{Factors of 10}} = \frac{\pm 1, \pm 2, \pm 3, \pm 4, \pm 6, \pm 12}{\pm 1, \pm 2, \pm 5, \pm 10}$$

With so many possibilities (32, in fact), it is worth your time to stop and sketch a graph. From Figure 4.26, it looks like three reasonable choices would be $x = -\frac{6}{5}$, $x = \frac{1}{2}$, and $x = 2$. Testing these by synthetic division shows that only $x = 2$ works. So, you have

$$f(x) = (x - 2)(10x^2 + 5x - 6).$$

Using the Quadratic Formula, you find that the two additional zeros are irrational numbers.

$$x = \frac{-5 + \sqrt{265}}{20} \approx 0.5639 \qquad \text{and} \qquad x = \frac{-5 - \sqrt{265}}{20} \approx -1.0639$$

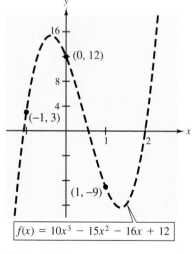

$f(x) = 10x^3 - 15x^2 - 16x + 12$

Figure 4.26

# The Intermediate Value Theorem

The next theorem, called the **Intermediate Value Theorem,** tells you of the existence of real zeros of polynomial functions. The theorem implies that if $(a, f(a))$ and $(b, f(b))$ are two points on the graph of a polynomial such that $f(a) \neq f(b)$, then for any number $d$ between $f(a)$ and $f(b)$ there must be a number $c$ between $a$ and $b$ such that $f(c) = d$. (See Figure 4.27.)

> ▶ **Intermediate Value Theorem**
>
> Let $a$ and $b$ be real numbers such that $a < b$. If $f$ is a polynomial function such that $f(a) \neq f(b)$, then, in the interval $[a, b]$, $f$ takes on every value between $f(a)$ and $f(b)$.

The Intermediate Value Theorem helps you locate the real zeros of a polynomial function in the following way. If you can find a value $x = a$ where a polynomial function is positive, and another value $x = b$ where it is negative, you can conclude that the function has at least one real zero between these two values. For example, the function

$$f(x) = x^3 + x^2 + 1$$

is negative when $x = -2$ and positive when $x = -1$. So, it follows from the Intermediate Value Theorem that $f$ must have a real zero somewhere between $-2$ and $-1$, as shown in Figure 4.28.

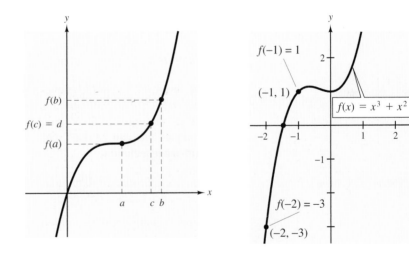

Figure 4.27    *If d lies between f(a) and f(b), there exists c between a and b such that f(c) = d.*

Figure 4.28    *f has a zero between $-2$ and $-1$.*

**Example 5** Approximating a Zero of a Polynomial Function

Use the Intermediate Value Theorem to approximate the real zero of

$$f(x) = x^3 - x^2 + 1.$$

Solution

Begin by computing a few function values, as follows.

| $x$ | $-2$ | $-1$ | 0 | 1 |
|-----|------|------|---|---|
| $f(x)$ | $-11$ | $-1$ | 1 | 1 |

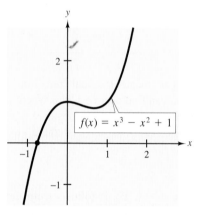

Because $f(-1)$ is negative and $f(0)$ is positive, you can apply the Intermediate Value Theorem to conclude that the function has a zero between $-1$ and 0. To pinpoint this zero more closely, divide the interval $[-1, 0]$ into tenths and evaluate the function at each point. When you do this, you will find that

$$f(-0.8) = -0.152$$

and

$$f(-0.7) = 0.167.$$

**Figure 4.29** *f has a zero between* $-0.8$ *and* $-0.7$.

So, $f$ must have a zero between $-0.8$ and $-0.7$, as shown in Figure 4.29. By continuing this process you can approximate this zero to any desired accuracy.

## Approximating Zeros of Polynomial Functions

There are several different techniques for approximating the zeros of a polynomial function. All such techniques are better suited to computers or graphing utilities than they are to "hand calculations." In this section, you will study two techniques that can be used with a graphing utility. The first is called the **zoom-and-trace** technique.

---

▶ **Zoom-and-Trace Technique**

To approximate a real zero of a function with a graphing utility, use the following steps.

1. Graph the function so that the real zero you want to approximate appears as an $x$-intercept on the screen.

2. Move the cursor near the $x$-intercept and use the zoom feature to zoom in to get a better look at the intercept.

3. Use the trace feature to find the $x$-values that occur just before and just after the $x$-intercept. If the difference in these values is sufficiently small, use their average as the approximation. If not, continue zooming in until the approximation reaches the desired accuracy.

To help you visually determine when you have zoomed in enough times to reach the desired accuracy, you can set the X-scale of the viewing window to the accuracy you need and zoom in repeatedly.

• To approximate the zero to the nearest hundredth, set the X-scale to 0.01.

• To approximate the zero to the nearest thousandth, set the X-scale to 0.001.

• To approximate the zero to the nearest ten-thousandth, set the X-scale to 0.0001.

The amount that a graphing utility zooms in is determined by the *zoom factor*. The zoom factor is an integer that gives the ratio of the larger screen to the smaller screen. For instance, if you zoom in with a zoom factor of 2, you will obtain a screen in which the *x*- and *y*-values are half their original values. In this text, we use a zoom factor of 4.

**Example 6**    Approximating the Zeros of a Polynomial Function

Approximate the real zeros of the polynomial function

$$f(x) = x^3 + 4x + 2.$$

Use an accuracy of 0.001.

**Solution**

To begin, use a graphing utility to graph the function, as shown in Figure 4.30(a). Set the X-scale to 0.001 and zoom in several times until the tick marks on the *x*-axis become visible. The final screen should be similar to the one shown in Figure 4.30(b). At this point, you can use the trace feature to determine that the *x*-values just to the left and right of the *x*-intercept are

$$x \approx -0.4735 \quad \text{and} \quad x \approx -0.4733.$$

So, to the nearest thousandth, you can approximate the zero of the function to be

$$x \approx -0.473.$$

To check this, try substituting $-0.473$ into the function. You should obtain a result that is approximately zero.

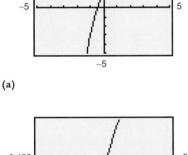
(a)

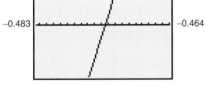
(b)

Figure 4.30

In Example 6, the cubic polynomial function has only one real zero. Remember that functions can have two or more real zeros. In such cases, you can use the zoom-and-trace technique for each zero separately. For instance, the function

$$f(x) = x^3 - 4x^2 + x + 2$$

has three real zeros, as shown in Figure 4.31. Using a zoom-and-trace approach for each real zero, you can approximate the real zeros to be

$$-0.562, \quad 1.000, \quad \text{and} \quad 3.562.$$

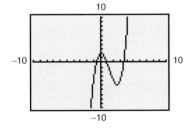

Figure 4.31

The second technique that can be used with some graphing utilities is to use the graphing utility's **solve, root,** or **zero key.** The name of this key or feature differs with different calculators. Consult your user's guide to determine if this feature is available.

### Example 7 Approximating the Zeros of a Polynomial Function

Approximate the real zeros of $f(x) = x^3 - 2x^2 - x + 1$.

### Solution

To begin, use a graphing utility to graph the function, as shown in Figure 4.32. Notice that the graph has three $x$-intercepts. To approximate the leftmost intercept, find an appropriate viewing window and use the root feature. After this has been done, the calculator should display an approximation of

$$x \approx -0.8019377$$

which is accurate to seven decimal places. By repeating this process, you can determine that the other two zeros are $x \approx 0.555$ and $x \approx 2.247$.

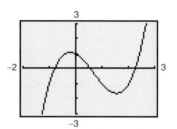

*Find an appropriate viewing window, then use the root feature.*

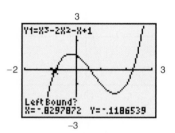

*Cursor to the left of the intercept and press "Enter."*

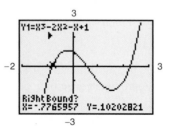

*Cursor to the right of the intercept and press "Enter."*

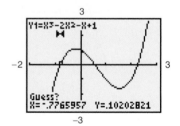

*Cursor near the intercept and press "Enter."*

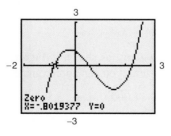

Figure 4.32

You may be wondering why we spend so much time in algebra trying to find the *zeros* of a function. The reason is that if you have a technique that will allow you to solve the equation $f(x) = 0$, you can use the same technique to solve the more general equation

$$f(x) = c$$

where $c$ is any real number. This procedure is demonstrated in Example 8.

**Example 8**    Solving the Equation $f(x) = c$

Find a value of $x$ such that $f(x) = 30$ for the function

$$f(x) = x^3 - 4x + 4.$$

**Solution**

The graph of $f(x) = x^3 - 4x + 4$ is shown in Figure 4.33. Note from the graph that $f(x) = 30$ when $x$ is about 3.5. To use the zoom-and-trace technique to approximate this $x$-value more closely, consider the equation

$$x^3 - 4x + 4 = 30$$
$$x^3 - 4x - 26 = 0.$$

So, the *solutions* of the equation $f(x) = 30$ are precisely the same $x$-values as the *zeros* of $g(x) = x^3 - 4x - 26$. Using the graph of $g$, as shown in Figure 4.34, you can approximate the zero of $g$ to be

$$x \approx 3.41.$$

You can check this value by substituting $x = 3.41$ into the original function.

$$f(3.41) = (3.41)^3 - 4(3.41) + 4$$
$$\approx 30.01 \checkmark$$

Remember that with decimal approximations, a check will usually not produce an exact value.

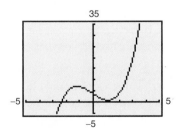

Figure 4.33

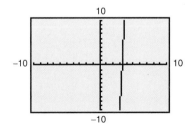

Figure 4.34

## Application

**Example 9** Profit and Advertising Expenses

A company that produces sports clothes estimates that the profit from selling a particular line of sportswear is given by

$$P = -140x^3 + 7520x^2 - 400{,}000, \qquad 0 \le x \le 50$$

where $P$ is the profit (in tens of thousands of dollars) and $x$ is the advertising expense (in tens of thousands of dollars). According to this model, how much money should the company spend to obtain a profit of $2,750,000?

### Solution

From Figure 4.35, it appears that there are two different values of $x$ between 0 and 50 that will produce a profit of $2,750,000. However, because of the context of the problem, it is clear that the better answer is the smaller of the two numbers. So, to solve the equation

$$-140x^3 + 7520x^2 - 400{,}000 = 2{,}750{,}000$$

$$-140x^3 + 7520x^2 - 3{,}150{,}000 = 0$$

you can find the zeros of the function $g(x) = -140x^3 + 7520x^2 - 3{,}150{,}000$. Using the zoom-and-trace technique, you can find that the leftmost zero is $x \approx 32.8$. You can check this solution by substituting $x = 32.8$ into the original function.

$$P = -140(32.8)^3 + 7520(32.8)^2 - 400{,}000 \approx \$2{,}750{,}060$$

So, the company should plan to spend about $328,000 in advertising for the line of sportswear.

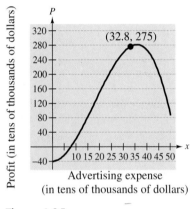

Figure 4.35

---

**Discussing the Concept** | **Comparing Real Zeros and Rational Zeros**

Discuss the meanings of *real zeros* and *rational zeros* of a polynomial function and compare these two types of zeros. Then answer the following questions.

**a.** Is it possible for a polynomial function to have no rational zeros but to have real zeros? If so, give an example.

**b.** If a polynomial function has three real zeros, and only one of them is a rational number, must the other two zeros be irrational numbers?

**c.** Consider a cubic polynomial function $f(x) = ax^3 + bx^2 + cx + d$ where $a \ne 0$. Is it possible that $f$ has no real zeros? Is it possible that $f$ has no rational zeros?

**d.** Is it possible that a second-degree polynomial function with integer coefficients has one rational zero and one irrational zero? If so, give an example.

# Warm Up

The following warm-up exercises involve skills that were covered in earlier sections. You will use these skills in the exercise set for this section.

In Exercises 1 and 2, find a polynomial function with integer coefficients having the given zeros.

**1.** $-1, \frac{2}{3}, 3$

**2.** $-2, 0, \frac{3}{4}, 2$

In Exercises 3 and 4, divide by synthetic division.

**3.** $\dfrac{x^5 - 9x^3 + 5x + 18}{x + 3}$

**4.** $\dfrac{3x^4 + 17x^3 + 10x^2 - 9x - 8}{x + \frac{2}{3}}$

In Exercises 5-8, use the given zero to find all the real zeros of $f$.

**5.** $f(x) = 2x^3 + 11x^2 + 2x - 4, \ x = \frac{1}{2}$

**6.** $f(x) = 6x^3 - 47x^2 - 124x - 60, \ x = 10$

**7.** $f(x) = 4x^3 - 13x^2 - 4x + 6, \ x = -\frac{3}{4}$

**8.** $f(x) = 10x^3 + 51x^2 + 48x - 28, \ x = \frac{2}{5}$

In Exercises 9 and 10, find all real solutions.

**9.** $x^4 - 3x^2 + 2 = 0$

**10.** $x^4 - 7x^2 + 12 = 0$

## 4.4   Exercises

In Exercises 1-4, use the Rational Zero Test to list all possible rational zeros of $f$. Then use a graphing utility to graph the function. Use the graph to help determine which of the possible rational zeros are actual zeros of the function.

**1.** $f(x) = x^3 + x^2 - 4x - 4$

**2.** $f(x) = -4x^3 + 15x^2 - 8x - 3$

**3.** $f(x) = 4x^4 - 17x^2 + 4$

**4.** $f(x) = -2x^4 + 13x^3 - 21x^2 + 2x + 8$

In Exercises 5-12, find the real zeros of the function.

**5.** $f(x) = x^3 - 6x^2 + 11x - 6$

**6.** $g(x) = x^3 - 4x^2 - x + 4$

**7.** $h(t) = t^3 + 12t^2 + 21t + 10$

**8.** $f(x) = x^3 - 4x^2 + 5x - 2$

**9.** $C(x) = 2x^3 + 3x^2 - 1$

**10.** $f(x) = 3x^3 - 19x^2 + 33x - 9$

**11.** $f(x) = x^4 - 3x^2 + 2$

**12.** $P(t) = t^4 - 7t^2 + 12$

In Exercises 13-18, find all real solutions of the polynomial equation.

**13.** $z^4 - z^3 - 2z - 4 = 0$

**14.** $x^4 - 13x^2 - 12x = 0$

**15.** $2y^4 + 7y^3 - 26y^2 + 23y - 6 = 0$

**16.** $2x^4 - 11x^3 - 6x^2 + 64x + 32 = 0$

**17.** $x^5 - x^4 - 3x^3 + 5x^2 - 2x = 0$

**18.** $x^5 - 7x^4 + 10x^3 + 14x^2 - 24x = 0$

In Exercises 19 and 20, (a) list the possible rational zeros of $f$, (b) sketch the graph of $f$ so that some of the possible zeros in part (a) can be discarded, and then (c) determine all real zeros of $f$.

**19.** $f(x) = 32x^3 - 52x^2 + 17x + 3$

**20.** $f(x) = 4x^3 + 7x^2 - 11x - 18$

In Exercises 21 and 22, find the rational zeros of the polynomial function.

**21.** $P(x) = x^4 - \frac{25}{4}x^2 + 9$

**22.** $f(x) = x^3 - \frac{3}{2}x^2 - \frac{23}{2}x + 6$

In Exercises 23–26, follow the procedure given in Example 5 to approximate the zero of $f(x)$ in the interval $[a, b]$. Give your approximation to the nearest tenth. (If you have a graphing utility, use it to help approximate the zero.)

**23.** $f(x) = x^3 + x - 1$, $\quad [0, 1]$

**24.** $f(x) = x^5 + x + 1$, $\quad [-1, 0]$

**25.** $f(x) = x^4 - 10x^2 - 11$, $\quad [3, 4]$

**26.** $f(x) = -x^3 + 3x^2 + 9x - 2$, $\quad [4, 5]$

In Exercises 27–32, match the function with its graph. Then approximate the real zeros of the function to three decimal places. [The graphs are labeled (a), (b), (c), (d), (e), and (f).]

(a)

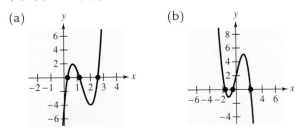

(b)

(c)

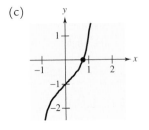

(d)

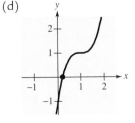

(e)

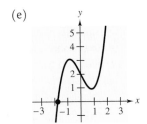

(f)

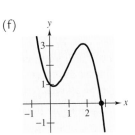

**27.** $f(x) = x^3 - 2x + 2$

**28.** $f(x) = x^5 + x - 1$

**29.** $f(x) = 2x^3 - 6x^2 + 6x - 1$

**30.** $f(x) = 5x^3 - 20x^2 + 20x - 4$

**31.** $f(x) = -x^3 + 3x^2 - x + 1$

**32.** $f(x) = -x^3 + 4x + 2$

In Exercises 33–40, use a graphing utility to approximate the real zeros of the function. Use an accuracy of 0.001.

**33.** $f(x) = x^4 - x - 3$

**34.** $f(x) = x^3 - 3.9x^2 + 4.79x - 1.881$

**35.** $f(x) = 4x^3 + 14x - 8$

**36.** $f(x) = -x^3 + 2x^2 + 4x + 5$

**37.** $f(x) = x^4 + x - 3$

**38.** $f(x) = -x^4 + 2x^3 + 4$

**39.** $f(x) = 7x^4 - 42x^3 + 43x^2 + 216x - 324$

**40.** $f(x) = 3x^4 - 12x^3 + 27x^2 + 4x - 4$

**41.** *Measurements of a Box* An open box is to be made from a rectangular piece of material, 12 inches by 10 inches, by cutting equal squares from the corners and turning up the sides.

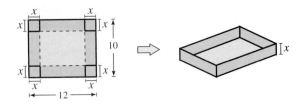

(a) Use the figure to write the volume $V$ of the box as a function of $x$. Determine the domain of the function.

(b) Sketch the graph of the function and approximate the measurements of the box that yield a maximum volume.

(c) Find values of $x$ such that $V = 96$. Which of these values is a physical impossibility in the construction of the box? Explain.

**42.** *Measurements of a Box*  An open box is to be made from a rectangular piece of material, 9 inches by 5 inches, by cutting equal squares from the corners and turning up the sides.

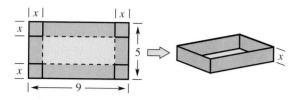

(a) Use the figure to write the volume $V$ of the box as a function of $x$. Determine the domain of the function.

(b) Sketch the graph of the function and approximate the measurements of the box that yield a maximum volume.

(c) Find values of $x$ such that $V = 56$. Which of these values is a physical impossibility in the construction of the box? Explain.

**43.** *Measurements of a Package*  A rectangular package to be sent by a postal service has a combined length and girth (perimeter of a cross section) of 108 inches (see figure). Find the measurements of the package, given that the volume is 11,664 cubic inches.

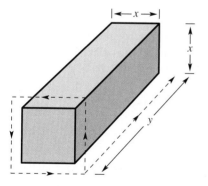

**44.** *Measurements of a Package*  A rectangular package to be sent by a postal service has a combined length and girth (perimeter of a cross section) of 120 inches. Find the measurements of the package, given that the volume is 16,000 cubic inches.

**45.** *Medicine*  The concentration of a certain chemical in the bloodstream $t$ hours after injection into muscle tissue is given by

$$C = \frac{3t^2 + t}{t^3 + 50}, \qquad t \ge 0.$$

The concentration is greatest when

$$3t^4 + 2t^3 - 300t - 50 = 0.$$

Approximate this time to the nearest tenth of an hour.

**46.** *Transportation Cost*  The transportation cost $C$ of the components used in manufacturing a certain product is given by

$$C = 100\left(\frac{200}{x^2} + \frac{x}{x + 30}\right), \qquad x \ge 1$$

where $C$ is measured in thousands of dollars and $x$ is the order size in hundreds. The cost is a minimum when

$$3x^3 - 40x^2 - 2400x - 36{,}000 = 0.$$

Approximate the optimal order size to the nearest hundred units.

**47.** *Imports to the United States*  The values $I$ (in billions of dollars) of the total goods imported into the United States from 1991 to 2000 are given in the table.  (Source: U.S. Statistical Abstract)

| Year | 1991 | 1992 | 1993 | 1994 | 1995 |
|------|------|------|------|------|------|
| $I$ | 488.5 | 532.7 | 580.7 | 663.3 | 743.4 |

| Year | 1996 | 1997 | 1998 | 1999 | 2000 |
|------|------|------|------|------|------|
| $I$ | 795.3 | 870.7 | 911.9 | 1024.6 | 1218.0 |

(a) Use a graphing utility to make a scatter plot of the data. Let $t = 1$ correspond to 1991.

(b) Use the regression capabilities of a graphing utility to find a linear model, a quadratic model, a cubic model, and a quartic model for the data.

(c) Graph each model separately with the data in the same viewing window.

(d) Choose the model that "best fits" the data. Explain why you chose that model. Use your model to estimate the year in which imports reached 1500 billion dollars.

**48.** *Cost of Dental Care*  The amount that $100 worth of dental care at 1982–1984 prices would cost at a different time is given by a CPI (Consumer Price Index). The CPIs for dental care in the United States from 1990 to 1999 are given in the table. (Source: U.S. Bureau of Labor Statistics)

| Year | 1990 | 1991 | 1992 | 1993 | 1994 |
|------|------|------|------|------|------|
| CPI | 155.8 | 167.4 | 178.7 | 188.1 | 197.1 |

| Year | 1995 | 1996 | 1997 | 1998 | 1999 |
|------|------|------|------|------|------|
| CPI | 206.8 | 216.5 | 226.6 | 236.2 | 247.2 |

(a) Use a graphing utility to make a scatter plot of the data. Let $t = 0$ correspond to 1990.

(b) Use the regression capabilities of a graphing utility to find a linear model, a quadratic model, a cubic model, and a quartic model for the data.

(c) Graph each model separately with the data in the same viewing window.

(d) Choose the model that "best fits" the data. Explain why you chose that model. Use your model to estimate the year in which the consumer price index for dental care reached 300.

**49.** *Lead Emissions*  The amount of lead (in tons) in air pollution in the United States from 1990 to 1998 is given in the table. (Source: U.S. Statistical Abstract)

| Year | 1990 | 1991 | 1992 | 1993 | 1994 |
|------|------|------|------|------|------|
| Lead | 4975 | 4169 | 3810 | 3916 | 4047 |

| Year | 1995 | 1996 | 1997 | 1998 |
|------|------|------|------|------|
| Lead | 3929 | 3899 | 3952 | 3973 |

(a) Use a graphing utility to make a scatter plot of the data. Let $t = 0$ correspond to 1990.

(b) Use the regression capabilities of a graphing utility to find a cubic model and a quartic model for the data.

(c) Graph each model separately with the data in the same viewing window. How well does each model fit the data?

(d) Use each model to estimate the amount of lead emitted in 2000 and 2005. Are the estimates from each model reasonable for years after 1998?

**50.** *Purchasing Power of the Dollar*  Based on the 1982–1984 value of a dollar, the purchasing power $p$ of a consumer dollar in the United States from 1990 to 1999 is given in the table. (Source: U.S. Bureau of Labor Statistics)

| Year | 1990 | 1991 | 1992 | 1993 | 1994 |
|------|------|------|------|------|------|
| $p$ | 0.766 | 0.734 | 0.713 | 0.692 | 0.675 |

| Year | 1995 | 1996 | 1997 | 1998 | 1999 |
|------|------|------|------|------|------|
| $p$ | 0.656 | 0.638 | 0.623 | 0.614 | 0.600 |

(a) Use a graphing utility to make a scatter plot of the data. Let $t = 0$ correspond to 1990.

(b) Use the regression capabilities of a graphing utility to find a linear model, a quadratic model, a cubic model, and a quartic model for the data.

(c) Graph each model separately with the data in the same viewing window. How well does each model fit the data?

(d) Use each model to estimate $p$ in 2000 and 2005. Are the estimates from each model reasonable?

(e) Discuss the appropriateness of each model for estimating the purchasing power of the dollar after 1999.

**51.** *Measurements of a Box*  An open box is to be made from a rectangular piece of material, 10 inches by 8 inches, by cutting equal squares from the corners and turning up the sides. Find the measurements of the box if the volume is to be 52.5 cubic inches.

**52.** *Demand Function*  A company that produces cellular pagers estimates that the demand for a particular model of pager is

$$D = -x^3 + 54x^2 - 140x - 3000, \quad 10 \leq x \leq 50$$

where $D$ is the demand for the pager and $x$ is the price of the pager.

(a) Use a graphing utility to graph $D$. Use the trace feature to determine for what values of $x$ the demand will be 14,400 pagers.

(b) You may also determine the price $x$ of the pagers that will yield a demand of 14,400 by setting $D$ equal to 14,400 and solving for $x$ with a graphing utility. Discuss this alternative method of solution. Of the actual solutions, what price would you recommend the company charge for the pagers?

# Mid-Chapter Quiz

Take this quiz as you would take a quiz in class. After you are done, check your work against the answers given in the back of the book.

In Exercises 1 and 2, sketch the graph of the quadratic function. Identify the vertex and the intercepts.

**1.** $f(x) = (x + 1)^2 - 2$        **2.** $f(x) = 16 - x^2$

In Exercises 3 and 4, determine the right- and left-hand behavior of the graph of the function. Verify with a graphing utility.

**3.** $f(x) = 5x^3 - 7x^2 + 2$        **4.** $f(x) = -x^4 + 5x^2 - 4$

**5.** Completely factor the polynomial $2x^3 - x^2 - 13x - 6$ using long division. One factor is $(x + 2)$.

**6.** Use synthetic division to evaluate $f(x) = 3x^3 - 5x^2 + 9$ at $x = 2$.

In Exercises 7 and 8, express the function in the form $f(x) = (x - k)q(x) + r$ for the given value of $k$, and demonstrate that $f(k) = r$.

**7.** $f(x) = x^4 - 5x^2 + 4, \quad k = 1$

**8.** $f(x) = x^3 + 5x^2 - 2x - 24, \quad k = -3$

**9.** Simplify: $\dfrac{2x^4 + 9x^3 - 32x^2 - 99x + 180}{x^2 + 2x - 15}$

**10.** The profit for a company is $P = -95x^3 + 5650x^2 - 250,000, 0 \le x \le 55$, where $x$ is the advertising expense (in tens of thousands of dollars) and $P$ is the profit in dollars. What is the profit for an advertising expense of $450,000? Use a graphing utility to approximate another advertising expense that would yield the same profit.

In Exercises 11–14, find all real zeros of the function.

**11.** $f(x) = -2x^3 - 7x^2 + 10x + 35$      **12.** $f(x) = 4x^4 - 37x^2 + 9$

**13.** $f(x) = 3x^4 + 4x^3 - 3x - 4$      **14.** $f(x) = 2x^3 - 3x^2 + 2x - 3$

| Year | $L$ |
|------|------|
| 1990 | 534 |
| 1991 | 667 |
| 1992 | 882 |
| 1993 | 1197 |
| 1994 | 1715 |
| 1995 | 1795 |
| 1996 | 1808 |
| 1997 | 2062 |
| 1998 | 1985 |

**15.** The numbers of consumer new passenger car leases $L$ (in thousands) in the United States from 1990 to 1998 are given in table at the left. (Source: U.S. Statistical Abstract)

(a) Use a graphing utility to make a scatter plot of the data. Let $t = 0$ correspond to 1990.

(b) Use the regression capabilities of a graphing utility to find a quadratic model, a cubic model, and a quartic model for the data.

(c) Graph each model separately with the data in the same viewing window. How well does each model fit the data?

(d) Use all three models to predict the number of consumer new passenger car leases in 2000 and 2001. Discuss the reasonableness of each model's prediction.

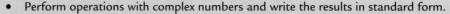

## 4.5 Complex Numbers

### Objectives

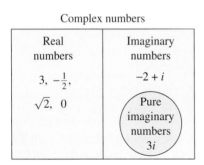

- Perform operations with complex numbers and write the results in standard form.
- Find the complex conjugate of a complex number.
- Solve a polynomial equation.
- Plot a complex number in the complex plane.
- Determine if a complex number is in the Mandelbrot Set.

## The Imaginary Unit *i*

In Section 1.4, you learned that some quadratic equations have no real solutions. For instance, the quadratic equation

$$x^2 + 1 = 0 \qquad \text{Equation with no real solution}$$

has no real solution because there is no real number $x$ that can be squared to produce $-1$. To overcome this deficiency, mathematicians created an expanded system of numbers using the **imaginary unit *i*,** defined as

$$i = \sqrt{-1} \qquad \text{Imaginary unit}$$

where $i^2 = -1$. By adding real numbers to real multiples of this imaginary unit, we obtain the set of **complex numbers.** Each complex number can be written in the **standard form, *a* + *bi*.**

Complex numbers

| Real numbers | Imaginary numbers |
|---|---|
| $3, -\frac{1}{2},$ $\sqrt{2}, \ 0$ | $-2 + i$ <br> Pure imaginary numbers <br> $3i$ |

Figure 4.36

### ▶ Definition of a Complex Number

If $a$ and $b$ are real numbers, the number $a + bi$ is a **complex number,** and it is said to be written in **standard form.** If $b = 0$, the number $a + bi = a$ is a real number. If $b \neq 0$, the number $a + bi$ is called an **imaginary number.** A number of the form $bi$, where $b \neq 0$, is called a **pure imaginary number.**

The set of real numbers is a subset of the set of complex numbers, as shown in Figure 4.36. This is true because every real number $a$ can be written as a complex number using $b = 0$. That is, for every real number $a$, we can write $a = a + 0i$.

### ▶ Equality of Complex Numbers

Two complex numbers $a + bi$ and $c + di$, written in standard form, are **equal** to each other

$$a + bi = c + di \qquad \text{Equality of two complex numbers}$$

if and only if $a = c$ and $b = d$.

## Operations with Complex Numbers

To add (or subtract) two complex numbers, you add (or subtract) the real and imaginary parts of the numbers separately.

---

▶ **Addition and Subtraction of Complex Numbers**

If $a + bi$ and $c + di$ are two complex numbers written in standard form, their sum and difference are defined as follows.

$$\text{Sum: } (a + bi) + (c + di) = (a + c) + (b + d)i$$
$$\text{Difference: } (a + bi) - (c + di) = (a - c) + (b - d)i$$

---

The **additive identity** in the complex number system is zero (the same as in the real number system). Furthermore, the **additive inverse** of the complex number $a + bi$ is

$$-(a + bi) = -a - bi. \qquad \text{Additive inverse}$$

So, you have

$$(a + bi) + (-a - bi) = 0 + 0i = 0.$$

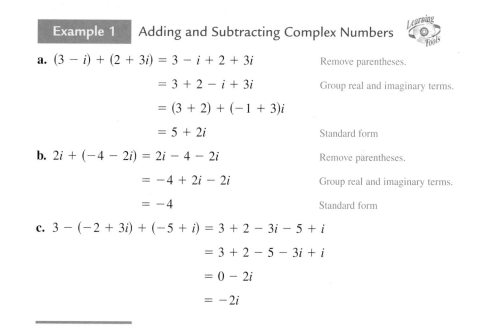

**Example 1**    Adding and Subtracting Complex Numbers

a. $(3 - i) + (2 + 3i) = 3 - i + 2 + 3i$    Remove parentheses.

$$= 3 + 2 - i + 3i \qquad \text{Group real and imaginary terms.}$$
$$= (3 + 2) + (-1 + 3)i$$
$$= 5 + 2i \qquad \text{Standard form}$$

b. $2i + (-4 - 2i) = 2i - 4 - 2i$    Remove parentheses.

$$= -4 + 2i - 2i \qquad \text{Group real and imaginary terms.}$$
$$= -4 \qquad \text{Standard form}$$

c. $3 - (-2 + 3i) + (-5 + i) = 3 + 2 - 3i - 5 + i$

$$= 3 + 2 - 5 - 3i + i$$
$$= 0 - 2i$$
$$= -2i$$

---

Note in Example 1(b) that the sum of two imaginary numbers can be a real number.

Complete the table:

$i^1 = i$     $i^5 =$        $i^9 =$

$i^2 = -1$ $i^6 =$      $i^{10} =$

$i^3 = -i$  $i^7 =$       $i^{11} =$

$i^4 = 1$    $i^8 =$       $i^{12} =$

What pattern do you see? Write a brief description of how you would find $i$ raised to a positive integer power.

Many of the properties of real numbers are valid for complex numbers as well. Here are some examples.

*Associative Property of Addition and Multiplication*

*Commutative Property of Addition and Multiplication*

*Distributive Property of Multiplication Over Addition*

Notice how these properties are used when two complex numbers are multiplied.

$$(a + bi)(c + di) = a(c + di) + bi(c + di) \qquad \text{Distributive}$$
$$= ac + (ad)i + (bc)i + (bd)i^2 \qquad \text{Distributive}$$
$$= ac + (ad)i + (bc)i + (bd)(-1) \qquad \text{Definition of } i$$
$$= ac - bd + (ad)i + (bc)i \qquad \text{Commutative}$$
$$= (ac - bd) + (ad + bc)i \qquad \text{Associative}$$

Rather than trying to memorize this multiplication rule, we suggest that you simply remember how the distributive property is used to multiply two complex numbers. The procedure is similar to multiplying two polynomials and combining like terms (as in the FOIL Method).

---

**Example 2**    Multiplying Complex Numbers    *Learning Tools*

**a.** $(i)(-3i) = -3i^2$      Multiply.

      $= -3(-1)$     $i^2 = -1$

       $= 3$       Simplify.

**b.** $(2 - i)(4 + 3i) = 8 + 6i - 4i - 3i^2$   Product of binomials

        $= 8 + 6i - 4i - 3(-1)$   $i^2 = -1$

        $= 8 + 3 + 6i - 4i$    Collect terms.

        $= 11 + 2i$      Standard form

**c.** $(3 + 2i)(3 - 2i) = 9 - 6i + 6i - 4i^2$   Product of binomials

        $= 9 - 4(-1)$     $i^2 = -1$

        $= 9 + 4$      Simplify.

        $= 13$       Simplify.

**d.** $(3 + 2i)^2 = 9 + 6i + 6i + 4i^2$   Product of binomials

       $= 9 + 4(-1) + 12i$   $i^2 = -1$

       $= 9 - 4 + 12i$    Simplify.

       $= 5 + 12i$     Simplify.

# Complex Conjugates

Notice in Example 2(c) that the product of two imaginary numbers can be a real number. This occurs with pairs of complex numbers of the form $a + bi$ and $a - bi$, called **complex conjugates.** In general, the product of two complex conjugates can be written as follows.

$$(a + bi)(a - bi) = a^2 - abi + abi - b^2i^2$$
$$= a^2 - b^2(-1)$$
$$= a^2 + b^2$$

Complex conjugates can be used to divide one complex number by another. That is, to find the quotient of $a + bi$ and $c + di$, multiply the numerator and denominator by the conjugate of the denominator to obtain

$$\frac{a + bi}{c + di} = \frac{a + bi}{c + di}\left(\frac{c - di}{c - di}\right) = \frac{(ac + bd) + (bc - ad)i}{c^2 + d^2}.$$

**Example 3**    Dividing Complex Numbers    *Learning Tools*

$$\frac{1}{1 + i} = \frac{1}{1 + i}\left(\frac{1 - i}{1 - i}\right) \qquad \text{Multiply by conjugate.}$$

$$= \frac{1 - i}{1^2 - i^2} \qquad \text{Product of conjugates}$$

$$= \frac{1 - i}{1 - (-1)} \qquad i^2 = -1$$

$$= \frac{1 - i}{2} \qquad \text{Simplify.}$$

$$= \frac{1}{2} - \frac{1}{2}i \qquad \text{Standard form}$$

**Example 4**    Dividing Complex Numbers    *Learning Tools*

$$\frac{2 + 3i}{4 - 2i} = \frac{2 + 3i}{4 - 2i}\left(\frac{4 + 2i}{4 + 2i}\right) \qquad \text{Multiply by conjugate.}$$

$$= \frac{8 + 4i + 12i + 6i^2}{16 - 4i^2} \qquad \text{Expand.}$$

$$= \frac{8 - 6 + 16i}{16 + 4} \qquad i^2 = -1$$

$$= \frac{2 + 16i}{20} \qquad \text{Simplify.}$$

$$= \frac{1}{10} + \frac{4}{5}i \qquad \text{Standard form}$$

## Complex Solutions

When using the Quadratic Formula to solve a quadratic equation, you often obtain a result such as $\sqrt{-3}$, which you know is not a real number. By factoring out $i = \sqrt{-1}$, you can write this number in standard form.

$$\sqrt{-3} = \sqrt{3(-1)} = \sqrt{3}\sqrt{-1} = \sqrt{3}\,i$$

The number $\sqrt{3}\,i$ is called the principal square root of $-3$.

---

▶ **Principal Square Root of a Negative Number**

If $a$ is a positive number, the **principal square root** of the negative number $-a$ is defined as

$$\sqrt{-a} = \sqrt{a}\,i.$$

---

### Study Tip

The definition of principal square root uses the rule

$$\sqrt{ab} = \sqrt{a}\sqrt{b}$$

for $a > 0$ and $b < 0$. This rule is not valid if *both* $a$ and $b$ are negative. For example,

$$\sqrt{-5}\sqrt{-5} = 5i^2 = -5$$

whereas

$$\sqrt{(-5)(-5)} = \sqrt{25} = 5.$$

To avoid problems with multiplying square roots of negative numbers, be sure to convert to standard form *before* multiplying.

**Example 5**   Writing Complex Numbers in Standard Form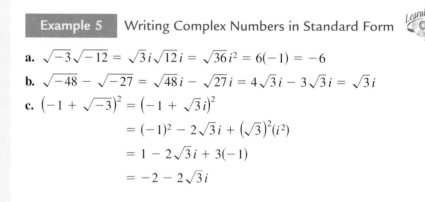

**a.** $\sqrt{-3}\sqrt{-12} = \sqrt{3}\,i\sqrt{12}\,i = \sqrt{36}\,i^2 = 6(-1) = -6$

**b.** $\sqrt{-48} - \sqrt{-27} = \sqrt{48}\,i - \sqrt{27}\,i = 4\sqrt{3}\,i - 3\sqrt{3}\,i = \sqrt{3}\,i$

**c.** $\left(-1 + \sqrt{-3}\right)^2 = \left(-1 + \sqrt{3}\,i\right)^2$

$$= (-1)^2 - 2\sqrt{3}\,i + \left(\sqrt{3}\right)^2(i^2)$$

$$= 1 - 2\sqrt{3}\,i + 3(-1)$$

$$= -2 - 2\sqrt{3}\,i$$

**Example 6**   Complex Solutions of a Quadratic Equation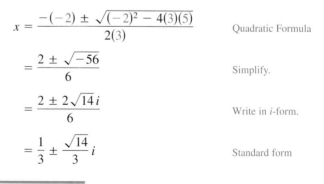

Solve the equation $3x^2 - 2x + 5 = 0$.

**Solution**

$$x = \frac{-(-2) \pm \sqrt{(-2)^2 - 4(3)(5)}}{2(3)} \qquad \text{Quadratic Formula}$$

$$= \frac{2 \pm \sqrt{-56}}{6} \qquad \text{Simplify.}$$

$$= \frac{2 \pm 2\sqrt{14}\,i}{6} \qquad \text{Write in } i\text{-form.}$$

$$= \frac{1}{3} \pm \frac{\sqrt{14}}{3}\,i \qquad \text{Standard form}$$

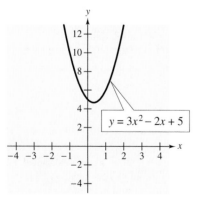

Figure 4.37

The graph of $f(x) = 3x^2 - 2x + 5$, shown in Figure 4.37, does not touch or cross the $x$-axis. This confirms that Example 6 has no real solutions.

Imaginary axis

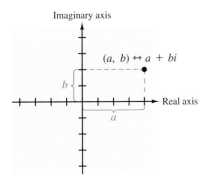

$(a, b) \leftrightarrow a + bi$

Real axis

**Figure 4.38**

## Applications

Most applications involving complex numbers are either theoretical (see the next section) or very technical, and so are not appropriate for inclusion in this text. However, to give you some idea of how complex numbers can be used in applications, we give a general description of their use in **fractal geometry.**

To begin, consider a coordinate system called the **complex plane.** Just as every real number corresponds to a point on the real line, every complex number corresponds to a point in the complex plane, as shown in Figure 4.38. In this figure, note that the vertical axis is the **imaginary axis** and the horizontal axis is the **real axis.** The point that corresponds to the complex number $a + bi$ is $(a, b)$.

Complex number
$a + bi$

Ordered pair
$(a, b)$

---

**Example 7**    Plotting Complex Numbers in the Complex Plane

Plot the following complex numbers in the complex plane.

**a.** $2 + 3i$

**b.** $-1 + 2i$

**c.** $4$

### Solution

**a.** To plot the complex number $2 + 3i$, move (from the origin) two units to the right on the real axis and then three units up, as shown in Figure 4.39. In other words, plotting the complex number $2 + 3i$ in the complex plane is comparable to plotting the point $(2, 3)$ in the Cartesian plane.

**b.** The complex number $-1 + 2i$ corresponds to the point $(-1, 2)$, as shown in Figure 4.39.

**c.** The complex number $4$ corresponds to the point $(4, 0)$, as shown in Figure 4.39.

Imaginary axis

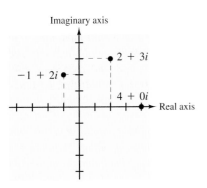

$2 + 3i$

$-1 + 2i$

$4 + 0i$

Real axis

**Figure 4.39**

In the hands of a person who understands "fractal geometry," the complex plane can become an easel on which stunning pictures, called **fractals,** can be drawn. The most famous such picture is called the **Mandelbrot Set,** named after the Polish-born mathematician Benoit Mandelbrot. To draw the Mandelbrot Set, consider the following sequence of numbers.

$$c, \quad c^2 + c, \quad (c^2 + c)^2 + c, \quad [(c^2 + c)^2 + c]^2 + c, \quad \cdots$$

The behavior of this sequence depends on the value of the complex number $c$. For some values of $c$ this sequence is **bounded,** and for other values it is **unbounded.** If the sequence is bounded, the complex number $c$ is in the Mandelbrot Set, and if the sequence is unbounded, the complex number $c$ is not in the Mandelbrot Set.

**Example 8**   Members of the Mandelbrot Set

**a.** The complex number $-2$ is in the Mandelbrot Set because for $c = -2$, the corresponding Mandelbrot sequence is $-2, 2, 2, 2, 2, 2, \ldots$, which is bounded.

**b.** The complex number $i$ is also in the Mandelbrot Set because for $c = i$, the corresponding Mandelbrot sequence is

$$i, \quad -1 + i, \quad -i, \quad -1 + i, \quad -i, \quad -1 + i, \quad \ldots$$

which is bounded.

**c.** The complex number $1 + i$ is *not* in the Mandelbrot Set because for $c = 1 + i$, the corresponding Mandelbrot sequence is

$$1 + i, \quad 1 + 3i, \quad -7 + 7i, \quad 1 - 97i, \quad -9407 - 193i,$$

$$88454401 + 3631103i, \ldots$$

which is unbounded.

With this definition, a picture of the Mandelbrot Set would have only two colors. One color for points that are in the set (the sequence is bounded) and one color for points that are outside the set (the sequence is unbounded). Figure 4.40 shows a black and yellow picture of the Mandelbrot Set. The points that are black are in the Mandelbrot Set and the points that are yellow are not.

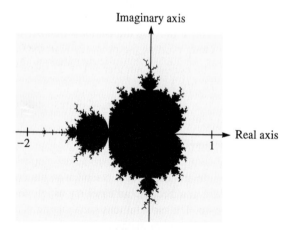

Figure 4.40   *Mandelbrot Set*

To add more interest to the picture, computer scientists discovered that the points that are not in the Mandelbrot Set can be assigned a variety of colors, depending on "how quickly" their sequences diverge. Figure 4.41 shows three different appendages of the Mandelbrot Set using a spectrum of colors. (The colored portions of the picture represent points that are *not* in the Mandelbrot Set.)

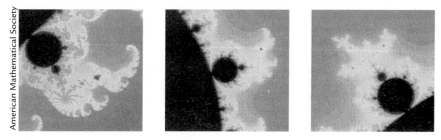

Figure 4.41

Figures 4.42, 4.43, and 4.44 show other types of fractal sets. From these pictures, you can see why fractals have fascinated people since their discovery (around 1980).

Figure 4.42                      Figure 4.43                      Figure 4.44

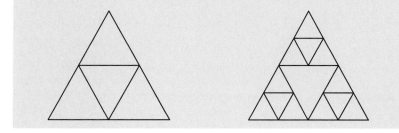

**Discussing the Concept**    **Building Your Own Fractal**

On a large sheet of paper, use a ruler to draw a large equilateral triangle. Find the midpoint of each side. Connect the midpoints as shown in the figure on the left. Repeat the process for each of the smaller upward-pointing triangles, as shown in the figure on the right. Continue the process, always connecting the midpoints of the upward-pointing triangles. This fractal triangle is called the Sierpinski triangle.

# Warm Up

The following warm-up exercises involve skills that were covered in earlier sections. You will use these skills in the exercise set for this section.

In Exercises 1–8, simplify the expression.

1. $\sqrt{12}$

2. $\sqrt{500}$

3. $\sqrt{20} - \sqrt{5}$

4. $\sqrt{27} - \sqrt{243}$

5. $\sqrt{24}\sqrt{6}$

6. $2\sqrt{18}\sqrt{32}$

7. $\dfrac{1}{\sqrt{3}}$

8. $\dfrac{2}{\sqrt{2}}$

In Exercises 9 and 10, solve the quadratic equation.

9. $x^2 + x - 1 = 0$

10. $x^2 + 2x - 1 = 0$

## 4.5    Exercises

1. Write out the first 16 positive integer powers of $i$ $(i, i^2, i^3, \ldots, i^{16})$, and express each as $i, -i, 1,$ or $-1$.

2. Express each of the powers of $i$ as $i, -i, 1,$ or $-1$.

   (a) $i^{28}$    (b) $i^{37}$    (c) $i^{127}$    (d) $i^{82}$

In Exercises 3–6, find the real numbers $a$ and $b$ such that the equation is true.

3. $a + bi = 3 + 12i$

4. $a + bi = -2 - 5i$

5. $(a + 3) + (b - 1)i = 7 - 4i$

6. $(a + 6) + 2bi = 6 - 5i$

In Exercises 7–18, write the complex number in standard form and find its complex conjugate.

7. $9 + \sqrt{-16}$

8. $2 + \sqrt{-25}$

9. $-3 - \sqrt{-12}$

10. $1 + \sqrt{-8}$

11. $-21$

12. $45$

13. $-6i + i^2$

14. $4i^2 - 2i^3$

15. $-5i^5$

16. $(-i)^3$

17. $\left(\sqrt{-6}\right)^2 + 3$

18. $\left(\sqrt{-4}\right)^2 - 5$

In Exercises 19–56, perform the indicated operation and write the result in standard form.

19. $(-4 + 3i) + (6 - 2i)$

20. $(13 - 2i) + (-5 + 6i)$

21. $(12 + 5i) - (7 - i)$

22. $(3 + 2i) - (6 + 13i)$

23. $\left(-2 + \sqrt{-8}\right) + \left(5 - \sqrt{-50}\right)$

24. $\left(8 + \sqrt{-18}\right) - \left(4 + 3\sqrt{2}\,i\right)$

25. $-\left(\frac{3}{2} + \frac{5}{2}i\right) + \left(\frac{5}{3} + \frac{11}{3}i\right)$

26. $(1.6 + 3.2i) + (-5.8 + 4.3i)$

27. $\sqrt{-3} \cdot \sqrt{-8}$

28. $\sqrt{-5} \cdot \sqrt{-10}$

29. $\left(\sqrt{-10}\right)^2$

30. $\left(\sqrt{-75}\right)^3$

31. $(1 + i)(3 - 2i)$

32. $(9 - 4i)(1 - i)$

33. $(4 + 5i)(4 - 5i)$

34. $(6 + 7i)(6 - 7i)$

35. $7i(6 - 3i)$

36. $-8i(9 + 4i)$

37. $\left(\sqrt{5} - \sqrt{3}\,i\right)\left(\sqrt{5} + \sqrt{3}\,i\right)$

38. $\left(\sqrt{14} + \sqrt{10}\,i\right)\left(\sqrt{14} - \sqrt{10}\,i\right)$

39. $\left(2 - \sqrt{-8}\right)\left(8 + \sqrt{-6}\right)$

40. $\left(3 + \sqrt{-5}\right)\left(7 - \sqrt{-10}\right)$

**41.** $(4 + 5i)^2$

**42.** $(2 - 3i)^3$

**43.** $\dfrac{2 + i}{2 - i}$

**44.** $\dfrac{3}{1 - i}$

**45.** $\dfrac{5}{4 - 2i}$

**46.** $\dfrac{8 - 7i}{1 - 2i}$

**47.** $\dfrac{6 - 7i}{i}$

**48.** $\dfrac{8 + 20i}{2i}$

**49.** $\dfrac{1}{(2i)^3}$

**50.** $\dfrac{1}{(4 - 5i)^2}$

**51.** $\dfrac{5}{(1 + i)^3}$

**52.** $\dfrac{(2 - 3i)(5i)}{2 + 3i}$

**53.** $\dfrac{(21 - 7i)(4 + 3i)}{2 - 5i}$

**54.** $\dfrac{1}{i^3}$

**55.** $(2 + 3i)^2 + (2 - 3i)^2$    **56.** $(1 - 2i)^2 - (1 + 2i)^2$

*Error Analysis*    In Exercises 57 and 58, a student has handed in the following problem. Find the error(s) and discuss how to explain each error to the student.

**57.** Write $\dfrac{5}{3 - 2i}$ in standard form.

$$\dfrac{5}{3 - 2i} \cdot \dfrac{3 + 2i}{3 + 2i} \; \dfrac{15 + 10i}{9 - 4} = 3 + 2i$$

**58.** Multiply $\left(\sqrt{-4} + 3\right)\left(i - \sqrt{-3}\right)$.

$$\left(\sqrt{-4} + 3\right)\left(i - \sqrt{-3}\right)$$
$$= i\sqrt{-4} - \sqrt{-4}\sqrt{-3} + 3i - 3\sqrt{-3}$$
$$= -2i - \sqrt{12} + 3i - 3i\sqrt{3}$$
$$= \left(1 - 3\sqrt{3}\right)i - 2\sqrt{3}$$

In Exercises 59–66, solve the equation.

**59.** $x^2 - 2x + 2 = 0$

**60.** $x^2 + 6x + 10 = 0$

**61.** $4x^2 + 16x + 17 = 0$

**62.** $9x^2 - 6x + 37 = 0$

**63.** $4x^2 + 16x + 15 = 0$

**64.** $9x^2 - 6x + 35 = 0$

**65.** $16t^2 - 4t + 3 = 0$

**66.** $5s^2 + 6s + 3 = 0$

In Exercises 67–72, plot the complex number.

**67.** $-2 + i$

**68.** $i$

**69.** $3$

**70.** $-2 - 3i$

**71.** $1 - 2i$

**72.** $-2i$

In Exercises 73–78, decide whether the number is in the Mandelbrot Set. Explain your reasoning.

**73.** $c = 0$

**74.** $c = 2$

**75.** $c = \frac{1}{2}i$

**76.** $c = -i$

**77.** $c = 1$

**78.** $c = -1$

## Math Matters • Acceptance of Imaginary Numbers

Imaginary numbers were given the name "imaginary" because many mathematicians had a difficult time accepting such numbers. The following excerpt, written by the famous mathematician Carl Gauss (1777–1855), shows that, even in the early 1800s, some people were hesitant to accept the use of imaginary numbers.

"Our general arithmetic, which far surpasses the extent of the geometry of the ancients, is entirely the creation of modern times. Starting with the notion of whole numbers, it has gradually enlarged its domain. To whole numbers have been added fractions; to rational numbers have been added irrational; to positive numbers have been added negative; and to real numbers have been added imaginary numbers.

This advance, however, has always been made with timid steps. The early algebraists called negative solutions of equations false solutions (and this is indeed the case when the problem to which they relate has been stated in such a way that negative solutions have no meaning). The reality of negative numbers is sufficiently justified since there are many cases where they have meaningful interpretations. This has long been admitted, but the imaginary numbers (formerly and occasionally now called impossible numbers) are still rather tolerated than fully accepted."

## 4.6     The Fundamental Theorem of Algebra

### Objectives

- Use the Fundamental Theorem of Algebra and the Linear Factorization Theorem to write a polynomial as the product of linear factors.
- Find a polynomial with integer coefficients whose zeros are given.
- Factor a polynomial over the real and complex numbers.
- Find all real and complex zeros of a polynomial function.

The Granger Collection

**Jean Le Rond d'Alembert**

**(1717–1783)**

Jean Le Rond d'Alembert worked independently of Carl Gauss trying to prove the Fundamental Theorem of Algebra. His efforts were such that in France, the Fundamental Theorem of Algebra is frequently known as the theorem of d'Alembert.

## The Fundamental Theorem of Algebra

You have been using the fact than an $n$th-degree polynomial can have at most $n$ real zeros. In the complex number system, this statement can be improved. That is, in the complex number system, every $n$th-degree polynomial function has *precisely* $n$ zeros. This important result is derived from the **Fundamental Theorem of Algebra,** first proved by the famous German mathematician Carl Friedrich Gauss (1777–1855).

▶ **The Fundamental Theorem of Algebra**

If $f(x)$ is a polynomial of degree $n$, where $n > 0$, then $f$ has at least one zero in the complex number system.

Using the Fundamental Theorem of Algebra and the equivalence of zeros and factors, you obtain the following theorem.

▶ **Linear Factorization Theorem**

If $f(x)$ is a polynomial of degree $n$

$$f(x) = a_n x^n + a_{n-1} x^{n-1} + \cdots + a_1 x + a_0$$

where $n > 0$, then $f$ has precisely $n$ linear factors

$$f(x) = a_n(x - c_1)(x - c_2) \cdots (x - c_n)$$

where $c_1, c_2, \ldots, c_n$ are complex numbers and $a_n$ is the leading coefficient of $f(x)$.

Note that neither the Fundamental Theorem of Algebra nor the Linear Factorization Theorem tells you *how* to find the zeros or factors of a polynomial. Such theorems are called **existence theorems.** To find the zeros of a polynomial function, you still rely on the techniques developed in the earlier parts of the text.

Remember that the $n$ zeros of a polynomial function can be real or complex, and they may be repeated. Example 1 illustrates several cases.

## CHAPTER SUMMARY

After studying this chapter, you should have acquired the following skills.
These skills are keyed to the Review Exercises that begin on page 292.
Answers to odd-numbered Review Exercises are given in the back of the book.

**4.1**
- Sketch the graph of a quadratic function and identify its vertex and intercepts.    Review Exercises 1–4
- Find a quadratic function given its vertex and a point on the graph.    Review Exercises 5, 6
- Construct and use a quadratic model to solve an application problem.    Review Exercises 7–12

**4.2**
- Apply the Leading Coefficient Test to determine right- and left-hand behavior of the graph of a polynomial function.    Review Exercises 13–16
- Find the real zeros of a polynomial function.    Review Exercises 17–20

**4.3**
- Divide one polynomial by a second polynomial using long division.    Review Exercises 21, 22
- Simplify a rational expression using long division.    Review Exercises 23, 24
- Use synthetic division to divide two polynomials.    Review Exercises 25–28, 31, 32
- Use the Remainder Theorem and synthetic division to evaluate a polynomial.    Review Exercises 29, 30

**4.4**
- Find all possible rational zeros of a function using the Rational Zero Test.    Review Exercises 33, 34
- Approximate the real zeros of a polynomial function using the Intermediate Value Theorem.    Review Exercises 35, 36
- Find and approximate the real zeros of a polynomial function.    Review Exercises 37–44
- Apply techniques for approximating real zeros to solve an application problem.    Review Exercises 45, 46

**4.5**
- Find the complex conjugate of a complex number.    Review Exercises 47–50
- Perform operations with complex numbers and write the results in standard form.    Review Exercises 51–66
- Solve a polynomial equation.    Review Exercises 67–70
- Plot a complex number in the complex plane.    Review Exercises 71, 72

**4.6**
- Use the Fundamental Theorem of Algebra and the Linear Factorization Theorem to write a polynomial as the product of linear factors.    Review Exercises 73–78
- Find a polynomial with integer coefficients whose zeros are given.    Review Exercises 79, 80
- Factor a polynomial over the real and complex numbers.    Review Exercises 81, 82
- Find all real and complex zeros of a polynomial function.    Review Exercises 83–86

**4.7**
- Find the domain of a rational function.    Review Exercises 87–90
- Find the vertical and horizontal asymptotes of the graph of a rational function.    Review Exercises 87–90
- Sketch the graph of a rational function.    Review Exercises 91–94
- Use a rational function model to solve an application problem.    Review Exercises 95–100

## REVIEW EXERCISES

In Exercises 1–4, sketch the graph of the quadratic function. Identify the vertex and intercepts.

**1.** $f(x) = (x + 3)^2 - 2$

**2.** $g(x) = -x^2 + 6x + 8$

**3.** $h(x) = 3x^2 - 12x + 11$

**4.** $f(x) = \frac{1}{3}(x^2 + 5x - 4)$

In Exercises 5 and 6, find the quadratic function that has the indicated vertex and whose graph passes through the given point.

**5.** Vertex: $(-4, -2)$; point: $(-3, -5)$

**6.** Vertex: $(2, 5)$; point: $(4, 7)$

**7.** *Maximum Area*   The perimeter of a rectangle is 200 feet. Let $x$ represent the width of the rectangle and write a quadratic function that expresses the area of the rectangle in terms of its width. Of all possible rectangles with perimeters of 200 feet, what are the measurements of the one that has the greatest area?

**8.** *Maximum Revenue*   Find the number of units that produces a maximum revenue $R$ for

$$R = 800x - 0.01x^2$$

where $R$ is the total revenue in dollars and $x$ is the number of units produced.

**9.** *Minimum Cost*   A manufacturer has daily production costs of

$$C = 20{,}000 - 40x + 0.055x^2$$

where $C$ is the total cost in dollars and $x$ is the number of units produced.

(a) Use a graphing utility to graph the cost function.

(b) Graphically estimate the number of units that should be produced to yield a minimum cost.

(c) Explain how to confirm the result of part (b) algebraically.

**10.** *Maximum Profit*   The profit for a company is given by

$$P = -0.0001x^2 + 130x - 300{,}000$$

where $P$ is the profit (in dollars) and $x$ is the number of units produced.

(a) Use a graphing utility to graph the profit function.

(b) Graphically estimate the number of units that should be produced to yield a maximum profit.

(c) Explain how to confirm the result of part (b) algebraically.

**11.** *Regression Problem*   Let $x$ be the angle (in degrees) at which a baseball is hit with a 30-hertz backspin at an initial speed of 40 meters per second and let $d(x)$ be the distance (in meters) the ball travels. The following table gives the distances for the different angles at which the ball is hit.   (Source: *The Physics of Sports*)

| $x$ | 10° | 15° | 30° | 36° | 42° | 43° |
|---|---|---|---|---|---|---|
| $d(x)$ | 61.2 | 83.0 | 130.4 | 139.4 | 143.2 | 143.3 |

| $x$ | 44° | 45° | 48° | 54° | 60° |
|---|---|---|---|---|---|
| $d(x)$ | 142.8 | 142.7 | 140.7 | 132.8 | 119.7 |

(a) Use a graphing utility to make a scatter plot using the data in the table.

(b) Use the regression capabilities of a graphing utility to find a quadratic model for $d(x)$.

(c) Use a graphing utility to graph your model for $d(x)$ and the scatter plot for the data.

(d) Use the model to approximate the angle that will result in a maximum distance.

**12.** *Regression Problem*   Let $x$ be the crude protein intake (in kilograms per day) for pigs and let $s(x)$ be the shoulder weight (in kilograms). The following table gives the shoulder weight for pigs given the crude protein intake.   (Source: *Livestock Research for Rural Development*)

| $x$ | 0.195 | 0.238 | 0.297 | 0.341 | 0.401 | 0.427 |
|---|---|---|---|---|---|---|
| $s(x)$ | 8.13 | 8.74 | 9.68 | 9.69 | 9.81 | 8.99 |

(a) Use a graphing utility to make a scatter plot using the data in the table.

(b) Use the regression capabilities of a graphing utility to find a quadratic model for $s(x)$.

(c) Use a graphing utility to graph your model for $s(x)$ and the scatter plot for the data.

(d) Use the model to approximate the amount of crude protein that will result in a maximum shoulder weight.

In Exercises 13–16, determine the right- and left-hand behavior of the graph of the function.

**13.** $f(x) = \frac{1}{2}x^3 + 2x$

**14.** $f(x) = 5 - 3x^2 + 4x^4 - x^6$

**15.** $f(x) = -x^5 + 3x^2 + 11$

**16.** $f(x) = \frac{3}{4}(x^4 + 3x^2 + 2)$

In Exercises 17–20, find all real zeros of the function.

**17.** $f(x) = 25 - x^2$

**18.** $f(x) = x^4 - 6x^2 + 8$

**19.** $f(x) = x^3 - 7x^2 + 10x$

**20.** $f(x) = x^3 - 6x^2 - 3x + 18$

In Exercises 21 and 22, divide by long division.

| Dividend | Divisor |
|----------|---------|
| **21.** $2x^3 - 5x^2 - x$ | $2x + 1$ |
| **22.** $x^4 - 5x^3 + 10x^2 - 12$ | $x^2 - 2x + 4$ |

In Exercises 23 and 24, simplify the expression.

**23.** $\dfrac{x^3 + 9x^2 + 2x - 48}{x - 2}$

**24.** $\dfrac{x^4 + 5x^3 - 20x - 16}{x^2 - 4}$

In Exercises 25 and 26, divide by synthetic division.

| Dividend | Divisor |
|----------|---------|
| **25.** $x^3 - 6x + 9$ | $x + 3$ |
| **26.** $x^5 - x^4 + x^3 - 13x^2 + x + 6$ | $x + 2$ |

In Exercises 27 and 28, use synthetic division to show that $x$ is a solution of the equation. Then factor the polynomial completely.

**27.** $x^3 - 4x^2 - 11x + 30 = 0, \quad x = 5$

**28.** $3x^3 + 23x^2 + 37x - 15 = 0, \quad x = \frac{1}{3}$

In Exercises 29 and 30, use synthetic division to evaluate the function.

**29.** $f(x) = 6 + 2x^2 - 3x^3$

   (a) $f(2)$     (b) $f(-1)$

**30.** $f(x) = 2x^4 + 3x^3 + 6$

   (a) $f\left(\frac{1}{2}\right)$     (b) $f(-1)$

**31.** *Measurements of a Room*    A rectangular room has a volume of

$$x^3 + 13x^2 + 50x + 56$$

cubic feet. The height of the room is $(x + 2)$ feet. Find the number of square feet of floor space in the room.

**32.** *Profit*    The profit for a product is given by

$$P = -130x^3 + 6500x^2 - 200,000, \quad 0 \le x \le 50$$

where $P$ is the profit (in dollars) and $x$ is the advertising expense (in tens of thousands of dollars). For this product, the advertising expense was \$400,000 ($x = 40$), and the profit was \$1,880,000. Use a graphing utility to graph the function and use the result to find another advertising expense that would have produced the same profit.

In Exercises 33 and 34, use the Rational Zero Test to list all possible rational zeros of $f$. Verify using a graphing utility.

**33.** $f(x) = -4x^3 + 8x^2 - 3x + 15$

**34.** $f(x) = 3x^4 + 4x^3 - 5x^2 + 10x - 8$

In Exercises 35 and 36, use the Intermediate Value Theorem to approximate the zero of $f(x)$ in the interval $[a, b]$. Give your approximation to the nearest tenth.

**35.** $f(x) = x^3 - 4x + 3, \quad [-3, -2]$

**36.** $f(x) = x^5 + 5x^2 + x - 1, \quad [0, 1]$

In Exercises 37–42, find and approximate all real zeros of the function.

**37.** $f(x) = x^3 + 2x^2 - 5x - 6$

**38.** $g(x) = 2x^3 - 15x^2 + 24x + 16$

**39.** $h(x) = 3x^4 - 27x^2 + 60$

**40.** $f(x) = x^5 - 4x^3 + 3x$

**41.** $C(x) = 3x^4 + 3x^3 - 7x^2 - x + 2$

**42.** $p(x) = x^4 - x^3 - 2x - 4$

In Exercises 43 and 44, approximate the real zeros of $f$ using a graphing utility. Use an accuracy of 0.001.

**43.** $f(x) = 5x^3 - 11x - 3$

**44.** $f(x) = 2x^4 - 9x^3 - 5x^2 + 10x + 12$

**45.** *Advertising Costs*    A company that manufactures motorcycles estimates that the profit from selling the top-of-the-line model is given by

$$P = -35x^3 + 2000x^2 - 27{,}500, \quad 10 \le x \le 55$$

where $P$ is the profit (in dollars) and $x$ is the advertising expense (in tens of thousands of dollars). According to this model, how much money should the company spend on advertising to obtain a profit of $538,000?

**46.** *Age of the Groom*    The average age of the groom in a marriage in the United States for a given age of the bride can be approximated by the model

$$y = -0.00428x^2 + 1.442x - 3.136, \quad 20 \le x \le 55$$

where $y$ is the groom's age and $x$ is the bride's age. For what age of the bride is the average age of the groom 30? (Source: U.S. National Center for Health Statistics)

In Exercises 47–50, write the complex number in standard form and find its complex conjugate.

**47.** $\sqrt{-32}$

**48.** $12$

**49.** $-3 + \sqrt{-16}$

**50.** $2 - \sqrt{-18}$

In Exercises 51–66, perform the indicated operation and write the result in standard form.

**51.** $(7 - 4i) + (-2 + 5i)$

**52.** $(14 + 6i) - (-1 - 2i)$

**53.** $\left(1 + \sqrt{-12}\right)\left(5 - \sqrt{-3}\right)$

**54.** $\left(3 - \sqrt{-4}\right)\left(4 - \sqrt{-49}\right)$

**55.** $(5 + 8i)(5 - 8i)$

**56.** $\left(\frac{1}{2} + \frac{3}{4}i\right)\left(\frac{1}{2} - \frac{3}{4}i\right)$

**57.** $-2i(4 - 5i)$

**58.** $-3(-2 + 4i)$

**59.** $(3 + 4i)^2$

**60.** $(2 - 5i)^2$

**61.** $\dfrac{8 - i}{2 + i}$

**62.** $\dfrac{3 - 4i}{1 - 5i}$

**63.** $\dfrac{4 - 3i}{i}$

**64.** $\dfrac{2}{(1 + i)^2}$

**65.** $(3 + 2i)^2 + (3 - 2i)^2$

**66.** $(1 + i)^2 - (1 - i)^2$

In Exercises 67–70, solve the equation.

**67.** $2x^2 - x + 3 = 0$

**68.** $3x^2 + 6x + 11 = 0$

**69.** $4x^2 + 11x + 3 = 0$

**70.** $9x^2 - 2x + 5 = 0$

In Exercises 71 and 72, plot the complex number.

**71.** $-3 + 2i$

**72.** $-1 - 4i$

In Exercises 73–78, write the polynomial as a product of linear factors.

**73.** $f(x) = x^4 - 81$

**74.** $h(x) = 2x^3 - 5x^2 + 4x - 10$

**75.** $f(t) = t^3 + 5t^2 + 3t + 15$

**76.** $h(x) = x^4 + 17x^2 + 16$

**77.** $g(x) = 4x^3 - 8x^2 + 9x - 18$

**78.** $f(x) = x^5 - 2x^4 + x^3 - x^2 + 2x - 1$

In Exercises 79 and 80, find a polynomial with integer coefficients that has the given zeros.

**79.** $3, 4i, -4i$

**80.** $2, -3, 1 - 2i, 1 + 2i$

In Exercises 81 and 82, write the polynomial (a) as the product of factors that are irreducible over the rationals, (b) as the product of linear and quadratic factors that are irreducible over the reals, and (c) in completely factored form.

**81.** $f(x) = x^4 + 5x^2 - 24$

**82.** $f(x) = x^4 - 3x^3 - 11x^2 + 15x + 30$

In Exercises 83–86, use the given zero of $f$ to find all the zeros of $f$.

**83.** $f(x) = 4x^3 - x^2 + 64x - 16, \quad -4i$

**84.** $f(x) = 50 - 75x + 2x^2 - 3x^3, \quad 5i$

**85.** $f(x) = x^4 + 7x^3 + 24x^2 + 58x + 40, \quad -1 + 3i$

**86.** $f(x) = x^4 + 4x^3 + 8x^2 + 4x + 7, \quad -2 - \sqrt{3}\,i$

In Exercises 87–90, find the domain of the function and identify any horizontal or vertical asymptotes.

**87.** $f(x) = \dfrac{5}{x - 6}$

**88.** $f(x) = \dfrac{2x^2 + 5x - 3}{x^2 + 2}$

**89.** $f(x) = \dfrac{x^2}{x^2 - 4}$

**90.** $f(x) = \dfrac{2x}{x^2 - 2x - 8}$

In Exercises 91–94, sketch the graph of the rational function. As sketching aids, check for intercepts, symmetry, vertical asymptotes, and horizontal asymptotes.

**91.** $P(x) = \dfrac{3 - x}{x + 2}$

**92.** $f(x) = \dfrac{4}{(x - 1)^2}$

**93.** $g(x) = \dfrac{1}{x + 3} + 2$

**94.** $h(x) = \dfrac{-3x}{2x^2 + 3x - 5}$

**95.** *Average Cost*   The cost in dollars of producing $x$ units is $C = 100{,}000 + 0.9x$, and so the average cost per unit is

$$\overline{C} = \frac{C}{x} = \frac{100{,}000 + 0.9x}{x}, \quad x > 0.$$

Sketch the graph of this average cost function and find the average cost of producing $x = 1000$, $x = 10{,}000$, and $x = 100{,}000$ units.

**96.** *Population of Fish*   The Parks and Wildlife Commission introduces 50,000 game fish into a large lake. The population of the fish (in thousands) is

$$N = \frac{20(4 + 3t)}{1 + 0.05t}, \quad t \geq 0$$

where $t$ is time in years.

(a) Find the population when $t = 5$, 10, and 25.

(b) What is the limiting number of this population of fish in the lake as time progresses?

**97.** *Human Memory Model*   Consider the learning curve

$$P = \frac{0.5 + 0.9(n - 1)}{1 + 0.9(n - 1)}, \quad n > 0$$

where $P$ is the percent of correct responses after $n$ trials. Complete the table for this model.

| $n$ | 1 | 2 | 3 | 4 | 5 | 6 | 7 | 8 | 9 | 10 |
|-----|---|---|---|---|---|---|---|---|---|----|
| $P$ |   |   |   |   |   |   |   |   |   |    |

**98.** *Smokestack Emission*   The cost in dollars of removing $p$ percent of the air pollutants in the stack emission of a utility company that burns coal to generate electricity is

$$C = \frac{95{,}000p}{100 - p}, \quad 0 \leq p < 100.$$

(a) Find the cost of removing 25%.

(b) Find the cost of removing 60%.

(c) Find the cost of removing 99%.

(d) According to this model, would it be possible to remove 100% of the pollutants? Explain.

**99.** *Average Recycling Cost*   The cost in dollars of recycling a waste product is

$$C = 450{,}000 + 6x, \quad x > 0$$

where $x$ is the number of pounds of waste. The average recycling cost is

$$\overline{C} = \frac{C}{x} = \frac{450{,}000 + 6x}{x}, \quad x > 0.$$

(a) Sketch the graph of $\overline{C}$.

(b) Find the average cost of recycling for $x = 1000$ pounds, $x = 10{,}000$ pounds, and $x = 100{,}000$ pounds. What can you conclude?

**100.** *Sound Recordings*   The total dollar values $V$ (in millions) for sound recordings shipped at manufacturers' suggested list prices in the United States from 1991 to 2000 are given in the table.   (Source: The Recording Industry Association of America)

| Year | 1991 | 1992 | 1993 | 1994 | 1995 |
|------|------|------|------|------|------|
| $V$ | 7834 | 9024 | 10,047 | 12,068 | 12,320 |

| Year | 1996 | 1997 | 1998 | 1999 | 2000 |
|------|------|------|------|------|------|
| $V$ | 12,534 | 12,237 | 13,724 | 14,585 | 14,323 |

(a) Use a graphing utility to make a scatter plot of the data. Let $t = 1$ correspond to 1991.

(b) Use the regression capabilities of a graphing utility to find a quadratic model, a cubic model, and a quartic model for the data.

(c) Graph each model separately with the data in the same viewing window. How well does each model fit the data?

(d) Use all three models to estimate the total values of recordings shipped in 2001 and 2002. Does each estimate seem reasonable?

## Chapter Test

Take this test as you would take a test in class. After you are done, check your work against the answers given in the back of the book.

1. Graph $f(x) = -(x + 1)^2 + 2$. Indicate the $x$- and $y$-intercepts (if any), the vertex, and the line of symmetry.

2. Describe the left- and right-hand behavior of the graph of $f(x)$.
   (a) $f(x) = 12x^3 - 5x^2 - 49x + 15$
   (b) $f(x) = 5x^4 - 3x^3 + 2x^2 + 11x + 12$

3. List all the possible rational zeros of $f(x) = 2x^3 - 5x^2 + 2x - 5$. Use synthetic division to show that $x = \frac{5}{2}$ is a zero of $f(x)$. Using this result, completely factor the polynomial.

4. Simplify: $\dfrac{x^4 + 4x^3 - 19x^2 - 106x - 120}{x^2 - 3x - 10}$.

In Exercises 5–12, perform the indicated operation and write the result in standard form.

5. $(12 + 3i) + (4 - 6i)$            6. $(10 - 2i) - (3 + 7i)$

7. $\left(5 + \sqrt{-12}\right)\left(3 - \sqrt{-12}\right)$     8. $(4 + 3i)(2 - 5i)$

9. $(2 - 3i)^2$                   10. $(5 + 2i)^2$

11. $\dfrac{1 + i}{1 - i}$                   12. $\dfrac{5 - 2i}{i}$

In Exercises 13 and 14, solve the quadratic equation.

13. $x^2 + 5x + 7 = 0$            14. $2x^2 - 5x + 11 = 0$

15. Find all zeros of $f(x) = x^3 + 2x^2 + 5x + 10$, given that $\sqrt{5}\,i$ is a zero.

16. Find a polynomial with integer coefficients that has $2, 5, 3i$, and $-3i$ as zeros.

17. Graph $f(x) = \dfrac{2x}{x + 1}$. Indicate all $x$- and $y$-intercepts and any asymptotes. What is the domain of $f(x)$?

18. The profit for a company is $P(x) = 300 + 10x - 0.002x^2$, where $x$ is the number of units produced. What production level will yield a maximum profit?

19. The average cost (in dollars) of recycling a waste product $x$ (in pounds) is
$$\overline{C}(x) = \dfrac{450{,}000 + 5x}{x}, x > 0.$$ Find the average cost of recycling $x = 10{,}000$ pounds, $x = 100{,}000$ pounds, and $x = 1{,}000{,}000$ pounds. What can you conclude?

20. A company estimates that the profit from selling its product is
$$P = -11x^3 + 900x^2 - 50{,}000, \quad x \ge 0$$
where $P$ is the profit (in dollars) and $x$ is the advertising expense (in tens of thousands of dollars). How much should the company spend on advertising to obtain a profit of $222{,}000? Explain your reasoning.

# 5

# Exponential and Logarithmic Functions

When a warm object is placed in a cool room, its temperature will gradually change to the temperature of the room. (The same is true for a cool object placed in a warm room.)

According to Newton's Law of Cooling, the rate at which the object's temperature $T$ changes is proportional to the difference between its temperature and the temperature of the room $L$. This statement can be translated as

$$T = L + Ce^{kt}$$

where $k$ is a constant and $L + C$ is the original temperature of the object.

The table and the scatter plot below show the temperature $T$ (in degrees Fahrenheit) for several times $t$ (in minutes) for 300 ml of hot water in a ceramic cup.

5.1 **Exponential Functions**

5.2 **Logarithmic Functions**

5.3 **Properties of Logarithms**

5.4 **Solving Exponential and Logarithmic Equations**

5.5 **Exponential and Logarithmic Models**

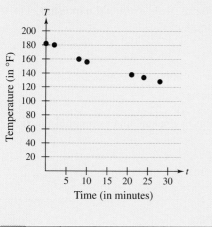

| $t$ | 0 | 2 | 8 | 10 | 21 | 24 | 28 |
|-----|-----|-----|-----|-----|-----|-----|-----|
| $T$ | 182 | 180 | 160 | 156 | 138 | 134 | 128 |

The chapter project related to this information is on page 351.

**297**

## 5.1    Exponential Functions

### Objectives

- Evaluate an exponential expression.
- Sketch the graph of an exponential function.
- Use the compound interest formula.
- Use an exponential model to solve an application problem.

## Exponential Functions

So far, this text has dealt only with **algebraic functions,** which include polynomial functions and rational functions. In this chapter you will study two types of nonalgebraic functions— *exponential* functions and *logarithmic* functions. These functions are examples of **transcendental functions.**

> ▶ Definition of Exponential Function
>
> The **exponential function** $f$ **with base** $a$ is denoted by
>
> $$f(x) = a^x$$
>
> where $a > 0$, $a \neq 1$, and $x$ is any real number.

The base $a = 1$ is excluded from the definition of exponential function because it yields $f(x) = 1^x = 1$, which is a constant function.

You already know how to evaluate $a^x$ for integer and rational values of $x$. For example, you know that $4^3 = 64$ and $4^{1/2} = 2$. However, to evaluate $4^x$ for any real number $x$, you need to interpret forms with *irrational* exponents. For the purposes of this text, it is sufficient to think of

$$a^{\sqrt{2}} \qquad (\text{where } \sqrt{2} \approx 1.414214)$$

as that value having the successively closer approximations

$$a^{1.4}, a^{1.41}, a^{1.414}, a^{1.4142}, a^{1.41421}, a^{1.414214}, \ldots$$

Example 1 shows how to use a calculator to evaluate an exponential function.

| Example 1 | Evaluating Exponential Expressions |  |

**Scientific Calculator**

| Number | Keystrokes | Display |
|--------|-----------|---------|
| $2^{-\pi}$ | 2 $\boxed{y^x}$ $\pi$ $\boxed{+/-}$ $\boxed{=}$ | 0.1133147 |

**Graphing Calculator**

| Number | Keystrokes | Display |
|--------|-----------|---------|
| $2^{-\pi}$ | 2 $\boxed{\wedge}$ $\boxed{(-)}$ $\pi$ $\boxed{\text{ENTER}}$ | 0.1133147 |

## Graphs of Exponential Functions

The graphs of all exponential functions have similar characteristics, as shown in Examples 2, 3, and 4.

**Example 2**    Graphs of $y = a^x$

In the same coordinate plane, sketch the graphs of the following functions.

**a.** $f(x) = 2^x$    **b.** $g(x) = 4^x$

**Solution**

The table below lists some values for each function, and Figure 5.1 shows their graphs. Note that both graphs are increasing. Moreover, the graph of $g(x) = 4^x$ is increasing more rapidly than the graph of $f(x) = 2^x$.

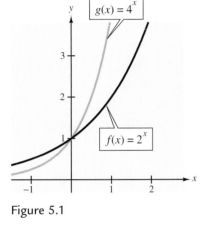

Figure 5.1

| $x$ | $-2$ | $-1$ | 0 | 1 | 2 | 3 |
|-----|------|------|---|---|---|---|
| $2^x$ | $\frac{1}{4}$ | $\frac{1}{2}$ | 1 | 2 | 4 | 8 |
| $4^x$ | $\frac{1}{16}$ | $\frac{1}{4}$ | 1 | 4 | 16 | 64 |

**Example 3**    Graphs of $y = a^{-x}$

In the same coordinate plane, sketch the graphs of the following functions.

**a.** $F(x) = 2^{-x}$    **b.** $G(x) = 4^{-x}$

**Solution**

The table below lists some values for each function, and Figure 5.2 shows their graphs. Note that both graphs are decreasing. Moreover, the graph of $G(x) = 4^{-x}$ is decreasing more rapidly than the graph of $F(x) = 2^{-x}$.

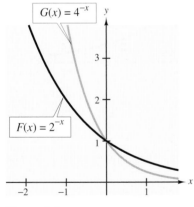

Figure 5.2

| $x$ | $-3$ | $-2$ | $-1$ | 0 | 1 | 2 |
|-----|------|------|------|---|---|---|
| $2^{-x}$ | 8 | 4 | 2 | 1 | $\frac{1}{2}$ | $\frac{1}{4}$ |
| $4^{-x}$ | 64 | 16 | 4 | 1 | $\frac{1}{4}$ | $\frac{1}{16}$ |

The tables in Examples 2 and 3 were evaluated by hand. You could, of course, use the table feature of a graphing utility to construct tables with even more values.

**·DISCOVERY·**

Use a graphing utility to graph

$y = a^x$

for $a = 2, 2.4,$ and $2.8$. What do you observe? What can you conclude about the value of $a$ and the behavior of the function?
Graph

$y = a^{-x}$

for $a = 2, 2.4,$ and $2.8$. What do you observe? What can you conclude about the value of $a$ and the behavior of the function?

Comparing the functions in Examples 2 and 3, observe that

$$F(x) = 2^{-x} = f(-x) \quad \text{and} \quad G(x) = 4^{-x} = g(-x).$$

Consequently, the graph of $F$ is a reflection (in the $y$-axis) of the graph of $f$. The graphs of $G$ and $g$ have the same relationship.

The graphs in Figures 5.1 and 5.2 are typical of the exponential functions $a^x$ and $a^{-x}$. They have one $y$-intercept and one horizontal asymptote (the $x$-axis), and they are continuous. The basic characteristics of these exponential functions are summarized in Figure 5.3.

Graph of $y = a^x, a > 1$

- Domain: $(-\infty, \infty)$
- Range: $(0, \infty)$
- Intercept: $(0, 1)$
- Increasing
- $x$-axis is a horizontal asymptote
  $(a^x \to 0$ as $x \to -\infty)$
- Continuous

Graph of $y = a^{-x}, a > 1$

- Domain: $(-\infty, \infty)$
- Range: $(0, \infty)$
- Intercept: $(0, 1)$
- Decreasing
- $x$-axis is a horizontal asymptote
  $(a^{-x} \to 0$ as $x \to \infty)$
- Continuous
- Reflection of graph of $y = a^x$
  about $y$-axis

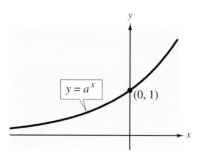 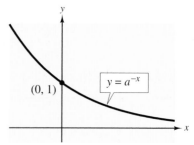

Figure 5.3

**·TECHNOLOGY·**

Try using a graphing utility to graph several exponential functions. For instance, the graph of $y = 2^x$ is shown below.

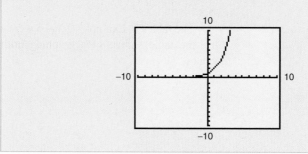

In the following example, notice how the graph of $y = a^x$ can be used to sketch the graphs of functions of the form $f(x) = b \pm a^{x+c}$.

**Example 4**    Sketching Graphs of Exponential Functions

Each of the following graphs is a transformation of the graph of $f(x) = 3^x$, as shown in Figure 5.4.

**a.** Because $g(x) = 3^{x+1} = f(x + 1)$, the graph of g can be obtained by shifting the graph of f one unit to the left.

**b.** Because $h(x) = 3^x - 2 = f(x) - 2$, the graph of h can be obtained by shifting the graph of f down two units.

**c.** Because $k(x) = -3^x = -f(x)$, the graph of k can be obtained by reflecting the graph of f in the x-axis.

**d.** Because $j(x) = 3^{-x} = f(-x)$, the graph of j can be obtained by reflecting the graph of f in the y-axis.

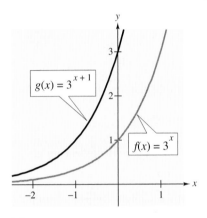

(a)

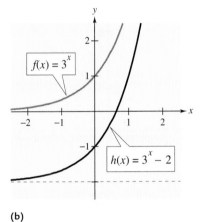

(b)

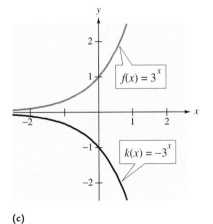

(c)

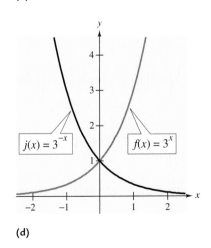

(d)

Figure 5.4

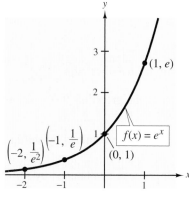

Figure 5.5

## The Natural Base $e$

In many applications, the most convenient choice for a base is the irrational number

$$e \approx 2.718281828 \ldots$$

called the **natural base.** The function $f(x) = e^x$ is called the **natural exponential function.** Its graph is shown in Figure 5.5. Be sure you see that for the exponential function $f(x) = e^x$, $e$ is the constant $2.718281828 \ldots$, whereas $x$ is the variable.

**Example 5** Evaluating the Natural Exponential Function

### Scientific Calculator

| Number | Keystrokes | Display |
|--------|------------|---------|
| $e^2$ | 2 $e^x$ $=$ | 7.3890561 |
| $e^{-1}$ | 1 $+/-$ $e^x$ $=$ | 0.3678794 |

### Graphing Calculator

| Number | Keystrokes | Display |
|--------|------------|---------|
| $e^2$ | $e^x$ 2 ENTER | 7.3890561 |
| $e^{-1}$ | $e^x$ $(-)$ 1 ENTER | 0.3678794 |

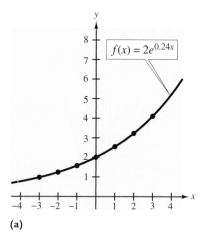

(a)

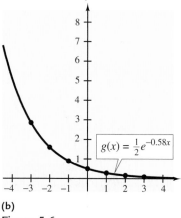

(b)

Figure 5.6

**Example 6** Graphing Natural Exponential Functions

Sketch the graphs of the following natural exponential functions.

**a.** $f(x) = 2e^{0.24x}$

**b.** $g(x) = \frac{1}{2}e^{-0.58x}$

### Solution

To sketch these two graphs, you can use a calculator to plot several points on each graph, as shown in the table. Then, connect the points with smooth curves, as shown in Figure 5.6. Note that the graph in part (a) is increasing whereas the graph in part (b) is decreasing.

| $x$ | $-3$ | $-2$ | $-1$ | 0 | 1 | 2 | 3 |
|-----|------|------|------|---|---|---|---|
| $f(x) = 2e^{0.24x}$ | 0.974 | 1.238 | 1.573 | 2.000 | 2.543 | 3.232 | 4.109 |
| $g(x) = \frac{1}{2}e^{-0.58x}$ | 2.849 | 1.595 | 0.893 | 0.500 | 0.280 | 0.157 | 0.088 |

## Compound Interest

One of the most familiar examples of exponential growth is that of an investment earning *continuously compounded interest*. The formula for the balance in an account that is compounded $n$ times per year is $A = P(1 + r/n)^{nt}$, where $A$ is the balance in the account, $P$ is the initial deposit, $r$ is the annual interest rate (in decimal form), and $t$ is the number of years. Using exponential functions, you will *develop* this formula and show how it leads to continuous compounding.

Suppose a principal $P$ is invested at an annual percentage rate $r$, compounded once a year. If the interest is added to the principal at the end of the year, the balance is

$$P_1 = P + Pr = P(1 + r).$$

This pattern of multiplying the previous principal by $1 + r$ is then repeated each successive year, as shown below.

| Year | Balance After Each Compounding |
|------|-------------------------------|
| 0 | $P = P$ |
| 1 | $P_1 = P(1 + r)$ |
| 2 | $P_2 = P_1(1 + r) = P(1 + r)(1 + r) = P(1 + r)^2$ |
| 3 | $P_3 = P_2(1 + r) = P(1 + r)^2(1 + r) = P(1 + r)^3$ |
| $\vdots$ | |
| $t$ | $P_t = P(1 + r)^t$ |

To accommodate more frequent (quarterly, monthly, or daily) compounding of interest, let $n$ be the number of compoundings per year and $t$ the number of years. Then the rate per compounding is $r/n$ and the account balance after $t$ years is

$$A = P\left(1 + \frac{r}{n}\right)^{nt}. \qquad \text{Amount with } n \text{ compoundings}$$

If you let the number of compoundings $n$ increase without bound, the process approaches what is called **continuous compounding.** In the formula for $n$ compoundings per year, let $m = n/r$. This produces

$$A = P\left(1 + \frac{r}{n}\right)^{nt}$$

$$= P\left(1 + \frac{1}{m}\right)^{mrt}$$

$$= P\left[\left(1 + \frac{1}{m}\right)^m\right]^{rt}.$$

As $m$ increases without bound, it can be shown that $\left[1 + (1/m)\right]^m$ approaches $e$. From this, you can conclude that the formula for continuous compounding is $A = Pe^{rt}$.

> ▶ **Formulas for Compound Interest**
>
> After $t$ years, the balance $A$ in an account with principal $P$ and annual percentage rate $r$ (in decimal form) is given by the following formulas.
>
> **1.** For $n$ compoundings per year: $A = P\left(1 + \dfrac{r}{n}\right)^{nt}$
>
> **2.** For continuous compounding: $A = Pe^{rt}$

Be sure that the annual percentage rate is expressed in decimal form. For instance, 6% should be expressed as 0.06 when using compound interest formulas.

**Example 7** Compounding $n$ Times and Continuously

A total of $12,000 is invested at an annual rate of 9%. Find the balance after 5 years if it is compounded

**a.** quarterly.    **b.** continuously.

**Solution**

**a.** For quarterly compoundings, you have $n = 4$. Thus, in 5 years at 9%, the balance is

$$A = P\left(1 + \frac{r}{n}\right)^{nt} \qquad \text{Formula for compound interest}$$

$$= 12{,}000\left(1 + \frac{0.09}{4}\right)^{4(5)} \qquad \text{Substitute for } P, r, n, \text{ and } t.$$

$$= \$18{,}726.11. \qquad \text{Use a calculator.}$$

**b.** Compounding continuously, the balance is

$$A = Pe^{rt} \qquad \text{Formula for continuous compounding}$$

$$= 12{,}000e^{0.09(5)} \qquad \text{Substitute for } P, r, \text{ and } t.$$

$$= \$18{,}819.75. \qquad \text{Use a calculator.}$$

Note that continuous compounding yields

$$\$18{,}819.75 - \$18{,}726.11 = \$93.64$$

more than quarterly compounding. This is typical of the two types of compounding. That is, for a given principal, interest rate, and time, continuous compounding will always yield a larger balance than compounding $n$ times a year.

## Another Application

 Example 8    Radioactive Decay

In 1957, a nuclear reactor accident occurred in the Soviet Union. The explosion spread radioactive chemicals over thousands of square miles, and the government set aside a large region as a "permanent preserve." To see why this area was permanently declared off-limits to people, consider the following model.

$$P = 10e^{-0.00002845t}$$

This model represents the amount of plutonium that remains (from an initial amount of 10 pounds) after $t$ years. Sketch the graph of this function over the interval from $t = 0$ to $t = 100{,}000$. How much of the 10 pounds will remain after 100,000 years?

### Solution

The graph of this function is shown in Figure 5.7. Note from this graph that plutonium has a *half-life* of about 24,360 years. That is, after 24,360 years, *half* of the original amount will remain. After another 24,360 years, one-quarter of the original amount will remain, and so on. After 100,000 years, there will still be

$$P = 10e^{-0.00002845(100{,}000)} = 10e^{-2.845} \approx 0.58 \text{ pounds}$$

of the original amount of plutonium remaining.

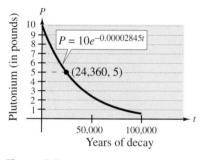

Figure 5.7

---

### Discussing the Concept    Exponential Growth

The pattern 3, 6, 9, 12, 15, . . . is given by $f(n) = 3n$ and is an example of **linear growth.** The pattern 3, 9, 27, 81, 243, . . . is given by $f(n) = 3^n$ and is an example of **exponential growth.** Explain the difference between these two types of growth. For each of the following patterns, indicate whether the pattern represents linear growth or exponential growth, and find a linear or exponential function that represents the pattern. Give several other examples of linear and exponential growth.

a. $\frac{1}{2}, \frac{1}{4}, \frac{1}{8}, \frac{1}{16}, \frac{1}{32}, \ldots$    b. 4, 8, 12, 16, 20, . . .

c. $\frac{2}{3}, \frac{4}{3}, 2, \frac{8}{3}, \frac{10}{3}, 4, \ldots$    d. 5, 25, 125, 625, . . .

# Warm Up

The following warm-up exercises involve skills that were covered in earlier sections. You will use these skills in the exercise set for this section.

In Exercises 1–10, use the properties of exponents to simplify the expression.

**1.** $5^{2x}(5^{-x})$

**2.** $3^{-x}(3^{3x})$

**3.** $\dfrac{4^{5x}}{4^{2x}}$

**4.** $\dfrac{10^{2x}}{10^{x}}$

**5.** $(4^{x})^{2}$

**6.** $(4^{2x})^{5}$

**7.** $\left(\dfrac{2^{x}}{3^{x}}\right)^{-1}$

**8.** $(4^{6x})^{1/2}$

**9.** $(2^{3x})^{-1/3}$

**10.** $(16^{x})^{1/4}$

## 5.1 Exercises

In Exercises 1–10, use a calculator to evaluate the expression. (Round the result to three decimal places.)

**1.** $(2.6)^{1.3}$

**2.** $(1.07)^{50}$

**3.** $100(1.03)^{-1.4}$

**4.** $1500(2^{-5/2})$

**5.** $6^{-\sqrt{2}}$

**6.** $1.3^{\sqrt{5}}$

**7.** $e^{4}$

**8.** $e^{-5}$

**9.** $e^{2/3}$

**10.** $e^{-2.7}$

In Exercises 11–18, match the function with its graph. [The graphs are labeled (a), (b), (c), (d), (e), (f), (g), and (h).]

(a)

(b)

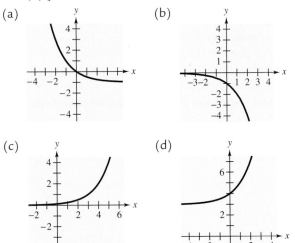

(e)

(f)

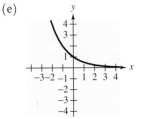
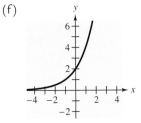

(g)

(h)

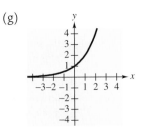
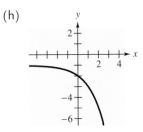

(c)

(d)

**11.** $f(x) = 2^{x}$

**12.** $f(x) = 2^{-x}$

**13.** $f(x) = -2^{x}$

**14.** $f(x) = -2^{x} - 1$

**15.** $f(x) = 2^{x} + 3$

**16.** $f(x) = 2^{-x} - 1$

**17.** $f(x) = 2^{x+1}$

**18.** $f(x) = 2^{x-3}$

In Exercises 19–36, sketch the graph of the function.

**19.** $g(x) = 4^{x}$

**20.** $f(x) = \left(\frac{3}{2}\right)^{x}$

**21.** $f(x) = 4^{x} + 2$

**22.** $h(x) = \left(\frac{3}{2}\right)^{-x}$

**23.** $h(x) = 4^{x-3}$

**24.** $g(x) = \left(\frac{3}{2}\right)^{x+2}$

**25.** $g(x) = 4^{-x} - 2$

**26.** $f(x) = \left(\frac{3}{2}\right)^{-x} + 2$

## Properties of Logarithms

You know from the preceding section that the logarithmic function with base $a$ is the *inverse* of the exponential function with base $a$. So, it makes sense that the properties of exponents should have corresponding properties involving logarithms. For instance, the exponential property $a^0 = 1$ has the corresponding logarithmic property $\log_a 1 = 0$.

---

▶ **Properties of Logarithms**

Let $a$ be a positive number such that $a \neq 1$, and let $n$ be a real number. If $u$ and $v$ are positive real numbers, the following properties are true.

**1.** $\log_a(uv) = \log_a u + \log_a v$   **1.** $\ln(uv) = \ln u + \ln v$

**2.** $\log_a \dfrac{u}{v} = \log_a u - \log_a v$   **2.** $\ln \dfrac{u}{v} = \ln u - \ln v$

**3.** $\log_a u^n = n \log_a u$   **3.** $\ln u^n = n \ln u$

---

There is no general property that can be used to rewrite $\log_a(u \pm v)$. Specifically, $\log_a(x + y)$ is not equal to $\log_a x + \log_a y$.

---

### Example 3   Using Properties of Logarithms

Write the logarithm in terms of $\ln 2$ and $\ln 3$.

**a.** $\ln 6$     **b.** $\ln \dfrac{2}{27}$

**Solution**

**a.** $\ln 6 = \ln(2 \cdot 3)$         Rewrite 6 as $2 \cdot 3$.

$\qquad = \ln 2 + \ln 3$         Property 1

**b.** $\ln \dfrac{2}{27} = \ln 2 - \ln 27$         Property 2

$\qquad = \ln 2 - \ln 3^3$         Rewrite 27 as $3^3$.

$\qquad = \ln 2 - 3 \ln 3$         Property 3

---

### Example 4   Using Properties of Logarithms

Use the properties of logarithms to verify that $-\ln \frac{1}{2} = \ln 2$.

**Solution**

$$-\ln \frac{1}{2} = -\ln(2^{-1}) = -(-1)\ln 2 = \ln 2$$

Try checking this result on your calculator.

---

## Rewriting Logarithmic Expressions

The properties of logarithms are useful for rewriting logarithmic expressions in forms that simplify the operations of algebra. This is true because they convert complicated products, quotients, and exponential forms into simpler sums, differences, and products, respectively. Examples 5 and 6 illustrate some cases.

### Example 5  Rewriting the Logarithm of a Product

$$\log_{10} 5x^3 y = \log_{10} 5 + \log_{10} x^3 y$$

$$= \log_{10} 5 + \log_{10} x^3 + \log_{10} y$$

$$= \log_{10} 5 + 3 \log_{10} x + \log_{10} y$$

### Example 6  Rewriting the Logarithm of a Quotient

$$\ln \frac{\sqrt{3x-5}}{7} = \ln(3x-5)^{1/2} - \ln 7 = \frac{1}{2}\ln(3x-5) - \ln 7$$

·DISCOVERY·

Use a calculator to find the approximate value of $\ln \sqrt[3]{e^2}$. Now find the exact value by rewriting $\ln \sqrt[3]{e^2}$ with a rational exponent using the properties of logarithms. How do the two values compare?

In Examples 5 and 6, the properties of logarithms were used to *expand* logarithmic expressions. In Examples 7 and 8, this procedure is reversed and the properties of logarithms are used to *condense* logarithmic expressions.

### Example 7  Condensing a Logarithmic Expression

$$\frac{1}{2}\log_{10} x - 3 \log_{10}(x+1) = \log_{10} x^{1/2} - \log_{10}(x+1)^3$$

$$= \log_{10} \frac{\sqrt{x}}{(x+1)^3}$$

### Example 8  Condensing a Logarithmic Expression

$$2 \ln(x+2) - \ln x = \ln(x+2)^2 - \ln x = \ln \frac{(x+2)^2}{x}$$

When applying the properties of logarithms to a logarithmic function, you should be careful to check the domain of the function. For example, the domain of $f(x) = \ln x^2$ is all real $x \neq 0$, whereas the domain of $g(x) = 2 \ln x$ is all real $x > 0$.

**25.** *Population*   The population $P$ of a city is given by

$$P = 105,300e^{0.015t}$$

where $t$ is the time in years, with $t = 0$ corresponding to 1990. Sketch the graph of this equation. According to this model, in what year will the city have a population of 150,000?

**26.** *Population*   The population $P$ of a city is given by

$$P = 240,360e^{0.012t}$$

where $t$ is the time in years, with $t = 0$ corresponding to 1990. Sketch the graph of this equation. According to this model, in what year will the city have a population of 300,000?

**27.** *Population*   The population $P$ of a city is given by

$$P = 25,000e^{kt}$$

where $t$ is the time in years, with $t = 0$ corresponding to 1990. In 1945, the population was 3350. Find the value of $k$ and use this result to predict the population in the year 2010.

**28.** *Population*   The population $P$ of a city is given by

$$P = 140,500e^{kt}$$

where $t = 0$ represents 1990. In 1960, the population was 100,250. Find the value of $k$ and use this result to predict the population in the year 2005.

**29.** *Bacteria Growth*   The number of bacteria $N$ in a culture is given by the model

$$N = 100e^{kt}$$

where $t$ is the time in hours, with $t = 0$ corresponding to the time when $N = 100$. When $t = 4$, there are 250 bacteria. How long does it take the population to double in size? To triple in size?

**30.** *Bacteria Growth*   The number of bacteria $N$ in a culture is given by the model

$$N = 250e^{kt}$$

where $t$ is the time in hours, with $t = 0$ corresponding to the time when $N = 250$. When $t = 10$, there are 280 bacteria. How long does it take the population to double in size? To triple in size?

**31.** *Radioactive Decay*   What percent of a present amount of radioactive radium ($^{226}$Ra) will remain after 100 years? Use the fact that radioactive radium has a half-life of 1620 years.

**32.** *Radioactive Decay*   Find the half-life of a radioactive material if, after 1 year, 99.57% of the initial amount remains.

**33.** *Learning Curve*   The management at a factory has found that the maximum number of units a worker can produce in a day is 30. The learning curve for the number of units $N$ produced per day after a new employee has worked $t$ days is given by

$$N = 30(1 - e^{kt}).$$

After 20 days on the job, a particular worker produced 19 units in 1 day.

(a) Find the learning curve for this worker (that is, find the value of $k$).

(b) How many days should pass before this worker is producing 25 units per day?

**34.** *Sales*   The number of sales units $S$ (in thousands) of a new product after it has been on the market for $t$ years is given by $S = 100(1 - e^{kt})$.

(a) Find $S$ as a function of $t$ if 15,000 units have been sold after 1 year.

(b) How many units will have been sold after 5 years?

**35.** *In-Line Skate Sales*   From 1990 to 1999, the sales $y$ (in millions of dollars) of in-line skating and wheel sports products can be approximated by the function

$$y = \begin{cases} 146.6(1.3777)^t, & 0 \le t \le 4 \\ 1106.44e^{-0.35531t} + 458.8, & 5 \le t \le 9 \end{cases}$$

where $t = 0$ represents 1990. (Source: National Sporting Good Association)

(a) Use a graphing utility to graph the function.

(b) Describe the change in the growth pattern that occurred in 1995.

**36.** *Textbook Sales*   The sales of a college textbook that was published in 1994 and revised in 2001 can be approximated by the function

$$y = \begin{cases} e^{6.96+1.619t-0.2199t^2}, & 4 \le t \le 7 \\ e^{-2.61+3.37t-0.22t^2}, & 8 \le t \le 11 \end{cases}$$

where $y$ is the number of copies sold and $t = 4$ represents 1994.

(a) Use a graphing utility to graph this model.

(b) The graph is typical of college textbook sales. How would you explain this sales pattern?

**37.** *Women's Heights*    The distribution for the heights of American women (between 30 and 39 years of age) can be approximated by the function

$$p = 0.148e^{-(x-64.9)^2/14.45}, \quad 56 \le x \le 72$$

where $x$ is the height in inches and $p$ is the percent (in decimal form).    (Source: U.S. National Center for Health Statistics)

(a) Use a graphing utility to graph this model.

(b) Use the graph to determine the average height of women in this age bracket.

**38.** *Men's Heights*    The distribution for the heights of American men (between 30 and 39 years of age) can be approximated by the function

$$p = 0.140e^{-(x-70)^2/16.27}, \quad 62 \le x \le 75$$

where $x$ is the height in inches and $p$ is the percent (in decimal form).    (Source: U.S. National Center for Health Statistics)

(a) Use a graphing utility to graph this model.

(b) Use the graph to determine the average height of men in this age bracket.

**39.** *Stocking a Lake with Fish*    A certain lake is stocked with 500 fish, and the fish population $P$ increases according to the logistics curve

$$P = \frac{10,000}{1 + 19e^{-t/5}}, \quad t \ge 0$$

where $t$ is measured in months.

(a) Use a graphing utility to graph this model.

(b) Find the fish population after 5 months.

(c) After how many months will the fish population be 2000?

**40.** *Endangered Species*    A conservation organization releases 100 animals of an endangered species into a game preserve. The organization believes that the preserve has a carrying capacity of 1000 animals and that the growth of the herd will be modeled by the logistics curve

$$p = \frac{1000}{1 + 9e^{-kt}}, \quad t \ge 0$$

where $p$ is the number of animals and $t$ is measured in years.

(a) Find $k$ if the herd size is 134 after 2 years.

(b) Find the population after 5 years.

**41.** *Recreational Vehicles*    The sales $S$ (in millions of dollars) of recreational vehicles in the United States from 1991 to 1999 are given in the table.    (Source: National Sporting Goods Association)

| Year | 1991 | 1992 | 1993 | 1994 | 1995 |
|------|------|------|------|------|------|
| $S$  | 3615 | 4412 | 4775 | 5690 | 5895 |

| Year | 1996 | 1997 | 1998 | 1999 |
|------|------|------|------|------|
| $S$  | 6327 | 6904 | 8364 | 9078 |

(a) Use a graphing utility to make a scatter plot of the data. Let $t = 0$ correspond to 1990. Would an exponential model give a good representation of the data?

(b) Use the regression capabilities of a graphing utility to find an exponential model for the data. Use the inverse property $b = e^{\ln b}$ to rewrite the model as an exponential model in base $e$.

(c) Use your graphing utility to graph the model.

(d) Use the model to estimate recreational vehicle sales in 2000 and 2001.

**42.** *Sports Clothes*    The sales $S$ (in millions of dollars) of athletic and sports clothing in the Untied States from 1992 to 1999 are given in the table.    (Source: National Sporting Goods Association)

| Year | 1992 | 1993 | 1994 | 1995 |
|------|------|------|------|------|
| $S$  | 8990 | 9096 | 9521 | 10,311 |

| Year | 1996 | 1997 | 1998 | 1999 |
|------|------|------|------|------|
| $S$  | 11,127 | 12,035 | 12,637 | 13,390 |

(a) Use a graphing utility to make a scatter plot of the data. Let $t = 2$ correspond to 1992. Would an exponential model give a good representation of the data?

(b) Use the regression capabilities of a graphing utility to find an exponential model for the data. Use the inverse property $b = e^{\ln b}$ to rewrite the model as an exponential model in base $e$.

(c) Use your graphing utility to graph the model.

(d) Use the model to estimate athletic and sports clothing sales in 2000 and 2001.

**43.** *Sales and Advertising*    After discontinuing all advertising for a certain product (in 1998), the manufacturer of the product found that the sales began to drop according to the model

$$S = \frac{500,000}{1 + 0.6e^{kt}}, \quad 0 \le t \le 6$$

where $S$ represents the number of units sold and $t$ represents the calendar year, with $t = 0$ corresponding to 1998.

(a) Find $k$ if the company sold 300,000 units in 2000.

(b) According to this model, what were the sales in 2002?

**44.** *Full-Length CDs*    The percents $P$ of sound recordings that were full-length CDs in the United States from 1991 to 2000 are given in the table.    (Source: Recording Industry Association of America)

| Year | 1991 | 1992 | 1993 | 1994 | 1995 |
|------|------|------|------|------|------|
| $P$ | 38.9 | 46.5 | 51.1 | 58.4 | 65.0 |

| Year | 1996 | 1997 | 1998 | 1999 | 2000 |
|------|------|------|------|------|------|
| $P$ | 68.4 | 70.2 | 74.8 | 83.2 | 89.3 |

(a) Use a graphing utility to make a scatter plot of the data. Let $t = 1$ correspond to 1991.

(b) Use the regression capabilities of a graphing utility to find an exponential model for the data. Use the inverse property $b = e^{\ln b}$ to rewrite the model as an exponential model in base $e$.

(c) Use the regression capabilities of a graphing utility to find a logistics model for the data.

(d) Graph the exponential and logistics models with the data. Use both models to estimate the percents of recordings that are full-length CDs in 2001 and 2002.

(e) Do both models give reasonable estimates?

**45.** *High School Graduates*    The table shows the percent $P$ of the U.S. population (who are 25 years old or older) who had not completed high school in each of the indicated years.    (Source: U.S. Bureau of the Census)

| Year | 1990 | 1991 | 1992 | 1993 | 1994 |
|------|------|------|------|------|------|
| $P$ | 22.4 | 21.6 | 20.6 | 19.8 | 19.1 |

| Year | 1995 | 1996 | 1997 | 1998 | 1999 |
|------|------|------|------|------|------|
| $P$ | 18.3 | 18.3 | 17.9 | 17.2 | 16.6 |

(a) Use a graphing utility to make a scatter plot of the data. Let $t = 0$ represent 1990.

(b) Use the regression capabilities of a graphing utility to find a quadratic model and an exponential model for the data. Use the inverse property $b = e^{\ln b}$ to rewrite the exponential model in base $e$.

(c) Graph the quadratic model and the exponential model (in base $e$) with the data.

(d) Use both models to predict the percents for the years 2005, 2006, and 2007. Do both models give reasonable predictions?

*Earthquake Magnitudes*    In Exercises 46 and 47, use the Richter scale (see Example 6) for measuring the magnitude of earthquakes.

**46.** Find the magnitude $R$ (on the Richter scale) of an earthquake of intensity $I$ (let $I_0 = 1$).

(a) $I = 80,500,000$

(b) $I = 48,275,000$

(c) $I = 40,000,000$

(d) $I = 20,000,000$

**47.** Find the intensity $I$ of an earthquake measuring $R$ on the Richter scale (let $I_0 = 1$).

(a) Turkmenistan in 2000, $R = 7.0$

(b) Near coast of Peru in 2001, $R = 8.4$

*Intensity of Sound*   In Exercises 48 and 49, use the information to determine the level of sound (in decibels) for the sound intensity. The level of sound $\beta$, in decibels, of a sound with an intensity of $I$ is given by

$$\beta(I) = 10 \log_{10} \frac{I}{I_0}$$

where $I_0$ is an intensity of $10^{-16}$ watts per square centimeter, corresponding roughly to the faintest sound that can be heard.

**48.** (a) $I = 10^{-14}$ watts per cm$^2$ (faint whisper)

    (b) $I = 10^{-9}$ watts per cm$^2$ (busy street corner)

    (c) $I = 10^{-6.5}$ watts per cm$^2$ (air hammer)

    (d) $I = 10^{-4}$ watts per cm$^2$ (threshold of pain)

**49.** (a) $I = 10^{-13}$ watts per cm$^2$ (whisper)

    (b) $I = 10^{-7.5}$ watts per cm$^2$ (DC-8 4 miles away)

    (c) $I = 10^{-7}$ watts per cm$^2$ (diesel truck at 25 feet)

    (d) $I = 10^{-4.5}$ watts per cm$^2$ (automobile horn at 3 feet)

**50.** *Noise Level*   Due to the installation of acoustical tiling, the noise level in an auditorium was reduced from 78 to 72 decibels. Find the percent decrease in the intensity level of the noise as a result of the installation of the tiling.

**51.** *Noise Level*   Due to the installation of a muffler, the noise level of an automobile engine was reduced from 88 to 76 decibels. Find the percent decrease in the intensity level of the noise as a result of the installation of the muffler.

**52.** *Estimating the Time of Death*   At 8:30 A.M., a coroner was called to the home of a person who had died during the night. In order to estimate the time of death, the coroner took the person's temperature twice. At 9:00 A.M. the temperature was 85.7°, and at 9:30 A.M. the temperature was 82.8°. From these two temperature readings, the coroner was able to determine that the time elapsed since death and the body temperature were related by the formula

$$t = -2.5 \ln \frac{T - 70}{98.6 - 70}$$

where $t$ is the time in hours that have elapsed since the person died and $T$ is the temperature (in degrees Fahrenheit) of the person's body. Assume the person had a normal body temperature of 98.6° at death, and the room temperature was a constant 70°. (This formula is derived from a general cooling principle called Newton's Law of Cooling.) Use this formula to estimate the time of death of the person.

**53.** *Thawing a Package of Steaks*   Suppose you take a 3-pound package of steaks out of the freezer at noon. Will the steaks be thawed in time to be grilled on the barbecue at 6 P.M.? To answer this question, assume that the room temperature is 70° Fahrenheit and that the freezer temperature is 24°. Use the formula

$$t = -3.8 \ln \frac{T - 70}{24 - 70}$$

where $t$ is the time (with $t = 0$ corresponding to noon) and $T$ is the temperature of the package of steaks.

**54.** *Population Growth*   The population of a city is given in the table, with $t = 4$ corresponding to 1994. Use a graphing utility to make a scatter plot of these data. Use the regression capabilities of your graphing utility to find an exponential model for the data. Graph the model with the data points. Use the model to predict the city's population in the years 2005 and 2006.

| $t$ | 4 | 6 | 8 | 10 | 12 |
|---|---|---|---|---|---|
| Population | 35,478 | 47,114 | 55,431 | 65,522 | 79,963 |

**55.** *Worker's Productivity*   The numbers of units per day $n$ that a new worker can produce after $t$ days on the job are listed in the table. Use a graphing utility to make a scatter plot of these data. Do the data fit an exponential model or a logarithmic model? Use the regression capabilities of your graphing utility to find the model. Graph the model with the data points. Is the model a good fit?

| $t$ | 5 | 10 | 12 | 15 | 21 |
|---|---|---|---|---|---|
| $n$ | 6 | 13 | 17 | 22 | 31 |

In Exercises 57–60, approximate the logarithm using the properties of logarithms given that $\log_b 2 \approx 0.3562$, $\log_b 3 \approx 0.5646$, and $\log_b 5 \approx 0.8271$.

**57.** $\log_b 6$

**58.** $\log_b\left(\frac{4}{25}\right)$

**59.** $\log_b \sqrt{3}$

**60.** $\log_b 30$

In Exercises 61–64, find the *exact* value of the logarithm without using a calculator.

**61.** $\log_7 49$

**62.** $\log_6 \frac{1}{36}$

**63.** $\ln e^{3.2}$

**64.** $\ln \sqrt[5]{e^3}$

In Exercises 65–74, solve the equation.

**65.** $e^x = 8$

**66.** $e^{2x} = 17$

**67.** $4e^{-3x} = 146$

**68.** $e^{2x} - 7e^x + 10 = 0$

**69.** $\ln 3x = 8.2$

**70.** $2 \ln 4x = 15$

**71.** $-2 + \ln 5x = 0$

**72.** $\ln 4x^2 = 21$

**73.** $\ln x - \ln 3 = 2$

**74.** $\ln \sqrt{x + 1} = 2$

**75.** *Demand Function*    The demand function for a certain product is given by

$$p = 600 - 0.3(e^{0.005x}).$$

Find the demand $x$ for a price of (a) $p = \$500$ and (b) $p = \$400$.

**76.** *Demand Function*    The demand function for a certain product is given by

$$p = 4000\left(1 - \frac{3}{3 + e^{-0.004x}}\right).$$

Find the demand $x$ for a price of (a) $p = \$700$ and (b) $p = \$400$.

*Radioactive Decay*    In Exercises 77 and 78, complete the table for the radioactive isotope.

| Isotope | Half-Life | Initial Quantity | Amount After 1000 Years |
|---------|-----------|------------------|-------------------------|
| **77.** $^{14}$C | 5730 | 12 g | |
| **78.** $^{230}$Pu | 24,360 | | 3.1 g |

**79.** *Population*    The population $P$ of a city is given by

$$P = 270,000e^{0.019t}$$

where $t$ is the time in years, with $t = 0$ corresponding to 2000.

(a) Use a graphing utility to graph this equation.

(b) According to this model, in what year will the city have a population of 300,000?

**80.** *Population*    The population $P$ of a city is given by

$$P = 50,000e^{kt}$$

where $t$ is the time in years, with $t = 0$ corresponding to 2000. In 1990, the population was 34,500.

(a) Find the value of $k$ and use this result to predict the population in the year 2030.

(b) Use a graphing utility to confirm the result of part (a).

**81.** *Curve Fitting*    Use the properties of logarithms and exponentiation to find an equation that relates $x$ and $y$. (*Hint:* See Example 9 in Section 5.3.)

| $x$ | 1 | 2 | 3 | 4 | 5 | 6 |
|-----|---|---|---|---|---|---|
| $y$ | 1 | 2.520 | 4.327 | 6.350 | 8.550 | 10.903 |

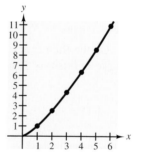

**82.** *Bacteria Growth*    The number of bacteria $N$ in a culture is given by the model

$$N = 200e^{kt}$$

where $t$ is the time in hours, with $t = 0$ corresponding to the time when $N = 200$. When $t = 6$, $N = 500$. How long does it take the population to double in size?

**83.** *Learning Curve* The management at a factory has found that the maximum number of units a worker can produce in a day is 50. The learning curve for the number of units $N$ produced per day after a new employee has worked $t$ days is given by $N = 50(1 - e^{kt})$. After 20 days on the job, a particular worker produced 31 units in 1 day.

(a) Find the learning curve for this worker.

(b) How many days should pass before this worker is producing 45 units per day?

**84.** *Test Scores* The scores on a general aptitude test roughly follow a normal distribution given by

$$y = 0.0028e^{-[(x-300)^2]/20,000}, \quad 100 \le x \le 500.$$

Sketch the graph of this function. Estimate the average score on this test.

**85.** *Wildlife Management* A state parks and wildlife department releases 100 deer into a wilderness area. The department believes that the carrying capacity of the area is 500 deer and that the growth of the herd will be modeled by the logistics curve

$$P = \frac{500}{1 + 4e^{-kt}}, \quad t \ge 0$$

where $t$ is measured in years.

(a) Find $k$ if the herd size is 170 after 2 years.

(b) Find the population after 5 years.

**86.** *Earthquake Magnitudes* On the Richter scale, the magnitude $R$ of an earthquake of intensity $I$ is given by

$$R = \log_{10} \frac{I}{I_0}$$

where $I_0 = 1$ is the minimum intensity used for comparison. Find the intensity per unit of area for the following values of $R$.

(a) $R = 8.4$    (b) $R = 6.85$    (c) $R = 9.1$

**87.** *Thawing Time* Suppose that you take a 5-pound package of steaks out of a freezer at 11 A.M. Will the steaks be thawed in time to be grilled at 6 P.M.? Assume that the room temperature is 70° Fahrenheit and the freezer temperature is 20°. Use the formula

$$t = -3.8 \ln \frac{T - 70}{20 - 70}$$

where $t$ is the time (with $t = 0$ corresponding to 11 A.M.) and $T$ is the temperature of the package of steaks.

**88.** *Sales and Advertising* In 2000, after discontinuing all advertising for a product, the manufacturer of the product found that the sales began to drop according to the model

$$S = \frac{600,000}{1 + 0.4e^{kt}}$$

where $S$ represents the number of units sold and $t$ represents the calendar year, with $t = 0$ corresponding to 2000.

(a) Find $k$ if the company sold 400,000 units in 2002.

(b) According to this model, what will sales be in 2004? In 2006?

**89.** *Bacteria Growth* The number of bacteria $N$ in a culture is given by the model $N = 200e^{kt}$, where $t$ is the time in hours, with $t = 0$ corresponding to the time when $N = 200$. When $t = 5$, there are 350 bacteria. How long does it take for the population to triple in size?

**90.** *Recording Sales* The percents $P$ of sound recordings purchased at stores other than record stores in the United States from 1991 to 2000 are given in the table. (Source: Recording Industry Association of America)

| Year | 1991 | 1992 | 1993 | 1994 | 1995 |
|------|------|------|------|------|------|
| $P$  | 23.4 | 24.9 | 26.1 | 26.7 | 28.2 |

| Year | 1996 | 1997 | 1998 | 1999 | 2000 |
|------|------|------|------|------|------|
| $P$  | 31.5 | 31.9 | 34.4 | 38.3 | 40.8 |

(a) Use a graphing utility to make a scatter plot of the data. Let $t = 1$ correspond to 1991.

(b) Use the regression capabilities of a graphing utility to find an exponential model for the data. Use the inverse property $b = e^{\ln b}$ to rewrite the model as an exponential model in base $e$.

(c) Use the regression capabilities of a graphing utility to find a linear and quadratic model for the data.

(d) Graph the exponential, linear, and quadratic models with the data. Use all three models to estimate the percents of sound recordings purchased at stores other than record stores in 2001 and 2002.

(e) Does each model give a reasonable estimate? Explain your reasoning.

# Chapter Test

Take this test as you would take a test in class. After you are done, check your work against the answers given in the back of the book.

In Exercises 1–4, sketch the graph of the function.

**1.** $y = 2^x$

**2.** $y = 3^{-x}$

**3.** $y = \ln x$

**4.** $y = \log_3(x - 1)$

**5.** You deposit $30,000 into a fund that pays 7.25% interest, compounded continuously. When will the balance be greater than $100,000?

In Exercises 6–8, expand the logarithmic expression.

**6.** $\ln \dfrac{x^2 y^3}{z}$

**7.** $\log_{10} 3xyz^2$

**8.** $\ln\left(x \sqrt[3]{x - 2}\right)$

In Exercises 9 and 10, condense the expression.

**9.** $\ln y + 2 \ln z - 3 \ln x$

**10.** $\frac{2}{3}(\log_{10} x + \log_{10} y)$

In Exercises 11–14, solve the equation.

**11.** $e^{4x} = 21$

**12.** $e^{2x} - 8e^x + 12 = 0$

**13.** $-3 + \ln 4x = 0$

**14.** $\ln \sqrt{x + 2} = 3$

In Exercises 15–17, students in a psychology class were given an exam and then were retested monthly with an equivalent exam. The average score for the class is given by the human memory model $f(t) = 87 - 15 \log_{10}(t + 1)$, $0 \le t \le 4$, where $t$ is the time in months.

**15.** What was the average score on the original exam?

**16.** What was the average score after 2 months? After 4 months?

**17.** Suppose the students in this psychology class participated in a study that required that they continue taking an equivalent exam every 6 months for 2 years. If the model remained valid, what would the average score be after 12 months? After 18 months? What could this indicate about human memory?

**18.** The population $P$ of a city is given by $P = 70,000e^{0.023t}$, where $t = 0$ represents 2000. When will the city have a population of 100,000? Explain.

**19.** The number of bacteria $N$ in a culture is given by $N = 100e^{kt}$, where $t$ is the time in hours, with $t = 0$ corresponding to the time when $N = 100$. When $t = 8$, $N = 500$. How long does it take the population to double?

**20.** Carbon 14 has a half-life of 5730 years. You have an initial quantity of 10 grams. How many grams will remain after 10,000 years? After 20,000 years?

# Cumulative Test: Chapters 3–5

Take this test as you would take a test in class. After you are done, check your work against the answers given in the back of the book.

In Exercises 1–6, given $f(x) = x^2 + 1$ and $g(x) = 3x - 5$, write the indicated function.

**1.** $(f + g)(x)$            **2.** $(f - g)(x)$            **3.** $(fg)(x)$

**4.** $\left(\dfrac{f}{g}\right)(x)$            **5.** $(f \circ g)(x)$            **6.** $(g \circ f)(x)$

In Exercises 7–11, sketch the graph of the function. Describe the domain and range of the function.

**7.** $f(x) = (x - 2)^2 + 3$     **8.** $g(x) = \dfrac{2}{x - 3}$     **9.** $h(x) = 2^{-x}$

**10.** $f(x) = \log_4 (x - 1)$     **11.** $g(x) = \begin{cases} x + 5, & x < 0 \\ 5, & x = 0 \\ x^2 + 5, & x > 0 \end{cases}$

**12.** The profit $P$ (in dollars) for a company is given by

$$P = 200 + 10x - 0.001x^2$$

where $x$ is the number of units produced. What production level will yield a maximum profit?

In Exercises 13–15, perform the indicated operation and write the result in standard form.

**13.** $(10 + 2i)(3 - 4i)$     **14.** $(4 + 5i)^2$     **15.** $\dfrac{1 + 2i}{3 - i}$

**16.** Use the Quadratic Formula to solve $3x^2 - 5x + 7 = 0$.

**17.** Find all the zeros of $f(x) = x^4 + 10x^2 + 9$ given that $3i$ is a zero. Explain your reasoning.

In Exercises 18 and 19, solve the equation.

**18.** $e^{2x} - 11e^x + 24 = 0$     **19.** $\frac{1}{2} \ln(x - 3) = 4$

**20.** The population $P$ of a city is given by

$$P = 80,000e^{0.035t}$$

where $t$ is the time in years, with $t = 6$ corresponding to 1996. According to this model, in what year will the city have a population of 140,000?

# 6 Systems of Equations and Inequalities

The amounts of U.S. consumer credit outstanding (in billions of dollars) held by finance companies $y_1$ and held by credit unions $y_2$ from 1991 to 2001 are shown in the table. Models for the data are given by

$$y_1 = 0.05854t^4 - 1.5680t^3 + 14.638t^2 - 46.01t + 160.0,$$
$$1 \leq t \leq 11 \quad \text{and}$$

$$y_2 = 0.05676t^4 - 1.4159t^3 + 11.900t^2 - 27.59t + 105.6,$$
$$1 \leq t \leq 11 \quad \text{where } t = 1 \text{ represents 1991.}$$

The graphs of these two models are shown above. If the graphs of the two models intersect, the amount of consumer credit outstanding for both would be the same. If the graphs do not intersect, what possible conclusion could you draw? (Source: Federal Reserve Statistical Release, Consumer Credit Table G.19, Oct. 5, 2001 [Internet])

**6.1 Systems of Equations**

**6.2 Linear Systems in Two Variables**

**6.3 Linear Systems in Three or More Variables**

**6.4 Systems of Inequalities**

**6.5 Linear Programming**

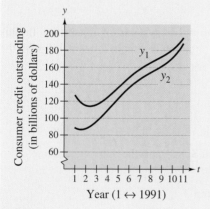

Year (1 ↔ 1991)

|       | 1991  | 1992  | 1993  | 1994  | 1995  | 1996  | 1997  | 1998  | 1999  | 2000  | 2001  |
|-------|-------|-------|-------|-------|-------|-------|-------|-------|-------|-------|-------|
| $y_1$ | 127.5 | 117.0 | 110.3 | 124.0 | 141.3 | 153.9 | 156.7 | 154.3 | 173.6 | 190.3 | 191.0 |
| $y_2$ | 88.5  | 88.3  | 95.4  | 108.2 | 125.7 | 136.1 | 147.6 | 152.4 | 158.2 | 176.0 | 187.0 |

The chapter project related to this information is on page 417.

## 6.1 Systems of Equations

### Objectives

- Solve a system of equations by the method of substitution.
- Solve a system of equations graphically.
- Construct and use a system of equations to solve an application problem.

## The Method of Substitution

Up to this point in the text, most problems have involved either a function of one variable or a single equation in two variables. However, many problems in science, business, and engineering involve two or more equations in two or more variables. To solve such problems, you need to find solutions of a **system of equations.** Here is an example of a system of two equations in $x$ and $y$.

$$2x + \phantom{2}y = 5 \qquad \text{Equation 1}$$
$$3x - 2y = 4 \qquad \text{Equation 2}$$

A **solution** of this system is an ordered pair that satisfies each equation in the system. Finding the set of all solutions is called **solving the system of equations.** For instance, the ordered pair $(2, 1)$ is a solution of this system. To check this, you can substitute 2 for $x$ and 1 for $y$ in *each* equation.

$$2x + y = 5 \qquad \text{Equation 1}$$
$$2(2) + 1 \stackrel{?}{=} 5 \qquad \text{Substitute 2 for } x \text{ and 1 for } y.$$
$$4 + 1 = 5 \qquad \text{Solution checks in Equation 1. } \checkmark$$
$$3x - 2y = 4 \qquad \text{Equation 2}$$
$$3(2) - 2(1) \stackrel{?}{=} 4 \qquad \text{Substitute 2 for } x \text{ and 1 for } y.$$
$$6 - 2 = 4 \qquad \text{Solution checks in Equation 2. } \checkmark$$

There are several different ways to solve systems of equations. In this chapter you will study three of the most common techniques, beginning with the **method of substitution.**

---

### ▶ Method of Substitution

1. *Solve* one of the equations for one variable in terms of the other.

2. *Substitute* the expression found in Step 1 into the other equation to obtain an equation in one variable.

3. *Solve* the equation obtained in Step 2.

4. *Back-substitute* the solution in Step 3 into the expression obtained in Step 1 to find the value of the other variable.

5. *Check* the solution to see that it satisfies *each* of the original equations.

---

| Example 1 | Solving a System of Two Equations |  |

Solve the following system of equations.

$$x + y = 4 \qquad \text{Equation 1}$$
$$x - y = 2 \qquad \text{Equation 2}$$

**Solution**

Begin by solving for $y$ in Equation 1.

$$y = 4 - x \qquad \text{Solve for } y \text{ in Equation 1.}$$

Next, substitute this expression for $y$ in Equation 2 and solve the resulting single-variable equation for $x$.

$$x - y = 2 \qquad \text{Equation 2}$$
$$x - (4 - x) = 2 \qquad \text{Substitute } 4 - x \text{ for } y.$$
$$x - 4 + x = 2 \qquad \text{Simplify.}$$
$$2x = 6 \qquad \text{Combine like terms.}$$
$$x = 3 \qquad \text{Divide both sides by 2.}$$

Finally, you can solve for $y$ by *back-substituting* $x = 3$ into the equation $y = 4 - x$, to obtain

$$y = 4 - x \qquad \text{Revised Equation 1}$$
$$y = 4 - 3 \qquad \text{Substitute 3 for } x.$$
$$y = 1 \qquad \text{Solve for } y.$$

The solution is the ordered pair $(3, 1)$. You can check this as follows.

*Check*

$$x + y = 4 \qquad \text{Equation 1}$$
$$3 + 1 \overset{?}{=} 4 \qquad \text{Substitute for } x \text{ and } y.$$
$$4 = 4 \qquad \text{Solution checks in Equation 1. } \checkmark$$
$$x - y = 2 \qquad \text{Equation 2}$$
$$3 - 1 \overset{?}{=} 2 \qquad \text{Substitute for } x \text{ and } y.$$
$$2 = 2 \qquad \text{Solution checks in Equation 2. } \checkmark$$

The term *back-substitution* implies that you work *backwards*. First you solve for one of the variables, and then you substitute that value *back* into one of the equations in the system to find the value of the other variable.

**Study Tip**

Because many steps are required to solve a system of equations, it is easy to make errors in arithmetic. We *strongly* suggest that you always *check your solution by substituting it into each equation in the original system.*

**Example 2** Solving a System by Substitution

A total of $12,000 is invested in two funds paying 9% and 11% simple interest. If the yearly interest is $1180, how much of the $12,000 is invested at each rate?

**Solution**

*Verbal Model:*  $\boxed{\begin{array}{c} 9\% \\ \text{fund} \end{array}} + \boxed{\begin{array}{c} 11\% \\ \text{fund} \end{array}} = \boxed{\begin{array}{c} \text{Total} \\ \text{investment} \end{array}}$

$\boxed{\begin{array}{c} 9\% \\ \text{interest} \end{array}} + \boxed{\begin{array}{c} 11\% \\ \text{interest} \end{array}} = \boxed{\begin{array}{c} \text{Total} \\ \text{interest} \end{array}}$

*Labels:*  Amount in 9% fund $= x$       (dollars)
Interest for 9% fund $= 0.09x$     (dollars)
Amount in 11% fund $= y$      (dollars)
Interest for 11% fund $= 0.11y$   (dollars)
Total investment $= \$12,000$    (dollars)
Total interest $= \$1180$      (dollars)

*System:* 
$$x + \quad y = 12,000 \qquad \text{Equation 1}$$
$$0.09x + 0.11y = \quad 1180 \qquad \text{Equation 2}$$

To begin, it is convenient to multiply both sides of the second equation by 100 to obtain $9x + 11y = 118,000$. This eliminates the need to work with decimals.

$$9x + 11y = 118,000 \qquad \text{Revised Equation 2}$$

To solve this system, you can solve for $x$ in Equation 1.

$$x = 12,000 - y \qquad \text{Solve for } x \text{ in Equation 1.}$$

Then, substitute this expression for $x$ into Equation 2, and solve the resulting equation for $y$.

$$9x + 11y = 118,000 \qquad \text{Revised Equation 2}$$
$$9(12,000 - y) + 11y = 118,000 \qquad \text{Substitute } 12,000 - y \text{ for } x.$$
$$108,000 - 9y + 11y = 118,000 \qquad \text{Distributive Property}$$
$$2y = 10,000 \qquad \text{Combine like terms.}$$
$$y = 5000 \qquad \text{Divide both sides by 2.}$$

Next, back-substitute the value $y = 5000$ to solve for $x$.

$$x = 12,000 - y \qquad \text{Revised Equation 1}$$
$$x = 12,000 - 5000 \qquad \text{Substitute 5000 for } y.$$
$$x = 7000 \qquad \text{Solve for } x.$$

The solution is (7000, 5000). Check this in the original statement of the problem.

The equations in Examples 1 and 2 are linear. The method of substitution can also be used to solve systems in which one or both of the equations are nonlinear.

**Example 3**    Solving a System by Substitution: Two-Solution Case

Solve the following system of equations.

$$x^2 - x - y = 1 \qquad \text{Equation 1}$$
$$-x + y = -1 \qquad \text{Equation 2}$$

**Solution**

Begin by solving for $y$ in Equation 2 to obtain $y = x - 1$. Next, substitute this expression for $y$ into Equation 1 and solve for $x$.

$$x^2 - x - y = 1 \qquad \text{Equation 1}$$
$$x^2 - x - (x - 1) = 1 \qquad \text{Substitute } x - 1 \text{ for } y.$$
$$x^2 - 2x + 1 = 1 \qquad \text{Simplify.}$$
$$x^2 - 2x = 0 \qquad \text{Standard form}$$
$$x(x - 2) = 0 \qquad \text{Factor.}$$
$$x = 0, 2 \qquad \text{Solve for } x.$$

Back-substituting these values of $x$ to solve for the corresponding values of $y$ produces the two solutions $(0, -1)$ and $(2, 1)$. Check these solutions in the original system.

**Example 4**    Solving a System by Substitution: No-Solution Case

Solve the following system of equations.

$$-x + y = 4 \qquad \text{Equation 1}$$
$$x^2 + y = 3 \qquad \text{Equation 2}$$

**Solution**

Begin by solving Equation 1 for $y$ to obtain $y = x + 4$. Next, substitute this expression for $y$ into Equation 2, and solve for $x$.

$$x^2 + y = 3 \qquad \text{Equation 2}$$
$$x^2 + (x + 4) = 3 \qquad \text{Substitute } x + 4 \text{ for } y.$$
$$x^2 + x + 1 = 0 \qquad \text{Simplify.}$$
$$x = \frac{-1 \pm \sqrt{1^2 - 4(1)(1)}}{2} \qquad \text{Quadratic Formula}$$

Because the discriminant is negative, the equation $x^2 + x + 1 = 0$ has no (real) solution. Hence, this system has no (real) solution.

# Graphical Approach to Finding Solutions

From Examples 2, 3, and 4, you can see that a system of two equations in two unknowns can have exactly one solution, more than one solution, or no solution. In practice, you can gain insight about the location and number of solutions of a system of equations by graphing each of the equations in the same coordinate plane. The solutions of the system correspond to the **points of intersection** of the graphs. For instance, in Figure 6.1(a) the two equations graph as two lines with a *single point* of intersection. The two equations in Example 3 graph as a parabola and a line with *two points* of intersection, as shown in Figure 6.1(b). Moreover, the two equations in Example 4 graph as a line and a parabola that happen to have *no points* of intersection, as shown in Figure 6.1(c).

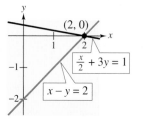

One point of intersection

**(a) One solution**

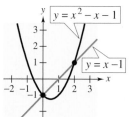

Two points of intersection

**(b) Two solutions**

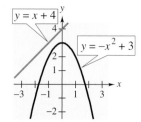

No points of intersection

**(c) No solution**

Figure 6.1

## Example 5  Solving a System of Equations

Solve the following system of equations.

$$y = \ln x \qquad \text{Equation 1}$$
$$x + y = 1 \qquad \text{Equation 2}$$

### Solution

The graph of each equation is shown in Figure 6.2. From this, it is clear that there is only one point of intersection. Also, it appears that $(1, 0)$ is the solution point. You can confirm this by substituting in *both* equations.

*Check*

$$0 = \ln 1 \qquad \text{Equation 1 checks.} \checkmark$$
$$1 + 0 = 1 \qquad \text{Equation 2 checks.} \checkmark$$

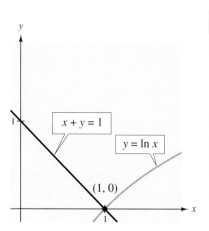

Figure 6.2

## Applications

The total cost $C$ of producing $x$ units of a product typically has two components—the initial cost and the cost per unit. When enough units have been sold so that the total revenue $R$ equals the total cost, the sales are said to have reached the **break-even point.** You will find that the break-even point corresponds to the point of intersection of the cost and revenue curves.

| Example 6 | An Application: Break-Even Analysis |
|---|---|

A small business invests $10,000 in equipment to produce a product. Each unit of the product costs $0.65 to produce and is sold for $1.20. How many items must be sold before the business breaks even?

### Solution

The total cost of producing $x$ units is

$$\boxed{\text{Total cost}} = \boxed{\text{Cost per unit}} \cdot \boxed{\text{Number of units}} + \boxed{\text{Initial cost}}$$

$$C = 0.65x + 10,000. \qquad \text{Equation 1}$$

The revenue obtained by selling $x$ units is

$$\boxed{\text{Total revenue}} = \boxed{\text{Price per unit}} \cdot \boxed{\text{Number of units}}$$

$$R = 1.2x. \qquad \text{Equation 2}$$

Because the break-even point occurs when $R = C$, you have

$$1.2x = 0.65x + 10,000 \qquad \text{Equate } R \text{ and } C.$$

$$0.55x = 10,000 \qquad \text{Subtract } 0.65x \text{ from both sides.}$$

$$x = \frac{10,000}{0.55} \qquad \text{Divide both sides by 0.55.}$$

$$x \approx 18,182 \text{ units.} \qquad \text{Use a calculator.}$$

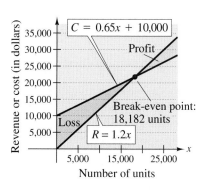

Figure 6.3

Note in Figure 6.3 that sales less than the break-even point correspond to an overall loss, whereas sales greater than the break-even point correspond to a profit.

---

Another way to view the solution in Example 6 is to consider the profit function

$$P = R - C.$$

The break-even point occurs when the profit is 0, which is the same as saying that $R = C$.

**Example 7** World Coal Production

From 1994 to 1998, the amount of coal produced each year in Australia was *increasing*, while the amount of coal produced each year in Germany was *decreasing*. Two models that approximate the values are

$$C = 184 + 15.8t \qquad \text{Australia}$$

$$C = 350 - 14.7t \qquad \text{Germany}$$

where $C$ is the amount of coal (in millions of tons) produced, and $t = 4$ represents 1994 (see Figure 6.4). According to these two models, when would you expect the amount of coal produced in Australia to have exceeded the amount produced in Germany? (Source: U.S. Energy Information Administration)

**Solution**

Because the first equation has already been solved for $C$ in terms of $t$, you substitute this value into the second equation and solve for $t$, as follows.

$$184 + 15.8t = 350 - 14.7t$$

$$15.8t + 14.7t = 350 - 184$$

$$30.5t = 166$$

$$t \approx 5.4$$

So, from the given models, you would expect that the amount of coal produced in Australia exceeded the amount produced in Germany sometime in 1995.

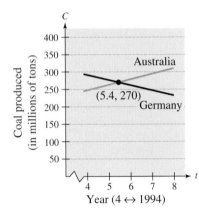

Figure 6.4

---

| Discussing the Concept | Interpreting Points of Intersection |
|---|---|

You plan to rent a 14-foot truck for a 2-day local move. At truck rental agency A, you can rent a truck for $29.95 per day plus $0.49 per mile. At agency B, you can rent a truck for $50 per day plus $0.25 per mile. The total cost y (in dollars) for the truck from agency A is

$$y = (\$29.95 \text{ per day})(2 \text{ days}) + 0.49x = 59.90 + 0.49x$$

where x is the total number of miles the truck is driven. Write a total cost equation in terms of x and y for the total cost for the truck from agency B. Use a graphing utility to graph the two equations and find the point of intersection. Interpret the meaning of the point of intersection in the context of the problem. Which agency should you choose if you plan to travel a total of 100 miles over the 2-day move? Why? How does the situation change if you plan to drive 200 miles per day instead?

# Warm Up

The following warm-up exercises involve skills that were covered in earlier sections. You will use these skills in the exercise set for this section.

In Exercises 1–4, sketch the graph of the equation.

**1.** $y = -\frac{1}{3}x + 6$

**2.** $y = 2(x - 3)$

**3.** $x^2 + y^2 = 4$

**4.** $y = 5 - (x - 3)^2$

In Exercises 5–8, perform the indicated operations and simplify.

**5.** $(3x + 2y) - 2(x + y)$

**6.** $(-10u + 3v) + 5(2u - 8v)$

**7.** $x^2 + (x - 3)^2 + 6x$

**8.** $y^2 - (y + 1)^2 + 2y$

In Exercises 9 and 10, solve the equation.

**9.** $3x + (x - 5) = 15 + 4$

**10.** $y^2 + (y - 2)^2 = 2$

---

## 6.1   Exercises

In Exercises 1 and 2, decide whether each ordered pair is a solution of the system.

**1.**  $x + 4y = -3$
    $5x - y = \phantom{-}6$
   (a) $(-1, -1)$
   (b) $(1, -1)$

**2.** $2x - \phantom{3}y = 2$
    $x + 3y = 8$
   (a) $(2, 1)$
   (b) $(2, 2)$

In Exercises 3–12, use substitution to solve the system. Then use the graph to confirm your solution.

**3.**  $x + y = -1$
    $-2x + y = -7$

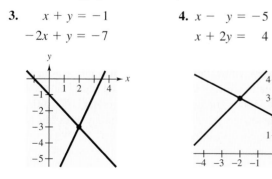

**4.** $x - \phantom{2}y = -5$
    $x + 2y = \phantom{-}4$

**5.**  $x - y = -3$
    $x^2 - y = -1$

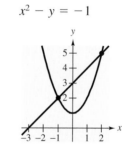

**6.** $x^2 - \phantom{4x +}y = 0$
    $x^2 - 4x + y = 0$

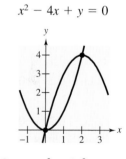

**7.**  $x - y = 0$
    $x^3 - 5x + y = 0$

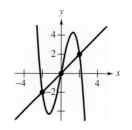

**8.** $y = x^3 - 3x^2 + 3$
    $2x + y = 3$

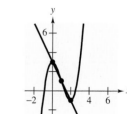

**9.** $3x + y = 4$
$x^2 + y^2 = 16$

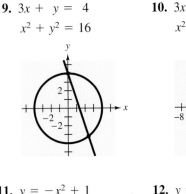

**10.** $3x - 4y = 18$
$x^2 + y^2 = 36$

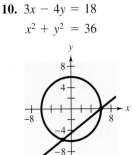

**11.** $y = -x^2 + 1$
$y = x^2 - 1$

**12.** $y = x^2 - 3x - 4$
$y = -x^2 + 3x + 4$

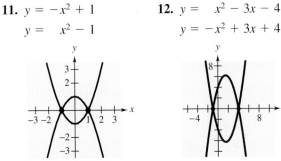

In Exercises 13–32, use substitution to solve the system.

**13.** $2x - y = -3$
$-3x - 4y = -1$

**14.** $x + 2y = 1$
$5x - 4y = -23$

**15.** $2x - y + 2 = 0$
$4x + y - 5 = 0$

**16.** $6x - 3y - 4 = 0$
$x + 2y - 4 = 0$

**17.** $0.3x - 0.4y - 0.33 = 0$
$0.1x + 0.2y - 0.21 = 0$

**18.** $1.5x + 0.8y = 2.3$
$0.3x - 0.2y = 0.1$

**19.** $\frac{1}{5}x + \frac{1}{2}y = 8$
$x + y = 20$

**20.** $\frac{1}{2}x + \frac{3}{4}y = 10$
$\frac{3}{2}x - y = 4$

**21.** $x - y = 7$
$2x + y = 23$

**22.** $x - 2y = -2$
$3x - y = 6$

**23.** $y = 2x$
$y = x^2 + 1$

**24.** $x + y = 4$
$x^2 - y = 2$

**25.** $3x - 7y + 6 = 0$
$x^2 - y^2 = 4$

**26.** $x^2 + y^2 = 25$
$2x + y = 10$

**27.** $x^2 + y^2 = 5$
$x - y = 1$

**28.** $x^2 + y = 4$
$2x - y = 1$

**29.** $y = x^4 - 2x^2 + 1$
$y = 1 - x^2$

**30.** $y = x^3 - 2x^2 + x - 1$
$y = -x^2 + 3x - 1$

**31.** $xy - 2 = 0$
$y = \sqrt{x - 1}$

**32.** $xy = 3$
$y = \sqrt{x - 2}$

In Exercises 33–40, use a graphing utility to find graphically the point of intersection of the graphs. Then confirm your solution algebraically.

**33.** $y = x^2 + 3x - 1$
$y = -x^2 - 2x + 2$

**34.** $y = -2x^2 + x - 1$
$y = x^2 - 2x - 1$

**35.** $x - y + 3 = 0$
$x^2 - 4x + 7 = y$

**36.** $x - y = 3$
$x - y^2 = 1$

**37.** $y = e^x$
$x - y + 1 = 0$

**38.** $y = \sqrt{x}$
$y = x$

**39.** $2x - y + 3 = 0$
$x^2 + y^2 - 4x = 0$

**40.** $x^2 + y^2 = 8$
$y = x^2 + 4$

***Break-Even Analysis*** In Exercises 41–44, find the sales necessary to break even ($R = C$) for the given cost $C$ of producing $x$ units and the given revenue $R$ obtained by selling $x$ units. (Round your answer to the nearest whole unit.)

**41.** $C = 8650x + 250{,}000;\ R = 9950x$

**42.** $C = 5.5\sqrt{x} + 10{,}000;\ R = 3.29x$

**43.** $C = 2.65x + 350{,}000;\ R = 4.15x$

**44.** $C = 0.08x + 50{,}000;\ R = 0.25x$

**45.** ***Break-Even Point*** Suppose you are setting up a small business and have invested $16,000 to produce an item that will sell for $5.95. If each unit can be produced for $3.45, how many units must you sell to break even?

**46.** ***Break-Even Point*** Suppose you are setting up a small business and have made an initial investment of $5000. The unit cost of the product is $21.60, and the selling price is $34.10. How many units must you sell to break even?

---

The symbol ⊞ indicates an exercise or parts of an exercise in which you are instructed to use a graphing utility.

**47.** *Comparing Populations* From 1990 to 1999, the population of New York grew more slowly than that of Texas. Two models that represent the populations of the two states are

$P = 0.911t^3 - 15.35t^2 + 88.1t + 17,976$   New York

$P = 339.5t + 16,983$   Texas

where $P$ is the population in thousands and $t = 0$ represents 1990. Use a graphing utility to determine when the population of Texas overtook the population of New York. (Source: U.S. Bureau of the Census)

**48.** *Comparing Populations* In 1998, Houston was the tenth most populated city in the United States and Atlanta was the eleventh. The populations of Houston and Atlanta from 1990 to 1998 can be approximated by the models

$P = -1.75t^2 + 94.3t + 3750$   Houston

$P = 1.31t^2 + 87.8t + 2960$   Atlanta

where $P$ is the population (within the city limits) in thousands and $t = 0$ represents 1990. Use a graphing utility to determine if, according to the models, the population of Atlanta will overtake the population of Houston. (Source: U.S. Bureau of the Census)

**49.** *Gross Domestic Product* The annual gross domestic products $G$ (in billions of dollars) of the United States and the European Union from 1995 to 1999 can be approximated by the models

$G = 465.6t + 4985$   United States

$G = -107.2917t^4 + 3038.417t^3 - 31,756.71t^2$
$\quad + 144,986.6t - 235,155$   European Union

where $t = 5$ represents 1995. Use a graphing utility to determine when the gross domestic products of the United States and the European Union were approximately equal. (Source: Organization for Economic Cooperation and Development)

**50.** *Advertising Expenditures* The amount of money $y$ (in millions of dollars) spent on advertising in newspapers and on television in the United States from 1990 to 1999 can be modeled by

$y = 31,035 - 87.7t + 212.88t^2$   Newspapers

$y = 26,068 + 802.7t + 103.45t^2$   Television

where $t = 0$ represents 1990. Use a graphing utility to determine whether, according to the models, television advertising expenditures will overtake newspaper advertising expenditures. (Source: McCann-Erickson, Inc.)

**51.** *Investment Portfolio* A total of $25,000 is invested in two funds paying 8% and 8.5% simple interest. If the yearly interest is $2060, how much of the $25,000 is invested at each rate?

**52.** *Investment Portfolio* A total of $18,000 is invested in two funds paying 7.75% and 8.25% simple interest. If the yearly interest is $1455, how much of the $18,000 is invested at each rate?

**53.** *Choice of Two Jobs* You are offered two different sales jobs. One company offers a straight commission of 6% of the sales. The other company offers a salary of $250 per week *plus* 3% of the sales. How much would you have to sell in a week in order to make the straight commission offer better?

**54.** *Choice of Two Jobs* You are offered two different sales jobs. One company offers an annual salary of $20,000 *plus* a year-end bonus of 1% of your total sales. The other company offers a salary of $15,000 *plus* a year-end bonus of 2% of your total sales. How much would you have to sell in a year in order to make the second offer better?

**55.** *Record Store or Other Stores?* The purchases $y$ (in percent) of sound recordings from record stores and from other stores by consumers in the United States from 1991 to 2000 can be approximated by the models

$y = 69.3 - 7.36t + 1.144t^2 - 0.0679t^3$   Record stores

$y = 0.138t^2 + 0.36t + 23.3$   Other stores

where $t = 1$ represents 1991. Use a graphing utility to determine when the purchases from other stores overtook record store purchases according to the models. (Source: Recording Industry Association of America, Inc.)

## 6.2    Linear Systems in Two Variables

### Objectives

* Solve a linear system by the method of elimination.
* Interpret the solution of a linear system graphically.
* Construct and use a linear system to solve an application problem.

## The Method of Elimination

In Section 6.1, you studied two methods for solving a system of equations: substitution and graphing. In this section, you will study a third method called the **method of elimination.** The key step in the method of elimination is to obtain, for one of the variables, coefficients that differ only in sign so that *adding* the two equations eliminates this variable. The following system provides an example.

$$3x + 5y = 7 \qquad \text{Equation 1}$$
$$\underline{-3x - 2y = -1} \qquad \text{Equation 2}$$
$$3y = 6 \qquad \text{Add equations.}$$

Note that by adding the two equations, you eliminate the variable $x$ and obtain a single equation in $y$. Solving this equation for $y$ produces $y = 2$, which you can then back-substitute into one of the original equations to solve for $x$.

| Example 1 | The Method of Elimination |

Solve the following system of linear equations.

$$3x + 2y = 4 \qquad \text{Equation 1}$$
$$5x - 2y = 8 \qquad \text{Equation 2}$$

### Solution

Because the coefficients for $y$ differ only in sign, you can eliminate $y$ by adding the two equations.

$$3x + 2y = 4 \qquad \text{Equation 1}$$
$$\underline{5x - 2y = 8} \qquad \text{Equation 2}$$
$$8x = 12 \qquad \text{Add equations.}$$

So, $x = \frac{3}{2}$. By back-substituting this value into the first equation, you can solve for $y$, as follows.

$$3x + 2y = 4 \qquad \text{Equation 1}$$
$$3\left(\tfrac{3}{2}\right) + 2y = 4 \qquad \text{Substitute } \tfrac{3}{2} \text{ for } x.$$
$$y = -\tfrac{1}{4} \qquad \text{Solve for } y.$$

The solution is $\left(\frac{3}{2}, -\frac{1}{4}\right)$. Check this in the original system.

### Study Tip

The method of substitution can also be used to solve the system in Example 1. Use substitution to solve the system. Which method do you think is easier? Many people find that the method of elimination is more efficient.

| Example 2 | The Method of Elimination  |
|---|---|

Solve the following system of linear equations.

$$2x - 3y = -7 \qquad \text{Equation 1}$$
$$3x + y = -5 \qquad \text{Equation 2}$$

**Solution**

For this system, you can obtain coefficients that differ only in sign by multiplying the second equation by 3.

$$2x - 3y = -7 \qquad\Longrightarrow\qquad 2x - 3y = -7 \qquad \text{Equation 1}$$
$$3x + y = -5 \qquad\Longrightarrow\qquad \underline{9x + 3y = -15} \qquad \text{Multiply Equation 2 by 3.}$$
$$11x \qquad\quad = -22 \qquad \text{Add equations.}$$

By dividing both sides by 11, you can see that $x = -2$. By back-substituting this value of $x$ into Equation 1, you can solve for $y$.

$$2x - 3y = -7 \qquad \text{Equation 1}$$
$$2(-2) - 3y = -7 \qquad \text{Substitute } -2 \text{ for } x.$$
$$-3y = -3 \qquad \text{Add 4 to both sides.}$$
$$y = 1 \qquad \text{Solve for } y.$$

The solution is $(-2, 1)$. Check this in the original system, as follows.

*Check*

$$2(-2) - 3(1) \overset{?}{=} -7 \qquad \text{Substitute into Equation 1.}$$
$$-4 - 3 = -7 \qquad \text{Equation 1 checks. } \checkmark$$
$$3(-2) + 1 \overset{?}{=} -5 \qquad \text{Substitute into Equation 2.}$$
$$-6 + 1 = -5 \qquad \text{Equation 2 checks. } \checkmark$$

In Example 2, the two systems of linear equations

$$2x - 3y = -7 \qquad \text{and} \qquad 2x - 3y = -7$$
$$3x + y = -5 \qquad\qquad\qquad 9x + 3y = -15$$

are called **equivalent** because they have precisely the same solution set. The operations that can be performed on a system of linear equations to produce an equivalent system are (1) interchanging any two equations, (2) multiplying an equation by a nonzero constant, and (3) adding a multiple of one equation to any other equation in the system.

> ▶ The Method of Elimination
>
> To use the **method of elimination** to solve a system of two linear equations in $x$ and $y$, use the following steps.
>
> 1. Obtain coefficients for $x$ (or $y$) that differ only in sign by multiplying all terms of one or both equations by suitably chosen constants.
>
> 2. Add the equations to eliminate one variable and solve the resulting equation.
>
> 3. Back-substitute the value obtained in Step 2 into either of the original equations and solve for the other variable.
>
> 4. Check your solution in both of the original equations.

**Example 3**    The Method of Elimination

Solve the following system of linear equations.

$$5x + 3y = 9 \qquad \text{Equation 1}$$

$$2x - 4y = 14 \qquad \text{Equation 2}$$

**Solution**

You can obtain coefficients that differ only in sign by multiplying Equation 1 by 4 and multiplying Equation 2 by 3.

| $5x + 3y = 9$ | ⟹ | $20x + 12y = 36$ | Multiply Equation 1 by 4. |
|---|---|---|---|
| $2x - 4y = 14$ | ⟹ | $6x - 12y = 42$ | Multiply Equation 2 by 3. |
| | | $26x \qquad\ = 78$ | Add equations. |

From this equation, you can see that $x = 3$. By back-substituting this value of $x$ into the second equation, you can solve for $y$, as follows.

$$2x - 4y = 14 \qquad \text{Equation 2}$$

$$2(3) - 4y = 14 \qquad \text{Substitute 3 for } x.$$

$$-4y = 8$$

$$y = -2 \qquad \text{Solve for } y.$$

The solution is $(3, -2)$. Check this in the original system.

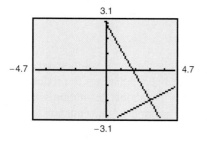

Figure 6.5

Remember that you can check the solution of a system of equations graphically. For instance, to check the solution found in Example 3, sketch the graphs of both equations on the same screen, as shown in Figure 6.5. Notice that the two lines intersect at $(3, -2)$.

## Graphical Interpretation of Solutions

It is possible for a *general* system of equations to have exactly one solution, two or more solutions, or no solution. If a system of *linear* equations has two different solutions, it must have an *infinite* number of solutions. To see why this is true, consider the following graphical interpretations of systems of two linear equations in two variables. (Remember that the graph of a linear equation in two variables is a straight line.)

▶ **Graphical Interpretation of Solutions** *Learning Tools*

For a system of two linear equations in two variables, the number of solutions is given by one of the following.

| *Number of Solutions* | *Graphical Interpretation* |
|---|---|
| **1.** Exactly one solution | The two lines intersect at one point. |
| **2.** Infinitely many solutions | The two lines are identical. |
| **3.** No solution | The two lines are parallel. |

These graphical interpretations are shown in Figure 6.6.

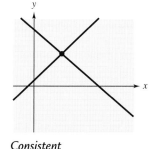

*Consistent*
*Two lines that intersect*
*Single point of intersection*
Figure 6.6

*Consistent*
*Two lines that coincide*
*Infinitely many points*

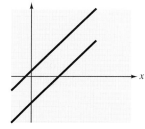

*Inconsistent*
*Two parallel lines*
*No point of intersection*

A system of linear equations is called **consistent** if it has at least one solution. A consistent system with exactly one solution is **independent.** A consistent system with infinitely many solutions is **dependent.** A system of linear equations is called **inconsistent** if it has no solution. In Examples 4 and 5, note how you can use the method of elimination to determine that a system of linear equations has no solution or infinitely many solutions.

| Example 4 | The Method of Elimination: No-Solution Case |  |

Solve the following system of linear equations.

$$x - 2y = 3 \qquad \text{Equation 1}$$
$$-2x + 4y = 1 \qquad \text{Equation 2}$$

### Solution

To obtain coefficients that differ only in sign, multiply Equation 1 by 2.

$$x - 2y = 3 \implies 2x - 4y = 6 \qquad \text{Multiply Equation 1 by 2.}$$
$$-2x + 4y = 1 \implies -2x + 4y = 1 \qquad \text{Equation 2}$$
$$0 = 7 \qquad \text{False statement}$$

Because there are no values of $x$ and $y$ for which $0 = 7$, you can conclude that the system is inconsistent and has no solution. The lines corresponding to the two equations given in this system are shown in Figure 6.7. Note that the two lines are parallel, and therefore have no point of intersection.

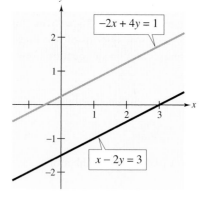
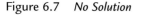

**Figure 6.7** *No Solution*

In Example 4, note that the occurrence of a false statement, such as $0 = 7$, indicates that the system has no solution. In the next example, note that the occurrence of a statement that is true for all values of the variables, such as $0 = 0$, indicates that the system has infinitely many solutions.

| Example 5 | The Method of Elimination: Many-Solutions Case |

Solve the following system of linear equations.

$$2x - y = 1 \qquad \text{Equation 1}$$
$$4x - 2y = 2 \qquad \text{Equation 2}$$

### Solution

To obtain coefficients that differ only in sign, multiply the second equation by $-\frac{1}{2}$.

$$2x - y = 1 \implies 2x - y = 1 \qquad \text{Equation 1}$$
$$4x - 2y = 2 \implies -2x + y = -1 \qquad \text{Multiply Equation 2 by } -\frac{1}{2}.$$
$$0 = 0 \qquad \text{Add equations.}$$

Because the two equations turn out to be equivalent (have the same solution set), you can conclude that the system has infinitely many solutions. The solution set consists of all points $(x, y)$ lying on the line $2x - y = 1$ as shown in Figure 6.8. To represent the solution set as an ordered pair, let $x = a$, where $a$ is any real number. Then $y = 2a - 1$ and the solution set can be written as $(a, 2a - 1)$.

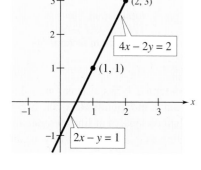

**Figure 6.8** *Infinite Number of Solutions*

**9.** $9x - 3y = -1$
$3x + 6y = -5$

**10.** $5x + 3y = 18$
$2x - 7y = -1$

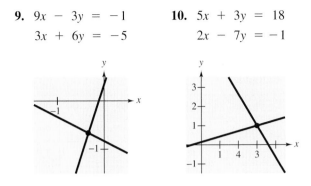

In Exercises 11–30, solve the system by elimination.

**11.** $4x - 3y = 11$
$-6x + 3y = 3$

**12.** $3x - 5y = 2$
$2x + 5y = 13$

**13.** $3x - y = 17$
$5x + 5y = -5$

**14.** $x + 7y = 12$
$3x - 5y = 10$

**15.** $3x + 2y = 10$
$2x + 5y = 3$

**16.** $8r + 16s = 20$
$16r + 50s = 55$

**17.** $2u + v = 120$
$u + 2v = 120$

**18.** $5u + 6v = 24$
$3u + 5v = 18$

**19.** $6r - 5s = 3$
$-1.2r + s = 0.5$

**20.** $1.8x + 1.2y = 4$
$9x + 6y = 3$

**21.** $\dfrac{x}{4} + \dfrac{y}{6} = 1$
$x - y = 3$

**22.** $\dfrac{2}{3}x + \dfrac{1}{6}y = \dfrac{2}{3}$
$4x + y = 4$

**23.** $\dfrac{x + 3}{4} + \dfrac{y - 1}{3} = 1$
$x - y = 3$

**24.** $\dfrac{x - 1}{2} + \dfrac{y + 2}{3} = 4$
$x - 2y = 5$

**25.** $2.5x - 3y = 1.5$
$10x - 12y = 6$

**26.** $3.5x - 2y = 2.6$
$7x - 2.5y = 4$

**27.** $0.05x - 0.03y = 0.21$
$0.07x + 0.02y = 0.16$

**28.** $0.02x - 0.05y = -0.19$
$0.03x + 0.04y = 0.52$

**29.** $4b + 3m = 3$
$3b + 11m = 13$

**30.** $3b + 3m = 7$
$3b + 5m = 3$

In Exercises 31 and 32, the graphs of the two equations appear to be parallel. Are they? Explain your reasoning.

**31.** $200y - x = 200$
$199y - x = -198$

**32.** $25x - 24y = 0$
$13x - 12y = 120$

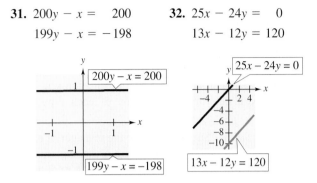

**33.** *Airplane Speed*   An airplane flying into a headwind travels the 1800-mile flying distance between two cities in 3 hours and 36 minutes. On the return flight, the distance is traveled in 3 hours. Find the air speed of the plane and the speed of the wind, assuming that both remain constant.

**34.** *Airplane Speed*   Two planes start from the same airport and fly in opposite directions. The second plane starts one-half hour after the first plane, but its speed is 50 miles per hour faster. Find the air speed of each plane if, 2 hours after the first plane starts its flight, the planes are 2000 miles apart.

**35.** *Acid Mixture*   Ten gallons of a 30% acid solution is obtained by mixing a 20% solution with a 50% solution. How much of each must be used?

**36.** *Fuel Mixture*   Five hundred gallons of 89-octane gasoline is obtained by mixing 87-octane gasoline with 90-octane gasoline. How much of each must be used? (Octane ratings can be interpreted as percents. A high octane rating indicates that the gasoline is knock resistant. A low octane rating indicates that the gasoline is knock prone.)

**37.** *Investment Portfolio*   A total of $12,000 is invested in two corporate bonds that pay 10.5% and 12% simple interest. The annual interest is $1380. How much is invested in each bond?

**38.** *Investment Portfolio*   A total of $32,000 is invested in two municipal bonds that pay 5.75% and 6.25% simple interest. The annual interest is $1930. How much is invested in each bond?

**39. Ticket Sales** You are the manager of a theater. On Saturday morning you are going over the ticket sales for Friday evening. A total of 750 tickets were sold. The tickets for adults and children sold for $7.50 and $4.00, respectively, and the receipts for the performance were $3808.50. However, your assistant manager did not record how many of each type of ticket were sold. From the information you have, can you determine how many of each type were sold?

**40. Shoe Sales** Suppose you are the manager of a shoe store. On Sunday morning you are going over the receipts for the previous week's sales. A total of 240 pairs of tennis shoes were sold. One style sold for $66.95 and the other sold for $84.95. The total receipts were $17,652. The cash register that was supposed to keep track of the number of each type of shoe sold malfunctioned. Can you recover the information? If so, how many of each type were sold?

**Supply and Demand** In Exercises 41–44, find the point of equilibrium for each pair of supply and demand equations.

| | Demand | Supply |
|---|---|---|
| **41.** | $p = 56 - 0.0001x$ | $p = 22 + 0.00001x$ |
| **42.** | $p = 60 - 0.00001x$ | $p = 15 + 0.00004x$ |
| **43.** | $p = 140 - 0.00002x$ | $p = 80 + 0.00001x$ |
| **44.** | $p = 400 - 0.0002x$ | $p = 225 + 0.0005x$ |

**45. U.S. Exports and General Imports** The amounts of U.S. exports and general imports from 1996 to 1999 are shown in the table. The numbers in the table represent billions of dollars. (Source: U.S. Bureau of the Census)

| Year | 1996 | 1997 | 1998 | 1999 |
|---|---|---|---|---|
| Exports | 624.8 | 689.2 | 682.1 | 695.0 |
| Imports | 791.4 | 870.7 | 911.9 | 1025.0 |

(a) Make a scatter plot for both sets of data and find a pair of linear equations that represent the exports and the imports. Let $t = 6$ represent 1996.

(b) Assuming that the amounts for the given 4 years are representative of future years, will the exports ever equal the imports?

**46. Students Per Computer** The number of students per computer in public and private elementary schools in the United States are given in the table for the school years shown. (Source: Market Data Retrieval)

| School year | 96–97 | 97–98 | 98–99 | 99–00 |
|---|---|---|---|---|
| Public | 8.0 | 6.9 | 6.3 | 5.5 |
| Private | 9.6 | 8.6 | 7.5 | 7.2 |

(a) Use a graphing utility to make a scatter plot of the data for public schools and use its regression capabilities to find a quadratic model. Let $x = 0$ represent the 96–97 school year. Repeat the procedure for the data for private schools.

(b) Use the equations to predict the numbers of students per computer in public and private elementary schools for the 2007–2008 school year.

(c) Discuss the reasonableness of your predictions.

**47. Delivery Service** A shipping department hires two delivery services to deliver 2400 packages. One service will deliver three times as many packages as the other. Find the number delivered by each.

**48. Reasoning** Design a system of two linear equations containing two variables that has (a) exactly one solution, (b) infinitely many solutions, and (c) no solutions. Solve each system algebraically and graph.

**Fitting a Line to Data** In Exercises 49–52, find the least squares regression line $y = ax + b$ for the points $(x_1, y_1), (x_2, y_2), \ldots, (x_n, y_n)$. To find the line, solve the system for $a$ and $b$. (If you are unfamiliar with summation notation, look at the discussion in Section 8.1.)

$$nb + \left( \sum_{i=1}^{n} x_i \right) a = \sum_{i=1}^{n} y_i$$

$$\left( \sum_{i=1}^{n} x_i \right) b + \left( \sum_{i=1}^{n} x_i^2 \right) a = \sum_{i=1}^{n} x_i y_i$$

**49.**  $5b + 10a = 20.2$      **50.**  $5b + 10a = 11.7$
$\phantom{49.}$  $10b + 30a = 50.1$      $\phantom{50.}$  $10b + 30a = 25.6$

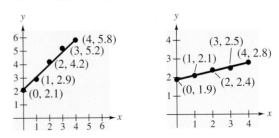

**51.**  $7b + 21a = 35.1$
  $21b + 91a = 114.2$

**52.**  $6b + 15a = 23.6$
  $15b + 55a = 48.8$

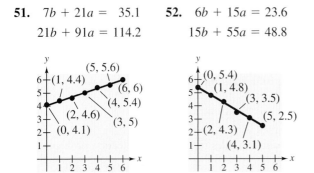

**53.** *Restaurant Sales*   The total sales (in billions of dollars) for full-service restaurants in the United States from 1995 to 2000 are shown in the table. (Source: National Restaurant Association)

(a) Solve the system below for $a$ and $b$ to find the least squares regression line $y = at + b$ for the data.

$$6b + 15a = 664.5$$

$$15b + 55a = 1773.5$$

(b) Graph the line on a graphing utility and estimate the total restaurant sales in 2001.

(c) Use the regression capabilities of a graphing utility to find a linear model. Compare this model with the one you found in part (a).

(d) Use the linear equation found in part (c) to estimate total restaurant sales in 2001. Compare this estimate with the one you found in part (b).

| Year | $t$ | Total Sales (in billions of dollars) |
|------|-----|--------------------------------------|
| 1995 | 0 | 96.4 |
| 1996 | 1 | 100.8 |
| 1997 | 2 | 106.4 |
| 1998 | 3 | 111.8 |
| 1999 | 4 | 121.0 |
| 2000 | 5 | 128.1 |

**54.** *SAT Scores*   The average math scores on the SAT for college-bound seniors in the United States from 1995 to 1999 are shown in the table. (Source: College Entrance Examination Board)

(a) Solve the system below for $a$ and $b$ to find the least squares regression line $y = at + b$ for the data.

$$5b + 10a = 2548$$

$$10b + 30a = 5110$$

(b) Graph the line on a graphing utility and estimate the average math score in 2001.

(c) Use the regression capabilities of a graphing utility to find a linear model. Compare this model with the one you found in part (a).

(d) Use the linear equation found in part (c) to estimate the average math score in 2001. Compare this estimate with the one you found in part (b).

| Year | $t$ | Average SAT Score in Math |
|------|-----|---------------------------|
| 1995 | 0 | 506 |
| 1996 | 1 | 508 |
| 1997 | 2 | 511 |
| 1998 | 3 | 512 |
| 1999 | 4 | 511 |

## 6.3 Linear Systems in Three or More Variables

### Objectives

- Solve a linear system in row-echelon form using back-substitution.
- Use Gaussian elimination to solve a linear system.
- Solve a nonsquare linear system.
- Construct and use a linear system in three or more variables to solve an application problem.
- Find the equation of a circle or a parabola using a linear system in three or more variables.

## Row-Echelon Form and Back-Substitution

### Chui-chang suan-shu

(250 B.C.)

One of the most influential Chinese mathematics books was the *Chui-chang suan-shu* or *Nine Chapters on the Mathematical Art* (written in approximately 250 B.C.). Chapter Eight of the *Nine Chapters* contained solutions of systems of linear equations using positive and negative numbers. One such system was

$$3x + 2y + z = 39$$
$$2x + 3y + z = 34$$
$$x + 2y + 3z = 26.$$

This system was solved using column operations on a matrix.

The method of elimination can be applied to a system of linear equations in more than two variables. In fact, this method easily adapts to computer use for solving linear systems with dozens of variables.

When elimination is used to solve a system of linear equations, the goal is to rewrite the system in a form to which back-substitution can be applied. To see how this works, consider the following two systems of linear equations.

$$\begin{aligned} x - 2y + 3z &= 9 \\ -x + 3y &= -4 \\ 2x - 5y + 5z &= 17 \end{aligned} \qquad \begin{aligned} x - 2y + 3z &= 9 \\ y + 3z &= 5 \\ z &= 2 \end{aligned}$$

The system on the right is said to be in **row-echelon form,** which means that it has a "stair-step" pattern with leading coefficients of 1. After comparing the two systems, it should be clear that it is easier to solve the system on the right.

### Example 1   Using Back-Substitution

Solve the following system of linear equations.

$$\begin{aligned} x - 2y + 3z &= 9 \qquad &\text{Equation 1} \\ y + 3z &= 5 \qquad &\text{Equation 2} \\ z &= 2 \qquad &\text{Equation 3} \end{aligned}$$

#### Solution

From Equation 3, you know the value of $z$. To solve for $y$, substitute $z = 2$ into Equation 2 to obtain

$$\begin{aligned} y + 3(2) &= 5 \qquad &\text{Substitute 2 for } z. \\ y &= -1. \qquad &\text{Solve for } y. \end{aligned}$$

Finally, substitute $y = -1$ and $z = 2$ into Equation 1 to obtain

$$\begin{aligned} x - 2(-1) + 3(2) &= 9 \qquad &\text{Substitute } -1 \text{ for } y \text{ and 2 for } z. \\ x &= 1. \qquad &\text{Solve for } x. \end{aligned}$$

The solution is $x = 1$, $y = -1$, and $z = 2$, which can be written as the **ordered triple** $(1, -1, 2)$. Check this in the original system of equations.

# Gaussian Elimination

Two systems of equations are **equivalent** if they have the same solution set. To solve a system that is not in row-echelon form, first convert it to an *equivalent* system that is in row-echelon form. To see how this is done, let's take another look at the method of elimination, as applied to a system of two linear equations.

Corbis-Bettmann

## Carl Friedrich Gauss

(1777–1855)

The process of rewriting a system of equations in row-echelon form by using the three basic row operations is called **Gaussian elimination,** after the German mathematician Carl Friedrich Gauss. Example 2 shows the chain of equivalent systems used to solve a linear system in two variables.

**Example 2**    The Method of Elimination

Solve the following system of linear equations.

$$3x - 2y = -1$$
$$x - y = 0$$

**Solution**

$$x - y = 0$$    You can interchange two equations in the system.
$$3x - 2y = -1$$

$$-3x + 3y = 0$$    Multiply the first equation by $-3$.

$$-3x + 3y = 0$$    You can add the multiple of the first equation to the second equation to obtain a new equation.
$$\underline{3x - 2y = -1}$$
$$y = -1$$

$$x - y = 0$$    New system in row-echelon form
$$y = -1$$

Now, using back-substitution, you can determine that the solution is $y = -1$ and $x = -1$, which can be written as the ordered pair $(-1, -1)$. Check this in the original system of equations.

---

▶ **Operations That Produce Equivalent Systems**

Each of the following **row operations** on a system of linear equations produces an *equivalent* system of linear equations.

**1.** Interchange two equations.

**2.** Multiply one of the equations by a nonzero constant.

**3.** Add a multiple of one of the equations to another equation to replace the latter equation.

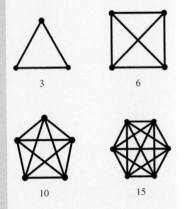
**Example 3** Using Elimination to Solve a System

Solve the following system of linear equations.

$$x - 2y + 3z = 9 \qquad \text{Equation 1}$$
$$-x + 3y \phantom{ + 3z} = -4 \qquad \text{Equation 2}$$
$$2x - 5y + 5z = 17 \qquad \text{Equation 3}$$

**Solution**

There are many ways to begin, but we suggest working from the upper left corner, saving the $x$ in the upper left position and eliminating the other $x$'s from the first column.

$$x - 2y + 3z = 9$$
$$y + 3z = 5$$
$$2x - 5y + 5z = 17$$

Adding the first equation to the second equation produces a new second equation.

$$x - 2y + 3z = 9$$
$$y + 3z = 5$$
$$-y - z = -1$$

Adding $-2$ times the first equation to the third equation produces a new third equation.

Now that all but the first $x$ have been eliminated from the first column, go to work on the second column. (You need to eliminate $y$ from the third equation.)

$$x - 2y + 3z = 9$$
$$y + 3z = 5$$
$$2z = 4$$

Adding the second equation to the third equation produces a new third equation.

Finally, you need a coefficient of 1 for $z$ in the third equation.

$$x - 2y + 3z = 9$$
$$y + 3z = 5$$
$$z = 2$$

Multiplying the third equation by $\frac{1}{2}$ produces a new third equation.

This is the same system that was solved in Example 1, and, as in that example, you can conclude that the solution is

$$x = 1, \quad y = -1, \quad \text{and} \quad z = 2.$$

In Example 3, you can check the solution by substituting $x = 1$, $y = -1$, and $z = 2$ into each original equation, as follows.

Equation 1: $(1) - 2(-1) + 3(2) = 9$ ✓

Equation 2: $-(1) + 3(-1) = -4$ ✓

Equation 3: $2(1) - 5(-1) + 5(2) = 17$ ✓

The next example involves an inconsistent system—one that has no solution. The key to recognizing an inconsistent system is that at some stage in the elimination process you obtain an absurdity such as $0 = -2$.

**Example 4**    An Inconsistent System

Solve the following system of linear equations.

$$x - 3y + z = 1 \qquad \text{Equation 1}$$
$$2x - y - 2z = 2 \qquad \text{Equation 2}$$
$$x + 2y - 3z = -1 \qquad \text{Equation 3}$$

**Solution**

$$x - 3y + z = 1$$
$$5y - 4z = 0$$
$$x + 2y - 3z = -1$$

Adding $-2$ times the first equation to the second equation produces a new second equation.

$$x - 3y + z = 1$$
$$5y - 4z = 0$$
$$5y - 4z = -2$$

Adding $-1$ times the first equation to the third equation produces a new third equation.

$$x - 3y + z = 1$$
$$5y - 4z = 0$$
$$0 = -2$$

Adding $-1$ times the second equation to the third equation produces a new third equation.

Because the third "equation" is impossible, you can conclude that this system is inconsistent and therefore has no solution. Moreover, because this system is equivalent to the original system, you can conclude that the original system also has no solution.

As with a system of linear equations in two variables, the solution(s) of a system of linear equations in more than two variables must fall into one of three categories. Because an equation in three variables represents a plane in space, the possible solutions can be shown graphically, as in Figure 6.11.

> ▶ **The Number of Solutions of a Linear System**
>
> For a system of linear equations, exactly one of the following is true.
>
> **1.** There is exactly one solution. [See Figure 6.11(a).]
>
> **2.** There are infinitely many solutions. [See Figures 6.11(b) and (c).]
>
> **3.** There is no solution. [See Figures 6.11(d) and (e).]

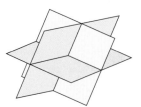

(a) Solution: one point

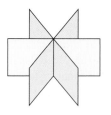

(b) Solution: one line

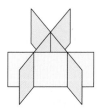

(c) Solution: one plane

(d) Solution: none

(e) Solution: none

Figure 6.11

> **Example 5**   A System with Infinitely Many Solutions

Solve the following system of linear equations.

$$
\begin{aligned}
x + y - 3z &= -1 \qquad \text{Equation 1} \\
y - z &= 0 \qquad\;\; \text{Equation 2} \\
-x + 2y \phantom{- 3z} &= 1 \qquad\;\; \text{Equation 3}
\end{aligned}
$$

**Solution**

$$
\begin{aligned}
x + y - 3z &= -1 \\
y - z &= 0 \\
3y - 3z &= 0
\end{aligned}
$$

> Adding the first equation to the third equation produces a new third equation.

$$
\begin{aligned}
x + y - 3z &= -1 \\
y - z &= 0 \\
0 &= 0
\end{aligned}
$$

> Adding $-3$ times the second equation to the third equation produces a new third equation.

This means that Equation 3 depends on Equations 1 and 2 in the sense that it gives us no additional information about the variables. So, the original system is equivalent to the system

$$
\begin{aligned}
x + y - 3z &= -1 \\
y - z &= 0.
\end{aligned}
$$

In this last equation, solve for $y$ in terms of $z$ to obtain $y = z$. Back-substituting for $y$ into the previous equation produces $x = 2z - 1$. Finally, letting $z = a$, the solutions to the given system are all of the form

$$
x = 2a - 1, \quad y = a, \quad \text{and} \quad z = a
$$

where $a$ is a real number. So, every ordered triple of the form

$$
(2a - 1, a, a), \qquad a \text{ is a real number}
$$

is a solution of the system.

In Example 5, there are other ways to write the same infinite set of solutions. For instance, the solutions could have been written as

$$
\left(b, \tfrac{1}{2}(b + 1), \tfrac{1}{2}(b + 1)\right), \qquad b \text{ is a real number.}
$$

Try convincing yourself of this by substituting $a = 0$, $a = 1$, $a = 2$, and $a = 3$ into the solution given in Example 5. Then substitute $b = -1$, $b = 1$, $b = 3$, and $b = 5$ into the solution given above. In both cases, you should obtain the same ordered triples. So, when comparing descriptions of an infinite solution set, keep in mind that there is more than one way to describe the set.

## Nonsquare Systems

So far, each system of linear equations has been **square,** which means that the number of equations is equal to the number of variables. In a **nonsquare** system, the number of equations differs from the number of variables. A system of linear equations cannot have a unique solution unless there are at least as many equations as there are variables in the system.

**Example 6**    A System with Fewer Equations than Variables

Solve the following system of linear equations.

$$x - 2y + z = 2 \qquad \text{Equation 1}$$
$$2x - y - z = 1 \qquad \text{Equation 2}$$

**Solution**

Begin by rewriting the system in row-echelon form, as follows.

$$\begin{aligned} x - 2y + z &= 2 \\ 3y - 3z &= -3 \end{aligned}$$

Adding $-2$ times the first equation to the second equation produces a new second equation.

$$\begin{aligned} x - 2y + z &= 2 \\ y - z &= -1 \end{aligned}$$

Multiplying the second equation by $\frac{1}{3}$ produces a new second equation.

Solving for $y$ in terms of $z$, you get $y = z - 1$, and back-substitution into Equation 1 yields

$$x - 2(z - 1) + z = 2$$
$$x - 2z + 2 + z = 2$$
$$x = z.$$

Finally, by letting $z = a$, you have the solution

$$x = a, \qquad y = a - 1, \qquad \text{and} \qquad z = a$$

where $a$ is a real number. So, every ordered triple of the form

$$(a, a - 1, a), \qquad a \text{ is a real number}$$

is a solution of the system.

In Example 6, try choosing some values of $a$ to obtain different solutions of the system, such as $(1, 0, 1)$, $(2, 1, 2)$, and $(3, 2, 3)$. Then check each of the solutions in the original system.

*In 1999, the amount Americans saved or invested increased by 18.1 billion dollars—an average increase of about $8 per person.* (Source: U.S. Bureau of Economic Analysis)

## Applications

| Example 7 | An Investment Portfolio

You have a portfolio totaling $450,000 and want to invest in (1) certificates of deposit, (2) municipal bonds, (3) blue-chip stocks, and (4) growth or speculative stocks. The certificates pay 9% annually, and the municipal bonds pay 6% annually. Over a 5-year period, you expect the blue-chip stocks to return 10% annually and the growth stocks to return 15% annually. You want a combined annual return of 8%, and you also want to have only one-third of the portfolio invested in stocks. How much should be allocated to each type of investment?

### Solution

To solve this problem, let $C$, $M$, $B$, and $G$ represent the amounts in the four types of investments. Because the total investment is $450,000, you can write the following equation.

$$C + M + B + G = 450,000$$

A second equation can be derived from the fact that the combined annual return should be 8%.

$$0.09C + 0.06M + 0.10B + 0.15G = 0.08(450,000)$$

Finally, because only one-third of the total investment should be allocated to stocks, you can write

$$B + G = \tfrac{1}{3}(450,000).$$

These three equations make up the following system.

$$C + M + B + G = 450,000$$
$$0.09C + 0.06M + 0.10B + 0.15G = 36,000$$
$$B + G = 150,000$$

Using elimination, you find that the system has infinitely many solutions, which can be written as follows.

$$C = -\tfrac{5}{3}a + 100,000, \qquad M = \tfrac{5}{3}a + 200,000, \qquad B = -a + 150,000,$$
$$G = a$$

So, you have many different options. One possible solution is to choose $a = 30,000$, which yields the following portfolio.

**1.** Certificates of deposit:            $50,000

**2.** Municipal bonds:                $250,000

**3.** Blue-chip stocks:                 $120,000

**4.** Growth or speculative stocks:      $30,000

### Example 8    Data Analysis: Curve-Fitting

Find a quadratic equation, $y = ax^2 + bx + c$, whose graph passes through the points $(-1, 3)$, $(1, 1)$, and $(2, 6)$.

#### Solution

Because the graph of $y = ax^2 + bx + c$ passes through the points $(-1, 3)$, $(1, 1)$, and $(2, 6)$, you can write the following.

When $x = -1$, $y = 3$:     $a(-1)^2 + b(-1) + c = 3$

When $x = 1$, $y = 1$:     $a(1)^2 + b(1) + c = 1$

When $x = 2$, $y = 6$:     $a(2)^2 + b(2) + c = 6$

This produces the following system of linear equations.

$$a - b + c = 3 \qquad \text{Equation 1}$$
$$a + b + c = 1 \qquad \text{Equation 2}$$
$$4a + 2b + c = 6 \qquad \text{Equation 3}$$

The solution of this system is $a = 2$, $b = -1$, and $c = 0$. So, the equation of the parabola is $y = 2x^2 - x$, as shown in Figure 6.12.

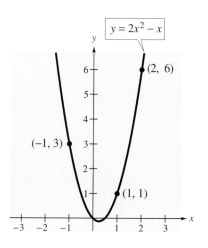

Figure 6.12

| $x$ | 1 | 3 | 5 |
|---|---|---|---|
| $y$ | 62 | 54 | 62 |

### Discussing the Concept        Mathematical Modeling

Suppose you work for an outerwear manufacturer, and the marketing department is concerned about sales trends in Arizona. Your manager has asked you to investigate climate data in hopes of explaining sales patterns and gives you the table at the left, which represents the average monthly temperature $y$ in degrees Fahrenheit for Phoenix, Arizona, for the month $x$, where $x = 1$ corresponds to November.    (Source: National Climatic Data Center)

Construct a scatter plot of the data; decide what type of mathematical model might be appropriate for the data; and use the methods you have learned thus far to fit an appropriate model. Your manager would like to know the average monthly temperatures for December and February. Explain to your manager how you found your model, what it represents, and how it can be used to find the December and February average temperatures. Investigate the usefulness of this model for the rest of the year. Would you recommend using the model to predict monthly average temperatures for the whole year or for just part of the year? Explain your reasoning.

# Warm Up

The following warm-up exercises involve skills that were covered in earlier sections. You will use these skills in the exercise set for this section.

In Exercises 1–4, solve the system of linear equations.

**1.** $x + y = 25$
$\phantom{x +}\ y = 10$

**2.** $2x - 3y = \phantom{-}4$
$\phantom{2}6x \phantom{- 3y}= -12$

**3.** $x + y = 32$
$x - y = 24$

**4.** $2r - \phantom{2}s = \phantom{1}5$
$\phantom{2}r + 2s = 10$

In Exercises 5–8, determine whether the ordered triple is a solution of the equation.

**5.** $5x - 3y + 4z = 2$
$(-1, -2, 1)$

**6.** $x - 2y + 12z = 9$
$(6, 3, 2)$

**7.** $2x - 5y + 3z = -9$
$(a - 2, a + 1, a)$

**8.** $-5x + y + z = 21$
$(a - 4, 4a + 1, a)$

In Exercises 9 and 10, solve for $x$ in terms of $a$.

**9.** $x + 2y - 3z = 4$
$y = 1 - a, z = a$

**10.** $x - 3y + 5z = 4$
$y = 2a + 3, z = a$

## 6.3   Exercises

In Exercises 1–26, solve the system of linear equations.

**1.** $x - \phantom{2}y + z = 4$
$x + 2y - z = 2$
$2x + 3y \phantom{- z}= 2$

**2.** $x + y + \phantom{3}z = 3$
$4x - y + 3z = 7$
$\phantom{4x -}\ 5y + 2z = 0$

**3.** $4x + \phantom{3}y - 3z = \phantom{1}11$
$2x - 3y + 2z = \phantom{11}9$
$\phantom{2}x + \phantom{3}y + \phantom{3}z = -3$

**4.** $\phantom{3x +}\ 6y + 4z = -12$
$3x + 3y \phantom{+ 4z}= \phantom{-1}9$
$2x \phantom{+ 3y}- 3z = \phantom{-1}10$

**5.** $3x \phantom{+ 2y}+ 2z = \phantom{-}13$
$\phantom{3}x + 2y \phantom{+ 2z}= -5$
$\phantom{3x +}-3y - \phantom{2}z = \phantom{-}10$

**6.** $2x + 3y + \phantom{3}z = -4$
$2x - 4y + 3z = \phantom{-}18$
$3x - 2y + 2z = \phantom{-1}9$

**7.** $3x - 2y + 4z = 1$
$\phantom{3}x + \phantom{2}y - 2z = 3$
$2x - 3y + 6z = 8$

**8.** $5x - \phantom{3}3y + 2z = 3$
$2x + \phantom{1}4y - \phantom{2}z = 7$
$\phantom{2}x - 11y + 4z = 3$

**9.** $3x + 3y + \phantom{1}5z = 1$
$3x + 5y + \phantom{1}9z = 0$
$5x + 9y + 17z = 0$

**10.** $2x + \phantom{2}y - \phantom{1}z = \phantom{-}13$
$\phantom{2}x + 2y + \phantom{1}z = \phantom{-1}2$
$8x - 3y + 4z = -2$

**11.**  $x + 2y - 7z = -4$
$2x + y + z = 13$
$3x + 9y - 36z = -33$

**12.**  $2x + y - 3z = 4$
$4x + 2z = 10$
$-2x + 3y - 13z = -8$

**13.**  $x + 4z = 13$
$4x - 2y + z = 7$
$2x - 2y - 7z = -19$

**14.**  $4x - y + 5z = 11$
$x + 2y - z = 5$
$5x - 8y + 13z = 7$

**15.**  $x - 2y + 5z = 2$
$3x + 2y - z = -2$

**16.**  $x - 3y + 2z = 18$
$5x - 13y + 12z = 80$

**17.**  $2x - 3y + z = -2$
$-4x + 9y = 7$

**18.**  $2x + 3y + 3z = 7$
$4x + 18y + 15z = 44$

**19.**  $x + 3w = 4$
$2y - z - w = 0$
$3y - 2w = 1$
$2x - y + 4z = 5$

**20.**  $x + y + z + w = 6$
$2x + 3y - w = 0$
$-3x + 4y + z + 2w = 4$
$x + 2y - z + w = 0$

**21.**  $x + 4z = 1$
$x + y + 10z = 10$
$2x - y + 2z = -5$

**22.**  $3x - 2y - 6z = -4$
$-3x + 2y + 6z = 1$
$x - y - 5z = -3$

**23.**  $4x + 3y + 5z = 10$
$5x + 2y + 10z = 13$
$3x + y - 2z = -9$

**24.**  $2x + 5y = 25$
$3x - 2y + 4z = 1$
$4x - 3y + z = 9$

**25.**  $5x + 5y - z = 0$
$10x + 5y + 2z = 0$
$5x + 15y - 9z = 0$

**26.**  $12x + 5y + z = 0$
$12x + 4y - z = 0$

**27.** One solution for Exercise 15 is $(-a, 2a - 1, a)$. A student gives $(b, -2b - 1, -b)$ as a solution to the same exercise. Explain how both solutions are correct.

In Exercises 28–30, write three ordered triples of the given form.

**28.** $\left(a, a - 5, \frac{2}{3}a + 1\right)$

**29.** $(3a, 5 - a, a)$

**30.** $\left(\frac{1}{2}a, 3a, 5\right)$

In Exercises 31 and 32, find the equation of the parabola that passes through the points.

**31.**                                              **32.**

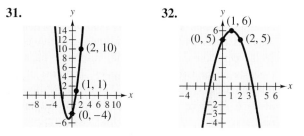

In Exercises 33 and 34, find the equation of the circle that passes through the points.

**33.**                                              **34.**

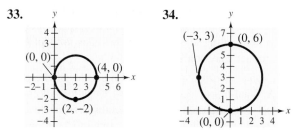

**35.** *Investment* An inheritance of $16,000 was divided among three investments yielding a total of $990 in interest per year. The interest rates for the three investments were 5%, 6%, and 7%. Find the amount placed in each investment if the 5% and 6% investments were $3000 and $2000 less than the 7% investment, respectively.

**36.** *Investment* Suppose you receive a total of $1520 a year in interest from three investments. The interest rates for the three investments are 5%, 7%, and 8%. The 5% investment is half of the 7% investment, and the 7% investment is $1500 less than the 8% investment. What is the amount of each investment?

**37.** *Investment* A company borrows $775,000. Some is borrowed at 8%, some at 9%, and some at 10%. How much is borrowed at each rate if the annual interest is $67,500 and the amount borrowed at 8% is four times the amount borrowed at 10%?

**38.** *Investment* A company borrows $800,000. Some is borrowed at 8%, some at 9%, and some at 10%. How much is borrowed at each rate if the annual interest is $67,000 and the amount borrowed at 8% is five times the amount borrowed at 10%?

**39.** *Grades of Paper* A manufacturer sells a 50-pound package of paper that consists of three grades of computer paper. Grade C costs $3.50 per pound, grade B costs $4.50 per pound, and grade A costs $6.00 per pound. Half of the 50-pound package consists of the two cheaper grades. The cost of the 50-pound package is $252.50. How many pounds of each grade of paper are there in the 50-pound package?

**40.** *Product Types* A company sells three types of products for $20, $10, and $5 per unit. In 1 year, the total revenue for the three products was $350,000, which corresponded to the sale of 32,500 units. The company sold half as many units of the $20 product as units of the $10 product. How many units of each product were sold?

**41.** *Crop Spraying* A mixture of 6 gallons of chemical A, 8 gallons of chemical B, and 13 gallons of chemical C is required to kill a certain destructive crop insect. Commercial spray X contains 1, 2, and 2 parts, respectively, of these chemicals. Commercial spray Y contains only chemical C. Commercial spray Z contains chemicals A, B, and C in equal amounts. How much of each type of commercial spray is needed to get the desired mixture?

**42.** *Chemistry* A chemist needs 10 liters of a 25% acid solution. The solution is to be mixed from three solutions whose concentrations are 10%, 20%, and 50%. How many liters of each solution should the chemist use to satisfy the following?

(a) Use as little as possible of the 50% solution.

(b) Use as much as possible of the 50% solution.

(c) Use 2 liters of the 50% solution.

*Investment Portfolio* In Exercises 43 and 44, you have a total of $500,000 that is to be invested in (1) certificates of deposit, (2) municipal bonds, (3) blue-chip stocks, and (4) growth or speculative stocks. How much should be put in each type of investment?

**43.** The certificates of deposit pay 10% annually, and the municipal bonds pay 8% annually. Over a 5-year period, you expect the blue-chip stocks to return 12% annually and the growth stocks to return 13% annually. You want a combined annual return of 10% and you also want to have only one-fourth of the portfolio in stocks.

**44.** The certificates of deposit pay 9% annually, and the municipal bonds pay 5% annually. Over a 5-year period, you expect the blue-chip stocks to return 12% annually and the growth stocks to return 14% annually. You want a combined annual return of 10% and you also want to have only one-fourth of the portfolio in stocks.

*Fitting a Parabola to Data* In Exercises 45–48, find the least squares regression parabola $y = ax^2 + bx + c$ for the points $(x_1, y_1)$, $(x_2, y_2)$, ..., $(x_n, y_n)$. To find the parabola, solve the system of linear equations for $a$, $b$, and $c$.

$$nc + \left( \sum_{i=1}^{n} x_i \right) b + \left( \sum_{i=1}^{n} x_i^2 \right) a = \sum_{i=1}^{n} y_i$$

$$\left( \sum_{i=1}^{n} x_i \right) c + \left( \sum_{i=1}^{n} x_i^2 \right) b + \left( \sum_{i=1}^{n} x_i^3 \right) a = \sum_{i=1}^{n} x_i y_i$$

$$\left( \sum_{i=1}^{n} x_i^2 \right) c + \left( \sum_{i=1}^{n} x_i^3 \right) b + \left( \sum_{i=1}^{n} x_i^4 \right) a = \sum_{i=1}^{n} x_i^2 y_i$$

**45.**
$$5c + 10a = 15.5$$
$$10b = 6.3$$
$$10c + 34a = 32.1$$

$(-1, 2.4)$, $(2, 4.5)$, $(1, 3.7)$, $(0, 2.9)$, $(-2, 2.0)$

**46.**  $5c \quad + 10a = 15.0$
$\qquad 10b \qquad = 17.3$
$\qquad 10c \quad + 34a = 34.5$

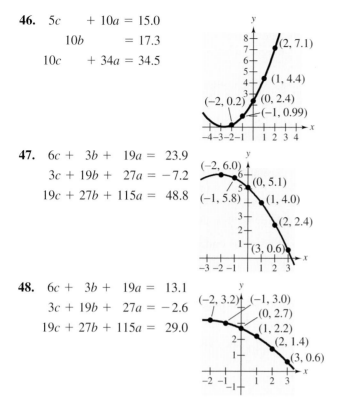

**47.**  $6c + 3b + 19a = 23.9$
$\qquad 3c + 19b + 27a = -7.2$
$\qquad 19c + 27b + 115a = 48.8$

**48.**  $6c + 3b + 19a = 13.1$
$\qquad 3c + 19b + 27a = -2.6$
$\qquad 19c + 27b + 115a = 29.0$

**49.**  *Computer Use in Schools*   The total numbers of personal computers used in grades K through 12 in the United States from 1996 to 2000 are shown in the accompanying table. In the table, $y$ represents the number of personal computers in millions and $x = 0$ represents 1998. Find the least squares regression parabola $y = ax^2 + bx + c$ that best fits these data by solving the following system. Then use the model to predict the number of personal computers in use in 2005.   (Source: Market Data Retrieval)

$5c \qquad + \quad 10a \ = \ 39.85$
$\qquad 10b \qquad\quad = \ 12.01$
$10c \qquad + \quad 34a \ = \ 79.89$

| $x$ | $-2$ | $-1$ | 0 | 1 | 2 |
|---|---|---|---|---|---|
| $y$ | 5.53 | 6.85 | 8.05 | 8.92 | 10.5 |

**50.**  *Prices of Mobile Homes*   The average prices of new mobile homes sold in the United States from 1995 to 1999 are shown in the accompanying table. In the table, $y$ represents the average price (in thousands of dollars) and $x = 0$ represents 1996. Find the least squares regression parabola

$$y = ax^2 + bx + c$$

that best fits these data by solving the system. Then use the model to predict the average price of a new mobile home in 2006.   (Source: U.S. Bureau of the Census)

$5c + \quad 5b + 15a = 203.3$
$5c + 15b + 35a = 223.8$
$15c + 35b + 99a = 646.8$

| $x$ | $-1$ | 0 | 1 | 2 | 3 |
|---|---|---|---|---|---|
| $y$ | 36.3 | 38.3 | 41.1 | 43.8 | 43.8 |

**51.**  *Stopping Distance*   In testing of the new braking system in an automobile, the speed in miles per hour and the stopping distance in feet were recorded in the table below.

| Speed ($x$) | 30 | 40 | 50 |
|---|---|---|---|
| Stopping distance ($y$) | 56 | 110 | 193 |

| Speed ($x$) | 60 | 70 |
|---|---|---|
| Stopping distance ($y$) | 305 | 425 |

(a)  Make a scatter plot of the data.

(b)  Use the regression capabilities of a graphing utility to find a quadratic model for the data.

(c)  Use the model to estimate the stopping distance if the speed is 75 miles per hour.

**52.** *New House Sales*   The numbers of new one-family houses sold (in thousands) in the southern United States from 1995 to 1999 are given in the table. (Source: U.S. Bureau of the Census)

| Year | 1995 | 1996 | 1997 | 1998 | 1999 |
|------|------|------|------|------|------|
| Houses | 300 | 337 | 363 | 398 | 408 |

(a) Make a scatter plot of the data. Let $x = 5$ correspond to 1995.

(b) Use the regression capabilities of a graphing utility to find a quadratic model for the data.

(c) Use the model to estimate home sales in the South in 2000 and 2001.

**53.** *Manufacturing*   The average monthly manufacturing sales (in billions of dollars) in the United States from 1994 to 1998 are given in the table at the top of the next column. (Source: U.S. Council of Economic Advisors)

| Year | 1994 | 1995 | 1996 | 1997 | 1998 |
|------|------|------|------|------|------|
| Sales | 279 | 300 | 310 | 327 | 338 |

(a) Make a scatter plot of the data. Let $x = 4$ correspond to 1994.

(b) Use the regression capabilities of a graphing utility to find a quadratic model for the data.

(c) Use the model to estimate average monthly manufacturing sales in 1999 and 2000.

In Exercises 54 and 55, find the position equation $s = \frac{1}{2}gt^2 + v_0t + s_0$, where $g = -32$ ft/sec$^2$.

**54.** At $t = 1$ second, $s = 134$ feet

    At $t = 2$ seconds, $s = 86$ feet

    At $t = 3$ seconds, $s = 6$ feet

**55.** At $t = 1$ second, $s = 184$ feet

    At $t = 2$ seconds, $s = 116$ feet

    At $t = 3$ seconds, $s = 16$ feet

## Math Matters • Regular Polygons and Regular Polyhedra

A regular polygon is a polygon that has $n$ sides of equal length and $n$ equal angles. For instance, a regular three-sided polygon is called an equilateral triangle, a regular four-sided polygon is called a square, and so on. There are infinitely many different types of regular polygons. For instance, it is possible to construct a regular polygon with 100, 1000, or even 1,000,000 sides (they would look very much like circles, but it is still possible to do).

For solid figures, the story is quite different. A regular polyhedron is a solid figure each of whose sides is a regular polygon (of the same size) and each of whose angles is formed by the same number of sides. At first, one might think that there are infinitely many different types of regular polyhedra, but in fact it can be shown that there are only five. The five different types are tetrahedron (4 triangular sides), cube (6 square sides), octahedron (8 triangular sides), dodecahedron (12 pentagonal sides), and icosahedron (20 triangular sides), as shown in the figures.

Tetrahedron     Cube (Hexahedron)     Octahedron     Dodecahedron     Icosahedron

## Mid-Chapter Quiz

Take this quiz as you would take a quiz in class. After you are done, check your work against the answers given in the back of the book.

In Exercises 1 and 2, write a system that has the given solution.

**1.** $(3, 2)$                                                **2.** $(5, -2, 3)$

In Exercises 3 and 4, solve algebraically. Use a graphing utility to verify the solution.

**3.** $y = 2\sqrt{x} + 1, y = -x + 4$                   **4.** $x^2 + y^2 = 9, y = 2x + 1$

In Exercises 5 and 6, find the number of sales necessary to break even.

**5.** $C = 12.50x + 10{,}000, R = 19.95x$

**6.** $C = 3.79x + 400{,}000, R = 4.59x$

In Exercises 7 and 8, solve the system by substitution or elimination. Verify the solution graphically.

**7.** $\begin{aligned} 2.5x - \phantom{0}y &= 6 \\ 3x + 4y &= 2 \end{aligned}$                        **8.** $\begin{aligned} \tfrac{1}{2}x + \tfrac{1}{3}y &= \phantom{-}1 \\ x - 2y &= -2 \end{aligned}$

In Exercises 9 and 10, find the point of equilibrium. Verify the solution graphically.

**9.** Demand: $p = 45 - 0.001x$;    supply: $p = 23 + 0.0002x$

**10.** Demand: $p = 95 - 0.0002x$;    supply: $p = 80 + 0.00001x$

In Exercises 11–13, solve the system of equations.

**11.** $\begin{aligned} 2x + 3y - \phantom{0}z &= -7 \\ x \phantom{+ 3y} + 3z &= \phantom{-}10 \\ 2y + \phantom{0}z &= -1 \end{aligned}$    **12.** $\begin{aligned} x + y - 2z &= 12 \\ 2x - y - \phantom{0}z &= \phantom{0}6 \\ y - \phantom{0}z &= \phantom{0}6 \end{aligned}$    **13.** $\begin{aligned} 3x + 2y + z &= 17 \\ -x + \phantom{0}y + z &= \phantom{0}4 \\ x - \phantom{0}y - z &= \phantom{0}3 \end{aligned}$

In Exercises 14 and 15, write three ordered triples of the given form.

**14.** $(a, a - 2, 3a)$                                 **15.** $(2a, a + 5, a)$

**16.** Find the position equation $s = \tfrac{1}{2}gt^2 + v_0 t + s_0$ given that

    $t = 1$ second, $s = 144$ feet

    $t = 2$ seconds, $s = 106$ feet

    $t = 3$ seconds, $s = 36$ feet.

**17.** How many solutions does the system $2x - 3y + z = 4$, $y - z = 7$ have?

**18.** Ten gallons of a 35% acid solution is obtained by mixing a 25% solution with a 40% solution. How much of each must be used?

**19.** Is it possible for a system of equations to have infinitely many solutions?

**20.** Is it true that a square linear system must have exactly one solution?

## 6.4    Systems of Inequalities

### Objectives

- Sketch the graph of an inequality in two variables.
- Solve a system of inequalities.
- Construct and use a system of inequalities to solve an application problem.

## The Graph of an Inequality

The following statements are inequalities in two variables:

$$3x - 2y < 6 \quad \text{and} \quad 2x^2 + 3y^2 \geq 6.$$

An ordered pair $(a, b)$ is a **solution of an inequality** in $x$ and $y$ if the inequality is true when $a$ and $b$ are substituted for $x$ and $y$, respectively. The **graph** of an inequality is the collection of all solutions of the inequality. To sketch the graph of an inequality, begin by sketching the graph of the *corresponding equation*. The graph of the equation will normally separate the plane into two or more regions. In each such region, one of the following must be true.

1. *All* points in the region are solutions of the inequality.

2. *No* point in the region is a solution of the inequality.

So, you can determine whether the points in an entire region satisfy the inequality by simply testing *one* point in the region.

▶ Sketching the Graph of an Inequality in Two Variables

1. Replace the inequality sign by an equal sign, and sketch the graph of the resulting equation. (Use a dashed line for $<$ or $>$ and a solid line for $\leq$ or $\geq$.)

2. Test one point in each of the regions formed by the graph in Step 1. If the point satisfies the inequality, shade the entire region to denote that every point in the region satisfies the inequality.

**Example 1**    Sketching the Graph of an Inequality

Sketch the graph of the inequality $y \geq x^2 - 1$.

### Solution

The graph of the corresponding *equation* $y = x^2 - 1$ is a parabola, as shown in Figure 6.13. By testing a point *above* the parabola $(0, 0)$ and a point *below* the parabola $(0, -2)$, you can see that the points that satisfy the inequality are those lying above (or on) the parabola.

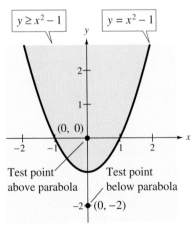

Figure 6.13

The inequality given in Example 1 is a nonlinear inequality in two variables. Most of the following examples involve **linear inequalities** such as $ax + by < c$. The graph of a linear inequality is a half-plane lying on one side of the line $ax + by = c$. The simplest linear inequalities are those corresponding to horizontal or vertical lines, as shown in Example 2.

## Study Tip

To graph a linear inequality, it can help to write the inequality in slope-intercept form. For instance, by writing $x - y < 2$ in the form

$$y > x - 2$$

you can see that the solution points lie *above* the line $x - y = 2$ (or $y = x - 2$), as shown in Figure 6.16.

**Example 2**   Sketching the Graph of a Linear Inequality

Sketch the graphs of the following linear inequalities.

**a.** $x > -2$     **b.** $y \le 3$

### Solution

**a.** The graph of the corresponding equation $x = -2$ is a vertical line. The points that satisfy the inequality $x > -2$ are those lying to the right of this line, as shown in Figure 6.14.

**b.** The graph of the corresponding equation $y = 3$ is a horizontal line. The points that satisfy the inequality $y \le 3$ are those lying below (or on) this line, as shown in Figure 6.15.

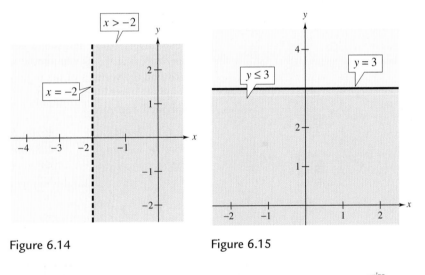

Figure 6.14                    Figure 6.15

**Example 3**   Sketching the Graph of a Linear Inequality

Sketch the graph of $x - y < 2$.

### Solution

The graph of the corresponding equation $x - y = 2$ is a line, as shown in Figure 6.16. Because the origin $(0, 0)$ satisfies the inequality, the graph consists of the half-plane lying above the line. (Try checking a point below the line. Regardless of which point you choose, you will see that it does not satisfy the inequality.)

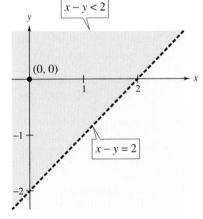

Figure 6.16

## Systems of Inequalities

Many practical problems in business, science, and engineering involve systems of linear inequalities. A **solution** of a system of inequalities in $x$ and $y$ is a point $(x, y)$ that satisfies each inequality in the system.

To sketch the graph of a system of inequalities in two variables, first sketch the graph of each individual inequality (on the same coordinate system) and then find the region that is *common* to every graph in the system. For systems of *linear* inequalities, it is helpful to find the vertices of the solution region.

**Example 4**    Solving a System of Inequalities

Sketch the graph (and label the vertices) of the solution set of the following system.

$$x - y < 2$$
$$x > -2$$
$$y \le 3$$

### Solution

The graphs of these inequalities are shown in Figures 6.16, 6.14, and 6.15 on page 397, respectively. The triangular region common to all three graphs can be found by superimposing the graphs on the same coordinate plane, as shown in Figure 6.17. To find the vertices of the region, solve the three systems of corresponding equations obtained by taking *pairs* of equations representing the boundaries of the individual regions.

| | | |
|---|---|---|
| *Vertex A:* $(-2, -4)$ | *Vertex B:* $(5, 3)$ | *Vertex C:* $(-2, 3)$ |
| *Obtained by solving the system* | *Obtained by solving the system* | *Obtained by solving the system* |

$$
\begin{aligned}
x - y &= 2 \\
x &= -2
\end{aligned}
\qquad
\begin{aligned}
x - y &= 2 \\
y &= 3
\end{aligned}
\qquad
\begin{aligned}
x &= -2 \\
y &= 3
\end{aligned}
$$

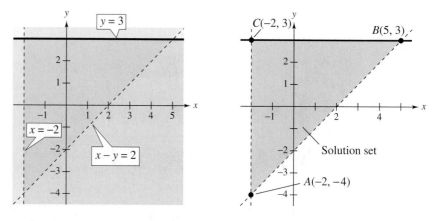

Figure 6.17

For the triangular region shown in Figure 6.17, each point of intersection of a pair of boundary lines corresponds to a vertex. With more complicated regions, two border lines can sometimes intersect at a point that is not a vertex of the region, as shown in Figure 6.18. In order to keep track of which points of intersection are actually vertices of the region, we suggest that you make a careful sketch of the region and refer to your sketch as you find each point of intersection.

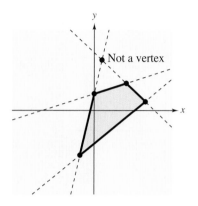

Not a vertex

**Figure 6.18**    *Boundary lines can intersect at a point that is not a vertex.*

**Example 5**    Solving a System of Inequalities

Sketch the region containing all points that satisfy the following system.

$$x^2 - y \le 1$$
$$-x + y \le 1$$

Solution

As shown in Figure 6.19, the points that satisfy the inequality $x^2 - y \le 1$ are the points lying above (or on) the parabola given by

$$y = x^2 - 1. \qquad \text{Parabola}$$

The points satisfying the inequality $-x + y \le 1$ are the points lying on or below the line given by

$$y = x + 1. \qquad \text{Line}$$

To find the points of intersection of the parabola and the line, solve the system of corresponding equations.

$$x^2 - y = 1$$
$$-x + y = 1$$

Using the method of substitution, you can find the solutions to be $(-1, 0)$ and $(2, 3)$, as shown in Figure 6.19.

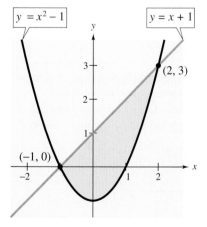

**Figure 6.19**

When solving a system of inequalities, you should be aware that the system might have no solution. For instance, the system

$$x + y > 3$$

$$x + y < -1$$

has no solution points, because the quantity $(x + y)$ cannot be both less than $-1$ and greater than 3, as shown in Figure 6.20.

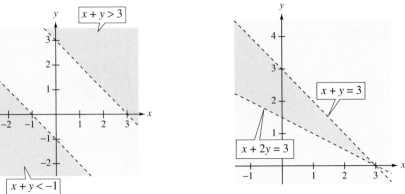

Figure 6.20  *No Solution*                    Figure 6.21  *Unbounded Region*

Another possibility is that the solution set of a system of inequalities can be unbounded. For instance, the solution set of

$$x + \phantom{2}y < 3$$

$$x + 2y > 3$$

forms an *infinite wedge*, as shown in Figure 6.21.

---

## •TECHNOLOGY•

**Inequalities and Graphing Utilities**    A graphing utility can be used to sketch the graph of an inequality. A sketch of the graph of $y \geq x^2 - 2$ is shown below. Consult the user's guide to determine how to graph an inequality in two variables (or a system of inequalities) on your graphing utility. Then use your graphing utility to graph the following inequalities.

**a.** $y \leq 2x + 2$    **b.** $y \geq \frac{1}{2}x^2 - 4$    **c.** $y \leq x^3 - 4x^2 + 4$

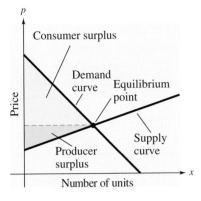

Figure 6.22

## Applications

Example 8 in Section 6.2 discussed the *point of equilibrium* for a demand and supply function. The next example discusses two related concepts that economists call **consumer surplus** and **producer surplus.** As shown in Figure 6.22, the consumer surplus is defined as the area of the region that lies *below* the demand curve, *above* the horizontal line passing through the equilibrium point, and to the right of the *y*-axis. Similarly, the producer surplus is defined as the area of the region that lies *above* the supply curve, *below* the horizontal line passing through the equilibrium point, and to the right of the *y*-axis. The consumer surplus is a measure of the amount that consumers would have been willing to pay *above what they actually paid*, whereas the producer surplus is a measure of the amount that producers would have been willing to receive *below what they actually received.*

**Example 6**    Consumer and Producer Surplus

The demand and supply functions for a certain type of calculator are given by

$$p = 150 - 0.00001x \qquad \text{Demand equation}$$

$$p = 60 + 0.00002x \qquad \text{Supply equation}$$

where $p$ is the price in dollars and $x$ represents the number of units. Find the consumer surplus and producer surplus for these two equations.

### Solution

Begin by finding the point of equilibrium by solving the equation

$$60 + 0.00002x = 150 - 0.00001x.$$

In Example 8 in Section 6.2, you saw that the solution is $x = 3,000,000$, which corresponds to an equilibrium price of $p = \$120$. Thus, the consumer surplus and producer surplus are the areas of the triangular regions given by the system of inequalities.

Figure 6.23

| *Consumer Surplus* | *Producer Surplus* |
|---|---|
| $p \le 150 - 0.00001x$ | $p \ge 60 + 0.00002x$ |
| $p \ge 120$ | $p \le 120$ |
| $x \ge 0$ | $x \ge 0$ |

In Figure 6.23, you can see that the consumer surplus is

$$\text{Consumer surplus} = \tfrac{1}{2}(\text{base})(\text{height}) = \tfrac{1}{2}(30)(3,000,000) = \$45,000,000$$

and the producer surplus is

$$\text{Producer surplus} = \tfrac{1}{2}(\text{base})(\text{height}) = \tfrac{1}{2}(60)(3,000,000) = \$90,000,000.$$

| Example 7 | Nutrition |
|---|---|

The liquid portion of a diet is to provide at least 300 calories, 36 units of vitamin A, and 90 units of vitamin C daily. A cup of dietary drink X provides 60 calories, 12 units of vitamin A, and 10 units of vitamin C. A cup of dietary drink Y provides 60 calories, 6 units of vitamin A, and 30 units of vitamin C. Set up a system of linear inequalities that describes the minimum daily requirements for calories and vitamins.

### Solution

Begin by letting $x$ and $y$ represent the following.

$$x = \text{number of cups of dietary drink X}$$

$$y = \text{number of cups of dietary drink Y}$$

To meet the minimum daily requirements, the following inequalities must be satisfied.

| For calories: | $60x + 60y \geq 300$ |
|---|---|
| For vitamin A: | $12x + 6y \geq 36$ |
| For vitamin C: | $10x + 30y \geq 90$ |
| | $x \geq 0$ |
| | $y \geq 0$ |

The last two inequalities are included because $x$ and $y$ cannot be negative. The graph of this system of inequalities is shown in Figure 6.24. (More is said about this application in Example 7 in Section 6.5.)

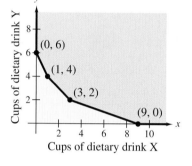

Figure 6.24

Cups of dietary drink Y

(0, 6)
(1, 4)
(3, 2)
(9, 0)

Cups of dietary drink X

---

| Discussing the Concept | Writing a System of Inequalities |
|---|---|

Write a system of inequalities that includes $x \geq 0$ and $y \geq 0$. Find the solution set by sketching a graph of the system and identifying the vertices. Exchange your system of inequalities for the system of inequalities written by another student. Find the solution set and identify the vertices of the system that you received. Does your answer agree with the answer of the student who wrote the problem?

# Warm Up

The following warm-up exercises involve skills that were covered in earlier sections. You will use these skills in the exercise set for this section.

In Exercises 1–6, identify the graph of the given equation.

1. $x + y = 3$
2. $4x - y = 8$
3. $y = x^2 - 4$
4. $y = -x^2 + 1$
5. $x^2 + y^2 = 9$
6. $\dfrac{x^2}{4} + \dfrac{y^2}{9} = 1$

In Exercises 7–10, solve the system of equations.

7. $x + 2y = 3$
   $4x - 7y = -3$
8. $2x - 3y = 4$
   $x + 5y = 2$
9. $x^2 + y = 5$
   $2x - 4y = 0$
10. $x^2 + y^2 = 13$
    $x + y = 5$

## 6.4   Exercises

In Exercises 1–6, match the inequality with its graph. [The graphs are labeled (a), (b), (c), (d), (e), and (f).]

(a)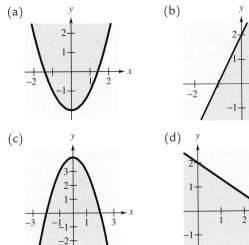

(b)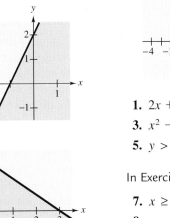

(c)

(d)

(e)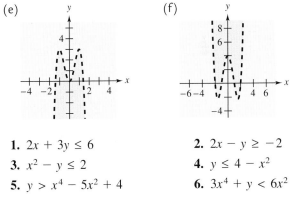

(f)

1. $2x + 3y \le 6$
2. $2x - y \ge -2$
3. $x^2 - y \le 2$
4. $y \le 4 - x^2$
5. $y > x^4 - 5x^2 + 4$
6. $3x^4 + y < 6x^2$

In Exercises 7–20, graph the inequality.

7. $x \ge 2$
8. $x < 4$
9. $y > -1$
10. $y \le 3$
11. $y < 2 - x$
12. $y > 2x - 4$
13. $2y - x \ge 4$
14. $5x + 3y \ge -15$

**15.** $y^2 + 2x > 0$

**16.** $y^2 - x < 0$

**17.** $y \le \dfrac{1}{1 + x^2}$

**18.** $y < \ln x$

**19.** $x^2 + y^2 \le 4$

**20.** $x^2 + y^2 > 4$

In Exercises 21–40, graph the solution set of the system of inequalities.

**21.** $\quad x + y \le 2$
$\quad -x + y \le 2$
$\qquad\quad y \ge 0$

**22.** $3x + 2y < 6$
$\quad x \qquad > 1$
$\qquad\quad y > 0$

**23.** $x + y \le 5$
$\quad x \qquad \ge 2$
$\qquad\quad y \ge 0$

**24.** $2x + y \ge 2$
$\quad x \qquad \le 2$
$\qquad\quad y \le 1$

**25.** $-3x + 2y < \quad 6$
$\quad x + 4y > -2$
$\quad 2x + \ y < \quad 3$

**26.** $\quad x - 7y > -36$
$\quad 5x + 2y > \quad 5$
$\quad 6x - 5y > \quad 6$

**27.** $2x + \ y < 2$
$\quad 6x + 3y > 2$

**28.** $5x - 3y < -6$
$\quad 5x - 3y > -9$

**29.** $y \ge -3$
$\quad y \le 1 - x^2$

**30.** $x - y^2 > 0$
$\quad y > (x - 3)^2 - 4$

**31.** $x^2 + y^2 \le 16$
$\quad x^2 + y^2 < \quad 1$

**32.** $x^2 + y^2 \le 25$
$\quad x^2 + y^2 \ge \quad 9$

**33.** $x > y^2$
$\quad x < y + 2$

**34.** $x < 2y - y^2$
$\quad 0 < x \ + y$

**35.** $y \le \sqrt{3x} + 1$
$\quad y \ge x + 1$

**36.** $y < \sqrt{2x} + 3$
$\quad y > \quad x + 3$

**37.** $y < x^3 - 2x + 1$
$\quad y > -2x$
$\quad x \le 1$

**38.** $\qquad\quad x \ge 1$
$\quad x - 2y \le 3$
$\quad 3x + 2y \ge 9$
$\quad x + \ y \le 6$

**39.** $y \le e^x$
$\quad y \ge \ln x$
$\quad x \ge \frac{1}{2}$
$\quad x \le 2$

**40.** $\qquad y \le e^{-x^2/2}$
$\qquad\quad y \ge 0$
$\quad -1 \le x \le 1$

In Exercises 41–46, write a system of inequalities that describes the region.

**41.** Rectangular region with vertices at $(1, 1)$, $(5, 1)$, $(5, 5)$, and $(1, 5)$

**42.** Parallelogram region with vertices at $(0, 0)$, $(5, 0)$, $(1, 6)$, and $(6, 6)$

**43.** Triangular region with vertices at $(0, 0)$, $(7, 0)$, and $(3, 5)$

**44.** Triangular region with vertices at $(-2, 0)$, $(6, 0)$, and $(0, 5)$

**45.** Sector of a circle

**46.** Sector of a circle

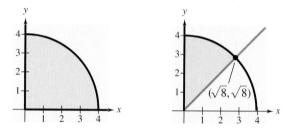

**47.** *Furniture Production* A furniture company can sell all the tables and chairs it produces. Each table requires 1 hour in the assembly center and $1\frac{1}{3}$ hours in the finishing center. Each chair requires $1\frac{1}{2}$ hours in the assembly center and $1\frac{1}{2}$ hours in the finishing center. The company's assembly center is available 12 hours per day, and its finishing center is available 15 hours per day. If $x$ is the number of tables produced per day and $y$ is the number of chairs, find a system of inequalities describing all possible production levels. Sketch the graph of the system.

**48.** *Computer Inventory* A store sells two models of a certain brand of computer. Because of the demand, it is necessary to stock three times as many units of model A as units of model B. The costs to the store for the two models are $900 and $1100, respectively. The management does not want more than $40,000 in computer inventory at any one time, and it wants at least six model A computers and two model B computers in inventory at all times. Devise a system of inequalities describing all possible inventory levels, and sketch the graph of the system.

**49.** *Investment*   A person plans to invest $20,000 in two different interest-bearing accounts. Each account is to contain at least $5000. Moreover, one account should have at least twice the amount that is in the other account.

(a) Find a system of inequalities that describes the amounts that can be deposited in each account.

(b) Sketch the graph of the system.

**50.** *Concert Ticket Sales*   Two types of tickets are to be sold for a concert. One type costs $15 and the other type costs $25. The promoter of the concert must sell at least 15,000 tickets, including at least 8000 of the $15 tickets and at least 4000 of the $25 tickets. Moreover, the gross receipts must total at least $275,000 in order for the concert to be held.

(a) Find a system of inequalities describing the different numbers of tickets that can be sold.

(b) Sketch the graph of the system.

**51.** *Diet Supplement*   A dietitian is asked to design a special diet supplement using two different foods. Each ounce of food X contains 20 units of calcium, 15 units of iron, and 10 units of vitamin B. Each ounce of food Y contains 10 units of calcium, 10 units of iron, and 20 units of vitamin B. The minimum daily requirements in the diet are 300 units of calcium, 150 units of iron, and 200 units of vitamin B.

(a) Find a system of inequalities describing the different amounts of food X and food Y that can be used in the diet.

(b) Sketch the graph of the system.

**52.** *Diet Supplement*   A dietitian is asked to design a special diet supplement using two different foods. Each ounce of food X contains 20 units of calcium, 15 units of iron, and 10 units of vitamin B. Each ounce of food Y contains 10 units of calcium, 10 units of iron, and 20 units of vitamin B. The minimum daily requirements in the diet are 280 units of calcium, 160 units of iron, and 180 units of vitamin B.

(a) Find a system of inequalities describing the different amounts of food X and food Y that can be used in the diet.

(b) Sketch the graph of the system.

*Consumer and Producer Surpluses*   In Exercises 53–56, find the consumer surplus and producer surplus for the pair of supply and demand equations.

|  | Demand | Supply |
|---|---|---|
| **53.** | $p = 56 - 0.0001x$ | $p = 22 + 0.00001x$ |
| **54.** | $p = 60 - 0.00001x$ | $p = 15 + 0.00004x$ |
| **55.** | $p = 140 - 0.00002x$ | $p = 80 + 0.00001x$ |
| **56.** | $p = 400 - 0.0002x$ | $p = 225 + 0.0005x$ |

**57.** *Births in District of Columbia*   The number of babies $y$ (in thousands) born to D.C. residents each year from 1994 to 1998 can be approximated by the linear model

$$y = 11.7 - 0.53t$$

where $t = 4$ represents 1994. The *total* number of babies born to D.C. residents during this 5-year period can be approximated by finding the area of the trapezoid represented by the system

$$y \leq 11.7 - 0.53t, \ y \geq 0, t \geq 3.5, t \leq 8.5.$$

(Source: U.S. National Center for Health Statistics)

(a) Graph this region using a graphing utility.

(b) Use the formula for the area of a trapezoid to approximate the total number of births.

**58.** *Spectator Amusements*   Annual consumer spending for admissions to spectator amusements in the United States from 1994 to 1998 can be approximated by the linear model

$$y = 1.42t + 12.3$$

where $y$ is consumer spending (in billions of dollars) and $t = 4$ represents 1994. The *total* amount of consumer spending during this 5-year period can be approximated by finding the area of the trapezoid represented by the system

$$y \leq 1.42t + 12.3, \ y \geq 0, t \geq 3.5, t \leq 8.5.$$

(Source: U.S. Bureau of Economic Analysis)

(a) Graph this region using a graphing utility.

(b) Use the formula for the area of a trapezoid to approximate the total consumer spending.

## 6.5    Linear Programming

### Objectives

- Solve a linear programming problem.
- Use linear programming to minimize or maximize an objective function.
- Use linear programming to optimize an application.

## Linear Programming: A Graphical Approach

Many applications in business and economics involve a process called **optimization,** in which you are asked to find the minimum cost, the maximum profit, or the minimum use of resources. In this section you will study an optimization strategy called **linear programming.**

A two-dimensional linear programming problem consists of a linear **objective function** and a system of linear inequalities called **constraints.** The objective function gives the quantity that is to be maximized (or minimized), and the constraints determine the set of **feasible solutions.** For example, consider a linear programming problem in which you are asked to maximize the value of

$$z = ax + by \qquad \text{Objective function}$$

subject to a set of constraints that determines the region in Figure 6.25. Because every point in the region satisfies each constraint, it is not clear how you should go about finding the point that yields a maximum value of $z$. Fortunately, it can be shown that if there is an optimal solution, it must occur at one of the vertices. This means that *you can find the maximum value by testing $z$ at each of the vertices.*

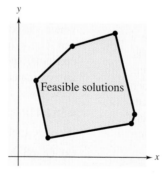

Figure 6.25

> ▶ Optimal Solution of a Linear Programming Problem
>
> If a linear programming problem has a solution, it must occur at a vertex of the set of feasible solutions. If the problem has more than one solution, then at least one of them must occur at a vertex of the set of feasible solutions. In either case, the value of the objective function is unique.

**Example 1**    Solving a Linear Programming Problem

Find the maximum value of

$$z = 3x + 2y \qquad \text{Objective function}$$

subject to the following constraints.

$$\left.\begin{array}{r} x \geq 0 \\ y \geq 0 \\ x + 2y \leq 4 \\ x - y \leq 1 \end{array}\right\} \qquad \text{Constraints}$$

### Solution

The constraints form the region shown in Figure 6.26. At the four vertices of this region, the objective function has the following values.

At $(0, 0)$:      $z = 3(0) + 2(0) = 0$

At $(1, 0)$:      $z = 3(1) + 2(0) = 3$

At $(2, 1)$:      $z = 3(2) + 2(1) = 8$      Maximum value of $z$

At $(0, 2)$:      $z = 3(0) + 2(2) = 4$

So, the maximum value of $z$ is 8, and this occurs when $x = 2$ and $y = 1$.

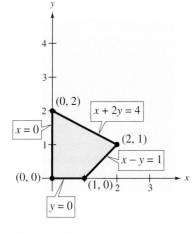

Figure 6.26

In Example 1, try testing some of the *interior* points in the region. You will see that the corresponding values of $z$ are less than 8. Here are some examples.

At $(1, 1)$:      $z = 3(1) + 2(1) = 5$

At $\left(1, \frac{1}{2}\right)$:      $z = 3(1) + 2\left(\frac{1}{2}\right) = 4$

At $\left(\frac{1}{2}, \frac{3}{2}\right)$:      $z = 3\left(\frac{1}{2}\right) + 2\left(\frac{3}{2}\right) = \frac{9}{2}$

To see why the maximum value of the objective function in Example 1 must occur at a vertex, consider writing the objective function in the form

$$y = -\frac{3}{2}x + \frac{z}{2} \qquad \text{Family of lines}$$

where $z/2$ is the $y$-intercept of the objective function. This equation represents a family of lines, each of slope $-\frac{3}{2}$. Of these infinitely many lines, you want the one that has the largest $z$-value while still intersecting the region determined by the constraints. In other words, of all the lines whose slope is $-\frac{3}{2}$, you want the one that has the largest $y$-intercept *and* intersects the given region, as shown in Figure 6.27. It should be clear that such a line will pass through one (or more) of the vertices of the region.

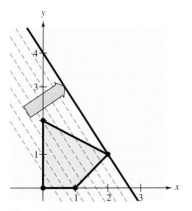

Figure 6.27

---

▶ Solving a Linear Programming Problem

To solve a linear programming problem involving two variables by the graphical method, use the following steps.

1. Sketch the region corresponding to the system of constraints. (The points inside or on the boundary of the region are called *feasible solutions.*)

2. Find the vertices of the region.

3. Test the objective function at each of the vertices and select the values of the variables that optimize the objective function. For a bounded region, both a minimum and a maximum value will exist. (For an unbounded region, *if* an optimal solution exists, it will occur at a vertex.)

---

## Study Tip

Remember that a vertex of the region can be found using a system of linear equations. The system will consist of the equations of the lines passing through the vertex.

These guidelines will work whether the objective function is to be maximized or minimized. For instance, the same test used in Example 1 to find the maximum value of $z$ can be used to conclude that the minimum value of $z$ is 0 and that this value occurs at the vertex $(0, 0)$.

| Example 2 | Solving a Linear Programming Problem |

Find the maximum value of the objective function

$$z = 4x + 6y \qquad \text{Objective function}$$

where $x \geq 0$ and $y \geq 0$, subject to the following constraints.

$$\left. \begin{array}{r} -x + \ y \leq 11 \\ x + \ y \leq 27 \\ 2x + 5y \leq 90 \end{array} \right\} \qquad \text{Constraints}$$

### Solution

The region bounded by the constraints is shown in Figure 6.28. By testing the objective function at each vertex, you obtain the following.

At $(0, 0)$:     $z = 4(0) \ + 6(0) \ = \ \ \ 0$

At $(0, 11)$:   $z = 4(0) \ + 6(11) = \ \ 66$

At $(5, 16)$:   $z = 4(5) \ + 6(16) = 116$

At $(15, 12)$:  $z = 4(15) + 6(12) = 132$    Maximum value of $z$

At $(27, 0)$:   $z = 4(27) + 6(0) \ = 108$

So, the maximum value of $z$ is 132, and this occurs when $x = 15$ and $y = 12$.

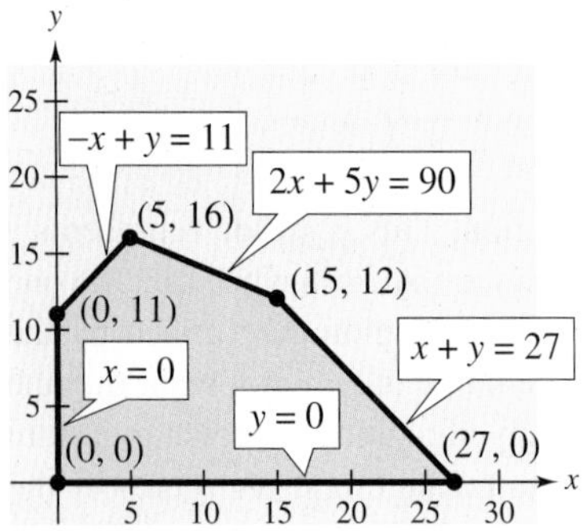

Figure 6.28

The next example shows that the same basic procedure can be used to solve a linear programming problem in which the objective function is to be *minimized*.

**Example 3** Minimizing an Objective Function

Find the minimum value of the objective function

$$z = 5x + 7y \qquad \text{Objection function}$$

where $x \geq 0$ and $y \geq 0$, subject to the following constraints.

$$\left. \begin{array}{rcl} 2x + 3y & \geq & 6 \\ 3x - y & \leq & 15 \\ -x + y & \leq & 4 \\ 2x + 5y & \leq & 27 \end{array} \right\} \quad \text{Constraints}$$

**Solution**

The region bounded by the constraints is shown in Figure 6.29. By testing the objective function at each vertex, you obtain the following.

At $(0, 2)$:   $z = 5(0) + 7(2) = 14$   Minimum value of $z$

At $(0, 4)$:   $z = 5(0) + 7(4) = 28$

At $(1, 5)$:   $z = 5(1) + 7(5) = 40$

At $(6, 3)$:   $z = 5(6) + 7(3) = 51$

At $(5, 0)$:   $z = 5(5) + 7(0) = 25$

At $(3, 0)$:   $z = 5(3) + 7(0) = 15$

So, the minimum value of $z$ is 14, and this occurs when $x = 0$ and $y = 2$.

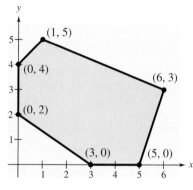

Figure 6.29

**Example 4** Maximizing an Objective Function

Find the maximum value of the objective function

$$z = 5x + 7y \qquad \text{Objective function}$$

where $x \geq 0$ and $y \geq 0$, subject to the following constraints.

$$\left. \begin{array}{rcl} 2x + 3y & \geq & 6 \\ 3x - y & \leq & 15 \\ -x + y & \leq & 4 \\ 2x + 5y & \leq & 27 \end{array} \right\} \quad \text{Constraints}$$

**Solution**

This linear programming problem is identical to that given in Example 3 above, *except* that the objective function is maximized instead of minimized. Using the values of $z$ at the vertices shown above, you can conclude that the maximum value of $z$ is 51, and that this value occurs when $x = 6$ and $y = 3$.

**Study Tip**

The steps used to find the minimum and maximum values in Examples 3 and 4 are precisely the same. In other words, once you have evaluated the objective function at the vertices of the feasible region, you simply choose the largest value as the maximum and the smallest value as the minimum.

It is possible for the maximum (or minimum) value in a linear programming problem to occur at *two* different vertices. For instance, at the vertices of the region shown in Figure 6.30, the objective function

$$z = 2x + 2y \qquad \text{Objective function}$$

has the following values.

At $(0, 0)$: $z = 2(0) + 2(0) = 0$

At $(0, 4)$: $z = 2(0) + 2(4) = 8$

At $(2, 4)$: $z = 2(2) + 2(4) = 12$     Maximum value of $z$

At $(5, 1)$: $z = 2(5) + 2(1) = 12$     Maximum value of $z$

At $(5, 0)$: $z = 2(5) + 2(0) = 10$

In this case, you can conclude that the objective function has a maximum value not only at the vertices $(2, 4)$ and $(5, 1)$; it also has a maximum value (of 12) at *any point on the line segment connecting these two vertices*. Note that the objective function

$$y = -x + \tfrac{1}{2}z$$

has the same slope as the line through the vertices $(2, 4)$ and $(5, 1)$.

Some linear programming problems have no optimal solution. This can occur if the region determined by the constraints is *unbounded*. Example 5 illustrates such a problem.

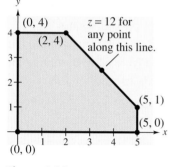

Figure 6.30

**Example 5**    An Unbounded Region

Find the maximum value of

$$z = 4x + 2y \qquad \text{Objective function}$$

where $x \geq 0$ and $y \geq 0$, subject to the following constraints.

$$\left. \begin{array}{r} x + 2y \geq 4 \\ 3x + y \geq 7 \\ -x + 2y \leq 7 \end{array} \right\} \quad \text{Constraints}$$

**Solution**

The region determined by the constraints is shown in Figure 6.31. For this unbounded region, there is no maximum value of $z$. To see this, note that the point $(x, 0)$ lies in the region for all values of $x \geq 4$. By choosing $x$ to be large, you can obtain values of

$$z = 4(x) + 2(0) = 4x$$

that are as large as you want. So, there is no maximum value of $z$. For this problem, there *is* a minimum value of $z = 10$, which occurs at the vertex $(2, 1)$.

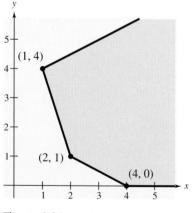

Figure 6.31

## Applications

Example 6 shows how linear programming can be used to find the maximum profit in a business application.

**Example 6**   Maximum Profit

A manufacturer wants to maximize the profit for two products. The first product yields a profit of $1.50 per unit, and the second product yields a profit of $2.00 per unit. Market tests and available resources have indicated the following constraints.

1. The combined production level should not exceed 1200 units per month.
2. The demand for product II is no more than half the demand for product I.
3. The production level of product I is less than or equal to 600 units plus three times the production level of product II.

### Solution

If you let $x$ be the number of units of product I and $y$ be the number of units of product II, the objective function (for the combined profit) is given by

$$P = 1.5x + 2y.$$   Objective function

The three constraints translate into the following linear inequalities.

1. $x + y \le 1200$    ⟹    $x + y \le 1200$
2. $y \le \frac{1}{2}x$    ⟹    $-x + 2y \le 0$
3. $x \le 3y + 600$    ⟹    $x - 3y \le 600$

Because neither $x$ nor $y$ can be negative, you also have the two additional constraints of $x \ge 0$ and $y \ge 0$. Figure 6.32 shows the region determined by the constraints. To find the maximum profit, test the value of $P$ at the vertices of the region.

At $(0, 0)$:          $P = 1.5(0)$      $+ 2(0)$   $=$   $0$

At $(800, 400)$:   $P = 1.5(800)$   $+ 2(400) = 2000$      Maximum profit

At $(1050, 150)$: $P = 1.5(1050) + 2(150) = 1875$

At $(600, 0)$:      $P = 1.5(600)$   $+ 2(0)$   $=$   $900$

So, the maximum profit is $2000, and it occurs when the monthly production consists of 800 units of product I and 400 units of product II.

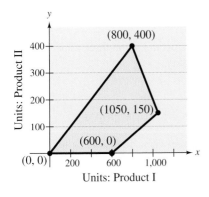

**Figure 6.32**

In Example 6, suppose the manufacturer improved the production of product I so that it yielded a profit of $2.50 per unit. How would this affect the number of units the manufacturer should sell to obtain a maximum profit?

Example 7    Minimum Cost

The liquid part of a diet is to provide at least 300 calories, 36 units of vitamin A, and 90 units of vitamin C daily. A cup of dietary drink X costs $0.12 and provides 60 calories, 12 units of vitamin A, and 10 units of vitamin C. A cup of dietary drink Y costs $0.15 and provides 60 calories, 6 units of vitamin A, and 30 units of vitamin C. How many cups of each drink should be consumed each day to minimize the cost and still meet the daily requirements?

**Solution**

As in Example 7 on page 402, let $x$ be the number of cups of dietary drink X and let $y$ be the number of cups of dietary drink Y.

$$
\left.
\begin{array}{lrcl}
\text{For calories:} & 60x + 60y & \geq & 300 \\
\text{For vitamin A:} & 12x + 6y & \geq & 36 \\
\text{For vitamin C:} & 10x + 30y & \geq & 90 \\
& x & \geq & 0 \\
& y & \geq & 0
\end{array}
\right\} \quad \text{Constraints}
$$

The cost $C$ is given by

$$C = 0.12x + 0.15y. \qquad \text{Objective function}$$

The graph of the region corresponding to the constraints is shown in Figure 6.33. To determine the minimum cost, test $C$ at each vertex of the region, as follows.

At $(0, 6)$:   $C = 0.12(0) + 0.15(6) = 0.90$

At $(1, 4)$:   $C = 0.12(1) + 0.15(4) = 0.72$

At $(3, 2)$:   $C = 0.12(3) + 0.15(2) = 0.66$    Minimum value of $C$

At $(9, 0)$:   $C = 0.12(9) + 0.15(0) = 1.08$

So, the minimum cost is $0.66 per day, and this occurs when three cups of drink X and two cups of drink Y are consumed each day.

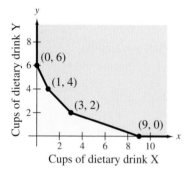

**Figure 6.33**

---

**Discussing the Concept**    **Analysis of Constraints**

Explain what difficulties you might encounter with the following sets of linear programming constraints.

**a.**   
$$
\begin{array}{rcl}
x - y & < & 0 \\
3x + y & > & 9 \\
-4x + y & > & -2
\end{array}
$$

**b.**  
$$
\begin{array}{rcl}
2x + y & > & 11 \\
x - y & > & 0 \\
x & < & 4 \\
y & < & 0
\end{array}
$$

# Warm Up

The following warm-up exercises involve skills that were covered in earlier sections. You will use these skills in the exercise set for this section.

In Exercises 1–4, sketch the graph of the linear equation.

**1.** $y + x = 3$         **2.** $y - x = 12$

**3.** $x = 0$         **4.** $y = 4$

In Exercises 5–8, find the point of intersection of the two lines.

**5.** $x + y = 4, x = 0$         **6.** $x + 2y = 12, y = 0$

**7.** $x + y = 4, 2x + 3y = 9$         **8.** $x + 2y = 12, 2x + y = 9$

In Exercises 9 and 10, sketch the graph of the inequality.

**9.** $2x + 3y \geq 18$         **10.** $4x + 3y \geq 12$

## 6.5   Exercises

In Exercises 1–8, find the minimum and maximum values of the objective function, subject to the indicated constraints. (For each exercise, the graph of the region determined by the constraints is provided.)

**1.** *Objective function:*

$z = 6x + 5y$

*Constraints:*

$$x \geq 0$$
$$y \geq 0$$
$$x + y \leq 6$$

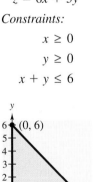

**2.** *Objective function:*

$z = 2x + 8y$

*Constraints:*

$$x \geq 0$$
$$y \geq 0$$
$$2x + y \leq 4$$

**3.** *Objective function:*

$z = 8x + 7y$

*Constraints:*

(See Exercise 1.)

**4.** *Objective function:*

$z = 7x + 3y$

*Constraints:*

(See Exercise 2.)

**5.** *Objective function:*

$z = 3x + 2y$

*Constraints:*

$$x \geq 0$$
$$y \geq 0$$
$$x + 3y \leq 15$$
$$4x + y \leq 16$$

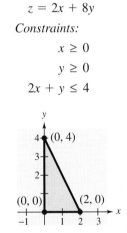

**6.** *Objective function:*

$z = 5x + 4y$

*Constraints:*

$$x \geq 0$$
$$2x + 3y \geq 6$$
$$3x - 2y \leq 9$$
$$x + 5y \leq 20$$

**7.** *Objective function:*

$z = 5x + 0.5y$

*Constraints:*

(See Exercise 5.)

**8.** *Objective function:*

$z = x + 6y$

*Constraints:*

(See Exercise 6.)

In Exercises 9–20, sketch the region determined by the constraints. Then find the minimum and maximum values of the objective function, subject to the indicated constraints.

**9.** *Objective function:*

$z = 6x + 10y$

*Constraints:*

$x \geq 0$

$y \geq 0$

$3x + 5y \leq 15$

**10.** *Objective function:*

$z = 7x + 8y$

*Constraints:*

$x \geq 0$

$y \geq 0$

$x + 2y \leq 8$

**11.** *Objective function:*

$z = 9x + 4y$

*Constraints:*

$x \geq 0$

$y \geq 0$

$3x + 5y \leq 15$

**12.** *Objective function:*

$z = 7x + 2y$

*Constraints:*

$x \geq 0$

$y \geq 0$

$x + 2y \leq 8$

**13.** *Objective function:*

$z = 4x + 5y$

*Constraints:*

$x \geq 0$

$y \geq 0$

$x + y \geq 8$

$3x + 5y \geq 30$

**14.** *Objective function:*

$z = 4x + 5y$

*Constraints:*

$x \geq 0$

$y \geq 0$

$2x + 2y \leq 10$

$x + 2y \leq 6$

**15.** *Objective function:*

$z = 2x + 7y$

*Constraints:*

(See Exercise 13.)

**16.** *Objective function:*

$z = 2x - y$

*Constraints:*

(See Exercise 14.)

**17.** *Objective function:*

$z = 5x + 2y$

*Constraints:*

$x \geq 0$

$y \geq 0$

$x + 2y \leq 40$

$x + y \geq 30$

$2x + 3y \geq 72$

**18.** *Objective function:*

$z = x$

*Constraints:*

$x \geq 0$

$y \geq 0$

$2x + 3y \leq 60$

$2x + y \leq 28$

$4x + y \leq 48$

**19.** *Objective function:*

$z = 2x + 5y$

*Constraints:*

(See Exercise 17.)

**20.** *Objective function:*

$z = y$

*Constraints:*

(See Exercise 18.)

In Exercises 21–24, maximize the objective function subject to the constraints

$3x + y \leq 15$   and   $4x + 3y \leq 30$

where $x \geq 0$ and $y \geq 0$.

**21.** $z = 2x + y$

**22.** $z = 5x + y$

**23.** $z = x + y$

**24.** $z = 3x + y$

In Exercises 25–28, maximize the objective function subject to the constraints

$x + 4y \leq 20$,   $x + y \leq 8$,   and   $3x + 2y \leq 21$

where $x \geq 0$ and $y \geq 0$.

**25.** $z = 2x + 5y$

**26.** $z = 3x + 5y$

**27.** $z = 10x + 7y$

**28.** $z = 10x + 4y$

In Exercises 29–32, find an objective function that has a maximum or a minimum value at the indicated vertex of the constraint region shown. (There are many correct answers.)

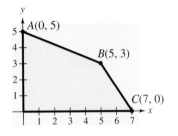

**29.** The maximum occurs at vertex $A$.

**30.** The maximum occurs at vertex $B$.

**31.** The maximum occurs at vertex $C$.

**32.** The minimum occurs at vertex $A$.

**33.** *Maximum Profit*    A store plans to sell two models of computers at costs of \$1100 and \$1500. The \$1100 model yields a profit of \$100 and the \$1500 model yields a profit of \$125. How many units of each model should be stocked to maximize profit, subject to the following constraints?

• The merchant estimates that the total monthly demand will not exceed 250 units.

• The merchant does not want to invest more than \$295,000 in computer inventory.

**34.** *Maximum Profit*   A fruit grower has 150 acres of land available to raise two crops, A and B. The profit is $140 per acre for crop A and $235 per acre for crop B. Find the number of acres of each fruit that should be planted to maximize profit, subject to the following constraints.

- It takes 1 day to trim an acre of crop A and 2 days to trim an acre of crop B, and there are 240 days per year available for trimming.
- It takes 0.3 day to pick an acre of crop A and 0.1 day to pick an acre of crop B, and there are 30 days per year available for picking.

**35.** *Minimum Cost*   A farming cooperative mixes two brands of cattle feed. Brand X costs $25 per bag, and brand Y costs $20 per bag. Find the number of bags of each brand that should be mixed to produce a mixture having a minimum cost per bag, subject to the following constraints.

- Brand X contains 2 units of nutritional element A, 2 units of element B, and 2 units of element C.
- Brand Y contains 1 unit of nutritional element A, 9 units of element B, and 3 units of element C.
- The minimum requirements of nutrients A, B, and C are 12 units, 36 units, and 24 units, respectively.

**36.** *Minimum Cost*   Two gasolines, type A and type B, have octane ratings of 89 and 93, respectively. Type A costs $0.99 per gallon and type B costs $1.25 per gallon. Determine the blend of minimum cost with an octane rating of at least 91. (*Hint:* Let $x$ be the fraction of each gallon that is type A and $y$ be the fraction that is type B.)

**37.** *Maximum Profit*   A manufacturer produces two models of bicycles. The times (in hours) required for assembling, painting, and packaging each model are as follows.

| Process | Model A | Model B |
|---|---|---|
| Assembling | 2 | 2.5 |
| Painting | 4 | 1 |
| Packaging | 1 | 0.75 |

The total times available for assembling, painting, and packaging are 4000 hours, 4800 hours, and 1500 hours, respectively. The profits per unit are $45 for model A and $50 for model B. How many of each should be made to obtain a maximum profit?

**38.** *Maximum Profit*   A company makes two models of a product. The times (in hours) required for assembling, painting, and packaging are as follows.

| Process | Model A | Model B |
|---|---|---|
| Assembling | 2.5 | 3 |
| Painting | 2 | 1 |
| Packaging | 0.75 | 1.25 |

The total times available for assembling, painting, and packaging are 4000 hours, 2500 hours, and 1500 hours, respectively. The profits per unit are $50 for model A and $52 for model B. How many of each should be made to obtain a maximum profit?

**39.** *Maximum Revenue*   An accounting firm has 900 hours of staff time and 100 hours of review time available each week. The firm charges $2000 for an audit and $300 for a tax return. What numbers of audits and tax returns will bring in a maximum revenue, subject to the following constraints?

- Each audit requires 100 hours of staff time and 10 hours of review time.
- Each tax return requires 12.5 hours of staff time and 2.5 hours of review time.

**40.** *Maximum Revenue*   The accounting firm in Exercise 39 lowers its charge for an audit to $1800. What numbers of audits and tax returns will bring in a maximum revenue?

**41.** *Media Selection*   A company has budgeted a maximum of $600,000 for advertising a product nationally. Each minute of television time costs $60,000 and each one-page newspaper ad costs $15,000. Each television ad is expected to be viewed by 15 million viewers, and each newspaper ad is expected to be seen by 3 million readers. The company's market research department recommends that at most 90% of the advertising budget be spent on television ads. How should the advertising budget be spent to maximize the total audience?

**42.** *Maximum Profit* A fruit juice company makes two drinks by blending apple and pineapple juices. The first uses 30% apple juice and 70% pineapple, and the second uses 60% apple and 40% pineapple. There are 1000 liters of apple juice and 1500 liters of pineapple juice available. The profit for the first drink is $0.60 per liter and the profit for the second drink is $0.50. Find the number of liters of each drink that should be produced in order to maximize profit.

**43.** *Investments* An investor has up to $250,000 to invest in two types of investments. Type A pays 8% annually and type B pays 10% annually. To have a well-balanced portfolio, the investor imposes the following conditions. At least one-fourth of the total portfolio is to be allocated to type A investments and at least one-fourth of the portfolio is to be allocated to type B investments. How much should be allocated to each type of investment to obtain a maximum return?

**44.** *Investments* An investor has up to $450,000 to invest in two types of investments. Type A pays 6% annually and type B pays 10% annually. To have a well-balanced portfolio, the investor imposes the following conditions. At least one-half of the total portfolio is to be allocated to type A investments and at least one-fourth of the portfolio is to be allocated to type B investments. How much should be allocated to each type of investment to obtain a maximum return?

In Exercises 45–50, the given linear programming problem has an unusual characteristic. Sketch a graph of the solution region for the problem and describe the unusual characteristic. (In each problem, the objective function is to be maximized.)

**45.** *Objective function:*

$z = 2.5x + y$

*Constraints:*

$x \geq 0$

$y \geq 0$

$3x + 5y \leq 15$

$5x + 2y \leq 10$

**46.** *Objective function:*

$z = x + y$

*Constraints:*

$x \geq 0$

$y \geq 0$

$-x + y \leq 1$

$-x + 2y \leq 4$

**47.** *Objective function:*

$z = -x + 2y$

*Constraints:*

$x \geq 0$

$y \geq 0$

$x \leq 10$

$x + y \leq 7$

**48.** *Objective function:*

$z = x + y$

*Constraints:*

$x \geq 0$

$y \geq 0$

$-x + y \leq 1$

$-3x + y \geq 3$

**49.** *Objective function:*

$z = 3x + 4y$

*Constraints:*

$x \geq 0$

$y \geq 0$

$x + y \leq 1$

$2x + y \leq 4$

**50.** *Objective function:*

$z = x + 2y$

*Constraints:*

$x \geq 0$

$y \geq 0$

$x + 2y \leq 4$

$2x + y \leq 4$

**51.** *Maximum Profit* A company makes two models of a patio furniture set. The times for assembling, finishing, and packaging model A are 3.0 hours, 2.5 hours, and 0.60 hour, respectively. The times for model B are 2.75 hours, 1.0 hour, and 1.25 hours. The total times available for assembling, finishing, and packaging are 3000 hours, 2400 hours, and 1200 hours, respectively. The profit per unit for model A is $80 and the profit per unit for model B is $65. How many of each model should be produced to obtain a maximum profit?

**52.** *Maximum Profit* A manufacturer produces two models of elliptical cross training exercise machines. The times for assembling, finishing, and packaging model A are 3.0 hours, 2.0 hours, and 1.2 hours, respectively. The times for model B are 4.0 hours, 2.5 hours, and 0.9 hour. The total times available for assembling, finishing, and packaging are 6000 hours, 4000 hours, and 900 hours, respectively. The profits per unit are $300 for model A and $375 for model B. How many of each model should be produced to obtain a maximum profit?

# Chapter Project

# Consumer Credit Outstanding

The following table gives the amounts $y_1$ of U.S. consumer credit outstanding (in billions of dollars) held by finance companies from 1991 to 2001. In the table, $t = 1$ represents 1991. From the scatter plot of the data in Figure A, what type of equation would you guess models the data?

| $t$ | 1 | 2 | 3 | 4 | 5 | 6 | 7 | 8 | 9 | 10 | 11 |
|-----|-----|-----|-----|-----|-----|-----|-----|-----|-----|-----|-----|
| $y_1$ | 127.5 | 117.0 | 110.3 | 124.0 | 141.3 | 153.9 | 156.7 | 154.3 | 173.6 | 190.3 | 191.0 |

The amounts $y_2$ of U.S. consumer credit outstanding (in billions of dollars) held by credit unions during the same time are listed in the table below. A scatter plot for this data is shown in Figure B. What is your suggested model for this data?

| $t$ | 1 | 2 | 3 | 4 | 5 | 6 | 7 | 8 | 9 | 10 | 11 |
|-----|-----|-----|-----|-----|-----|-----|-----|-----|-----|-----|-----|
| $y_2$ | 88.5 | 88.3 | 95.4 | 108.2 | 125.7 | 136.1 | 147.6 | 152.4 | 158.2 | 176.0 | 187.0 |

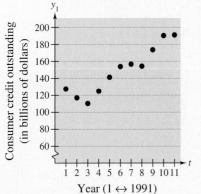

**Figure A**

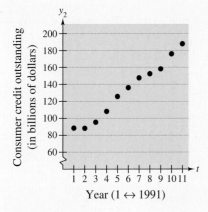

**Figure B**

Use this information to investigate the following questions. (Source: Federal Reserve Statistical Release, Consumer Credit Table G.19, Oct. 5, 2001. [Internet])

1. *Fitting a Model to Data*   Use the regression capabilities of a graphing utility to find a quadratic model and a quartic model for the data in the first table. Graph the two models along with the scatter plot of the data. Which model is a better fit?

2. *Comparing a Model to Actual Data*   Make a numerical comparison by constructing a table of the values given by the best-fitting model found in Question 1 and the actual data. Does the numerical comparison confirm that the model you found is valid?

3. *Fitting a Model to Data*   Use the regression capabilities of a graphing utility to find a quadratic model and a quartic model for the data in the second table. Graph the two models along with the scatter plot of the data. Which model is a better fit?

4. *Comparing a Model to Actual Data*   Make a numerical comparison by constructing a table of the values given by the best-fitting model found in Question 3 and the actual data. Does the numerical comparison confirm that the model you found is valid?

5. *Comparing Two Models*   Use a graphing utility to graph both best-fitting models in the same viewing window, where $1 \le t \le 11$. Do the graphs intersect? Does that agree with your data? Modify your viewing window so that $1 \le t \le 14$. Do the graphs intersect? Are the models likely to be valid for $11 < t \le 14$? Explain your reasoning.

## CHAPTER SUMMARY

After studying this chapter, you should have acquired the following skills.
These skills are keyed to the Review Exercises that begin on page 419.
Answers to odd-numbered Review Exercises are given in the back of the book.

| 6.1 | · Solve a system of equations by the method of substitution. | Review Exercises 1–6 |
| --- | --- | --- |
| | · Solve a system of equations graphically. | Review Exercises 7, 8, 10 |
| | · Construct and use a system of equations to solve an application problem. | Review Exercises 9, 11, 12, 21, 22, 52, 53 |
| 6.2 | · Solve a linear system by the method of elimination. | Review Exercises 13–18 |
| | · Interpret the solution of a linear system graphically. | Review Exercises 19, 20 |
| | · Construct and use a linear system to solve an application problem. | Review Exercises 23, 24, 40, 72 |
| 6.3 | · Solve a linear system in row-echelon form using back-substitution. | Review Exercises 25, 26 |
| | · Use Gaussian elimination to solve a linear system. | Review Exercises 27–32 |
| | · Construct and use a linear system in three or more variables to solve an application problem. | Review Exercises 37–39, 72 |
| | · Find the equation of a circle or a parabola using a linear system in three or more variables. | Review Exercises 33–36, 41 |
| 6.4 | · Sketch the graph of an inequality in two variables. | Review Exercises 42–47 |
| | · Solve a system of inequalities. | Review Exercises 50, 51 |
| | · Use a system of inequalities to model and solve an application problem. | Review Exercises 48–51, 54, 55 |
| 6.5 | · Use linear programming to minimize or maximize an objective function. | Review Exercises 56–63 |
| | · Use linear programming to optimize an application. | Review Exercises 64–71 |

## REVIEW EXERCISES

In Exercises 1–6, solve the system by the method of substitution.

**1.**  $x + 3y = 10$
     $4x - 5y = -28$

**2.** $3x - y - 13 = 0$
     $4x + 3y - 26 = 0$

**3.** $\frac{1}{2}x + \frac{3}{5}y = -2$
     $2x + y = 6$

**4.** $1.3x + 0.9y = 7.5$
     $0.4x - 0.5y = -0.8$

**5.** $x^2 + y^2 = 100$
     $x + 2y = 20$

**6.** $y = x^3 - 2x^2 - 2x - 3$
     $y = -x^2 + 4x - 3$

In Exercises 7 and 8, use a graphing utility to find all points of intersection of the graphs of the equations.

**7.** $y = x^2 - 3x + 11$
     $y = -x^2 + 2x + 8$

**8.** $y = \sqrt{9 - x^2}$
     $y = e^x + 1$

**9.** *Break-Even Point*  You are setting up a business and have made an initial investment of $20,000. The unit cost of the product is $3.25 and the selling price is $6.95. How many units must you sell to break even? (Round to the nearest whole unit.)

**10.** *Comparing Product Sales*  Your company produces three types of CD players. The research and development department has designed a new CD player that is predicted to sell better than the three players combined. The sales equations for the CD players are

$S = 2150.78 - 56.8t + 0.45t^2$   Three players combined

$S = 121.27 + 11.12t + 3.98t^2$   New player

where $S$ is the sales in thousands and $t = 8$ represents 1998. Use a graphing utility to determine when the sales of the new CD player will overtake the combined sales of the other players.

**11.** *Investment Portfolio*  A total of $50,000 is invested in two funds paying 6.75% and 7.5% simple interest. If the yearly interest is $3637.50, how much of the $50,000 is invested at each rate?

**12.** *Choice of Two Jobs*  You are offered two different jobs selling computers. One company offers an annual salary of $28,000 plus a year-end bonus of 1.5% of your total sales. The other company offers a salary of $22,000 plus a year-end bonus of 3.5% of your total sales. How much would you have to sell in order to make the second offer better?

In Exercises 13–18, solve the system by elimination.

**13.** $2x - 3y = 21$
     $3x + y = 4$

**14.** $3u + 5v = 9$
     $12u + 10v = 22$

**15.** $1.25x - 2y = 3.5$
     $5x - 8y = 14$

**16.** $\dfrac{x - 2}{3} + \dfrac{y + 3}{4} = 5$
     $2x - y = 7$

**17.** $1.5x + 2.5y = 8.5$
     $6x + 10y = 24$

**18.** $\frac{3}{5}x + \frac{2}{7}y = 10$
     $x + 2y = 38$

In Exercises 19 and 20, describe the graph of the solution of the linear system.

**19.** $2x + y = -1$
     $3x - 2y = -5$

**20.** $x - 2y = -1$
     $-2x + 4y = 2$

**21.** *Acid Mixture*  Twelve gallons of a 25% acid solution are obtained by mixing a 10% solution with a 50% solution.

(a) Write a system of equations that represents the problem and use a graphing utility to graph both equations.

(b) How much of each solution must be used?

**22.** *Compact Disc Sales*  Suppose you are the manager of a music store. At the end of the week you are going over receipts for the previous week's sales. Eight hundred and fifty compact discs were sold. One type of CD sold for $11.95 and a second type sold for $15.95. The total CD receipts were $11,257.50. The cash register that was supposed to keep track of the number of each type of CD sold malfunctioned. Can you recover the information? If so, how many of each type of CD were sold?

*Supply and Demand* In Exercises 23 and 24, find the point of equilibrium for the given pair of demand and supply equations.

|  *Demand*  |  *Supply*  |
|---|---|
| **23.** $p = 37 - 0.0002x$ | $p = 22 + 0.00001x$ |
| **24.** $p = 120 - 0.0001x$ | $p = 45 + 0.0002x$ |

In Exercises 25–32, solve the system of equations.

**25.** $\begin{aligned} 4x - 3y + 2z &= 1 \\ 2y - 4z &= 2 \\ 3z &= 6 \end{aligned}$

**26.** $\begin{aligned} 2x + y - 4z &= 6 \\ 3y + z &= 2 \\ 2z &= -8 \end{aligned}$

**27.** $\begin{aligned} 2x + y + z &= 6 \\ x - 4y - z &= 3 \\ x + y + z &= 4 \end{aligned}$

**28.** $\begin{aligned} x + 3y - z &= 13 \\ 2x - 5z &= 23 \\ 4x - y - 2z &= 14 \end{aligned}$

**29.** $\begin{aligned} x + y + z &= 10 \\ -2x + 3y + 4z &= 22 \end{aligned}$

**30.** $\begin{aligned} x + y + z &= 1 \\ 2x - 4y + 3z &= -20 \\ 2x + 3y + 2z &= 5 \end{aligned}$

**31.** $\begin{aligned} 2x + 6y - z &= 1 \\ x - 3y + z &= 2 \\ \tfrac{3}{2}x + \tfrac{3}{2}y &= 6 \end{aligned}$

**32.** $\begin{aligned} x + y + z + w &= 8 \\ 4y + 5z - 2w &= 3 \\ 2x + 3y - z &= -2 \\ 3x + 2y - 4w &= -20 \end{aligned}$

In Exercises 33 and 34, find the equation of the parabola $y = ax^2 + bx + c$ that passes through the points.

**33.** $(0, -6), (1, -3), (2, 4)$

**34.** $(-5, 0), (1, -6), (2, 14)$

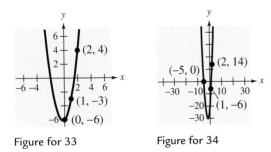

Figure for 33      Figure for 34

In Exercises 35 and 36, find the equation of the circle $x^2 + y^2 + Dx + Ey + F = 0$ that passes through the points.

**35.** $(2, 2), (5, -1), (-1, -1)$

**36.** $(4, 2), (1, 3), (-2, -6)$

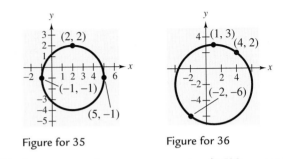

Figure for 35      Figure for 36

**37.** *Investment* Suppose you receive $1520 a year in interest from three investments. The interest rates for the three investments are 6%, 7%, and 8%. The 7% investment is six times the 6% investment, and the 8% investment is $1000 more than the 6% investment. What is the amount of each investment?

**38.** *Product Types* A company sells three types of products for $20, $15, and $10 per unit. In 1 year, the total revenue for the three products was $1,500,000, which corresponded to the sale of 110,000 units. The company sold half as many units of the $20 product as units of the $15 product. How many units of each product were sold?

**39.** *Investment Portfolio* Suppose an investor has a portfolio totaling $500,000 that is to be allocated among the following types of investments: (1) certificates of deposit, (2) municipal bonds, (3) blue-chip stocks, and (4) growth or speculative stocks. The certificates of deposit pay 8% annually, and the municipal bonds pay 6% annually. Over a 5-year period, the investor expects the blue-chip stocks to return 10% annually and the growth stocks to return 15% annually. The investor wishes a combined return of $8\frac{1}{2}\%$ and also wants to have only two-fifths of the portfolio invested in stocks. How much should be allocated to each type of investment if the amount invested in certificates of deposit is twice that invested in municipal bonds?

**40.** *Fitting a Line to Data*  Find the least squares regression line $y = ax + b$ for the points

$(0, 1.6)$, $(1, 2.4)$, $(2, 3.6)$, $(3, 4.7)$, $(4, 5.5)$.

To find the equation of the line, solve the system of linear equations for $a$ and $b$.

$5b + 10a = 17.8$

$10b + 30a = 45.7$

**41.** *Fitting a Parabola to Data*  Find the least squares regression parabola $y = ax^2 + bx + c$ for the points

$(-2, 0.4)$, $(-1, 0.9)$, $(0, 1.9)$, $(1, 2.1)$, $(2, 3.8)$.

To find the parabola, solve the system of linear equations for $a$, $b$, and $c$.

$$5c \quad + 10a = \quad 9.1$$
$$10b \qquad = \quad 8.0$$
$$10c \quad + 34a = \quad 19.8$$

In Exercises 42–47, sketch the graph of the solution set of the system of inequalities.

**42.** $2x + 3y < 9$

$\qquad x > 0$

$\qquad y > 0$

**43.** $2x - y > 6$

$\qquad x < 5$

$\qquad y \leq 8$

**44.** $3x - y > -4$

$\quad 2x + y > -1$

$\quad 7x + y < \quad 4$

**45.** $\quad x + y > \quad 4$

$\quad 3x + y < 10$

**46.** $\quad x^2 + y^2 \leq 9$

$\quad x^2 - x - 2 \leq y$

**47.** $\ln x < y$

$\qquad y > -1$

$\qquad x < 4$

In Exercises 48 and 49, derive a set of inequalities that describes the region.

**48.** Parallelogram

**49.** Triangle

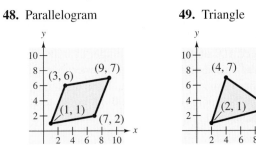

**50.** *Videocassette Recorder Inventory*  A store sells two models of a certain brand of VCR. Because of the demand, it is necessary to stock twice as many units of model A as units of model B. The costs to the store for the two models are $200 and $300, respectively. The management does not want more than $4000 in VCR inventory at any one time, and it wants at least four model A VCRs and two model B VCRs in inventory at all times. Derive a system of inequalities that describes all possible inventory levels, and sketch the graph of the system.

**51.** *Concert Ticket Sales*  Two types of tickets are to be sold for a concert. One type costs $20 per ticket and the other type costs $30 per ticket. The promoter of the concert must sell at least 16,000 tickets, including at least 9000 of the $20 tickets and at least 4000 of the $30 tickets. Moreover, the gross receipts must total at least $400,000 in order for the concert to be held. Find a system of inequalities describing the different numbers of tickets that can be sold, and sketch the graph of the system.

**52.** *Morning or Evening Paper?*  The total circulations of morning and evening newspapers in the United States from 1980 to 1999 can be approximated by the models

$y = 31.7 + 0.84t$      Morning

$y = 32.4 - 1.21t$      Evening

where $y$ is the circulation (in millions) and $t$ represents the year, with $t = 0$ corresponding to 1980. During which year did the circulation of morning papers overtake the circulation of evening papers? (Source: Editor and Publisher Company)

**53.** *News Programs*  A local television station tracked the numbers of viewers of its two prime news shows for 1 year. The numbers of viewers for the two news shows can be approximated by the models

$y = 9500 + 350t$      6 P.M. news

$y = 15,000 - 275t$      10 P.M. news

where $y$ is the number of viewers and $t = 1$ corresponds to the month of January. During what month did the number of viewers of the 6 P.M. news program surpass the number of viewers of the 10 P.M. news program?

In Exercises 54 and 55, find the consumer surplus and producer surplus for the given pair of demand and supply equations.

| Demand | Supply |
|---|---|
| **54.** $p = 160 - 0.0001x$ | $p = 70 + 0.0002x$ |
| **55.** $p = 130 - 0.0002x$ | $p = 30 + 0.0003x$ |

In Exercises 56–59, find the minimum and maximum values of the objective function, subject to the indicated constraints. (For each exercise, the graph of the constraints is provided.)

**56.** *Objective function:*

$z = 5x + 6y$

*Constraints:*

$$x \geq 0$$
$$y \geq 0$$
$$x + y \leq 8$$

**57.** *Objective function:*

$z = 15x + 12y$

*Constraints:*

$$x \geq 0$$
$$y \geq 0$$
$$x + 3y \leq 12$$
$$3x + 2y \leq 15$$

**58.** *Objective function:*

$z = 8x + 3y$

*Constraints:*

$$0 \leq x \leq 50$$
$$0 \leq y \leq 35$$
$$4x + 5y \leq 275$$

**59.** *Objective function:*

$z = 50x + 60y$

*Constraints:*

$$x \geq 0$$
$$y \geq 0$$
$$3x + 4y \geq 1200$$
$$5x + 6y \leq 3000$$

In Exercises 60–63, sketch the region determined by the constraints. Find the minimum and maximum values of the objective function, subject to the indicated constraints.

**60.** *Objective function:*

$z = 6x + 8y$

*Constraints:*

$$x \geq 0$$
$$y \geq 0$$
$$x + 4y \leq 16$$
$$3x + 2y \leq 18$$

**61.** *Objective function:*

$z = 5x + 8y$

*Constraints:*

$$0 \leq x \leq 5$$
$$y \geq 0$$
$$x + 2y \leq 12$$
$$2x + 3y \leq 19$$

**62.** *Objective function:*

$z = 8x + 3y$

*Constraints:*

$$0 \leq x \leq 5$$
$$0 \leq y \leq 7$$
$$x + y \leq 9$$
$$3x + y \leq 17$$

**63.** *Objective function:*

$z = 10x + 11y$

*Constraints:*

$$x \geq 0$$
$$y \geq 0$$
$$2x + 5y \leq 30$$
$$x + y \geq 3$$
$$2x + y \leq 14$$

**64.** *Maximum Profit* A merchant plans to sell two models of camcorders at prices of $800 and $1000. The $800 model yields a profit of $75 and the $1000 model yields a profit of $125. The merchant estimates that the total monthly demand will not exceed 300 units. Find the number of units of each model that should be stocked in order to maximize profit. Assume that the merchant does not want to invest more than $280,000 in camcorder inventory.

**65.** *Maximum Profit* A merchant plans to sell two models of color televisions at prices of $300 and $500. The $300 model yields a profit of $30 and the $500 model yields a profit of $45. The merchant estimates that the total monthly demand will not exceed 100 units. Find the number of units of each model that should be stocked in order to maximize profit. Assume that the merchant does not want to invest more than $40,000 in color television inventory.

**66.** *Maximum Profit*   A factory manufactures two color television set models: a basic model that yields $100 profit and a deluxe model that yields a profit of $200. The times (in hours) required for assembling, finishing, and packaging are as follows.

| Process | Basic Model | Deluxe Model |
|---|---|---|
| Assembling | 2 | 5 |
| Finishing | 1 | 2 |
| Packaging | 1 | 1 |

The total times available for assembling, finishing, and packaging are 3000 hours, 1400 hours, and 1000 hours, respectively. How many of each model should be produced for maximum profit?

**67.** *Maximum Profit*   A company makes two models of a product. The times (in hours) required for assembling, finishing, and packaging each model are as follows.

| Process | Model A | Model B |
|---|---|---|
| Assembling | 3.5 | 8 |
| Finishing | 2.5 | 2 |
| Packaging | 1.3 | 0.7 |

The total times available for assembling, finishing, and packaging are 5600 hours, 2000 hours, and 910 hours, respectively. The profits per unit are $100 for model A and $150 for model B. How many of each model should be produced to obtain a maximum profit?

**68.** *Minimum Cost*   A pet supply company mixes two brands of dry dog food. Brand X costs $15 per bag and contains 8 units of nutritional element A, 1 unit of nutritional element B, and 2 units of nutritional element C. Brand Y costs $30 per bag and contains 2 units of nutritional element A, 1 unit of nutritional element B, and 7 units of nutritional element C. Each bag of mixed dog food must contain at least 16 units of element A, 5 units of element B, and 20 units of element C. Find the number of bags of each brand that should be mixed to produce a mixture having a minimum cost per bag.

**69.** *Minimum Cost*   Two gasolines, type A and type B, have octane ratings of 85 and 89, respectively. Type A costs $1.01 per gallon and type B costs $1.13 per gallon. Determine the blend of minimum cost that has an octane rating of at least 87. (Hint: Let $x$ be the fraction of each gallon that is type A and let $y$ be the fraction of each gallon that is type B.)

**70.** *Maximum Revenue*   An accounting firm has 800 hours of staff time and 90 hours of review time available each week. The firm charges $2400 for an audit and $300 for a tax return. Each audit requires 100 hours of staff time and 10 hours of review time. Each tax return requires 10 hours of staff time and 2 hours of review time. What numbers of audits and tax returns will bring in maximum revenue?

**71.** *Maximum Profit*   Suppose the accounting firm in Exercise 70 realizes a profit of $800 for each audit and $120 for each tax return. What combination of audits and tax returns will yield a maximum profit?

**72.** *Median Price*   The median prices $y$ (in thousands of dollars) of new one-family homes sold in the United States from 1995 to 1999 are given in the table, where $t = 5$ represents 1995.   (Source: U.S. Bureau of the Census, U.S. Dept. of Housing and Urban Development)

| $t$ | 5 | 6 | 7 | 8 | 9 |
|---|---|---|---|---|---|
| $y$ | 133.9 | 140.0 | 146.0 | 152.5 | 159.8 |

(a) Make a scatter plot of the data.

(b) Solve the system for $a$ and $b$ to find the least squares regression line $y = at + b$ for the data.

$$5b + 35a = 732.2$$
$$35b + 255a = 5189.7$$

(c) Solve the system to find the least squares regression parabola for the data.

$$5c + 35b + 255a = 732.2$$
$$35c + 255b + 1925a = 5189.7$$
$$255c + 1925b + 14{,}979a = 38{,}245.3$$

(d) Use the regression capabilities of your graphing utility to find linear and quadratic models for the data. Compare them to the least squares regression models found in parts (b) and (c).

(e) Graph the linear and quadratic models. Use the trace feature with each model to estimate the median prices in 2000 and 2001. How close are the estimates to each other for each year?

## Chapter Test

Take this test as you would take a test in class. After you are done, check your work against the answers given in the back of the book.

In Exercises 1–8, solve the system of equations using the indicated method.

**1.** *Substitution*

$$3x - 2y = -2$$
$$4x + 3y = 20$$

**2.** *Substitution*

$$x + y = 3$$
$$x^2 + y = 9$$

**3.** *Substitution*

$$2x - y = 13$$
$$5x + 4y = 13$$

**4.** *Graphing*

$$5x - y = 6$$
$$2x^2 + y = 8$$

**5.** *Graphing*

$$1.5x - 2.25y = 8$$
$$2.5x + 2y = 5.75$$

**6.** *Elimination*

$$2x - 4y + z = 11$$
$$x + 2y + 3z = 9$$
$$3y + 5z = 12$$

**7.** *Elimination*

$$3x - 2y + z = 16$$
$$5x - z = 6$$
$$2x - y - z = 3$$

**8.** *Elimination*

$$-x + y - 2z = 3$$
$$x - 4y - 2z = 1$$
$$x + 2y + 6z = 5$$

**9.** Find the point of equilibrium for a system that has a demand function of $p = 45 - 0.0003x$ and a supply function of $p = 29 + 0.00002x$.

**10.** A system of linear equations reduces to $-23 = 0$. What can you conclude?

**11.** A system of linear equations reduces to $0 = 0$. What can you conclude?

**12.** Find an equation of the parabola passing through $(1, -5), (3, 1)$, and $(0, -2)$.

**13.** A total of $50,000 is invested in two funds paying 8% and 8.5% simple interest. The annual interest is $4150. How much is invested in each fund?

In Exercises 14–17, sketch the inequality.

**14.** $x \geq 0$     **15.** $y \geq 0$     **16.** $x + 3y \leq 12$     **17.** $3x + 2y \leq 15$

**18.** Sketch the solution of the system of inequalities comprised of the inequalities in Exercises 14–17.

**19.** Find the minimum and maximum values of the objective function $z = 6x + 7y$, subject to the constraints given in Exercises 14–17.

**20.** A manufacturer produces two models of stair climbers. The times (in hours) required for assembling, painting, and packaging each model are as follows.

- Assembling: 3.5 hours for model A; 8 hours for model B
- Painting: 2.5 hours for model A; 2 hours for model B
- Packaging: 1.3 hours for model A; 0.9 hour for model B

The total times available for assembling, painting, and packaging are 5600 hours, 2000 hours, and 900 hours, respectively. The profits per unit are $200 for model A and $275 for model B. How many of each type should be produced to obtain a maximum profit? Explain your reasoning.

# 7

# Matrices and Determinants

Your advertising group has been hired by Company C, a cable/internet service provider, to develop an advertising campaign to increase its market share. Currently your client competes with two other service providers, Companies A and B. With this new campaign, you predict that each month (1) 92% of A's subscribers will remain with A, 3% will switch to B, and 5% will switch to C; (2) 91% of B's subscribers will remain with B, 7% will switch to A, and 2% will change to C; and (3) 93% of C's subscribers will remain with C, 6% will change to B, and 1% will change to A. The current shares of the market are given in the matrix.

$$x_0 = \begin{bmatrix} 0.2 \\ 0.7 \\ 0.1 \end{bmatrix} \begin{matrix} \text{Company A} \\ \text{Company B} \\ \text{Company C} \end{matrix}$$

The market shares for Company C for the first six months of the campaign are shown in the table and bar graph.

**7.1 Matrices and Linear Systems**

**7.2 Operations with Matrices**

**7.3 The Inverse of a Square Matrix**

**7.4 The Determinant of a Square Matrix**

**7.5 Applications of Matrices and Determinants**

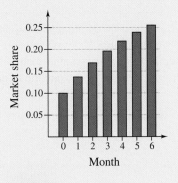

| Month | 0 | 1 | 2 | 3 | 4 | 5 | 6 |
|-------|-----|-------|-------|-------|-------|-------|-------|
| Share | 0.1 | 0.137 | 0.169 | 0.196 | 0.219 | 0.239 | 0.255 |

The chapter project related to this information is on page 484.

425

## 7.1 Matrices and Linear Systems

### Objectives

- Determine the order of a matrix.
- Perform elementary row operations on a matrix.
- Write a matrix in row-echelon form or reduced row-echelon form.
- Solve a system of linear equations using Gaussian elimination or Gauss-Jordan elimination.

## Matrices

In this section you will study a streamlined technique for solving systems of linear equations. This technique involves the use of a rectangular array of real numbers called a **matrix.** The plural of matrix is *matrices.*

> ▶ **Definition of a Matrix**
>
> If $m$ and $n$ are positive integers, an $m \times n$ **matrix** (read "$m$ by $n$") is a rectangular array
>
> $$\begin{bmatrix} a_{11} & a_{12} & a_{13} & \cdots & a_{1n} \\ a_{21} & a_{22} & a_{23} & \cdots & a_{2n} \\ a_{31} & a_{32} & a_{33} & \cdots & a_{3n} \\ \vdots & \vdots & \vdots & & \vdots \\ a_{m1} & a_{m2} & a_{m3} & \cdots & a_{mn} \end{bmatrix} \Big\} \; m \text{ rows}$$
>
> $\underbrace{\qquad\qquad\qquad}_{n \text{ columns}}$
>
> in which each **entry,** $a_{ij}$, of the matrix is a number. An $m \times n$ matrix has $m$ **rows** (horizontal lines) and $n$ **columns** (vertical lines). A matrix that has only one row is called a **row matrix,** and a matrix that has only one column is called a **column matrix.**

The entry in the $i$th row and $j$th column is denoted by the *double subscript* notation $a_{ij}$. That is, $a_{21}$ refers to the entry in row 2, column 1. A matrix having $m$ rows and $n$ columns is said to be of **order** $m \times n$. If $m = n$, the matrix is **square** of order $n$. For a square matrix, the entries $a_{11}, a_{22}, a_{33}, \ldots$ are the **main diagonal** entries.

**Example 1** Examples of Matrices

The following matrices have the indicated orders.

**a.** *Order:* $1 \times 1$

$$[2]$$

**b.** *Order:* $1 \times 4$

$$\begin{bmatrix} 1 & -3 & 0 & \frac{1}{2} \end{bmatrix}$$

**c.** *Order:* $2 \times 2$

$$\begin{bmatrix} 0 & 0 \\ 0 & 0 \end{bmatrix}$$

**d.** *Order:* $3 \times 2$

$$\begin{bmatrix} 5 & 0 \\ 2 & -2 \\ -7 & 4 \end{bmatrix}$$

A matrix derived from a system of linear equations (each written in standard form with the constant term on the right) is the **augmented matrix** of the system. Moreover, the matrix derived from the coefficients of the system (but that does not include the constant terms) is the **coefficient matrix** of the system. Note in the matrices the use of 0 for the missing $y$-variable in the third equation, and also note the fourth column of constant terms in the augmented matrix.

$$
\begin{array}{ccc}
\textit{System} & \textit{Augmented Matrix} & \textit{Coefficient Matrix} \\[4pt]
\begin{aligned}
x - 4y + 3z &= 5 \\
-x + 3y - z &= -3 \\
2x \phantom{{}+3y} - 4z &= 6
\end{aligned}
&
\left[\begin{array}{ccc:c}
1 & -4 & 3 & 5 \\
-1 & 3 & -1 & -3 \\
2 & 0 & -4 & 6
\end{array}\right]
&
\left[\begin{array}{ccc}
1 & -4 & 3 \\
-1 & 3 & -1 \\
2 & 0 & -4
\end{array}\right]
\end{array}
$$

When forming either the coefficient matrix or the augmented matrix of a system, you should begin by vertically aligning the variables in the equations and using 0's for the missing variables.

$$
\begin{array}{ccc}
\textit{Given System} & \textit{Line Up Variables} & \textit{Form Augmented Matrix} \\[4pt]
\begin{aligned}
x + 3y &= 9 \\
-y + 4z &= -2 \\
x - 5z &= 0
\end{aligned}
&
\begin{aligned}
x + 3y \phantom{{}+4z} &= 9 \\
-y + 4z &= -2 \\
x \phantom{{}+3y} - 5z &= 0
\end{aligned}
&
\left[\begin{array}{ccc:c}
1 & 3 & 0 & 9 \\
0 & -1 & 4 & -2 \\
1 & 0 & -5 & 0
\end{array}\right]
\end{array}
$$

## Elementary Row Operations

In Section 6.3, you studied three operations that can be used on a system of linear equations to produce an equivalent system.

**1.** Interchange two equations.

**2.** Multiply an equation by a nonzero constant.

**3.** Add a multiple of an equation to another equation.

In matrix terminology, these three operations correspond to **elementary row operations.** An elementary row operation on an augmented matrix of a given system of linear equations produces a new augmented matrix corresponding to a new (but equivalent) system of linear equations. Two matrices are **row-equivalent** if one can be obtained from the other by a sequence of elementary row operations.

---

▶ Elementary Row Operations

**1.** Interchange two rows.

**2.** Multiply a row by a nonzero constant.

**3.** Add a multiple of a row to another row.

---

## Study Tip

Although elementary row operations are simple to perform, they involve a lot of arithmetic. Because it is easy to make a mistake, we suggest that you get in the habit of noting, next to the row you are changing, the elementary row operation performed in each step so that you can go back and check your work.

The next example demonstrates each of the elementary row operations that can be performed on a matrix to produce an equivalent matrix.

**Example 2** Elementary Row Operations

**a.** Interchange the first and second rows.

*Original Matrix*

$$\begin{bmatrix} 0 & 1 & 3 & 4 \\ -1 & 2 & 0 & 3 \\ 2 & -3 & 4 & 1 \end{bmatrix}$$

*New Row-Equivalent Matrix*

$$\begin{matrix} R_2 \\ R_1 \end{matrix} \begin{bmatrix} -1 & 2 & 0 & 3 \\ 0 & 1 & 3 & 4 \\ 2 & -3 & 4 & 1 \end{bmatrix}$$

**b.** Multiply the first row by $\frac{1}{2}$.

*Original Matrix*

$$\begin{bmatrix} 2 & -4 & 6 & -2 \\ 1 & 3 & -3 & 0 \\ 5 & -2 & 1 & 2 \end{bmatrix}$$

*New Row-Equivalent Matrix*

$$\frac{1}{2}R_1 \rightarrow \begin{bmatrix} 1 & -2 & 3 & -1 \\ 1 & 3 & -3 & 0 \\ 5 & -2 & 1 & 2 \end{bmatrix}$$

**c.** Add $-2$ times the first row to the third row.

*Original Matrix*

$$\begin{bmatrix} 1 & 2 & -4 & 3 \\ 0 & 3 & -2 & -1 \\ 2 & 1 & 5 & -2 \end{bmatrix}$$

*New Row-Equivalent Matrix*

$$\begin{matrix} \\ \\ -2R_1 + R_3 \rightarrow \end{matrix} \begin{bmatrix} 1 & 2 & -4 & 3 \\ 0 & 3 & -2 & -1 \\ 0 & -3 & 13 & -8 \end{bmatrix}$$

Note that we write the elementary row operation beside the row that we are *changing*.

---

## •TECHNOLOGY•

Most graphing utilities can perform elementary row operations on matrices. Consult your user's guide to determine how to enter a matrix and perform elementary row operations. The screen on the left shows how one graphing utility displays the original matrix in Example 2(a). The screen on the right shows the new row-equivalent matrix in Example 2(a).

```
[A]
[ 0   1   3   4]
[ -1  2   0   3]
[ 2  -3   4   1]
```

```
[A]
[ -1  2   0   3]
[ 0   1   3   4]
[ 2  -3   4   1]
```

In Example 3 of Section 6.3, you used Gaussian elimination with back-substitution to solve a system of linear equations. The next example demonstrates the matrix version of Gaussian elimination. The two methods are essentially the same. The basic difference is that with matrices you do not need to keep writing the variables.

**Example 3**    Using Elementary Row Operations

*Linear System*                                    *Associated Augmented Matrix*

$$\begin{aligned} x - 2y + 3z &= 9 \\ -x + 3y\phantom{{} + 3z} &= -4 \\ 2x - 5y + 5z &= 17 \end{aligned}$$
$$\left[ \begin{array}{ccc:c} 1 & -2 & 3 & 9 \\ -1 & 3 & 0 & -4 \\ 2 & -5 & 5 & 17 \end{array} \right]$$

Add the first equation to the second equation.

Add the first row to the second row $(R_1 + R_2)$.

$$\begin{aligned} x - 2y + 3z &= 9 \\ y + 3z &= 5 \\ 2x - 5y + 5z &= 17 \end{aligned}$$
$$R_1 + R_2 \longrightarrow \left[ \begin{array}{ccc:c} 1 & -2 & 3 & 9 \\ 0 & 1 & 3 & 5 \\ 2 & -5 & 5 & 17 \end{array} \right]$$

Add $-2$ times the first equation to the third equation.

Add $-2$ times the first row to the third row $(-2R_1 + R_3)$.

$$\begin{aligned} x - 2y + 3z &= 9 \\ y + 3z &= 5 \\ -y - z &= -1 \end{aligned}$$
$$-2R_1 + R_3 \longrightarrow \left[ \begin{array}{ccc:c} 1 & -2 & 3 & 9 \\ 0 & 1 & 3 & 5 \\ 0 & -1 & -1 & -1 \end{array} \right]$$

Add the second equation to the third equation.

Add the second row to the third row $(R_2 + R_3)$.

$$\begin{aligned} x - 2y + 3z &= 9 \\ y + 3z &= 5 \\ 2z &= 4 \end{aligned}$$
$$R_2 + R_3 \longrightarrow \left[ \begin{array}{ccc:c} 1 & -2 & 3 & 9 \\ 0 & 1 & 3 & 5 \\ 0 & 0 & 2 & 4 \end{array} \right]$$

Multiply the third equation by $\frac{1}{2}$.

Multiply the third row by $\frac{1}{2}$.

$$\begin{aligned} x - 2y + 3z &= 9 \\ y + 3z &= 5 \\ z &= 2 \end{aligned}$$
$$\tfrac{1}{2}R_3 \longrightarrow \left[ \begin{array}{ccc:c} 1 & -2 & 3 & 9 \\ 0 & 1 & 3 & 5 \\ 0 & 0 & 1 & 2 \end{array} \right]$$

At this point, you can use back-substitution to find that the solution is $x = 1$, $y = -1$, and $z = 2$, as was done in Example 3 of Section 6.3.

Remember that you can check a solution by substituting the values of $x$, $y$, and $z$ into each equation in the original system.

The last matrix in Example 3 is said to be in **row-echelon form.** The term *echelon* refers to the stair-step pattern formed by the nonzero elements of the matrix. To be in this form, a matrix must have the following properties.

---

▶ **Row-Echelon Form and Reduced Row-Echelon Form**

A matrix in **row-echelon form** has the following properties.

1. All rows consisting entirely of zeros occur at the bottom of the matrix.

2. For each row that does not consist entirely of zeros, the first nonzero entry is 1 (called a **leading 1**).

3. For two successive (nonzero) rows, the leading 1 in the higher row is farther to the left than the leading 1 in the lower row.

A matrix in *row-echelon form* is in **reduced row-echelon form** if every column that has a leading 1 has zeros in every position above and below its leading 1.

---

**Example 4**   Row-Echelon Form   *Learning Tools*

The following matrices are in row-echelon form.

**a.** $\begin{bmatrix} 1 & 2 & -1 & 4 \\ 0 & 1 & 0 & 3 \\ 0 & 0 & 1 & -2 \end{bmatrix}$   **b.** $\begin{bmatrix} 0 & 1 & 0 & 5 \\ 0 & 0 & 1 & 3 \\ 0 & 0 & 0 & 0 \end{bmatrix}$

**c.** $\begin{bmatrix} 1 & -5 & 2 & -1 & 3 \\ 0 & 0 & 1 & 3 & -2 \\ 0 & 0 & 0 & 1 & 4 \\ 0 & 0 & 0 & 0 & 1 \end{bmatrix}$   **d.** $\begin{bmatrix} 1 & 0 & 0 & -1 \\ 0 & 1 & 0 & 2 \\ 0 & 0 & 1 & 3 \\ 0 & 0 & 0 & 0 \end{bmatrix}$

The matrices in (b) and (d) also happen to be in *reduced* row-echelon form. The following matrices are not in row-echelon form.

**e.** $\begin{bmatrix} 1 & 2 & -3 & 4 \\ 0 & 2 & 1 & -1 \\ 0 & 0 & 1 & -3 \end{bmatrix}$   **f.** $\begin{bmatrix} 1 & 2 & -1 & 2 \\ 0 & 0 & 0 & 0 \\ 0 & 1 & 2 & -4 \end{bmatrix}$

---

Every matrix is row-equivalent to a matrix in row-echelon form. For instance, in Example 4, you can change the matrix in part (e) to row-echelon form by multiplying its second row by $\frac{1}{2}$. What elementary row operation could you perform on the matrix in part (f) so that it would be in row-echelon form?

## Gaussian Elimination with Back-Substitution

**Example 5**   Gaussian Elimination with Back-Substitution

Solve the following system.

$$
\begin{aligned}
y + z - 2w &= -3 \\
x + 2y - z \phantom{-3w} &= 2 \\
2x + 4y + z - 3w &= -2 \\
x - 4y - 7z - w &= -19
\end{aligned}
$$

### Study Tip

Gaussian elimination with back-substitution works well for solving systems of linear equations by hand or with a computer. For this algorithm, the order in which the elementary row operations are performed is important. We suggest operating from *left to right by columns*, using elementary row operations to obtain zeros in all entries directly below the leading 1's.

**Solution**

$$
\begin{matrix} R_2 \\ R_1 \end{matrix}
\left[\begin{array}{cccc:c}
1 & 2 & -1 & 0 & 2 \\
0 & 1 & 1 & -2 & -3 \\
2 & 4 & 1 & -3 & -2 \\
1 & -4 & -7 & -1 & -19
\end{array}\right]
$$

First column has leading 1 in upper left corner.

$$
\begin{matrix} \\ \\ -2R_1 + R_3 \rightarrow \\ -R_1 + R_4 \rightarrow \end{matrix}
\left[\begin{array}{cccc:c}
1 & 2 & -1 & 0 & 2 \\
0 & 1 & 1 & -2 & -3 \\
0 & 0 & 3 & -3 & -6 \\
0 & -6 & -6 & -1 & -21
\end{array}\right]
$$

First column has zeros below its leading 1.

$$
\begin{matrix} \\ \\ \\ 6R_2 + R_4 \rightarrow \end{matrix}
\left[\begin{array}{cccc:c}
1 & 2 & -1 & 0 & 2 \\
0 & 1 & 1 & -2 & -3 \\
0 & 0 & 3 & -3 & -6 \\
0 & 0 & 0 & -13 & -39
\end{array}\right]
$$

Second column has zeros below its leading 1.

$$
\begin{matrix} \\ \\ \frac{1}{3}R_3 \rightarrow \\ \\ \end{matrix}
\left[\begin{array}{cccc:c}
1 & 2 & -1 & 0 & 2 \\
0 & 1 & 1 & -2 & -3 \\
0 & 0 & 1 & -1 & -2 \\
0 & 0 & 0 & -13 & -39
\end{array}\right]
$$

Third column has zeros below its leading 1.

$$
\begin{matrix} \\ \\ \\ -\frac{1}{13}R_4 \rightarrow \end{matrix}
\left[\begin{array}{cccc:c}
1 & 2 & -1 & 0 & 2 \\
0 & 1 & 1 & -2 & -3 \\
0 & 0 & 1 & -1 & -2 \\
0 & 0 & 0 & 1 & 3
\end{array}\right]
$$

Fourth column has a leading 1.

The matrix is now in row-echelon form, and the corresponding system is

$$
\begin{aligned}
x + 2y - z \phantom{- 2w} &= 2 \\
y + z - 2w &= -3 \\
z - w &= -2 \\
w &= 3.
\end{aligned}
$$

Using back-substitution, you can determine that the solution is $x = -1$, $y = 2$, $z = 1$, and $w = 3$. Check this in the original system of equations.

> ▶ **Gaussian Elimination with Back-Substitution**
>
> 1. Write the augmented matrix of the system of linear equations.
>
> 2. Use elementary row operations to rewrite the augmented matrix in row-echelon form.
>
> 3. Write the system of linear equations corresponding to the matrix in row-echelon form, and use back-substitution to find the solution.

When solving a system of linear equations, remember that it is possible for the system to have no solution. If, in the elimination process, you obtain a row with zeros except for the last entry, it is unnecessary to continue the elimination process. You can simply conclude that the system is inconsistent.

**Example 6** A System with No Solution

Solve the following system.

$$x - y + 2z = 4$$
$$x \qquad + z = 6$$
$$2x - 3y + 5z = 4$$
$$3x + 2y - z = 1$$

**Solution**

$$\begin{bmatrix} 1 & -1 & 2 & \vdots & 4 \\ 1 & 0 & 1 & \vdots & 6 \\ 2 & -3 & 5 & \vdots & 4 \\ 3 & 2 & -1 & \vdots & 1 \end{bmatrix} \quad \begin{matrix} \\ -R_1 + R_2 \rightarrow \\ -2R_1 + R_3 \rightarrow \\ -3R_1 + R_4 \rightarrow \end{matrix} \begin{bmatrix} 1 & -1 & 2 & \vdots & 4 \\ 0 & 1 & -1 & \vdots & 2 \\ 0 & -1 & 1 & \vdots & -4 \\ 0 & 5 & -7 & \vdots & -11 \end{bmatrix}$$

$$R_2 + R_3 \rightarrow \begin{bmatrix} 1 & -1 & 2 & \vdots & 4 \\ 0 & 1 & -1 & \vdots & 2 \\ 0 & 0 & 0 & \vdots & -2 \\ 0 & 5 & -7 & \vdots & -11 \end{bmatrix}$$

Note that the third row of this matrix consists of zeros except for the last entry. This means that the original system of linear equations is *inconsistent.* You can see why this is true by converting back to a system of linear equations.

$$x - y + 2z = 4$$
$$y - z = 2$$
$$0 = -2$$
$$5y - 7z = -11$$

Because the third equation is not possible, the system has no solution.

# Gauss-Jordan Elimination

With Gaussian elimination, we apply elementary row operations to a matrix to obtain a (row-equivalent) row-echelon form. A second method of elimination called **Gauss-Jordan elimination,** after Carl Friedrich Gauss and Wilhelm Jordan (1842–1899), continues the reduction process until a *reduced* row-echelon form is obtained. We demonstrate this procedure in the following example.

**Example 7**    Gauss-Jordan Elimination

Use Gauss-Jordan elimination to solve the following system.

$$x - 2y + 3z = 9$$
$$-x + 3y \phantom{+3z} = -4$$
$$2x - 5y + 5z = 17$$

**Solution**

In Example 3 we used Gaussian elimination to obtain the row-echelon form

$$\begin{bmatrix} 1 & -2 & 3 & \vdots & 9 \\ 0 & 1 & 3 & \vdots & 5 \\ 0 & 0 & 1 & \vdots & 2 \end{bmatrix}.$$

Now, rather than using back-substitution, apply elementary row operations until you obtain a matrix in reduced row-echelon form. To do this, you must produce zeros above each of the leading 1's, as follows.

$$2R_2 + R_1 \longrightarrow \begin{bmatrix} 1 & 0 & 9 & \vdots & 19 \\ 0 & 1 & 3 & \vdots & 5 \\ 0 & 0 & 1 & \vdots & 2 \end{bmatrix}$$

Second column has zeros above its leading 1.

$$\begin{matrix} -9R_3 + R_1 \longrightarrow \\ -3R_3 + R_2 \longrightarrow \end{matrix} \begin{bmatrix} 1 & 0 & 0 & \vdots & 1 \\ 0 & 1 & 0 & \vdots & -1 \\ 0 & 0 & 1 & \vdots & 2 \end{bmatrix}$$

Third column has zeros above its leading 1.

Now, converting back to a system of linear equations, you have

$$x = 1$$
$$y = -1$$
$$z = 2.$$

The beauty of Gauss-Jordan elimination is that, from the reduced row-echelon form, you can simply read the solution.

The elimination procedures described in this section employ an algorithmic approach that is easily adapted to computer use. However, the procedure makes no effort to avoid fractional coefficients. For instance, if the system given in Example 7 had been listed as

$$2x - 5y + 5z = 17$$
$$x - 2y + 3z = 9$$
$$-x + 3y = -4$$

the procedure would have required multiplication of the first row by $\frac{1}{2}$, which would have introduced fractions in the first row. For hand computations, fractions can sometimes be avoided by judiciously choosing the order in which the elementary row operations are applied.

## Study Tip

Remember that the solution set for a system with an infinite number of solutions can be written in several ways. For example, the solution set in Example 8 could have been written as

$$\left(\frac{1 - 5b}{3}, b, \frac{b + 1}{3}\right)$$

where $b$ is a real number.

**Example 8**  A System with an Infinite Number of Solutions

Solve the following system.

$$2x + 4y - 2z = 0$$
$$3x + 5y = 1$$

**Solution**

$$\begin{bmatrix} 2 & 4 & -2 & \vdots & 0 \\ 3 & 5 & 0 & \vdots & 1 \end{bmatrix} \qquad \frac{1}{2}R_1 \rightarrow \begin{bmatrix} 1 & 2 & -1 & \vdots & 0 \\ 3 & 5 & 0 & \vdots & 1 \end{bmatrix}$$

$$-3R_1 + R_2 \rightarrow \begin{bmatrix} 1 & 2 & -1 & \vdots & 0 \\ 0 & -1 & 3 & \vdots & 1 \end{bmatrix}$$

$$-R_2 \rightarrow \begin{bmatrix} 1 & 2 & -1 & \vdots & 0 \\ 0 & 1 & -3 & \vdots & -1 \end{bmatrix}$$

$$-2R_2 + R_1 \rightarrow \begin{bmatrix} 1 & 0 & 5 & \vdots & 2 \\ 0 & 1 & -3 & \vdots & -1 \end{bmatrix}$$

The corresponding system of equations is

$$x + 5z = 2$$
$$y - 3z = -1.$$

Solving for $x$ and $y$ in terms of $z$, you have $x = -5z + 2$ and $y = 3z - 1$. Then, letting $z = a$, the solution set has the form

$$(-5a + 2, 3a - 1, a)$$

where $a$ is a real number. Try substituting values for $a$ to obtain a few solutions. Then check each solution in the original system of equations.

It is worth noting that the row-echelon form of a matrix is not unique. That is, two different sequences of elementary row operations may yield different row-echelon forms. For instance, the following sequence of elementary row operations on the matrix in Example 3 produces a slightly different row-echelon form.

$$\begin{bmatrix} 1 & -2 & 3 & \vdots & 9 \\ -1 & 3 & 0 & \vdots & -4 \\ 2 & -5 & 5 & \vdots & 17 \end{bmatrix} \quad \begin{matrix} R_2 \\ R_1 \end{matrix} \quad \begin{bmatrix} -1 & 3 & 0 & \vdots & -4 \\ 1 & -2 & 3 & \vdots & 9 \\ 2 & -5 & 5 & \vdots & 17 \end{bmatrix}$$

$$-R_1 \rightarrow \begin{bmatrix} 1 & -3 & 0 & \vdots & 4 \\ 1 & -2 & 3 & \vdots & 9 \\ 2 & -5 & 5 & \vdots & 17 \end{bmatrix}$$

$$\begin{matrix} -R_1 + R_2 \rightarrow \\ -2R_1 + R_3 \rightarrow \end{matrix} \begin{bmatrix} 1 & -3 & 0 & \vdots & 4 \\ 0 & 1 & 3 & \vdots & 5 \\ 0 & 1 & 5 & \vdots & 9 \end{bmatrix}$$

$$-R_2 + R_3 \rightarrow \begin{bmatrix} 1 & -3 & 0 & \vdots & 4 \\ 0 & 1 & 3 & \vdots & 5 \\ 0 & 0 & 2 & \vdots & 4 \end{bmatrix}$$

$$\tfrac{1}{2}R_3 \rightarrow \begin{bmatrix} 1 & -3 & 0 & \vdots & 4 \\ 0 & 1 & 3 & \vdots & 5 \\ 0 & 0 & 1 & \vdots & 2 \end{bmatrix}$$

The corresponding system of linear equations is

$$x - 3y \qquad = 4$$
$$y + 3z = 5$$
$$z = 2 .$$

Try using back-substitution on this system to see that you obtain the same solution that was obtained in Example 3.

| Discussing the Concept | **Unique Solutions Versus Infinite Solutions** |
|---|---|

Construct one linear system of three variables that has a unique solution and another linear system of three variables that has an infinite number of solutions. Exchange systems with another student. Solve the systems you receive using the methods of this section. Compare and discuss your solutions with each other. Did you both write the infinite solution in the same form? If not, verify that your answers represent the same solution set.

# Warm Up

The following warm-up exercises involve skills that were covered in earlier sections. You will use these skills in the exercise set for this section.

In Exercises 1–4, evaluate the expression.

**1.** $2(-1) - 3(5) + 7(2)$

**2.** $-4(-3) + 6(7) + 8(-3)$

**3.** $11\left(\frac{1}{2}\right) - 7\left(-\frac{3}{2}\right) - 5(2)$

**4.** $\frac{2}{3}\left(\frac{1}{2}\right) + \frac{4}{3}\left(-\frac{1}{3}\right)$

In Exercises 5 and 6, decide whether $x = 1$, $y = 3$, and $z = -1$ is a solution of the system.

**5.** $\begin{aligned} 4x - 2y + 3z &= -5 \\ x + 3y - z &= 11 \\ -x + 2y &= 5 \end{aligned}$

**6.** $\begin{aligned} -x + 2y + z &= 4 \\ 2x - 3z &= 5 \\ 3x + 5y - 2z &= 21 \end{aligned}$

In Exercises 7–10, use back-substitution to solve the system of linear equations.

**7.** $\begin{aligned} 2x - 3y &= 4 \\ y &= 2 \end{aligned}$

**8.** $\begin{aligned} 5x + 4y &= 0 \\ y &= -3 \end{aligned}$

**9.** $\begin{aligned} x - 3y + z &= 0 \\ y - 3z &= 8 \\ z &= 2 \end{aligned}$

**10.** $\begin{aligned} 2x - 5y + 3z &= -2 \\ y - 4z &= 0 \\ z &= 1 \end{aligned}$

## 7.1    Exercises

In Exercises 1–6, determine the order of the matrix.

**1.** $\begin{bmatrix} 5 & -1 & 6 \\ 3 & 1 & -2 \end{bmatrix}$

**2.** $\begin{bmatrix} 4 & -1 \end{bmatrix}$

**3.** $\begin{bmatrix} 6 & 4 & 1 \\ 8 & 3 & 0 \\ -1 & 2 & 1 \\ 1 & 5 & 4 \end{bmatrix}$

**4.** $\begin{bmatrix} 1 \\ 0 \\ 3 \\ 5 \\ 6 \end{bmatrix}$

**5.** $\begin{bmatrix} 33 & 45 \\ -9 & 20 \\ 12 & 15 \\ 16 & -2 \end{bmatrix}$

**6.** $\begin{bmatrix} 12 & -2 & 4 \\ -3 & 4 & 0 \\ -8 & 12 & 2 \end{bmatrix}$

In Exercises 7–10, determine whether the matrix is in row-echelon form. If it is, determine if it is also in reduced row-echelon form.

**7.** $\begin{bmatrix} 1 & 0 & 0 & 0 \\ 0 & 1 & 1 & 5 \\ 0 & 0 & 0 & 0 \end{bmatrix}$

**8.** $\begin{bmatrix} 1 & 0 & 2 & 1 \\ 0 & 1 & -3 & 10 \\ 0 & 0 & 1 & 0 \end{bmatrix}$

**9.** $\begin{bmatrix} 2 & 0 & 4 & 0 \\ 0 & -1 & 3 & 6 \\ 0 & 0 & 1 & 5 \end{bmatrix}$

**10.** $\begin{bmatrix} 1 & 3 & 0 & 0 & 0 & 0 \\ 0 & 0 & 1 & 8 & 1 & 0 \\ 0 & 0 & 0 & 0 & 1 & 1 \\ 0 & 0 & 0 & 0 & 1 & 1 \end{bmatrix}$

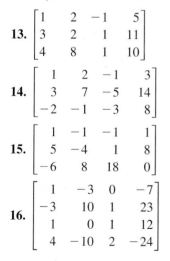 **11.** Use a graphing utility to perform the row operations to reduce the matrix to row-echelon form.

$$\begin{bmatrix} 1 & 1 & 2 \\ 3 & 4 & -3 \\ 2 & -1 & 1 \end{bmatrix}$$

(a) Add $-3$ times Row 1 to Row 2.

(b) Add $-2$ times Row 1 to Row 3.

(c) Add 3 times Row 2 to Row 3.

(d) Multiply Row 3 by $-\frac{1}{30}$.

**12.** Use a graphing utility to perform the operations to reduce the matrix to *reduced* row-echelon form.

$$\begin{bmatrix} 7 & 1 \\ 0 & 2 \\ -3 & 4 \\ 4 & 1 \end{bmatrix}$$

(a) Add Row 3 to Row 4.

(b) Interchange Rows 1 and 4.

(c) Add 3 times Row 1 to Row 3.

(d) Add $-7$ times Row 1 to Row 4.

(e) Multiply Row 2 by $\frac{1}{2}$.

(f) Add the appropriate multiple of Row 2 to Rows 1, 3, and 4.

In Exercises 13–16, write the matrix in row-echelon form. (*Note:* Row-echelon forms are not unique.)

**13.** $\begin{bmatrix} 1 & 2 & -1 & 5 \\ 3 & 2 & 1 & 11 \\ 4 & 8 & 1 & 10 \end{bmatrix}$

**14.** $\begin{bmatrix} 1 & 2 & -1 & 3 \\ 3 & 7 & -5 & 14 \\ -2 & -1 & -3 & 8 \end{bmatrix}$

**15.** $\begin{bmatrix} 1 & -1 & -1 & 1 \\ 5 & -4 & 1 & 8 \\ -6 & 8 & 18 & 0 \end{bmatrix}$

**16.** $\begin{bmatrix} 1 & -3 & 0 & -7 \\ -3 & 10 & 1 & 23 \\ 1 & 0 & 1 & 12 \\ 4 & -10 & 2 & -24 \end{bmatrix}$

In Exercises 17–20, write the matrix in *reduced* row-echelon form.

**17.** $\begin{bmatrix} 4 & 4 & 8 \\ 1 & 2 & 2 \\ -3 & 6 & -9 \end{bmatrix}$

**18.** $\begin{bmatrix} 1 & 3 & 2 \\ 5 & 15 & 9 \\ 2 & 6 & 10 \end{bmatrix}$

**19.** $\begin{bmatrix} 1 & 2 & 3 & -5 \\ 1 & 2 & 4 & -9 \\ -2 & -4 & -4 & 3 \\ 4 & 8 & 11 & -14 \end{bmatrix}$

**20.** $\begin{bmatrix} 1 & -3 \\ -1 & 8 \\ 0 & 4 \\ -2 & 10 \end{bmatrix}$

In Exercises 21–24, write the system of linear equations represented by the augmented matrix. (Use variables $w$, $x$, $y$, and $z$.)

**21.** $\begin{bmatrix} 2 & 4 & \vdots & 6 \\ -1 & 3 & \vdots & -8 \end{bmatrix}$

**22.** $\begin{bmatrix} 7 & -2 & \vdots & 7 \\ -8 & 3 & \vdots & -3 \end{bmatrix}$

**23.** $\begin{bmatrix} 1 & 0 & 2 & \vdots & -10 \\ 0 & 3 & -1 & \vdots & 5 \\ 4 & 2 & 0 & \vdots & 3 \end{bmatrix}$

**24.** $\begin{bmatrix} 5 & 8 & 2 & 0 & \vdots & -1 \\ -2 & 15 & 5 & 1 & \vdots & 9 \\ 1 & 6 & -7 & 0 & \vdots & -3 \end{bmatrix}$

In Exercises 25–28, write the system of equations represented by the augmented matrix. Use back-substitution to find the solution. (Use $w$, $x$, $y$, and $z$.)

**25.** $\begin{bmatrix} 1 & -5 & \vdots & 6 \\ 0 & 1 & \vdots & -2 \end{bmatrix}$

**26.** $\begin{bmatrix} 1 & 2 & -1 & \vdots & 3 \\ 0 & 1 & -2 & \vdots & -3 \\ 0 & 0 & 1 & \vdots & 4 \end{bmatrix}$

**27.** $\begin{bmatrix} 1 & 3 & -1 & \vdots & 15 \\ 0 & 1 & 4 & \vdots & -12 \\ 0 & 0 & 1 & \vdots & -5 \end{bmatrix}$

**28.** $\begin{bmatrix} 1 & 2 & -2 & 0 & \vdots & -1 \\ 0 & 1 & 1 & 2 & \vdots & 9 \\ 0 & 0 & 1 & 0 & \vdots & 2 \\ 0 & 0 & 0 & 1 & \vdots & -3 \end{bmatrix}$

The symbol ▦ indicates an exercise or parts of an exercise in which you are instructed to use a graphing utility.

In Exercises 29–32, an augmented matrix that represents a system of linear equations (in variables $x, y$, and $z$) has been reduced using Gauss-Jordan elimination. Write the solution represented by the augmented matrix.

**29.** $\begin{bmatrix} 1 & 0 & \vdots & -4 \\ 0 & 1 & \vdots & 6 \end{bmatrix}$
**30.** $\begin{bmatrix} 1 & 0 & \vdots & 9 \\ 0 & 1 & \vdots & -3 \end{bmatrix}$

**31.** $\begin{bmatrix} 1 & 0 & 0 & \vdots & -4 \\ 0 & 1 & 0 & \vdots & -8 \\ 0 & 0 & 1 & \vdots & 2 \end{bmatrix}$

**32.** $\begin{bmatrix} 1 & 0 & 0 & \vdots & 3 \\ 0 & 1 & 0 & \vdots & -1 \\ 0 & 0 & 1 & \vdots & 0 \end{bmatrix}$

In Exercises 33–36, write the augmented matrix for the system of linear equations.

**33.** $2x - y = 3$
$5x + 7y = 12$

**34.** $8x + 3y = 25$
$3x - 9y = 12$

**35.** $x + 10y - 3z = 2$
$5x - 3y + 4z = 0$
$2x + 4y = 6$

**36.** $9w - 3x + 20y + z = 13$
$12w - 8y = 5$
$w + 2x + 3y - 4z = -2$
$-w - x + y + z = 1$

In Exercises 37–58, solve the system of equations. Use Gaussian elimination with back-substitution or Gauss-Jordan elimination.

**37.** $x + 2y = 7$
$2x + y = 8$

**38.** $2x + 6y = 16$
$2x + 3y = 7$

**39.** $-3x + 5y = -22$
$3x + 4y = 4$
$4x - 8y = 32$

**40.** $x + 2y = 0$
$x + y = 6$
$3x - 2y = 8$

**41.** $8x - 4y = 7$
$5x + 2y = 1$

**42.** $2x - y = -0.1$
$3x + 2y = 1.6$

**43.** $-x + 2y = 1.5$
$2x - 4y = 3$

**44.** $x - 3y = 5$
$-2x + 6y = -10$

**45.** $2x + 3z = 3$
$4x - 3y + 7z = 5$
$8x - 9y + 15z = 9$

**46.** $2x - y + 3z = 24$
$2y - z = 14$
$7x - 5y = 6$

**47.** $x + y - 5z = 3$
$x - 2z = 1$
$2x - y - z = 0$

**48.** $x - 3z = -2$
$3x + y - 2z = 5$
$2x + 2y + z = 4$

**49.** $x + 2y + z = 8$
$3x + 7y + 6z = 26$

**50.** $4x + 12y - 7z - 20w = 22$
$3x + 9y - 5z - 28w = 30$

**51.** $3x + 3y + 12z = 6$
$x + y + 4z = 2$
$2x + 5y + 20z = 10$
$-x + 2y + 8z = 4$

**52.** $2x + 10y + 2z = 6$
$x + 5y + 2z = 6$
$x + 5y + z = 3$
$-3x - 15y - 3z = -9$

**53.** $2x + y - z + 2w = -16$
$3x + 4y + w = 1$
$x + 5y + 2z + 6w = -3$
$5x + 2y - z - w = 3$

**54.** $x + 2y + 2z + 4w = 11$
$3x + 6y + 5z + 12w = 30$

**55.** $x + 2y = 0$
$-x - y = 0$

**56.** $x + 2y = 0$
$2x + 4y = 0$

**57.** $x + y + z = 0$
$2x + 3y + z = 0$
$3x + 5y + z = 0$

**58.** $x + 2y + z + 3w = 0$
$x - y + w = 0$
$y - z + 2w = 0$

**59.** *Borrowing Money* A small corporation borrowed $500,000 to expand its product line. Some of the money was borrowed at 9%, some at 10%, and some at 12%. How much was borrowed at each rate if the annual interest was $52,000 and the amount borrowed at 10% was two and one-half times the amount borrowed at 9%?

**60.** *Borrowing Money*    A company borrowed $800,000 to expand its product line. Some of the money was borrowed at 10%, some at 10.5%, and some at 11.5%. How much was borrowed at each rate if the annual interest was $83,500 and the amount borrowed at 10% was four times the amount borrowed at 11.5%?

**61.** You and a friend solve the following system of equations independently.

$$2x - 4y - 3z = 3$$
$$x + 3y + z = -1$$
$$5x + y - 2z = 2$$

You write your solution set as $(a, -a, 2a - 1)$, where $a$ is any real number. Your friend's solution set is $\left(\frac{1}{2}b + \frac{1}{2}, -\frac{1}{2}b - \frac{1}{2}, b\right)$, where $b$ is any real number. Are you both correct? Explain. If you let $a = 3$, what value of $b$ should be selected so that you have the same ordered triple?

**62.** Describe how you would explain to another student that the augmented matrix below represents a dependent system of equations. Describe a way to write the infinitely many solutions to this system.

$$\begin{bmatrix} 1 & -2 & 3 & \vdots & -6 \\ 0 & 1 & 2 & \vdots & 5 \\ 0 & 0 & 0 & \vdots & 0 \end{bmatrix}$$

**63.** *Average Salary*    From 1990 to 1999, the average annual salary for professors at public colleges and universities in the United States increased in a pattern that was approximately linear (see figure). Find the least squares regression line $y = at + b$ for the data shown in the figure by solving the following system using a graphing utility. ($y$ represents the average salary in thousands of dollars and $t = 0$ represents 1990.)

$$10b + 45a = 615.6$$
$$45b + 285a = 2924.7$$

Use the result to predict the average salary for a professor at a public college in 2005. (Source: American Association of University Professors)

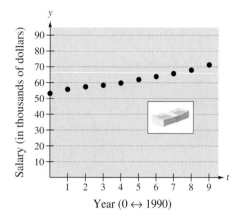

Figure for 63

**64.** *Average Salary*    From 1990 to 1999, the average annual salary for professors at private (non-church-related) colleges and universities in the United States increased in a pattern that was approximately linear (see figure). Find the least squares regression line $y = at + b$ for the data shown in the figure by solving the following system using a graphing utility. ($y$ represents the average salary in thousands of dollars and $t = 0$ represents 1990.)

$$10b + 45a = 719.9$$
$$45b + 285a = 3461.2$$

Use the result to predict the average salary for a professor at a private college in 2005. (Source: American Association of University Professors)

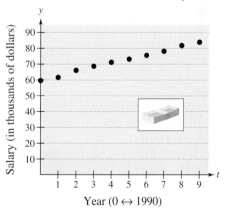

## 7.2    Operations with Matrices

### Objectives

- Determine if two matrices are equal.
- Add or subtract two matrices.
- Multiply a matrix by a scalar.
- Find the product of two matrices.
- Solve a matrix equation.

**Arthur Cayley**

**1821–1895**

A British mathematician, Arthur Cayley, invented matrices around 1858. Cayley was a Cambridge University graduate and a lawyer by profession. His groundbreaking work on matrices was begun as he studied the theory of transformations. Cayley also was instrumental in the development of determinants. Cayley and two American mathematicians, Benjamin Peirce (1809–1880) and his son Charles S. Peirce (1839–1914), are credited with developing "matrix algebra."

## Equality of Matrices

In Section 7.1, you used matrices to solve systems of linear equations. Matrices, however, can do much more than this. There is a rich mathematical theory of matrices, and its applications are numerous. This section and the next introduce some fundamentals of matrix theory. It is standard mathematical convention to represent matrices in any of the following three ways.

**1.** A matrix can be denoted by an uppercase letter such as $A$, $B$, or $C$.

**2.** A matrix can be denoted by a representative element enclosed in brackets, such as $[a_{ij}]$, $[b_{ij}]$, or $[c_{ij}]$.

**3.** A matrix can be denoted by a rectangular array of numbers such as

$$A = [a_{ij}] = \begin{bmatrix} a_{11} & a_{12} & a_{13} & \cdots & a_{1n} \\ a_{21} & a_{22} & a_{23} & \cdots & a_{2n} \\ a_{31} & a_{32} & a_{33} & \cdots & a_{3n} \\ \vdots & \vdots & \vdots & & \vdots \\ a_{m1} & a_{m2} & a_{m3} & \cdots & a_{mn} \end{bmatrix}.$$

Two matrices $A = [a_{ij}]$ and $B = [b_{ij}]$ are **equal** if they have the same order $(m \times n)$ and

$$a_{ij} = b_{ij}$$

for $1 \leq i \leq m$ and $1 \leq j \leq n$. In other words, two matrices are equal if their corresponding entries are equal.

**Example 1**    Equality of Matrices

Solve for $a_{11}$, $a_{12}$, $a_{21}$, and $a_{22}$ in the following matrix equation.

$$\begin{bmatrix} a_{11} & a_{12} \\ a_{21} & a_{22} \end{bmatrix} = \begin{bmatrix} 2 & -1 \\ -3 & 0 \end{bmatrix}$$

### Solution

Because two matrices are equal only if their corresponding entries are equal, you can conclude that

$$a_{11} = 2, \quad a_{12} = -1, \quad a_{21} = -3, \quad \text{and} \quad a_{22} = 0.$$

The general pattern for matrix multiplication is as follows. To obtain the entry in the *i*th row and the *j*th column of the product $AB$, use the *i*th row of $A$ and the *j*th column of $B$.

$$\begin{bmatrix} a_{11} & a_{12} & a_{13} & \cdots & a_{1n} \\ a_{21} & a_{22} & a_{23} & \cdots & a_{2n} \\ a_{31} & a_{32} & a_{33} & \cdots & a_{3n} \\ \vdots & \vdots & \vdots & & \vdots \\ a_{i1} & a_{i2} & a_{i3} & \cdots & a_{in} \\ \vdots & \vdots & \vdots & & \vdots \\ a_{m1} & a_{m2} & a_{m3} & \cdots & a_{mn} \end{bmatrix} \begin{bmatrix} b_{12} & b_{11} & \cdots & b_{1j} & \cdots & b_{1p} \\ b_{22} & b_{21} & \cdots & b_{2j} & \cdots & b_{2p} \\ b_{32} & b_{31} & \cdots & b_{3j} & \cdots & b_{3p} \\ \vdots & \vdots & & \vdots & & \vdots \\ b_{n2} & b_{n1} & \cdots & b_{nj} & \cdots & b_{np} \end{bmatrix} = \begin{bmatrix} c_{11} & c_{12} & \cdots & c_{1j} & \cdots & c_{1p} \\ c_{21} & c_{22} & \cdots & c_{2j} & \cdots & c_{2p} \\ \vdots & \vdots & & \vdots & & \vdots \\ c_{i1} & c_{i2} & \cdots & c_{ij} & \cdots & c_{ip} \\ \vdots & \vdots & & \vdots & & \vdots \\ c_{m1} & c_{m2} & \cdots & c_{mj} & \cdots & c_{mp} \end{bmatrix}$$

$$a_{i1}b_{1j} + a_{i2}b_{2j} + a_{i3}b_{3j} + \cdots + a_{in}b_{nj} = c_{ij}$$

---

▶ **Properties of Matrix Multiplication**

Let $A$, $B$, and $C$ be matrices and let $c$ be a scalar.

**1.** $A(BC) = (AB)C$        Associative Property of Multiplication

**2.** $A(B + C) = AB + AC$        Left Distributive Property

**3.** $(A + B)C = AC + BC$        Right Distributive Property

**4.** $c(AB) = (cA)B = A(cB)$

---

The $n \times n$ matrix that consists of 1's on its main diagonal and 0's elsewhere is called the **identity matrix of order *n*** and is denoted by

$$I_n = \begin{bmatrix} 1 & 0 & 0 & \cdots & 0 \\ 0 & 1 & 0 & \cdots & 0 \\ 0 & 0 & 1 & \cdots & 0 \\ \vdots & \vdots & \vdots & & \vdots \\ 0 & 0 & 0 & \cdots & 1 \end{bmatrix}.$$      Identity matrix

Note that an identity matrix must be *square*. When the order is understood to be $n$, you can denote $I_n$ simply by $I$. If $A$ is an $n \times n$ matrix, the identity matrix has the property that $AI_n = A$ and $I_nA = A$. For example,

$$\begin{bmatrix} 3 & -2 & 5 \\ 1 & 0 & 4 \\ -1 & 2 & -3 \end{bmatrix} \begin{bmatrix} 1 & 0 & 0 \\ 0 & 1 & 0 \\ 0 & 0 & 1 \end{bmatrix} = \begin{bmatrix} 3 & -2 & 5 \\ 1 & 0 & 4 \\ -1 & 2 & -3 \end{bmatrix}$$

and

$$\begin{bmatrix} 1 & 0 & 0 \\ 0 & 1 & 0 \\ 0 & 0 & 1 \end{bmatrix} \begin{bmatrix} 3 & -2 & 5 \\ 1 & 0 & 4 \\ -1 & 2 & -3 \end{bmatrix} = \begin{bmatrix} 3 & -2 & 5 \\ 1 & 0 & 4 \\ -1 & 2 & -3 \end{bmatrix}.$$

## Applications

One application of matrix multiplication is representation of a system of linear equations. Note how the system

$$a_{11}x_1 + a_{12}x_2 + a_{13}x_3 = b_1$$

$$a_{21}x_1 + a_{22}x_2 + a_{23}x_3 = b_2$$

$$a_{31}x_1 + a_{32}x_2 + a_{33}x_3 = b_3$$

can be written as the matrix equation $AX = B$, where $A$ is the *coefficient matrix* of the system, and $X$ and $B$ are column matrices.

$$\begin{bmatrix} a_{11} & a_{12} & a_{13} \\ a_{21} & a_{22} & a_{23} \\ a_{31} & a_{32} & a_{33} \end{bmatrix} \begin{bmatrix} x_1 \\ x_2 \\ x_3 \end{bmatrix} = \begin{bmatrix} b_1 \\ b_2 \\ b_3 \end{bmatrix}$$

$$A \qquad \times X = B$$

### Example 8    Solving a System of Linear Equations

Solve the matrix equation $AX = B$ for $X$, where

$$\underset{\text{Coefficient matrix}}{A = \begin{bmatrix} 1 & -2 & 1 \\ 0 & 1 & 2 \\ 2 & 3 & -2 \end{bmatrix}} \quad \text{and} \quad \underset{\text{Constant matrix}}{B = \begin{bmatrix} -4 \\ 4 \\ 2 \end{bmatrix}}.$$

### Solution

As a system of linear equations, $AX = B$ is as follows.

$$x_1 - 2x_2 + x_3 = -4$$

$$x_2 + 2x_3 = 4$$

$$2x_1 + 3x_2 - 2x_3 = 2$$

Using Gauss-Jordan elimination on the augmented matrix of this system, you obtain the following reduced row-echelon matrix.

$$\begin{bmatrix} 1 & 0 & 0 & \vdots & -1 \\ 0 & 1 & 0 & \vdots & 2 \\ 0 & 0 & 1 & \vdots & 1 \end{bmatrix}$$

So, the solution of the system of linear equations is $x_1 = -1$, $x_2 = 2$, and $x_3 = 1$, and the solution of the matrix equation is

$$X = \begin{bmatrix} x_1 \\ x_2 \\ x_3 \end{bmatrix} = \begin{bmatrix} -1 \\ 2 \\ 1 \end{bmatrix}.$$

Check this solution in the original system of equations.

Example 9   Softball Team Expenses

Two softball teams submit equipment lists to their sponsors.

|  | Women's Team | Men's Team |
|---|---|---|
| Bats | 12 | 15 |
| Balls | 45 | 38 |
| Gloves | 15 | 17 |

Each bat costs $55, each ball costs $4, and each glove costs $80. Use matrices to find the total cost of equipment for each team.

Solution

The equipment lists and the costs per item can be written in matrix form as

$$E = \begin{bmatrix} 12 & 15 \\ 45 & 38 \\ 15 & 17 \end{bmatrix} \quad \text{and} \quad C = \begin{bmatrix} 55 & 4 & 80 \end{bmatrix}.$$

The total cost of equipment for each team is given by the product

$$CE = \begin{bmatrix} 55 & 4 & 80 \end{bmatrix} \begin{bmatrix} 12 & 15 \\ 45 & 38 \\ 15 & 17 \end{bmatrix} = \begin{bmatrix} 2040 & 2337 \end{bmatrix}.$$

So, the total cost of equipment for the women's team is $2040, and the total cost of equipment for the men's team is $2337.

---

## Discussing the Concept   Matrix Multiplication

Discuss the requirements for matrix order $m \times n$ for multiplication of two matrices. Determine which of the following matrix multiplications $AB$ is (are) defined. For each case in which $AB$ is defined, what is the order of the resulting matrix?

**a.** $A$ is of order $1 \times 3$
   $B$ is of order $2 \times 1$

**b.** $A$ is of order $2 \times 3$
   $B$ is of order $2 \times 3$

**c.** $A$ is of order $3 \times 4$
   $B$ is of order $4 \times 2$

**d.** $A$ is of order $3 \times 1$
   $B$ is of order $3 \times 3$

Discuss why matrix multiplication is not, in general, commutative. Give an example of two $2 \times 2$ matrices such that $AB \neq BA$. Find an example of two $2 \times 2$ matrices such that $AB = BA$.

# Warm Up

The following warm-up exercises involve skills that were covered in earlier sections. You will use these skills in the exercise set for this section.

In Exercises 1 and 2, evaluate the expression.

**1.** $-3\left(-\frac{5}{6}\right) + 10\left(-\frac{3}{4}\right)$

**2.** $-22\left(\frac{5}{2}\right) + 6(8)$

In Exercises 3 and 4, determine whether the matrix is in *reduced* row-echelon form.

**3.** $\begin{bmatrix} 0 & 1 & 0 & -5 \\ 1 & 0 & 3 & 2 \\ 0 & 0 & 1 & 0 \end{bmatrix}$

**4.** $\begin{bmatrix} 1 & 0 & 0 & 2 & 3 \\ 0 & 0 & 0 & 0 & 0 \\ 0 & 1 & 1 & 3 & 10 \end{bmatrix}$

In Exercises 5 and 6, write the augmented matrix for the system of linear equations.

**5.** $\begin{aligned} -5x + 10y &= 12 \\ 7x - 3y &= 0 \\ -x + 7y &= 25 \end{aligned}$

**6.** $\begin{aligned} 10x + 15y - 9z &= 42 \\ 6x - 5y &= 0 \end{aligned}$

In Exercises 7–10, solve the system of linear equations represented by the augmented matrix.

**7.** $\left[\begin{array}{cc:c} 1 & 0 & 0 \\ 0 & 1 & 2 \end{array}\right]$

**8.** $\left[\begin{array}{ccc:c} 1 & 0 & -1 & 2 \\ 0 & 1 & 1 & 3 \end{array}\right]$

**9.** $\left[\begin{array}{ccc:c} 1 & 2 & 1 & 0 \\ 0 & 0 & 1 & -1 \\ 0 & 0 & 0 & 0 \end{array}\right]$

**10.** $\left[\begin{array}{ccc:c} 1 & -1 & 0 & 3 \\ 0 & 1 & -2 & 1 \\ 0 & 0 & 1 & -1 \end{array}\right]$

## 7.2    Exercises

In Exercises 1–4, find $x$ and $y$.

**1.** $\begin{bmatrix} 4 & x \\ -1 & y \end{bmatrix} = \begin{bmatrix} 4 & -3 \\ -1 & 2 \end{bmatrix}$

**2.** $\begin{bmatrix} x & -7 \\ 9 & y \end{bmatrix} = \begin{bmatrix} 5 & -7 \\ 9 & -8 \end{bmatrix}$

**3.** $\begin{bmatrix} -4 & 3 \\ 6 & -1 \\ 8 & 2 \\ 5 & 9 \end{bmatrix} = \begin{bmatrix} x-2 & 3 \\ 6 & -1 \\ 8 & -x \\ 5 & 2y-1 \end{bmatrix}$

**4.** $\begin{bmatrix} x+2 & 8 & -3 \\ 1 & 2y & 2x \\ 7 & -2 & y+2 \end{bmatrix} = \begin{bmatrix} 2x+6 & 8 & -3 \\ 1 & 18 & -8 \\ 7 & -2 & 11 \end{bmatrix}$

In Exercises 5–10, find (a) $A + B$, (b) $A - B$, (c) $3A$, and (d) $3A - 2B$.

**5.** $A = \begin{bmatrix} 5 & -2 \\ 3 & 1 \end{bmatrix}$, $B = \begin{bmatrix} 3 & 1 \\ -2 & 6 \end{bmatrix}$

**6.** $A = \begin{bmatrix} 7 & 4 \\ -4 & 5 \end{bmatrix}$, $B = \begin{bmatrix} -3 & 1 \\ 8 & -4 \end{bmatrix}$

**7.** $A = \begin{bmatrix} 6 & -1 \\ 2 & 4 \\ -3 & 5 \end{bmatrix}$, $B = \begin{bmatrix} 1 & 4 \\ -1 & 5 \\ 1 & 10 \end{bmatrix}$

**8.** $A = \begin{bmatrix} 6 & 8 & -3 & 2 & 1 \\ -4 & 2 & 1 & 5 & -2 \end{bmatrix}$,

$B = \begin{bmatrix} 6 & 0 & 4 & -1 & 3 \\ 4 & 5 & -2 & 1 & 2 \end{bmatrix}$

**9.** $A = \begin{bmatrix} 2 & 2 & -1 & 0 \\ 1 & 1 & -2 & 0 \\ 1 & -1 & 3 & 2 \end{bmatrix}$,

$B = \begin{bmatrix} 1 & 1 & -1 & 1 \\ -3 & 4 & 9 & -6 \\ 0 & -7 & 8 & 2 \end{bmatrix}$

**10.** $A = \begin{bmatrix} 3 \\ 2 \\ -1 \end{bmatrix}$, $B = \begin{bmatrix} -4 \\ 6 \\ 2 \end{bmatrix}$

In Exercises 11–16, find (a) $AB$, (b) $BA$, and, if possible, (c) $A^2$. (*Note:* $A^2 = AA$.)

**11.** $A = \begin{bmatrix} 1 & 2 \\ 4 & 2 \end{bmatrix}$, $B = \begin{bmatrix} 2 & -1 \\ -1 & 8 \end{bmatrix}$

**12.** $A = \begin{bmatrix} 2 & -1 \\ 1 & 4 \end{bmatrix}$, $B = \begin{bmatrix} 0 & 0 \\ 3 & -3 \end{bmatrix}$

**13.** $A = \begin{bmatrix} -1 & 2 & 3 \\ 4 & 1 & -1 \end{bmatrix}$, $B = \begin{bmatrix} 1 & 3 \\ -1 & -2 \\ 2 & 4 \end{bmatrix}$

**14.** $A = \begin{bmatrix} 1 & -1 & 7 \\ 2 & -1 & 8 \\ 3 & 1 & -1 \end{bmatrix}$, $B = \begin{bmatrix} 1 & 1 & 2 \\ 2 & 1 & 1 \\ 1 & -3 & 2 \end{bmatrix}$

**15.** $A = \begin{bmatrix} -4 & 2 & 3 \end{bmatrix}$, $B = \begin{bmatrix} 1 \\ 0 \\ 5 \end{bmatrix}$

**16.** $A = \begin{bmatrix} 3 & 2 & 1 & 0 \end{bmatrix}$, $B = \begin{bmatrix} 2 \\ 3 \\ 1 \\ 0 \end{bmatrix}$

In Exercises 17–24, find $AB$, if possible.

**17.** $A = \begin{bmatrix} 3 & -2 \\ 4 & 5 \\ 1 & -1 \end{bmatrix}$, $B = \begin{bmatrix} -1 & 4 & -2 & 5 \\ 2 & 1 & 3 & -1 \end{bmatrix}$

**18.** $A = \begin{bmatrix} 0 & -1 & 0 \\ 4 & 0 & 2 \\ 8 & -1 & 7 \end{bmatrix}$, $B = \begin{bmatrix} 2 & 1 \\ -3 & 4 \\ 1 & 6 \end{bmatrix}$

**19.** $A = \begin{bmatrix} -1 & 3 \\ 4 & -5 \\ 0 & 2 \end{bmatrix}$, $B = \begin{bmatrix} 1 & 2 \\ 0 & 7 \end{bmatrix}$

**20.** $A = \begin{bmatrix} 1 & 0 & 0 \\ 0 & 4 & 0 \\ 0 & 0 & -2 \end{bmatrix}$, $B = \begin{bmatrix} 3 & 0 & 0 \\ 0 & -1 & 0 \\ 0 & 0 & 5 \end{bmatrix}$

**21.** $A = \begin{bmatrix} 5 & 0 & 0 \\ 0 & -8 & 0 \\ 0 & 0 & 7 \end{bmatrix}$, $B = \begin{bmatrix} \frac{1}{5} & 0 & 0 \\ 0 & -\frac{1}{8} & 0 \\ 0 & 0 & \frac{1}{2} \end{bmatrix}$

**22.** $A = \begin{bmatrix} 0 & 1 & 0 \\ 3 & 0 & 2 \\ 5 & 0 & 0 \end{bmatrix}$, $B = \begin{bmatrix} 4 \\ -2 \\ 0 \\ 1 \end{bmatrix}$

**23.** $A = \begin{bmatrix} 6 \\ -2 \\ 1 \\ 6 \end{bmatrix}$, $B = \begin{bmatrix} 10 & 12 \end{bmatrix}$

**24.** $A = \begin{bmatrix} 1 & 0 & 3 & -2 & 4 \\ 6 & 13 & 8 & -17 & 10 \end{bmatrix}$, $B = \begin{bmatrix} 1 & 6 \\ 4 & 2 \end{bmatrix}$

In Exercises 25–28, solve for $X$ given

$A = \begin{bmatrix} -2 & -1 \\ 1 & 0 \\ 3 & -4 \end{bmatrix}$ and $B = \begin{bmatrix} 0 & 3 \\ 2 & 0 \\ -4 & -1 \end{bmatrix}$.

**25.** $X = 3A - 2B$  **26.** $2X = 2A - B$
**27.** $2X + 3A = B$  **28.** $2A + 4B = -2X$

In Exercises 29–34, find matrices $A$, $X$, and $B$ such that the system of linear equations can be written as the matrix equation $AX = B$. Solve the system of equations.

**29.** $\begin{aligned} -x + y &= 4 \\ -2x + y &= 0 \end{aligned}$  **30.** $\begin{aligned} 2x + 3y &= 5 \\ x + 4y &= 10 \end{aligned}$

**31.** $\begin{aligned} x + 2y &= 3 \\ 3x - y &= 2 \end{aligned}$  **32.** $\begin{aligned} 2x - 4y + z &= 0 \\ -x + 3y + z &= 1 \\ x + y &= 3 \end{aligned}$

**33.** $\begin{aligned} x - 2y + 3z &= 9 \\ -x + 3y - z &= -6 \\ 2x - 5y + 5z &= 17 \end{aligned}$

**34.** $\begin{aligned} x + y - 3z &= -1 \\ -x + 2y &= 1 \\ -y + z &= 0 \end{aligned}$

**35.** If $a, b,$ and $c$ are real numbers such that $c \neq 0$ and $ac = bc$, then $a = b$. However, if $A, B,$ and $C$ are matrices such that $AC = BC$, then $A$ is *not* necessarily equal to $B$. Illustrate this using the following matrices.

$$A = \begin{bmatrix} 1 & 2 & 3 \\ 0 & 5 & 4 \\ 3 & -2 & 1 \end{bmatrix}, \quad B = \begin{bmatrix} 4 & -6 & 3 \\ 5 & 4 & 4 \\ -1 & 0 & 1 \end{bmatrix},$$

and $C = \begin{bmatrix} 0 & 0 & 0 \\ 0 & 0 & 0 \\ 4 & -2 & 3 \end{bmatrix}$

**36.** If $a$ and $b$ are real numbers such that $ab = 0$, then $a = 0$ or $b = 0$. However, if $A$ and $B$ are matrices such that $AB = O$, then it is *not* necessarily true that $A = O$ or $B = O$. Illustrate this using the following matrices.

$$A = \begin{bmatrix} 3 & 3 \\ 4 & 4 \end{bmatrix} \quad \text{and} \quad B = \begin{bmatrix} 1 & -1 \\ -1 & 1 \end{bmatrix}$$

Find another example of two nonzero matrices whose product is the zero matrix.

**37.** *Factory Production*    A certain corporation has four factories, each of which manufactures two products. The number of units of product $i$ produced at factory $j$ in one day is represented by $a_{ij}$ in the matrix

$$A = \begin{bmatrix} 100 & 90 & 70 & 30 \\ 40 & 20 & 60 & 60 \end{bmatrix}.$$

Find the output if production is increased by 10%. (*Hint:* Because an increase of 10% corresponds to 100% + 10%, multiply the matrix by 1.10.)

**38.** *Factory Production*    A certain corporation has three factories, each of which manufactures two products. The number of units of product $i$ produced at factory $j$ in one day is represented by $a_{ij}$ in the matrix

$$A = \begin{bmatrix} 60 & 40 & 20 \\ 30 & 90 & 60 \end{bmatrix}.$$

Find the output if production is decreased by 20%. (*Hint:* Because a decrease of 20% corresponds to 100% − 20%, multiply the matrix by 0.80.)

**39.** *Hotel Pricing*    A convention planning service has identified three suitable hotels for a convention. The quoted room rates are for single, double, triple, and quadruple occupancy. The current cost for each type of room by hotel is represented by the matrix $A$.

$$A = \begin{bmatrix} 79 & 85 & 109 \\ 87 & 93 & 120 \\ 95 & 110 & 129 \\ 95 & 125 & 135 \end{bmatrix} \begin{matrix} \text{Single} \\ \text{Double} \\ \text{Triple} \\ \text{Quadruple} \end{matrix}$$

with columns labeled *Hotel* $x$, *Hotel* $y$, *Hotel* $z$ and the rows grouped as Occupancy.

If room rates were guaranteed not to increase more than 10% by the time of the convention, what would be the maximum rate per room per hotel?

**40.** *Vacation Packages*    A vacation service has identified four resort hotels with a special all-inclusive package (room and meals included) at a popular travel destination. The quoted room rates are for a double or family (maximum of four people) occupancy for 5 days and 4 nights. The current cost for each type of room by hotel is represented by the matrix $A$.

$$A = \begin{bmatrix} 595 & 650 & 727 & 983 \\ 985 & 1020 & 1105 & 1097 \end{bmatrix} \begin{matrix} \text{Double} \\ \text{Family} \end{matrix}$$

with columns labeled *Hotel* $w$, *Hotel* $x$, *Hotel* $y$, *Hotel* $z$ and the rows grouped as Occupancy.

If room rates were guaranteed not to increase more than 15% by next season, what would be the maximum rate per package per hotel?

**41.** *Inventory Levels*    A company sells five different models of computers through three retail outlets. The inventories of the five models at the three outlets are given by the matrix $S$.

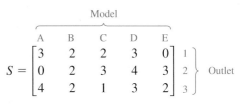

$$S = \begin{bmatrix} 3 & 2 & 2 & 3 & 0 \\ 0 & 2 & 3 & 4 & 3 \\ 4 & 2 & 1 & 3 & 2 \end{bmatrix} \begin{matrix} 1 \\ 2 \\ 3 \end{matrix}$$

with columns labeled Model A, B, C, D, E and the rows grouped as Outlet.

The wholesale and retail prices for each model are given by the matrix $T$.

$$
T = \begin{array}{c} \\ \\ \\ \\ \\ \\ \end{array}
\overbrace{\begin{bmatrix} \$840 & \$1100 \\ \$1200 & \$1350 \\ \$1450 & \$1650 \\ \$2650 & \$3000 \\ \$3050 & \$3200 \end{bmatrix}}^{\substack{\text{Price} \\ \text{Wholesale} \quad \text{Retail}}}
\begin{array}{l} A \\ B \\ C \\ D \\ E \end{array} \Bigg\} \; \text{Model}
$$

(a) What is the total retail price of the inventory at Outlet 1?

(b) What is the total wholesale price of the inventory at Outlet 3?

(c) Compute the product $ST$ and use the context of the problem to interpret the result.

**42.** *Labor/Wage Requirements*    A company that manufactures boats has the following labor-hour and wage requirements.

*Labor-Hour Requirements (per boat)*

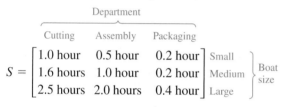

$$
S = \begin{array}{c} \\ \\ \\ \end{array}
\overbrace{\begin{bmatrix} 1.0 \text{ hour} & 0.5 \text{ hour} & 0.2 \text{ hour} \\ 1.6 \text{ hours} & 1.0 \text{ hour} & 0.2 \text{ hour} \\ 2.5 \text{ hours} & 2.0 \text{ hours} & 0.4 \text{ hour} \end{bmatrix}}^{\substack{\text{Department} \\ \text{Cutting} \quad \text{Assembly} \quad \text{Packaging}}}
\begin{array}{l} \text{Small} \\ \text{Medium} \\ \text{Large} \end{array} \Bigg\} \; \substack{\text{Boat} \\ \text{size}}
$$

*Wage Requirements (per hour)*

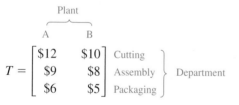

$$
T = \begin{array}{c} \\ \\ \\ \end{array}
\overbrace{\begin{bmatrix} \$12 & \$10 \\ \$9 & \$8 \\ \$6 & \$5 \end{bmatrix}}^{\substack{\text{Plant} \\ A \qquad B}}
\begin{array}{l} \text{Cutting} \\ \text{Assembly} \\ \text{Packaging} \end{array} \Bigg\} \; \text{Department}
$$

(a) What is the labor cost for a medium-sized boat at Plant B?

(b) What is the labor cost for a large-sized boat at Plant A?

(c) Compute $ST$ and interpret the result.

*Think About It*    In Exercises 43 and 44, find a matrix $B$ such that $AB$ is the identity matrix. Is there more than one correct result?

**43.** $A = \begin{bmatrix} 1 & 3 \\ 1 & 2 \end{bmatrix}$    **44.** $A = \begin{bmatrix} 2 & 1 \\ 5 & 2 \end{bmatrix}$

**45.** *Voting Preference*    The matrix

$$
P = \begin{array}{c} \\ \\ \\ \end{array}
\overbrace{\begin{bmatrix} 0.6 & 0.1 & 0.1 \\ 0.2 & 0.7 & 0.1 \\ 0.2 & 0.2 & 0.8 \end{bmatrix}}^{\substack{\text{From} \\ R \quad D \quad I}}
\begin{array}{l} R \\ D \\ I \end{array} \Bigg\} \; \text{To}
$$

is called a stochastic matrix. Each entry $p_{ij}$ $(i \neq j)$ represents the proportion of the voting population that changes from party $i$ to party $j$, and $p_{ii}$ represents the proportion that remains loyal to the party from one election to the next. Use a graphing utility to find $P^2$. (This matrix gives the transition probabilities from the first election to the third.)

**46.** *Voting Preference*    Use a graphing utility to find $P^3$, $P^4$, $P^5$, $P^6$, $P^7$, and $P^8$ for the matrix given in Exercise 45. Can you detect a pattern as $P$ is raised to higher and higher powers?

**47.** *Contract Bonuses*    Professional athletes frequently have bonus or incentive clauses in their contracts. For example, a defensive football player might receive a bonus for a sack, interception, and/or key tackle. Suppose that in a contract a sack is worth $2000, an interception is worth $1000, and a key tackle is worth $700. Use matrices to calculate the bonuses for defensive players A, B, and C if the following matrix describes the numbers of sacks, interceptions, and key tackles in a game.

| Player | Sacks | Interceptions | Key Tackles |
|--------|-------|---------------|-------------|
| Player A | 2 | 0 | 4 |
| Player B | 0 | 1 | 5 |
| Player C | 1 | 3 | 2 |

## 7.3    The Inverse of a Square Matrix

### Objectives

- Verify that a matrix is the inverse of a given matrix.
- Find the inverse of a matrix.
- Use an inverse matrix to solve a system of linear equations.

## The Inverse of a Matrix

This section further develops the algebra of matrices. To begin, consider the real number equation $ax = b$. To solve this equation for $x$, multiply both sides of the equation by $a^{-1}$ (provided $a \neq 0$).

$$ax = b$$
$$(a^{-1}a)x = a^{-1}b$$
$$(1)x = a^{-1}b$$
$$x = a^{-1}b$$

The number $a^{-1}$ is called the *multiplicative inverse of a* because it has the property that $a^{-1}a = 1$. The definition of a multiplicative inverse of a matrix is similar.

> ▶ **Definition of an Inverse of a Square Matrix**
>
> Let $A$ be an $n \times n$ matrix. If there exists a matrix $A^{-1}$ such that
>
> $$AA^{-1} = I_n = A^{-1}A$$
>
> $A^{-1}$ is called the **inverse** of $A$. (The symbol $A^{-1}$ is read "$A$ inverse.")

### Study Tip

Recall that it is not always true that $AB = BA$, even if both products are defined. However, if $A$ and $B$ are both square matrices and $AB = I_n$, it can be shown that $BA = I_n$. So, in Example 1, you need only to check that $AB = I_2$.

**Example 1**    The Inverse of a Matrix     *Learning Tools*

Show that $B$ is the inverse of $A$, where

$$A = \begin{bmatrix} -1 & 2 \\ -1 & 1 \end{bmatrix} \quad \text{and} \quad B = \begin{bmatrix} 1 & -2 \\ 1 & -1 \end{bmatrix}.$$

**Solution**

To show that $B$ is the inverse of $A$, show that $AB = I = BA$, as follows.

$$AB = \begin{bmatrix} -1 & 2 \\ -1 & 1 \end{bmatrix} \begin{bmatrix} 1 & -2 \\ 1 & -1 \end{bmatrix} = \begin{bmatrix} -1+2 & 2-2 \\ -1+1 & 2-1 \end{bmatrix} = \begin{bmatrix} 1 & 0 \\ 0 & 1 \end{bmatrix}$$

$$BA = \begin{bmatrix} 1 & -2 \\ 1 & -1 \end{bmatrix} \begin{bmatrix} -1 & 2 \\ -1 & 1 \end{bmatrix} = \begin{bmatrix} -1+2 & 2-2 \\ -1+1 & 2-1 \end{bmatrix} = \begin{bmatrix} 1 & 0 \\ 0 & 1 \end{bmatrix}$$

If a matrix $A$ has an inverse, $A$ is called **invertible** (or **nonsingular**); otherwise, $A$ is called **singular.** A nonsquare matrix cannot have an inverse. To see this, note that if $A$ is of order $m \times n$ and $B$ is of order $n \times m$ (where $m \neq n$), the products $AB$ and $BA$ are of different orders and therefore cannot be equal to each other. Not all square matrices have inverses (see the matrix at the bottom of page 457). If, however, a matrix does have an inverse, that inverse is unique. The following example shows how to use a system of equations to find the inverse.

---

**Example 2**    Finding the Inverse of a Matrix

Find the inverse of the matrix

$$A = \begin{bmatrix} 1 & 4 \\ -1 & -3 \end{bmatrix}.$$

**Solution**

To find the inverse of $A$, try to solve the matrix equation $AX = I$ for $X$.

$$\begin{matrix} A & & X & & I \end{matrix}$$

$$\begin{bmatrix} 1 & 4 \\ -1 & -3 \end{bmatrix} \begin{bmatrix} x_{11} & x_{12} \\ x_{21} & x_{22} \end{bmatrix} = \begin{bmatrix} 1 & 0 \\ 0 & 1 \end{bmatrix}$$

$$\begin{bmatrix} x_{11} + 4x_{21} & x_{12} + 4x_{22} \\ -x_{11} - 3x_{21} & -x_{12} - 3x_{22} \end{bmatrix} = \begin{bmatrix} 1 & 0 \\ 0 & 1 \end{bmatrix}$$

Equating corresponding entries, you obtain the following two systems of linear equations.

$$\begin{aligned} x_{11} + 4x_{21} &= 1 & x_{12} + 4x_{22} &= 0 \\ -x_{11} - 3x_{21} &= 0 & -x_{12} - 3x_{22} &= 1 \end{aligned}$$

From the first system you can determine that $x_{11} = -3$ and $x_{21} = 1$, and from the second system you can determine that $x_{12} = -4$ and $x_{22} = 1$. Therefore, the inverse of $A$ is

$$X = A^{-1} = \begin{bmatrix} -3 & -4 \\ 1 & 1 \end{bmatrix}.$$

You can use matrix multiplication to check this result.

**Check**

$$AA^{-1} = \begin{bmatrix} 1 & 4 \\ -1 & -3 \end{bmatrix} \begin{bmatrix} -3 & -4 \\ 1 & 1 \end{bmatrix} = \begin{bmatrix} 1 & 0 \\ 0 & 1 \end{bmatrix} \checkmark$$

$$A^{-1}A = \begin{bmatrix} -3 & -4 \\ 1 & 1 \end{bmatrix} \begin{bmatrix} 1 & 4 \\ -1 & -3 \end{bmatrix} = \begin{bmatrix} 1 & 0 \\ 0 & 1 \end{bmatrix} \checkmark$$

## Finding Inverse Matrices

In Example 2, note that the two systems of linear equations have the *same coefficient matrix A*. Rather than solve the two systems represented by

$$\begin{bmatrix} 1 & 4 & \vdots & 1 \\ -1 & -3 & \vdots & 0 \end{bmatrix} \quad \text{and} \quad \begin{bmatrix} 1 & 4 & \vdots & 0 \\ -1 & -3 & \vdots & 1 \end{bmatrix}$$

separately, you can solve them *simultaneously* by **adjoining** the identity matrix to the coefficient matrix to obtain

$$\overset{A}{\phantom{x}} \qquad \overset{I}{\phantom{x}}$$
$$\begin{bmatrix} 1 & 4 & \vdots & 1 & 0 \\ -1 & -3 & \vdots & 0 & 1 \end{bmatrix}.$$

Then, applying Gauss-Jordan elimination to this matrix, you can solve *both* systems with a single elimination process, as follows.

$$\begin{bmatrix} 1 & 4 & \vdots & 1 & 0 \\ -1 & -3 & \vdots & 0 & 1 \end{bmatrix}$$

$$R_1 + R_2 \rightarrow \begin{bmatrix} 1 & 4 & \vdots & 1 & 0 \\ 0 & 1 & \vdots & 1 & 1 \end{bmatrix}$$

$$-4R_2 + R_1 \rightarrow \begin{bmatrix} 1 & 0 & \vdots & -3 & -4 \\ 0 & 1 & \vdots & 1 & 1 \end{bmatrix}$$

Thus, from the "doubly augmented" matrix $[A : I]$, you obtained the matrix $[I : A^{-1}]$.

$$\overset{A}{\phantom{x}} \qquad \overset{I}{\phantom{x}} \qquad\qquad \overset{I}{\phantom{x}} \qquad \overset{A^{-1}}{\phantom{x}}$$
$$\begin{bmatrix} 1 & 4 & \vdots & 1 & 0 \\ -1 & -3 & \vdots & 0 & 1 \end{bmatrix} \implies \begin{bmatrix} 1 & 0 & \vdots & -3 & -4 \\ 0 & 1 & \vdots & 1 & 1 \end{bmatrix}$$

This procedure (or algorithm) works for an arbitrary square matrix that has an inverse.

---

▶ **Finding an Inverse Matrix**

Let $A$ be a square matrix of order $n$.

1. Write the $n \times 2n$ matrix that consists of the given matrix $A$ on the left and the $n \times n$ identity matrix $I$ on the right to obtain $[A : I]$. Note that we separate the matrices $A$ and $I$ by a dotted line. We call this process **adjoining** the matrices $A$ and $I$.

2. If possible, row reduce $A$ to $I$ using elementary row operations on the *entire* matrix $[A : I]$. The result will be the matrix $[I : A^{-1}]$. If this is not possible, $A$ is not invertible.

3. Check your work by multiplying to see that $AA^{-1} = I = A^{-1}A$.

**Example 3**    Finding the Inverse of a Matrix

Find the inverse of the matrix

$$A = \begin{bmatrix} 1 & -1 & 0 \\ 1 & 0 & -1 \\ 6 & -2 & -3 \end{bmatrix}.$$

**Solution**

Begin by adjoining the identity matrix to $A$ to form the matrix

$$[A \ \vdots \ I] = \begin{bmatrix} 1 & -1 & 0 & \vdots & 1 & 0 & 0 \\ 1 & 0 & -1 & \vdots & 0 & 1 & 0 \\ 6 & -2 & -3 & \vdots & 0 & 0 & 1 \end{bmatrix}.$$

Use elementary row operations to obtain the form $[I \ \vdots \ A^{-1}]$, as follows.

$$\begin{bmatrix} 1 & -1 & 0 & \vdots & 1 & 0 & 0 \\ 1 & 0 & -1 & \vdots & 0 & 1 & 0 \\ 6 & -2 & -3 & \vdots & 0 & 0 & 1 \end{bmatrix}$$

$$\begin{matrix} \\ -R_1 + R_2 \rightarrow \\ -6R_1 + R_3 \rightarrow \end{matrix} \begin{bmatrix} 1 & -1 & 0 & \vdots & 1 & 0 & 0 \\ 0 & 1 & -1 & \vdots & -1 & 1 & 0 \\ 0 & 4 & -3 & \vdots & -6 & 0 & 1 \end{bmatrix}$$

$$\begin{matrix} R_2 + R_1 \rightarrow \\ \\ -4R_2 + R_3 \rightarrow \end{matrix} \begin{bmatrix} 1 & 0 & -1 & \vdots & 0 & 1 & 0 \\ 0 & 1 & -1 & \vdots & -1 & 1 & 0 \\ 0 & 0 & 1 & \vdots & -2 & -4 & 1 \end{bmatrix}$$

$$\begin{matrix} R_3 + R_1 \rightarrow \\ R_3 + R_2 \rightarrow \\ \\ \end{matrix} \begin{bmatrix} 1 & 0 & 0 & \vdots & -2 & -3 & 1 \\ 0 & 1 & 0 & \vdots & -3 & -3 & 1 \\ 0 & 0 & 1 & \vdots & -2 & -4 & 1 \end{bmatrix}$$

So, the matrix $A$ is invertible and its inverse is

$$A^{-1} = \begin{bmatrix} -2 & -3 & 1 \\ -3 & -3 & 1 \\ -2 & -4 & 1 \end{bmatrix}.$$

Try confirming this result by multiplying $A$ and $A^{-1}$ to obtain $I$.

---

The process shown in Example 3 applies to any $n \times n$ matrix $A$. If $A$ has an inverse, this process will find it. On the other hand, if $A$ does not have an inverse, this process will tell us so. For instance, the following matrix has no inverse.

$$A = \begin{bmatrix} 1 & 2 & 0 \\ 3 & -1 & 2 \\ -2 & 3 & -2 \end{bmatrix}$$

Explain how the elimination process shows that this matrix is singular.

## The Inverse of a $2 \times 2$ Matrix (Quick Method)

·DISCOVERY·

Use a graphing utility with matrix operations to find the inverse of the matrix

$$A = \begin{bmatrix} 1 & -3 \\ -2 & 6 \end{bmatrix}.$$

What message appears on the screen? Why does the graphing utility display an error message?

Using Gauss-Jordan elimination to find the inverse of a matrix works well (even as a computer technique) for matrices of order $3 \times 3$ or greater. For $2 \times 2$ matrices, however, many people prefer to use a formula for the inverse rather than Gauss-Jordan elimination. This simple formula, which works *only* for $2 \times 2$ matrices, is explained as follows. If $A$ is a $2 \times 2$ matrix given by

$$A = \begin{bmatrix} a & b \\ c & d \end{bmatrix}$$

then $A$ is invertible if and only if $ad - bc \neq 0$. Moreover, if $ad - bc \neq 0$, the inverse is given by

$$A^{-1} = \frac{1}{ad - bc} \begin{bmatrix} d & -b \\ -c & a \end{bmatrix}.$$

Try verifying this inverse by multiplication.

The denominator $ad - bc$ is called the **determinant** of the $2 \times 2$ matrix $A$. You will study determinants in the next section.

**Example 4** Finding the Inverse of a $2 \times 2$ Matrix

If possible, find the inverses of the following matrices.

**a.** $A = \begin{bmatrix} 3 & -1 \\ -2 & 2 \end{bmatrix}$

**b.** $B = \begin{bmatrix} 3 & -1 \\ -6 & 2 \end{bmatrix}$

**Solution**

**a.** For the matrix $A$, apply the formula for the inverse of a $2 \times 2$ matrix to obtain

$$ad - bc = (3)(2) - (-1)(-2) = 4.$$

Because this quantity is not zero, the inverse is formed by interchanging the entries on the main diagonal and changing the signs of the other two entries, as follows.

$$A^{-1} = \frac{1}{4} \begin{bmatrix} 2 & 1 \\ 2 & 3 \end{bmatrix} = \begin{bmatrix} \frac{1}{2} & \frac{1}{4} \\ \frac{1}{2} & \frac{3}{4} \end{bmatrix}$$

**b.** For the matrix $B$, you have

$$ad - bc = (3)(2) - (-1)(-6) = 0$$

which means that $B$ is not invertible.

## Systems of Linear Equations

We know that a system of linear equations can have exactly one solution, infinitely many solutions, or no solution. If the coefficient matrix $A$ of a *square* system (a system that has the same number of equations as variables) is invertible, the system has a unique solution, which is given as follows.

> ▶ **A System of Equations with a Unique Solution**
>
> If $A$ is an invertible matrix, the system of linear equations represented by $AX = B$ has a unique solution given by
>
> $$X = A^{-1}B.$$

•**TECHNOLOGY**•

The formula $X = A^{-1}B$ is used in most graphing utilities to solve linear systems that have invertible coefficient matrices. That is, you enter the $n \times n$ coefficient matrix $[A]$ and the $n \times 1$ column matrix $[B]$. The solution $X$ is given by $[A]^{-1}[B]$. Use your graphing utility to verify the solution in Example 5.

**Example 5**    Solving a System of Equations Using an Inverse Matrix

Use an inverse matrix to solve the following system.

$$2x + 3y + z = -1$$

$$3x + 3y + z = 1$$

$$2x + 4y + z = -2$$

**Solution**

$$X = A^{-1}B = \begin{bmatrix} -1 & 1 & 0 \\ -1 & 0 & 1 \\ 6 & -2 & -3 \end{bmatrix} \begin{bmatrix} -1 \\ 1 \\ -2 \end{bmatrix} = \begin{bmatrix} 2 \\ -1 \\ -2 \end{bmatrix}$$

So, the solution is $x = 2$, $y = -1$, and $z = -2$.

---

| Discussing the Concept | Methods of Problem Solving |
|---|---|

Describe how the method of solving Example 5 is similar to the method used to solve a simple equation such as

$$2x = 10.$$

How are the two methods different?

# Warm Up

The following warm-up exercises involve skills that were covered in earlier sections. You will use these skills in the exercise set for this section.

In Exercises 1–8, perform the indicated matrix operations.

1. $4\begin{bmatrix} 1 & 6 \\ 0 & -4 \\ 12 & 2 \end{bmatrix}$

2. $\dfrac{1}{2}\begin{bmatrix} 11 & 10 & 48 \\ 1 & 0 & 16 \\ 0 & 2 & 8 \end{bmatrix}$

3. $\begin{bmatrix} 1 & -10 & 3 \\ 4 & 1 & 0 \end{bmatrix} - 2\begin{bmatrix} 3 & -4 & 8 \\ 0 & 7 & 1 \end{bmatrix}$

4. $\begin{bmatrix} 5 & 20 \\ -7 & 15 \end{bmatrix} - 3\begin{bmatrix} 6 & 3 \\ 4 & -2 \end{bmatrix}$

5. $\begin{bmatrix} 1 & -2 \\ -1 & 3 \end{bmatrix}\begin{bmatrix} 3 & 2 \\ 1 & 1 \end{bmatrix}$

6. $\begin{bmatrix} 1 & 0 \\ 0 & 1 \end{bmatrix}\begin{bmatrix} 6 & 5 \\ 3 & -2 \end{bmatrix}$

7. $\begin{bmatrix} 2 & 0 & 0 \\ 0 & -1 & 0 \\ 0 & 0 & 3 \end{bmatrix}\begin{bmatrix} \frac{1}{2} & 0 & 0 \\ 0 & -1 & 0 \\ 0 & 0 & \frac{1}{3} \end{bmatrix}$

8. $\begin{bmatrix} 1 & -1 & 0 \\ 1 & 0 & -1 \\ 6 & -2 & -3 \end{bmatrix}\begin{bmatrix} -2 & -3 & 1 \\ -3 & -3 & 1 \\ -2 & -4 & 1 \end{bmatrix}$

In Exercises 9 and 10, rewrite the matrix in reduced row-echelon form.

9. $\begin{bmatrix} 3 & -2 & 1 & 0 \\ 4 & -3 & 0 & 1 \end{bmatrix}$

10. $\begin{bmatrix} 1 & 1 & 2 & 1 & 0 & 0 \\ -1 & 0 & 3 & 0 & 1 & 0 \\ 1 & 2 & 8 & 0 & 0 & 1 \end{bmatrix}$

## 7.3 Exercises

In Exercises 1–10, show that $B$ is the inverse of $A$.

1. $A = \begin{bmatrix} 7 & 4 \\ 5 & 3 \end{bmatrix}$, $B = \begin{bmatrix} 3 & -4 \\ -5 & 7 \end{bmatrix}$

2. $A = \begin{bmatrix} -4 & 1 \\ -9 & 2 \end{bmatrix}$, $B = \begin{bmatrix} 2 & -1 \\ 9 & -4 \end{bmatrix}$

3. $A = \begin{bmatrix} 2 & -1 \\ 5 & -4 \end{bmatrix}$, $B = \begin{bmatrix} \frac{4}{3} & -\frac{1}{3} \\ \frac{5}{3} & -\frac{2}{3} \end{bmatrix}$

4. $A = \begin{bmatrix} 1 & -2 \\ 3 & -10 \end{bmatrix}$, $B = \begin{bmatrix} \frac{5}{2} & -\frac{1}{2} \\ \frac{3}{4} & -\frac{1}{4} \end{bmatrix}$

5. $A = \begin{bmatrix} -2 & 2 & 3 \\ 1 & -1 & 0 \\ 0 & 1 & 4 \end{bmatrix}$, $B = \dfrac{1}{3}\begin{bmatrix} -4 & -5 & 3 \\ -4 & -8 & 3 \\ 1 & 2 & 0 \end{bmatrix}$

6. $A = \begin{bmatrix} 2 & -17 & 11 \\ -1 & 11 & -7 \\ 0 & 3 & -2 \end{bmatrix}$, $B = \begin{bmatrix} 1 & 1 & 2 \\ 2 & 4 & -3 \\ 3 & 6 & -5 \end{bmatrix}$

7. $A = \begin{bmatrix} -1 & 0 & 2 \\ 1 & -2 & 0 \\ 1 & 0 & 3 \end{bmatrix}$, $B = \dfrac{1}{10}\begin{bmatrix} -6 & 0 & 4 \\ -3 & -5 & 2 \\ 2 & 0 & 2 \end{bmatrix}$

8. $A = \begin{bmatrix} -1 & 1 & -3 \\ 2 & -1 & 4 \\ -1 & 1 & -2 \end{bmatrix}$, $B = \begin{bmatrix} 2 & 1 & -1 \\ 0 & 1 & 2 \\ -1 & 0 & 1 \end{bmatrix}$

9. $A = \begin{bmatrix} 2 & 0 & 1 & 1 \\ 3 & 0 & 0 & 1 \\ -1 & 1 & -2 & 1 \\ 4 & -1 & 1 & 0 \end{bmatrix}$,

$B = \begin{bmatrix} -1 & 2 & -1 & -1 \\ -4 & 9 & -5 & -6 \\ 0 & 1 & -1 & -1 \\ 3 & -5 & 3 & 3 \end{bmatrix}$

**10.** $A = \begin{bmatrix} -1 & 1 & 0 & -1 \\ 1 & -1 & 2 & 0 \\ -1 & 1 & 2 & 0 \\ 0 & -1 & 1 & 1 \end{bmatrix}$,

$B = \dfrac{1}{4}\begin{bmatrix} -4 & 1 & 1 & -4 \\ -4 & -1 & 3 & -4 \\ 0 & 1 & 1 & 0 \\ -4 & -2 & 2 & 0 \end{bmatrix}$

In Exercises 11–36, find the inverse of the matrix (if it exists).

**11.** $\begin{bmatrix} 8 & 4 \\ -2 & -2 \end{bmatrix}$

**12.** $\begin{bmatrix} 1 & 2 \\ 3 & 7 \end{bmatrix}$

**13.** $\begin{bmatrix} 2 & 3 \\ 6 & 9 \end{bmatrix}$

**14.** $\begin{bmatrix} -7 & 33 \\ 4 & -19 \end{bmatrix}$

**15.** $\begin{bmatrix} -1 & 1 \\ -2 & 1 \end{bmatrix}$

**16.** $\begin{bmatrix} 2 & 3 \\ 1 & 4 \end{bmatrix}$

**17.** $\begin{bmatrix} 0 & 4 \\ -3 & 6 \end{bmatrix}$

**18.** $\begin{bmatrix} 11 & 1 \\ -1 & 0 \end{bmatrix}$

**19.** $\begin{bmatrix} 2 & 7 & 1 \\ -3 & -9 & 2 \end{bmatrix}$

**20.** $\begin{bmatrix} -2 & 5 \\ 6 & -15 \\ 0 & 1 \end{bmatrix}$

**21.** $\begin{bmatrix} 1 & 1 & 1 \\ 3 & 5 & 4 \\ 3 & 6 & 5 \end{bmatrix}$

**22.** $\begin{bmatrix} 1 & 2 & 2 \\ 3 & 7 & 9 \\ -1 & -4 & -7 \end{bmatrix}$

**23.** $\begin{bmatrix} 1 & 2 & -1 \\ 3 & 7 & -10 \\ -5 & -7 & -15 \end{bmatrix}$

**24.** $\begin{bmatrix} 10 & 5 & -7 \\ -5 & 1 & 4 \\ 3 & 2 & -2 \end{bmatrix}$

**25.** $\begin{bmatrix} 1 & 1 & 2 \\ 3 & 1 & 0 \\ -2 & 0 & 3 \end{bmatrix}$

**26.** $\begin{bmatrix} 3 & 2 & 2 \\ 2 & 2 & 2 \\ -4 & 4 & 3 \end{bmatrix}$

**27.** $\begin{bmatrix} 3 & 0 & 0 \\ 0 & -2 & 0 \\ 0 & 0 & 4 \end{bmatrix}$

**28.** $\begin{bmatrix} 2 & 0 & 0 \\ 0 & 3 & 0 \\ 0 & 0 & 5 \end{bmatrix}$

**29.** $\begin{bmatrix} 1 & 0 & 0 \\ 3 & 4 & 0 \\ 2 & 5 & 5 \end{bmatrix}$

**30.** $\begin{bmatrix} 1 & 0 & 0 \\ 3 & 0 & 0 \\ 2 & 5 & 5 \end{bmatrix}$

**31.** $\begin{bmatrix} 1 & 0 & 3 & 0 \\ 0 & 2 & 0 & 4 \\ 1 & 0 & 3 & 0 \\ 0 & 2 & 0 & 4 \end{bmatrix}$

**32.** $\begin{bmatrix} 1 & 3 & -2 & 0 \\ 0 & 2 & 4 & 6 \\ 0 & 0 & -2 & 1 \\ 0 & 0 & 0 & 5 \end{bmatrix}$

**33.** $\begin{bmatrix} -8 & 0 & 0 & 0 \\ 0 & 1 & 0 & 0 \\ 0 & 0 & 4 & 0 \\ 0 & 0 & 0 & -5 \end{bmatrix}$

**34.** $\begin{bmatrix} -1 & 0 & 1 & 0 \\ 0 & 2 & 0 & -1 \\ 2 & 0 & -1 & 0 \\ 0 & -1 & 0 & 1 \end{bmatrix}$

**35.** $\begin{bmatrix} 1 & -2 & -1 & -2 \\ 3 & -5 & -2 & -3 \\ 2 & -5 & -2 & -5 \\ -1 & 4 & 4 & 11 \end{bmatrix}$

**36.** $\begin{bmatrix} 4 & 8 & -7 & 14 \\ 2 & 5 & -4 & 6 \\ 0 & 2 & 1 & -7 \\ 3 & 6 & -5 & 10 \end{bmatrix}$

In Exercises 37–40, use an inverse matrix to solve the system of linear equations. (Use the inverse matrix found in Exercise 15.)

**37.** $-x + y = 4$
$-2x + y = 0$

**38.** $-x + y = -3$
$-2x + y = 5$

**39.** $-x + y = 20$
$-2x + y = 10$

**40.** $-x + y = 0$
$-2x + y = 7$

In Exercises 41–44, use an inverse matrix to solve the system of linear equations. (Use the inverse matrix found in Exercise 16.)

**41.** $2x + 3y = 5$
$x + 4y = 10$

**42.** $2x + 3y = 0$
$x + 4y = 3$

**43.** $2x + 3y = 4$
$x + 4y = 2$

**44.** $2x + 3y = 1$
$x + 4y = -2$

In Exercises 45 and 46, use an inverse matrix to solve the system of linear equations. (Use the inverse matrix found in Exercise 26.)

**45.** $3x + 2y + 2z = 0$
$2x + 2y + 2z = 5$
$-4x + 4y + 3z = 2$

**46.** $3x + 2y + 2z = -1$
$2x + 2y + 2z = 2$
$-4x + 4y + 3z = 0$

In Exercises 47 and 48, use an inverse matrix to solve the system of linear equations. (Use the inverse matrix found in Exercise 35.)

**47.**
$$\begin{aligned} x_1 - 2x_2 - x_3 - 2x_4 &= 0 \\ 3x_1 - 5x_2 - 2x_3 - 3x_4 &= 1 \\ 2x_1 - 5x_2 - 2x_3 - 5x_4 &= -1 \\ -x_1 + 4x_2 + 4x_3 + 11x_4 &= 2 \end{aligned}$$

**48.**
$$\begin{aligned} x_1 - 2x_2 - x_3 - 2x_4 &= 1 \\ 3x_1 - 5x_2 - 2x_3 - 3x_4 &= -2 \\ 2x_1 - 5x_2 - 2x_3 - 5x_4 &= 0 \\ -x_1 + 4x_2 + 4x_3 + 11x_4 &= -3 \end{aligned}$$

**49.** *Research and Development Expenditures* The amounts of their own funds that universities and colleges in the United States spent on research and development for the years 1990 to 1999 are shown in the figure. The least squares regression parabola $y = at^2 + bt + c$ for these data is found by solving the system

$$\begin{aligned} 10c + 45b + 285a &= 42.45 \\ 45c + 285b + 2025a &= 213.899 \\ 285c + 2025b + 15{,}333a &= 1429.095 \end{aligned}$$

where $y$ is the expenditures (in billions of dollars) and $t = 0$ represents 1990. (Source: U.S. National Science Foundation)

(a) Use a graphing utility to find an inverse matrix to solve this system and find the equation of the least squares regression parabola.

(b) Use the result of part (a) to approximate the expenditures in 2005.

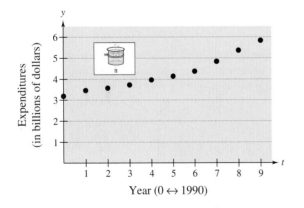

Year (0 ↔ 1990)

**50.** *Service Industry in the United States* The total numbers of people employed in the "service industry" (excluding self-employed people) from 1995 to 1999 are shown in the figure. The least squares regression parabola $y = at^2 + bt + c$ for these data is found by solving the system

$$\begin{aligned} 5c + 10b + 30a &= 231.3 \\ 10c + 30b + 100a &= 474.2 \\ 30c + 100b + 354a &= 1434.6 \end{aligned}$$

where $y$ is the number of workers (in millions) and $t = 0$ represents 1995. (Source: U.S. Bureau of Labor Statistics)

(a) Use a graphing utility to find an inverse matrix to solve this system and write the equation of the least squares regression line.

(b) Use the result of part (a) to approximate the number of workers in the service industry in 2005.

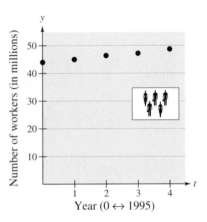

Year (0 ↔ 1995)

In Exercises 51 and 52, develop for the given matrix a system of equations that will have the given solution. Use an inverse matrix to verify that the system of equations gives the desired solution.

**51.** $\begin{bmatrix} 2 & 1 & 3 \\ 4 & 0 & -2 \\ 0 & 3 & 2 \end{bmatrix}$    $\begin{aligned} x &= 2 \\ y &= -3 \\ z &= 5 \end{aligned}$

**52.** $\begin{bmatrix} 1 & 0 & 2 \\ 1 & 1 & 1 \\ 2 & -1 & 0 \end{bmatrix}$    $\begin{aligned} x &= 5 \\ y &= -2 \\ z &= 1 \end{aligned}$

# Triangular Matrices

Evaluating determinants of matrices of order 4 or higher can be tedious. There is, however, an important exception: the determinant of a **triangular** matrix. A square matrix is **upper triangular** if it has all zero entries below its main diagonal and **lower triangular** if it has all zero entries above its main diagonal. A matrix that is both upper and lower triangular is called **diagonal.** That is, a diagonal matrix is one in which all entries above and below the main diagonal are zero.

*Upper Triangular Matrix*

$$\begin{bmatrix} a_{11} & a_{12} & a_{13} & \cdots & a_{1n} \\ 0 & a_{22} & a_{23} & \cdots & a_{2n} \\ 0 & 0 & a_{33} & \cdots & a_{3n} \\ \vdots & \vdots & \vdots & & \vdots \\ 0 & 0 & 0 & \cdots & a_{nn} \end{bmatrix}$$

*Lower Triangular Matrix*

$$\begin{bmatrix} a_{11} & 0 & 0 & \cdots & 0 \\ a_{21} & a_{22} & 0 & \cdots & 0 \\ a_{31} & a_{32} & a_{33} & \cdots & 0 \\ \vdots & \vdots & \vdots & & \vdots \\ a_{n1} & a_{n2} & a_{n3} & \cdots & a_{nn} \end{bmatrix}$$

To find the determinant of a triangular matrix of any order, simply form the product of the entries on the main diagonal.

---

**Example 6**    The Determinant of a Triangular Matrix

**a.**
$$\begin{vmatrix} 2 & 0 & 0 & 0 \\ 4 & -2 & 0 & 0 \\ -5 & 6 & 1 & 0 \\ 1 & 5 & 3 & 3 \end{vmatrix} = (2)(-2)(1)(3) = -12$$

**b.**
$$\begin{vmatrix} -1 & 0 & 0 & 0 & 0 \\ 0 & 3 & 0 & 0 & 0 \\ 0 & 0 & 2 & 0 & 0 \\ 0 & 0 & 0 & 4 & 0 \\ 0 & 0 & 0 & 0 & -2 \end{vmatrix} = (-1)(3)(2)(4)(-2) = 48$$

---

**Discussing the Concept**    **The Determinant of a Triangular Matrix**

Explain why the determinant of a 3 × 3 triangular matrix is the product of its main-diagonal entries.

$$\begin{vmatrix} a_{11} & a_{12} & a_{13} \\ 0 & a_{22} & a_{23} \\ 0 & 0 & a_{33} \end{vmatrix} = a_{11}a_{22}a_{33}$$

Is this true of higher-order triangular matrices? Explain.

# Warm Up

The following warm-up exercises involve skills that were covered in earlier sections. You will use these skills in the exercise set for this section.

In Exercises 1–4, perform the indicated matrix operations.

**1.** $\begin{bmatrix} 1 & -2 \\ 0 & 3 \end{bmatrix} + \begin{bmatrix} 2 & 7 \\ 4 & -3 \end{bmatrix}$

**2.** $\begin{bmatrix} -2 & 5 \\ 3 & -2 \end{bmatrix} - \begin{bmatrix} 0 & -3 \\ 1 & 2 \end{bmatrix}$

**3.** $3\begin{bmatrix} 3 & -4 & 2 \\ 1 & 0 & -1 \\ 0 & 1 & -2 \end{bmatrix}$

**4.** $4\begin{bmatrix} 0 & 2 & 3 \\ -1 & 2 & 3 \\ -2 & 1 & -2 \end{bmatrix}$

In Exercises 5–10, perform the indicated arithmetic operations.

**5.** $[(1)(3) + (-3)(2)] - [(1)(4) + (3)(5)]$

**6.** $[(4)(4) + (-1)(-3)] - [(-1)(2) + (-2)(7)]$

**7.** $\dfrac{4(7) - 1(-2)}{(-5)(-2) - 3(4)}$

**8.** $\dfrac{3(6) - 2(7)}{6(-5) - 2(1)}$

**9.** $-5(-1)^2[6(-2) - 7(-3)]$

**10.** $4(-1)^3[3(6) - 2(7)]$

## 7.4  Exercises

In Exercises 1–24, find the determinant of the matrix.

**1.** $\begin{bmatrix} -2 \end{bmatrix}$

**2.** $\begin{bmatrix} 6 \end{bmatrix}$

**3.** $\begin{bmatrix} 1 & 3 \\ 2 & 7 \end{bmatrix}$

**4.** $\begin{bmatrix} -3 & 4 \\ -2 & 1 \end{bmatrix}$

**5.** $\begin{bmatrix} 5 & 6 \\ 2 & 3 \end{bmatrix}$

**6.** $\begin{bmatrix} -7 & -4 \\ 8 & 7 \end{bmatrix}$

**7.** $\begin{bmatrix} 9 & 3 \\ 12 & 4 \end{bmatrix}$

**8.** $\begin{bmatrix} -5 & -2 \\ 10 & 4 \end{bmatrix}$

**9.** $\begin{bmatrix} \frac{2}{3} & 0 \\ -1 & 6 \end{bmatrix}$

**10.** $\begin{bmatrix} 9 & -\frac{1}{4} \\ 8 & 0 \end{bmatrix}$

**11.** $\begin{bmatrix} 2 & -1 & 0 \\ 4 & 2 & 1 \\ 4 & 2 & 1 \end{bmatrix}$

**12.** $\begin{bmatrix} -2 & 2 & 3 \\ 1 & -1 & 0 \\ 0 & 1 & 4 \end{bmatrix}$

**13.** $\begin{bmatrix} 0.3 & 0.2 & 0.2 \\ 0.2 & 0.2 & 0.2 \\ -0.4 & 0.4 & 0.3 \end{bmatrix}$

**14.** $\begin{bmatrix} 0.1 & 0.2 & 0.3 \\ -0.3 & 0.2 & 0.2 \\ 0.5 & 0.4 & 0.4 \end{bmatrix}$

**15.** $\begin{bmatrix} 1 & 4 & -2 \\ 3 & 6 & -6 \\ -2 & 1 & 4 \end{bmatrix}$

**16.** $\begin{bmatrix} -1 & 3 & 1 \\ 4 & 2 & 5 \\ -2 & 1 & 6 \end{bmatrix}$

**17.** $\begin{bmatrix} 6 & 3 & -7 \\ 0 & 0 & 0 \\ 4 & -6 & 3 \end{bmatrix}$

**18.** $\begin{bmatrix} 5 & 0 & 3 \\ -4 & 0 & 8 \\ 3 & 0 & -6 \end{bmatrix}$

**19.** $\begin{bmatrix} 2 & 0 & 0 \\ 4 & -3 & 0 \\ 6 & 5 & 1 \end{bmatrix}$

**20.** $\begin{bmatrix} 1 & 0 & 0 \\ -4 & -1 & 0 \\ 5 & 1 & 5 \end{bmatrix}$

**21.** $\begin{bmatrix} 2 & 3 & -1 & -1 \\ 0 & -1 & -3 & 5 \\ 0 & 0 & -2 & 7 \\ 0 & 0 & 0 & -4 \end{bmatrix}$

**22.** $\begin{bmatrix} 4 & 0 & 0 & 0 \\ 1 & -4 & 0 & 0 \\ 2 & 1 & -1 & 0 \\ 6 & -2 & 3 & -1 \end{bmatrix}$

**23.** $\begin{bmatrix} 1 & 0 & 0 & 0 & 0 \\ 0 & 2 & 0 & 0 & 0 \\ 0 & 0 & 3 & 0 & 0 \\ 0 & 0 & 0 & 4 & 0 \\ 0 & 0 & 0 & 0 & 5 \end{bmatrix}$

**24.** $\begin{bmatrix} -2 & 0 & 0 & 0 & 0 \\ 0 & 3 & 0 & 0 & 0 \\ 0 & 0 & -1 & 0 & 0 \\ 0 & 0 & 0 & 2 & 0 \\ 0 & 0 & 0 & 0 & -4 \end{bmatrix}$

In Exercises 25–28, find (a) all minors and (b) all cofactors of the given matrix.

**25.** $\begin{bmatrix} 3 & 4 \\ 2 & -5 \end{bmatrix}$

**26.** $\begin{bmatrix} 11 & 0 \\ -3 & 2 \end{bmatrix}$

**27.** $\begin{bmatrix} 3 & -2 & 8 \\ 3 & 2 & -6 \\ -1 & 3 & 6 \end{bmatrix}$

**28.** $\begin{bmatrix} -2 & 9 & 4 \\ 7 & -6 & 0 \\ 6 & 7 & -6 \end{bmatrix}$

In Exercises 29–34, find the determinant of the matrix by the method of expansion by cofactors. Expand using the indicated row or column.

**29.** $\begin{bmatrix} 4 & 1 & -3 \\ 6 & 5 & -2 \\ -1 & 3 & -4 \end{bmatrix}$

(a) Row 3

(b) Column 2

**30.** $\begin{bmatrix} -3 & 4 & 2 \\ 6 & 3 & 1 \\ 4 & -7 & -8 \end{bmatrix}$

(a) Row 2

(b) Column 3

**31.** $\begin{bmatrix} 7 & 0 & -4 \\ 2 & -3 & 0 \\ 5 & 8 & 1 \end{bmatrix}$

(a) Row 1

(b) Column 3

**32.** $\begin{bmatrix} 10 & -5 & 5 \\ 30 & 0 & 10 \\ 0 & 10 & 1 \end{bmatrix}$

(a) Row 3

(b) Column 1

**33.** $\begin{bmatrix} 6 & 0 & -3 & 5 \\ 4 & 13 & 6 & -8 \\ -1 & 0 & 7 & 4 \\ 8 & 6 & 0 & 2 \end{bmatrix}$

(a) Row 2

(b) Column 2

**34.** $\begin{bmatrix} 10 & 8 & 3 & -7 \\ 4 & 0 & 5 & -6 \\ 0 & 3 & 2 & 7 \\ 1 & 0 & -3 & 2 \end{bmatrix}$

(a) Row 3

(b) Column 1

In Exercises 35–44, find the determinant of the matrix. Use a graphing utility to confirm your result.

**35.** $\begin{bmatrix} 1 & 4 & -2 \\ 3 & 2 & 0 \\ -1 & 4 & 3 \end{bmatrix}$

**36.** $\begin{bmatrix} 2 & -1 & 3 \\ 1 & 4 & 4 \\ 1 & 0 & 2 \end{bmatrix}$

**37.** $\begin{bmatrix} 2 & 4 & 6 \\ 0 & 3 & 1 \\ 0 & 0 & -5 \end{bmatrix}$

**38.** $\begin{bmatrix} -3 & 0 & 0 \\ 7 & 11 & 0 \\ 1 & 2 & 2 \end{bmatrix}$

**39.** $\begin{bmatrix} 3 & 6 & -5 & 4 \\ -2 & 0 & 6 & 0 \\ 1 & 1 & 2 & 2 \\ 0 & 3 & -1 & -1 \end{bmatrix}$

**40.** $\begin{bmatrix} 2 & 6 & 6 & 2 \\ 2 & 7 & 3 & 6 \\ 1 & 5 & 0 & 1 \\ 3 & 7 & 0 & 7 \end{bmatrix}$

**41.** $\begin{bmatrix} 5 & 3 & 0 & 6 \\ 4 & 6 & 4 & 12 \\ 0 & 2 & -3 & 4 \\ 0 & 1 & -2 & 2 \end{bmatrix}$

**42.** $\begin{bmatrix} 1 & 4 & 3 & 2 \\ -5 & 6 & 2 & 1 \\ 0 & 0 & 0 & 0 \\ 3 & -2 & 1 & 5 \end{bmatrix}$

**43.** $\begin{bmatrix} 3 & 2 & 4 & -1 & 5 \\ -2 & 0 & 1 & 3 & 2 \\ 1 & 0 & 0 & 4 & 0 \\ 6 & 0 & 2 & -1 & 0 \\ 3 & 0 & 5 & 1 & 0 \end{bmatrix}$

**44.** $\begin{bmatrix} 5 & 2 & 0 & 0 & -2 \\ 0 & 1 & 4 & 3 & 2 \\ 0 & 0 & 2 & 6 & 3 \\ 0 & 0 & 3 & 4 & 1 \\ 0 & 0 & 0 & 0 & 2 \end{bmatrix}$

In Exercises 45 and 46, find a 4 × 4 *upper* triangular matrix whose determinant is equal to the given value and a 4 × 4 *lower* triangular matrix whose determinant is equal to the given value. Use a graphing utility to confirm your results.

**45.** $-18$

**46.** 28

In Exercises 47 and 48, explain why the determinant of the matrix is equal to zero.

**47.** $\begin{bmatrix} 3 & 4 & -2 & 7 \\ 1 & 3 & -1 & 2 \\ 0 & 5 & 7 & 1 \\ 1 & 3 & -1 & 2 \end{bmatrix}$

**48.** $\begin{bmatrix} 3 & 2 & -1 \\ -6 & -4 & 2 \\ 5 & -7 & 9 \end{bmatrix}$

## Math Matters • Guess the Number

Here is a guessing game that can be made using the four cards shown in the figure (note that the fourth card has numbers on the front *and* back). To play the game, ask someone to think of a number between 1 and 15. Ask the person if the number is on the first card. If it is, place the card face up with the "YES" on top. If it isn't, place the card face up with the "NO" on top. Repeat this with each of the four cards (using the same number), stacking the cards one on top of another. Be sure that the fourth card is used last. After all four cards are in a stack, turn the stack over, The person's number will be the number that is showing through a window in the cards. Can you explain why this card game works?

Card 1

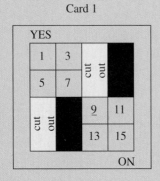

Card 2

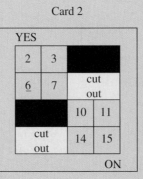

Card 3

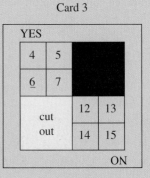

Card 4 (front)

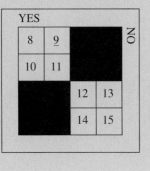

Card 4 (back)

## 7.5    Applications of Matrices and Determinants

### Objectives

- Find the area of a triangle using a determinant.
- Determine whether three points are collinear using a determinant.
- Use a determinant to find the equation of a line.
- Encode and decode a cryptogram using a matrix.

## Area of a Triangle

In this section, you will study some additional applications of matrices and determinants. The first involves a formula for finding the area of a triangle whose vertices are given by three points on a rectangular coordinate system.

> ▶ **Area of a Triangle**
>
> The area of a triangle with vertices $(x_1, y_1)$, $(x_2, y_2)$, and $(x_3, y_3)$ is given by
>
> $$\text{Area} = \pm \frac{1}{2} \begin{vmatrix} x_1 & y_1 & 1 \\ x_2 & y_2 & 1 \\ x_3 & y_3 & 1 \end{vmatrix}$$
>
> where the symbol $(\pm)$ indicates that the appropriate sign should be chosen to yield a positive area.

**Example 1**    Finding the Area of a Triangle

Find the area of the triangle whose vertices are $(1, 0)$, $(2, 2)$, and $(4, 3)$, as shown in Figure 7.1.

**Solution**

Let $(x_1, y_1) = (1, 0)$, $(x_2, y_2) = (2, 2)$, and $(x_3, y_3) = (4, 3)$. Then, to find the area of the triangle, evaluate the determinant

$$\begin{vmatrix} x_1 & y_1 & 1 \\ x_2 & y_2 & 1 \\ x_3 & y_3 & 1 \end{vmatrix} = \begin{vmatrix} 1 & 0 & 1 \\ 2 & 2 & 1 \\ 4 & 3 & 1 \end{vmatrix} = 1 \begin{vmatrix} 2 & 1 \\ 3 & 1 \end{vmatrix} - 0 \begin{vmatrix} 2 & 1 \\ 4 & 1 \end{vmatrix} + 1 \begin{vmatrix} 2 & 2 \\ 4 & 3 \end{vmatrix}$$

$$= 1(-1) - 0(-2) + 1(-2)$$

$$= -3.$$

Using this value, you can conclude that the area of the triangle is

$$\text{Area} = -\frac{1}{2} \begin{vmatrix} 1 & 0 & 1 \\ 2 & 2 & 1 \\ 4 & 3 & 1 \end{vmatrix} = -\frac{1}{2}(-3) = \frac{3}{2}.$$

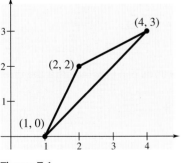

Figure 7.1

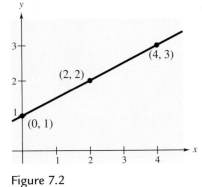

Figure 7.2

## Lines in the Plane

Suppose the three points in Example 1 had been on the same line. What would have happened had the area formula been applied to three such points? The answer is that the determinant would have been zero. Consider, for instance, the three collinear points $(0, 1)$, $(2, 2)$, and $(4, 3)$, as shown in Figure 7.2. The area of the "triangle" that has these three points as vertices is

$$\frac{1}{2}\begin{vmatrix} 0 & 1 & 1 \\ 2 & 2 & 1 \\ 4 & 3 & 1 \end{vmatrix} = \frac{1}{2}\left(0\begin{vmatrix} 2 & 1 \\ 3 & 1 \end{vmatrix} - 1\begin{vmatrix} 2 & 1 \\ 4 & 1 \end{vmatrix} + 1\begin{vmatrix} 2 & 2 \\ 4 & 3 \end{vmatrix}\right)$$

$$= \frac{1}{2}[0(-1) - 1(-2) + 1(-2)] = 0.$$

This result is generalized as follows.

▶ **Test for Collinear Points**

Three points $(x_1, y_1)$, $(x_2, y_2)$, and $(x_3, y_3)$ are collinear (lie on the same line) if and only if

$$\begin{vmatrix} x_1 & y_1 & 1 \\ x_2 & y_2 & 1 \\ x_3 & y_3 & 1 \end{vmatrix} = 0.$$

**Example 2**   Testing for Collinear Points

Determine whether the points $(-2, -2)$, $(1, 1)$, and $(7, 5)$ lie on the same line. (See Figure 7.3.)

**Solution**

Letting $(x_1, y_1) = (-2, -2)$, $(x_2, y_2) = (1, 1)$, and $(x_3, y_3) = (7, 5)$, you have

$$\begin{vmatrix} x_1 & y_1 & 1 \\ x_2 & y_2 & 1 \\ x_3 & y_3 & 1 \end{vmatrix} = \begin{vmatrix} -2 & -2 & 1 \\ 1 & 1 & 1 \\ 7 & 5 & 1 \end{vmatrix}$$

$$= -2\begin{vmatrix} 1 & 1 \\ 5 & 1 \end{vmatrix} - (-2)\begin{vmatrix} 1 & 1 \\ 7 & 1 \end{vmatrix} + 1\begin{vmatrix} 1 & 1 \\ 7 & 5 \end{vmatrix}$$

$$= -2(-4) - (-2)(-6) + 1(-2)$$

$$= -6.$$

Because the value of this determinant is *not* zero, you can conclude that the three points do not lie on the same line.

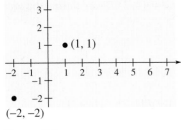

Figure 7.3

The test for collinear points can be adapted to another use. That is, if you are given two points on a rectangular coordinate system, you can find an equation of the line passing through the two points, as follows.

> ▶ **Two-Point Form of the Equation of a Line**
>
> An equation of the line passing through the distinct points $(x_1, y_1)$ and $(x_2, y_2)$ is given by
>
> $$\begin{vmatrix} x & y & 1 \\ x_1 & y_1 & 1 \\ x_2 & y_2 & 1 \end{vmatrix} = 0.$$

**Example 3**    Finding an Equation of a Line

Find an equation of the line passing through the two points $(2, 4)$ and $(-1, 3)$, as shown in Figure 7.4.

**Solution**

Applying the determinant formula for the equation of a line produces

$$\begin{vmatrix} x & y & 1 \\ 2 & 4 & 1 \\ -1 & 3 & 1 \end{vmatrix} = 0.$$

To evaluate this determinant, you can expand by cofactors along the first row to obtain the following.

$$x \begin{vmatrix} 4 & 1 \\ 3 & 1 \end{vmatrix} - y \begin{vmatrix} 2 & 1 \\ -1 & 1 \end{vmatrix} + 1 \begin{vmatrix} 2 & 4 \\ -1 & 3 \end{vmatrix} = x - 3y + 10 = 0$$

So, an equation of the line is

$$x - 3y + 10 = 0.$$

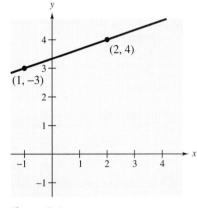

(2, 4)

(1, −3)

**Figure 7.4**

Note that this method of finding the equation of a line works for all lines, including horizontal and vertical lines. For instance, the equation of the vertical line through $(2, 0)$ and $(2, 2)$ is

$$\begin{vmatrix} x & y & 1 \\ 2 & 0 & 1 \\ 2 & 2 & 1 \end{vmatrix} = 0$$

$$4 - 2x = 0$$

$$x = 2.$$

Library of Congress

*Both governments and private industry use codes to transmit messages. Many different systems are used. The first mechanical device for coding and decoding was invented in 1917 by an American engineer, Gilbert Vernam.*

# Cryptography

A **cryptogram** is a message written according to a secret code. (The Greek word *kryptos* means "hidden.") Matrix multiplication can be used to **encode** and **decode** messages. To begin, you need to assign a number to each letter in the alphabet (with 0 assigned to a blank space), as follows.

| | | |
|---|---|---|
| 0 = __ | 9 = I | 18 = R |
| 1 = A | 10 = J | 19 = S |
| 2 = B | 11 = K | 20 = T |
| 3 = C | 12 = L | 21 = U |
| 4 = D | 13 = M | 22 = V |
| 5 = E | 14 = N | 23 = W |
| 6 = F | 15 = O | 24 = X |
| 7 = G | 16 = P | 25 = Y |
| 8 = H | 17 = Q | 26 = Z |

Then the message is converted to numbers and partitioned into **uncoded row matrices,** each having $n$ entries, as demonstrated in Example 4.

### Example 4    Forming Uncoded Row Matrices

Write the uncoded row matrices of order $1 \times 3$ for the message

MEET ME MONDAY.

#### Solution

Partitioning the message (including blank spaces, but ignoring punctuation) into groups of three produces the following uncoded row matrices.

$$\begin{bmatrix} 13 & 5 & 5 \end{bmatrix} \begin{bmatrix} 20 & 0 & 13 \end{bmatrix} \begin{bmatrix} 5 & 0 & 13 \end{bmatrix} \begin{bmatrix} 15 & 14 & 4 \end{bmatrix} \begin{bmatrix} 1 & 25 & 0 \end{bmatrix}$$
   M    E    E    T        M    E        M    O    N    D    A    Y

Note that a blank space is used to fill out the last uncoded row matrix.

---

To **encode** a message, choose an $n \times n$ invertible matrix $A$ and multiply the uncoded row matrices by $A$ to obtain **coded row matrices.** The uncoded matrix should be on the left whereas the encoding matrix $A$ should be on the right. Here is an example.

*Uncoded Matrix*   *Encoding Matrix A*   *Coded Matrix*

$$\begin{bmatrix} 13 & 5 & 5 \end{bmatrix} \begin{bmatrix} 1 & -2 & 2 \\ -1 & 1 & 3 \\ 1 & -1 & -4 \end{bmatrix} = \begin{bmatrix} 13 & -26 & 21 \end{bmatrix}$$

This technique is further illustrated in Example 5.

**Example 5**    Encoding a Message

Use the following matrix to encode the message MEET ME MONDAY.

$$A = \begin{bmatrix} 1 & -2 & 2 \\ -1 & 1 & 3 \\ 1 & -1 & -4 \end{bmatrix}$$

### Solution

The coded row matrices are obtained by multiplying each of the uncoded row matrices found in Example 4 by the matrix $A$, as follows.

| *Uncoded Matrix* | *Encoding Matrix A* | *Coded Matrix* |
|---|---|---|

$$\begin{bmatrix} 13 & 5 & 5 \end{bmatrix} \begin{bmatrix} 1 & -2 & 2 \\ -1 & 1 & 3 \\ 1 & -1 & -4 \end{bmatrix} = \begin{bmatrix} 13 & -26 & 21 \end{bmatrix}$$

$$\begin{bmatrix} 20 & 0 & 13 \end{bmatrix} \begin{bmatrix} 1 & -2 & 2 \\ -1 & 1 & 3 \\ 1 & -1 & -4 \end{bmatrix} = \begin{bmatrix} 33 & -53 & -12 \end{bmatrix}$$

$$\begin{bmatrix} 5 & 0 & 13 \end{bmatrix} \begin{bmatrix} 1 & -2 & 2 \\ -1 & 1 & 3 \\ 1 & -1 & -4 \end{bmatrix} = \begin{bmatrix} 18 & -23 & -42 \end{bmatrix}$$

$$\begin{bmatrix} 15 & 14 & 4 \end{bmatrix} \begin{bmatrix} 1 & -2 & 2 \\ -1 & 1 & 3 \\ 1 & -1 & -4 \end{bmatrix} = \begin{bmatrix} 5 & -20 & 56 \end{bmatrix}$$

$$\begin{bmatrix} 1 & 25 & 0 \end{bmatrix} \begin{bmatrix} 1 & -2 & 2 \\ -1 & 1 & 3 \\ 1 & -1 & -4 \end{bmatrix} = \begin{bmatrix} -24 & 23 & 77 \end{bmatrix}$$

So, the sequence of coded row matrices is

$$\begin{bmatrix} 13 & -26 & 21 \end{bmatrix} \begin{bmatrix} 33 & -53 & -12 \end{bmatrix} \begin{bmatrix} 18 & -23 & -42 \end{bmatrix} \begin{bmatrix} 5 & -20 & 56 \end{bmatrix} \begin{bmatrix} -24 & 23 & 77 \end{bmatrix}.$$

Finally, removing the matrix notation produces the following cryptogram.

$$13 \ -26 \ 21 \ 33 \ -53 \ -12 \ 18 \ -23 \ -42 \ 5 \ -20 \ 56 \ -24 \ 23 \ 77$$

---

For those who do not know the matrix $A$, decoding the cryptogram found in Example 5 is difficult. But for an authorized receiver who knows the matrix $A$, decoding is simple. The receiver need only multiply the coded row matrices by $A^{-1}$ (on the right) to retrieve the uncoded row matrices. Here is an example.

$$\underbrace{\begin{bmatrix} 13 & -26 & 21 \end{bmatrix}}_{\text{Coded}} A^{-1} = \underbrace{\begin{bmatrix} 13 & 5 & 5 \end{bmatrix}}_{\text{Uncoded}}$$

Example 6    Decoding a Message

Use the inverse of the matrix $A = \begin{bmatrix} 1 & -2 & 2 \\ -1 & 1 & 3 \\ 1 & -1 & -4 \end{bmatrix}$ to decode the cryptogram

13  −26  21  33  −53  −12  18  −23  −42  5  −20  56  −24  23  77.

Solution

Partition the message into groups of three to form the coded row matrices. Multiply each coded row matrix on the right by $A^{-1}$ to obtain the decoded row matrices.

| Coded Matrix | Decoding Matrix $A^{-1}$ | Decoded Matrix |
|---|---|---|

$$\begin{bmatrix} 13 & -26 & 21 \end{bmatrix} \begin{bmatrix} -1 & -10 & -8 \\ -1 & -6 & -5 \\ 0 & -1 & -1 \end{bmatrix} = \begin{bmatrix} 13 & 5 & 5 \end{bmatrix}$$

$$\begin{bmatrix} 33 & -53 & -12 \end{bmatrix} \begin{bmatrix} -1 & -10 & -8 \\ -1 & -6 & -5 \\ 0 & -1 & -1 \end{bmatrix} = \begin{bmatrix} 20 & 0 & 13 \end{bmatrix}$$

$$\begin{bmatrix} 18 & -23 & -42 \end{bmatrix} \begin{bmatrix} -1 & -10 & -8 \\ -1 & -6 & -5 \\ 0 & -1 & -1 \end{bmatrix} = \begin{bmatrix} 5 & 0 & 13 \end{bmatrix}$$

$$\begin{bmatrix} 5 & -20 & 56 \end{bmatrix} \begin{bmatrix} -1 & -10 & -8 \\ -1 & -6 & -5 \\ 0 & -1 & -1 \end{bmatrix} = \begin{bmatrix} 15 & 14 & 4 \end{bmatrix}$$

$$\begin{bmatrix} -24 & 23 & 77 \end{bmatrix} \begin{bmatrix} -1 & -10 & -8 \\ -1 & -6 & -5 \\ 0 & -1 & -1 \end{bmatrix} = \begin{bmatrix} 1 & 25 & 0 \end{bmatrix}$$

So, the message is as follows.

$$\begin{bmatrix} 13 & 5 & 5 \end{bmatrix} \begin{bmatrix} 20 & 0 & 13 \end{bmatrix} \begin{bmatrix} 5 & 0 & 13 \end{bmatrix} \begin{bmatrix} 15 & 14 & 4 \end{bmatrix} \begin{bmatrix} 1 & 25 & 0 \end{bmatrix}$$

M  E  E    T    M  E    M    O  N  D    A  Y

---

**Discussing the Concept    Decoding**

Show how to use a graphing utility with matrix operations to decode the following cryptogram. (Use the matrix in Example 6.)

12  −25  15    28  −32  −89    10  −10  −49    12

−12  −51  17  −31    10    10  −28    55    4  −8  8

# Warm Up

The following warm-up exercises involve skills that were covered in earlier sections. You will use these skills in the exercise set for this section.

In Exercises 1–6, evaluate the determinant.

**1.** $\begin{vmatrix} 4 & 3 \\ -3 & -2 \end{vmatrix}$

**2.** $\begin{vmatrix} 10 & -20 \\ -1 & 2 \end{vmatrix}$

**3.** $\begin{vmatrix} 4 & 0 \\ -3 & -2 \end{vmatrix}$

**4.** $\begin{vmatrix} x & x^2 \\ 1 & 2x \end{vmatrix}$

**5.** $\begin{vmatrix} 4 & 0 & -2 \\ 3 & 1 & 2 \\ -8 & 0 & 6 \end{vmatrix}$

**6.** $\begin{vmatrix} 3 & 2 & 5 \\ 0 & 0 & -4 \\ -6 & 1 & 1 \end{vmatrix}$

In Exercises 7 and 8, find the inverse of the matrix.

**7.** $A = \begin{bmatrix} 1 & 3 \\ 2 & 7 \end{bmatrix}$

**8.** $A = \begin{bmatrix} 10 & 5 & -2 \\ -4 & -2 & 1 \\ 1 & 1 & 0 \end{bmatrix}$

In Exercises 9 and 10, perform the indicated matrix multiplication.

**9.** $\begin{bmatrix} 0.1 & 0.2 & 0.2 \\ 0.4 & 0.3 & 0.5 \\ 0.5 & 0.5 & 0.3 \end{bmatrix} \begin{bmatrix} 0.4 \\ 0.5 \\ 0.1 \end{bmatrix}$

**10.** $\begin{bmatrix} 2 & 5 & 8 \end{bmatrix} \begin{bmatrix} 1 & 2 & -1 \\ 1 & 2 & 2 \\ 2 & 5 & 0 \end{bmatrix}$

## 7.5   Exercises

In Exercises 1–10, use a determinant to find the area of the triangle with the given vertices.

**1.**

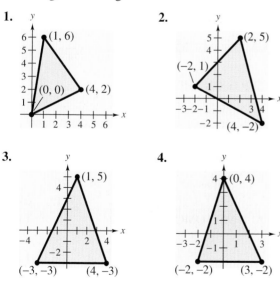

**2.**

**3.**

**4.**

**5.**

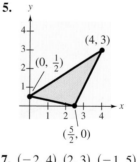

**6.**

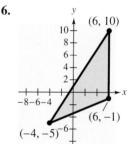

**7.** $(-2, 4), (2, 3), (-1, 5)$

**8.** $(0, -2), (-1, 4), (3, 5)$

**9.** $(-3, 5), (2, 6), (3, -5)$

**10.** $(-2, 4), (1, 5), (3, -2)$

In Exercises 11 and 12, find a value of $x$ such that the triangle has an area of 4.

**11.** $(-5, 1), (0, 2), (-2, x)$

**12.** $(-4, 2), (-3, 5), (-1, x)$

**13.** *Area of a Region* A large region of forest has been infected with gypsy moths. The region is roughly triangular, as shown in the figure. From the northernmost vertex *A* of the region, the distances to the other vertices are 25 miles south and 10 miles east (for vertex *B*), and 20 miles south and 28 miles east (for vertex *C*). Use a graphing utility to approximate the number of square miles in this region.

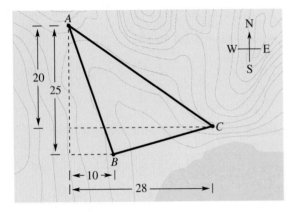

**14.** *Area of a Region* You own a triangular tract of land, as shown in the figure. To estimate the number of square feet in the tract, you start at one vertex, walk 65 feet east and 50 feet north to the second vertex, and then walk 85 feet west and 30 feet north to the third vertex. Use a graphing utility to determine how many square feet there are in the tract of land.

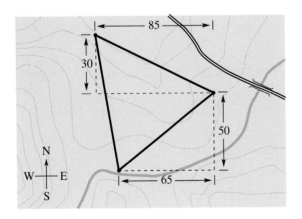

In Exercises 15–20, use a determinant to determine whether the points are collinear.

**15.** $(-4, -7), (0, -4), (4, -1)$
**16.** $(2, 4), (4, 5), (-2, 2)$
**17.** $(-1, -7), (0, -3), (1, 2)$
**18.** $(-2, -11), (4, 13), (2, 5)$
**19.** $(1, 7), (0, 4), (-1, 2)$
**20.** $(4, 3), (3, 1), (2, -1)$

In Exercises 21–26, use a determinant to find an equation of the line through the points.

**21.** $(0, 0), (5, 3)$
**22.** $(0, 0), (-2, 2)$
**23.** $(-4, 3), (2, 1)$
**24.** $(10, 7), (-2, -7)$
**25.** $\left(-\frac{1}{2}, 3\right), \left(\frac{5}{2}, 1\right)$
**26.** $\left(\frac{2}{3}, 4\right), (6, 12)$

In Exercises 27 and 28, find *x* such that the points are collinear.

**27.** $(2, -5), (4, x), (5, -2)$
**28.** $(-6, 2), (-5, x), (-3, 5)$

In Exercises 29 and 30, find the uncoded $1 \times 2$ row matrices for the message. Then encode the message using the matrix.

|  *Message* | *Matrix* |
|---|---|
| **29.** COME HOME SOON | $\begin{bmatrix} 1 & 2 \\ 3 & 5 \end{bmatrix}$ |
| **30.** HELP IS ON THE WAY | $\begin{bmatrix} -2 & 3 \\ -1 & 1 \end{bmatrix}$ |

In Exercises 31 and 32, find the uncoded $1 \times 3$ row matrices for the message. Then encode the message using the matrix.

|  *Message* | *Matrix* |
|---|---|
| **31.** CALL ME TOMORROW | $\begin{bmatrix} 1 & -1 & 0 \\ 1 & 0 & -1 \\ -6 & 2 & 3 \end{bmatrix}$ |
| **32.** PLEASE SEND MONEY | $\begin{bmatrix} 4 & 2 & 1 \\ -3 & -3 & -1 \\ 3 & 2 & 1 \end{bmatrix}$ |

In Exercises 33–36, write a cryptogram for the message using the matrix

$$A = \begin{bmatrix} 1 & 2 & 2 \\ 3 & 7 & 9 \\ -1 & -4 & -7 \end{bmatrix}.$$

**33.** LANDING SUCCESSFUL

**34.** BEAM ME UP SCOTTY

**35.** HAPPY BIRTHDAY

**36.** OPERATION OVERLORD

In Exercises 37–40, use $A^{-1}$ to decode the cryptogram.

**37.** $A = \begin{bmatrix} 1 & 2 \\ 3 & 5 \end{bmatrix}$

11, 21, 64, 112, 25, 50, 29, 53, 23, 46, 40, 75, 55, 92

**38.** $A = \begin{bmatrix} 2 & 3 \\ 3 & 4 \end{bmatrix}$

85, 120, 6, 8, 10, 15, 84, 117, 42, 56, 90, 125, 60, 80, 30, 45, 19, 26

**39.** $A = \begin{bmatrix} 4 & 2 & 1 \\ -3 & -3 & -1 \\ 3 & 2 & 1 \end{bmatrix}$

110, 59, 32, 79, 40, 20, 95, 50, 25, 99, 53, 29, -22, -32, -9

**40.** $A = \begin{bmatrix} 1 & -1 & 0 \\ 1 & 0 & -1 \\ -6 & 2 & 3 \end{bmatrix}$

9, -1, -9, 38, -19, -19, 28, -9, -19, -80, 25, 41, -64, 21, 31, -7, -4, 7

In Exercises 41 and 42, decode the cryptogram by using the inverse of the matrix

$$A = \begin{bmatrix} 1 & 2 & 2 \\ 3 & 7 & 9 \\ -1 & -4 & -7 \end{bmatrix}.$$

**41.** 20, 17, -15, -9, -44, -83, 64, 136, 157, 24, 31, 12, 4, -37, -102

**42.** 13, -9, -59, 61, 112, 106, -17, -73, -131, 11, 24, 29, 65, 144, 172

**43.** The following cryptogram was encoded with a $2 \times 2$ matrix.

8, 21, -15, -10, -13, -13, 5, 10, 5, 25, 5, 19, -1, 6, 20, 40, -18, -18, 1, 16

The last word of the message is __RON. What is the message?

**44.** The following cryptogram was encoded with a $2 \times 2$ matrix.

5, 2, 25, 11, -2, -7, -15, -15, 32, 14, -8, -13, 38, 19, -19, -19, 37, 16

The last word of the message is __SUE. What is the message?

## Chapter Project

## Market Share

Your advertising group has been hired by Company C, a cable/internet service provider, to develop an advertising campaign that will increase its market share from 10% to at least 30% in one year. Currently, your client competes with two other cable/internet service providers, Company A and Company B. Your group has designed a campaign that you believe will meet the 30% objective. With this new campaign, you predict that each month

(1) 92% of A's subscribers will remain with A, 3% will switch to B, and 5% will switch to C;

(2) 91% of B's subscribers will remain with B, 7% will switch to A, and 2% will change to C; and

(3) 93% of C's subscribers will remain with C, 6% will change to B, and 1% will change to A.

$$x_0 = \begin{bmatrix} 0.2 \\ 0.7 \\ 0.1 \end{bmatrix} \begin{matrix} \text{Company A} \\ \text{Company B} \\ \text{Company C} \end{matrix}$$

The current shares of the market are given in the matrix at the left. After one month, the shares of the three companies will be given by

$$x_1 = px_0 = \begin{bmatrix} 0.92 & 0.03 & 0.05 \\ 0.07 & 0.91 & 0.02 \\ 0.01 & 0.06 & 0.93 \end{bmatrix} \begin{bmatrix} 0.2 \\ 0.7 \\ 0.1 \end{bmatrix} = \begin{bmatrix} 0.210 \\ 0.653 \\ 0.137 \end{bmatrix} \begin{matrix} \text{Company A} \\ \text{Company B} \\ \text{Company C} \end{matrix}.$$

Use this information to investigate the following questions.

1. *Making a Table*   Construct a table that shows the market share for each of the three companies during each of the first 12 months. The matrix operations on your graphing utility can help you with the calculations.

2. *Fitting a Model to the Data*   Enter the results for Company C from Question 1 into the statistical package of your graphing utility.

   (a) Make a scatter plot of the data.

   (b) What type of model do you think best fits the data? Select a regression program on your graphing utility to model the data.

   (c) Graph the model you found in part (b) for the data for Company C.

   (d) Repeat (a), (b), and (c) using the data for Company A and Company B from Question 1.

3. *Comparing Models*   Graph the models for Companies A, B, and C in the same viewing window. Will Company C's market share reach 30% within 12 months? Based on the model for Company C's market share, will Company C maintain a 30% or better market share?

4. *Making Predictions*   Are the models you found good predictors over long periods of time? Would you expect this type of market to fluctuate or be static? As Company C begins to take more of the market share, what action do you predict Companies A and B will take? What possible effect on this system could their actions have?

## CHAPTER SUMMARY

After studying this chapter, you should have acquired the following skills.
These skills are keyed to the Review Exercises that begin on page 486.
Answers to odd-numbered Review Exercises are given in the back of the book.

**7.1** · Determine the order of a matrix. — Review Exercises 1, 2

· Perform elementary row operations on a matrix. — Review Exercises 3–6

· Write a matrix in row-echelon form or reduced row-echelon form. — Review Exercises 3–6

· Solve a system of linear equations using Gaussian elimination or Gauss-Jordan elimination. — Review Exercises 7–14, 17, 18

**7.2** · Add or subtract two matrices. — Review Exercises 19–22

· Multiply a matrix by a scalar. — Review Exercises 19–22, 37, 38

· Find the product of two matrices. — Review Exercises 23–32, 39, 40

· Solve a matrix equation. — Review Exercises 35, 36

**7.3** · Verify that a matrix is the inverse of a given matrix. — Review Exercises 41, 42

· Find the inverse of a matrix. — Review Exercises 43–46

· Use an inverse matrix to solve a system of linear equations. — Review Exercises 47–54

**7.4** · Evaluate the determinant of a $2 \times 2$ matrix. — Review Exercises 55–58

· Find the determinant of a square matrix. — Review Exercises 59, 60, 65

· Find the determinant of a triangular matrix. — Review Exercises 61–64

**7.5** · Find the area of a triangle using a determinant. — Review Exercises 67, 68

· Determine whether three points are collinear using a determinant. — Review Exercises 69–72

· Use a determinant to find the equation of a line. — Review Exercises 73, 74

· Encode and decode a cryptogram using a matrix. — Review Exercises 75–78

In Exercises 1 and 2, determine the order of the matrix.

**1.** $\begin{bmatrix} 3 & 7 & 4 & -2 & -3 \\ -1 & 8 & 6 & 1 & 2 \end{bmatrix}$

**2.** $\begin{bmatrix} 5 \\ -1 \\ 2 \\ 4 \end{bmatrix}$

In Exercises 3 and 4, write the matrix in row-echelon form.

**3.** $\begin{bmatrix} 1 & 3 & 0 & 2 \\ 3 & 10 & 1 & 8 \\ 2 & 3 & 3 & 10 \end{bmatrix}$

**4.** $\begin{bmatrix} 1 & 2 & -1 & 0 \\ -2 & -3 & 3 & 4 \\ 4 & 0 & 1 & 3 \end{bmatrix}$

In Exercises 5 and 6, write the matrix in reduced row-echelon form.

**5.** $\begin{bmatrix} 1 & 2 & 3 \\ -2 & 0 & 2 \\ 2 & 1 & 2 \end{bmatrix}$

**6.** $\begin{bmatrix} 2 & 3 & 1 & -5 \\ 1 & 0 & 5 & 2 \\ -1 & 4 & 3 & 6 \\ 0 & -2 & 6 & -8 \end{bmatrix}$

In Exercises 7–14, use matrices to solve the system.

**7.** $4x - 3y = 18$
$x + y = 1$

**8.** $2x + 4y = 16$
$-x + 3y = 17$

**9.** $2x + 3y - z = 13$
$3x + z = 8$
$x - 2y + 3z = -4$

**10.** $3x + 4y + 2z = 5$
$2x + 3y = 7$
$2y - 3z = 12$

**11.** $x + 2y + 2z = 10$
$2x + 3y + 5z = 20$

**12.** $3x + 10y + 4z = 20$
$x + 3y - 2z = 8$

**13.** $2x + y - 3z = 4$
$x + 2y + 2z = 10$
$x - 2z = 12$
$x + y + z = 6$

**14.** $2x + 4y + 2z = 10$
$x + 3z = 9$
$3x - 2y = 4$
$x + y + z = 8$

**15.** *Reasoning* Discuss how you know that the systems of equations in Exercises 11 and 12 will have infinitely many solutions.

**16.** *Reasoning* Discuss the three possible outcomes for the solution of a system of linear equations.

**17.** *Borrowing Money* A company borrowed $600,000 to expand its product line. Some of the money was borrowed at 9%, some at 10%, and some at 12%. How much was borrowed at each rate if the annual interest was $63,000 and the amount borrowed at 10% was three times the amount borrowed at 9%?

**18.** *Borrowing Money* A company borrowed $700,000 to expand its product line. Some of the money was borrowed at 9.5%, some at 10.5%, and some at 11%. How much was borrowed at each rate if the annual interest was $70,000 and the amount borrowed at 9.5% was four times the amount borrowed at 11%?

In Exercises 19–22, find (a) $A + B$, (b) $A - B$, (c) $4A$, and (d) $4A - 3B$.

**19.** $A = \begin{bmatrix} -1 & 5 \\ 2 & 1 \end{bmatrix}$, $B = \begin{bmatrix} 4 & 2 \\ -6 & 3 \end{bmatrix}$

**20.** $A = \begin{bmatrix} 1 & 0 & 2 \\ -1 & 3 & 5 \\ 2 & -2 & 3 \end{bmatrix}$, $B = \begin{bmatrix} 2 & 0 & 1 \\ 3 & -4 & 6 \\ 1 & 2 & -3 \end{bmatrix}$

**21.** $A = \begin{bmatrix} 1 & 3 & -2 & 6 \\ 0 & 1 & 3 & 2 \end{bmatrix}$,
$B = \begin{bmatrix} 2 & 1 & 4 & -5 \\ 3 & -6 & 3 & -2 \end{bmatrix}$

**22.** $A = \begin{bmatrix} 3 \\ -2 \\ 3 \end{bmatrix}$, $B = \begin{bmatrix} -1 \\ 4 \\ 5 \end{bmatrix}$

In Exercises 23–26, find (a) $AB$, (b) $BA$, and, if possible, (c) $A^2$. (*Note:* $A^2 = AA$.)

**23.** $A = \begin{bmatrix} 1 & -3 & 4 \end{bmatrix}$, $B = \begin{bmatrix} 2 \\ -2 \\ -1 \end{bmatrix}$

**24.** $A = \begin{bmatrix} 2 & -3 \\ 4 & 5 \end{bmatrix}$, $B = \begin{bmatrix} -1 & 0 \\ 2 & -1 \end{bmatrix}$

**25.** $A = \begin{bmatrix} 1 & 0 & 2 \\ 3 & 1 & -2 \\ 1 & 1 & 1 \end{bmatrix}$, $B = \begin{bmatrix} 2 & 0 & 0 \\ 1 & -2 & 1 \\ 5 & 4 & -2 \end{bmatrix}$

**26.** $A = \begin{bmatrix} 0 & 2 & 3 \\ 1 & -1 & 0 \\ 1 & 2 & 1 \end{bmatrix}$, $B = \begin{bmatrix} 1 & 1 & 4 \\ 2 & 3 & 0 \\ 4 & 3 & 2 \end{bmatrix}$

In Exercises 27–32, find $AB$, if possible.

**27.** $A = \begin{bmatrix} 1 & 4 \\ -2 & -1 \\ 3 & 2 \end{bmatrix}$, $B = \begin{bmatrix} -4 \\ 3 \end{bmatrix}$

**28.** $A = \begin{bmatrix} 3 \\ 2 \\ 4 \\ 6 \end{bmatrix}$, $B = \begin{bmatrix} 2 & 0 & -1 \end{bmatrix}$

**29.** $A = \begin{bmatrix} 4 & 0 & 0 \\ 0 & 3 & 0 \\ 0 & 0 & -2 \end{bmatrix}$, $B = \begin{bmatrix} \frac{1}{4} & 0 & 0 \\ 0 & \frac{1}{3} & 0 \\ 0 & 0 & -\frac{1}{2} \end{bmatrix}$

**30.** $A = \begin{bmatrix} 3 & 1 \\ 4 & 7 \\ 1 & 1 \end{bmatrix}$, $B = \begin{bmatrix} 1 & 2 & -2 \\ 3 & 4 & 0 \\ 0 & 1 & 0 \end{bmatrix}$

**31.** $A = \begin{bmatrix} 1 & 2 & 3 & 6 & -1 \\ 2 & 8 & 0 & 0 & 2 \end{bmatrix}$,

$B = \begin{bmatrix} 3 & 2 \\ 4 & -1 \end{bmatrix}$

**32.** $A = \begin{bmatrix} 0 & 0 & 2 \\ 1 & 0 & 6 \\ 0 & 2 & 2 \end{bmatrix}$, $B = \begin{bmatrix} 3 & 4 & 0 & 1 \\ 2 & 1 & 0 & 0 \\ 0 & 0 & 1 & 1 \end{bmatrix}$

**33.** *Think About It*    Give an example of two matrices $A$ and $B$ such that the products of $AB$ and $BA$ exist.

**34.** *Think About It*    Give an example of two matrices $A$ and $B$ such that the product of $AB$ exists but $BA$ does not exist.

In Exercises 35 and 36, solve for $X$ given

$A = \begin{bmatrix} 1 & -2 \\ 0 & 1 \\ 2 & 3 \end{bmatrix}$    and    $B = \begin{bmatrix} 0 & 1 \\ 1 & 1 \\ 3 & 5 \end{bmatrix}$.

**35.** $X = 5A - 3B$        **36.** $3X = 2A + 3B$

**37.** *Factory Production*    A corporation has four factories, each of which manufactures three products. The number of units of product $i$ produced at factory $j$ in one day is represented by $a_{ij}$ in the matrix

$A = \begin{bmatrix} 80 & 70 & 90 & 40 \\ 50 & 30 & 80 & 20 \\ 90 & 60 & 100 & 50 \end{bmatrix}$.

Find the production levels if production is increased by 20%.

**38.** *Factory Production*    A corporation has three factories, each of which manufactures four products. The number of units of product $j$ produced at factory $i$ in one day is represented by $a_{ij}$ in the matrix

$A = \begin{bmatrix} 30 & 80 & 70 & 20 \\ 50 & 100 & 90 & 90 \\ 60 & 70 & 80 & 100 \end{bmatrix}$.

Find the production levels if production is decreased by 10%. (*Hint:* Because a 10% decrease corresponds to $100\% - 10\%$, multiply the given matrix by 0.90.)

**39.** *Inventory Levels*    A company sells four different models of car sound systems through three retail outlets. The inventories of the four models at the three outlets are given by matrix $S$.

Model

$S = \begin{array}{c} \\ \\ \\ \end{array} \begin{matrix} A & B & C & D \\ \begin{bmatrix} 3 & 2 & 1 & 4 \\ 1 & 3 & 4 & 3 \\ 5 & 3 & 2 & 2 \end{bmatrix} & & & \end{matrix} \begin{array}{c} 1 \\ 2 \\ 3 \end{array} \text{Outlet}$

The wholesale and retail prices of the four models are given by matrix $T$.

Price

Wholesale    Retail

$T = \begin{bmatrix} 300 & 500 \\ 400 & 650 \\ 200 & 350 \\ 800 & 1200 \end{bmatrix} \begin{array}{c} A \\ B \\ C \\ D \end{array} \text{Model}$

Use a graphing utility to compute $ST$ and interpret the result.

**40.** *Labor/Wage Requirements* A company that manufactures racing bicycles has the following labor-hour and wage requirements.

*Labor-Hour Requirements (per bicycle)*

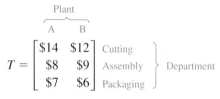

$$S = \begin{bmatrix} 0.9 \text{ hour} & 0.8 \text{ hour} & 0.2 \text{ hour} \\ 1.5 \text{ hours} & 1.0 \text{ hour} & 0.4 \text{ hour} \\ 3.5 \text{ hours} & 3.0 \text{ hours} & 0.5 \text{ hour} \end{bmatrix} \begin{matrix} \text{Basic} \\ \text{Light} \\ \text{Ultra-} \\ \text{Light} \end{matrix} \Big\} \text{Models}$$

*Wage Requirements (per hour)*

$$T = \begin{bmatrix} \$14 & \$12 \\ \$8 & \$9 \\ \$7 & \$6 \end{bmatrix} \begin{matrix} \text{Cutting} \\ \text{Assembly} \\ \text{Packaging} \end{matrix} \Big\} \text{Department}$$

(a) What is the labor cost for a light racing bicycle at Plant A?

(b) What is the labor cost for an ultralight racing bicycle at Plant B?

(c) Use a graphing utility to compute $ST$ and interpret the result.

In Exercises 41 and 42, show that $B$ is the inverse of $A$.

**41.** $A = \begin{bmatrix} 1 & 2 & 1 \\ 3 & 6 & 4 \\ 0 & 1 & 3 \end{bmatrix}, B = \begin{bmatrix} -14 & 5 & -2 \\ 9 & -3 & 1 \\ -3 & 1 & 0 \end{bmatrix}$

**42.** $A = \begin{bmatrix} 2 & 0 & 1 & 2 \\ 3 & 0 & 0 & 1 \\ -1 & 1 & 2 & 0 \\ 0 & -1 & 2 & 2 \end{bmatrix}$,

$B = \dfrac{1}{9}\begin{bmatrix} -4 & 6 & 1 & 1 \\ 10 & -6 & 2 & -7 \\ -7 & 6 & 4 & 4 \\ 12 & -9 & -3 & -3 \end{bmatrix}$

In Exercises 43–46, find the inverse of the matrix (if it exists).

**43.** $\begin{bmatrix} 1 & 3 \\ 2 & 5 \end{bmatrix}$  **44.** $\begin{bmatrix} -2 & 1 \\ 4 & 3 \end{bmatrix}$

**45.** $\begin{bmatrix} -1 & 0 & 0 & 0 \\ 0 & 2 & 0 & 0 \\ 0 & 0 & 4 & 0 \\ 0 & 0 & 0 & 6 \end{bmatrix}$  **46.** $\begin{bmatrix} 3 & 2 & 2 \\ 0 & 2 & 1 \\ 1 & 0 & 1 \end{bmatrix}$

In Exercises 47 and 48, use an inverse matrix to solve the system of linear equations. (Use the inverse matrix found in Exercise 43.)

**47.** $\begin{aligned} x + 3y &= 15 \\ 2x + 5y &= 26 \end{aligned}$  **48.** $\begin{aligned} x + 3y &= 7 \\ 2x + 5y &= 11 \end{aligned}$

In Exercises 49 and 50, use an inverse matrix to solve the system of linear equations. (Use the inverse matrix found in Exercise 46.)

**49.** $\begin{aligned} 3x + 2y + 2z &= 13 \\ 2y + z &= 4 \\ x + z &= 5 \end{aligned}$  **50.** $\begin{aligned} 3x + 2y + 2z &= 12 \\ 2y + z &= 13 \\ x + z &= 3 \end{aligned}$

*Computer Models* In Exercises 51 and 52, consider a company that produces three computer models. Model A requires 2 units of plastic, 2 units of computer chips, and 1 unit of computer "cards." Model B requires 1 unit of plastic, 3 units of computer chips, and 2 units of computer "cards." Model C requires 3 units of plastic, 2 units of computer chips, and 2 units of computer "cards." A system of linear equations (where $x$, $y$, and $z$ represent models A, B, and C, respectively) is as follows.

$2x + y + 3z = $ (units of plastic)

$2x + 3y + 2z = $ (units of computer chips)

$x + 2y + 2z = $ (units of computer "cards")

Use the inverse of the coefficient matrix of this system to find the numbers of models A, B, and C that the company can produce with the given amounts of components.

**51.** 700 units of plastic

900 units of computer chips

600 units of computer "cards"

**52.** 600 units of plastic

800 units of computer chips

500 units of computer "cards"

*Investment Portfolio*    In Exercises 53 and 54, consider a person who invests in blue-chip stocks, common stocks, and municipal bonds. The average yields are 8% on blue-chip stocks, 6% on common stocks, and 7% on municipal bonds. Twice as much is invested in municipal bonds as in common stocks. A system of linear equations (where $x$, $y$, and $z$ represent the amounts invested in blue-chip stocks, common stocks, and municipal bonds, respectively) is as follows.

$$x + \quad y + \quad z = \text{(total investment)}$$
$$0.08x + 0.06y + 0.07z = \text{(annual return)}$$
$$2y - \quad z = 0$$

Use the inverse of the coefficient matrix of this system to find the amount invested in each type of stock or bond for the indicated total investment.

**53.** Total investment = \$50,000; annual return = \$3720
**54.** Total investment = \$45,000; annual return = \$3380

In Exercises 55–66, find the determinant of the matrix.

**55.** $\begin{bmatrix} 8 & 4 \\ 3 & 2 \end{bmatrix}$    **56.** $\begin{bmatrix} 7 & 2 \\ 9 & -3 \end{bmatrix}$

**57.** $\begin{bmatrix} 5 & 2 \\ 0 & 0 \end{bmatrix}$    **58.** $\begin{bmatrix} 3 & 0 \\ 0 & -7 \end{bmatrix}$

**59.** $\begin{bmatrix} 1 & 2 & 3 \\ 8 & 6 & 7 \\ 0 & 2 & -1 \end{bmatrix}$    **60.** $\begin{bmatrix} -2 & 3 & 3 \\ -1 & 0 & 5 \\ 1 & 2 & -1 \end{bmatrix}$

**61.** $\begin{bmatrix} 1 & 3 & 2 & 4 \\ 0 & -1 & 2 & 2 \\ 0 & 0 & 3 & 0 \\ 0 & 0 & 0 & 4 \end{bmatrix}$    **62.** $\begin{bmatrix} 2 & 0 & 0 & 0 \\ 3 & 4 & 0 & 0 \\ 5 & 1 & -2 & 0 \\ 6 & 3 & 1 & 1 \end{bmatrix}$

**63.** $\begin{bmatrix} 3 & 0 & 0 & 0 \\ 0 & 2 & 0 & 0 \\ 0 & 0 & -1 & 0 \\ 0 & 0 & 0 & -10 \end{bmatrix}$    **64.** $\begin{bmatrix} 2 & 0 & 0 & 0 \\ 0 & 5 & 0 & 0 \\ 0 & 0 & 4 & 0 \\ 0 & 0 & 0 & 3 \end{bmatrix}$

**65.** $\begin{bmatrix} 1 & 3 & 0 & 7 \\ 5 & 2 & 4 & 2 \\ 0 & 1 & 0 & 1 \\ 2 & 1 & 0 & 0 \end{bmatrix}$

**66.** $\begin{bmatrix} 1 & 0 & 0 & 0 & 0 \\ 0 & 2 & 0 & 0 & 0 \\ 0 & 0 & 0 & 8 & 0 \\ 0 & 0 & 0 & 0 & \frac{1}{2} \end{bmatrix}$

In Exercises 67 and 68, use a determinant to find the area of the triangle with the given vertices.

**67.** $(3, 4), (2, -3), (-1, -4)$
**68.** $(1, -4), (-2, 3), (0, 6)$

In Exercises 69–72, use a determinant to ascertain whether the points are collinear.

**69.** $(0, 3), (1, 5), (2, 8)$
**70.** $(2, 6), (-2, 3), (0, 5)$
**71.** $(-4, 1), (6, 6), (0, 3)$
**72.** $(-3, -1), (0, 5), (-4, -3)$

In Exercises 73 and 74, use a determinant to find an equation of the line through the points $(x_1, y_1)$ and $(x_2, y_2)$.

**73.** $(-7, 3), (8, 2)$    **74.** $(5, -4), (-3, 2)$

In Exercises 75 and 76, use the matrix to encode the message.

| *Message* | *Matrix* |
|---|---|
| **75.** TRANSMIT NOW | $\begin{bmatrix} 2 & 3 \\ 3 & 4 \end{bmatrix}$ |
| **76.** CALL AT MIDNIGHT | $\begin{bmatrix} 1 & 2 & 2 \\ 3 & 7 & 9 \\ -1 & -4 & -7 \end{bmatrix}$ |

In Exercises 77 and 78, use $A^{-1}$ to decode the cryptogram.

**77.** $A = \begin{bmatrix} 1 & 2 \\ -1 & 3 \end{bmatrix}$

14, 53, $-17$, 96, 5, 10, 12, 64, 5, 10, 3, 11, 25, 50

**78.** $A = \begin{bmatrix} 1 & -1 & 0 \\ 1 & 0 & -1 \\ -6 & 2 & 3 \end{bmatrix}$

$-14$, $-1$, 10, $-38$, 2, 27, $-94$, 18, 57, 7, $-11$, $-1$, $-96$, 20, 57, $-74$, 23, 35, 17, $-12$, $-5$

## Chapter Test

Take this test as you would take a test in class. After you are done, check your work against the answers given in the back of the book.

In Exercises 1 and 2, write an augmented matrix that represents the system.

**1.** $3x + y = 1$
$\phantom{3}x - y = 7$

**2.** $3x + 4y + 2z = \phantom{-}4$
$\phantom{3}2x + 3y \phantom{+ 2z} = -2$
$\phantom{3x + }2y - 3z = -13$

In Exercises 3–5, use matrices to solve the system.

**3.** $\phantom{5}x + 2y + 3z = 16$
$5x + 4y - \phantom{3}z = 22$

**4.** $x - 2y + \phantom{3}z = 14$
$\phantom{x - 2}y - 3z = \phantom{1}2$
$\phantom{x - 2y + 3}z = -6$

**5.** $2x - 3y + z = \phantom{-}14$
$\phantom{2}x + 2y \phantom{+ z} = -4$
$\phantom{2x + }y - z = -4$

In Exercises 6–9, use the matrices to find the indicated matrix.

$$A = \begin{bmatrix} 1 & 3 \\ 2 & 4 \end{bmatrix}, \ B = \begin{bmatrix} 2 & -1 & 3 \\ 4 & 0 & 1 \end{bmatrix}, \ C = \begin{bmatrix} 0 & -2 \\ 3 & 5 \end{bmatrix}, \ D = \begin{bmatrix} 3 \\ 2 \\ -1 \end{bmatrix}$$

**6.** $2A + C$    **7.** $AB$    **8.** $BD$    **9.** $A^2$

In Exercises 10–12, find the inverse of the matrix.

**10.** $A = \begin{bmatrix} 2 & -1 \\ -3 & 4 \end{bmatrix}$    **11.** $A = \begin{bmatrix} 1 & 0 \\ 0 & 1 \end{bmatrix}$    **12.** $A = \begin{bmatrix} 3 & 4 & 2 \\ 2 & 3 & 0 \\ 0 & 2 & -3 \end{bmatrix}$

In Exercises 13–15, find the determinant of the matrix.

**13.** $\begin{bmatrix} 3 & -1 \\ 4 & 7 \end{bmatrix}$    **14.** $\begin{bmatrix} 3 & 2 & -1 \\ 1 & 0 & 2 \\ 4 & 5 & 2 \end{bmatrix}$    **15.** $\begin{bmatrix} 2 & 0 & 0 \\ 0 & 5 & 0 \\ 0 & 0 & -2 \end{bmatrix}$

**16.** Use the inverse found in Exercise 12 to solve the system in Exercise 2.

**17.** Find two nonzero matrices whose product is a zero matrix.

**18.** Use a determinant to decide whether $(2, 9)$, $(-2, 1)$, and $(3, 11)$ are collinear.

**19.** Find the area of the triangle whose vertices are $(-2, 4)$, $(0, 5)$, and $(3, -1)$.

**20.** A manufacturer produces three models of a product, which are shipped to two warehouses. The number of units $i$ that are shipped to warehouse $j$ is represented by $a_{ij}$ in matrix $A$ at the left. The price per unit is represented by matrix $B$. Find the product $BA$ and interpret the result.

$$A = \begin{bmatrix} 1000 & 3000 \\ 2000 & 4000 \\ 5000 & 8000 \end{bmatrix}$$

$$B = \begin{bmatrix} \$50 & \$35 & \$28 \end{bmatrix}$$

Matrices for 20

**Example 7**    Writing a Sum in Summation Notation

Write the partial sum in summation notation.

**a.** $\dfrac{3}{1+2} + \dfrac{3}{2+2} + \dfrac{3}{3+2} + \dfrac{3}{4+2} + \dfrac{3}{5+2}$

**b.** $2 - 4 + 8 - 16$

Solution

**a.** Begin by looking for similarities and differences in each of the terms. Each term has 3 in the numerator. In the denominator, 2 is added to a number that is increasing by 1 for each term. Let $i$ be the index of summation, let 1 be the lower limit of summation, and let 5 be the upper limit of summation. Then, using summation notation, you can write

$$\frac{3}{1+2} + \frac{3}{2+2} + \frac{3}{3+2} + \frac{3}{4+2} + \frac{3}{5+2} = \sum_{i=1}^{5} \frac{3}{i+2}.$$

**b.** This series has terms with alternating signs, and each term can be written as a power of 2. Let $n$ be the index of summation, let 1 be the lower limit of summation, and let 4 be the upper limit of summation. Then, using summation notation, you can write

$$2 - 4 + 8 - 16 = (-1)^{1+1}(2) + (-1)^{2+1}(4) + (-1)^{3+1}(8) + (-1)^{4+1}(16)$$

$$= (-1)^{1+1}(2)^1 + (-1)^{2+1}(2)^2 + (-1)^{3+1}(2)^3 + (-1)^{4+1}(2)^4$$

$$= \sum_{n=1}^{4} (-1)^{n+1}(2)^n.$$

---

▶ **Properties of Sums**

**1.** $\displaystyle\sum_{i=1}^{n} ca_i = c\sum_{i=1}^{n} a_i, \qquad c$ is any constant

**2.** $\displaystyle\sum_{i=1}^{n} (a_i + b_i) = \sum_{i=1}^{n} a_i + \sum_{i=1}^{n} b_i$

**3.** $\displaystyle\sum_{i=1}^{n} (a_i - b_i) = \sum_{i=1}^{n} a_i - \sum_{i=1}^{n} b_i$

The sum of the terms of any *finite* sequence must be a finite number. Variations in the upper and lower limits of summation can produce quite different-looking summation notations for *the same sum*. For example, consider the following two sums.

$$\sum_{i=1}^{5} 3(2^i) = 3\sum_{i=1}^{5} 2^i = 3(2^1 + 2^2 + 2^3 + 2^4 + 2^5)$$

$$\sum_{i=0}^{4} 3(2^{i+1}) = 3\sum_{i=0}^{4} 2^{i+1} = 3(2^1 + 2^2 + 2^3 + 2^4 + 2^5)$$

## Application

Sequences have many applications in business and science. One is illustrated in Example 8.

### Example 8 Population of the United States

From 1956 to 1999, the resident population of the United States can be approximated by the model

$$a_n = \sqrt{29{,}283 + 830.1n + 4.74n^2}, \qquad n = 0, 1, \ldots, 43$$

where $a_n$ is the population in millions and $n$ represents the calendar year, with $n = 0$ corresponding to 1956. Find the last five terms of this finite sequence. (Source: U.S. Bureau of the Census)

#### Solution

The last five terms of this finite sequence are as follows.

$$a_{39} = \sqrt{29{,}283 + 830.1(39) + 4.74(39)^2} \approx 262.4 \qquad \text{1995 population}$$

$$a_{40} = \sqrt{29{,}283 + 830.1(40) + 4.74(40)^2} \approx 264.7 \qquad \text{1996 population}$$

$$a_{41} = \sqrt{29{,}283 + 830.1(41) + 4.74(41)^2} \approx 267.0 \qquad \text{1997 population}$$

$$a_{42} = \sqrt{29{,}283 + 830.1(42) + 4.74(42)^2} \approx 269.3 \qquad \text{1998 population}$$

$$a_{43} = \sqrt{29{,}283 + 830.1(43) + 4.74(43)^2} \approx 271.6 \qquad \text{1999 population}$$

The bar graph in Figure 8.1 graphically represents the population given by this sequence for the entire 44-year period from 1956 to 1999.

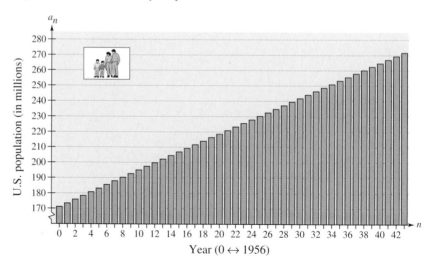

Figure 8.1

## Discussing the Concept     A Summation Program

A graphing calculator can be programmed to calculate the sum of the first $n$ terms of a sequence. Consult your user's guide for specific instructions on programming your calculator. Use the algorithm below to write the summation program, or access the website for this text at *college.hmco.com* for programs written for selected graphing calculators.

### Summation Program

To find the sum of a finite sequence, write the program for your graphing calculator from the algorithm given below and enter it into your calculator. Then enter the $n$th term of the sequence in $Y_1$ and run the program. For instance, to find the sum

$$\sum_{i=1}^{10} 2^i$$

enter 2^X into $Y_1$. Then run the program, entering 1 as the lower limit and 10 as the upper limit. The last sum displayed is the final sum. If you wish to view each partial sum, add a pause command after displaying the value of $S$. For the example above, 10 numbers will be displayed, with the final sum being 2046.

### Summation Algorithm

1. Enter $M$ as the lower limit of summation.
2. Enter $N$ as the upper limit of summation.
3. Set $S$ equal to 0 so that $S$ can be used to store the sums.
4. Use a FOR loop to sum from the lower limit $M$ to the upper limit $N$ incrementing by $X$.
5. Add the sum of $Y_1$ and $S$ and store the result in $S$.
6. Display the value of $S$.
7. End the FOR loop when $Y_1$ has been calculated and summed for the upper limit.

Try using this program to find the following sums.

a. $\displaystyle\sum_{i=1}^{15} (i - 2)^2$     b. $\displaystyle\sum_{i=1}^{20} 3^{(i-1)}$     c. $\displaystyle\sum_{i=1}^{10} \sqrt{e^i}$

# Warm Up

The following warm-up exercises involve skills that were covered in earlier sections. You will use these skills in the exercise set for this section.

**1.** Find $f(2)$ for $f(n) = \dfrac{2n}{n^2 + 1}$.

**2.** Find $f(3)$ for $f(n) = \dfrac{4}{3(n + 1)}$.

In Exercises 3–6, factor the expression.

**3.** $4n^2 - 1$

**4.** $4n^2 - 8n + 3$

**5.** $n^2 - 3n + 2$

**6.** $n^2 + 3n + 2$

In Exercises 7–10, perform the indicated operations and/or simplify.

**7.** $\left(\dfrac{2}{3}\right)\left(\dfrac{3}{4}\right)\left(\dfrac{4}{5}\right)\left(\dfrac{5}{6}\right)$

**8.** $\dfrac{2 \cdot 4 \cdot 6 \cdot 8}{2^4}$

**9.** $\dfrac{1}{2 \cdot 2} + \dfrac{1}{2 \cdot 3} + \dfrac{1}{2 \cdot 4}$

**10.** $\dfrac{1}{1 \cdot 2} + \dfrac{1}{2 \cdot 3} + \dfrac{1}{3 \cdot 4}$

## 8.1     Exercises

In Exercises 1–22, write the first five terms of the sequence. (Assume that $n$ begins with 1.)

**1.** $a_n = 3n - 2$

**2.** $a_n = 4 - 2n$

**3.** $a_n = 2^n$

**4.** $a_n = (-2)^n$

**5.** $a_n = \left(\dfrac{1}{2}\right)^n$

**6.** $a_n = \left(-\dfrac{1}{2}\right)^n$

**7.** $a_n = \dfrac{2}{n + 1}$

**8.** $a_n = \dfrac{n + 1}{3}$

**9.** $a_n = \dfrac{(-1)^n}{n}$

**10.** $a_n = \dfrac{(-1)^{n+1}}{n}$

**11.** $a_n = 3 - \dfrac{1}{2^n}$

**12.** $a_n = \dfrac{3^n}{4^n}$

**13.** $a_n = \dfrac{3^n}{n!}$

**14.** $a_n = \dfrac{n!}{n}$

**15.** $a_n = \dfrac{n^2 - 1}{n + 1}$

**16.** $a_n = \dfrac{n^2 - 1}{n^2 + 2}$

**17.** $a_n = \dfrac{(-1)^n}{n^2}$

**18.** $a_n = \dfrac{(-1)^n n}{n + 1}$

**19.** $a_n = \dfrac{3n!}{(n - 1)!}$

**20.** $a_n = \dfrac{(n + 1)!}{n!}$

**21.** $a_n = \dfrac{5n^2 - n + 1}{n^2}$

**22.** $a_n = \dfrac{3n^2 - n + 4}{2n^2 + 1}$

In Exercises 23–26, evaluate the expression.

**23.** $\dfrac{8!}{6!}$

**24.** $\dfrac{10!}{9!}$

**25.** $\dfrac{28!}{25!}$

**26.** $\dfrac{49!}{50!}$

In Exercises 27–30, simplify the expression. Then evaluate the expression when $n = 0$ and $n = 1$.

**27.** $\dfrac{(n + 1)!}{n!}$

**28.** $\dfrac{(n + 2)!}{n!}$

**29.** $\dfrac{(2n - 1)!}{(2n + 1)!}$

**30.** $\dfrac{(2n + 2)!}{(2n)!}$

In Exercises 31–44, write an expression for the most apparent $n$th term of the sequence. (Assume that $n$ begins with 1.)

**31.** $1, 3, 5, 7, 9, \ldots$

**32.** $5, 8, 11, 14, 17, \ldots$

**33.** $2, \frac{1}{2}, \frac{2}{9}, \frac{1}{8}, \frac{2}{25}, \ldots$

**34.** $1, \frac{1}{4}, \frac{1}{9}, \frac{1}{16}, \frac{1}{25}, \ldots$

**35.** $\frac{2}{3}, \frac{3}{4}, \frac{4}{5}, \frac{5}{6}, \frac{6}{7}, \ldots$

**36.** $\frac{2}{1}, \frac{3}{3}, \frac{4}{5}, \frac{5}{7}, \frac{6}{9}, \ldots$

**37.** $\frac{1}{2}, -\frac{1}{4}, \frac{1}{8}, -\frac{1}{16}, \frac{1}{32}, \ldots$

**38.** $\frac{1}{3}, \frac{2}{9}, \frac{4}{27}, \frac{8}{81}, \frac{16}{243}, \ldots$

**39.** $1 + \frac{1}{1}, 1 + \frac{1}{2}, 1 + \frac{1}{3}, 1 + \frac{1}{4}, 1 + \frac{1}{5}, \ldots$

**40.** $1 + \frac{1}{2}, 1 + \frac{3}{4}, 1 + \frac{7}{8}, 1 + \frac{15}{16}, 1 + \frac{31}{32}, \ldots$

**41.** $1, \frac{1}{2}, \frac{1}{6}, \frac{1}{24}, \frac{1}{120}, \ldots$

**42.** $2, -4, 6, -8, 10, \ldots$

**43.** $1, -1, 1, -1, 1, \ldots$

**44.** $1, 2, \dfrac{2^2}{2}, \dfrac{2^3}{6}, \dfrac{2^4}{24}, \dfrac{2^5}{120}, \ldots$

In Exercises 45–58, find the sum.

**45.** $\displaystyle\sum_{i=1}^{4} (6i + 3)$

**46.** $\displaystyle\sum_{i=1}^{6} (4i - 2)$

**47.** $\displaystyle\sum_{i=1}^{7} -3i$

**48.** $\displaystyle\sum_{i=1}^{5} -3(i - 2)$

**49.** $\displaystyle\sum_{k=1}^{4} 10$

**50.** $\displaystyle\sum_{k=1}^{5} 4$

**51.** $\displaystyle\sum_{i=0}^{4} i^2$

**52.** $\displaystyle\sum_{i=0}^{5} 3i^2$

**53.** $\displaystyle\sum_{k=2}^{6} \frac{1}{2k}$

**54.** $\displaystyle\sum_{j=3}^{5} \frac{1}{j}$

**55.** $\displaystyle\sum_{i=1}^{4} (i - 1)^2$

**56.** $\displaystyle\sum_{k=2}^{5} (k + 1)(k - 3)$

**57.** $\displaystyle\sum_{i=1}^{4} (-1)^i (2i + 4)$

**58.** $\displaystyle\sum_{i=1}^{4} (-2)^i$

In Exercises 59–68, use summation notation to write the sum.

**59.** $\dfrac{1}{3(1)} + \dfrac{1}{3(2)} + \dfrac{1}{3(3)} + \cdots + \dfrac{1}{3(9)}$

**60.** $\dfrac{5}{1 + 1} + \dfrac{5}{1 + 2} + \dfrac{5}{1 + 3} + \cdots + \dfrac{5}{1 + 15}$

**61.** $\left[2\left(\frac{1}{8}\right) + 3\right] + \left[2\left(\frac{2}{8}\right) + 3\right] + \cdots + \left[2\left(\frac{8}{8}\right) + 3\right]$

**62.** $\left[1 - \left(\frac{1}{6}\right)^2\right] + \left[1 - \left(\frac{2}{6}\right)^2\right] + \cdots + \left[1 - \left(\frac{6}{6}\right)^2\right]$

**63.** $3 - 9 + 27 - 81 + 243 - 729$

**64.** $1 - \frac{1}{2} + \frac{1}{4} - \frac{1}{8} + \cdots - \frac{1}{128}$

**65.** $\dfrac{1}{1^2} - \dfrac{1}{2^2} + \dfrac{1}{3^2} - \dfrac{1}{4^2} + \cdots - \dfrac{1}{20^2}$

**66.** $\dfrac{1}{1 \cdot 3} + \dfrac{1}{2 \cdot 4} + \dfrac{1}{3 \cdot 5} + \cdots + \dfrac{1}{10 \cdot 12}$

**67.** $\frac{1}{4} + \frac{3}{8} + \frac{7}{16} + \frac{15}{32} + \frac{31}{64}$

**68.** $\frac{1}{2} + \frac{2}{4} + \frac{6}{8} + \frac{24}{16} + \frac{120}{32} + \frac{720}{64}$

**69.** *Compound Interest*   A deposit of \$5000 is made in an account that earns 8% interest compounded quarterly. The balance in the account after $n$ quarters is given by

$$A_n = 5000\left(1 + \frac{0.08}{4}\right)^n, \quad n = 1, 2, 3, \ldots.$$

(a) Compute the first eight terms of this sequence.

(b) Find the balance in this account after 10 years by computing the 40th term of the sequence.

(c) Is the balance after 20 years twice the balance after 10 years? Explain.

**70.** *Compound Interest*   A deposit of \$100 is made *each* month in an account that earns 12% interest compounded monthly. The balance in the account after $n$ months is given by

$$A_n = 100(101)[(1.01)^n - 1], \quad n = 1, 2, 3, \ldots.$$

(a) Compute the first six terms of this sequence.

(b) Find the balance after 5 years by computing the 60th term of the sequence.

(c) Find the balance after 20 years by computing the 240th term of the sequence.

**71.** *Ratio of Men to Women* In 1940, the population of the United States had about the same number of men as women. After that, there were more women than men. The ratio of men to women is approximated by the model

$$a_n = 1.04 - 0.032n + 0.0028n^2,$$

$$n = 1, 2, \ldots, 8$$

where $a_n$ is the ratio of men to women and $n$ is the year, with $n = 1, 2, 3, \ldots, 8$ corresponding to 1940, 1950, 1960, ..., 2010. (Source: U.S. Bureau of the Census)

(a) Use a graphing utility to find the terms of this finite sequence.

(b) Construct a bar graph that represents the sequence.

(c) In 2000, the population of the United States was 275 million. How many of these were women? How many were men?

**72.** *Annual Payroll* The annual payrolls for new car dealerships in the United States from 1994 to 1999 can be approximated by the model

$$a_n = 29.7 + 2.20n + 0.079n^2,$$

$$n = 0, 1, 2, 3, 4, 5$$

where $a_n$ is the annual payroll (in billions of dollars) and $n = 0$ represents 1994. Find the total payroll from 1994 to 1999 by evaluating the sum

$$\sum_{n=0}^{5} (29.7 + 2.20n + 0.079n^2).$$

(Source: National Automobile Dealers Association)

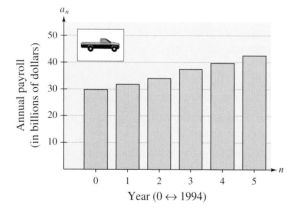

Annual payroll (in billions of dollars)

Year (0 ↔ 1994)

**73.** *Hospital Costs* The average cost of a day in a hospital in the United States from 1990 to 1998 is approximated by the model

$$a_n = 686 + 73.5n - 3.31n^2,$$

$$n = 0, 1, \ldots, 8$$

where $a_n$ is the average cost in dollars and $n = 0$ represents 1990. Use a graphing utility to find the terms of this finite sequence and construct a bar graph that represents the sequence. Predict the average cost of a day in a hospital in 2005. (Source: American Hospital Association)

**74.** *Total Revenue* The total annual revenues for WorldCom, Inc. from 1996 to 2000 can be approximated by the model

$$a_n = 5151 - 8822.7n + 11{,}761.86n^2 - 1853.25n^3,$$

$$n = 0, 1, \ldots, 4$$

where $a_n$ is the annual revenue (in millions of dollars) and $n = 0$ represents 1996. Find the total revenue from 1996 to 2000 by evaluating the sum

$$\sum_{n=0}^{4} (5151 - 8822.7n + 11{,}761.86n^2 - 1853.25n^3).$$

(Source: WorldCom, Inc.)

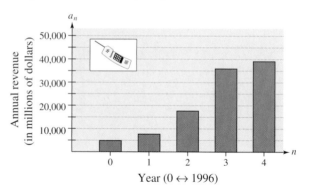

Annual revenue (in millions of dollars)

Year (0 ↔ 1996)

**75.** *Dividends* A company declares dividends per share of common stock that can be approximated by the model $a_n = 0.35n + 1.25$, $n = 0, 1, 2, 3, 4$, where $a_n$ is the dividend in dollars and $n = 0$ represents 1995. Approximate the sum of the dividends per share of common stock for the years 1995 through 1999 by evaluating

$$\sum_{n=0}^{4} (0.35n + 1.25).$$

The symbol ▦ indicates an exercise or parts of an exercise in which you are instructed to use a graphing utility.

## 8.2    Arithmetic Sequences and Series

### Objectives

- Determine whether a sequence is arithmetic.
- Find the *n*th term of an arithmetic sequence.
- Find a formula for an arithmetic sequence.
- Find the sum of an arithmetic sequence.
- Use an arithmetic sequence to solve an application problem.

## Arithmetic Sequences

A sequence whose consecutive terms have a common difference is called an **arithmetic sequence.**

·DISCOVERY·

Determine the arithmetic rule used to generate the terms of each sequence.

**a.** $1, 4, 7, 10, 13, \ldots$

**b.** $6, 12, 18, 24, \ldots$

**c.** $100, 90, 80, 70, \ldots$

▶ Definition of Arithmetic Sequence

A sequence is **arithmetic** if the differences between consecutive terms are the same. Thus, the sequence

$$a_1, a_2, a_3, a_4, \ldots, a_n, \ldots$$

is arithmetic if there is a number $d$ such that

$$a_2 - a_1 = d, \quad a_3 - a_2 = d, \quad a_4 - a_3 = d,$$

and so on. The number $d$ is the **common difference** of the arithmetic sequence.

---

**Example 1**    Examples of Arithmetic Sequences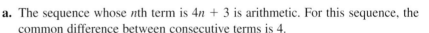

**a.** The sequence whose *n*th term is $4n + 3$ is arithmetic. For this sequence, the common difference between consecutive terms is 4.

$$\underbrace{7, 11}, 15, 19, \ldots, 4n + 3, \ldots$$
$$11 - 7 = 4$$

**b.** The sequence whose *n*th term is $7 - 5n$ is arithmetic. For this sequence, the common difference between consecutive terms is $-5$.

$$\underbrace{2, -3}, -8, -13, \ldots, 7 - 5n, \ldots$$
$$-3 - 2 = -5$$

**c.** The sequence whose *n*th term is $\frac{1}{4}(n + 3)$ is arithmetic. For this sequence, the common difference between consecutive terms is $\frac{1}{4}$.

$$\underbrace{1, \frac{5}{4}}, \frac{3}{2}, \frac{7}{4}, \ldots, \frac{n + 3}{4}, \ldots$$
$$\tfrac{5}{4} - 1 = \tfrac{1}{4}$$

In Example 1, notice that each of the arithmetic sequences has an $n$th term that is of the form $dn + c$, where the common difference of the sequence is $d$. This result is summarized as follows.

---

▶ **The $n$th Term of an Arithmetic Sequence**

The $n$th term of an arithmetic sequence has the form

$$a_n = dn + c$$

where $d$ is the common difference between consecutive terms of the sequence and $c = a_1 - d$.

---

**Example 2** Finding the $n$th Term of an Arithmetic Sequence

Find a formula for the $n$th term of the arithmetic sequence whose common difference is 3 and whose first term is 2.

**Solution**

Because the sequence is arithmetic, you know that the formula for the $n$th term is of the form $a_n = dn + c$. Moreover, because the common difference is $d = 3$, the formula must have the form

$$a_n = 3n + c.$$

Because $a_1 = 2$, it follows that

$$c = a_1 - d = 2 - 3 = -1.$$

So, the formula for the $n$th term is

$$a_n = 3n - 1.$$

The sequence therefore has the following form.

$$2, 5, 8, 11, 14, \ldots, 3n - 1, \ldots$$

---

Another way to find a formula for the $n$th term of the sequence in Example 2 is to begin by writing the terms of the sequence.

| $a_1$ | $a_2$ | $a_3$ | $a_4$ | $a_5$ | $a_6$ | $a_7$ | |
|-------|-------|-------|-------|-------|-------|-------|-----|
| 2 | $2 + 3$ | $5 + 3$ | $8 + 3$ | $11 + 3$ | $14 + 3$ | $17 + 3$ | $\ldots$ |
| 2 | 5 | 8 | 11 | 14 | 17 | 20 | $\ldots$ |

From these terms, you can reason that the $n$th term is of the form

$$a_n = dn + c = 3n - 1.$$

| Example 3 | Finding the $n$th Term of an Arithmetic Sequence |
|---|---|

The fourth term of an arithmetic sequence is 20, and the 13th term is 65. Write the first several terms of this sequence.

### Solution

The fourth and 13th terms of the sequence are related by

$$a_{13} = a_4 + 9d.$$

Using $a_4 = 20$ and $a_{13} = 65$, you can conclude that $d = 5$, which implies that the sequence is as follows.

| $a_1$ | $a_2$ | $a_3$ | $a_4$ | $a_5$ | $a_6$ | $a_7$ | $a_8$ | $a_9$ | $a_{10}$ | $a_{11}$ | |
|---|---|---|---|---|---|---|---|---|---|---|---|
| 5, | 10, | 15, | 20, | 25, | 30, | 35, | 40, | 45, | 50, | 55, | . . . |

## Study Tip

If you substitute $a_1 - d$ for $c$ in the formula $a_n = dn + c$, then the $n$th term of an arithmetic sequence has the alternative recursion formula

$$a_n = a_1 + (n - 1)d.$$

Use this formula to solve Example 4 and Example 9.

If you know the $n$th term of an arithmetic sequence *and* you know the common difference of the sequence, you can find the $(n + 1)$th term by using the **recursion formula**

$$a_{n+1} = a_n + d.$$

With this formula, you can find any term of an arithmetic sequence, *provided* that you know the previous term. For instance, if you know the first term, you can find the second term. Then, knowing the second term, you can find the third term, and so on.

| Example 4 | Using a Recursion Formula |
|---|---|

Find the ninth term of the arithmetic sequence whose first two terms are 2 and 9.

### Solution

For this sequence, the common difference is $d = 9 - 2 = 7$. There are two ways to find the ninth term. One way is simply to write out the first nine terms (by repeatedly adding 7).

2, 9, 16, 23, 30, 37, 44, 51, 58

Another way to find the ninth term is first to find a formula for the $n$th term. Because the first term is 2, it follows that

$$c = a_1 - d = 2 - 7 = -5.$$

So, a formula for the $n$th term is $a_n = 7n - 5$, which implies that the ninth term is $a_9 = 7(9) - 5 = 58$.

## The Sum of an Arithmetic Series

The *sum* of the terms of an arithmetic sequence is called an **arithmetic series.** There is a simple formula for the sum of a finite arithmetic series. Be sure you see that this formula works only for *arithmetic* series.

> ▶ **The Sum of a Finite Arithmetic Series**
>
> The sum of a finite arithmetic series with $n$ terms is given by
>
> $$S = \frac{n}{2}(a_1 + a_n).$$

**Example 5**  Finding the Sum of an Arithmetic Series

Find the sum: $1 + 3 + 5 + 7 + 9 + 11 + 13 + 15 + 17 + 19$.

**Solution**

To begin, notice that the series is arithmetic (with a common difference of 2). Moreover, the series has 10 terms. So, the sum of the series is

$$S = 1 + 3 + 5 + 7 + 9 + 11 + 13 + 15 + 17 + 19$$

$$= \frac{n}{2}(a_1 + a_n)$$

$$= \frac{10}{2}(1 + 19)$$

$$= 5(20)$$

$$= 100.$$

**Example 6**  Finding the Sum of an Arithmetic Series

Find the sum of the integers from 1 to 100.

**Solution**

The integers from 1 to 100 form an arithmetic series that has 100 terms. So, you can use the formula for the sum of an arithmetic series, as follows.

$$S = 1 + 2 + 3 + 4 + 5 + 6 + \cdots + 99 + 100$$

$$= \frac{n}{2}(a_1 + a_n)$$

$$= \frac{100}{2}(1 + 100)$$

$$= 50(101)$$

$$= 5050$$

Example 7    Finding the Sum of an Arithmetic Series

Find the sum of the first 150 terms of the arithmetic sequence

$$5, 16, 27, 38, 49, \ldots .$$

**Solution**

For this arithmetic series, you have $a_1 = 5$ and $d = 16 - 5 = 11$. Thus, $c = a_1 - d = 5 - 11 = -6$, and the $n$th term is

$$a_n = 11n - 6.$$

So, $a_{150} = 11(150) - 6 = 1644$, and the sum of the first 150 terms is as follows.

$$S = \frac{n}{2}(a_1 + a_n)$$

$$= \frac{150}{2}(5 + 1644)$$

$$= 75(1649)$$

$$= 123{,}675$$

## Applications

Example 8    Seating Capacity

An auditorium has 20 rows of seats. There are 20 seats in the first row, 22 seats in the second row, 24 seats in the third row, and so on (see Figure 8.2). How many seats are there in all 20 rows?

**Solution**

The number of seats in the rows forms an arithmetic sequence in which the common difference is $d = 2$. Because $c = a_1 - d = 20 - 2 = 18$, you can determine that the formula for the $n$th term in the sequence is $a_n = 2n + 18$. So, the 20th term in the sequence is $a_{20} = 2(20) + 18 = 58$, and the total number of seats is

$$S = 20 + 22 + 24 + \cdots + 58$$

$$= \frac{n}{2}(a_1 + a_{20})$$

$$= \frac{20}{2}(20 + 58)$$

$$= 10(78)$$

$$= 780.$$

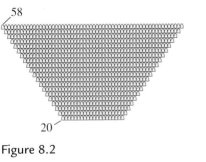

58

20

Figure 8.2

| Example 9 | Total Sales  |

A small business sells $10,000 worth of products during its first year. The owner of the business has set a goal of increasing annual sales by $7500 each year for 9 years. Assuming that this goal is met, find the total sales during the first 10 years this business is in operation.

### Solution

The annual sales form an arithmetic sequence in which $a_1 = 10{,}000$ and $d = 7500$. So, $c = a_1 - d = 10{,}000 - 7500 = 2500$, and the $n$th term of the sequence is

$$a_n = 7500n + 2500.$$

This implies that the 10th term of the sequence is $a_{10} = 77{,}500$. So, the total sales for the first 10 years are as follows.

$$S = \frac{n}{2}(a_1 + a_{10})$$

$$= \frac{10}{2}(10{,}000 + 77{,}500)$$

$$= 5(87{,}500)$$

$$= \$437{,}500$$

---

### Discussing the Concept    Numerical Relationships

Decide whether it is possible to fill in the blanks in each of the following such that the resulting sequence is arithmetic. If so, find a recursion formula for the sequence.

a. $-7$, ___, ___, ___, ___, ___, 11

b. 17, ___, ___, ___, ___, ___, ___, ___, ___, 71

c. 2, 6, ___, ___, 162

d. 4, 7.5, ___, ___, ___, ___, ___, ___, ___, ___, 39

e. 8, 12, ___, ___, ___, 60.75

# Warm Up

The following warm-up exercises involve skills that were covered in earlier sections. You will use these skills in the exercise set for this section.

In Exercises 1 and 2, find the sum.

**1.** $\displaystyle\sum_{i=1}^{6} (2i - 1)$

**2.** $\displaystyle\sum_{i=1}^{10} (4i + 2)$

In Exercises 3 and 4, find the distance between the two real numbers.

**3.** $\frac{5}{2}, 8$

**4.** $\frac{4}{3}, \frac{14}{3}$

In Exercises 5 and 6, evaluate the function as indicated.

**5.** Find $f(3)$ for $f(n) = 10 + (n - 1)4$.

**6.** Find $f(10)$ for $f(n) = 1 + (n - 1)\frac{1}{3}$.

In Exercises 7–10, evaluate the expression.

**7.** $\frac{11}{2}(1 + 25)$

**8.** $\frac{16}{2}(4 + 16)$

**9.** $\frac{20}{2}[2(5) + (12 - 1)3]$

**10.** $\frac{8}{2}[2(-3) + (15 - 1)5]$

## 8.2   Exercises

In Exercises 1–10, determine whether the sequence is arithmetic. If it is, find the common difference.

**1.** $3, 7, 11, 15, 19, \ldots$

**2.** $20, 17, 14, 11, 8, \ldots$

**3.** $-3, -2, 0, 3, 7, \ldots$

**4.** $-12, -8, -4, 0, 4, 8, \ldots$

**5.** $3, \frac{5}{2}, 2, \frac{3}{2}, 1, \frac{1}{2}, \ldots$

**6.** $\frac{9}{4}, 2, \frac{7}{4}, \frac{3}{2}, \frac{5}{4}, 1, \ldots$

**7.** $\frac{1}{3}, \frac{2}{3}, \frac{4}{3}, \frac{8}{3}, \frac{16}{3}, \frac{32}{3}, \ldots$

**8.** $\ln 1, \ln 2, \ln 3, \ln 4, \ln 5, \ldots$

**9.** $5.3, 5.7, 6.1, 6.5, 6.9, \ldots$

**10.** $1^2, 2^2, 3^2, 4^2, 5^2, \ldots$

In Exercises 11–18, write the first five terms of the specified sequence. Determine whether the sequence is arithmetic. If it is, find the common difference.

**11.** $a_n = 5 + 3n$

**12.** $a_n = 100 - 3n$

**13.** $a_n = (2 + n) - (1 + n)$

**14.** $a_n = 1 + (n - 1)4$

**15.** $a_n = (2^n)n$

**16.** $a_n = 2^{n-1}$

**17.** $a_n = \dfrac{1}{n + 1}$

**18.** $a_n = (-1)^n$

In Exercises 19–28, find a formula for $a_n$ for the arithmetic sequence.

**19.** $a_1 = 4, d = 5$

**20.** $a_1 = -2, d = 3$

**21.** $a_1 = 50, d = -12$

**22.** $a_1 = 120, d = -15$

**23.** $a_1 = 7, a_5 = 27$
**24.** $a_1 = -5, a_7 = -17$
**25.** $a_3 = 8, a_9 = 32$
**26.** $a_6 = 29, a_{18} = 95$
**27.** 2, 8, 14, 20, 26, . . .
**28.** 3, 13, 23, 33, 43, . . .

In Exercises 29–36, write the first five terms of the arithmetic sequence.

**29.** $a_1 = 10, d = 4$
**30.** $a_1 = 2, d = \frac{1}{3}$
**31.** $a_1 = 5, a_{n+1} = a_n + 9$
**32.** $a_1 = 7, a_{n+1} = a_n - 2$
**33.** $a_1 = 2, a_{12} = 46$
**34.** $a_5 = 28, a_{10} = 53$
**35.** $a_8 = 26, a_{12} = 42$
**36.** $a_4 = 16, a_{10} = 46$

In Exercises 37–44, find the sum of the first $n$ terms of the arithmetic sequence.

**37.** 4, 8, 12, 16, . . . , $n = 6$
**38.** 3, 11, 19, 27, . . . , $n = 10$
**39.** $-10, -8, -6, -4, \ldots , n = 12$
**40.** $-23, -16, -9, -2, \ldots , n = 10$
**41.** 0.5, 0.9, 1.3, 1.7, . . . , $n = 10$
**42.** 1.50, 1.45, 1.40, 1.35, . . . , $n = 20$
**43.** $a_1 = 100, a_{25} = 220, n = 25$
**44.** $a_1 = 15, a_{100} = 307, n = 100$

In Exercises 45–54, find the indicated sum.

**45.** $\displaystyle\sum_{n=1}^{50} n$
**46.** $\displaystyle\sum_{n=1}^{100} 2n$
**47.** $\displaystyle\sum_{n=1}^{100} 5n$
**48.** $\displaystyle\sum_{n=51}^{100} 7n$
**49.** $\displaystyle\sum_{n=0}^{50} (1000 - 5n)$
**50.** $\displaystyle\sum_{n=1}^{250} (1000 - n)$
**51.** $\displaystyle\sum_{n=1}^{20} (2n + 5)$
**52.** $\displaystyle\sum_{n=1}^{500} (n + 3)$
**53.** $\displaystyle\sum_{n=1}^{100} \frac{n + 4}{2}$
**54.** $\displaystyle\sum_{n=0}^{100} \frac{8 - 3n}{16}$

**55.** Find the sum of the first 100 odd integers.

**56.** Find the sum of the integers from $-10$ to 50.

**57.** *Job Offer* A person accepts a position with a company and will receive a salary of $27,500 for the first year. The person is guaranteed a raise of $1500 per year for the first 5 years.

(a) What will the salary be during the sixth year of employment?

(b) Use a graphing utility to find how much the company will have paid the person by the end of the sixth year.

**58.** *Job Offer* A person accepts a position with a company and will receive a salary of $26,800 for the first year. The person is guaranteed a raise of $1750 per year for the first 5 years.

(a) What will the salary be during the sixth year of employment?

(b) Use a graphing utility to find how much the company will have paid the person by the end of the sixth year.

**59.** *Seating Capacity* Determine the seating capacity of an auditorium with 30 rows of seats if there are 15 seats in the first row, 20 seats in the second row, 25 seats in the third row, and so on.

**60.** *Baling Hay* As a farmer bales a field of hay, each trip around the field gets shorter. Suppose that on the first round there are 267 bales and on the second round there are 253 bales. Assume that the decrease will be the same on each round and that there will be 11 more trips. How many bales of hay will the farmer get from the field?

**61.** *Total Sales* The annual sales for Wal-Mart Stores, Inc. from 1995 to 2000 can be approximated by the model

$$a_n = 85.881 + 19.6757n, \quad n = 0, 1, \ldots , 5$$

where $a_n$ is the total annual sales (in billions of dollars) and $n = 0$ represents 1995. (Source: Wal-Mart Stores, Inc.)

(a) Sketch a bar graph showing the annual sales for Wal-Mart Stores, Inc. from 1995 to 2000.

(b) Find the total sales from 1995 to 2000.

**62.** *Total Revenue*   The annual revenues for SBC Communications, Inc. from 1995 to 2000 can be modeled by

$$a_n = 8.149 + 8.9183n, \quad n = 0, 1, \ldots, 5$$

where $a_n$ is the annual revenue (in billions of dollars) and $n = 0$ represents 1995.   (Source: SBC Communications, Inc.)

(a) Sketch a bar graph showing the annual revenues for SBC Communications, Inc. from 1995 to 2000.

(b) Find the total revenue from 1995 to 2000.

**63.** *Falling Object*   A heavy object (with negligible air resistance) is dropped from a plane. During the first second of fall, the object falls 16 feet; during the second second, it falls 48 feet; during the third second, it falls 80 feet; during the fourth second, it falls 112 feet. If this pattern continues, how many feet will the object fall in 8 seconds?

**64.** *Falling Object*   A heavy object (with negligible air resistance) is dropped from a plane. During the first second of fall, the object falls 4.9 meters; during the second second, it falls 14.7 meters; during the third second, it falls 24.5 meters; during the fourth second, it falls 34.3 meters. How many meters will the object fall in 10 seconds?

**65.** *Number of Logs*   Logs are stacked in a pile, as shown in the figure. The top row has 15 logs and the bottom row has 21 logs. How many logs are in the stack?

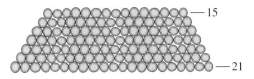
— 15
— 21

**66.** *Brick Pattern*   A brick patio is roughly the shape of a trapezoid (see figure). The patio has 20 rows of bricks. The first row has 14 bricks and the 20th row has 33 bricks. How many bricks are in the patio?

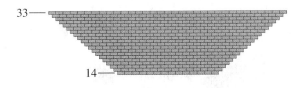
33 —
14 —

**67.** *Total Profit*   A small company makes a profit of $20,000 during its first year. The company president sets a goal of increasing profit by $5000 each year for 4 years. Assuming that this goal is met, find the total profit during the first 5 years of business.

**68.** *Total Sales*   An entrepreneur sells $16,000 worth of gumballs in the first year of operation. The entrepreneur has set a goal of increasing annual sales by $4000 each year for 9 years. Assuming that this goal is met, find the total sales during the first 10 years of this gumball business.

**69.** *Total Revenue*   The annual revenues for McDonald's Corporation from 1995 to 2000 are given in the table, where $a_n$ is the annual revenue (in billions of dollars) and $n = 0$ represents 1995. (Source: McDonald's Corp.)

| $n$ | 0 | 1 | 2 | 3 | 4 | 5 |
|---|---|---|---|---|---|---|
| $a_n$ | 9.795 | 10.687 | 11.409 | 12.421 | 13.259 | 14.243 |

(a) Sketch a bar graph showing the annual revenues for McDonald's Corp. from 1995 to 2000.

(b) Use a linear regression program to find an arithmetic sequence that approximates the annual revenues for McDonald's Corp. from 1995 to 2000. Let $n = 0$ represent 1995.

(c) Use the sequence to write an arithmetic series for the *total* revenue of McDonald's Corp. from 1995 to 2000. Find the total revenue.

**70.** *Total Sales*   The annual sales for Home Depot, Inc. from 1995 to 2000 are given in the table, where $a_n$ is the amount of sales (in billions of dollars) and $n = 0$ represents 1995.   (Source: Home Depot, Inc.)

| $n$ | 0 | 1 | 2 | 3 | 4 | 5 |
|---|---|---|---|---|---|---|
| $a_n$ | 15.470 | 19.536 | 24.156 | 30.219 | 38.434 | 45.738 |

(a) Sketch a bar graph showing the annual sales for Home Depot, Inc. from 1995 to 2000.

(b) Use a linear regression program to find an arithmetic sequence that approximates the annual sales for Home Depot, Inc. from 1995 to 2000. Let $n = 0$ represent 1995.

(c) Use the sequence to write an arithmetic series for the *total* sales from 1995 to 2000. Find the total sales.

## 8.3    Geometric Sequences and Series

### Objectives

- Determine whether a sequence is geometric.
- Find the $n$th term of a geometric sequence.
- Find the sum of a finite or infinite geometric sequence.
- Use a geometric sequence to solve an application problem.

## Geometric Sequences

In Section 8.2, you learned that a sequence whose consecutive terms have a common *difference* is an arithmetic sequence. In this section, you will study another important type of sequence called a **geometric sequence.** Consecutive terms of a geometric sequence have a common *ratio*, as indicated in the following definition.

---

▶ **Definition of Geometric Sequence**

A sequence is **geometric** if the ratios of consecutive terms are the same. Thus, a sequence is geometric if there is a number $r$, $r \neq 0$, such that

$$\frac{a_2}{a_1} = r, \quad \frac{a_3}{a_2} = r, \quad \frac{a_4}{a_3} = r,$$

and so on. The number $r$ is the **common ratio** of the geometric sequence.

---

**Example 1**    Examples of Geometric Sequences

**a.** The sequence whose $n$th term is $2^n$ is geometric. For this sequence, the common ratio between consecutive terms is 2.

$$2, 4, 8, 16, \ldots, 2^n, \ldots$$

$$\frac{4}{2} = 2$$

**b.** The sequence whose $n$th term is $4(3^n)$ is geometric. For this sequence, the common ratio between consecutive terms is 3.

$$12, 36, 108, 324, \ldots, 4(3^n), \ldots$$

$$\frac{36}{12} = 3$$

**c.** The sequence whose $n$th term is $\left(-\frac{1}{3}\right)^n$ is geometric. For this sequence, the common ratio between consecutive terms is $-\frac{1}{3}$.

$$-\frac{1}{3}, \frac{1}{9}, -\frac{1}{27}, \frac{1}{81}, \ldots, \left(-\frac{1}{3}\right)^n, \ldots$$

$$\frac{1/9}{-1/3} = -\frac{1}{3}$$

In Example 1, notice that each of the geometric sequences has an $n$th term that is of the form $ar^n$, where the common ratio of the sequence is $r$.

---

▶ **The $n$th Term of a Geometric Sequence**

The $n$th term of a geometric sequence has the form

$$a_n = a_1 r^{n-1}$$

where $r$ is the common ratio of consecutive terms of the sequence. Thus, every geometric sequence can be written in the following form.

$$
\begin{array}{ccccccc}
a_1, & a_2, & a_3, & a_4, & a_5, & \ldots, & a_n, \ldots \\
\downarrow & \downarrow & \downarrow & \downarrow & \downarrow & \ldots, & \downarrow \ldots \\
a_1, & a_1 r, & a_1 r^2, & a_1 r^3, & a_1 r^4, & \ldots, & a_1 r^{n-1}, \ldots
\end{array}
$$

---

If you know the $n$th term of a geometric sequence, you can find the $(n+1)$th term by multiplying by $r$. That is, $a_{n+1} = r a_n$.

**Example 2**    Finding the Terms of a Geometric Sequence

Write the first five terms of the geometric sequence whose first term is $a_1 = 3$ and whose common ratio is $r = 2$.

**Solution**

Starting with 3, repeatedly multiply by 2 to obtain the following.

$$a_1 = 3 \qquad\qquad\qquad \text{1st term}$$
$$a_2 = 3(2^1) = 6 \qquad\quad \text{2nd term}$$
$$a_3 = 3(2^2) = 12 \qquad\quad \text{3rd term}$$
$$a_4 = 3(2^3) = 24 \qquad\quad \text{4th term}$$
$$a_5 = 3(2^4) = 48 \qquad\quad \text{5th term}$$

**Example 3**    Finding a Term of a Geometric Sequence

Find the 15th term of the geometric sequence whose first term is 20 and whose common ratio is 1.05.

**Solution**

$$a_{15} = a_1 r^{n-1} \qquad\qquad \text{Formula for geometric sequence}$$
$$\quad = 20(1.05)^{15-1} \qquad \text{Substitute for } a_1, r, \text{ and } n.$$
$$\quad \approx 39.599 \qquad\qquad\quad \text{Use a calculator.}$$

| Example 4 | Finding a Term of a Geometric Sequence |

Find the 12th term of the geometric sequence

$$5, 15, 45, \ldots .$$

**Solution**

The common ratio of this sequence is $r = \frac{15}{5} = 3$. Because the first term is $a_1 = 5$, you can determine the 12th term ($n = 12$) to be

$$a_{12} = a_1 r^{n-1} \qquad \text{Formula for geometric sequence}$$

$$= 5(3)^{12-1} \qquad \text{Substitute 12 for } n.$$

$$= 5(177,147) \qquad \text{Use a calculator.}$$

$$= 885,735. \qquad \text{Simplify.}$$

## Study Tip

Remember that $r$ is the common ratio of consecutive terms of a sequence. So in Example 5,

$$a_{10} = a_1 r^9$$

$$= a_1 \cdot r \cdot r \cdot r \cdot r^6$$

$$= a_1 \cdot \frac{a_2}{a_1} \cdot \frac{a_3}{a_2} \cdot \frac{a_4}{a_3} \cdot r^6$$

$$= a_4 r^6.$$

If you know any two terms of a geometric sequence, you can use that information to find a formula for the $n$th term of the sequence.

| Example 5 | Finding a Term of a Geometric Sequence |  |

The fourth term of a geometric sequence is 125, and the 10th term is $\frac{125}{64}$. Find the 14th term. (Assume that the terms of the sequence are positive.)

**Solution**

The 10th term is related to the fourth term by the equation

$$a_{10} = a_4 r^6.$$

Because $a_{10} = 125/64$ and $a_4 = 125$, you can solve for $r$ as follows.

$$\frac{125}{64} = 125 r^6$$

$$\frac{1}{64} = r^6$$

$$\frac{1}{2} = r$$

You can obtain the 14th term by multiplying the 10th term by $r^4$.

$$a_{14} = a_{10} r^4$$

$$= \frac{125}{64}\left(\frac{1}{2}\right)^4 = \frac{125}{1024}$$

## The Sum of a Geometric Series

The sum of the terms of a geometric sequence is called a **geometric series.** The formula for the sum of a *finite* geometric series is as follows.

> ▶ **The Sum of a Finite Geometric Series**
>
> The sum of the geometric series
>
> $$a_1 + a_1 r + a_1 r^2 + a_1 r^3 + a_1 r^4 + \cdots + a_1 r^{n-1}$$
>
> with common ratio $r \neq 1$ is given by
>
> $$S = a_1\left(\frac{1 - r^n}{1 - r}\right).$$

·**DISCOVERY**·

To *develop* the formula for the sum of a finite geometric series, consider the following two equations.

$$S = a_1 + a_1 r + a_1 r^2$$
$$+ \cdots + a_1 r^{n-2}$$
$$+ a_1 r^{n-1}$$
$$rS = a_1 r + a_1 r^2 + a_1 r^3$$
$$+ \cdots + a_1 r^{n-1}$$
$$+ a_1 r^n$$

Can you discover a way to simplify the difference between these two equations to obtain the formula for the sum of a finite geometric series?

**Example 6**  Finding the Sum of a Finite Geometric Series

Find the sum $\displaystyle\sum_{n=1}^{12} 4(0.3)^n$.

Solution

By writing out a few terms, you have

$$\sum_{n=1}^{12} 4(0.3)^n = 4(0.3) + 4(0.3)^2 + 4(0.3)^3 + \cdots + 4(0.3)^{12}.$$

Now, because $a_1 = 4(0.3)$, $r = 0.3$, and $n = 12$, you can apply the formula for the sum of a finite geometric series to obtain

$$\sum_{n=1}^{12} 4(0.3)^n = a_1\left(\frac{1 - r^n}{1 - r}\right)$$

$$= 4(0.3)\left[\frac{1 - (0.3)^{12}}{1 - 0.3}\right]$$

$$\approx 1.714.$$

When using the formula for the sum of a finite geometric series, be careful to check that the index begins at $i = 1$. If the index begins at $i = 0$, you must adjust the formula for the $n$th partial sum. For instance, if the index in Example 6 had begun with $n = 0$, the sum would have been

$$\sum_{n=0}^{12} 4(0.3)^n = 4 + \sum_{n=1}^{12} 4(0.3)^n \approx 4 + 1.714 = 5.714.$$

·DISCOVERY·

Use a calculator to evaluate $r^n$ for $r = \frac{1}{2}$ and $n = 1, 2, 5,$ 10, 50, and 100. What happens to the value of $r^n$ as $n$ increases? Make a conjecture about the value of $r^n$ as $n \to \infty$. Use a graphing utility to graph $y = \left(\frac{1}{2}\right)^x$ to verify your conjecture.

The formula for the sum of a *finite* geometric series can, depending on the value of $r$, be extended to produce a formula for the sum of an *infinite* geometric series. Specifically, if the common ratio $r$ has the property that $|r| < 1$, it can be shown that $r^n$ becomes arbitrarily close to zero as $n$ increases without bound. Consequently,

$$a_1\left(\frac{1 - r^n}{1 - r}\right) \to a_1\left(\frac{1 - 0}{1 - r}\right) \quad \text{as} \quad n \to \infty.$$

This result is summarized as follows.

---

▶ **The Sum of an Infinite Geometric Series**

If $|r| < 1$, the infinite geometric series

$$a_1 + a_1 r + a_1 r^2 + a_1 r^3 + \cdots + a_1 r^{n-1} + \cdots$$

has the sum

$$S = \frac{a_1}{1 - r}.$$

---

**Example 7**    Finding the Sum of an Infinite Geometric Series

Find the sum of the infinite geometric series.

**a.** $\displaystyle\sum_{n=1}^{\infty} 4(0.6)^{n-1}$    **b.** $\displaystyle\sum_{n=1}^{\infty} 3(0.1)^{n-1}$

**Solution**

**a.** $\displaystyle\sum_{n=1}^{\infty} 4(0.6)^{n-1} = 4 + 4(0.6) + 4(0.6)^2 + 4(0.6)^3 + \cdots + 4(0.6)^{n-1} + \cdots$

$$= \frac{4}{1 - (0.6)} \qquad \frac{a_1}{1 - r}$$

$$= 10$$

**b.** $\displaystyle\sum_{n=1}^{\infty} 3(0.1)^{n-1} = 3 + 3(0.1) + 3(0.1)^2 + 3(0.1)^3 + \cdots + 3(0.1)^{n-1} + \cdots$

$$= \frac{3}{1 - (0.1)} \qquad \frac{a_1}{1 - r}$$

$$= \frac{10}{3}$$

$$\approx 3.33$$

# Application

Example 8  Compound Interest

A deposit of $50 is made on the first day of each month in a savings account that pays 6% compounded monthly. What is the balance at the end of 2 years?

Solution

The first deposit will gain interest for 24 months, and its balance will be

$$A_{24} = 50\left(1 + \frac{0.06}{12}\right)^{24} = 50(1.005)^{24}.$$

The second deposit will gain interest for 23 months, and its balance will be

$$A_{23} = 50\left(1 + \frac{0.06}{12}\right)^{23} = 50(1.005)^{23}.$$

The last deposit will gain interest for only 1 month, and its balance will be

$$A_{1} = 50\left(1 + \frac{0.06}{12}\right)^{1} = 50(1.005).$$

The total balance in the account will be the sum of the balances of the 24 deposits. Using the formula for the sum of a finite geometric series, with $A_1 = 50(1.005)$ and $r = 1.005$, you have

$$S = 50(1.005)\left[\frac{1 - (1.005)^{24}}{1 - 1.005}\right] = \$1277.96.$$

---

## Discussing the Concept    An Experiment

You will need a piece of string or yarn, a pair of scissors, and a tape measure. Measure out any length of string at least 5 feet long. Double over the string and cut it in half. Take one of the resulting halves, double it over, and cut it in half. Continue this process until you are no longer able to cut a length of string in half. How many cuts were you able to make? Construct a sequence of the resulting string lengths after each cut, starting with the original length of the string. Find a formula for the $n$th term of this sequence. How many cuts could you theoretically make? Discuss why you were not able to make that many cuts.

# Warm Up

The following warm-up exercises involve skills that were covered in earlier sections. You will use these skills in the exercise set for this section.

In Exercises 1–4, evaluate the expression.

**1.** $\left(\frac{4}{5}\right)^3$

**2.** $\left(\frac{3}{4}\right)^2$

**3.** $2^{-4}$

**4.** $4(3^4)$

In Exercises 5–8, simplify the expression.

**5.** $(2n)(3n^2)$

**6.** $n(3n)^3$

**7.** $\dfrac{4n^5}{n^2}$

**8.** $\dfrac{(2n)^3}{8n}$

In Exercises 9 and 10, use sigma notation to write the sum.

**9.** $2 + 2(3) + 2(3^2) + 2(3^3)$

**10.** $3 + 3(2) + 3(2^2) + 3(2^3)$

## 8.3  Exercises

In Exercises 1–10, determine whether the sequence is geometric. If it is, find its common ratio and write a formula for $a_n$.

**1.** $1, 4, 16, 64, \ldots$

**2.** $2, 10, 50, 250, \ldots$

**3.** $4, -12, 36, -108, \ldots$

**4.** $1, -2, 4, -8, \ldots$

**5.** $1, -\frac{1}{2}, \frac{1}{4}, -\frac{1}{8}, \ldots$

**6.** $\frac{3}{2}, 1, \frac{2}{3}, \frac{4}{9}, \ldots$

**7.** $\frac{1}{2}, \frac{2}{3}, \frac{3}{4}, \frac{4}{5}, \ldots$

**8.** $9, -6, 4, -\frac{8}{3}, \ldots$

**9.** $1, \frac{1}{2}, \frac{1}{3}, \frac{1}{4}, \ldots$

**10.** $\frac{1}{5}, \frac{2}{3}, \frac{3}{9}, \frac{4}{11}, \ldots$

In Exercises 11–18, write the first five terms of the geometric sequence.

**11.** $a_1 = 4, r = 2$

**12.** $a_1 = 6, r = 3$

**13.** $a_1 = 1, r = \frac{1}{3}$

**14.** $a_1 = 1, r = \frac{2}{3}$

**15.** $a_1 = 7, r = -\frac{1}{5}$

**16.** $a_1 = \frac{1}{2}, r = -\frac{1}{3}$

**17.** $a_1 = 1, r = e$

**18.** $a_1 = 2, r = e^{0.1}$

In Exercises 19–30, find the indicated term of the geometric sequence.

**19.** $a_1 = 16, r = \frac{1}{4}$, 5th term

**20.** $a_1 = 9, r = \frac{2}{3}$, 7th term

**21.** $a_1 = 6, r = -\frac{1}{3}$, 12th term

**22.** $a_1 = 1, r = -\frac{2}{3}$, 10th term

**23.** $a_1 = 100, r = e$, 9th term

**24.** $a_1 = 6, r = e^{0.1}$, 9th term

**25.** $a_1 = 500, r = 1.02$, 40th term

**26.** $a_1 = 1000, r = 1.005$, 60th term

**27.** $a_1 = 16, a_4 = \frac{27}{4}$, 3rd term

**28.** $a_2 = 3, a_5 = \frac{3}{64}$, 1st term

**29.** $a_2 = -18, a_5 = \frac{2}{3}$, 6th term

**30.** $a_3 = \frac{16}{3}, a_5 = \frac{64}{27}$, 7th term

In Exercises 31–38, find the indicated sum.

**31.** $\displaystyle\sum_{n=1}^{10} 8(2^n)$

**32.** $\displaystyle\sum_{n=1}^{6} 3(4^n)$

**33.** $\displaystyle\sum_{i=1}^{10} 8\left(\frac{1}{4}\right)^{i-1}$

**34.** $\displaystyle\sum_{i=1}^{10} 5\left(\frac{1}{3}\right)^{i-1}$

**35.** $\displaystyle\sum_{n=0}^{8} 2^n$

**36.** $\displaystyle\sum_{n=0}^{8} (-2)^n$

**37.** $\displaystyle\sum_{n=0}^{20} 3\left(\frac{3}{2}\right)^n$

**38.** $\displaystyle\sum_{n=0}^{15} 2\left(\frac{4}{3}\right)^n$

In Exercises 39–46, find the sum, if it exists.

**39.** $\displaystyle\sum_{n=0}^{\infty} \left(\frac{1}{2}\right)^n = 1 + \frac{1}{2} + \frac{1}{4} + \frac{1}{8} + \cdots$

**40.** $\displaystyle\sum_{n=0}^{\infty} 2\left(\frac{2}{3}\right)^n = 2 + \frac{4}{3} + \frac{8}{9} + \frac{16}{27} + \cdots$

**41.** $\displaystyle\sum_{n=1}^{\infty} \left(-\frac{1}{2}\right)^{n-1} = 1 - \frac{1}{2} + \frac{1}{4} - \frac{1}{8} + \cdots$

**42.** $\displaystyle\sum_{n=1}^{\infty} 2\left(-\frac{2}{3}\right)^{n-1} = 2 - \frac{4}{3} + \frac{8}{9} - \frac{16}{27} + \cdots$

**43.** $\displaystyle\sum_{n=0}^{\infty} 4\left(\frac{1}{4}\right)^n = 4 + 1 + \frac{1}{4} + \frac{1}{16} + \cdots$

**44.** $\displaystyle\sum_{n=0}^{\infty} 3\left(\frac{1}{10}\right)^n = 3 + 0.3 + 0.03 + 0.003 + \cdots$

**45.** $\displaystyle\sum_{n=0}^{\infty} \left(\frac{3}{2}\right)^n = 1 + \frac{3}{2} + \frac{9}{4} + \frac{27}{8} + \cdots$

**46.** $\displaystyle\sum_{n=0}^{\infty} 5\left(\frac{4}{3}\right)^n = 5 + \frac{20}{3} + \frac{80}{9} + \frac{320}{27} + \cdots$

**47.** *Compound Interest*    A deposit of $100 is made at the beginning of each month for 5 years in an account that pays 6% compounded monthly. The balance $A$ in the account at the end of 5 years is

$$A = 100\left(1 + \frac{0.06}{12}\right)^1 + \cdots + 100\left(1 + \frac{0.06}{12}\right)^{60}.$$

(a) Find the balance after 5 years.

(b) How much would the balance increase if the interest rate were raised to 7%?

(c) At 7% interest, find the balance after 10 years.

**48.** *Compound Interest*    A deposit of $50 is made at the beginning of each month for 5 years in an account that pays 5% compounded monthly. Use a graphing utility to find the balance $A$ in the account at the end of 5 years.

$$A = 50\left(1 + \frac{0.05}{12}\right)^1 + \cdots + 50\left(1 + \frac{0.05}{12}\right)^{60}$$

**49.** *Compound Interest*    Suppose you deposit $75 in an account at the beginning of each month for 10 years. The account pays 8% compounded monthly. Use a graphing utility to find your balance at the end of 10 years. If the interest were compounded continuously, what would the balance be?

**50.** *Compound Interest*    Suppose you deposit $200 in an account at the beginning of each month for 40 years. The account pays 6% compounded monthly. What would your balance be at the end of 40 years? If the interest were compounded continuously, what would the balance be?

**51.** *Profit*    The annual profits for the Procter & Gamble Company from 1995 to 2000 can be approximated by the model

$$a_n = 2.747e^{0.096n}, \quad n = 0, 1, \ldots, 5$$

where $a_n$ is the annual profit (in billions of dollars) and $n$ represents the year, with $n = 0$ corresponding to 1995 (see figure). Use the formula for the sum of a finite geometric series to approximate the total profit earned during this 6-year period. (Source: Procter & Gamble Company)

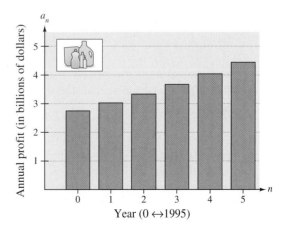

**52.** *Profit* The annual profits for Microsoft Corporation from 1991 to 2000 can be approximated by the model

$$a_n = 456.31e^{0.336n}, \quad n = 0, 1, 2, 3, \ldots, 9$$

where $a_n$ is the annual profit (in millions of dollars) and $n$ represents the year, with $n = 0$ corresponding to 1991 (see figure). Use the formula for the sum of a finite geometric series to approximate the total profit earned during this 10-year period. (Source: Microsoft Corp.)

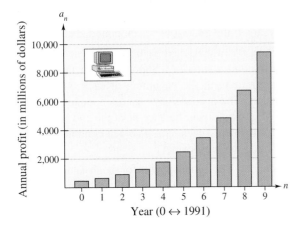

Year (0 ↔ 1991)

**53.** *Sales* The annual sales for La-Z-Boy, Inc. from 1997 to 2001 can be modeled by

$$a_n = 0.981e^{0.197n}, \quad n = 0, 1, 2, 3, 4$$

where $a_n$ is the annual sales (in billions of dollars) and $n$ represents the year, with $n = 0$ corresponding to 1997. Use the formula for the sum of a finite geometric series to approximate the total sales earned during this 5-year period. (Source: La-Z-Boy, Inc.)

**54.** *Sales* The annual sales for Newell Rubbermaid from 1990 to 2000 can be approximated by the model

$$a_n = 2.93e^{0.095n}, \quad n = 0, 1, 2, 3, \ldots, 10$$

where $a_n$ is the annual sales (in billions of dollars) and $n$ represents the year, with $n = 0$ corresponding to 1990. Use the formula for the sum of a finite geometric series to approximate the total sales earned during this 11-year period. (Source: Newell Rubbermaid, Inc.)

**55.** *Would You Take This Job?* Suppose you go to work at a company that pays $0.01 for the first day, $0.02 for the second day, $0.04 for the third day, and so on. If the daily wage keeps doubling, what will your total income be for working 29 days? 30 days? 31 days?

**56.** *Gambler's Ruin* Suppose you were planning to visit a gambling casino and a friend suggested the following gambling strategy. Find a game that pays $2 for each dollar bet (such as the pass line at craps). Start by betting $2, and every time you win, return the $2 bet. If you lose the $2 bet, then bet $4. If you lose the $4 bet, then bet $8. For instance, if you lose four times in a row and then win, you will have bet 2 + 4 + 8 + 16 + 32 and you will have won $64 on the last bet (which puts you $2 ahead). What is wrong with this gambling strategy? How much money would you have to take to cover the possibility of losing 12 times in a row?

**57.** *Area* The sides of a square are 16 inches in length. A new square is formed by connecting the midpoints of the sides of the original square, and two of the triangles are shaded (see figure). If this process is repeated five more times, what will the total area of the shaded region be?

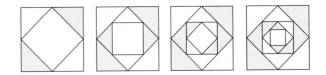

**58.** *Area* The sides of a square are 16 inches in length. The square is divided into nine smaller squares and the center square is shaded (see figure). Each of the eight unshaded squares is then divided into nine smaller squares and the center square of each is shaded. If this process is repeated four more times, what will the total area of the shaded region be?

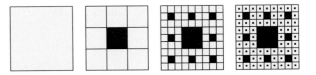

## Math Matters • The Snowflake Curve

To create a **snowflake curve,** begin with an equilateral triangle (called Stage 1). To form Stage 2, trisect each side of the triangle and at each of the middle sections construct an equilateral triangle pointing outward. To form each additional stage, repeat this process. The following diagram shows the first six stages of the snowflake curve, which is the curve obtained by continuing this process infinitely many times.

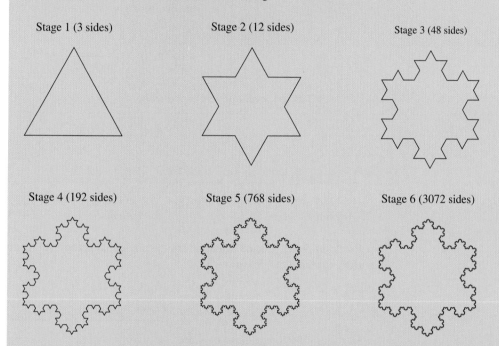

Stage 1 (3 sides)    Stage 2 (12 sides)    Stage 3 (48 sides)

Stage 4 (192 sides)    Stage 5 (768 sides)    Stage 6 (3072 sides)

In this figure, note that the $n$th stage of the snowflake curve has $3(4^{n-1})$ sides. If each side in Stage 1 has a length of 1, then each side in Stage 2 will have a length of $\frac{1}{3}$, each side in Stage 3 will have a length of $\left(\frac{1}{3}\right)^2 = \frac{1}{9}$, and so on. Thus, the total perimeter of the $n$th stage is

$$3(4^{n-1}) \cdot \left(\frac{1}{3}\right)^{n-1} = 3\left(\frac{4}{3}\right)^{n-1}.$$

Because the value of this expression is unbounded as $n$ approaches infinity, you can conclude that the perimeter of the snowflake curve is infinite. This seems to be a paradox because it is difficult to conceive of a curve that is infinitely long being drawn on a piece of paper that has a finite area.

## Mid-Chapter Quiz

Take this quiz as you would take a quiz in class. After you are done, check your work against the answers given in the back of the book.

In Exercises 1 and 2, write the first five terms of the sequence. (Begin with $n = 1$.)

**1.** $a_n = 2n + 1$

**2.** $a_n = \dfrac{n}{(n + 1)^2}$

In Exercises 3 and 4, evaluate the expression.

**3.** $\dfrac{5!}{3!}$

**4.** $\dfrac{30!}{27!}$

In Exercises 5–8, write an expression for $a_n$. (Begin with $n = 1$.)

**5.** $2, \frac{1}{2}, \frac{2}{9}, \frac{1}{8}, \frac{2}{25}, \ldots$

**6.** $\frac{1}{2}, \frac{1}{3}, \frac{1}{4}, \frac{1}{5}, \frac{1}{6}, \ldots$

**7.** *Arithmetic:* $a_1 = 2, d = 5$

**8.** *Arithmetic:* $a_1 = 7, d = 3$

In Exercises 9–12, decide whether the sequence is arithmetic, geometric, or neither. Explain your reasoning.

**9.** $a_n = 2 + 3(n - 1)$

**10.** $3, -6, 18, -72, \ldots$

**11.** $1, \frac{1}{2}, \frac{1}{4}, \frac{1}{8}, \ldots$

**12.** $a_n = 3(2)^n$

In Exercises 13–18, evaluate the sum.

**13.** $\displaystyle\sum_{i=1}^{5} (2i - 1)$

**14.** $\displaystyle\sum_{i=1}^{4} 2i^2$

**15.** $\displaystyle\sum_{n=1}^{25} 2n$

**16.** $\displaystyle\sum_{n=1}^{6} 2^n$

**17.** $\displaystyle\sum_{n=0}^{\infty} \left(\tfrac{1}{3}\right)^n$

**18.** $\displaystyle\sum_{n=0}^{\infty} 5\left(\tfrac{1}{5}\right)^n$

**19.** A person accepts a position with a company and will receive a salary of $28,500 for the first year. The person is guaranteed a raise of $2000 per year for the first 5 years.

   (a) What will the salary be during the sixth year of employment?

   (b) How much will the company have paid the person by the end of the sixth year?

**20.** Suppose that you deposit $75 in an account at the beginning of each month for 20 years. The account pays 7% compounded monthly. What will your balance be at the end of the 20 years?

## 8.4    The Binomial Theorem

### Objectives

- Find a binomial coefficient.
- Use Pascal's Triangle to determine a binomial coefficient.
- Expand a binomial using binomial coefficients.
- Find a specific term in a binomial expansion.

## Binomial Coefficients

Recall that a **binomial** is a polynomial that has two terms. In this section, you will study a formula that gives a quick method of raising a binomial to a power. To begin, let's look at the expansion of $(x + y)^n$ for several values of $n$.

$$(x + y)^0 = 1$$
$$(x + y)^1 = x + y$$
$$(x + y)^2 = x^2 + 2xy + y^2$$
$$(x + y)^3 = x^3 + 3x^2y + 3xy^2 + y^3$$
$$(x + y)^4 = x^4 + 4x^3y + 6x^2y^2 + 4xy^3 + y^4$$
$$(x + y)^5 = x^5 + 5x^4y + 10x^3y^2 + 10x^2y^3 + 5xy^4 + y^5$$

There are several observations you can make about these expansions.

1. In each expansion, there are $n + 1$ terms.

2. In each expansion, $x$ and $y$ have symmetrical roles. The powers of $x$ decrease by 1 in successive terms, whereas the powers of $y$ increase by 1.

3. The sum of the powers of each term is $n$. For instance, in the expansion of $(x + y)^5$, the sum of the powers of each term is 5.

$$4 + 1 = 5 \qquad 3 + 2 = 5$$
$$(x + y)^5 = x^5 + 5x^4y^1 + 10x^3y^2 + 10x^2y^3 + 5xy^4 + y^5$$

4. The coefficients increase and then decrease in a symmetric pattern.

The coefficients of a binomial expansion are called **binomial coefficients.** To find them, you can use the following theorem.

---

▶ **The Binomial Theorem**

In the expansion of $(x + y)^n$

$$(x + y)^n = x^n + nx^{n-1}y + \cdots + {}_nC_r x^{n-r}y^r + \cdots + nxy^{n-1} + y^n$$

the coefficient of $x^{n-r}y^r$ is given by

$${}_nC_r = \frac{n!}{(n - r)!r!}.$$

---

**Example 1**    Finding Binomial Coefficients

Find the binomial coefficients.

**a.** $_8C_2$    **b.** $_{10}C_3$    **c.** $_7C_0$    **d.** $_8C_8$

Solution

**a.** $_8C_2 = \dfrac{8!}{6! \cdot 2!} = \dfrac{(8 \cdot 7) \cdot 6!}{6! \cdot 2!} = \dfrac{8 \cdot 7}{2 \cdot 1} = 28$

**b.** $_{10}C_3 = \dfrac{10!}{7! \cdot 3!} = \dfrac{(10 \cdot 9 \cdot 8) \cdot 7!}{7! \cdot 3!} = \dfrac{10 \cdot 9 \cdot 8}{3 \cdot 2 \cdot 1} = 120$

**c.** $_7C_0 = \dfrac{7!}{7! \cdot 0!} = 1$

**d.** $_8C_8 = \dfrac{8!}{0! \cdot 8!} = 1$

When $r \neq 0$ and $r \neq n$, as in parts (a) and (b) above, there is a simple pattern for evaluating binomial coefficients.

$$_8C_2 = \overbrace{\dfrac{8 \cdot 7}{\underbrace{2 \cdot 1}_{\text{2 factorial}}}}^{\text{2 factors}} \quad \text{and} \quad _{10}C_3 = \overbrace{\dfrac{10 \cdot 9 \cdot 8}{\underbrace{3 \cdot 2 \cdot 1}_{\text{3 factorial}}}}^{\text{3 factors}}$$

**Example 2**    Finding Binomial Coefficients

Find the binomial coefficients.

**a.** $_7C_3$    **b.** $_7C_4$    **c.** $_{12}C_1$    **d.** $_{12}C_{11}$

Solution

**a.** $_7C_3 = \dfrac{7 \cdot 6 \cdot 5}{3 \cdot 2 \cdot 1} = 35$

**b.** $_7C_4 = \dfrac{7 \cdot 6 \cdot 5 \cdot 4}{4 \cdot 3 \cdot 2 \cdot 1} = 35$

**c.** $_{12}C_1 = \dfrac{12}{1} = 12$

**d.** $_{12}C_{11} = \dfrac{12!}{1! \cdot 11!} = \dfrac{(12) \cdot 11!}{1! \cdot 11!} = \dfrac{12}{1} = 12$

It is not a coincidence that the results in parts (a) and (b) of Example 2 are the same and that the results in parts (c) and (d) are the same. In general, it is true that

$$_nC_r = _nC_{n-r}.$$

This shows the symmetric property of binomial coefficients that was identified earlier.

# Pascal's Triangle

There is a convenient way to remember a pattern for binomial coefficients. By arranging the coefficients in a triangular pattern, you obtain the following array, which is called **Pascal's Triangle.** This triangle is named after the famous French mathematician Blaise Pascal (1623–1662).

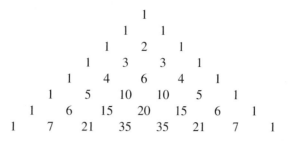

The first and last number in each row of Pascal's Triangle is 1. Every other number in each row is formed by adding the two numbers immediately above the number. Pascal noticed that numbers in this triangle are precisely the same numbers that are the coefficients of binomial expansions, as follows.

$$(x + y)^0 = 1$$

$$(x + y)^1 = 1x + 1y$$

$$(x + y)^2 = 1x^2 + 2xy + 1y^2$$

$$(x + y)^3 = 1x^3 + 3x^2y + 3xy^2 + 1y^3$$

$$(x + y)^4 = 1x^4 + 4x^3y + 6x^2y^2 + 4xy^3 + 1y^4$$

$$(x + y)^5 = 1x^5 + 5x^4y + 10x^3y^2 + 10x^2y^3 + 5xy^4 + 1y^5$$

$$(x + y)^6 = 1x^6 + 6x^5y + 15x^4y^2 + 20x^3y^3 + 15x^2y^4 + 6xy^5 + 1y^6$$

$$(x + y)^7 = 1x^7 + 7x^6y + 21x^5y^2 + 35x^4y^3 + 35x^3y^4 + 21x^2y^5 + 7xy^6 + 1y^7$$

The top row in Pascal's Triangle is called the *zero row* because it corresponds to the binomial expansion $(x + y)^0 = 1$.

Similarly, the next row is called the *first row* because it corresponds to the binomial expansion $(x + y)^1 = 1x + 1y$. In general, the *nth row* in Pascal's Triangle gives the coefficients of $(x + y)^n$.

**Example 3**   Using Pascal's Triangle

Use the seventh row of Pascal's Triangle to find the binomial coefficients.

$$_8C_0, \, _8C_1, \, _8C_2, \, _8C_3, \, _8C_4, \, _8C_5, \, _8C_6, \, _8C_7, \, _8C_8$$

Solution

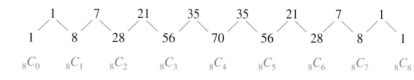

# Binomial Expansions

As mentioned at the beginning of this section, when you write out the coefficients for a binomial that is raised to a power, you are **expanding a binomial.** The formulas for binomial coefficients give you an easy way to expand binomials, as demonstrated in the next four examples.

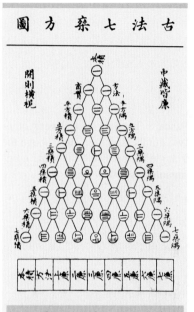

圖 方 蔡 七 法 古

開則橫視

中藏皆庶

**Pascal's Triangle**

"Pascal's" Triangle and forms of the Binomial Theorem were known in Eastern cultures prior to the Western "discovery" of the theorem. A Chinese text *Precious Mirror* contains a triangle of binomial expansions through the eighth power.

**Example 4**    Expanding a Binomial

Write the expansion for the expression

$$(x + 1)^3.$$

Solution

The binomial coefficients from the third row of Pascal's Triangle are

$$1, 3, 3, 1.$$

So, the expansion is as follows.

$$(x + 1)^3 = (1)x^3 + (3)x^2(1) + (3)x(1^2) + (1)(1^3)$$

$$= x^3 + 3x^2 + 3x + 1$$

To expand binomials representing *differences,* rather than sums, you alternate signs. Here are two examples.

$$(x - 1)^3 = x^3 - 3x^2 + 3x - 1$$

$$(x - 1)^4 = x^4 - 4x^3 + 6x^2 - 4x + 1$$

**Example 5**    Expanding a Binomial

Write the expansion for the expression

$$(x - 3)^4.$$

Solution

The binomial coefficients from the fourth row of Pascal's Triangle are

$$1, 4, 6, 4, 1.$$

So, the expansion is as follows.

$$(x - 3)^4 = (1)x^4 - (4)x^3(3) + (6)x^2(3^2) - (4)x(3^3) + (1)(3^4)$$

$$= x^4 - 12x^3 + 54x^2 - 108x + 81$$

**Example 6**   Expanding a Binomial   *Learning Tools*

Write the expansion for $(x - 2y)^4$.

**Solution**

Use the fourth row of Pascal's Triangle, as follows.

$$(x - 2y)^4 = (1)x^4 - (4)x^3(2y) + (6)x^2(2y)^2 - (4)x(2y)^3 + (1)(2y)^4$$

$$= x^4 - 8x^3y + 24x^2y^2 - 32xy^3 + 16y^4$$

**Example 7**   Finding a Term in a Binomial Expansion   *Learning Tools*

Find the sixth term of $(a + 2b)^8$.

**Solution**

From the Binomial Theorem, you can see that the $(r + 1)$th term is $_nC_r x^{n-r}y^r$. So in this case, $6 = r + 1$ means that $r = 5$. Because $n = 8$, $x = a$, and $y = 2b$, the sixth term in the binomial expansion is

$$_8C_5a^{8-5}(2b)^5 = 56 \cdot a^3 \cdot (2b)^5$$

$$= 56(2^5)a^3b^5$$

$$= 1792a^3b^5.$$

---

**Discussing the Concept**   **Finding a Pattern**

By adding the terms in each of the rows of Pascal's Triangle, you obtain the following.

Row 0:   $1 = 1$

Row 1:   $1 + 1 = 2$

Row 2:   $1 + 2 + 1 = 4$

Row 3:   $1 + 3 + 3 + 1 = 8$

Row 4:   $1 + 4 + 6 + 4 + 1 = 16$

Find a pattern for this sequence. Then use the pattern to find the sum of the terms in the 10th row of Pascal's Triangle. Check your answer by actually adding the terms of the 10th row.

# Warm Up

The following warm-up exercises involve skills that were covered in earlier sections. You will use these skills in the exercise set for this section.

In Exercises 1–6, perform the indicated operations and/or simplify.

**1.** $5x^2(x^3 + 3)$

**2.** $(x + 5)(x^2 - 3)$

**3.** $(x + 4)^2$

**4.** $(2x - 3)^2$

**5.** $x^2y(3xy^{-2})$

**6.** $(-2z)^5$

In Exercises 7–10, evaluate the expression.

**7.** $5!$

**8.** $\dfrac{8!}{5!}$

**9.** $\dfrac{10!}{7!}$

**10.** $\dfrac{6!}{3!3!}$

## 8.4 Exercises

In Exercises 1–10, evaluate $_nC_r$.

**1.** $_6C_2$

**2.** $_9C_4$

**3.** $_{12}C_0$

**4.** $_8C_8$

**5.** $_{20}C_{15}$

**6.** $_{20}C_5$

**7.** $_{100}C_{98}$

**8.** $_{10}C_4$

**9.** $_{100}C_2$

**10.** $_{10}C_6$

In Exercises 11–30, use the Binomial Theorem to expand the expression. Simplify your answer.

**11.** $(x + 1)^4$

**12.** $(x + 1)^6$

**13.** $(x + 2)^3$

**14.** $(x + 2)^4$

**15.** $(x - 2)^4$

**16.** $(x - 2)^5$

**17.** $(x + y)^5$

**18.** $(x + y)^6$

**19.** $(s + 2)^4$

**20.** $(s + 3)^5$

**21.** $(2 + 3s)^6$

**22.** $(2 + 3s)^5$

**23.** $(x - y)^5$

**24.** $(2x - y)^5$

**25.** $(1 - 2x)^3$

**26.** $(1 - 3x)^4$

**27.** $(x^2 - y^2)^5$

**28.** $(x^2 + y^2)^6$

**29.** $\left(\dfrac{1}{x} + y\right)^5$

**30.** $\left(\dfrac{x}{2} - 3y\right)^3$

**31.** How many terms are in the expanded form of $(x + y)^5$? $(x + y)^6$? $(x + y)^7$? $(x + y)^8$? How many terms are in $(x + y)^n$?

**32.** How are the expansions of the binomials $(x + y)^n$ and $(x - y)^n$ alike? How are they different?

In Exercises 33–38, use the Binomial Theorem to expand the complex number. Simplify your answer by using the fact that $i^2 = -1$.

**33.** $(1 + i)^4$

**34.** $(2 - i)^5$

**35.** $\left(4 - \sqrt{-4}\right)^6$

**36.** $\left(5 + \sqrt{-9}\right)^3$

**37.** $\left(\dfrac{-1}{2} + \dfrac{\sqrt{3}}{2}i\right)^3$

**38.** $\left(\dfrac{3}{4} - \dfrac{\sqrt{2}}{4}i\right)^4$

In Exercises 39–42, expand the expression using Pascal's Triangle to determine the coefficients.

**39.** $(2t - 1)^5$

**40.** $(x - 4)^5$

**41.** $(3 + 2z)^4$

**42.** $(3y + 2)^5$

In Exercises 43–50, find the indicated term in the expansion of the binomial expression.

**43.** $(x + 2)^{10}$      $ax^6$

**44.** $(x^2 + 3)^{12}$      $ax^8$

**45.** $(x - 2y)^{10}$      $ax^8y^2$

**46.** $(4x - y)^{10}$      $ax^3y^7$

**47.** $(3x - 2y)^9$      $ax^4y^5$

**48.** $(2x - 3y)^{11}$      $ax^3y^8$

**49.** $(x^2 - 1)^8$      $ax^8$

**50.** $(x^2 + 2)^{12}$      $ax^{10}$

In Exercises 51–56, use the Binomial Theorem to expand the expression. [In the study of probability, it is sometimes necessary to use the expansion of $(p + q)^n$, where $p + q = 1$.]

**51.** $\left(\frac{1}{2} + \frac{1}{2}\right)^7$

**52.** $\left(\frac{1}{4} + \frac{3}{4}\right)^{10}$

**53.** $\left(\frac{1}{3} + \frac{2}{3}\right)^8$

**54.** $(0.3 + 0.7)^{12}$

**55.** $(0.6 + 0.4)^5$

**56.** $(0.35 + 0.65)^6$

**57.** *Finding a Pattern*   Describe the pattern formed by the sums of the numbers along the diagonal segments of Pascal's Triangle (see figure).

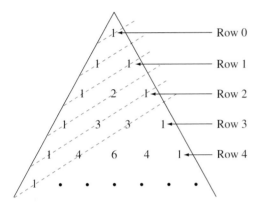

**58.** *Finding a Pattern*   Use each of the encircled groups of numbers in the figure to form a $2 \times 2$ matrix. Find the determinant of the matrix. Then describe the pattern.

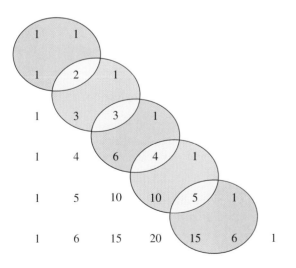

**59.** *Finding a Pattern*   Each of the "flowers" in the figure is taken from Pascal's Triangle. Copy the flowers and fill in the missing numbers. Which rows did the flowers come from?

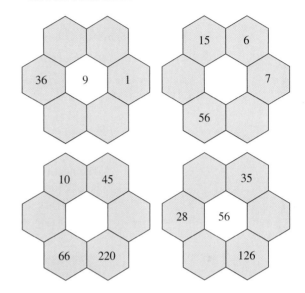

**60.** *Life Insurance*   The average amount of life insurance per household $f(t)$ (in thousands of dollars) in the United States from 1970 through 1998 can be approximated by

$$f(t) = 0.096t^2 + 1.78t + 18.2, \quad 0 \le t \le 28$$

where $t = 0$ represents 1970 (see figure). You want to adjust this model so that $t = 0$ corresponds to 1980 rather than 1970. To do this, you shift the graph of $f$ 10 units *to the left* and obtain

$$g(t) = f(t + 10).$$

(Source: American Council of Life Insurance)

(a) Write $g(t)$ in standard form.

(b) Use a graphing utility to graph $f$ and $g$ in the same viewing window.

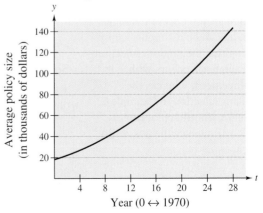

## 8.5    Counting Principles

### Objectives

- Solve simple counting problems.
- Use the Fundamental Counting Principle.
- Determine the number of permutations of $n$ elements taken $r$ at a time.
- Determine the number of combinations of $n$ elements taken $r$ at a time.

## Simple Counting Problems

The next two sections of this chapter contain a brief introduction to some of the basic counting principles and their application to probability. In the next section, you will see that much of probability has to do with counting the number of ways an event can occur.

 **Example 1**    Selecting Pairs of Numbers at Random

Eight pieces of paper are numbered from 1 to 8 and placed in a box. One piece of paper is drawn from the box, its number is written down, and the piece of paper is replaced in the box. Then, a second piece of paper is drawn from the box, and its number is written down. Finally, the two numbers are added together. How many different ways can a total of 12 be obtained?

### Solution

To solve this problem, count the different ways that a total of 12 can be obtained using two numbers from 1 to 8.

| *First number* | 4 | 5 | 6 | 7 | 8 |
|---|---|---|---|---|---|
| *Second number* | 8 | 7 | 6 | 5 | 4 |

From this list, you can see that a total of 12 can occur in five different ways.

**Example 2**    Selecting Pairs of Numbers at Random

Eight pieces of paper are numbered from 1 to 8 and placed in a box. Two pieces of paper are drawn from the box, and the numbers on the pieces of paper are written down and totaled. How many different ways can a total of 12 be obtained?

### Solution

To solve this problem, you can count the different ways that a total of 12 can be obtained *using two different numbers* from 1 to 8.

| *First number* | 4 | 5 | 7 | 8 |
|---|---|---|---|---|
| *Second number* | 8 | 7 | 5 | 4 |

So, a total of 12 can be obtained in four different ways.

# Warm Up

The following warm-up exercises involve skills that were covered in earlier sections. You will use these skills in the exercise set for this section.

In Exercises 1–4, evaluate the expression.

**1.** $13 \cdot 8^2 \cdot 2^3$

**2.** $10^2 \cdot 9^3 \cdot 4$

**3.** $\dfrac{12!}{2!(7!)(3!)}$

**4.** $\dfrac{25!}{22!}$

In Exercises 5 and 6, find the binomial coefficient.

**5.** $_{12}C_7$

**6.** $_{25}C_{22}$

In Exercises 7–10, simplify the expression.

**7.** $\dfrac{n!}{(n-4)!}$

**8.** $\dfrac{(2n)!}{4(2n-3)!}$

**9.** $\dfrac{2 \cdot 4 \cdot 6 \cdot 8 \cdots (2n)}{2^n}$

**10.** $\dfrac{3 \cdot 6 \cdot 9 \cdot 12 \cdots (3n)}{3^n}$

## 8.5    Exercises

**1.** *Job Applicants*  A small college needs two additional faculty members: a chemist and a statistician. In how many ways can these positions be filled if there are four applicants for the chemistry position and five for the position in statistics?

**2.** *Computer Systems*  A customer in a computer store can choose one of five monitors, one of four keyboards, and one of six computers. If all the choices are compatible, how many different systems could be chosen?

**3.** *Toboggan Ride*  Six people are lining up for a ride on a toboggan, but only two of the six are willing to take the first position. With that constraint, in how many ways can the six people be seated on the toboggan? Draw a diagram to illustrate the number of ways that people can sit.

**4.** *Course Schedule*  A college student is preparing her course schedule for the next semester. She may select one of five mathematics courses, one of three science courses, and one of eight courses from the social sciences and humanities. In how many ways can she select her schedule?

**5.** *License Plate Numbers*  In a certain state, each automobile license plate number consists of two letters followed by a four-digit number. How many distinct license plate numbers can be formed?

**6.** *License Plate Numbers*  In a certain state, each automobile license plate number consists of two letters followed by a four-digit number. To avoid confusion between "O" and "zero" and "I" and "one," the letters "O" and "I" are not used. How many distinct license plate numbers can be formed?

**7.** *True-False Exam*  In how many ways can a six-question true-false exam be answered? (Assume that no questions are omitted.)

**8.** *True-False Exam*  In how many ways can a 10-question true-false exam be answered? (Assume that no questions are omitted.)

**9.** *Three-Digit Numbers*  How many three-digit numbers can be formed under the following conditions?

(a) Leading digits cannot be zero.

(b) Leading digits cannot be zero and no repetition of digits is allowed.

(c) Leading digits cannot be zero and the number must be a multiple of 5.

**10.** *Four-Digit Numbers*    How many four-digit numbers can be formed under the following conditions?

(a) Leading digits cannot be zero.

(b) Leading digits cannot be zero and no repetition of digits is allowed.

(c) Leading digits cannot be zero and the number must be a multiple of 5.

**11.** *Combination Lock*    A combination lock will open when the right choice of three numbers (from 1 to 40, inclusive) is selected. How many different lock combinations are possible?

**12.** *Combination Lock*    A combination lock will open when the right choice of three numbers (from 1 to 50, inclusive) is selected. How many different lock combinations are possible?

**13.** *Concert Seats*    Four couples have reserved seats in a given row for a concert. In how many different ways can they be seated, given the following conditions?

(a) There are no restrictions.

(b) The two members of each couple wish to sit together.

**14.** *Single File*    In how many orders can five girls and three boys walk through a doorway single-file, given the following conditions?

(a) There are no restrictions.

(b) The boys go before the girls.

(c) The girls go before the boys.

In Exercises 15–24, evaluate $_nP_r$.

**15.** $_4P_3$

**16.** $_5P_4$

**17.** $_8P_3$

**18.** $_{10}P_5$

**19.** $_{12}P_4$

**20.** $_{20}P_2$

**21.** $_{100}P_2$

**22.** $_{48}P_3$

**23.** $_6P_6$

**24.** $_7P_7$

**25.** *Posing for a Photograph*    In how many ways can five children line up in one row to have their picture taken?

**26.** *Riding in a Car*    In how many ways can eight people sit in an eight-passenger van?

**27.** *Choosing Officers*    From a pool of 12 candidates, the offices of president, vice-president, secretary, and treasurer will be filled. In how many different ways can the offices be filled if each of the 12 candidates can hold any office?

**28.** *Assembly Line Production*    There are four processes involved in assembling a certain product, and these processes can be performed in any order. The management wants to test each order to determine which is the least time consuming. How many different orders will have to be tested?

**29.** *Forming an Experimental Group*    In order to conduct a certain experiment, four students are randomly selected from a class of 26. How many different groups of four students are possible?

**30.** *Test Questions*    A student may answer any 12 questions from a total of 15 questions on an exam. How many different ways can the student select the questions?

**31.** *Lottery Choices*    There are 40 numbers in a particular state lottery. In how many ways can a player select six of the numbers? (The order of selection is not important.)

**32.** *Lottery Choices*    There are 50 numbers in a particular state lottery. In how many ways can a player select eight of the numbers? (The order of selection is not important.)

**33.** *Number of Subsets*    How many subsets of four elements can be formed from a set of 100 elements?

**34.** *Number of Subsets*    How many subsets of five elements can be formed from a set of 80 elements?

**35.** *Forming a Committee*    A committee composed of three graduate students and two undergraduate students is to be selected from a group of 10 graduates and six undergraduates. How many different committees can be formed?

**36.** *Defective Units*    A shipment of 12 microwave ovens contains three defective units. In how many ways can a vending company purchase four of these units and receive (a) all good units, (b) two good units, and (c) at least two good units?

**37.** *Job Applicants*    An employer interviews 10 people for four openings in the company. Four of the 10 people are women. If all 10 are qualified, in how many ways can the employer fill the four positions if (a) the selection is random and (b) exactly two selections are women?

**38.** *Forming a Committee*    Four people are randomly chosen from a group of four couples. In how many ways can this happen, given the following conditions?

(a) There are no restrictions.

(b) The selection must include one member from each couple.

**39.** *Poker Hand*    Five cards are selected from an ordinary deck of 52 playing cards. In how many ways can you get a full house? (A full house consists of three of one kind and two of another. For example, A-A-A-5-5 and K-K-K-10-10 are full houses.)

**40.** *Poker Hand*    Five cards are selected from an ordinary deck of 52 playing cards. In how many ways can you get a straight flush? (A straight flush consists of five cards that are in order and of the same suit. For example, A♥, 2♥, 3♥, 4♥, 5♥ and 10♠, J♠, Q♠, K♠, A♠ are straight flushes.)

*Diagonals of a Polygon*    In Exercises 41–44, find the number of diagonals of the polygon. (A line segment connecting any two nonadjacent vertices is called a *diagonal* of the polygon.)

**41.** Pentagon        **42.** Hexagon

**43.** Octagon        **44.** Decagon (10 sides)

In Exercises 45–50, find the number of distinguishable permutations of the group of letters.

**45.** A, S, S, E, S, S

**46.** P, A, P, A, Y, A

**47.** R, E, F, E, R, E, E

**48.** P, A, R, A, L, L, E, L

**49.** A, L, G, E, B, R, A

**50.** M, I, S, S, I, S, S, I, P, P, I

In Exercises 51–53, use the example of the tree diagram to determine the possible arrangements needed to answer the question.

A college theater company is planning a tour with performances in Phoenix, Dallas, and New Orleans. If there are no restrictions on the order of the performances, how many ways can the tour be arranged?

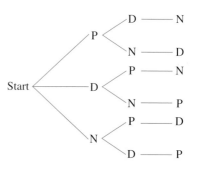

To determine the possible arrangements, we count the paths. A path moves from the start to the end of the last branch. In this case, there are six paths, so there are six possible ways to arrange the itinerary of the tour. The tree diagram verifies visually what we know by using counting principles. That is, the number of possible arrangements would be $3 \cdot 2 \cdot 1 = 6$.

**51.** Use a tree diagram to show the possible arrangements for the college theater company if the city of Houston is added to its tour.

**52.** Use a tree diagram to describe visually the possible arrangements for the genders of three children within a family.

**53.** Use a tree diagram to describe visually the possible arrangements for the genders of four children within a family.

**54.** Complete the following table.

| $n$ | $_nP_n$ | $n!$ |
|---|---|---|
| 4 | | |
| 5 | | |
| 6 | | |
| 7 | | |

What do you observe about the values in the second and third columns? Explain.

## 8.6    Probability

### Objectives

- Determine the sample space of an experiment.
- Find the probability of an event.
- Find the probability of mutually exclusive events and independent events.
- Find the probability of the complement of an event.

## The Probability of an Event

In measuring the uncertainties in everyday life, we often use ambiguous terminology, such as "fairly certain" and "highly unlikely." In mathematics we attempt to assign a number to the likelihood of the occurrence of an event. We call this measurement the **probability** that the event will occur. For example, if we toss a fair coin, we say that the probability that it will land face up is $\frac{1}{2}$.

Any happening whose result is uncertain is called an **experiment.** The possible results of the experiment are **outcomes,** the set of all possible outcomes of an experiment is the **sample space** of the experiment, and any subcollection of a sample space is an **event.**

For instance, when a six-sided die is tossed, the sample space can be represented by the numbers from 1 through 6, each having *equally likely* outcomes.

To describe sample spaces in such a way that each outcome is equally likely, you must sometimes distinguish between various outcomes in ways that appear artificial. Example 1 illustrates such a situation.

| Example 1 | Finding the Sample Space |  |

Find the sample spaces for the following.

**a.** One coin is tossed.    **b.** Two coins are tossed.    **c.** Three coins are tossed.

Solution

**a.** Because the coin will land either heads up (denoted by *H*) or tails up (denoted by *T*), the sample space is $S = \{H, T\}$.

**b.** Because either coin can land heads up or tails up, the possible outcomes are

$HH$ = heads up on both coins

$HT$ = heads up on first coin and tails up on second coin

$TH$ = tails up on first coin and heads up on second coin

$TT$ = tails up on both coins.

The sample space is $S = \{HH, HT, TH, TT\}$. This list distinguishes between the two cases *HT* and *TH*, even though these two outcomes appear to be similar.

**c.** Following the notation of part (b), the sample space is

$$S = \{HHH, HHT, HTH, HTT, THH, THT, TTH, TTT\}.$$

To calculate the probability of an event, count the number of outcomes in the event and in the sample space. The *number of outcomes* in event $E$ is denoted by $n(E)$, and the number of outcomes in the sample space $S$ is denoted by $n(S)$. The probability that event $E$ will occur is given by $n(E)/n(S)$.

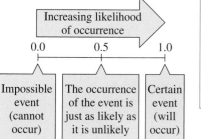

Figure 8.3

▶ **The Probability of an Event**

If an event $E$ has $n(E)$ equally likely outcomes and its sample space $S$ has $n(S)$ equally likely outcomes, the **probability** of event $E$ is

$$P(E) = \frac{n(E)}{n(S)}.$$

Because the number of outcomes in an event must be less than or equal to the number of outcomes in the sample space, the probability of an event must be a number between 0 and 1. That is, for any event $E$, it must be true that $0 \le P(E) \le 1$, as indicated in Figure 8.3.

▶ **Properties of the Probability of an Event**

Let $E$ be an event that is a subset of a finite sample space $S$.

**1.** $0 \le P(E) \le 1$

**2.** If $P(E) = 0$, $E$ *cannot occur* and is called an **impossible event.**

**3.** If $P(E) = 1$, $E$ *must occur* and is called a **certain event.**

**Example 2**    Finding the Probability of an Event

**a.** Two coins are tossed. What is the probability that both land heads up?

**b.** A card is drawn from a standard deck of playing cards. What is the probability that it is an ace?

**Solution**

**a.** Following the procedure in Example 1(b), let $E = \{HH\}$ and $S = \{HH, HT, TH, TT\}$. The probability of getting two heads is

$$P(E) = \frac{n(E)}{n(S)} = \frac{1}{4}.$$

**b.** Because there are 52 cards in a standard deck of playing cards and there are four aces (one in each suit), the probability of drawing an ace is

$$P(E) = \frac{n(E)}{n(S)} = \frac{4}{52} = \frac{1}{13}.$$

Figure 8.4

**Example 3**    Finding the Probability of an Event

Two six-sided dice are tossed. What is the probability that the total of the two dice is 7? (See Figure 8.4.)

**Solution**

Because there are six possible outcomes on each die, you can use the Fundamental Counting Principle to conclude that there are 6 · 6 or 36 different outcomes when two dice are tossed. To find the probability of rolling a total of 7, you must first count the number of ways this can occur.

| First Die | 1 | 2 | 3 | 4 | 5 | 6 |
|---|---|---|---|---|---|---|
| Second Die | 6 | 5 | 4 | 3 | 2 | 1 |

So, a total of 7 can be rolled in six ways, which means that the probability of rolling a 7 is

$$P(E) = \frac{n(E)}{n(S)} = \frac{6}{36} = \frac{1}{6}.$$

You could have written out each sample space in Examples 2 and 3 and simply counted the outcomes in the desired events. For larger sample spaces, however, you must make more use of the counting principles discussed in Section 8.5.

**Example 4**    Finding the Probability of an Event

Twelve-sided dice can be constructed (in the shape of regular dodecahedrons) so that each of the numbers from 1 to 6 appears twice on each die, as shown in Figure 8.5. Prove that these dice can be used in any game requiring ordinary six-sided dice without changing the probabilities of different outcomes.

**Solution**

For an ordinary six-sided die, each of the numbers 1, 2, 3, 4, 5, and 6 occurs only once, so the probability of any particular number coming up is

$$P(E) = \frac{n(E)}{n(S)} = \frac{1}{6}.$$

For one of the twelve-sided dice, each number occurs twice, so the probability of any particular number coming up is

$$P(E) = \frac{n(E)}{n(S)} = \frac{2}{12} = \frac{1}{6}.$$

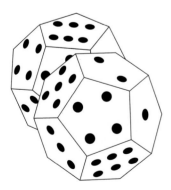

Figure 8.5

*Although popular in the early 1800s, lotteries were banned in more and more states until, by 1894, no state allowed lotteries. In 1964, New Hampshire became the first state to reinstitute a state lottery. Today, lotteries are conducted by almost all states.*

**Example 5**     The Probability of Winning a Lottery

In a state lottery, a player chooses six numbers from 1 to 40. If these six numbers match the six numbers drawn by the lottery commission, the player wins (or shares) the top prize. What is the probability of winning the top prize?

Solution

To find the number of elements in the sample space, use the formula for the number of combinations of 40 elements taken six at a time.

$$n(S) = {}_{40}C_6 = \frac{40 \cdot 39 \cdot 38 \cdot 37 \cdot 36 \cdot 35}{6 \cdot 5 \cdot 4 \cdot 3 \cdot 2 \cdot 1} = 3{,}838{,}380.$$

If a person buys only one ticket, the probability of winning is

$$P(E) = \frac{n(E)}{n(S)} = \frac{1}{3{,}838{,}380}.$$

**Example 6**    Random Selection

The total number of colleges and universities in the United States in 1999 is shown in Figure 8.6. One institution is selected at random. What is the probability that the institution is in one of the three southern regions?    (Source: U.S. National Center for Education Statistics)

Solution

From the figure, the total number of colleges and universities is 4065. Because there are 643 + 275 + 377 = 1295 colleges and universities in the three southern regions, the probability that the institution is in one of these regions is

$$P(E) = \frac{n(E)}{n(S)} = \frac{1295}{4065} \approx 0.319.$$

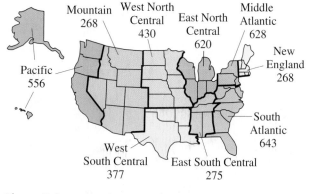

Figure 8.6

## Mutually Exclusive Events

Two events $A$ and $B$ (from the same sample space) are **mutually exclusive** if $A$ and $B$ have no outcomes in common. In the terminology of sets, the **intersection of $A$ and $B$** is the empty set and

$$P(A \cap B) = 0.$$

For instance, if two dice are tossed, the event $A$ of rolling a total of 6 and the event $B$ of rolling a total of 9 are mutually exclusive. To find the probability that one or the other of two mutually exclusive events will occur, you can *add* their individual probabilities.

---

▶ **Probability of the Union of Two Events**

If $A$ and $B$ are events in the same sample space, the probability of $A$ *or* $B$ occurring is given by

$$P(A \cup B) = P(A) + P(B) - P(A \cap B).$$

If $A$ and $B$ are mutually exclusive, then

$$P(A \cup B) = P(A) + P(B).$$

---

**Example 7**   The Probability of a Union

One card is selected from a standard deck of 52 playing cards. What is the probability that the card is either a heart or a face card?

**Solution**

Because the deck has 13 hearts, the probability of selecting a heart (event $A$) is $P(A) = \frac{13}{52}$. Similarly, because the deck has 12 face cards, the probability of selecting a face card (event $B$) is $P(B) = \frac{12}{52}$. Because three of the cards are hearts and face cards (see Figure 8.7), it follows that $P(A \cap B) = \frac{3}{52}$. Finally, applying the formula for the probability of the union of two events, you can conclude that the probability of selecting a heart or a face card is

$$P(A \cup B) = P(A) + P(B) - P(A \cap B)$$

$$= \frac{13}{52} + \frac{12}{52} - \frac{3}{52}$$

$$= \frac{22}{52}$$

$$\approx 0.423.$$

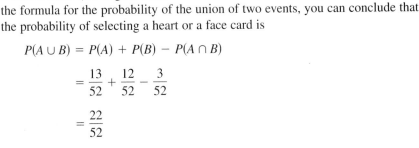

Figure 8.7

**Example 8**    Probability of Mutually Exclusive Events

The personnel department of a company has compiled data showing the number of years of service for each employee. The results are shown in the table.

| Years of Service | Number of Employees |
|---|---|
| 0–4 | 157 |
| 5–9 | 89 |
| 10–14 | 74 |
| 15–19 | 63 |
| 20–24 | 42 |
| 25–29 | 38 |
| 30–34 | 37 |
| 35–39 | 21 |
| 40–44 | 8 |

If an employee is chosen at random, what is the probability that the employee has had nine or fewer years of service?

Solution

To begin, add the number of employees and find that the total is 529. Next, let event $A$ represent choosing an employee with 0 to 4 years of service and let event $B$ represent choosing an employee with 5 to 9 years of service. Then

$$P(A) = \frac{157}{529} \quad \text{and} \quad P(B) = \frac{89}{529}.$$

Because $A$ and $B$ have no outcomes in common, you can conclude that these two events are mutually exclusive and that

$$P(A \cup B) = P(A) + P(B) = \frac{157}{529} + \frac{89}{529}$$

$$= \frac{246}{529}$$

$$\approx 0.465.$$

So, the probability of choosing an employee who has nine or fewer years of service is about 0.465.

## Independent Events

Two events are **independent** if the occurrence of one has no effect on the occurrence of the other. To find the probability that two independent events will occur, *multiply* the probabilities of each. For instance, rolling a total of 12 with two six-sided dice has no effect on the outcome of future rolls of the dice.

---

▶ **Probability of Independent Events**

If $A$ and $B$ are independent events, the probability that both $A$ and $B$ will occur is

$$P(A \text{ and } B) = P(A) \cdot P(B).$$

---

**Example 9**    Probability of Independent Events

A random number generator on a computer selects three integers from 1 to 20. What is the probability that all three numbers are less than or equal to 5?

**Solution**

The probability of selecting a number from 1 to 5 is

$$P(A) = \frac{5}{20} = \frac{1}{4}.$$

So, the probability that all three numbers are less than or equal to 5 is

$$P(A) \cdot P(A) \cdot P(A) = \left(\frac{1}{4}\right)\left(\frac{1}{4}\right)\left(\frac{1}{4}\right)$$

$$= \frac{1}{64}.$$

**Example 10**    Probability of Independent Events

In 2000, 58.2% of the population of the United States was 30 years old or older. Suppose that in a survey, 10 people were chosen at random from the population. What is the probability that all 10 were 30 years old or older?    (Source: U.S. Bureau of the Census)

**Solution**

Let $A$ represent choosing a person who was 30 years old or older. Because the probability of choosing a person who was 30 years old or older is 0.582, you can conclude that the probability that all 10 people were 30 years old or older is

$$[P(A)]^{10} = (0.582)^{10}$$

$$\approx 0.0045.$$

# The Complement of an Event

The **complement of an event** $A$ is the collection of all outcomes in the sample space that are *not* in $A$. The complement of event $A$ is denoted by $A'$. Because $P(A \text{ or } A') = 1$ and because $A$ and $A'$ are mutually exclusive, it follows that $P(A) + P(A') = 1$. So, the probability of $A'$ is given by

$$P(A') = 1 - P(A).$$

For instance, if the probability of *winning* a certain game is

$$P(A) = \frac{1}{4}$$

the probability of *losing* the game is

$$P(A') = 1 - \frac{1}{4} = \frac{3}{4}.$$

---

·**DISCOVERY**·

You are in a class with 22 other people. What is the probability that at least two out of the 23 people will have a birthday on the same day of the year?

The complement of the probability that at least two people have the same birthday is the probability that all 23 birthdays are different. So, first find the probability that all 23 people have different birthdays and then find the complement.

Now, determine the probability that in a room with 50 people at least two people have the same birthday.

---

▶ **Probability of a Complement**

Let $A$ be an event and let $A'$ be its complement. If the probability of $A$ is $P(A)$, then the probability of the complement is

$$P(A') = 1 - P(A).$$

---

**Example 11**    Finding the Probability of a Complement

A manufacturer has determined that a certain machine averages one faulty unit for every 1000 it produces. What is the probability that an order of 200 units will have one or more faulty units?

### Solution

To solve this problem as stated, you would need to find the probability of having exactly one faulty unit, exactly two faulty units, exactly three faulty units, and so on. However, using complements, you can simply find the probability that all units are perfect and then subtract this value from 1. Because the probability that any given unit is perfect is $999/1000$, the probability that all 200 units are perfect is

$$P(A) = \left(\frac{999}{1000}\right)^{200} \approx 0.8186.$$

So, the probability that at least one unit is faulty is

$$P(A') = 1 - P(A) \approx 0.1814.$$

## Discussing the Concept    An Experiment in Probability

In this section you have been finding probabilities from a *theoretical* point of view. Another way to find probabilities is from an *experimental* point of view. For instance, suppose you wanted to find the probability of obtaining a given total when two six-sided dice are tossed. The following BASIC program simulates the tossing of a pair of dice 5000 times.

*BASIC Program:*

```
10      RANDOMIZE
20      DIM TALLY (12)
30      FOR I=1 TO 5000
40      ROLLONE=INT (6*RND)+1
50      ROLLTWO=INT (6*RND)+1
60      DICETOTAL=ROLLONE+ROLLTWO
70      TALLY(DICETOTAL)=TALLY(DICETOTAL)+1
80      NEXT
90      FOR I=2 TO 12
100     PRINT "TOTAL OF", I,"OCCURRED", TALLY(I), "TIMES"
110     NEXT
120     END
```

When we ran this program, the printout was as follows.

```
TOTAL OF     2   OCCURRED    139   TIMES
TOTAL OF     3   OCCURRED    264   TIMES
TOTAL OF     4   OCCURRED    443   TIMES
TOTAL OF     5   OCCURRED    553   TIMES
TOTAL OF     6   OCCURRED    691   TIMES
TOTAL OF     7   OCCURRED    810   TIMES
TOTAL OF     8   OCCURRED    715   TIMES
TOTAL OF     9   OCCURRED    557   TIMES
TOTAL OF    10   OCCURRED    398   TIMES
TOTAL OF    11   OCCURRED    270   TIMES
TOTAL OF    12   OCCURRED    160   TIMES
```

In Example 3 you found that the theoretical probability of tossing a total of 7 on a pair of dice is $\frac{1}{6} \approx 0.167$. From this experiment, calculate the experimental probability of tossing a total of 7 on a pair of dice. If you increase the number of trials from 5000 to 10,000, do you expect the experimental result to get closer to the theoretical result? Explain.

# Warm Up

The following warm-up exercises involve skills that were covered in earlier sections. You will use these skills in the exercise set for this section.

In Exercises 1–8, evaluate the expression.

**1.** $\frac{1}{4} + \frac{5}{8} - \frac{5}{16}$

**2.** $\frac{4}{15} + \frac{3}{5} - \frac{1}{3}$

**3.** $\frac{5 \cdot 4}{5!}$

**4.** $\frac{5!22!}{27!}$

**5.** $\frac{4!}{8!12!}$

**6.** $\frac{9 \cdot 8 \cdot 7 \cdot 6 \cdot 5}{9!}$

**7.** $\frac{{}_5C_3}{{}_{10}C_3}$

**8.** $\frac{{}_{10}C_2 \cdot {}_{10}C_2}{{}_{20}C_4}$

In Exercises 9 and 10, evaluate the expression. (Round your answer to three decimal places.)

**9.** $\left(\frac{99}{100}\right)^{100}$

**10.** $1 - \left(\frac{89}{100}\right)^{50}$

## 8.6    Exercises

*Heads or Tails?*    In Exercises 1–4, a coin is tossed three times. Find the probability of the event.

**1.** Getting exactly one head

**2.** Getting a tail on the second toss

**3.** Getting at least one tail

**4.** Getting at least two heads

*Tossing a Die*    In Exercises 5–10, two six-sided dice are tossed. Find the probability of the event.

**5.** The sum is 5.

**6.** The sum is less than 10.

**7.** The sum is at least 7.

**8.** The total is 2, 3, or 8.

**9.** The sum is odd and no more than 7.

**10.** The sum is odd or a prime number.

*Drawing a Card*    In Exercises 11–14, a card is selected from a standard deck of 52 cards. Find the probability of the event.

**11.** Getting a face card

**12.** Not getting a face card

**13.** Getting a black card that is not a face card

**14.** Getting a card that is a 6 or less (aces are low)

*Drawing Marbles*    In Exercises 15–18, two marbles are drawn (without replacement) from a bag containing one green, two yellow, and three red marbles. Find the probability of the event.

**15.** Drawing exactly one red marble

**16.** Drawing two green marbles

**17.** Drawing neither of the yellow marbles

**18.** Drawing marbles of different colors

In Exercises 19 and 20, you are given the probability that an event *will* happen. Find the probability that the event *will not* happen.

**19.** $p = 0.4$

**20.** $p = 0.72$

In Exercises 21 and 22, you are given the probability that an event *will not* happen. Find the probability that the event *will* happen.

**21.** $p = 0.2$

**22.** $p = 0.68$

**23.** Use information from news sources, the Internet, or an original experiment to write an example of two mutually exclusive events.

**24.** Use information from news sources, the Internet, or an original experiment to write an example of two independent events.

**25.** *Winning an Election*   Three people have been nominated for president of a college class. From a small poll, it is estimated that the probability of Jane winning the election is 0.46, and the probability of Larry winning the election is 0.32. What is the probability of the third candidate winning the election?

**26.** *Winning an Election*   Taylor, Moore, and Jenkins are candidates for a public office. It is estimated that Moore and Jenkins have about the same probability of winning, and Taylor is believed to be twice as likely to win as either of the others. Find each candidate's probability of winning the election.

**27.** *College Bound*   In a high school graduating class of 198 students, 43 are on the honor roll. Of these, 37 are going on to college, and of the other 155 students, 102 are going on to college. If a student is selected at random from the class, what are the probabilities that the person chosen is

(a) going to college?

(b) not going to college?

(c) on the honor roll, but not going to college?

**28.** *Alumni Association*   The alumni office of a college is sending a survey to selected members of the class of 1995. Of the 1254 people who graduated that year, 672 are women, and of those, 124 went to graduate school. Of the 582 male graduates, 198 went to graduate school. If an alumnus is selected at random, what are the probabilities that the person is

(a) female?      (b) male?

(c) female and did not attend graduate school?

**29.** *Random Number Generator*   Two integers (from 1 to 30, inclusive) are chosen by a random number generator on a computer. What are the probabilities that (a) the numbers are both even, (b) one number is even and one is odd, (c) both numbers are less than 10, and (d) the same number is chosen twice?

**30.** *Random Number Generator*   Two integers (from 1 to 40, inclusive) are chosen by a random number generator on a computer. What are the probabilities that (a) the numbers are both even, (b) one number is even and one is odd, (c) both numbers are less than 30, and (d) the same number is chosen twice?

**31.** *Preparing for a Test*   An instructor gives her class a list of eight study problems, from which she will select five to be answered on an exam. If a given student knows how to solve six of the problems, find the probability that the student will be able to answer all five questions on the exam.

**32.** *Preparing for a Test*   An instructor gives his class a list of 20 study problems, from which he will select 10 to be answered on an exam. If a given student knows how to solve 15 of the problems, find the probability that the student will be able to answer all 10 questions on the exam.

**33.** *Drawing Cards from a Deck*   Two cards are selected at random from an ordinary deck of 52 playing cards. Find the probabilities that two aces are selected under the following conditions.

(a) The cards are drawn in sequence, with the first card being replaced and the deck reshuffled prior to the second drawing.

(b) The two cards are drawn consecutively, without replacement.

**34.** *Drawing Cards from a Deck*   Two cards are selected at random from an ordinary deck of 52 playing cards. Find the probabilities that two hearts are selected under the following conditions.

(a) The first card is replaced and the deck reshuffled prior to the second drawing.

(b) The two cards are drawn, without replacement.

**35.** *Game Show*   On a game show, you are given five digits to arrange in the proper order to represent the price of a car. If you are correct, you win the car. Find the probabilities of winning if

(a) You must guess the position of each digit.

(b) You know the first digit, but must guess the remaining four.

(c) You know the first and last digits, but must guess the remaining three.

**36.** *Game Show*   On a game show, you are given six digits to arrange in the proper order to represent the price of a house. If you are correct, you win the house. Find the probabilities of winning if

(a) You guess the position of each digit.

(b) You know the first digit, but must guess the remaining five.

(c) You know the first and last digits, but must guess the remaining four.

**37.** *Letter Mix-Up*   Five letters and envelopes are addressed to five different people. If the letters are randomly inserted into the envelopes, what are the probabilities that (a) exactly one will be inserted in the correct envelope and (b) at least one will be inserted in the correct envelope?

**38.** *Payroll Mix-Up*  Three paychecks and envelopes are addressed to three different people. If the paychecks get mixed up and are randomly inserted into the envelopes, what are the probabilities that (a) exactly one will be inserted in the correct envelope and (b) at least one will be inserted in the correct envelope?

**39.** *Poker Hand*  Five cards are drawn from a standard deck of 52 cards. What is the probability of getting a full house? (See Exercise 39 in Section 8.5.)

**40.** *Poker Hand*  Five cards are drawn from a standard deck of 52 cards. What is the probability of getting a straight flush? (See Exercise 40 in Section 8.5.)

**41.** *Defective Units*  A shipment of 1000 compact disc players contains four defective units. A retail outlet has ordered 20 units. (a) What is the probability that all 20 units are good? (b) What is the probability that at least one unit is defective?

**42.** *Defective Units*  A shipment of 12 stereos contains three defective units. Four of the units are shipped to a retail store. What are the probabilities that (a) all four units are good, (b) exactly two units are good, and (c) at least two units are good?

**43.** *Backup System*  A space vehicle has an independent backup system for one of its communication networks. The probability that either system will function satisfactorily for the duration of a flight is 0.985. What are the probabilities that during a given flight (a) both systems function satisfactorily, (b) at least one system functions satisfactorily, and (c) both systems fail?

**44.** *Backup Vehicle*  A fire company keeps two rescue vehicles to serve the community. Because of the demand on the company's time and the chance of mechanical failure, the probability that a specific vehicle is available when needed is 90%. If the availability of one vehicle is *independent* of the other, find the probabilities that (a) both vehicles are available at a given time, (b) neither vehicle is available at a given time, and (c) at least one vehicle is available at a given time.

**45.** *Making a Sale*  A sales representative makes a sale at approximately one-third of the offices she calls on. If, on a given day, she goes to four offices, what are the probabilities that she will make a sale at (a) all four offices, (b) none of the offices, and (c) at least one office?

**46.** *Making a Sale*  A sales representative makes a sale at approximately one-fourth of the businesses he calls on. If, on a given day, he goes to five businesses, what

are the probabilities that he will make a sale at (a) all five businesses, (b) none of the businesses, and (c) at least one of the businesses?

**47.** *A Boy or a Girl?*  Assume that the probability of the birth of a child of a particular gender is 50%. In a family with six children, what are the probabilities that

(a) all six children are girls?

(b) all six children are of the same gender?

(c) there is at least one girl?

**48.** *A Boy or a Girl?*  Assume that the probability of the birth of a child of a particular gender is 50%. In a family with four children, what are the probabilities that

(a) all four children are boys?

(b) all four children are of the same gender?

(c) there is at least one boy?

*Probability*  In Exercises 49–52, consider $n$ independent trials of an experiment in which each trial has two possible outcomes, called success and failure. The probability of a success on each trial is $p$, and the probability of a failure is $q = 1 - p$. In this context, the term $_nC_k\, p^k\, q^{n-k}$ in the expansion of $(p + q)^n$ gives the probability of $k$ successes in the $n$ trials of the experiment.

**49.** A fair coin is tossed eight times. To find the probability of obtaining five heads, evaluate the term

$$_8C_5\left(\frac{1}{2}\right)^5\left(\frac{1}{2}\right)^3$$

in the expansion of $\left(\frac{1}{2} + \frac{1}{2}\right)^8$.

**50.** The probability of a baseball player getting a hit during any given time at bat is $\frac{1}{5}$. To find the probability that the player gets four hits during the next 10 times at bat, evaluate the term

$$_{10}C_4\left(\frac{1}{5}\right)^4\left(\frac{4}{5}\right)^6$$

in the expansion of $\left(\frac{1}{5} + \frac{4}{5}\right)^{10}$.

**51.** The probability of a sales representative making a sale with any one customer is $\frac{1}{4}$. The sales representative makes 10 contacts a day. To find the probability of making four sales, evaluate the term

$$_{10}C_4\left(\frac{1}{4}\right)^4\left(\frac{3}{4}\right)^6$$

in the expansion of $\left(\frac{1}{4} + \frac{3}{4}\right)^{10}$.

**52.** To find the probability that the baseball player in Exercise 50 makes five hits during the next 10 times at bat if the probability of a hit is $\frac{1}{3}$, evaluate the term

$$_{10}C_5\left(\frac{1}{3}\right)^5\left(\frac{2}{3}\right)^5$$

in the expansion of $\left(\frac{1}{3} + \frac{2}{3}\right)^{10}$.

**53.** *Is That Cash or Charge?*    According to a survey by *USA Today*, the methods used by Christmas shoppers to pay for gifts are as shown in the pie graph. Suppose two Christmas shoppers are chosen at random. What is the probability that both shoppers paid for their gifts in cash only?

How Shoppers Pay for Gifts

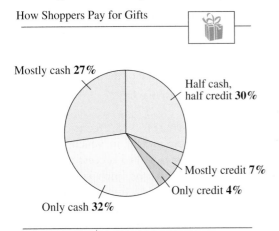

**54.** *Watching the Ads*    According to a survey by Video Storyboard Tests, Inc., television viewers pay varying amounts of attention to television advertisements as compared with the television programs (see figure). Suppose two television viewers are chosen at random. What is the probability that both viewers pay no attention to the television ads they see?

Watching the Ads

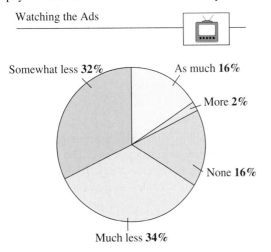

**55.** *Number of Telephones*    According to a survey by Maritz Ameripoll, the numbers of telephones in American households are as shown in the figure. Suppose three households are chosen at random. What is the probability that all three households have at least four telephones?

Telephones in the Home

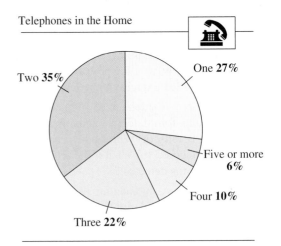

**56.** *Flexible Work Hours*    In a survey by Robert Hall International, people were asked if they would prefer to work flexible hours—even if it meant slower career advancement—so that they could spend more time with their families. The results of the survey are shown in the figure. Suppose three people are chosen at random. What is the probability that all three people prefer flexible work hours?

Flexible Work Hours

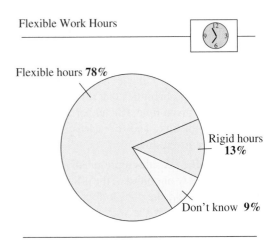

## 8.7 Mathematical Induction

### Objectives

- Use mathematical induction to prove a formula.
- Find the sums of powers of integers.
- Find a formula for a finite sum.
- Use finite differences to find a quadratic model.

## Introduction

In this section you will study a form of mathematical proof called **mathematical induction.** It is important that you clearly see the logical need for it, so let's look at the following problem.

$$S_1 = 1 = 1^2$$

$$S_2 = 1 + 3 = 2^2$$

$$S_3 = 1 + 3 + 5 = 3^2$$

$$S_4 = 1 + 3 + 5 + 7 = 4^2$$

$$S_5 = 1 + 3 + 5 + 7 + 9 = 5^2$$

Judging from the pattern formed by these first five sums, it appears that the sum of the first $n$ integers is

$$S_n = 1 + 3 + 5 + 7 + 9 + \cdots + (2n - 1) = n^2.$$

Although this particular formula *is* valid, it is important for you to see that recognizing a pattern and then simply *jumping to the conclusion* that the pattern must be true for all values of $n$ is *not* a logically valid method of proof. There are many examples in which a pattern appears to be developing for small values of $n$ and then at some point the pattern fails. One of the most famous cases of this was the conjecture by the French mathematician Pierre de Fermat (1601–1665), who speculated that all numbers of the form

$$F_n = 2^{2^n} + 1, \quad n = 0, 1, 2, \ldots$$

are prime. For $n = 0$, 1, 2, 3, and 4, the conjecture is true.

$$F_0 = 3, F_1 = 5, F_2 = 17, F_3 = 257, F_4 = 65{,}537$$

The size of the next Fermat number ($F_5 = 4{,}294{,}967{,}297$) is so great that it was difficult for Fermat to determine whether it was prime or not. However, another well-known mathematician, Leonhard Euler (1707–1783), later found the factorization

$$F_5 = 4{,}294{,}967{,}297 = 641(6{,}700{,}417)$$

which proved that $F_5$ is not prime and therefore Fermat's conjecture was false.

Just because a rule, pattern, or formula seems to work for several values of $n$, you cannot simply decide that it is valid for all values of $n$ without going through a *legitimate proof.*

▶ **The Principle of Mathematical Induction**

Let $P_n$ be a statement involving the positive integer $n$. If

**1.** $P_1$ is true, and

**2.** the truth of $P_k$ implies the truth of $P_{k+1}$ for every positive $k$,

then $P_n$ must be true for all positive integers $n$.

It is important to recognize that both parts of the Principle of Mathematical Induction are necessary. To apply the Principle of Mathematical Induction, you need to be able to determine the statement $P_{k+1}$ for a given statement $P_k$.

| **Example 1** | A Preliminary Example |
|---|---|

Find $P_{k+1}$ for the following.

**a.** $P_k$: $S_k = \dfrac{k^2(k+1)^2}{4}$

**b.** $P_k$: $S_k = 1 + 5 + 9 + \cdots + [4(k-1) - 3] + (4k - 3)$

**c.** $P_k$: $3^k \geq 2k + 1$

**Solution**

**a.** $P_{k+1}$: $S_{k+1} = \dfrac{(k+1)^2(k+1+1)^2}{4}$      Replace $k$ by $k+1$.

$\qquad\qquad = \dfrac{(k+1)^2(k+2)^2}{4}.$      Simplify.

**b.** $P_{k+1}$: $S_{k+1} = 1 + 5 + 9 + \cdots + \{4[(k+1) - 1] - 3\} + [4(k+1) - 3]$

$\qquad\qquad = 1 + 5 + 9 + \cdots + (4k - 3) + (4k + 1).$

**c.** $P_{k+1}$: $3^{k+1} \geq 2(k+1) + 1$

$\qquad\qquad 3^{k+1} \geq 2k + 3.$

A well-known illustration used to explain why the Principle of Mathematical Induction works is the unending line of dominoes shown in Figure 8.8. If the line actually contains infinitely many dominoes, it is clear that you could not knock the entire line down by knocking down only *one domino* at a time. However, suppose it were true that each domino would knock down the next one as it fell. Then you could knock them all down simply by pushing the first one and starting a chain reaction. Mathematical induction works in the same way. If the truth of $P_k$ implies the truth of $P_{k+1}$ and if $P_1$ is true, the chain reaction proceeds as follows: $P_1$ implies $P_2$, $P_2$ implies $P_3$, $P_3$ implies $P_4$, and so on.

**Figure 8.8**

| Example 2 | Using Mathematical Induction |

Use mathematical induction to prove the following formula.

$$S_n = 1 + 3 + 5 + 7 + \cdots + (2n - 1)$$

$$= n^2$$

**Solution**

Mathematical induction consists of two distinct parts. First, you must show that the formula is true when $n = 1$.

**1.** When $n = 1$, the formula is valid, because

$$S_1 = 1 = 1^2.$$

The second part of mathematical induction has two steps. The first step is to assume that the formula is valid for *some* integer $k$. The second step is to use this assumption to prove that the formula is valid for the next integer, $k + 1$.

**2.** Assuming that the formula

$$S_k = 1 + 3 + 5 + 7 + \cdots + (2k - 1)$$

$$= k^2$$

is true, you must show that the formula $S_{k+1} = (k + 1)^2$ is true.

$$S_{k+1} = 1 + 3 + 5 + 7 + \cdots + (2k - 1) + [2(k + 1) - 1]$$

$$= [1 + 3 + 5 + 7 + \cdots + (2k - 1)] + (2k + 2 - 1)$$

$$= S_k + (2k + 1) \qquad \text{Group terms to form } S_k.$$

$$= k^2 + 2k + 1 \qquad \text{Replace } S_k \text{ by } k^2.$$

$$= (k + 1)^2$$

Combining the results of parts (1) and (2), you can conclude by mathematical induction that the formula is valid for *all* positive integer values of $n$.

---

## Study Tip

When using mathematical induction to prove a *summation* formula (such as the one in Example 2), it is helpful to think of $S_{k+1}$ as $S_{k+1} = S_k + a_{k+1}$, where $a_{k+1}$ is the $(k + 1)$th term of the original sum.

---

It occasionally happens that a statement involving natural numbers is not true for the first $k - 1$ positive integers but is true for all values of $n \geq k$. In these instances, you use a slight variation of the Principle of Mathematical Induction in which you verify $P_k$ rather than $P_1$. This variation is called the **extended principle of mathematical induction.** To see the validity of this variation, note from Figure 8.8 that all but the first $k - 1$ dominoes can be knocked down by knocking over the $k$th domino. This suggests that you can prove a statement $P_n$ to be true for $n \geq k$ by showing that $P_k$ is true and that $P_k$ implies $P_{k+1}$. In Exercises 35–38 of this section you are asked to apply this extension of mathematical induction.

Example 3   Using Mathematical Induction

Use mathematical induction to prove the following formula.

$$S_n = 1^2 + 2^2 + 3^2 + 4^2 + \cdots + n^2$$

$$= \frac{n(n + 1)(2n + 1)}{6}$$

### Solution

1. When $n = 1$, the formula is valid, because

$$S_1 = 1^2 = \frac{1(2)(3)}{6}.$$

2. Assuming that

$$S_k = 1^2 + 2^2 + 3^2 + 4^2 + \cdots + k^2$$

$$= \frac{k(k + 1)(2k + 1)}{6}$$

you must show that

$$S_{k+1} = \frac{(k + 1)(k + 2)(2k + 3)}{6}.$$

To do this, write the following.

$$S_{k+1} = S_k + a_{k+1}$$

$$= (1^2 + 2^2 + 3^2 + 4^2 + \cdots + k^2) + (k + 1)^2$$

$$= \frac{k(k + 1)(2k + 1)}{6} + (k + 1)^2$$

$$= \frac{k(k + 1)(2k + 1) + 6(k + 1)^2}{6}$$

$$= \frac{(k + 1)[k(2k + 1) + 6(k + 1)]}{6}$$

$$= \frac{(k + 1)(2k^2 + 7k + 6)}{6}$$

$$= \frac{(k + 1)(k + 2)(2k + 3)}{6}$$

Combining the results of parts (1) and (2), you can conclude by mathematical induction that the formula is valid for *all* $n \geq 1$.

When proving a formula with mathematical induction, the only statement that you *need* to verify is $P_1$. As a check, however, it is good to try verifying other statements. For instance, in Example 3, try verifying $S_2$ and $S_3$.

## Sums of Powers of Integers

The formula in Example 3 is one of a collection of useful summation formulas. We summarize this and other formulas dealing with the sums of various powers of the first $n$ positive integers as follows.

---

▶ **Sums of Powers of Integers**

**1.** $1 + 2 + 3 + 4 + \cdots + n = \dfrac{n(n+1)}{2}$

**2.** $1^2 + 2^2 + 3^2 + 4^2 + \cdots + n^2 = \dfrac{n(n+1)(2n+1)}{6}$

**3.** $1^3 + 2^3 + 3^3 + 4^3 + \cdots + n^3 = \dfrac{n^2(n+1)^2}{4}$

**4.** $1^4 + 2^4 + 3^4 + 4^4 + \cdots + n^4 = \dfrac{n(n+1)(2n+1)(3n^2+3n-1)}{30}$

**5.** $1^5 + 2^5 + 3^5 + 4^5 + \cdots + n^5 = \dfrac{n^2(n+1)^2(2n^2+2n-1)}{12}$

---

Each of these formulas for sums can be proven by mathematical induction.

**Example 4**    Finding a Sum of Powers of Integers

Find $\displaystyle\sum_{n=1}^{7} n^3 = 1^3 + 2^3 + 3^3 + 4^3 + 5^3 + 6^3 + 7^3$.

**Solution**

Using the formula for the sum of the cubes of the first $n$ positive integers, you obtain the following.

$$\sum_{n=1}^{7} n^3 = 1^3 + 2^3 + 3^3 + 4^3 + 5^3 + 6^3 + 7^3$$

$$= \frac{7^2(7+1)^2}{4}$$

$$= \frac{49(64)}{4}$$

$$= 784$$

Check this sum by adding the numbers 1, 8, 27, 64, 125, 216, and 343.

| Example 5 | Proving an Inequality by Mathematical Induction |

Prove that $n < 2^n$ for all positive integers $n$.

**Solution**

1. For $n = 1$, the formula is true, because

   $$1 < 2^1.$$

2. Assuming that

   $$k < 2^k$$

   you need to show that $k + 1 < 2^{k+1}$. For $n = k$, you have

   $$2^{k+1} = 2(2^k) > 2(k) = 2k. \qquad \text{By assumption}$$

   Because $2k = k + k > k + 1$ for all $k > 1$, it follows that

   $$2^{k+1} > 2k > k + 1$$

   or

   $$k + 1 < 2^{k+1}.$$

   So, $n < 2^n$ for all integers $n \geq 1$.

## Pattern Recognition

Although choosing a formula on the basis of a few observations does *not* guarantee the validity of the formula, pattern recognition *is* important. Once you have a pattern that you think works, you can try using mathematical induction to prove your formula.

---

▶ **Finding a Formula for the *n*th Term of a Sequence**

To find a formula for the *n*th term of a sequence, consider the following guidelines.

1. Calculate the first several terms of the sequence. It is often a good idea to write the terms in both simplified and factored forms.

2. Try to find a recognizable pattern for the terms and write a formula for the *n*th term of the sequence. This is your *hypothesis* or *conjecture*. You might try computing one or two more terms in the sequence to test your hypothesis.

3. Use mathematical induction to prove your hypothesis.

---

**Example 6**    Finding a Formula for a Finite Sum

Find a formula for the following finite sum.

$$\frac{1}{1 \cdot 2} + \frac{1}{2 \cdot 3} + \frac{1}{3 \cdot 4} + \frac{1}{4 \cdot 5} + \cdots + \frac{1}{n(n+1)}$$

**Solution**

Begin by writing out the first few sums.

$$S_1 = \frac{1}{1 \cdot 2} = \frac{1}{2} = \frac{1}{1+1}$$

$$S_2 = \frac{1}{1 \cdot 2} + \frac{1}{2 \cdot 3} = \frac{4}{6} = \frac{2}{3} = \frac{2}{2+1}$$

$$S_3 = \frac{1}{1 \cdot 2} + \frac{1}{2 \cdot 3} + \frac{1}{3 \cdot 4} = \frac{9}{12} = \frac{3}{4} = \frac{3}{3+1}$$

$$S_4 = \frac{1}{1 \cdot 2} + \frac{1}{2 \cdot 3} + \frac{1}{3 \cdot 4} + \frac{1}{4 \cdot 5} = \frac{48}{60} = \frac{4}{5} = \frac{4}{4+1}$$

From this sequence, it appears that the formula for the $k$th sum is

$$S_k = \frac{1}{1 \cdot 2} + \frac{1}{2 \cdot 3} + \frac{1}{3 \cdot 4} + \frac{1}{4 \cdot 5} + \cdots + \frac{1}{k(k+1)} = \frac{k}{k+1}.$$

To prove the validity of this hypothesis, use mathematical induction, as follows. Note that you have already verified the formula for $n = 1$, so you can begin by assuming that the formula is valid for $n = k$ and trying to show that it is valid for $n = k + 1$.

$$S_{k+1} = \left[ \frac{1}{1 \cdot 2} + \frac{1}{2 \cdot 3} + \frac{1}{3 \cdot 4} + \frac{1}{4 \cdot 5} + \cdots + \frac{1}{k(k+1)} \right] + \frac{1}{(k+1)(k+2)}$$

$$= \frac{k}{k+1} + \frac{1}{(k+1)(k+2)}$$

$$= \frac{k(k+2) + 1}{(k+1)(k+2)}$$

$$= \frac{k^2 + 2k + 1}{(k+1)(k+2)}$$

$$= \frac{(k+1)^2}{(k+1)(k+2)}$$

$$= \frac{k+1}{k+2}$$

So, the hypothesis is valid.

## Finite Differences

The **first differences** of a sequence are found by subtracting consecutive terms. The **second differences** are found by subtracting consecutive first differences. The first and second differences of the sequence 3, 5, 8, 12, 17, 23, . . . are as follows.

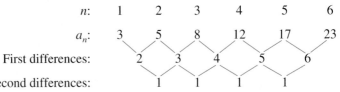

For this sequence, the second differences are all the same. When this happens, the sequence has a perfect quadratic model. If the first differences are all the same, the sequence has a linear model. That is, it is arithmetic.

---

**Example 7**  Finding a Quadratic Model

Find the quadratic model for the sequence

3, 5, 8, 12, 17, 23, . . . .

**Solution**

You know that the model has the form

$$a_n = an^2 + bn + c.$$

By substituting 1, 2, and 3 for $n$, you can obtain a system of three linear equations in three variables.

$a_1 = a(1)^2 + b(1) + c = 3$     Substitute 1 for $n$.

$a_2 = a(2)^2 + b(2) + c = 5$     Substitute 2 for $n$.

$a_3 = a(3)^2 + b(3) + c = 8$     Substitute 3 for $n$.

You now have a system of three equations in $a$, $b$, and $c$.

$a + b + c = 3$     Equation 1

$4a + 2b + c = 5$     Equation 2

$9a + 3b + c = 8$     Equation 3

Using the techniques discussed in Chapter 6, you can find the solution to be $a = \frac{1}{2}$, $b = \frac{1}{2}$, and $c = 2$. Thus, the quadratic model is

$$a_n = \frac{1}{2}n^2 + \frac{1}{2}n + 2.$$

Try checking the values of $a_1$, $a_2$, and $a_3$.

## Discussing the Concept    Mathematical Modeling

Use finite differences to determine whether each sequence in the table can be represented by either a linear or quadratic model.

| $n$ | 1 | 2 | 3 | 4 | 5 | 6 |
|---|---|---|---|---|---|---|
| $a_n$ | 12 | 14 | 22 | 36 | 56 | 82 |
| $b_n$ | $-23.5$ | $-20.0$ | $-16.5$ | $-13.0$ | $-9.5$ | $-6.0$ |
| $c_n$ | 7 | 13 | 20 | 26 | 33 | 39 |
| $d_n$ | 0.8 | 4.2 | 9.2 | 15.8 | 24.0 | 33.8 |

If the sequence can be represented by a linear or quadratic model, find the appropriate model, and predict the value of the tenth term. Discuss how the finite differences technique complements other modeling techniques you have learned thus far.

# Warm Up

The following warm-up exercises involve skills that were covered in previous courses and earlier sections. You will use these skills in the exercise set for this section.

In Exercises 1–4, find the sum.

**1.** $\displaystyle\sum_{k=3}^{6} (2k - 3)$

**2.** $\displaystyle\sum_{j=1}^{5} (j^2 - j)$

**3.** $\displaystyle\sum_{k=2}^{5} \frac{1}{k}$

**4.** $\displaystyle\sum_{i=1}^{2} \left(1 + \frac{1}{i}\right)$

In Exercises 5–10, simplify the expression.

**5.** $\dfrac{2(k + 1) + 3}{5}$

**6.** $\dfrac{3(k + 1) - 2}{6}$

**7.** $2 \cdot 2^{2(k+1)}$

**8.** $\dfrac{3^{2k}}{3^{2(k+1)}}$

**9.** $\dfrac{k + 1}{k^2 + k}$

**10.** $\dfrac{\sqrt{32}}{\sqrt{50}}$

## 8.7    Exercises

In Exercises 1–4, find $P_{k+1}$ for the given $P_k$.

**1.** $P_k = \dfrac{5}{k(k + 1)}$

**2.** $P_k = \dfrac{1}{(k + 1)(k + 3)}$

**3.** $P_k = \dfrac{k^2(k + 1)^2}{4}$

**4.** $P_k = \dfrac{k}{2}(3k - 1)$

In Exercises 5–18, use mathematical induction to prove the formula for every positive integer $n$.

**5.** $2 + 4 + 6 + 8 + \cdots + 2n = n(n + 1)$

**6.** $3 + 7 + 11 + 15 + \cdots + (4n - 1) = n(2n + 1)$

**7.** $2 + 7 + 12 + 17 + \cdots + (5n - 3) = \dfrac{n}{2}(5n - 1)$

**8.** $1 + 4 + 7 + 10 + \cdots + (3n - 2) = \dfrac{n}{2}(3n - 1)$

**9.** $1 + 2 + 2^2 + 2^3 + \cdots + 2^{n-1} = 2^n - 1$

**10.** $2(1 + 3 + 3^2 + 3^3 + \cdots + 3^{n-1}) = 3^n - 1$

**11.** $1 + 2 + 3 + 4 + \cdots + n = \dfrac{n(n + 1)}{2}$

**12.** $1^2 + 2^2 + 3^2 + 4^2 + \cdots + n^2 = \dfrac{n(n + 1)(2n + 1)}{6}$

**13.** $1^3 + 2^3 + 3^3 + 4^3 + \cdots + n^3 = \dfrac{n^2(n + 1)^2}{4}$

**14.** $\left(1 + \dfrac{1}{1}\right)\left(1 + \dfrac{1}{2}\right)\left(1 + \dfrac{1}{3}\right) \cdots \left(1 + \dfrac{1}{n}\right) = n + 1$

**15.** $\displaystyle\sum_{i=1}^{n} i^5 = \dfrac{n^2(n + 1)^2(2n^2 + 2n - 1)}{12}$

**16.** $\displaystyle\sum_{i=1}^{n} i^4 = \dfrac{n(n + 1)(2n + 1)(3n^2 + 3n - 1)}{30}$

**17.** $\displaystyle\sum_{i=1}^{n} i(i + 1) = \dfrac{n(n + 1)(n + 2)}{3}$

**18.** $\displaystyle\sum_{i=1}^{n} \dfrac{1}{(2i - 1)(2i + 1)} = \dfrac{n}{2n + 1}$

In Exercises 19–28, find the sum using the formulas for the sums of powers of integers.

**19.** $\displaystyle\sum_{n=1}^{20} n$

**20.** $\displaystyle\sum_{n=1}^{50} n$

**21.** $\displaystyle\sum_{n=1}^{6} n^2$

**22.** $\displaystyle\sum_{n=1}^{10} n^3$

**23.** $\displaystyle\sum_{n=1}^{5} n^4$

**24.** $\displaystyle\sum_{n=1}^{8} n^5$

**25.** $\displaystyle\sum_{n=1}^{6} (n^2 - n)$

**26.** $\displaystyle\sum_{n=1}^{10} (n^3 - n^2)$

**27.** $\displaystyle\sum_{i=1}^{6} (6i - 8i^3)$

**28.** $\displaystyle\sum_{j=1}^{4} \left(2 + \tfrac{5}{2}j - \tfrac{3}{2}j^2\right)$

In Exercises 29–34, find a formula for the sum of the first $n$ terms of the sequence.

**29.** $1, 5, 9, 13, \ldots$

**30.** $25, 22, 19, 16, \ldots$

**31.** $1, \frac{9}{10}, \frac{81}{100}, \frac{729}{1000}, \ldots$

**32.** $3, -\frac{9}{2}, \frac{27}{4}, -\frac{81}{8}, \ldots$

**33.** $\frac{1}{4}, \frac{1}{12}, \frac{1}{24}, \frac{1}{40}, \ldots, \frac{1}{2n(n+1)}, \ldots$

**34.** $\frac{1}{2 \cdot 3}, \frac{1}{3 \cdot 4}, \frac{1}{4 \cdot 5}, \frac{1}{5 \cdot 6}, \ldots, \frac{1}{(n+1)(n+2)}, \ldots$

In Exercises 35–38, prove the inequality for the indicated integer values of $n$.

**35.** $n! > 2^n, \quad n \geq 4$

**36.** $\left(\frac{4}{3}\right)^n > n, \quad n \geq 7$

**37.** $\frac{1}{\sqrt{1}} + \frac{1}{\sqrt{2}} + \frac{1}{\sqrt{3}} + \cdots + \frac{1}{\sqrt{n}} > \sqrt{n}, \quad n \geq 2$

**38.** $\left(\frac{x}{y}\right)^{n+1} < \left(\frac{x}{y}\right)^n$, if $n \geq 1$ and $0 < x < y$.

In Exercises 39–46, use mathematical induction to prove the given property for all positive integers $n$.

**39.** $(ab)^n = a^n b^n$

**40.** $\left(\frac{a}{b}\right)^n = \frac{a^n}{b^n}$

**41.** If $x_1 \neq 0, x_2 \neq 0, \ldots, x_n \neq 0$, then
$(x_1 x_2 x_3 \cdots x_n)^{-1} = x_1^{-1} x_2^{-1} x_3^{-1} \cdots x_n^{-1}$.

**42.** If $x_1 > 0, x_2 > 0, \ldots, x_n > 0$, then
$\ln(x_1 x_2 x_3 \cdots x_n) = \ln x_1 + \ln x_2 + \ln x_3 + \cdots + \ln x_n$.

**43.** Generalized Distributive Law:
$x(y_1 + y_2 + \cdots + y_n) = xy_1 + xy_2 + \cdots + xy_n$

**44.** $(a + bi)^n$ and $(a - bi)^n$ are complex conjugates for all $n \geq 1$.

**45.** A factor of $(n^3 + 3n^2 + 2n)$ is 3.

**46.** A factor of $(2^{2n-1} + 3^{2n-1})$ is 5.

**47.** In your own words, explain what is meant by a proof by mathematical induction.

**48.** *Think About It*   What conclusion can be drawn from the given information about the sequence of statements $P_n$?

(a) $P_3$ is true and $P_k$ implies $P_{k+1}$.

(b) $P_1, P_2, P_3, \ldots, P_{50}$ are all true.

(c) $P_1, P_2,$ and $P_3$ are all true, but the truth of $P_k$ does not imply that $P_{k+1}$ is true.

(d) $P_2$ is true and $P_{2k}$ implies $P_{2k+2}$.

In Exercises 49–52, write the first five terms of the sequence.

**49.** $a_0 = 1$
$a_n = a_{n-1} + 2$

**50.** $a_0 = 10$
$a_n = 4a_{n-1}$

**51.** $a_0 = 4$
$a_1 = 2$
$a_n = a_{n-1} - a_{n-2}$

**52.** $a_0 = 0$
$a_1 = 2$
$a_n = a_{n-1} + 2a_{n-2}$

In Exercises 53–62, write the first five terms of the sequence where $a_1 = f(1)$. Then calculate the first and second differences of the sequence. Does the sequence have a linear model, a quadratic model, or neither?

**53.** $f(1) = 0$
$a_n = a_{n-1} + 3$

**54.** $f(1) = 2$
$a_n = n - a_{n-1}$

**55.** $f(1) = 3$
$a_n = a_{n-1} - n$

**56.** $f(2) = -3$
$a_n = -2a_{n-1}$

**57.** $a_0 = 0$
$a_n = a_{n-1} + n$

**58.** $a_0 = 2$
$a_n = (a_{n-1})^2$

**59.** $f(1) = 2$
$a_n = a_{n-1} + 2$

**60.** $f(1) = 0$
$a_n = a_{n-1} + 2n$

**61.** $a_0 = 1$
$a_n = a_{n-1} + n^2$

**62.** $a_0 = 0$
$a_n = a_{n-1} - 1$

In Exercises 63–66, find a quadratic model for the sequence with the indicated terms.

**63.** $a_0 = 3, a_1 = 3, a_4 = 15$

**64.** $a_0 = 7, a_1 = 6, a_3 = 10$

**65.** $a_0 = -3, a_2 = 1, a_4 = 9$

**66.** $a_0 = 3, a_2 = 0, a_6 = 36$

# Chapter Project

# The Multiplier Effect

| Cycle, $x$ | Total spending (in dollars), $y$ |
|---|---|
| 1 | 92,000,000 |
| 2 | 73,460,000 |
| 3 | 57,598,000 |
| 4 | 45,122,000 |
| 5 | 37,020,000 |
| 6 | 29,491,000 |
| 7 | 23,592,960 |
| 8 | 18,854,365 |
| 9 | 15,099,500 |
| 10 | 12,080,600 |
| 11 | 9,578,675 |
| 12 | 7,740,950 |
| 13 | 6,185,752 |
| 14 | 4,950,800 |
| 15 | 3,980,200 |
| 16 | 3,150,600 |
| 17 | 2,540,300 |
| 18 | 2,025,700 |
| 19 | 1,600,300 |
| 20 | 1,300,050 |

The ideas of economist John Maynard Keynes have been influential in the development of economic policy at the national, state, and local levels. Keynesian economics brought the term "multiplier effect" into popular use. The "multiplier effect" is based on the idea that the expenditures of one person become the income of another. As the second person spends his or her income, then his/her expenditures become a third person's income. This pattern continues as one person's expenditures become another person's income. If individuals in an economy spend 75% of their income (some amount will go to income taxes, forced retirement contributions, voluntary savings/investments, and so forth), then a person spends $0.75 of every dollar earned. For one dollar earned the first person would spend $0.75. The next person spends only $0.56 of the original dollar ($0.75 × 75%). Each subsequent person would spend progressively smaller and smaller amounts of the original dollar, so the income "leaks" out of the spending system. The greater the leakage, the smaller the "multiplier effect" will be.

The table shows the cycle of income in a community that has hosted a large event such as the Super Bowl. In the table, $x$ corresponds to the cycle number of a dollar and $y$ is the total spending in dollars for that cycle. Use this information to investigate the following questions.

1. *Fitting the Model to Data* Use a graphing utility to make a scatter plot of the data in the table. What type or types of graphs does the scatter plot suggest might fit the data? Use the statistics package of a graphing utility to find a quadratic model ($y = ax^2 + bx + c$), a power series model ($y = ax^b$), and an exponential model ($y = a \cdot b^x$).

2. *Comparing a Model to Actual Data* Numerically compare the values given by the models found in Question 1 with the actual data in the table by constructing a table. Graphically compare the models with the data by graphing each model over the scatter plot.

3. *Discussing the Models* Which model fits the data best? Explain. Find the total economic impact of this event on a community by summing the total dollars spent after 20 cycles and the total according to the model you selected. Did the model fit the data well? Explain.

4. *Further Explorations* Use a reference to further investigate the "multiplier effect." Can you find examples of situations where the "multiplier effect" could apply?

## CHAPTER SUMMARY

After studying this chapter, you should have acquired the following skills.
These skills are keyed to the Review Exercises that begin on page 566.
Answers to odd-numbered Review Exercises are given in the back of the book.

**8.1**  · Find the terms of a sequence. ..................................... Review Exercises 1–6

· Evaluate a factorial expression. ................................. Review Exercises 7, 8

· Write an expression for the $n$th term of a sequence. ...... Review Exercises 9–14

· Find the sum of a finite sequence. ............................. Review Exercises 15–18

· Use summation notation to write the sum of a sequence. ... Review Exercises 19–22

· Use a sequence to solve an application problem. ............ Review Exercises 23, 24

**8.2**  · Determine whether a sequence is arithmetic. ............... Review Exercises 25–30

· Find the $n$th term of an arithmetic sequence. ............. Review Exercises 31, 32

· Write the first five terms of an arithmetic sequence. ...... Review Exercises 33, 34

· Find the sum of an arithmetic sequence. ..................... Review Exercises 35–38

· Use an arithmetic sequence to solve an application problem. ... Review Exercises 39, 40

**8.3**  · Determine whether a sequence is geometric. ............... Review Exercises 41–44

· Write the first five terms of a geometric sequence. ........ Review Exercises 45–48

· Find the $n$th term of a geometric sequence. .............. Review Exercises 49–52

· Find the sum of a finite or infinite geometric sequence. ... Review Exercises 53–58

· Use a geometric sequence to solve an application problem. ... Review Exercises 59–62

**8.4**  · Find a binomial coefficient. ..................................... Review Exercises 63–66

· Use Pascal's Triangle to determine a binomial coefficient. ... Review Exercises 67, 68

· Expand a binomial using binomial coefficients. ............. Review Exercises 69–74, 77, 78

· Find a specific term in a binomial expansion. ............... Review Exercises 75, 76

**8.5**  · Use the Fundamental Counting Principle. ................... Review Exercises 79–85

· Determine the number of permutations of $n$ elements taken $r$ at a time. ... Review Exercises 86–89

· Determine the number of combinations of $n$ elements taken $r$ at a time. ... Review Exercises 90–93

**8.6**  · Find the probability of an event. ............................. Review Exercises 94–97

· Find the probability of mutually exclusive events and independent events. ... Review Exercises 100–106

· Find the probability of the complement of an event. ...... Review Exercises 98, 99, 102, 104–106

**8.7**  · Use mathematical induction to prove a formula. .......... Review Exercises 107–110

· Find the sum of powers of integers. .......................... Review Exercises 111–114

· Find a formula for a finite sum. ............................... Review Exercises 115, 116

· Use finite differences to find a quadratic model. .......... Review Exercises 117, 118

## REVIEW EXERCISES

In Exercises 1–6, write the first five terms of the sequence. (Begin with $n = 1$.)

**1.** $a_n = 4n + 1$

**2.** $a_n = 2n - 3$

**3.** $a_n = \dfrac{2n}{2n + 1}$

**4.** $a_n = \dfrac{5n - 2}{n}$

**5.** $a_n = (-1)^{n+1} \dfrac{2^n}{n + 1}$

**6.** $a_n = (-1)^n \dfrac{3^n - 1}{3^n + 1}$

In Exercises 7 and 8, evaluate the fraction.

**7.** $\dfrac{16!}{12!}$

**8.** $\dfrac{24!}{22!}$

In Exercises 9–14, write an expression for the apparent $n$th term of the sequence. (Begin with $n = 1$.)

**9.** $1, 2, 4, 8, \ldots$

**10.** $4, -4, 4, -4, \ldots$

**11.** $1, \frac{1}{2}, \frac{1}{3}, \frac{1}{4}, \frac{1}{5}, \ldots$

**12.** $-\frac{2}{3}, \frac{3}{4}, -\frac{4}{5}, \frac{5}{6}, -\frac{6}{7}, \ldots$

**13.** $3, -9, 27, -81, 243, \ldots$

**14.** $-2, 4, -8, 16, -32, \ldots$

In Exercises 15–18, find the sum.

**15.** $\displaystyle\sum_{i=1}^{6} (2i - 3)$

**16.** $\displaystyle\sum_{i=1}^{6} (3i + 2)$

**17.** $\displaystyle\sum_{k=1}^{6} (k - 1)(k + 2)$

**18.** $\displaystyle\sum_{k=1}^{8} k^2 + 1$

In Exercises 19–22, use summation notation to write the sum.

**19.** $\dfrac{3}{1 + 1} + \dfrac{3}{1 + 2} + \dfrac{3}{1 + 3} + \dfrac{3}{1 + 4} + \cdots + \dfrac{3}{1 + 12}$

**20.** $1 + \frac{1}{2} + \frac{1}{3} + \frac{1}{4} + \frac{1}{5} + \cdots + \frac{1}{60}$

**21.** $2 - 4 + 8 - 16 + 32 - 64 + \cdots + 8192$

**22.** $2 + 8 + 18 + 32 + \cdots + 98$

**23.** *Compound Interest*  Suppose on your next birthday you deposit $2000 in an account that earns 8% compounded quarterly. The balance in the account after $n$ quarters is given by

$$A_n = 2000\left(1 + \frac{0.08}{4}\right)^n, \qquad n = 1, 2, 3, \ldots$$

(a) Compute the first eight terms of the sequence.

(b) Find the balance in this account 10 years from the date of deposit by computing the 40th term of the sequence.

(c) Assuming you do not withdraw money from the account, find the balance 30 years from the date of deposit by computing the 120th term of the sequence.

**24.** *Compound Interest*  A deposit of $20,000 is made in an account that earns 6% compounded monthly. The balance in the account after $n$ months is given by

$$A_n = 20,000\left(1 + \frac{0.06}{12}\right)^n, \qquad n = 1, 2, 3, \ldots$$

(a) Compute the first 10 terms of the sequence.

(b) Find the balance in this account after 10 years by computing the 120th term of the sequence.

In Exercises 25–28, determine whether the sequence is arithmetic. If it is, find the common difference.

**25.** $3, 8, 13, 18, \ldots$

**26.** $1, 3, 9, 27, \ldots$

**27.** $4, \frac{7}{2}, 3, \frac{5}{2}, 2, \ldots$

**28.** $3.2, 4.0, 4.8, 5.6, 6.4, \ldots$

In Exercises 29 and 30, write the first five terms of the specified sequence. Determine whether the sequence is arithmetic, and, if it is, find the common difference.

**29.** $a_n = 4 + 5n$

**30.** $a_n = 2 + (n - 3)5$

In Exercises 31 and 32, find a formula for $a_n$ for the arithmetic sequence.

**31.** $a_1 = 2, d = 3$

**32.** $a_1 = 15, d = -3$

In Exercises 33 and 34, write the first five terms of the arithmetic sequence.

**33.** $a_1 = 4, a_8 = 25$

**34.** $a_6 = 28, a_{14} = 16$

In Exercises 35–38, find the sum of the first $n$ terms of the arithmetic sequence.

**35.** $3, 9, 15, 21, 27, \ldots, n = 8$

**36.** $-8, -4, 0, 4, 8, \ldots, n = 12$

**37.** $a_1 = 27, a_{40} = 300, n = 20$

**38.** $a_1 = 9, a_{15} = 93, n = 15$

**39.** *Job Offers*   A person is offered a job by two companies. The position with Company A has a salary of $26,500 for the first year with a guaranteed annual raise of $1600 per year for the first 5 years. The position with Company B has a salary of $23,000 for the first year with a guaranteed annual raise of $2500 per year for the first 5 years.

(a) What will the salary be during the sixth year of employment at Company A? At Company B?

(b) How much will company A have paid the person at the end of 6 years?

(c) How much will Company B have paid the person at the end of 6 years?

(d) Which job should the person accept, and why?

**40.** *Seating Capacity*   Determine the seating capacity of an auditorium with 20 rows of seats if there are 25 seats in the first row, 28 seats in the second row, and so on, in arithmetic progression.

In Exercises 41–44, determine whether the sequence is geometric. If it is, find its common ratio.

**41.** $1, -3, 9, -27, \ldots$    **42.** $3, 9, 15, 21, \ldots$

**43.** $16, 8, 4, 2, 1, \frac{1}{2}, \frac{1}{4}, \frac{1}{8}, \ldots$

**44.** $1, -\frac{1}{3}, \frac{1}{9}, -\frac{1}{27}, \frac{1}{81}, -\frac{1}{243}, \ldots$

In Exercises 45–48, write the first five terms of the geometric sequence.

**45.** $a_1 = 2, r = 3$    **46.** $a_1 = 2, r = e$

**47.** $a_1 = 10, r = -\frac{1}{5}$    **48.** $a_1 = 1, r = \frac{3}{4}$

In Exercises 49–52, find the indicated term of the geometric sequence.

**49.** $a_1 = 8, r = \frac{1}{2}$, 40th term

**50.** $a_1 = 100, r = -\frac{1}{10}$, 5th term

**51.** $a_1 = 200, r = -\frac{1}{2}$, 10th term

**52.** $a_1 = 10, r = 1.05$, 20th term

In Exercises 53 and 54, find the sum.

**53.** $\displaystyle\sum_{n=1}^{10} 4(2^n)$    **54.** $\displaystyle\sum_{n=0}^{20} 2\left(\frac{2}{3}\right)^n$

In Exercises 55–58, find the sum of the infinite geometric series, if it exists.

**55.** $\displaystyle\sum_{n=0}^{\infty} \left(\frac{1}{3}\right)^n$    **56.** $\displaystyle\sum_{n=0}^{\infty} \left(\frac{3}{2}\right)^n$

**57.** $\displaystyle\sum_{n=0}^{\infty} \left(\frac{4}{3}\right)^n$    **58.** $\displaystyle\sum_{n=0}^{\infty} 2\left(\frac{1}{10}\right)^n$

**59.** *Compound Interest*   A deposit of $100 is made at the beginning of each month for 10 years in an account that pays 8% compounded monthly. The balance $A$ at the end of 10 years is as follows.

$$A = 100\left(1 + \frac{0.08}{12}\right)^1 + \cdots + 100\left(1 + \frac{0.08}{12}\right)^{120}$$

(a) Find the balance.

(b) Is the balance after 20 years twice what it is after 10 years? Explain your reasoning.

**60.** *Compound Interest*   Suppose you deposit $200 in an account at the beginning of each month for 10 years. The account pays 6% compounded monthly. What will your balance be at the end of 10 years? If the interest were compounded continuously, what would the balance be?

**61.** *Profit*   The annual profit for a company from 1986 to 2001 can be approximated by the model

$$a_n = 156.4e^{0.16n}, \quad n = 0, 1, 2, 3, \ldots, 15$$

where $a_n$ is the annual profit (in millions of dollars) and $n$ represents the year, with $n = 0$ corresponding to 1986.

(a) Sketch a bar graph that represents the company's profit during the 16-year period.

(b) Find the total profit during the 16-year period.

**62.** *Sales*   The annual sales for a company from 1986 to 2001 can be approximated by the model

$$a_n = 221e^{0.146n}, \quad n = 0, 1, 2, 3, \ldots, 15$$

where $a_n$ is the annual sales (in millions of dollars) and $n$ represents the year, with $n = 0$ corresponding to 1986. Use the formula for the sum of a geometric series to approximate the total sales earned during the 16-year period.

In Exercises 63–66, evaluate $_nC_r$.

**63.** $_7C_4$

**64.** $_{12}C_8$

**65.** $_{30}C_{30}$

**66.** $_{20}C_9$

In Exercises 67 and 68, expand the binomial using Pascal's Triangle to determine the coefficients.

**67.** $(3 - 2y)^4$

**68.** $(3x + 4y)^5$

In Exercises 69–74, use the Binomial Theorem to expand the binomial. Simplify your answer.

**69.** $(x + 1)^6$

**70.** $(x - 1)^7$

**71.** $(x + y)^5$

**72.** $(2x - y)^6$

**73.** $(x - 2y)^8$

**74.** $(x^2 + 4)^5$

In Exercises 75 and 76, find the required term in the expansion of the binomial.

| *Binomial* | *Term* |
| --- | --- |
| **75.** $(x + 2)^{12}$ | $ax^7$ |
| **76.** $(2x - 5y)^9$ | $ax^4y^5$ |

In Exercises 77 and 78, use the Binomial Theorem to expand the expression. [In the study of probability, it is sometimes necessary to use the expansion of $(p + q)^n$, where $p + q = 1$.]

**77.** $\left(\frac{1}{3} + \frac{2}{3}\right)^5$

**78.** $(0.2 + 0.8)^{10}$

**79.** *Computer Systems* A customer in a computer store can choose one of five monitors, one of four keyboards, and one of five computers. If all of the choices are compatible, how many different systems can be chosen?

**80.** *Sound System* A customer in an electronics store can choose one of five CD players, one of six speaker systems, and one of three radio/cassette players to design a sound system. How many different systems can be designed?

**81.** *Roller Coaster Ride* Six people are lining up for a ride on a roller coaster, but only three of the six are willing to sit in the front two seats of the coaster car, which has six seats. With that constraint, in how many ways can the six people be seated in the car?

**82.** *True-False Exam* In how many ways can a 15-question true-false exam be answered? (Assume that no questions are omitted.)

**83.** *True-False Exam* In how many ways can a 20-question true-false exam be answered? (Assume that no questions are omitted.)

**84.** *Four-Digit Numbers* How many four-digit numbers can be formed under the following conditions?

(a) Leading digits cannot be zero.

(b) Leading digits cannot be zero and no repetition of digits is allowed.

(c) Leading digits cannot be zero and the number must be divisible by 2.

**85.** *Five-Digit Numbers* How many five-digit numbers can be formed under the following conditions?

(a) Leading digits cannot be zero.

(b) Leading digits cannot be zero and no repetition of digits is allowed.

(c) Leading digits cannot be zero and the number must be odd.

In Exercises 86–89, evaluate $_nP_r$.

**86.** $_9P_3$

**87.** $_{30}P_1$

**88.** $_8P_8$

**89.** $_{15}P_2$

**90.** *Test Questions* A student may answer any 15 questions from a total of 20 questions on an exam. In how many different ways can the student select the 15 questions?

**91.** *Starting Lineup* In how many ways can the starting lineup for a coed, six-member flag football team be selected from a team of eight men and eight women? League rules require that every team have three men and three women on the field.

**92.** *Defective Units* A shipment of 30 VCRs contains three defective units. In how many ways can a vending company purchase five of these units and receive (a) all good units, (b) two good units, and (c) at least two good units?

**93.** *Job Applicants* An employer interviews 10 people for five openings in the company. Six of the 10 people are women. If all 10 are qualified, in how many ways can the employer fill the five positions if (a) the selection is random, (b) exactly three of those hired are women, and (c) all five of those hired are women?

*Drawing a Card* In Exercises 94 and 95, find the indicated probability in the experiment of selecting one card from a standard deck of 52 playing cards.

**94.** The probability of getting a ten, jack, or queen

**95.** The probability of getting a red card that is not a face card

*Tossing a Die*   In Exercises 96 and 97, find the indicated probability in the experiment of tossing a six-sided die twice.

**96.** The probability that the sum is 7

**97.** The probability that the sum is 7 or 11

In Exercises 98 and 99, you are given the probability that an event will happen. Find the probability that the event will not happen.

**98.** $p = 0.57$          **99.** $p = 0.23$

**100.** *Preparing for a Test*   An instructor gives a class a list of 12 study problems, from which eight will be selected to construct an exam. If a given student knows how to solve 10 of the problems, find the probabilities that the student will be able to answer (a) all eight questions on the exam, (b) exactly seven questions on the exam, and (c) at least seven questions on the exam.

**101.** *Game Show*   On a game show, you are given five digits to arrange in the proper order to represent the price of a car. If you are correct, you win the car. What are the probabilities of winning under the following conditions?

(a) You must guess the position of each digit.

(b) You know the first two digits, but must guess the remaining three.

**102.** *Payroll Mix-Up*   Four paychecks and envelopes are addressed to four different people. If the paychecks get mixed up and are randomly inserted into the envelopes, what are the probabilities that (a) exactly one will be inserted in the correct envelope and (b) at least one will be inserted in the correct envelope?

**103.** *Poker Hand*   Five cards are drawn from an ordinary deck of 52 playing cards. What is the probability of getting two pair? (Two pair consists of two of one kind, two of another kind, and one of a third kind. For example, KK995 is two pair.)

**104.** *A Boy or a Girl?*   Assume that the probability of the birth of a child of a particular gender is 50%. In a family with five children, what are the probabilities that (a) all the children are boys, (b) all the children are of the same gender, and (c) there is at least one boy?

**105.** *A Boy or a Girl?*   Assume that the probability of the birth of a child of a particular gender is 50%. In a family with 10 children, what are the probabilities that (a) all the children are girls, (b) all the children are of the same gender, and (c) there is at least one girl?

**106.** *Defective Units*   A shipment of 1000 radar detectors contains five defective units. A retail outlet has ordered 50 units.

(a) What is the probability that all 50 units are good?

(b) What is the probability that at least one unit is defective?

In Exercises 107–110, use mathematical induction to prove the formula for every positive integer $n$.

**107.** $1 + 4 + \cdots + (3n - 2) = \dfrac{n}{2}(3n - 1)$

**108.** $1 + \dfrac{3}{2} + 2 + \dfrac{5}{2} + \cdots + \dfrac{1}{2}(n + 1) = \dfrac{n}{4}(n + 3)$

**109.** $\displaystyle\sum_{i=0}^{n-1} ar^i = \dfrac{a(1 - r^n)}{1 - r}$

**110.** $\displaystyle\sum_{k=0}^{n-1} (a + kd) = \dfrac{n}{2}[2a + (n - 1)d]$

In Exercises 111–114, find the sum of the integers using the formulas for the sums of powers of integers.

**111.** $\displaystyle\sum_{n=1}^{30} n$          **112.** $\displaystyle\sum_{n=1}^{5} n^2$

**113.** $\displaystyle\sum_{n=1}^{8} n^3$          **114.** $\displaystyle\sum_{n=1}^{6} (n^2 + n)$

In Exercises 115 and 116, find a formula for the sum of the first $n$ terms of the sequence.

**115.** $1, \dfrac{3}{5}, \dfrac{9}{25}, \dfrac{27}{125}, \ldots$          **116.** $5, -\dfrac{5}{2}, \dfrac{5}{4}, -\dfrac{5}{8}, \ldots$

In Exercises 117 and 118, write the first five terms of the sequence where $a_1 = f(1)$. Then calculate the first and second differences of the sequence. Does the sequence have a linear model, a quadratic model, or neither?

**117.** $a_0 = 2;\ a_n = a_{n-1} + n$

**118.** $f(1) = 1;\ a_n = a_{n-1} + 2n$

## Chapter Test

Take this test as you would take a test in class. After you are done, check your work against the answers given in the back of the book.

In Exercises 1–4, write the first five terms of the sequence. (Begin with $n = 1$.)

**1.** $a_n = 2n + 1$ **2.** $a_n = (-1)^n n^2$ **3.** $a_n = (n + 1)!$ **4.** $a_n = \left(\frac{1}{2}\right)^n$

In Exercises 5–8, decide whether the sequence is arithmetic, geometric, or neither. If possible, find its common difference or common ratio.

**5.** $a_n = 5n - 2$ **6.** $a_n = 2^n + 2$ **7.** $a_n = 5(2^n)$ **8.** $a_n = 3n^2 - 2$

In Exercises 9–11, evaluate the sum.

**9.** $\displaystyle\sum_{n=1}^{50} (2n + 1)$ **10.** $\displaystyle\sum_{n=1}^{20} 3\left(\frac{3}{2}\right)^n$ **11.** $\displaystyle\sum_{n=1}^{\infty} \left(\frac{1}{3}\right)^n$

**12.** A deposit of $10,000 is made in an account that pays 8% compounded monthly. The balance in the account after $n$ months is given by

$$A = 10,000\left(1 + \frac{0.08}{12}\right)^n, \quad n = 1, 2, 3, \ldots .$$

Find the balance in the account after 10 years.

**13.** You deposit $250 in an account at the beginning of each month for 10 years. The account pays 6% compounded monthly. What is the balance at the end of 10 years?

**14.** Use the Binomial Theorem to expand $(x + 2)^5$.

**15.** Determine the numerical coefficient of the $x^7$ term of $(x - 3)^{12}$.

**16.** A customer in an electronics store can choose one of five CD players, one of six speaker systems, and one of four radio/cassette players to design a sound system. How many different systems can be designed?

**17.** In how many ways can a 10-question true-false exam be answered? (Assume that no question is omitted.) Explain your reasoning.

**18.** How many five-digit numbers can be formed if the leading digit cannot be zero and the number must be odd?

**19.** On a game show, you are given five different digits to arrange in the proper order to represent the price of a car. You know only the first digit. What is the probability that you will arrange the other four digits correctly?

**20.** Five coins are tossed. What is the probability that all are heads?

# Cumulative Test: Chapters 6–8

Take this test as you would take a test in class. After you are done, check your work against the answers given in the back of the book.

In Exercises 1 and 2, solve the system by the indicated method.

**1.** *Elimination*

$$2x - 3y + z = 18$$
$$3x \quad\quad - 2z = -4$$
$$x - y + 3z = 20$$

**2.** *Matrices*

$$3x - 4y + 2z = -32$$
$$2x + 3y \quad\quad = 8$$
$$y - 3z = 19$$

$$A = \begin{bmatrix} 3 & 2 & 2 \\ 1 & 2 & 2 \\ 1 & 0 & 1 \end{bmatrix}$$

$$B = \begin{bmatrix} 4 & 1 \\ -1 & 2 \\ 3 & 1 \end{bmatrix}$$

$$C = \begin{bmatrix} 5 & 0 & -4 \\ 3 & 0 & 1 \\ 2 & -1 & -3 \end{bmatrix}$$

$$D = \begin{bmatrix} 1 & 3 \\ -2 & 0 \end{bmatrix}$$

Matrices for 3–6

In Exercises 3–6, use the matrices to find the indicated matrix.

**3.** $2A - C$    **4.** $AB$    **5.** $BD$    **6.** $A^{-1}$

In Exercises 7 and 8, evaluate the determinant of the matrix.

**7.** $\begin{bmatrix} 1 & -2 & 4 \\ 3 & 7 & -5 \\ 6 & 1 & 4 \end{bmatrix}$    **8.** $\begin{bmatrix} 1 & 0 & 0 \\ 0 & 7 & 0 \\ 0 & 0 & -1 \end{bmatrix}$

**9.** Find the point of equilibrium for a system with demand function $p = 75 - 0.0005x$ and supply function $p = 30 + 0.002x$.

**10.** A total of $45,000 is invested at 7.5% and 8.5% simple interest. If the yearly interest is $3625, how much of the $45,000 is invested at each rate?

**11.** A factory produces three different models of a product, which are shipped to three different warehouses. The number of units $i$ that are shipped to warehouse $j$ is represented by $a_{ij}$ in matrix $A$ at the left. The price per unit is represented by matrix $B$. Find the product $BA$ and state what each entry of the product represents.

$$A = \begin{bmatrix} 1000 & 3000 & 2000 \\ 2000 & 4000 & 5000 \\ 3000 & 1000 & 1000 \end{bmatrix}$$

$$B = \begin{bmatrix} \$25 & \$32 & \$27 \end{bmatrix}$$

Matrices for 11

**12.** Write the first five terms of the sequence given by $a_n = 2 + 3n$.

**13.** Write the first five terms of the geometric sequence with $a_1 = 3$ and $r = \frac{1}{3}$.

**14.** Use the Binomial Theorem to expand $(y - 4)^7$.

**15.** In how many ways can the letters in MATRIX be arranged?

**16.** How many five-digit numbers are multiples of 5?

**17.** A shipment of 2000 microwave ovens contains 15 defective units. You order three units from the shipment. What is the probability that all three are good?

**18.** On a game show, the digits 7, 8, and 9 must be arranged in the proper order to form the price of an appliance. If the digits are arranged correctly, the contestant wins the appliance. What is the probability of winning if the contestant knows that the price is at least $800?

**19.** Write the first five terms of the sequence where $a_1 = 2$ and $a_n = a_{n-1} + n$. Then calculate the first and second differences of the sequence. Does the sequence have a linear model, a quadratic model, or neither?

## Appendix A

# Review of Fundamental Concepts of Algebra

---

### A.1    Real Numbers:  Order and Absolute Value

Real Numbers  ■  The Real Number Line and Ordering  ■  Absolute Value and Distance  ■
Application

## Real Numbers

The formal term that is used in mathematics to refer to a collection of objects is the word **set.** For instance, the set

$$\{1, 2, 3\}$$

contains the three numbers 1, 2, and 3. Note that a pair of braces { } is used to enclose the members of the set. In this text, a *pair* of braces will always indicate the members of a set. Parentheses ( ) and brackets [ ] are used to represent other ideas.

The set of numbers that is used in arithmetic is the **set of real numbers.** The term *real* distinguishes real numbers from *imaginary* or *complex* numbers.

A set $A$ is called a **subset** of a set $B$ if every member of $A$ is also a member of $B$. Here are two examples.

- $\{1, 2, 3\}$ is a subset of $\{1, 2, 3, 4\}$.

- $\{0, 4\}$ is a subset of $\{0, 1, 2, 3, 4\}$.

One of the most commonly used subsets of real numbers is the set of **natural numbers** or **positive integers**

$$\{1, 2, 3, 4, \ldots\}. \qquad \text{Set of positive integers}$$

Note that the three dots indicate that the pattern continues. For instance, the set also contains the numbers 5, 6, 7, and so on.

Positive integers can be used to describe many quantities that you encounter in everyday life—for instance, you might be taking four classes this term, or you might be paying $240 dollars a month for rent. But even in everyday life, positive integers cannot describe some concepts accurately. For instance, you could have a zero balance in your checking account, or the temperature could be $-10°$ (10 degrees below zero). To describe such quantities you need to expand the set of positive integers to include **zero** and the **negative integers.** The expanded set is called the set of **integers,** which can be written as follows.

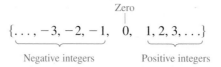

The set of integers is a **subset** of the set of real numbers—this means that every integer is a real number.

Even with the set of integers, there are still many quantities in everyday life that you cannot describe accurately. The costs of many items are not in whole-dollar amounts, but in parts of dollars, such as $1.19 or $39.98. You might work $8\frac{1}{2}$ hours, or you might miss the first *half* of a movie. To describe such quantities, the set of integers is expanded to include **fractions.** The expanded set is called the set of **rational numbers.** Formally, a real number is called **rational** if it can be written as the ratio $p/q$ of two integers, where $q \neq 0$. (The symbol $\neq$ means **not equal to.**) For instance,

$$2 = \frac{2}{1}, \quad 0.333\ldots = \frac{1}{3}, \quad 0.125 = \frac{1}{8}, \quad \text{and} \quad 1.126126\ldots = \frac{125}{111}$$

are rational numbers. Real numbers that cannot be written as ratios of two integers are called **irrational.** For instance, the numbers

$$\sqrt{2} = 1.4142135\ldots \quad \text{and} \quad \pi = 3.1415926\ldots$$

are irrational. The decimal representation of a rational number is either *terminating* or *repeating.* For instance, the decimal representation of $\frac{1}{4} = 0.25$ is terminating, and the decimal representation of $\frac{4}{11} = 0.363636\ldots = 0.\overline{36}$ is repeating. (The line over "36" indicates which digits repeat.)

The decimal representation of an irrational number neither terminates nor repeats. When you perform calculations using decimal representations of nonterminating decimals, you usually use a decimal approximation that has been **rounded** to a certain number of decimal places. For instance, rounded to four decimal places, the decimal approximations of $\frac{2}{3}$ and $\pi$ are

$$\frac{2}{3} \approx 0.6667 \quad \text{and} \quad \pi \approx 3.1416.$$

The symbol $\approx$ means **approximately equal to.**

Figure A.1 shows several commonly used subsets of real numbers and their relationships to each other.

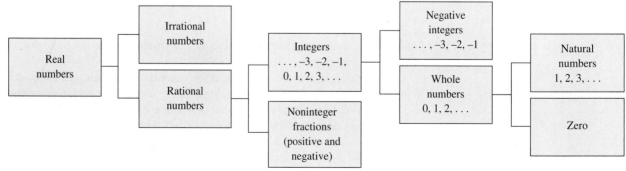

Figure A.1

# The Real Number Line and Ordering

The picture that is used to represent the real numbers is the **real number line.** It consists of a horizontal line with a point (the **origin**) labeled as 0. Points to the left of 0 are associated with **negative numbers,** and points to the right of 0 are associated with **positive numbers,** as shown in Figure A.2. The real number zero is neither positive nor negative. Thus, when you want to talk about real numbers that might be positive *or* zero, you can use the term **nonnegative real numbers.**

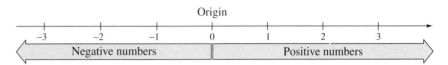

**Figure A.2**    *The Real Number Line*

Each point on the real number line corresponds to exactly one real number, and each real number corresponds to exactly one point on the real number line, as shown in Figure A.3. The number associated with a point on the real number line is the **coordinate** of the point.

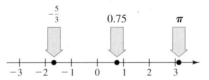

**Figure A.3**    *Every real number corresponds to a point on the real number line.*

The real number line provides you with a way of comparing any two real numbers. For instance, if you choose any two (different) numbers on the real number line, one of the numbers must be to the left of the other number. The number to the left is **less than** the number to the right, and the number to the right is **greater than** the number to the left.

---

▶ Definition of Order on the Real Number Line

If the real number $a$ lies to the left of the real number $b$ on the real number line, $a$ is **less than** $b$, which is denoted by

$$a < b$$

as shown in Figure A.4. This relationship can also be described by saying that $b$ is **greater than** $a$ and writing $b > a$. The symbol $a \leq b$ means that $a$ is **less than or equal to** $b$, and the symbol $b \geq a$ means that $b$ is **greater than or equal to** $a$.

---

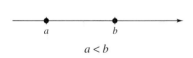

$a < b$

**Figure A.4**    *a is to the left of b.*

The symbols $<$, $>$, $\leq$, and $\geq$ are called **inequality symbols.** Inequalities are useful in denoting subsets of real numbers, as shown in Examples 1 and 2.

**Example 1** Interpreting Inequalities

**a.** The inequality $x \leq 2$ denotes all real numbers that are less than or equal to 2, as shown in Figure A.5 (a).

**b.** The inequality $-2 \leq x < 3$ means that $x \geq -2$ *and* $x < 3$. This **double inequality** denotes all real numbers between $-2$ and 3, including $-2$ but *not* including 3, as shown in Figure A.5 (b).

**c.** The inequality $x > -5$ denotes all real numbers that are greater than $-5$, as shown in Figure A.5 (c).

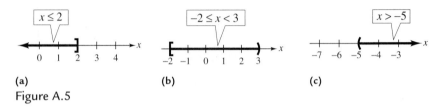

(a)  (b)  (c)

Figure A.5

In Figure A.5, notice that a bracket is used to *include* the endpoint of an interval and a parenthesis is used to *exclude* the endpoint.

**Example 2** Inequalities and Sets of Real Numbers

**a.** "$c$ is nonnegative" means that $c$ is greater than or equal to zero, which you can write as $c \geq 0$.

**b.** "$b$ is at most 5" can be written as $b \leq 5$.

**c.** "$d$ is negative" can be written as $d < 0$, and "$d$ is greater than $-3$" can be written as $-3 < d$. Combining these two inequalities produces $-3 < d < 0$.

**d.** "$x$ is positive" can be written as $0 < x$, and "$x$ is not more than 6" can be written as $x \leq 6$. Combining these two inequalities produces $0 < x \leq 6$.

The following property of real numbers is the **Law of Trichotomy.** As the "tri" in its name suggests, this law tells you that for any two real numbers $a$ and $b$, precisely one of *three* relationships is possible.

$$a < b, \qquad a = b, \qquad \text{or} \qquad a > b \qquad \text{Law of Trichotomy}$$

## Absolute Value and Distance

The **absolute value** of a real number is its *magnitude*, or its value disregarding its sign. For instance, the absolute value of $-3$ is $|-3|$, which has the value of 3.

▶ **Definition of Absolute Value**

Let $a$ be a real number. The **absolute value** of $a$, denoted by $|a|$, is

$$|a| = \begin{cases} a, \text{ if } a \geq 0 \\ -a, \text{ if } a < 0. \end{cases}$$

The absolute value of any real number is either positive or zero. Moreover, 0 is the only real number whose absolute value is zero. That is, $|0| = 0$.

Be sure you see from the definition on the preceding page that the absolute value of a real number is never negative. For instance, if $a = -5$, then $|-5| = -(-5) = 5$.

### Example 3    Finding Absolute Value

**a.** $|-7| = 7$        **b.** $\left|\frac{1}{2}\right| = \frac{1}{2}$

**c.** $|-4.8| = 4.8$        **d.** $-|-9| = -(9) = -9$

### Example 4    Comparing Real Numbers

Place the correct symbol ($<$, $>$, or $=$) between the real numbers.

**a.** $|-4|$ ____ $|4|$        **b.** $|-5|$ ____ $3$        **c.** $-|-1|$ ____ $|-1|$

**Solution**

**a.** $|-4| = |4|$, because both are equal to 4.

**b.** $|-5| > 3$, because $|-5| = 5$ and 5 is greater than 3.

**c.** $-|-1| < |-1|$, because $-|-1| = -1$ and $|-1| = 1$.

---

> ▶ **Properties of Absolute Value**
>
> Let $a$ and $b$ be real numbers. Then the following properties are true.
>
> **1.** $|a| \geq 0$        **2.** $|-a| = |a|$
>
> **3.** $|ab| = |a|\,|b|$        **4.** $\left|\dfrac{a}{b}\right| = \dfrac{|a|}{|b|}$

Absolute value can be used to define the distance between two numbers on the real number line. To see how this is done, consider the numbers $-3$ and 4, as shown in Figure A.6. To find the distance between these two numbers, subtract *either* number from the other and then take the absolute value of the difference. For instance,

(Distance between $-3$ and 4) $= |-3 - 4| = |-7| = 7$.

**Figure A.6**    *The distance between $-3$ and 4 is 7.*

> ▶ **Distance Between Two Numbers**
>
> Let $a$ and $b$ be real numbers. The **distance between $a$ and $b$** is given by
>
> Distance $= |b - a| = |a - b|$.

**Example 5**  Finding the Distance Between Two Numbers

**a.** The distance between 2 and 7 is

$$\text{Distance} = |2 - 7| = |-5| = 5.$$

**b.** The distance between 0 and $-4$ is

$$\text{Distance} = |0 - (-4)| = |4| = 4.$$

**c.** The statement "the distance between $x$ and 2 is at least 3" can be written as

$$|x - 2| \geq 3.$$

## Application

**Example 6**  Budget Variance

You work in the accounting department of a company. One of your jobs is to check whether the monthly expenses of the other departments vary too much from their budgets. Your company's definition of "too much" is that the difference between actual and budgeted expenses must be less than or equal to $500 *and* less than or equal to 5% of the budgeted expenses. By letting $a$ represent the actual expenses and $b$ the budgeted expenses, you can translate these requirements as follows.

$$|a - b| \leq 500 \quad \text{and} \quad |a - b| \leq 0.05b$$

For travel, the budgeted expense was $12,500 and the actual expense was $12,872.56. For office supplies, the budgeted expense was $750 and the actual expense was $704.15. For wages, the budgeted expense was $84,600 and the actual expense was $85,143.95. Are these amounts within budget restrictions?

**Solution**

One way to determine whether these three expenses are within budget restrictions is to create a table, as follows.

|  | Budgeted Expense, b | Actual Expense, a | $\|a - b\|$ | 0.05b |
|---|---|---|---|---|
| Travel | $12,500.00 | $12,872.56 | $372.56 | $625.00 |
| Office supplies | $750.00 | $704.15 | $45.85 | $37.50 |
| Wages | $84,600.00 | $85,143.95 | $543.95 | $4230.00 |

From this table, you can see that the expense for travel passes both tests, but that each of the other two expenses fails one of the tests. Can you see why?

## A.1  Exercises

In Exercises 1–4, determine which numbers in the set are (a) natural numbers, (b) integers, (c) rational numbers, and (d) irrational numbers.

**1.** $\left\{-9, -\frac{7}{2}, 5, \frac{2}{3}, \sqrt{2}, 0.1\right\}$

**2.** $\left\{12, -13, 1, \sqrt{4}, \sqrt{6}, \frac{3}{2}\right\}$

**3.** $\left\{3, -1, \frac{1}{3}, \frac{6}{3}, -\frac{1}{2}, \sqrt{2}, -7.5\right\}$

**4.** $\left\{25, -17, \frac{12}{5}, \sqrt{9}, \sqrt{8}, -\sqrt{8}\right\}$

In Exercises 5–8, plot the two real numbers on the real number line and place the appropriate inequality symbol (< or >) between them.

**5.** $\frac{3}{2}, 7$

**6.** $-4, -8$

**7.** $\frac{5}{6}, \frac{2}{3}$

**8.** $-\frac{8}{7}, -\frac{3}{7}$

In Exercises 9–14, describe the subset of real numbers that is represented by the inequality, and sketch the subset on the real number line.

**9.** $x \le 5$

**10.** $x \ge -2$

**11.** $x > 3$

**12.** $x \ge 4$

**13.** $-2 < x < 2$

**14.** $0 \le x \le 5$

In Exercises 15–18, use inequality notation to describe the set of real numbers.

**15.** $x$ is negative.

**16.** $y$ is greater than 5 and less than or equal to 12.

**17.** The annual rate of inflation $r$ is expected to be at least 3.5%, but no more than 6%.

**18.** The price $p$ of unleaded gasoline is not expected to go above \$1.45 per gallon during the coming year.

In Exercises 19–24, write the expression without using absolute value signs.

**19.** $|-10|$

**20.** $|3 - \pi|$

**21.** $\dfrac{-5}{|-5|}$

**22.** $-3 - |-3|$

**23.** $-3|-3|$

**24.** $|-1| - |-2|$

In Exercises 25–28, place the correct symbol (<, >, or =) between the two real numbers.

**25.** $|-4| \quad\quad |4|$

**26.** $-5 \quad\quad -|5|$

**27.** $-|-6| \quad\quad |-6|$

**28.** $-|-2| \quad\quad -|2|$

In Exercises 29–34, find the distance between $a$ and $b$.

**29.**

**30.**

**31.** $a = -\frac{7}{2}, b = 0$

**32.** $a = \frac{3}{4}, b = \frac{9}{4}$

**33.** $a = 126, b = 75$

**34.** $a = -126, b = -75$

In Exercises 35–38, use absolute value notation to describe the expression.

**35.** The distance between $z$ and $\frac{3}{2}$ is greater than 1.

**36.** The distance between $z$ and 0 is less than 8.

**37.** $y$ is at least 6 units from 0.

**38.** $y$ is at most 2 units from $a$.

In Exercises 39 and 40, use a calculator to order the numbers from smallest to largest.

**39.** $\frac{7071}{5000}, \frac{584}{413}, \sqrt{2}, \frac{47}{33}, \frac{127}{90}$

**40.** $\frac{26}{15}, \sqrt{3}, 1.73\overline{20}, \frac{381}{220}, \sqrt{10} - \sqrt{2}$

In Exercises 41 and 42, use a calculator to find the decimal form of the rational number. If it is a nonterminating decimal, write the repeating pattern.

**41.** $\frac{5}{8}$

**42.** $\frac{6}{11}$

*Budget Variance*   In Exercises 43–48, the accounting department of a company is checking to see whether the actual expenses of a department differ from the budgeted expenses by more than \$500 or 5%. Complete the missing parts of the table and determine whether each actual expense passes the "budget variance test."

| | Budgeted Expense, $b$ | Actual Expense, $a$ | $|a - b|$ | $0.05b$ |
|---|---|---|---|---|
| **43.** | \$25,000.00 | \$24,872.12 | | |
| **44.** | \$112,700.00 | \$113,356.52 | | |
| **45.** | \$9400.00 | \$9972.59 | | |
| **46.** | \$7500.00 | \$7104.68 | | |
| **47.** | \$37,640.00 | \$36,968.25 | | |
| **48.** | \$2575.00 | \$2613.15 | | |

*Median Incomes* In Exercises 49–54, use the bar graph, which shows the median income for American households that purchase equipment for skiing, tennis, golf, fishing, camping, and hunting. In each exercise you are given a household income. Find the amount by which the income differs from the median income for the particular sport. (Source: National Sporting Goods Association)

| | Median Income, y | Household Income, x | \|y − x\| |
|---|---|---|---|
| **49.** Skiing | | $65,400 | |
| **50.** Hunting | | $75,400 | |
| **51.** Tennis | | $37,300 | |
| **52.** Golf | | $18,760 | |
| **53.** Camping | | $21,300 | |
| **54.** Fishing | | $26,370 | |

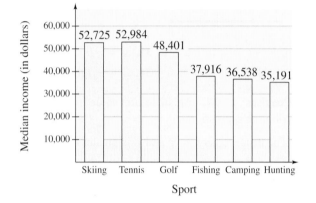

---

## A.2 The Basic Rules of Algebra

Algebraic Expressions ■ Basic Rules of Algebra ■ Equations ■ Calculators and Rounding

## Algebraic Expressions

One of the basic characteristics of algebra is the use of letters (or combinations of letters) to represent numbers. The letters used to represent numbers are called **variables,** and combinations of letters and numbers are called **algebraic expressions.** Here are a few examples of algebraic expressions.

$$5x, \qquad 2x - 3, \qquad \frac{4}{x^2 + 2}, \qquad 7x + y$$

> ▶ **Algebraic Expression**
>
> A collection of letters (called **variables**) and real numbers (called **constants**) that are combined using the operations of addition, subtraction, multiplication, and division is an **algebraic expression.** (Other operations can also be used to form an algebraic expression.)

The **terms** of an algebraic expression are those parts that are separated by addition. For example, the algebraic expression $x^2 - 5x + 8$ has three terms: $x^2$, $-5x$, and 8. Note that $-5x$, rather than $5x$, is a term, because

$$x^2 - 5x + 8 = x^2 + (-5x) + 8.$$

The terms $x^2$ and $-5x$ are the **variable terms** of the expression, and 8 is the **constant term** of the expression. The numerical factor of a variable term  is the **coefficient** of the variable term. For instance, the coefficient of the variable term $-5x$ is $-5$, and the coefficient of the variable term $x^2$ is 1.

### Example 1   Identifying the Terms of an Algebraic Expression

| *Algebraic Expression* | *Terms* |
|---|---|
| **a.** $4x - 3$ | $4x, -3$ |
| **b.** $2x + 4y - 5$ | $2x, 4y, -5$ |

To **evaluate** an algebraic expression, substitute numerical values for each of the variables in the expression. Here are some examples.

| *Expression* | *Value of Variable* | *Substitute* | *Value of Expression* |
|---|---|---|---|
| $-3x + 5$ | $x = 3$ | $-3(3) + 5$ | $-9 + 5 = -4$ |
| $3x^2 + 2x - 1$ | $x = -1$ | $3(-1)^2 + 2(-1) - 1$ | $3 - 2 - 1 = 0$ |
| $-2x(x + 4)$ | $x = -2$ | $-2(-2)(-2 + 4)$ | $-2(-2)(2) = 8$ |
| $\dfrac{1}{x - 2}$ | $x = 2$ | $\dfrac{1}{2 - 2}$ | Undefined |

### Example 2   Evaluating Algebraic Expressions

Evaluate the following algebraic expressions when $x = -2$ and $y = 3$.

**a.** $4y - 2x$    **b.** $5 + x^2$    **c.** $5 - x^2$

Solution

**a.** When $x = -2$ and $y = 3$, the expression $4y - 2x$ has a value of

$$4(3) - 2(-2) = 12 + 4 = 16.$$

**b.** When $x = -2$, the expression $5 + x^2$ has a value of

$$5 + (-2)^2 = 5 + 4 = 9.$$

**c.** When $x = -2$, the expression $5 - x^2$ has a value of

$$5 - (-2)^2 = 5 - 4 = 1.$$

## Basic Rules of Algebra

The four basic arithmetic operations are **addition, multiplication, subtraction,** and **division,** denoted by the symbols $+$, $\times$ or $\cdot$, $-$, and $\div$. Of these, addition and multiplication are considered to be the two primary arithmetic operations. Subtraction and division are defined as the inverse operations of addition and multiplication, as follows.

*Subtraction*                    *Division*

$$a - b = a + (-b) \qquad \text{If } b \neq 0, \text{ then } a \div b = a\left(\frac{1}{b}\right) = \frac{a}{b}.$$

In these definitions, $-b$ is called the **opposite** (or additive inverse) of $b$, and $1/b$ is called the **reciprocal** (or multiplicative inverse) of $b$. In place of $a \div b$, you can use the fraction symbol $a/b$. In this fractional form, $a$ is called the **numerator** of the fraction and $b$ is called the **denominator.**

Be sure you see that the **basic rules of algebra,** listed below, are true for variables and algebraic expressions as well as for real numbers.

---

▶ **Basic Rules of Algebra**

Let $a$, $b$, and $c$ be real numbers, variables, or algebraic expressions.

| *Property* | *Example* |
|---|---|
| Commutative Property of Addition | |
| $a + b = b + a$ | $4x + x^2 = x^2 + 4x$ |
| Commutative Property of Multiplication | |
| $ab = ba$ | $(4 - x)x^2 = x^2(4 - x)$ |
| Associative Property of Addition | |
| $(a + b) + c = a + (b + c)$ | $(-x + 5) + 2x^2 =$ $-x + (5 + 2x^2)$ |
| Associative Property of Multiplication | |
| $(ab)c = a(bc)$ | $(2x \cdot 3y)(8) = (2x)(3y \cdot 8)$ |
| Distributive Property | |
| $a(b + c) = ab + ac$ | $3x(5 + 2x) = 3x \cdot 5 + 3x \cdot 2x$ |
| $(a + b)c = ac + bc$ | $(y + 8)y = y \cdot y + 8 \cdot y$ |
| Additive Identity Property | |
| $a + 0 = a$ | $5y^2 + 0 = 5y^2$ |
| Multiplicative Identity Property | |
| $a \cdot 1 = 1 \cdot a = a$ | $(4x^2)(1) = (1)(4x^2) = 4x^2$ |
| Additive Inverse Property | |
| $a + (-a) = 0$ | $5x^3 + (-5x^3) = 0$ |
| Multiplicative Inverse Property | |
| $a \cdot \dfrac{1}{a} = 1, \quad a \neq 0$ | $(x^2 + 4)\left(\dfrac{1}{x^2 + 4}\right) = 1$ |

---

Because subtraction is defined as "adding the opposite," the Distributive Properties are also true for subtraction. For instance, the "subtraction form" of $a(b + c) = ab + ac$ is

$$a(b - c) = a[b + (-c)] = ab + a(-c) = ab - ac.$$

**Example 3**    Identifying the Basic Rules of Algebra

Identify the rule of algebra illustrated in each of the following.

**a.** $(4x^2)5 = 5(4x^2)$

**b.** $(2y^3 + y) - (2y^3 + y) = 0$

**c.** $(4 + x^2) + 3x^2 = 4 + (x^2 + 3x^2)$

**d.** $(x - 5)7 + (x - 5)x = (x - 5)(7 + x)$

**e.** $2x \cdot \dfrac{1}{2x} = 1, \quad x \neq 0$

**Solution**

**a.** This equation illustrates the Commutative Property of Multiplication.

**b.** This equation illustrates the Additive Inverse Property.

**c.** This equation illustrates the Associative Property of Addition. In other words, to form the sum $4 + x^2 + 3x^2$, it doesn't matter whether 4 and $x^2$ are added first or $x^2$ and $3x^2$ are added first.

**d.** This equation illustrates the Distributive Property in reverse order.

$$ab + ac = a(b + c) \qquad \text{Distributive Property}$$

$$(x - 5)7 + (x - 5)x = (x - 5)(7 + x)$$

**e.** This equation illustrates the Multiplicative Inverse Property. Note that it is important that $x$ be a nonzero number. If $x$ were allowed to be zero, you would be in trouble because the reciprocal of zero is undefined.

---

The following three lists summarize the basic properties of negation, zero, and fractions. When you encounter such lists, we suggest that you not only *memorize* a verbal description of each property, but that you also try to gain an *intuitive feeling* for the validity of each.

> ▶ **Properties of Negation**
>
> Let $a$ and $b$ be real numbers, variables, or algebraic expressions.
>
> | *Property* | *Example* |
> |---|---|
> | **1.** $(-1)a = -a$ | $(-1)7 = -7$ |
> | **2.** $-(-a) = a$ | $-(-6) = 6$ |
> | **3.** $(-a)b = -(ab) = a(-b)$ | $(-5)3 = -(5 \cdot 3) = 5(-3)$ |
> | **4.** $(-a)(-b) = ab$ | $(-2)(-6) = 12$ |
> | **5.** $-(a + b) = (-a) + (-b)$ | $-(3 + 8) = (-3) + (-8)$ |

Be sure you see the difference between the *negative* (or *opposite*) *of a number* and a *negative number*. If $a$ is already negative, then its additive inverse, $-a$, is positive. For instance, if $a = -5$, then $-a = -(-5) = 5$.

▶ Properties of Zero

Let $a$ and $b$ be real numbers, variables, or algebraic expressions. Then the following properties are true.

**1.** $a + 0 = a$    and    $a - 0 = a$

**2.** $a \cdot 0 = 0$

**3.** $\dfrac{0}{a} = 0, \quad a \neq 0$

**4.** $\dfrac{a}{0}$ is undefined.

**5.** Zero-Factor Property: If $ab = 0$, then $a = 0$ or $b = 0$.

The "or" in the Zero-Factor Property includes the possibility that both factors are zero. This is called an **inclusive or,** and it is the way the word "or" is always used in mathematics.

▶ Properties of Fractions

Let $a$, $b$, $c$, and $d$ be real numbers, variables, or algebraic expressions such that $b \neq 0$ and $d \neq 0$. Then the following properties are true.

**1.** *Equivalent fractions:* $\dfrac{a}{b} = \dfrac{c}{d}$ if and only if $ad = bc$.

**2.** *Rules of signs:* $-\dfrac{a}{b} = \dfrac{-a}{b} = \dfrac{a}{-b}$    and    $\dfrac{-a}{-b} = \dfrac{a}{b}$

**3.** *Generate equivalent fractions:* $\dfrac{a}{b} = \dfrac{ac}{bc}, \quad c \neq 0$

**4.** *Add or subtract with like denominators:* $\dfrac{a}{b} \pm \dfrac{c}{b} = \dfrac{a \pm c}{b}$

**5.** *Add or subtract with unlike denominators:* $\dfrac{a}{b} \pm \dfrac{c}{d} = \dfrac{ad \pm bc}{bd}$

**6.** *Multiply fractions:* $\dfrac{a}{b} \cdot \dfrac{c}{d} = \dfrac{ac}{bd}$

**7.** *Divide fractions:* $\dfrac{a}{b} \div \dfrac{c}{d} = \dfrac{a}{b} \cdot \dfrac{d}{c} = \dfrac{ad}{bc}, \quad c \neq 0$

In Property 1 (equivalent fractions) the phrase "if and only if" implies two statements. One statement is: If $a/b = c/d$, then $ad = bc$. The other statement is: If $ad = bc$, where $b \neq 0$ and $d \neq 0$, then $a/b = c/d$.

**Example 4**   Properties of Zero and Properties of Fractions

**a.** $x - \dfrac{0}{5} = x - 0 = x$     Properties 3 and 1 of zero

**b.** $\dfrac{x}{5} = \dfrac{3 \cdot x}{3 \cdot 5} = \dfrac{3x}{15}$     Generate equivalent fractions.

**c.** $\dfrac{x}{3} + \dfrac{2x}{5} = \dfrac{x \cdot 5 + 3 \cdot 2x}{15}$     Add fractions with unlike denominators.

**d.** $\dfrac{7}{x} \div \dfrac{3}{2} = \dfrac{7}{x} \cdot \dfrac{2}{3} = \dfrac{14}{3x}$     Divide fractions.

If $a$, $b$, and $c$ are integers such that $ab = c$, then $a$ and $b$ are **factors** or **divisors** of $c$. For example, 2 and 3 are factors of 6 because $2 \cdot 3 = 6$. A **prime number** is a positive integer that has exactly two factors: itself and 1. For example, 2, 3, 5, 7, and 11 are prime numbers, whereas 1, 4, 6, 8, 9, and 10 are not. The numbers 4, 6, 8, 9, and 10 are **composite** because they can be written as the products of two or more prime numbers. The number 1 is neither prime nor composite. The **Fundamental Theorem of Arithmetic** states that every positive integer greater than 1 can be written as the product of prime numbers in precisely one way (disregarding order). For instance, the prime factorization of 24 is

$$24 = 2 \cdot 2 \cdot 2 \cdot 3.$$

When adding or subtracting fractions with unlike denominators, you can use Property 4 of fractions by rewriting both fractions so that they have the same denominator. This is called the **least common denominator** method.

**Example 5**   Adding and Subtracting Fractions

Evaluate $\dfrac{2}{15} - \dfrac{5}{9} + \dfrac{4}{5}$.

Solution

By prime factoring the denominators ($15 = 3 \cdot 5$, $9 = 3 \cdot 3$, and $5 = 5$) you can see that the least common denominator is $3 \cdot 3 \cdot 5 = 45$. Thus, it follows that

$$\frac{2}{15} - \frac{5}{9} + \frac{4}{5} = \frac{2 \cdot 3}{15 \cdot 3} - \frac{5 \cdot 5}{9 \cdot 5} + \frac{4 \cdot 9}{5 \cdot 9}$$

$$= \frac{6 - 25 + 36}{45}$$

$$= \frac{17}{45}.$$

## Equations

An **equation** is a statement of equality between two expressions. Thus, the statement

$$a + b = c + d$$

means that the expressions $a + b$ and $c + d$ represent the same number. For instance, because $1 + 4$ and $3 + 2$ both represent the number 5, you can write $1 + 4 = 3 + 2$. Three important properties of equality are as follows.

---

▶ **Properties of Equality**

Let $a$, $b$, and $c$ be real numbers, variables, or algebraic expressions.

**1.** Reflexive: $a = a$

**2.** Symmetric: If $a = b$, then $b = a$.

**3.** Transitive: If $a = b$ and $b = c$, then $a = c$.

---

In algebra, you often rewrite expressions by making substitutions that are permitted under the **Substitution Principle.** "If $a = b$, then $a$ can be replaced by $b$ in any expression involving $a$." Two important consequences of the Substitution Principle are the following rules.

**1.** If $a = b$, then $a + c = b + c$.         Add $c$ to both sides.

**2.** If $a = b$, then $ac = bc$.              Multiply both sides by $c$.

The first rule allows us to add the same number to both sides of an equation. The second allows us to multiply both sides of an equation by the same number. The converses of these two rules are listed below.

**1.** If $a + c = b + c$, then $a = b$.         Subtract $c$ from both sides.

**2.** If $ac = bc$ and $c \neq 0$ then $a = b$.     Divide both sides by $c$.

When adding, subtracting, multiplying, or dividing more than two numbers, it is important to use symbols of grouping (such as parentheses) to indicate the order of operations.

---

**Example 6**    Symbols of Grouping

**a.** $7 - 3(4 - 2) = 7 - 3(2) = 7 - 6 = 1$

**b.** $(4 - 5) - (3 - 6) = (-1) - (-3) = -1 + 3 = 2$

---

## Calculators and Rounding

The specific keystrokes that are listed here are those that correspond to a standard scientific calculator and to a graphing calculator. The keystrokes listed here may not be the same as those for your graphing calculator. Consult your user's guide for specific keystrokes. Here are some comparisons between graphing calculator keys and scientific calculator keys.

**1.** The key marked $\boxed{\text{ENTER}}$ is similar to $\boxed{=}$.

**2.** The key marked $\boxed{(-)}$ is similar to $\boxed{+/-}$.

**3.** The key marked $\boxed{\wedge}$ is similar to $\boxed{y^x}$.

**4.** The key marked $\boxed{x^{-1}}$ is similar to $\boxed{1/x}$.

For example, you can evaluate $13^3$ on a graphing calculator or a scientific calculator as follows.

| Graphing Calculator | Scientific Calculator |
|---|---|
| 13 $\boxed{\wedge}$ 3 $\boxed{\text{ENTER}}$ | 13 $\boxed{y^x}$ 3 $\boxed{=}$ |

### Example 7    Using a Calculator

**Scientific Calculator**

| Expression | Keystrokes | Display |
|---|---|---|
| **a.** $7 - (5 \cdot 3)$ | 7 $\boxed{-}$ 5 $\boxed{\times}$ 3 $\boxed{=}$ | $-8$ |
| **b.** $-12^2 - 100$ | 12 $\boxed{x^2}$ $\boxed{+/-}$ $\boxed{-}$ 100 $\boxed{=}$ | $-244$ |
| **c.** $24 \div 2^3$ | 24 $\boxed{\div}$ 2 $\boxed{y^x}$ 3 $\boxed{=}$ | 3 |
| **d.** $3(10 - 4^2) \div 2$ | 3 $\boxed{\times}$ $\boxed{(}$ 10 $\boxed{-}$ 4 $\boxed{x^2}$ $\boxed{)}$ $\boxed{\div}$ 2 $\boxed{=}$ | $-9$ |
| **e.** 37% of 40 | .37 $\boxed{\times}$ 40 $\boxed{=}$ | 14.8 |

**Graphing Calculator**

| Expression | Keystrokes | Display |
|---|---|---|
| **a.** $7 - (5 \cdot 3)$ | 7 $\boxed{-}$ 5 $\boxed{\times}$ 3 $\boxed{\text{ENTER}}$ | $-8$ |
| **b.** $-12^2 - 100$ | $\boxed{(-)}$ 12 $\boxed{x^2}$ $\boxed{-}$ 100 $\boxed{\text{ENTER}}$ | $-244$ |
| **c.** $24 \div 2^3$ | 24 $\boxed{\div}$ 2 $\boxed{\wedge}$ 3 $\boxed{\text{ENTER}}$ | 3 |
| **d.** $3(10 - 4^2) \div 2$ | 3 $\boxed{(}$ 10 $\boxed{-}$ 4 $\boxed{x^2}$ $\boxed{)}$ $\boxed{\div}$ 2 $\boxed{\text{ENTER}}$ | $-9$ |
| **e.** 37% of 40 | .37 $\boxed{\times}$ 40 $\boxed{\text{ENTER}}$ | 14.8 |

Be sure you see the difference between the change sign keys $\boxed{+/-}$ or $\boxed{(-)}$ and the subtraction key $\boxed{-}$, as used in Example 7(b).

For all their usefulness, calculators do have a problem representing some numbers because they are limited to a finite number of digits. For instance, what does your calculator display when you compute $2 \div 3$? Some calculators simply truncate (drop) the digits that exceed their display range and display .66666666. Others round the number and display .66666667. Although the second display is more accurate, *both* of these decimal representations of $\frac{2}{3}$ contain rounding errors. When rounding decimals in this text, use the following rule: *round up on 5 or greater, round down on 4 or less.*

### Example 8    Rounding Decimal Numbers

| Number | Rounded to Three Decimal Places | |
|---|---|---|
| **a.** $\sqrt{2} = 1.4142135\ldots$ | 1.414 | Round down. |
| **b.** $\pi = 3.1415926\ldots$ | 3.142 | Round up. |
| **c.** $\dfrac{7}{9} = 0.7777777\ldots$ | 0.778 | Round up. |

One of the best ways to minimize error due to rounding is to leave numbers in your calculator until your calculations are complete. If you want to save a number for future use, store it in your calculator's memory.

# Warm Up

The following warm-up exercises involve skills that were covered in earlier sections. You will use these skills in the exercise set for this section.

In Exercises 1–4, place the correct inequality symbol ($<$ or $>$) between the two numbers.

**1.** $-4 \quad -2$  **2.** $0 \quad -3$  **3.** $\sqrt{3} \quad 1.73$  **4.** $-\pi \quad -3$

In Exercises 5–8, find the distance between the two numbers.

**5.** $4, 6$  **6.** $-2, 2$  **7.** $0, -5$  **8.** $-1, 3$

In Exercises 9 and 10, evaluate the expression.

**9.** $|-7| + |7|$  **10.** $-|8 - 10|$

## A.2    Exercises

In Exercises 1–4, identify the terms of the algebraic expression.

**1.** $7x + 4$  **2.** $-5 + 3x$
**3.** $x^2 - 4x + 8$  **4.** $4x^3 + x - 5$

In Exercises 5–8, evaluate the expression for the values of $x$. (If not possible, state the reason.)

| Expression | Values |
|---|---|
| **5.** $4x - 6$ | (a) $x = -1$  (b) $x = 0$ |
| **6.** $x^2 - 3x + 4$ | (a) $x = -2$  (b) $x = 2$ |
| **7.** $-x^3 + 4x$ | (a) $x = 0$  (b) $x = 2$ |
| **8.** $\dfrac{x + 1}{x - 1}$ | (a) $x = 1$  (b) $x = -1$ |

In Exercises 9–23, identify the rule(s) of algebra illustrated by the equation.

**9.** $3 + 4 = 4 + 3$  **10.** $x + 9 = 9 + x$
**11.** $-15 + 15 = 0$  **12.** $2(x + 3) = 2x + 6$
**13.** $2\left(\frac{1}{2}\right) = 1$
**14.** $(5 + 11) \cdot 6 = 5 \cdot 6 + 11 \cdot 6$
**15.** $\dfrac{1}{h + 6}(h + 6) = 1, \quad h \neq -6$
**16.** $h + 0 = h$
**17.** $57 \cdot 1 = 57$

**18.** $(z - 2) + 0 = z - 2$
**19.** $1 \cdot (1 + x) = 1 + x$
**20.** $6 + (7 + 8) = (6 + 7) + 8$
**21.** $x + (y + 10) = (x + y) + 10$
**22.** $x(3y) = (x \cdot 3)y = (3x)y$
**23.** $\frac{1}{7}(7 \cdot 12) = \left(\frac{1}{7} \cdot 7\right)12 = 1 \cdot 12 = 12$

In Exercises 24 and 25, write the prime factorization of the integer.

**24.** $36$  **25.** $30$

In Exercises 26–37, perform the operations. (Write fractional answers in reduced form.)

**26.** $(8 - 17) + 3$  **27.** $-3(5 - 2)$
**28.** $(4 - 7)(-2)$  **29.** $(-5)(-8)$
**30.** $2\left(\dfrac{77}{-11}\right)$  **31.** $\dfrac{27 - 35}{4}$
**32.** $8(-6)(-2)$  **33.** $\frac{10}{11} + \frac{6}{33} - \frac{13}{66}$
**34.** $\frac{2}{5} \cdot \frac{7}{8}$  **35.** $\frac{2}{3} \cdot \frac{5}{8} \cdot \frac{3}{4}$
**36.** $\frac{2}{3} \div 8$  **37.** $\left(\frac{3}{5} \div 3\right) - \left(6 \cdot \frac{4}{8}\right)$

In Exercises 38–41, use a calculator to evaluate the expression. (Round to two decimal places.)

**38.** $3\left(-\frac{5}{12} + \frac{3}{8}\right)$  **39.** $2\left(-7 + \frac{1}{6}\right)$

**40.** $\dfrac{11.46 - 5.37}{3.91}$      **41.** $\dfrac{-8.31 + 4.83}{7.65}$

In Exercises 42 and 43, use a calculator to solve the percent problem.

**42.** 33% of 57      **43.** 121% of 34

In Exercises 44 and 45, find the percent that corresponds to the unlabeled portion of the circle graph.

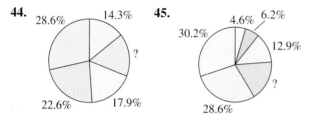

**44.**  28.6%   14.3%   ?   22.6%   17.9%

**45.**  30.2%   4.6%   6.2%   12.9%   ?   28.6%

**46.** *Cookie Sales*   The circle graph shows the shares for the best-selling Girl Scout cookies for 1998. What percent of the total cookie sales did Thin Mints have? If 175 million boxes of Girl Scout cookies were sold in 1998, find the numbers of boxes of Thin Mints, Samoas, Tagalongs, Do-Si-Dos, and Trefoils that were sold. (Round your answers to the nearest tenth of a million boxes.)   (Source: Girl Scouts of the USA)

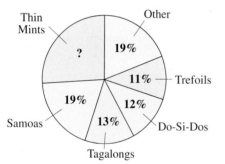

Thin Mints   Other   ?   19%   11% — Trefoils   19%   12%   13%   Do-Si-Dos   Samoas   Tagalongs

**47.** *Federal Government Expenses*   The circle graph shows the types of expenses for the federal government in 2000. What percent of the total budget was spent by Social Security? If the total expenses were 1,789,600,000,000 dollars, find the expense for each of the indicated categories. (Round your answers to the nearest billion dollars.)   (Source: Office of Management and Budget)

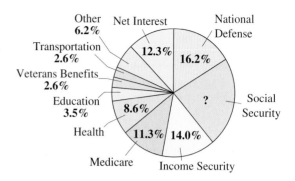

Other 6.2%   Net Interest   National Defense   Transportation 2.6%   12.3%   16.2%   Veterans Benefits 2.6%   Education 3.5%   8.6%   ?   Social Security   Health   11.3%   14.0%   Medicare   Income Security

**Figure for 47**

**48.** *Calculator Keystrokes*   Write the algebraic expression that corresponds to the following keystrokes.

5 ⊠ ⦅ 2.7 ⊟ 9.4 ⦆ ⊜      Scientific

5 ⦅ 2.7 ⊟ 9.4 ⦆ ⌷ENTER⌷      Graphing

**49.** *Calculator Keystrokes*   Write the algebraic expression that corresponds to the following keystrokes.

2 ⊠ ⦅ 4 ⊬ ⊞ 2 ⦆ ⊜      Scientific

2 ⦅ ⊟ 4 ⊞ 2 ⦆ ⌷ENTER⌷      Graphing

**50.** *Calculator Keystrokes*   Write the keystrokes used to evaluate the following algebraic expression. (Base your description on either a scientific or a graphing calculator.)

$5(18 - 2^3) \div 10$

**51.** *Calculator Keystrokes*   Write the keystrokes used to evaluate the following algebraic expression. (Base your description on either a scientific or a graphing calculator.)

$-6^2 - [7 + (-2)^3]$

**52.** *College Enrollment*   The percent of students 24 years old or older at a college is 44.7%. If the college enrollment for the 1998–1999 academic year was 13,385 students, how many students were under 24 years old?

**53.** *College Enrollment*   The percent of students who attend evening classes at a college is 44.1%. If the total enrollment for the college is 11,875 students, how many students attend evening classes?

## A.3    Integer Exponents

Properties of Exponents ■ Scientific Notation ■ Applications

# Properties of Exponents

Repeated multiplication of a real number by itself can be written in **exponential form.** Here are some examples.

| Repeated Multiplication | Exponential Form |
|---|---|
| $7 \cdot 7$ | $7^2$ |
| $a \cdot a \cdot a \cdot a \cdot a$ | $a^5$ |
| $(-4)(-4)(-4)$ | $(-4)^3$ |
| $(2x)(2x)(2x)(2x)$ | $(2x)^4$ |

## Study Tip

It is important to recognize the difference between exponential forms such as $(-2)^4$ and $-2^4$. In $(-2)^4$, the parentheses indicate that the exponent applies to the negative sign as well as to the 2, but in $-2^4 = -(2^4)$, the exponent applies only to the 2. Similarly, in $(5x)^3$, the parentheses indicate that the exponent applies to the 5 as well as to the $x$, whereas in $5x^3 = 5(x^3)$, the exponent applies only to the $x$.

▶ **Exponential Notation**

Let $a$ be a real number, a variable, or an algebraic expression, and let $n$ be a positive integer. Then

$$a^n = \underbrace{a \cdot a \cdot a \cdots a}_{n \text{ factors}}$$

where $n$ is the **exponent** and $a$ is the **base.** The expression $a^n$ is read as "$a$ to the $n$th **power**" or simply "$a$ to the $n$th."

When multiplying exponential expressions with the same base, *add* exponents.

$$a^m \cdot a^n = a^{m+n} \qquad \text{Add exponents when multiplying.}$$

For instance, to multiply $2^2$ and $2^3$, you can write

$$2^2 \cdot 2^3 = \overbrace{(2 \cdot 2)}^{\substack{\text{Two} \\ \text{factors}}} \cdot \overbrace{(2 \cdot 2 \cdot 2)}^{\substack{\text{Three} \\ \text{factors}}} = \overbrace{2 \cdot 2 \cdot 2 \cdot 2 \cdot 2}^{\substack{\text{Five} \\ \text{factors}}}$$

$$= 2^{2+3} = 2^5.$$

On the other hand, when dividing exponential expressions, *subtract* exponents. That is,

$$\frac{a^m}{a^n} = a^{m-n}, \qquad a \neq 0. \qquad \text{Subtract exponents when dividing.}$$

These and other properties of exponents are summarized in the list on the following page.

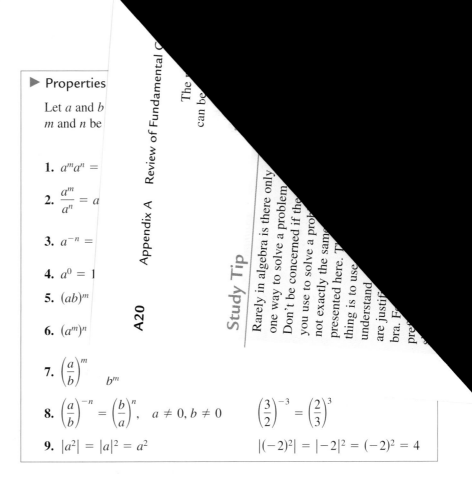

▶ Properties

Let $a$ and $b$
$m$ and $n$ be

1. $a^m a^n =$

2. $\dfrac{a^m}{a^n} = a$

3. $a^{-n} =$

4. $a^0 = 1$

5. $(ab)^m$

6. $(a^m)^n$

7. $\left(\dfrac{a}{b}\right)^m$    $b^m$

8. $\left(\dfrac{a}{b}\right)^{-n} = \left(\dfrac{b}{a}\right)^n,\quad a \neq 0, b \neq 0$    $\left(\dfrac{3}{2}\right)^{-3} = \left(\dfrac{2}{3}\right)^3$

9. $|a^2| = |a|^2 = a^2$    $|(-2)^2| = |-2|^2 = (-2)^2 = 4$

Be sure you see that these properties of exponents apply for *all* integers $m$ and $n$, not just positive ones. For instance, by Property 2, you have

$$\frac{3^4}{3^{-5}} = 3^{4-(-5)}$$

$$= 3^{4+5}$$

$$= 3^9.$$

**Example 1**    Using Properties of Exponents

a. $3^4 \cdot 3^{-1} = 3^{4-1} = 3^3 = 27$

b. $\dfrac{5^6}{5^4} = 5^{6-4} = 5^2 = 25$

c. $5\left(\dfrac{2}{5}\right)^3 = 5 \cdot \dfrac{2^3}{5^3} = 5 \cdot 5^{-3} \cdot 2^3 = 5^{-2}2^3 = \dfrac{2^3}{5^2} = \dfrac{8}{25}$

d. $(-5 \cdot 2^3)^2 = (-5)^2 \cdot (2^3)^2 = 25 \cdot 2^6 = 25 \cdot 64 = 1600$

e. $(-3ab^4)(4ab^{-3}) = -12a \cdot a(b^4)(b^{-3}) = -12a^2b$

f. $3a(-4a^2)^0 = 3a(1) = 3a,\quad a \neq 0$

g. $\left(\dfrac{5x^3}{y}\right)^2 = \dfrac{5^2(x^3)^2}{y^2} = \dfrac{25x^6}{y^2}$

ext two examples show how expressions involving negative exponents rewritten using positive exponents.

steps lem are as the steps he important steps that you *and* that, of course, ed by the rules of alge- or instance, you might er the following steps to implify Example 3.

$$\left(\frac{3x^2}{y}\right)^{-2} = \left(\frac{y}{3x^2}\right)^2 = \frac{y^2}{9x^4}$$

**Example 2** Rewriting with Positive Exponents

**a.** $x^{-1} = \dfrac{1}{x}$ — Property 3: $a^{-n} = \dfrac{1}{a^n}$

**b.** $\dfrac{1}{3x^{-2}} = \dfrac{1(x^2)}{3} = \dfrac{x^2}{3}$ — $-2$ exponent does not apply to 3.

**c.** $\dfrac{12a^3b^{-4}}{4a^{-2}b} = \dfrac{12a^3 \cdot a^2}{4b \cdot b^4} = \dfrac{3a^5}{b^5}$

**Example 3** Quotients Raised to Negative Powers

Rewrite $\left(\dfrac{3x^2}{y}\right)^{-2}$ and simplify.

Solution

$$\left(\frac{3x^2}{y}\right)^{-2} = \frac{3^{-2}(x^2)^{-2}}{y^{-2}} = \frac{3^{-2}x^{-4}}{y^{-2}} = \frac{y^2}{3^2x^4} = \frac{y^2}{9x^4}$$

**Example 4** Ratio of Volume to Surface Area

The volume $V$ and surface area $S$ of a sphere are given by

$$V = \frac{4}{3}\pi r^3 \qquad \text{and} \qquad S = 4\pi r^2$$

where $r$ is the radius of the sphere. A spherical weather balloon has a radius of 2 feet, as shown in Figure A.7. Find the quotient of the volume and the surface area.

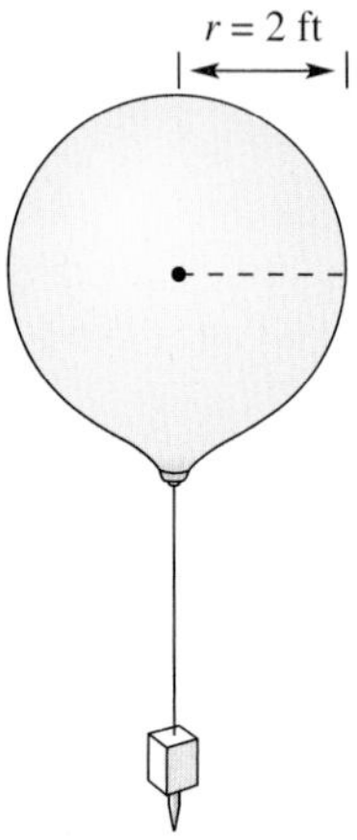


Figure A.7

Solution

Form the quotient $V/S$, and simplify, as follows.

$$\frac{V}{S} = \frac{\frac{4}{3}\pi r^3}{4\pi r^2} = \frac{\frac{4}{3}\pi 2^3}{4\pi 2^2} = \frac{1}{3}(2) = \frac{2}{3}$$

## Scientific Notation

Exponents provide an efficient way of writing and computing with very large (or very small) numbers. For instance, a drop of water contains more than 33 billion billion molecules—that is, 33 followed by 18 zeros.

33,000,000,000,000,000,000

It is convenient to write such numbers in **scientific notation.** This notation has the form $c \times 10^n$, where $1 \le c < 10$ and $n$ is an integer. Thus, the number of molecules in a drop of water can be written in scientific notation as

$$3.3 \times 10,000,000,000,000,000,000 = 3.3 \times 10^{19}.$$

The *positive* exponent 19 indicates that the number is large (10 or more) and that the decimal point has been moved 19 places. A *negative* exponent in scientific notation indicates that the number is *small* (less than 1). For instance, the mass (in grams) of one electron is approximately

$$9.0 \times 10^{-28} = 0.00000000000000000000000000009.$$

28 decimal places

---

**Example 5**    Converting to Scientific Notation

**a.** $0.0000572 = 5.72 \times 10^{-5}$          Number is less than 1.

**b.** $149,400,000 = 1.494 \times 10^8$          Number is greater than 10.

---

**Example 6**    Converting to Decimal Notation

**a.** $3.125 \times 10^2 = 312.5$          Number is greater than 10.

**b.** $3.73 \times 10^{-6} = 0.00000373$          Number is less than 1.

---

Most calculators automatically switch to scientific notation when they are showing large (or small) numbers that exceed the display range. Try multiplying $86,500,000 \times 6000$. If your calculator follows standard conventions, its display should be

5.19   11     or     5.19   E   11 .

This means that $c = 5.19$ and the exponent of 10 is $n = 11$, which implies that the number is $5.19 \times 10^{11}$. To *enter* numbers in scientific notation, your calculator should have an exponential entry key labeled $\boxed{\text{EE}}$.

---

**Example 7**    The Speed of Light

The distance between Earth and the sun is approximately 93 million miles, as shown in Figure A.8. How long does it take for light to travel from the sun to Earth? Use the fact that light travels at a rate of approximately 186,000 miles per second.

Solution

Using the formula Distance = (rate)(time), you find the time as follows.

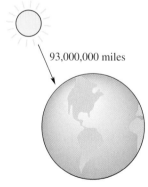

93,000,000 miles

Figure A.8

$$\text{Time} = \frac{\text{distance}}{\text{rate}} = \frac{93 \text{ million miles}}{186,000 \text{ miles per second}}$$

$$= \frac{9.3 \times 10^7 \text{ miles}}{1.86 \times 10^5 \text{ miles/second}}$$

$$= 5 \times 10^2 \text{ seconds}$$

$$\approx 8.33 \text{ minutes}$$

Note that to convert 500 seconds to 8.33 minutes, you divide by 60 because there are 60 seconds in a minute.

## Applications

One of the most useful features of a calculator is its ability to evaluate exponential expressions. Consult your user's guide for specific keystrokes.

**Example 8**   Using a Calculator to Raise a Number to a Power

**Scientific Calculator**

| Expression | Keystrokes | Display |
|---|---|---|
| **a.** $13^4 + 5$ | 13 [$y^x$] 4 [+] 5 [=] | 28566 |
| **b.** $3^{-2} + 4^{-1}$ | 3 [$y^x$] 2 [+/−] [+] 4 [$y^x$] 1 [+/−] [=] | .3611111111 |
| **c.** $\dfrac{3^5 + 1}{3^5 - 1}$ | [(] 3 [$y^x$] 5 [+] 1 [)] [÷]  [(] 3 [$y^x$] 5 [−] 1 [)] [=] | 1.008264463 |

**Graphing Calculator**

| Expression | Keystrokes | Display |
|---|---|---|
| **a.** $13^4 + 5$ | 13 [^] 4 [+] 5 [ENTER] | 28566 |
| **b.** $3^{-2} + 4^{-1}$ | 3 [^] [(−)] 2 [+] 4 [^] [(−)] 1 [ENTER] | .3611111111 |
| **c.** $\dfrac{3^5 + 1}{3^5 - 1}$ | [(] 3 [^] 5 [+] 1 [)] [÷]  [(] 3 [^] 5 [−] 1 [)] [ENTER] | 1.008264463 |

The following box gives a formula for finding the balance in a savings account.

▶ **Balance in an Account**

The balance $A$ in an account that earns an annual interest rate of $r$ (in decimal form) for $t$ years is given by one of the following.

$A = P(1 + rt)$    Simple interest

$A = P\left(1 + \dfrac{r}{n}\right)^{nt}$    Compound interest

In both formulas, $P$ is the principal (or the initial deposit). In the formula for compound interest, $n$ is the number of compoundings *per year*.

Example 9    Finding the Balance in an Account

**a.** If a friend borrows $250 at 8% annual interest for 6 months, the amount due after 6 months (at simple interest) is

$$A = P(1 + rt) = 250\left[1 + 0.08\left(\frac{1}{2}\right)\right] = \$260.00.$$

**b.** If you deposit $1000 at 6% annual interest compounded quarterly for 5 years, the balance after 5 years is

$$A = P\left(1 + \frac{r}{n}\right)^{nt} = 1000\left(1 + \frac{0.06}{4}\right)^{(4)(5)} = \$1346.86.$$

In addition to finding the balance in an account, the compound interest formula can also be used to determine the rate of inflation. To apply the formula, you must know the cost of an item for two different years, as demonstrated in Example 10.

Example 10    Finding the Rate of Inflation

In 1976, the cost of a first-class postage stamp was $0.13. By 2001, the cost had risen to $0.34, as shown in Figure A.9. Find the average annual rate of inflation for first-class postage over this 25-year period.    (Source: U.S. Postal Service)

**Solution**

To find the average annual rate of inflation, use the formula for compound interest with *annual* compounding. Thus, you need to find the value of $r$ that will make the following equation true.

$$A = P\left(1 + \frac{r}{n}\right)^{nt}$$

$$0.34 = 0.13(1 + r)^{25}$$

You can begin by guessing that the average annual rate of inflation was 5%. Entering $r = 0.05$ in the formula, you find that $0.13(1 + 0.05)^{25} \approx 0.4402$. Because this result is more than 0.34, try some smaller values of $r$. Finally, you can discover that

$$0.13(1 + 0.039)^{25} \approx 0.34.$$

So, the average annual rate of inflation for first-class postage from 1976 to 2001 was about 3.9%.

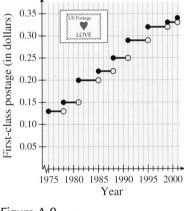

**Figure A.9**

# Warm Up

The following warm-up exercises involve skills that were covered in earlier sections. You will use these skills in the exercise set for this section.

In Exercises 1–10, perform the indicated operation(s) and simplify.

1. $\left(\frac{2}{3}\right)\left(\frac{3}{2}\right)$

2. $\left(\frac{1}{4}\right)(5)(4)$

3. $3\left(\frac{2}{7}\right) + 11\left(\frac{2}{7}\right)$

4. $\frac{1}{2} \div 2$

5. $\frac{1}{3} + \frac{1}{2} - \frac{5}{6}$

6. $\frac{1}{3} \div \frac{1}{3}$

7. $\frac{1}{7} + \frac{1}{3} - \frac{1}{21}$

8. $11\left(\frac{1}{4}\right) + \frac{5}{4}$

9. $\frac{1}{12} - \frac{1}{3} + \frac{1}{8}$

10. $\left(\frac{1}{2} - \frac{1}{3}\right) \div \frac{1}{6}$

## A.3    Exercises

In Exercises 1–16, evaluate the expression. Write fractional answers in simplified form.

1. $2^2 \cdot 2^4$

2. $\dfrac{2^6}{2^3}$

3. $(3^3)^2$

4. $(2^4)^2$

5. $-3^4$

6. $(-3)^4$

7. $\left(\frac{1}{2}\right)^{-3}$

8. $5^{-1} - 2^{-1}$

9. $\left(\frac{3}{4}\right)^2$

10. $\left(\frac{3}{4}\right)^{-2}$

11. $(2^3 \cdot 3^2)^2$

12. $\left(-\frac{3}{5}\right)^3\left(\frac{5}{3}\right)^2$

13. $6^5 \cdot 6^{-3}$

14. $6 \cdot 2^{-3} \cdot 3^{-1}$

15. $3^0$

16. $(-2)^0$

In Exercises 17–20, evaluate the expression for the indicated value of x.

| Expression | Value |
|---|---|
| 17. $\dfrac{x^2}{2}$ | $x = 6$ |
| 18. $4x^{-3}$ | $x = 2$ |
| 19. $7x^{-2}$ | $x = 4$ |
| 20. $6x^0 - (6x)^0$ | $x = 10$ |

In Exercises 21–38, simplify the expression.

21. $(-5z)^3$

22. $(8x^4)(2x^3)$

23. $5x^4(x^2)$

24. $10(x^2)^2$

25. $(-z)^3(3z^4)$

26. $\dfrac{25y^8}{10y^4}$

27. $\dfrac{3x^5}{x^3}$

28. $\left(\dfrac{4}{y}\right)^3\left(\dfrac{3}{y}\right)^4$

29. $\dfrac{7x^2}{x^3}$

30. $(2x^5)^0, \quad x \neq 0$

31. $(x + 5)^0, \quad x \neq -5$

32. $(4y^{-2})(8y^4)$

33. $(-2x^2)^3(4x^3)^{-1}$

34. $\left(\dfrac{x^{-3}y^4}{5}\right)^{-3}$

35. $\left(\dfrac{x}{10}\right)^{-1}$

36. $\dfrac{x^2 \cdot x^n}{x^3 \cdot x^n}$

37. $3^n \cdot 3^{2n}$

38. $2^m \cdot 2^{3m}$

In Exercises 39–43, write the number in scientific notation.

39. *Land area of Earth:*   57,500,000 square miles

40. *Ocean area of Earth:*   139,400,000 square miles

41. *Light year:*   9,461,000,000,000,000 kilometers

42. *Relative density of hydrogen:*   0.0000899

43. *One point (printer's measure):*   0.013837 inch

In Exercises 44–48, write the number in decimal form.

44. *Number of air sacs in lungs:*   $3.5 \times 10^8$

45. *Temperature of sun:*   $1.3 \times 10^7$ degrees Celsius

46. *Number of insect species:*   $9.5 \times 10^5$

47. *Electron charge:*   $4.8 \times 10^{-10}$ electrostatic unit

48. *Width of human hair:*   $9.0 \times 10^{-4}$ meter

In Exercises 49 and 50, use a calculator to evaluate the expression. (Round to three decimal places.)

49. (a) $2400(1 + 0.06)^{20}$

    (b) $750\left(1 + \dfrac{0.11}{365}\right)^{800}$

50. (a) $5000(1 + 0.02)^{60}$

    (b) $2500\left(1 + \dfrac{0.085}{12}\right)^{60}$

**51.** *Balance in an Account*    Ten thousand dollars is deposited in an account with an annual percentage rate of 6.5% for 10 years. What is the balance in the account if the interest is compounded (a) quarterly and (b) monthly?

**52.** *Balance in an Account*    Two thousand dollars is deposited in an account with an annual percentage rate of 7.75% for 15 years. What is the balance in the account if the interest is compounded (a) quarterly and (b) monthly?

**53.** *Reasoning*    A student calculates the amount in an account after 5 years to be $10,616.78 when $2000 was invested at 6% interest compounded monthly. Determine the error the student made in the calculation if the problem was set up correctly. Explain to the student how to correct the error.

**54.** *Reasoning*    A student tells you that $1000 invested at 8% for 10 years compounded quarterly will yield $2208.04. The student then concludes that $2000 invested at 8% for 10 years compounded quarterly will yield $4416.08. Do you agree? Explain.

**55.** *College Tuition*    The bar chart shows the average tuition at private four-year colleges in the United States from 1989 to 1999. Estimate the average annual rate of inflation for tuition over this 10-year period.    (Source: U.S. National Center for Education Statistics)

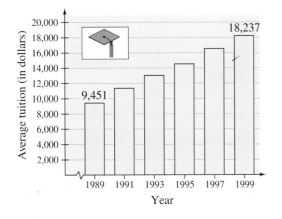

**56.** *Minimum Wage*    The bar chart shows the minimum wage in the United States from 1978 to 1998. Estimate the average annual rate of inflation for minimum wages over this 20-year period.    (Source: U.S. Employment Standards Administration)

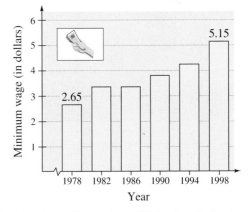

**57.** *Calculator Keystrokes*    Write the algebraic expression that corresponds to the following keystrokes.

( 5.1 − 3.6 ) $y^x$ 5 =          Scientific

( 5.1 − 3.6 ) ^ 5 ENTER          Graphing

**58.** *Calculator Keystrokes*    Write the algebraic expression that corresponds to the following keystrokes.

1 + 3 × 2 = $y^x$ 2 +/− =          Scientific

1 + 3 × 2 ENTER ^ (−) 2 ENTER          Graphing

**59.** *Think About It*    Because $-2^3 = -8$ and $(-2)^3 = -8$, a student concludes that $-a^n = (-a)^n$ where $n$ is an integer. Do you agree? Can you find an example where $-a^n \neq (-a)^n$?

**60.** *Think About It*    Because $2^0 = 1$ and $3^0 = 1$, a student concludes that $0^0 = 1$. Do you agree? What argument would you give to show that $0^0 \neq 1$?

## A.4   Radicals and Rational Exponents

Radicals and Properties of Radicals ■ Simplifying Radicals ■ Rational Exponents ■ Radicals and Calculators

## Radicals and Properties of Radicals

A **square root** of a number is defined as one of its two equal factors. For example, 5 is a square root of 25 because 5 is one of the two equal factors of 25. In a similar way, a **cube root** of a number is one of its three equal factors. Here are some examples.

| *Number* | *Equal Factors* | *Root* | |
|---|---|---|---|
| $25 = (-5)^2$ | $(-5)(-5)$ | $-5$ | (square root) |
| $-64 = (-4)^3$ | $(-4)(-4)(-4)$ | $-4$ | (cube root) |
| $81 = 3^4$ | $3 \cdot 3 \cdot 3 \cdot 3$ | $3$ | (fourth root) |

In general, an ***n*th root** of a number is defined as follows.

---

▶ **Definition of *n*th Root of a Number**

Let $a$ and $b$ be real numbers and let $n$ be a positive integer. If

$$a = b^n$$

then $b$ is an ***n*th root of *a*.** If $n = 2$, the root is a **square root,** and if $n = 3$, the root is a **cube root.**

---

From this definition, you can see that some numbers have more than one *n*th root. For example, both 5 and $-5$ are square roots of 25. The following definition distinguishes between these two roots.

---

▶ **Principal *n*th Root of a Number**

Let $a$ be a real number that has at least one real *n*th root. The **principal *n*th root of *a*** is the *n*th root that has the same sign as $a$, and it is denoted by the **radical symbol**

$$\sqrt[n]{a}. \qquad \text{Principal } n\text{th root}$$

The positive integer $n$ is the **index** (the plural of index is *indexes* or *indices*) of the radical, and the number $a$ is the **radicand.** If $n = 2$, omit the index and write $\sqrt{a}$ rather than $\sqrt[2]{a}$.

---

### Example 1   Evaluating Expressions Involving Radicals

**a.** The principal square root of 121 is $\sqrt{121} = 11$ because $11^2 = 121$.

**b.** The principal cube root of $\frac{125}{64}$ is $\sqrt[3]{\frac{125}{64}} = \frac{5}{4}$ because $\left(\frac{5}{4}\right)^3 = \frac{5^3}{4^3} = \frac{125}{64}$.

**c.** The principal fifth root of

**d.** $-\sqrt{49} = -7$ because $7^2$

**e.** $\sqrt[4]{-81}$ is not a real numbe
to the fourth power to prod

From Example 1, you can n
of a real number.

**1.** If $a$ is a positive real number $\imath$
ly two real $n$th roots, which a

**2.** If $a$ is any real number and $n$ i
root. It is the principal $n$th root

**3.** If $a$ is negative and $n$ is an *even*

Integers such as 1, 4, 9, 16, 4 ... $\ldots$ are called **perfect squares** because
they have integer square roots. Similarly, integers such as 1, 8, 27, 64, and 125
are called **perfect cubes** because they have integer cube roots.

---

▶ **Properties of Radicals**

Let $a$ and $b$ be real numbers, variables, or algebraic expressions such
that the indicated roots are real numbers, and let $m$ and $n$ be positive
integers. Then the following properties are true.

| *Property* | *Example* |
|---|---|
| **1.** $\sqrt[n]{a^m} = \left(\sqrt[n]{a}\right)^m$ | $\sqrt[3]{8^2} = \left(\sqrt[3]{8}\right)^2 = (2)^2 = 4$ |
| **2.** $\sqrt[n]{a} \cdot \sqrt[n]{b} = \sqrt[n]{ab}$ | $\sqrt{5} \cdot \sqrt{7} = \sqrt{5 \cdot 7} = \sqrt{35}$ |
| **3.** $\dfrac{\sqrt[n]{a}}{\sqrt[n]{b}} = \sqrt[n]{\dfrac{a}{b}}, \quad b \neq 0$ | $\dfrac{\sqrt[4]{27}}{\sqrt[4]{9}} = \sqrt[4]{\dfrac{27}{9}} = \sqrt[4]{3}$ |
| **4.** $\sqrt[m]{\sqrt[n]{a}} = \sqrt[mn]{a}$ | $\sqrt[3]{\sqrt{10}} = \sqrt[6]{10}$ |
| **5.** $\left(\sqrt[n]{a}\right)^n = a$ | $\left(\sqrt{3}\right)^2 = 3$ |
| **6.** For $n$ even, $\sqrt[n]{a^n} = \lvert a \rvert$. | $\sqrt{(-12)^2} = \lvert -12 \rvert = 12$ |
| For $n$ odd, $\sqrt[n]{a^n} = a$. | $\sqrt[3]{(-12)^3} = -12$ |

A common special case of Property 6 is $\sqrt{a^2} = \lvert a \rvert$.

## Simplifying Radicals

An expression involving radicals is in **simplest form** when the following conditions are satisfied.

**1.** All possible factors have been removed from the radical.

**2.** All fractions have radical-free denominators (accomplished by a process
called *rationalizing the denominator*).

**3.** The index of the radical has been reduced as far as possible.

To simplify a radical, factor the radicand into factors whose powers are multiples of the index. The roots of these factors are written outside the radical, and the "leftover" factors make up the new radicand.

### Example 2    Simplifying Even Roots

**a.** $\sqrt[4]{48} = \sqrt[4]{16 \cdot 3}$                       Find largest fourth-power factor.

       $= \sqrt[4]{2^4 \cdot 3}$                     Rewrite.

       $= 2\sqrt[4]{3}$                       Find fourth root.

**b.** $\sqrt{75x^3} = \sqrt{25x^2 \cdot 3x}$            Find largest square factor.

       $= \sqrt{(5x)^2 \cdot 3x}$               Rewrite.

         $= 5x\sqrt{3x}, \quad x \geq 0$       Find root of perfect square.

**c.** $\sqrt[4]{(5x)^4} = |5x| = 5|x|.$

---

In Example 2(c), note that absolute value is included in the answer because $\sqrt[4]{x^4} = |x|.$

### Example 3    Simplifying Odd Roots

**a.** $\sqrt[3]{24} = \sqrt[3]{8 \cdot 3}$                     Find largest cube factor.

       $= \sqrt[3]{2^3 \cdot 3}$                     Rewrite.

       $= 2\sqrt[3]{3}$                       Find root of perfect cube.

**b.** $\sqrt[3]{24a^4} = \sqrt[3]{8a^3 \cdot 3a}$           Find largest cube factor.

       $= \sqrt[3]{(2a)^3 \cdot 3a}$           Rewrite.

        $= 2a\sqrt[3]{3a}$             Find root of perfect cube.

**c.** $\sqrt[3]{-40x^6} = \sqrt[3]{(-8x^6) \cdot 5}$        Find largest cube factor.

       $= \sqrt[3]{(-2x^2)^3 \cdot 5}$       Rewrite.

       $= -2x^2\sqrt[3]{5}$          Find root of perfect cube.

---

To **rationalize the denominator** of a fraction, multiply the numerator and denominator by an appropriate factor. To find the "appropriate factor," make use of the form $a + b\sqrt{m}$ and its **conjugate** $a - b\sqrt{m}$. The product of this conjugate pair has no radical. For instance,

$$\left(4 + \sqrt{3}\right)\left(4 - \sqrt{3}\right) = 4^2 - \left(\sqrt{3}\right)^2 = 16 - 3 = 13.$$

Therefore, to rationalize a denominator of the form $a - b\sqrt{m}$ (or $a + b\sqrt{m}$), you can multiply the numerator and denominator by the conjugate factor. If $a = 0$, the rationalizing factor of $\sqrt{m}$ is itself, $\sqrt{m}$.

**Example 4**    Rationalizing Single-Term Denominators

**a.** To rationalize the denominator of the following fraction, multiply *both* the numerator and the denominator by $\sqrt{3}$ to obtain

$$\frac{5}{2\sqrt{3}} = \frac{5}{2\sqrt{3}} \cdot \frac{\sqrt{3}}{\sqrt{3}} = \frac{5\sqrt{3}}{2\sqrt{3^2}} = \frac{5\sqrt{3}}{2(3)} = \frac{5\sqrt{3}}{6}.$$

**b.** To rationalize the denominator of the following fraction, multiply *both* the numerator and the denominator by $\sqrt[3]{5^2}$. Note how this eliminates the radical from the denominator by producing a perfect *cube* in the radicand.

$$\frac{2}{\sqrt[3]{5}} = \frac{2}{\sqrt[3]{5}} \cdot \frac{\sqrt[3]{5^2}}{\sqrt[3]{5^2}} = \frac{2\sqrt[3]{5^2}}{\sqrt[3]{5^3}} = \frac{2\sqrt[3]{25}}{5}$$

**Example 5**    Rationalizing a Denominator with Two Terms

$$\frac{2}{3 + \sqrt{7}} = \frac{2}{3 + \sqrt{7}} \cdot \frac{3 - \sqrt{7}}{3 - \sqrt{7}} \qquad \text{Multiply numerator and denominator by conjugate.}$$

$$= \frac{2(3 - \sqrt{7})}{3^2 - (\sqrt{7})^2} \qquad \text{Multiply fractions.}$$

$$= \frac{2(3 - \sqrt{7})}{9 - 7} \qquad \text{Simplify.}$$

$$= \frac{2(3 - \sqrt{7})}{2} \qquad \text{Divide out like factors.}$$

$$= 3 - \sqrt{7} \qquad \text{Simplify.}$$

Don't confuse an expression such as $\sqrt{2} + \sqrt{7}$ with $\sqrt{2 + 7}$. In general, $\sqrt{x + y} \neq \sqrt{x} + \sqrt{y}$.

## Rational Exponents

The following definition shows how radicals are used to define **rational exponents**. Until now, work with exponents has been restricted to integer exponents.

▶ **Definition of Rational Exponents**

If $a$ is a real number and $n$ is a positive integer such that the principal $n$th root of $a$ exists, then $a^{1/n}$ is defined to be

$$a^{1/n} = \sqrt[n]{a}.$$

If $m$ is a positive integer that has no common factor with $n$, then

$$a^{m/n} = (a^{1/n})^m = \left(\sqrt[n]{a}\right)^m$$

and

$$a^{m/n} = (a^m)^{1/n} = \sqrt[n]{a^m}.$$

The properties of exponents that were listed in Section A.3 also apply to rational exponents (provided the roots indicated by the denominators exist). Some of those properties are relisted here, with different examples.

---

▶ **Properties of Exponents**

Let $r$ and $s$ be rational numbers, and let $a$ and $b$ be real numbers, variables, or algebraic expressions. If the roots indicated by the rational exponents exist, then the following properties are true.

| *Property* | *Example* |
|---|---|
| **1.** $a^r a^s = a^{r+s}$ | $4^{1/2}(4^{1/3}) = 4^{5/6}$ |
| **2.** $\dfrac{a^r}{a^s} = a^{r-s}, \quad a \neq 0$ | $\dfrac{x^2}{x^{1/2}} = x^{2-(1/2)} = x^{3/2}$ |
| **3.** $a^{-r} = \dfrac{1}{a^r}, \quad a \neq 0$ | $4^{-1/2} = \dfrac{1}{4^{1/2}} = \dfrac{1}{2}$ |
| **4.** $\left(\dfrac{a}{b}\right)^{-r} = \left(\dfrac{b}{a}\right)^r, \quad a \neq 0, \quad b \neq 0$ | $\left(\dfrac{x}{4}\right)^{-1/2} = \left(\dfrac{4}{x}\right)^{1/2} = \dfrac{2}{x^{1/2}}$ |
| **5.** $(ab)^r = a^r b^r$ | $(2x)^{1/2} = 2^{1/2}(x^{1/2})$ |
| **6.** $(a^r)^s = a^{rs}$ | $(x^3)^{1/3} = x$ |
| **7.** $\left(\dfrac{a}{b}\right)^r = \dfrac{a^r}{b^r}, \quad b \neq 0$ | $\left(\dfrac{x}{3}\right)^{1/3} = \dfrac{x^{1/3}}{3^{1/3}}$ |

---

Rational exponents are particularly useful for evaluating roots of numbers on a calculator, for reducing the index of a radical, and for simplifying (and factoring) algebraic expressions. Examples 6 and 7 demonstrate some of these uses.

**Study Tip**

Rational exponents can be tricky, and you must remember that the expression $b^{m/n}$ is not defined unless $\sqrt[n]{b}$ is a real number. This restriction produces some unusual-looking results. For instance, the number $(-8)^{1/6}$ is not defined because $\sqrt[6]{-8}$ is not a real number. And yet, $(-8)^{1/3}$ is defined because $\sqrt[3]{-8} = -2$.

**Example 6**    Simplifying with Rational Exponents

**a.** $(27)^{1/3} = \sqrt[3]{27} = 3$

**b.** $(-32)^{-4/5} = \left(\sqrt[5]{-32}\right)^{-4} = (-2)^{-4} = \dfrac{1}{(-2)^4} = \dfrac{1}{16}$

**c.** $(-5x^{2/3})(3x^{-1/3}) = -15x^{(2/3)-(1/3)} = -15x^{1/3}, \quad x \neq 0$

**Example 7**    Reducing the Index of a Radical

**a.** $\sqrt[6]{a^4} = a^{4/6} = a^{2/3} = \sqrt[3]{a^2}$

**b.** $\sqrt[3]{\sqrt{125}} = (125^{1/2})^{1/3}$      Rewrite with rational exponents.

$\qquad\qquad = (125)^{1/6}$      Multiply exponents.

$\qquad\qquad = (5^3)^{1/6}$      Rewrite base as perfect cube.

$\qquad\qquad = 5^{3/6}$      Multiply exponents.

$\qquad\qquad = 5^{1/2}$      Reduce exponent.

$\qquad\qquad = \sqrt{5}$      Rewrite as radical.

Radical expressions can be combined (added or subtracted) if they are **like radicals**—that is, if they have the same index and radicand. For instance, $2\sqrt{3x}$ and $\frac{1}{2}\sqrt{3x}$ are like radicals, but $\sqrt[3]{3x}$ and $2\sqrt{3x}$ are not like radicals. To determine whether two expressions involve like radicals, you should first simplify each radical.

**Example 8**    Simplifying and Combining Like Radicals

**a.** $2\sqrt{48} + 3\sqrt{27} = 2\sqrt{16 \cdot 3} + 3\sqrt{9 \cdot 3}$     Find square factors.

$\qquad\qquad\qquad\quad = 8\sqrt{3} + 9\sqrt{3}$     Find square roots.

$\qquad\qquad\qquad\quad = 17\sqrt{3}$     Combine like terms.

**b.** $\sqrt[3]{16x} - \sqrt[3]{54x} = \sqrt[3]{8 \cdot 2x} - \sqrt[3]{27 \cdot 2x}$     Find cube factors.

$\qquad\qquad\qquad\quad = 2\sqrt[3]{2x} - 3\sqrt[3]{2x}$     Find cube roots.

$\qquad\qquad\qquad\quad = -\sqrt[3]{2x}$     Combine like terms.

## Radicals and Calculators

There are two methods of evaluating radicals on most calculators. For square roots, you can use the *square root key*. For other roots, you should first convert the radical to exponential form and then use the *exponential key*.

**Example 9**    Evaluating a Cube Root with a Calculator

Use a calculator to evaluate $\sqrt[3]{25}$. Round to three decimal places.

**Solution**

First write $\sqrt[3]{25}$ as $25^{1/3}$. Then use the following keystrokes.

$\quad$ 25 $\boxed{y^x}$ $\boxed{(}$ 1 $\boxed{\div}$ 3 $\boxed{)}$ $\boxed{=}$     Scientific

$\quad$ 25 $\boxed{\wedge}$ $\boxed{(}$ 1 $\boxed{\div}$ 3 $\boxed{)}$ $\boxed{\text{ENTER}}$     Graphing

For either of these keystroke sequences, the calculator display should read 2.9240177. Thus, you have

$$\sqrt[3]{25} \approx 2.924.$$

**Example 10**    Evaluating Radicals with a Calculator

Evaluate the following radicals. Round to three decimal places.

**a.** $\sqrt[3]{-4}$     **b.** $(1.4)^{-2/5}$

**Solution**

**a.** Because

$$\sqrt[3]{-4} = \sqrt[3]{(-1)(4)} = \sqrt[3]{-1} \cdot \sqrt[3]{4} = -\sqrt[3]{4}$$

you can attach the negative sign of the radicand as follows.

4 $\boxed{y^x}$ $\boxed{(}$ 1 $\boxed{\div}$ 3 $\boxed{)}$ $\boxed{=}$ $\boxed{+/-}$    *Scientific*

$\boxed{(-)}$ $\boxed{(}$ 4 $\boxed{\wedge}$ $\boxed{(}$ 1 $\boxed{\div}$ 3 $\boxed{)}$ $\boxed{)}$ $\boxed{\text{ENTER}}$    *Graphing*

The calculator display is $-1.5874011$, which implies that

$$\sqrt[3]{-4} \approx -1.587.$$

**b.** Using the following keystroke sequence

1.4 $\boxed{y^x}$ $\boxed{(}$ 2 $\boxed{\div}$ 5 $\boxed{+/-}$ $\boxed{)}$ $\boxed{=}$    *Scientific*

1.4 $\boxed{\wedge}$ $\boxed{(}$ $\boxed{(-)}$ 2 $\boxed{\div}$ 5 $\boxed{)}$ $\boxed{\text{ENTER}}$    *Graphing*

you obtain a calculator display of 0.8740752. Thus, you have

$$(1.4)^{-2/5} \approx 0.874.$$

# Warm Up

The following warm-up exercises involve skills that were covered in earlier sections. You will use these skills in the exercise set for this section.

In Exercises 1–10, simplify the expression.

**1.** $\left(\frac{1}{3}\right)\left(\frac{2}{3}\right)^2$

**2.** $3(-4)^2$

**3.** $(-2x)^3$

**4.** $(-2x^3)(-3x^4)$

**5.** $(7x^5)(4x)$

**6.** $(5x^4)(25x^2)^{-1}$, $x \neq 0$

**7.** $\dfrac{12z^6}{4z^2}$, $z \neq 0$

**8.** $\left(\dfrac{2x}{5}\right)^2\left(\dfrac{2x}{5}\right)^{-4}$, $x \neq 0$

**9.** $\left(\dfrac{3y^2}{x}\right)^0$, $x \neq 0,\ y \neq 0$

**10.** $[(x + 2)^2(x + 2)^3]^2$

## A.4    Exercises

In Exercises 1–12, fill in the missing form.

| Radical Form | Rational Exponent Form |
|---|---|
| **1.** $\sqrt{9} = 3$ | |
| **2.** $\sqrt[3]{64} = 4$ | |
| **3.** | $32^{1/5} = 2$ |
| **4.** | $-(144^{1/2}) = -12$ |
| **5.** | $196^{1/2} = 14$ |
| **6.** $\sqrt[3]{614.125} = 8.5$ | |
| **7.** $\sqrt[3]{-216} = -6$ | |
| **8.** | $(-243)^{1/5} = -3$ |
| **9.** | $27^{2/3} = 9$ |
| **10.** $\left(\sqrt[4]{81}\right)^3 = 27$ | |
| **11.** $\sqrt[4]{81^3} = 27$ | |
| **12.** | $16^{5/4} = 32$ |

In Exercises 13–28, evaluate the expression.

**13.** $\sqrt{9}$

**14.** $\sqrt[3]{64}$

**15.** $-\sqrt[3]{-27}$

**16.** $\sqrt[3]{0}$

**17.** $\dfrac{4}{\sqrt{64}}$

**18.** $\dfrac{\sqrt[4]{81}}{3}$

**19.** $\left(\sqrt[3]{-125}\right)^3$

**20.** $\sqrt[4]{562^4}$

**21.** $16^{1/2}$

**22.** $27^{1/3}$

**23.** $36^{3/2}$

**24.** $16^{3/2}$

**25.** $\left(\frac{16}{81}\right)^{-3/4}$

**26.** $\left(\frac{9}{4}\right)^{-1/2}$

**27.** $\left(-\frac{1}{64}\right)^{-1/3}$

**28.** $\left(-\frac{125}{27}\right)^{-1/3}$

In Exercises 29–32, simplify the expression.

**29.** $\sqrt[3]{16x^5}$

**30.** $\sqrt[4]{(3x^2)^4}$

**31.** $\sqrt{75x^2y^{-4}}$

**32.** $\sqrt[5]{96x^5}$

In Exercises 33–40, rewrite the expression by rationalizing the denominator. Simplify the result.

**33.** $\dfrac{1}{\sqrt{3}}$          **34.** $\dfrac{5}{\sqrt{10}}$

**35.** $\dfrac{8}{\sqrt[3]{2}}$          **36.** $\dfrac{5}{\sqrt[3]{(5x)^2}}$

**37.** $\dfrac{2x}{5 - \sqrt{3}}$          **38.** $\dfrac{5}{\sqrt{14} - 2}$

**39.** $\dfrac{3}{\sqrt{5} + \sqrt{6}}$          **40.** $\dfrac{5}{2\sqrt{10} - 5}$

In Exercises 41–44, simplify the expression.

**41.** $5\sqrt{x} - 3\sqrt{x}$          **42.** $3\sqrt{x+1} + 10\sqrt{x+1}$

**43.** $2\sqrt{4y} - 2\sqrt{9y}$          **44.** $2\sqrt{80} + \sqrt{125}$

In Exercises 45–50, simplify the expression.

**45.** $\sqrt{2}\sqrt{3}$          **46.** $\sqrt{2}\sqrt{5}$

**47.** $5^{1/2} \cdot 5^{3/2}$          **48.** $\dfrac{2^{3/2}}{2}$

**49.** $\dfrac{x^2}{x^{1/2}}$          **50.** $\sqrt{\sqrt{\sqrt{x^4}}}$

In Exercises 51 and 52, use rational exponents to reduce the index of the radical.

**51.** $\sqrt[4]{3^2}$          **52.** $\sqrt[4]{(3x^2)^4}$

In Exercises 53–56, fill in the blank with <, =, or >.

**53.** $\sqrt{5} + \sqrt{3}$     $\sqrt{5+3}$

**54.** $5$     $\sqrt{3^2 + 4^2}$

**55.** $\sqrt{3} \cdot \sqrt[4]{3}$     $\sqrt[8]{3}$      **56.** $\sqrt{\dfrac{3}{11}}$     $\dfrac{\sqrt{3}}{\sqrt{11}}$

**57.** *Geometry*   Find the measurements of a cube that has a volume of 13,824 cubic inches (see figure).

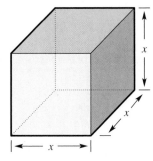

**58.** *Geometry*   Find the measurements of a square classroom that has 800 square feet of floor space (see figure).

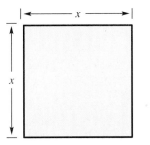

*Declining Balances Depreciation*   In Exercises 59 and 60, find the annual depreciation rate $r$. To find the annual depreciation rate by the declining balances method, use the formula

$$r = 1 - \left(\dfrac{S}{C}\right)^{1/n}$$

where $n$ is the useful life of the item (in years), $S$ is the salvage value (in dollars), and $C$ is the original cost (in dollars).

**59.** A truck whose original cost is $75,000 is depreciated over an 8-year period, as shown in the bar graph.

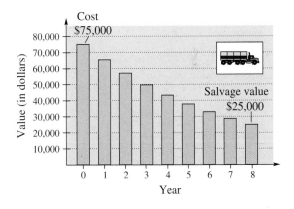

**60.** A printing press whose original cost is $125,000 is depreciated over a 10-year period, as shown in the bar graph.

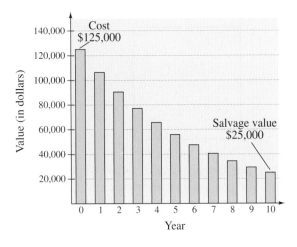

**61.** *Period of a Pendulum* The period $T$ of a pendulum in seconds is given by

$$T = 2\pi\sqrt{\frac{L}{32}}$$

where $L$ is the length of the pendulum in feet. Find the period of a pendulum whose length is 2 feet.

**62.** *Period of a Pendulum* Use the formula given in Exercise 61 to find the period of a pendulum whose length is 1.5 feet.

*Notes on a Musical Scale* In Exercises 63 and 64, find the frequency of the indicated note on a piano (see figure). The musical note A above middle C has a frequency of 440 vibrations per second. If we denote this frequency by $F_1$, then the frequency of the next higher note is given by $F_2 = F_1 \cdot 2^{1/12}$. Similarly, the frequency of the next note is given by $F_3 = F_2 \cdot 2^{1/12}$.

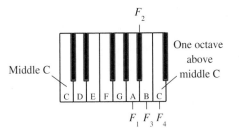

**63.** Find the frequency of the musical note B above middle C.

**64.** Find the frequency of the musical note C that is one octave above middle C.

**65.** *Calculator Experiment* Enter any positive real number in your calculator and repeatedly take the square root. What real number does the display appear to be approaching?

**66.** *Calculator Experiment* Square the real number $2/\sqrt{5}$ and note that the radical is eliminated from the denominator. Is this equivalent to rationalizing the denominator? Why or why not?

## A.5 Polynomials and Special Products

Polynomials ■ Operations with Polynomials ■ Special Products ■ Applications

### Polynomials

One of the simplest and most common types of algebraic expressions is a **polynomial.** Here are some examples.

$$2x + 5, \qquad 3x^4 - 7x^2 + 2x + 4, \qquad 5x^2y^2 - xy + 3$$

The first two are *polynomials in x* and the third is a *polynomial in x and y.* The terms of a polynomial in $x$ have the form $ax^k$, where $a$ is the **coefficient** and $k$ is the **degree** of the term. Because a polynomial is defined as an algebraic sum, the coefficients take on the signs between the terms. For instance, the polynomial

$$2x^3 - 5x^2 + 1 = 2x^3 + (-5)x^2 + (0)x + 1$$

has coefficients 2, $-5$, 0, and 1.

> ▶ **Definition of a Polynomial in *x***
>
> Let $a_n, \ldots, a_2, a_1, a_0$ be real numbers and let $n$ be a *nonnegative integer*. A **polynomial in *x*** is an expression of the form
>
> $$a_n x^n + \cdots + a_2 x^2 + a_1 x + a_0$$
>
> where $a_n \neq 0$. The polynomial is of **degree** $n$, and the number $a_n$ is the **leading coefficient.** The number $a_0$ is the **constant term.** The constant term is considered to have a degree of zero.

Note in the definition of a polynomial in $x$ that the polynomial is written with descending powers of $x$. This is called the **standard form** of a polynomial.

### Example 1    Rewriting a Polynomial in Standard Form

| Polynomial | Standard Form | Degree |
|---|---|---|
| **a.** $4x^2 - 5x^3 - 2 + 3x$ | $-5x^3 + 4x^2 + 3x - 2$ | 3 |
| **b.** $4 - 9x^2$ | $-9x^2 + 4$ | 2 |
| **c.** $8$ | $8 \ (8 = 8x^0)$ | 0 |

Polynomials with one, two, and three terms are called **monomials, binomials,** and **trinomials,** respectively.

A polynomial that has all zero coefficients is called the **zero polynomial,** denoted by 0. This particular polynomial is not considered to have a degree.

### Example 2    Identifying a Polynomial and Its Degree

**a.** $-2x^3 + x^2 + 3x - 2$ is a polynomial of degree 3.

**b.** $\sqrt{x^2 - 3x}$ is not a polynomial because the radical sign indicates a noninteger power of $x$.

**c.** $x^2 + 5x^{-1}$ is not a polynomial because of the negative exponent.

For polynomials in more than one variable, the *degree of a term* is the sum of the powers of the variables in the term. The *degree of the polynomial* is the highest degree of all its terms. For instance, the polynomial

$$5x^3y - x^2y^2 + 2xy - 5$$

has two terms of degree 4, one term of degree 2, and one term of degree 0. The degree of the polynomial is 4.

## Operations with Polynomials

You can **add** and **subtract** polynomials in much the same way that you add and subtract real numbers—you simply add or subtract the *like terms* (terms having the same variables to the same powers) by adding their coefficients. For instance, $-3x^2$ and $5x^2$ are like terms and their sum is given by

$$-3x^2 + 5x^2 = (-3 + 5)x^2 = 2x^2.$$

### Example 3    Sums and Differences of Polynomials

**a.** $(5x^3 - 7x^2 - 3) + (x^3 + 2x^2 - x + 8)$

$\qquad = (5x^3 + x^3) + (2x^2 - 7x^2) - x + (8 - 3)$      Group like terms.

$\qquad = 6x^3 - 5x^2 - x + 5$      Combine like terms.

**b.** $(7x^4 - x^2 - 4x + 2) - (3x^4 - 4x^2 + 3x)$

$\qquad = 7x^4 - x^2 - 4x + 2 - 3x^4 + 4x^2 - 3x$      Distribute sign.

$\qquad = (7x^4 - 3x^4) + (4x^2 - x^2) + (-3x - 4x) + 2$      Group like terms.

$\qquad = 4x^4 + 3x^2 - 7x + 2$      Combine like terms.

---

A common mistake is to fail to change the sign of *each* term inside parentheses preceded by a minus sign. For instance, note the following.

$$-(3x^4 - 4x^2 + 3x) = -3x^4 + 4x^2 - 3x$$      Correct

$$-(3x^4 - 4x^2 + 3x) = -3x^4 - 4x^2 + 3x$$      Common mistake

To find the **product** of two polynomials, you can use the left and right Distributive Properties. For example, if you treat $(5x + 7)$ as a single quantity, you can multiply $(3x - 2)$ by $(5x + 7)$ as follows.

$$(3x - 2)(5x + 7) = 3x(5x + 7) - 2(5x + 7)$$

$$= (3x)(5x) + (3x)(7) - (2)(5x) - (2)(7)$$

$$= 15x^2 + 21x - 10x - 14$$

| Product of First terms | Product of Outer terms | Product of Inner terms | Product of Last terms |

$$= 15x^2 + 11x - 14$$

With practice you should be able to multiply two binomials without writing all of the steps above. In fact, the four products in the boxes above suggest that you can put the product of two binomials in the FOIL form in just one step. This is called the **FOIL Method.**

When multiplying two polynomials, be sure to multiply *each* term of one polynomial by *each* term of the other. The following vertical pattern is a convenient way to multiply two polynomials.

### Example 4    Using a Vertical Format to Multiply Polynomials

Multiply $(x^2 - 2x + 2)$ by $(x^2 + 2x + 2)$.

**Solution**

$$x^2 - 2x + 2$$

$$\underline{x^2 + 2x + 2}$$

$$x^4 - 2x^3 + 2x^2$$

$$2x^3 - 4x^2 + 4x$$

$$\underline{2x^2 - 4x +}$$

$$x^4 + 0x^3 + 0x^2 + 0x$$

Thus, $(x^2 - 2x + 2)(x^2 + 2x$ ┐ ．

## Special Products

▶ **Special Products**

Let $u$ and $v$ be real numbers, variables, or algebraic expressions.

| *Special Product* | *Example* |
|---|---|
| **Sum and Difference of Two Terms** | |
| $(u + v)(u - v) = u^2 - v^2$ | $(x + 4)(x - 4) = x^2 - 16$ |
| **Square of a Binomial** | |
| $(u + v)^2 = u^2 + 2uv + v^2$ | $(x + 3)^2 = x^2 + 6x + 9$ |
| $(u - v)^2 = u^2 - 2uv + v^2$ | $(3x - 2)^2 = 9x^2 - 12x + 4$ |
| **Cube of a Binomial** | |
| $(u + v)^3 = u^3 + 3u^2v + 3uv^2 + v^3$ | $(x + 2)^3 = x^3 + 6x^2 + 12x + 8$ |
| $(u - v)^3 = u^3 - 3u^2v + 3uv^2 - v^3$ | $(x - 1)^3 = x^3 - 3x^2 + 3x - 1$ |

**Example 5**    Sum and Difference of Two Terms

$$(5x + 9)(5x - 9) = (5x)^2 - 9^2 = 25x^2 - 81$$

**Example 6**    Square of a Binomial

$$(6x - 5)^2 = (6x)^2 - 2(6x)(5) + 5^2$$
$$= 36x^2 - 60x + 25$$

**Example 7**    Cube of a Binomial

$$(3x + 2)^3 = (3x)^3 + 3(3x)^2(2) + 3(3x)(2)^2 + 2^3$$
$$= 27x^3 + 54x^2 + 36x + 8$$

**Example 8**   The Product of Two Trinomials

$$(x + y - 2)(x + y + 2) = [(x + y) - 2][(x + y) + 2]$$
$$= (x + y)^2 - 2^2$$
$$= x^2 + 2xy + y^2 - 4$$

## Applications

**Example 9**   A Savings Plan

During the summers prior to your four college years you are able to save $1500, $1800, $2400, and $2600. During the summer after graduation you save $3000. This money is deposited at the end of each summer in an account that pays 7% interest, compounded annually. Find the balance in the account at the end of the summer after graduation.

Solution

Using the formula for compound interest, for *each* deposit you have

$$\text{Balance} = P\left(1 + \frac{r}{n}\right)^{nt} = P(1 + 0.07)^t = P(1.07)^t.$$

For the first deposit, $P = 1500$ and $t = 4$. For the second deposit, $P = 1800$ and $t = 3$, and so on. The balances for the five deposits are as follows.

| Date | Deposit | Time in Account | Balance in Account |
|------|---------|-----------------|--------------------|
| First Summer | $1500 | 4 years | $1500(1.07)^4$ |
| Second Summer | $1800 | 3 years | $1800(1.07)^3$ |
| Third Summer | $2400 | 2 years | $2400(1.07)^2$ |
| Fourth Summer | $2600 | 1 year | $2600(1.07)$ |
| Fifth Summer | $3000 | 0 years | 3000 |

By adding these five balances, you can find the total balance in the account to be

$$1500(1.07)^4 + 1800(1.07)^3 + 2400(1.07)^2 + 2600(1.07) + 3000.$$

Note that this expression is in polynomial form. By evaluating the expression, you can find the balance to be $12,701.03, as shown in Figure A.10.

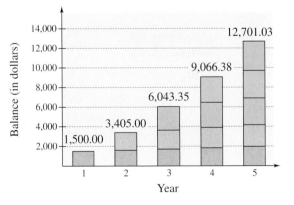

Figure A.10

**Example 10**  Geometry:  Volume of a Box

An open box is made by cutting squares from the corners of a piece of metal that is 16 inches by 20 inches, and turning up the sides, as shown in Figure A.11. If the edge of each cut-out square is $x$ inches, what is the volume of the box? Find the volume when $x = 1$, $x = 2$, and $x = 3$.

**Solution**

*Verbal Model:*

Volume = Length · Width · Height

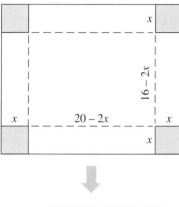

*Labels:*

Height = $x$                                          (inches)

Width = $16 - 2x$                              (inches)

Length = $20 - 2x$                            (inches)

*Equation:*

$$\text{Volume} = (20 - 2x)(16 - 2x)(x)$$
$$= (320 - 72x + 4x^2)(x)$$
$$= 320x - 72x^2 + 4x^3$$

When $x = 1$ inch, the volume of the box is

$$\text{Volume} = 320(1) - 72(1)^2 + 4(1)^3 = 252 \text{ cubic inches.}$$

When $x = 2$ inches, the volume of the box is

$$\text{Volume} = 320(2) - 72(2)^2 + 4(2)^3 = 384 \text{ cubic inches.}$$

When $x = 3$ inches, the volume of the box is

$$\text{Volume} = 320(3) - 72(3)^2 + 4(3)^3 = 420 \text{ cubic inches.}$$

**Figure A.11**

# Warm Up

The following warm-up exercises involve skills that were covered in earlier sections. You will use these skills in the exercise set for this section.

In Exercises 1–10, perform the indicated operation(s).

**1.** $(7x^2)(6x)$

**2.** $(10z^3)(-2z^{-1})$

**3.** $(-3x^2)^3$

**4.** $-3(x^2)^3$

**5.** $\dfrac{27z^5}{12z^2}$

**6.** $\sqrt{24} \cdot \sqrt{2}$

**7.** $\left(\dfrac{2x}{3}\right)^{-2}$

**8.** $16^{3/4}$

**9.** $\dfrac{4}{\sqrt{8}}$

**10.** $\sqrt[3]{-27x^3}$

## A.5    Exercises

In Exercises 1–4, find the degree and leading coefficient of the polynomial.

**1.** $2x^2 - x + 1$

**2.** $-3x^4 + 2x^2 - 5$

**3.** $x^5 - 1$

**4.** 3

In Exercises 5–8, determine whether the algebraic expression is a polynomial. If it is, write the polynomial in standard form and state its degree.

**5.** $2x - 3x^3 + 8$

**6.** $2x^3 + x - 3x^{-1}$

**7.** $\dfrac{3x + 4}{x}$

**8.** $\sqrt{y^2 - y^4}$

In Exercises 9 and 10, evaluate the polynomial for the indicated values of $x$.

**9.** $4x + 3$     (a) $x = -1$  (b) $x = 0$
                    (c) $x = 1$   (d) $x = 2$

**10.** $-2x^2 + 3x + 4$   (a) $x = -2$  (b) $x = -1$
                          (c) $x = 0$   (d) $x = 1$

In Exercises 11–20, perform the indicated operations and write the resulting polynomial in standard form.

**11.** $(6x + 5) - (8x + 15)$

**12.** $-(5x^2 - 1) - (-3x^2 + 5)$

**13.** $(15x^2 - 6) - (-8x^3 - 14x^2 - 17)$

**14.** $(15x^4 - 18x - 19) - (13x^4 - 5x + 15)$

**15.** $5z - [3z - (10z + 8)]$

**16.** $(y^3 + 1) - [(y^2 + 1) + (3y - 7)]$

**17.** $3x(x^2 - 2x + 1)$       **18.** $-4x(3 - x^3)$

**19.** $(-2x)(-3x)(5x + 2)$     **20.** $(1 - x^3)(4x)$

In Exercises 21–44, find the product.

**21.** $(x + 3)(x + 4)$         **22.** $(x - 5)(x + 10)$

**23.** $(3x - 5)(2x + 1)$       **24.** $(7x - 2)(4x - 3)$

**25.** $(x + 6)^2$              **26.** $(3x - 2)^2$

**27.** $(2x - 5y)^2$            **28.** $(5 - 8x)^2$

**29.** $[(x - 3) + y]^2$        **30.** $[(x + 1) - y]^2$

**31.** $(x + 2y)(x - 2y)$       **32.** $(2x + 3y)(2x - 3y)$

**33.** $(m - 3 + n)(m - 3 - n)$

**34.** $(x + y + 1)(x + y - 1)$

**35.** $(x + 1)^3$              **36.** $(x - 2)^3$

**37.** $(2x - y)^3$             **38.** $(3x + 2y)^3$

**39.** $\left(\sqrt{x} + \sqrt{y}\right)\left(\sqrt{x} - \sqrt{y}\right)$

**40.** $\left(5 + \sqrt{x}\right)\left(5 - \sqrt{x}\right)$

**41.** $(x^2 - x + 1)(x^2 + x + 1)$

**42.** $(x^2 + 9)(x^2 - x - 4)$

**43.** $5x(x + 1) - 3x(x + 1)$

**44.** $(2x - 1)(x + 3) + 3(x + 3)$

**45.** *Savings Plan*   A person who is attending 4 years of college has a summer job. During the summers prior to each of the four college years, the person is able to save $1200, $1400, $800, and $2300. During the summer after graduation the person saves $2900. This money is deposited in an account that pays 6% interest, compounded annually. Find the balance in the account at the end of the summer after graduation.

**46.** *Savings Plan*   Suppose you have an investment that pays an annual dividend in July. Each year for 6 years, you reinvest this dividend in an account that pays 6.5% interest, compounded annually. For the dividends shown in the table, find the balance in the account at the end of 6 years. (*Note:* The first deposit will earn 5 years of interest, and the sixth deposit will earn no interest.)

| Year     | 1     | 2     | 3     |
|----------|-------|-------|-------|
| Dividend | $840  | $930  | $760  |

| Year     | 4      | 5     | 6      |
|----------|--------|-------|--------|
| Dividend | $1020  | $980  | $1130  |

**47.** *Compound Interest*   After 2 years, an investment of $500 compounded annually at an interest rate of $r$ will yield an amount of

$$500(1 + r)^2.$$

Write this polynomial in standard form.

**48.** *Compound Interest*   After 3 years, an investment of $1200 compounded annually at an interest rate of *r* will yield an amount of

$1200(1 + r)^3$.

Write this polynomial in standard form.

**49.** *Geometry*   An open box is made by cutting squares from the corners of a piece of metal that is 18 inches by 26 inches, and folding up the sides (see figure). If the edge of each cut-out square is *x* inches, what is the volume of the box? Find the volume when $x = 1$, $x = 2$, and $x = 3$.

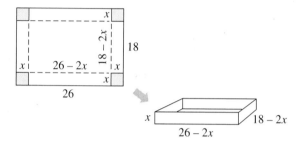

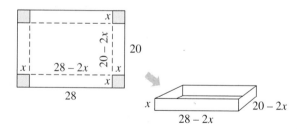

**50.** *Geometry*   An open box is made by cutting squares from the corners of a piece of cardboard that is 20 inches by 28 inches, and turning up the sides (see figure). If the edge of each cut-out square is *x* inches, what is the volume of the box? Find the volume when $x = 1$, $x = 2$, and $x = 3$.

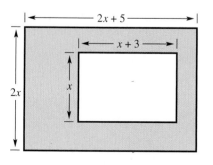

**51.** *Geometry*   Find the area of the shaded region in the figure. Write your answer as a polynomial in standard form.

**52.** *Geometry*   Find the area of the shaded region in the figure. Write your answer as a polynomial in standard form.

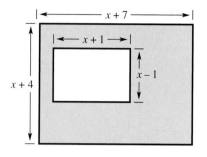

**53.** *Geometry*   Find a polynomial that represents the total number of square feet for the following floor plan.

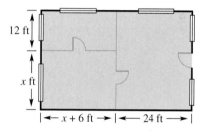

**54.** *Geometry*   Find a polynomial that represents the total number of square feet for the following floor plan.

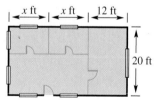

## A.6    Factoring

Common Factors    ■    Factoring Special Polynomial Forms    ■
Trinomials with Binomial Factors    ■    Factoring by Grouping

## Common Factors

The process of writing a polynomial as a product is called **factoring.** It is an important tool for solving equations and reducing fractional expressions.

A polynomial that cannot be factored using integer coefficients is called **prime** or **irreducible over the integers.** For instance, the polynomial $x^2 - 3$ is irreducible over the integers. [Over the *real numbers*, this polynomial can be factored as

$$x^2 - 3 = (x + \sqrt{3})(x - \sqrt{3}).]$$

A polynomial is **completely factored** when each of its factors is prime. For instance,

$$x^3 - x^2 + 4x - 4 = (x - 1)(x^2 + 4)    \qquad \text{Completely factored}$$

is completely factored, but

$$x^3 - x^2 - 4x + 4 = (x - 1)(x^2 - 4)    \qquad \text{Not completely factored}$$

is not completely factored. Its complete factorization would be

$$x^3 - x^2 - 4x + 4 = (x - 1)(x + 2)(x - 2).$$

The simplest type of factoring involves a polynomial that is the product of a monomial and another polynomial. To factor such a polynomial, you can use the Distributive Property in the *reverse* direction.

$$ab + ac = a(b + c)    \qquad \text{$a$ is a common factor.}$$

### Example 1    Removing Common Factors

Factor each of the following.

**a.** $6x^3 - 4x$    **b.** $(x - 2)(2x) + (x - 2)(3)$

**Solution**

**a.** Each term of this polynomial has $2x$ as a common factor.

$$6x^3 - 4x = 2x(3x^2) - 2x(2)$$
$$= 2x(3x^2 - 2)$$

**b.** The binomial factor $(x - 2)$ is common to both terms.

$$(x - 2)(2x) + (x - 2)(3) = (x - 2)(2x + 3)$$

## Factoring Special Polynomial Forms

> ▶ Factoring Special Polynomial Forms
>
> | *Factored Form* | *Example* |
> |---|---|
>
> **Difference of Two Squares**
>
> $u^2 - v^2 = (u + v)(u - v)$        $9x^2 - 4 = (3x + 2)(3x - 2)$
>
> **Perfect Square Trinomial**
>
> $u^2 + 2uv + v^2 = (u + v)^2$        $x^2 + 6x + 9 = (x + 3)^2$
>
> $u^2 - 2uv + v^2 = (u - v)^2$        $x^2 - 6x + 9 = (x - 3)^2$
>
> **Sum or Difference of Two Cubes**
>
> $u^3 + v^3 = (u + v)(u^2 - uv + v^2)$    $x^3 + 8 = (x + 2)(x^2 - 2x + 4)$
>
> $u^3 - v^3 = (u - v)(u^2 + uv + v^2)$    $27x^3 - 1 = (3x - 1)(9x^2 + 3x + 1)$

### Study Tip

In Example 2, note that the first step in factoring a polynomial is to check for common factors. Once the common factor is removed, it is often possible to recognize patterns that were not obvious at first glance.

**Example 2**    Removing a Common Factor First

Factor $3 - 12x^2$.

Solution

$$3 - 12x^2 = 3(1 - 4x^2) \qquad \text{Common factor}$$
$$= 3[1^2 - (2x)^2] \qquad \text{Difference of squares}$$
$$= 3(1 + 2x)(1 - 2x) \qquad \text{Completely factored}$$

**Example 3**    Factoring the Difference of Two Squares

**a.** $(x + 2)^2 - y^2 = [(x + 2) + y][(x + 2) - y]$
$$= (x + 2 + y)(x + 2 - y)$$
$$= (x + y + 2)(x - y + 2)$$

**b.** Factor $16x^4 - 81$ by applying the difference of two squares formula twice.

$$16x^4 - 81 = (4x^2)^2 - 9^2$$
$$= (4x^2 + 9)(4x^2 - 9) \qquad \text{First application}$$
$$= (4x^2 + 9)[(2x)^2 - 3^2]$$
$$= (4x^2 + 9)(2x + 3)(2x - 3). \qquad \text{Second application}$$

A perfect square trinomial is the square of a binomial, and it has the following form.

$$u^2 + 2uv + v^2 = (u + v)^2 \qquad \text{or} \qquad u^2 - 2uv + v^2 = (u - v)^2$$

        Same sign                                    Same sign

Note that the first and last terms are squares and the middle term is twice the product of $u$ and $v$.

**Example 4**    Factoring Perfect Square Trinomials

**a.** $16x^2 + 8x + 1 = (4x)^2 + 2(4x)(1) + 1^2$

$$= (4x + 1)^2$$

**b.** $x^2 - 10x + 25 = x^2 - 2(x)(5) + 5^2$

$$= (x - 5)^2$$

The next two formulas show that sums and differences of cubes factor easily. Pay special attention to the signs of the terms.

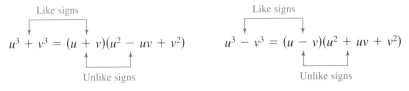

**Example 5**    Factoring the Sum and Difference of Cubes

Factor the following.

**a.** $x^3 - 27$     **b.** $3x^3 + 192$

**Solution**

**a.** $x^3 - 27 = x^3 - 3^3$

$$= (x - 3)(x^2 + 3x + 9)$$

**b.** $3x^3 + 192 = 3(x^3 + 64)$                    Common factor

$$= 3(x^3 + 4^3)$$

$$= 3(x + 4)(x^2 - 4x + 16)$$

## Trinomials with Binomial Factors

To factor a trinomial of the form $ax^2 + bx + c$, use the following pattern.

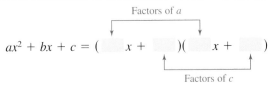

The goal is to find a combination of factors of $a$ and $c$ such that the outer and inner products add up to the middle term $bx$. For instance, in the trinomial $6x^2 + 17x + 5$, you have $a = 6$, $c = 5$, and $b = 17$. After some experimentation, you can find that the factorization is $(2x + 5)(3x + 1)$.

**Example 6**   Factoring a Trinomial:  Leading Coeffi

Factor the trinomial $x^2 - 7x + 12$.

**Solution**

For this trinomial, you have $a = 1$, $b = -7$, and $c = 12$. Because $b$ is nega
and $c$ is positive, both factors of 12 must be negative. That is, $12 = (-2)(-6)$
$12 = (-1)(-12)$, or $12 = (-3)(-4)$. Therefore, the possible factorizations of
$x^2 - 7x + 12$ are

$$(x - 2)(x - 6), \qquad (x - 1)(x - 12), \qquad \text{and} \qquad (x - 3)(x - 4).$$

Testing the middle term, you can find the correct factorization to be

$$x^2 - 7x + 12 = (x - 3)(x - 4).$$

**Example 7**   Factoring a Trinomial:  Leading Coefficient Is Not 1

Factor the trinomial $2x^2 + x - 15$.

**Solution**

For this trinomial, you have $a = 2$ and $c = -15$, which means that the factors of
$-15$ must have unlike signs. The eight possible factorizations are as follows.

$$(2x - 1)(x + 15) \qquad (2x + 1)(x - 15)$$
$$(2x - 3)(x + 5) \qquad (2x + 3)(x - 5)$$
$$(2x - 5)(x + 3) \qquad (2x + 5)(x - 3)$$
$$(2x - 15)(x + 1) \qquad (2x + 15)(x - 1)$$

Testing the middle term, you can find the correct factorization to be
$$2x^2 + x - 15 = (2x - 5)(x + 3).$$

## Factoring by Grouping

Sometimes polynomials with more than three terms can be **factored by grouping.**

**Example 8**   Factoring by Grouping

$$x^3 - 2x^2 - 3x + 6 = (x^3 - 2x^2) - (3x - 6) \qquad \text{Group terms.}$$
$$= x^2(x - 2) - 3(x - 2) \qquad \text{Factor groups.}$$
$$= (x - 2)(x^2 - 3) \qquad \text{Common factor}$$

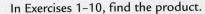

In Exercises 1–10, find the product.

1. $3x(5x - 2)$
2. $-2y(y + 1)$
3. $(2x + 3)^2$
4. $(3x - 8)^2$
5. $(2x - 3)(x + 8)$
6. $(4 - 5z)(1 + z)$
7. $(2y + 1)(2y - 1)$
8. $(x + a)(x - a)$
9. $(x + 4)^3$
10. $(2x - 3)^3$

## A.6   Exercises

In Exercises 1–4, factor out the common factor.

1. $3x + 6$
2. $4x^3 - 6x^2 + 12x$
3. $(x - 1)^2 + 6(x - 1)$
4. $3x(x + 2) - 4(x + 2)$

In Exercises 5–8, factor the difference of two squares.

5. $x^2 - 36$
6. $49 - 9y^2$
7. $(x - 1)^2 - 4$
8. $25 - (z + 5)^2$

In Exercises 9–12, factor the perfect square trinomial.

9. $x^2 - 4x + 4$
10. $x^2 + 10x + 25$
11. $25y^2 - 10y + 1$
12. $z^2 + z + \frac{1}{4}$

In Exercises 13–22, factor the trinomial.

13. $x^2 + x - 2$
14. $x^2 + 5x + 6$
15. $y^2 + y - 20$
16. $z^2 - 5z - 24$
17. $x^2 - 30x + 200$
18. $x^2 - 13x + 42$
19. $3x^2 - 5x + 2$
20. $2x^2 - x - 1$
21. $5x^2 + 26x + 5$
22. $5u^2 + 13u - 6$

In Exercises 23–26, factor the sum or difference of cubes.

23. $x^3 - 8$
24. $x^3 - 27$
25. $8t^3 - 1$
26. $27x^3 + 8$

In Exercises 27–30, factor by grouping.

27. $x^3 - x^2 + 2x - 2$
28. $x^3 + 5x^2 - 5x - 25$
29. $2x^3 - x^2 - 6x + 3$
30. $5x^3 - 10x^2 + 3x - 6$

In Exercises 31–46, completely factor the expression.

31. $4x^2 - 8x$
32. $12x^2 - 48$

33. $x^3 - 4x^2$
34. $16 + 6x - x^2$
35. $1 - 4x + 4x^2$
36. $2y^3 - 7y^2 - 15y$
37. $3x^3 + x^2 + 15x + 5$
38. $5 - x + 5x^2 - x^3$
39. $x^4 - 4x^3 + x^2 - 4x$
40. $3u - 2u^2 + 6 - u^3$
41. $25 - (x + 5)^2$
42. $(t - 1)^2 - 49$
43. $(x^2 + 1)^2 - 4x^2$
44. $(x^2 + 8)^2 - 36x^2$
45. $2t^3 - 16$
46. $5x^3 + 40$

47. *Geometry*   The room that is shown in the figure has a floor space of $2x^2 + 3x + 1$ square feet. If the width of the room is $(x + 1)$ feet, what is the length?

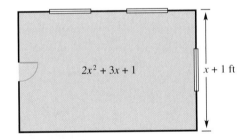

$2x^2 + 3x + 1$

$x + 1$ ft

48. *Geometry*   The room that is shown in the figure has a floor space of $3x^2 + 8x + 4$ square feet. If the width of the room is $(x + 2)$ feet, what is the length?

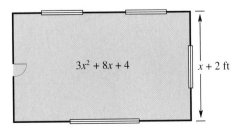

$3x^2 + 8x + 4$

$x + 2$ ft

## A.7    Fractional Expressions and Probability

Domain of an Expression ■ Rational Expressions ■ Complex Fractions ■
Introduction to Probability

## Domain of an Expression

The set of all real numbers for which an algebraic expression is defined is called the **domain** of the expression. For instance, the domain of

$$\frac{1}{x}$$

is all real numbers other than $x = 0$. Two algebraic expressions are said to be **equivalent** if they yield the same value for all numbers in their domain. For instance, the expressions

$$[(x + 1) + (x + 2)] \quad \text{and} \quad 2x + 3$$

are equivalent.

### Study Tip

For all algebraic expressions, exclude from the domain all values that could create *division by zero* or *the square root of a negative number.*

**Example 1**    Finding the Domain of an Algebraic Expression

**a.** The domain of the polynomial

$$2x^3 + 3x + 4$$

is the set of all real numbers. In fact, the domain of any polynomial is the set of all real numbers (unless the domain is specifically restricted).

**b.** The domain of the polynomial

$$x^2 + 5x + 2, \qquad x > 0$$

is the set of positive real numbers, because the polynomial is specifically restricted to that set.

**c.** The domain of the radical expression

$$\sqrt{x}$$

is the set of nonnegative real numbers, because the square root of a negative number is not a real number.

**d.** The domain of the expression

$$\frac{x + 2}{x - 3}$$

is the set of all real numbers except $x = 3$, because the value of $x = 3$ would produce an undefined division by zero.

## Rational Expressions

The quotient of two algebraic expressions is a **fractional expression.** Moreover, the quotient of two *polynomials* such as

$$\frac{1}{x}, \qquad \frac{2x-1}{x+1}, \qquad \text{or} \qquad \frac{x^2-1}{x^2+1}$$

is a **rational expression.** Recall that a fraction is in reduced form if its numerator and denominator have no factors in common aside from $\pm 1$. To write a fraction in reduced form, divide out common factors.

$$\frac{a \cdot \cancel{c}}{b \cdot \cancel{c}} = \frac{a}{b}, \qquad b \neq 0, \qquad c \neq 0$$

The key to success in simplifying rational expressions lies in your ability to *factor* polynomials. For example,

$$\frac{18x^2-18}{6x-6} = \frac{3(6)(x+1)\,\cancel{(x-1)}}{6\,\cancel{(x-1)}} = 3(x+1), \quad x \neq 1.$$

Note that the original expression is undefined when $x = 1$ (because division by zero is undefined). Thus, in order to make sure that the reduced expression is *equivalent* to the original expression, we must restrict the domain of the reduced expression by excluding the value $x = 1$.

---

**Example 2**    Reducing a Rational Expression

$$\frac{x^2+4x-12}{3x-6} = \frac{(x+6)\cancel{(x-2)}}{3\cancel{(x-2)}} \qquad \text{Factor completely.}$$

$$= \frac{x+6}{3}, \quad x \neq 2 \qquad \text{Divide out common factors.}$$

---

In Example 2, do not make the mistake of trying to reduce further by canceling *terms*.

$$\frac{x+6}{3} = \frac{\cancel{x}+\overset{2}{\cancel{6}}}{\cancel{3}} = x+2$$

Remember that to reduce fractions, you cancel *factors*, not terms.

When simplifying rational expressions, be sure to factor each polynomial completely before concluding that the numerator and denominator have no factors in common. Moreover, changing the sign of a factor may allow further reduction, as demonstrated in part (b) of the next example.

**Example 3**    Reducing Rational Expressions

**a.** $\dfrac{x^3-4x}{x^2+x-2} = \dfrac{x(x+2)(x-2)}{(x+2)(x-1)}$ \qquad Factor completely.

$$= \frac{x(x-2)}{x-1}, \qquad x \neq -2 \qquad \text{Divide out common factors.}$$

**b.** $\dfrac{12 + x - x^2}{2x^2 - 9x + 4} = \dfrac{(4 - x)(3 + x)}{(2x - 1)(x - 4)}$          Factor completely.

$\qquad\qquad = \dfrac{-(x - 4)(3 + x)}{(2x - 1)(x - 4)}$          $4 - x = -(x - 4)$

$\qquad\qquad = -\dfrac{3 + x}{2x - 1}, \quad x \neq 4$          Divide out common factors.

To multiply or divide rational expressions, use the properties of fractions (see Section A.2). Recall that to divide fractions you invert the divisor and multiply.

### Example 4    Multiplying Rational Expressions

$\qquad \dfrac{6x^2 - 6x}{x^2 + 2x - 3} \cdot \dfrac{x^2 + x - 6}{2x}$

$\qquad = \dfrac{6x(x - 1)(x + 3)(x - 2)}{(x - 1)(x + 3)(2x)}$          Factor and multiply.

$\qquad = \dfrac{3(2x)(x - 1)(x + 3)(x - 2)}{(x - 1)(x + 3)(2x)}$          Divide out common factors.

$\qquad = 3(x - 2), \quad x \neq 0, x \neq 1, x \neq -3$          Simplify.

### Example 5    Dividing Rational Expressions

$\qquad \dfrac{2x}{3x - 12} \div \dfrac{x^2 - 2x}{x^2 - 6x + 8} = \dfrac{2x}{3x - 12} \cdot \dfrac{x^2 - 6x + 8}{x^2 - 2x}$          Invert and multiply.

$\qquad\qquad = \dfrac{(2x)(x - 2)(x - 4)}{(3)(x - 4)(x)(x - 2)}$          Factor and multiply.

$\qquad\qquad = \dfrac{(2x)(x - 2)(x - 4)}{(3)(x - 4)(x)(x - 2)}$          Divide out common factors.

$\qquad\qquad = \dfrac{2}{3}, \quad x \neq 0, x \neq 2, x \neq 4$          Simplify.

To add or subtract rational expressions, use the least common denominator method or the following basic definition for adding two fractions.

$$\frac{a}{b} \pm \frac{c}{d} = \frac{ad \pm bc}{bd}, \qquad b \neq 0, d \neq 0$$

This definition is efficient for adding or subtracting *two* fractions that have no common factors in their denominators.

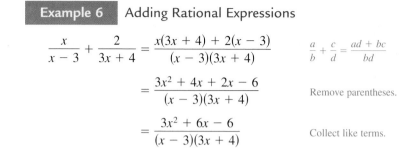

**Example 6**    Adding Rational Expressions

$$\frac{x}{x-3} + \frac{2}{3x+4} = \frac{x(3x+4) + 2(x-3)}{(x-3)(3x+4)} \qquad \frac{a}{b} + \frac{c}{d} = \frac{ad+bc}{bd}$$

$$= \frac{3x^2 + 4x + 2x - 6}{(x-3)(3x+4)} \qquad \text{Remove parentheses.}$$

$$= \frac{3x^2 + 6x - 6}{(x-3)(3x+4)} \qquad \text{Collect like terms.}$$

**Example 7**    Combining Rational Expressions

Perform the given operations and simplify.

$$\frac{3}{x-1} - \frac{2}{x} + \frac{x+3}{x^2-1}$$

**Solution**

Using the factored denominators $(x-1)$, $x$, and $(x+1)(x-1)$, you can see that the least common denominator is $x(x+1)(x-1)$.

$$\frac{3}{x-1} - \frac{2}{x} + \frac{x+3}{x^2-1}$$

$$= \frac{3(x)(x+1)}{x(x+1)(x-1)} - \frac{2(x+1)(x-1)}{x(x+1)(x-1)} + \frac{(x+3)(x)}{x(x+1)(x-1)}$$

$$= \frac{3(x)(x+1) - 2(x+1)(x-1) + (x+3)(x)}{x(x+1)(x-1)}$$

$$= \frac{3x^2 + 3x - 2x^2 + 2 + x^2 + 3x}{x(x+1)(x-1)}$$

$$= \frac{2x^2 + 6x + 2}{x(x+1)(x-1)}$$

$$= \frac{2(x^2 + 3x + 1)}{x(x+1)(x-1)}$$

## Complex Fractions

Problems involving the division of two rational expressions are sometimes written as **complex fractions.** The rules for dividing fractions still apply in such cases: invert the denominator and multiply, as follows.

$$\frac{\left(\dfrac{x-1}{5}\right)}{\left(\dfrac{x+3}{x}\right)} = \frac{x-1}{5} \cdot \frac{x}{x+3} = \frac{x(x-1)}{5(x+3)}$$

**Example 8** Simplifying a Complex Fraction

$$\frac{\left(\dfrac{x^2 + 2x - 3}{x - 3}\right)}{(4x + 12)} = \frac{\left(\dfrac{x^2 + 2x - 3}{x - 3}\right)}{\left(\dfrac{4x + 12}{1}\right)} \qquad \text{Rewrite denominator.}$$

$$= \frac{x^2 + 2x - 3}{x - 3} \cdot \frac{1}{4x + 12} \qquad \text{Invert and multiply.}$$

$$= \frac{(x - 1)(x + 3)}{(x - 3)(4)(x + 3)} \qquad \text{Factor and multiply.}$$

$$= \frac{(x - 1)(x + 3)}{(x - 3)(4)(x + 3)} \qquad \text{Divide out common factors.}$$

$$= \frac{x - 1}{4(x - 3)}, \quad x \neq -3 \qquad \text{Simplify.}$$

**Example 9** Simplifying a Complex Fraction

$$\frac{\left(\dfrac{x}{x + 1}\right)}{\left(\dfrac{2x}{x^2 - 1}\right)} = \frac{x}{x + 1} \cdot \frac{x^2 - 1}{2x} \qquad \text{Invert and multiply.}$$

$$= \frac{x(x - 1)(x + 1)}{(x + 1)(2)(x)} \qquad \text{Multiply and factor.}$$

$$= \frac{x - 1}{2}, \quad x \neq 0, x \neq -1 \qquad \text{Divide out common factors.}$$

## Introduction to Probability

As a member of a complex society, you are used to living with varying amounts of uncertainty. For example, you may be questioning the likelihood of getting a good job after graduation, or winning a state lottery, or the probability of rain tomorrow.

In assigning measurements to uncertainties in everyday life, we often use ambiguous terminology, such as "fairly certain," "probable," or "highly unlikely." In mathematics, we attempt to remove this ambiguity by assigning a number to the likelihood of the occurrence of an event. We call this measurement the **probability** that the event will occur. For example, if we toss a fair coin, we say that the probability that it will land heads up is one-half, or 0.5.

In the study of probability, we call any happening whose result is uncertain an **experiment.** The possible results of the experiment are called **outcomes,** the set of all possible outcomes of an experiment is called the **sample space** of the experiment, and any subcollection of a sample space is called an **event.**

To calculate the probability of an event, we count the number of outcomes in the event and in the sample space. The *number of outcomes* in event $E$ is denoted by $n(E)$ and the number of outcomes in the sample space $S$ is denoted by $n(S)$.

▶ The Probability of an Event

If an event $E$ has $n(E)$ equally likely outcomes and its sample space has $n(S)$ equally likely outcomes, the **probability** of event $E$ is

$$P(E) = \frac{n(E)}{n(S)}.$$

Because the number of outcomes in an event must be *less than or equal to* the number of outcomes in the sample space, we can see that the probability of an event must be a number between 0 and 1. That is, for any event $E$, it must be true that $0 \le P(E) \le 1$, as indicated in Figure A.12.

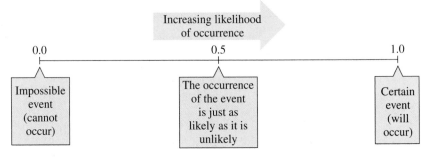

Figure A.12

▶ Properties of the Probability of an Event

Let $E$ be an event that is a subset of a finite sample space $S$.

**1.** $0 \le P(E) \le 1$

**2.** If $P(E) = 0$, $E$ *cannot occur* and is called an **impossible event.**

**3.** If $P(E) = 1$, $E$ *must occur* and is called a **certain event.**

Example 10    Finding the Probability of an Event   

A card is drawn at random from a standard deck of 52 playing cards.

**a.** What is the probability that the card is the ace of hearts?

**b.** What is the probability that the card is an ace?

Solution

**a.** Because there are 52 cards in the deck, the number of possible outcomes in the sample space is 52. Moreover, because only one of these cards is the ace of hearts, the number of outcomes in the event "drawing the ace of hearts" is 1. Thus, the probability of drawing the ace of hearts is

$$P(\text{drawing the ace of hearts}) = \frac{1}{52}.$$

**b.** Because there are 52 cards in the deck, the number of possible outcomes in the sample space is 52. Moreover, because four of these cards are aces, the number of outcomes in the event "drawing an ace" is 4. Thus, the probability of drawing an ace is

$$P(\text{drawing an ace}) = \frac{4}{52} = \frac{1}{13}.$$

This probability is shown graphically in Figure A.13.

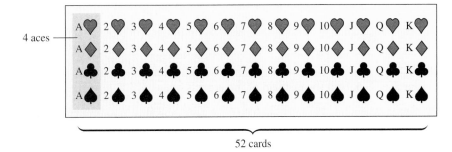

4 aces

52 cards

Figure A.13

**Example 11**    Finding the Probability of an Event

A parachutist plans to land in the middle of a football stadium during a halftime show. The parachutist is certain of landing in the stadium field, but cannot guarantee a landing in a square area in the center of the field, as shown in Figure A.14. Assuming that the probability of landing at any given point in the field is the same, what is the probability that the parachutist will land in the center square?

Solution

The area of the entire field is given by

$$\text{Area of field} = (\text{length})(\text{width}) = (6x)(12x + 25).$$

The area of the center square is given by

$$\text{Area of center square} = (\text{length})(\text{width}) = x^2.$$

Therefore, the probability of landing in the center square is

$$P(\text{landing in center square}) = \frac{x^2}{(6x)(12x + 25)} = \frac{x}{6(12x + 25)}.$$

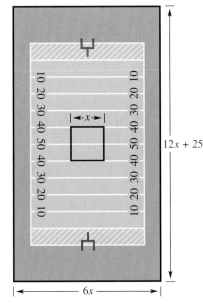

$12x + 25$

$6x$

Figure A.14

# Warm Up

The following warm-up exercises involve skills that were covered in earlier sections. You will use these skills in the exercise set for this section.

In Exercises 1–10, completely factor the polynomials.

**1.** $5x^2 - 15x^3$

**2.** $16x^2 - 9$

**3.** $9x^2 - 6x + 1$

**4.** $9 + 12y + 4y^2$

**5.** $z^2 + 4z + 3$

**6.** $x^2 - 15x + 50$

**7.** $3 + 8x - 3x^2$

**8.** $3x^2 - 46x + 15$

**9.** $s^3 + s^2 - 4s - 4$

**10.** $y^3 + 64$

## A.7    Exercises

In Exercises 1–6, find the domain of the expression.

**1.** $3x^2 - 4x + 7$

**2.** $6x^2 + 7x - 9, \ x > 0$

**3.** $\dfrac{1}{x - 2}$

**4.** $\dfrac{x + 1}{2x + 1}$

**5.** $\dfrac{x - 1}{x^2 - 4x}$

**6.** $\dfrac{1}{\sqrt{x + 1}}$

In Exercises 7–12, find the missing factor in the numerator such that the two fractions will be equivalent.

**7.** $\dfrac{5}{2x} = \dfrac{5(\quad)}{6x^2}$

**8.** $\dfrac{3}{4} = \dfrac{3(\quad)}{4(x + 1)}$

**9.** $\dfrac{x + 1}{x} = \dfrac{(x + 1)(\quad)}{x(x - 2)}$

**10.** $\dfrac{3y - 4}{y + 1} = \dfrac{(3y - 4)(\quad)}{y^2 - 1}$

**11.** $\dfrac{3x}{x - 3} = \dfrac{3x(\quad)}{x^2 - x - 6}$

**12.** $\dfrac{1 - z}{z^2} = \dfrac{(1 - z)(\quad)}{z^3 + z^2}$

In Exercises 13–24, write the rational expression in reduced form.

**13.** $\dfrac{15x^2}{10x}$

**14.** $\dfrac{18y^2}{60y^5}$

**15.** $\dfrac{2x}{4x + 4}$

**16.** $\dfrac{9x^2 + 9x}{2x + 2}$

**17.** $\dfrac{x - 5}{10 - 2x}$

**18.** $\dfrac{x^2 - 25}{5 - x}$

**19.** $\dfrac{x^3 + 5x^2 + 6x}{x^2 - 4}$

**20.** $\dfrac{x^2 + 8x - 20}{x^2 + 11x + 10}$

**21.** $\dfrac{y^2 - 7y + 12}{y^2 + 3y - 18}$

**22.** $\dfrac{x + 1}{x^2 + 11x + 10}$

**23.** $\dfrac{z^3 - 8}{z^2 + 2z + 4}$

**24.** $\dfrac{x^2 - 9}{x^3 + x^2 - 9x - 9}$

In Exercises 25–46, perform the indicated operations and simplify.

**25.** $\dfrac{5}{x - 1} \cdot \dfrac{x - 1}{25(x - 2)}$

**26.** $\dfrac{x + 13}{x^3(3 - x)} \cdot \dfrac{x(x - 3)}{5}$

**27.** $\dfrac{(x - 9)(x + 7)}{x + 1} \cdot \dfrac{x}{9 - x}$

**28.** $\dfrac{(x + 5)(x - 3)}{x + 2} \cdot \dfrac{1}{(x + 5)(x + 2)}$

**29.** $\dfrac{r}{r - 1} \cdot \dfrac{r^2 - 1}{r^2}$

**30.** $\dfrac{4y - 16}{5y + 15} \cdot \dfrac{2y + 6}{4 - y}$

**31.** $\dfrac{t^2 - t - 6}{t^2 + 6t + 9} \cdot \dfrac{t + 3}{t^2 - 4}$

**32.** $\dfrac{y^3 - 8}{2y^3} \cdot \dfrac{4y}{y^2 - 5y + 6}$

**33.** $\dfrac{x^2 + x - 2}{x^3 + x^2} \cdot \dfrac{x}{x^2 + 3x + 2}$

**34.** $\dfrac{x^3 - 1}{x + 1} \cdot \dfrac{x^2 + 1}{x^2 - 1}$

**35.** $\dfrac{3(x + y)}{4} \div \dfrac{x + y}{2}$

**36.** $\dfrac{x + 2}{5(x - 3)} \div \dfrac{x - 2}{5(x - 3)}$

**37.** $\dfrac{\left[\dfrac{x^2}{(x + 1)^2}\right]}{\left[\dfrac{x}{(x + 1)^3}\right]}$

**38.** $\dfrac{\left(\dfrac{x^2 - 1}{x}\right)}{\left[\dfrac{(x - 1)^2}{x}\right]}$

**39.** $\dfrac{5}{x-1} + \dfrac{x}{x-1}$

**40.** $\dfrac{2x-1}{x+3} + \dfrac{1-x}{x+3}$

**41.** $6 - \dfrac{5}{x+3}$

**42.** $\dfrac{2x}{x-5} - \dfrac{5}{5-x}$

**43.** $\dfrac{2}{x^2-4} - \dfrac{1}{x^2-3x+2}$

**44.** $\dfrac{x}{x^2+x-2} - \dfrac{1}{x+2}$

**45.** $-\dfrac{1}{x} + \dfrac{2}{x^2+1} + \dfrac{1}{x^3+x}$

**46.** $\dfrac{2}{x+1} + \dfrac{2}{x-1} + \dfrac{1}{x^2-1}$

In Exercises 47–52, simplify the complex fraction.

**47.** $\dfrac{\left(\dfrac{x}{2} - 1\right)}{(x-2)}$

**48.** $\dfrac{(x-3)}{\left(\dfrac{x}{4} - \dfrac{4}{x}\right)}$

**49.** $\dfrac{\left(\dfrac{1}{x} - \dfrac{1}{x+1}\right)}{\left(\dfrac{1}{x+1}\right)}$

**50.** $\dfrac{\left(\dfrac{5}{y} - \dfrac{6}{2y+1}\right)}{\left(\dfrac{5}{y} + 4\right)}$

**51.** $\dfrac{\left(\sqrt{x} - \dfrac{1}{2\sqrt{x}}\right)}{\sqrt{x}}$

**52.** $\dfrac{\left(\dfrac{x+4}{x+5} - \dfrac{x}{x+1}\right)}{4}$

**53.** *Random Selection*   One of the 35 students in your class is to be chosen at random. What is the probability that it will be you?

**54.** *Misplaced Test*   Thirty-eight of the students in your class took a test. Your instructor lost two of the tests. What is the probability that yours got lost?

**55.** *Lottery*   The "Daily Number" lottery in many states involves choosing an integer from 000 to 999. Suppose that you have purchased a lottery ticket that has one number. What is the probability that your number will be selected?

**56.** *Defective Product*   Suppose that you just bought a new camera. The manufacturer of your camera made 250,000 cameras of your model, and 5000 of them had defective lenses. What is the probability that your new camera has a defective lens?

**57.** *Playing Cards*   Suppose that you draw a single card from a standard deck of playing cards. What is the probability that the card is a heart?

*Monthly Payment*   In Exercises 58 and 59, use the following formula, which gives the approximate annual percentage rate $r$ of a monthly installment loan:

$$r = \dfrac{\left[\dfrac{24(NM - P)}{N}\right]}{\left(P + \dfrac{NM}{12}\right)}$$

where $N$ is the total number of payments, $M$ is the monthly payment, and $P$ is the amount financed.

**58.** (a) Approximate the rate $r$ for a 4-year car loan of $15,000 that has monthly payments of $400.

   (b) Simplify the expression for the annual percentage rate $r$, and then rework part (a).

**59.** (a) Approximate the rate $r$ for a 5-year car loan of $18,000 that has monthly payments of $400.

   (b) Simplify the expression for the annual percentage rate $r$, and then rework part (a).

**60.** *Refrigeration*   When food (at room temperature) is placed in a refrigerator, the time required for the food to cool depends on the amount of food, the air circulation in the refrigerator, the original temperature of the food, and the temperature of the refrigerator. Consider the following model, which gives the temperature of food that is at 75°F and is placed in a 40°F refrigerator:

$$T = 10\left(\dfrac{4t^2 + 16t + 75}{t^2 + 4t + 10}\right), \; t \geq 0$$

where $T$ is the temperature in degrees Fahrenheit and $t$ is the time in hours. Sketch a bar graph showing the temperature of the food when $t = 0, 1, 2, 3, 4,$ and 5 hours.

**61.** *Oxygen Level*   The mathematical model

$$O = \dfrac{t^2 - t + 1}{t^2 + 1}, \; t \geq 0$$

measures the percentage of the normal level of oxygen in a pond, where $t$ is the time in weeks after organic waste is dumped into the pond. Sketch a bar graph showing the oxygen level of the pond when $t = 0, 1, 2, 3, 4,$ and 5 weeks.

## Appendix B
# Conic Sections

### B.1    Conic Sections

Introduction to Conic Sections    ■    Parabolas    ■    Ellipses    ■    Hyperbolas

## Introduction to Conic Sections

Conic sections were discovered during the classical Greek period, which lasted from 600 to 300 B.C. By the beginning of the Alexandrian period, enough was known of conics for Apollonius (262–190 B.C.) to produce an eight-volume work on the subject.

This early Greek study was largely concerned with the geometric properties of conics. It was not until the early seventeenth century that the broad applicability of conics became apparent.

A **conic section** (or simply **conic**) can be described as the intersection of a plane and a double-napped cone. Notice from Figure B.1 that in the formation of the four basic conics, the intersecting plane does not pass through the vertex of the cone. When the plane does pass through the vertex, we call the resulting figure a **degenerate conic,** as shown in Figure B.2.

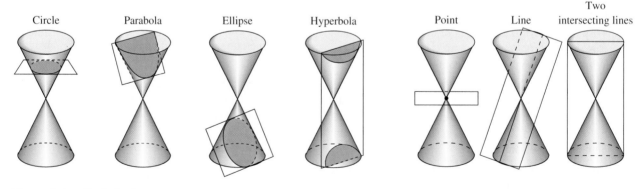

Figure B.1    *Conic Sections*

Figure B.2    *Degenerate Conics*

There are several ways to approach the study of conics. We could begin by defining conics in terms of the intersections of planes and cones, as the Greeks did, or we could define them algebraically, in terms of the general second-degree equation

$$Ax^2 + Bxy + Cy^2 + Dx + Ey + F = 0.$$

However, we will use a third approach, in which each of the conics is defined as a *locus*, or collection of points satisfying a certain geometric property. For example, in Section 2.2 we saw how the definition of a circle as *the collection of all points (x, y) that are equidistant from a fixed point (h, k)* led easily to the standard equation of a circle, $(x - h)^2 + (y - k)^2 = r^2$.

We will restrict our study of conics in this section to parabolas with vertices at the origin and ellipses and hyperbolas with centers at the origin. In the following section, we will look at the more general cases.

## Parabolas

In Section 4.1 we determined that the graph of the quadratic function $f(x) = ax^2 + bx + c$ is a parabola that opens upward or downward. The following definition of a parabola is more general in the sense that it is independent of the orientation of the parabola.

▶ **Definition of a Parabola**

A **parabola** is the set of all points $(x, y)$ that are equidistant from a fixed line called the **directrix** and a fixed point called the **focus** (not on the line).

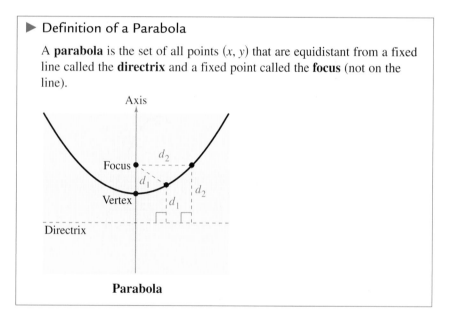

**Parabola**

The midpoint between the focus and the directrix is called the **vertex,** and the line passing through the focus and the vertex is called the **axis** of the parabola.

Using this definition, we can derive the following standard form of the equation of a parabola.

▶ **Standard Equation of a Parabola (Vertex at Origin)**

The **standard form of the equation of a parabola** with a vertex at $(0, 0)$ and directrix $y = -p$ is

$$x^2 = 4py, \quad p \neq 0. \qquad \text{Vertical axis}$$

For directrix $x = -p$, the equation is

$$y^2 = 4px, \quad p \neq 0. \qquad \text{Horizontal axis}$$

The focus is on the axis $p$ units (directed distance) from the vertex.

### Example 1    Finding the Focus of a Parabola

Find the focus of the parabola whose equation is $y = -2x^2$.

### Solution

Since the squared term in the equation involves $x$, we know that the axis is vertical, and we should use the standard form

$$x^2 = 4py. \qquad \text{Standard form, vertical axis}$$

Writing the given equation in this form, we have

$$-2x^2 = y \qquad \text{Original equation}$$

$$x^2 = -\frac{1}{2}y \qquad \text{Divide both sides by } -2.$$

$$x^2 = 4\left(-\frac{1}{8}\right)y. \qquad \text{Standard form}$$

Thus, $p = -\frac{1}{8}$. Since $p$ is negative, the parabola opens downward and the focus of the parabola is $(0, p) = \left(0, -\frac{1}{8}\right)$, as shown in Figure B.3.

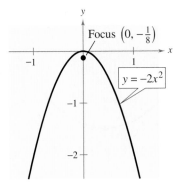

Focus $\left(0, -\frac{1}{8}\right)$

$y = -2x^2$

Figure B.3

### Example 2    A Parabola with a Horizontal Axis

Write the standard form of the equation of the parabola with the vertex at the origin and the focus at $(2, 0)$.

### Solution

The axis of the parabola is horizontal, passing through $(0, 0)$ and $(2, 0)$, as shown in Figure B.4. So, we consider the standard form

$$y^2 = 4px. \qquad \text{Standard form, horizontal axis}$$

Since the focus is $p = 2$ units from the vertex, the equation is

$$y^2 = 4(2)x \qquad \text{Standard form}$$

$$y^2 = 8x.$$

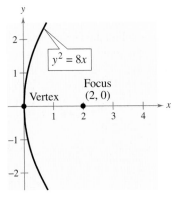

$y^2 = 8x$

Vertex

Focus
$(2, 0)$

Figure B.4

The term *parabola* is a technical term used in mathematics and does not simply refer to *any* U-shaped curve.

Parabolas occur in a wide variety of applications. For instance, a parabolic reflector can be formed by revolving a parabola about its axis. The resulting surface has the property that all incoming rays parallel to the axis are reflected through the focus of the parabola—this is the principal behind the construction of the parabolic mirrors used in reflecting telescopes. Conversely, the light rays emanating from the focus of a parabolic reflector used in a flashlight are all reflected parallel to one another, as shown in Figure B.5.

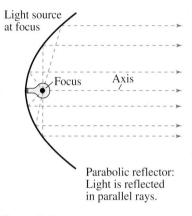

Figure B.5

## Ellipses

The second basic type of conic is called an **ellipse,** and it is defined as follows.

► Definition of an Ellipse

An **ellipse** is the set of all points $(x, y)$ the sum of whose distances from two distinct fixed points, called **foci,** is constant.

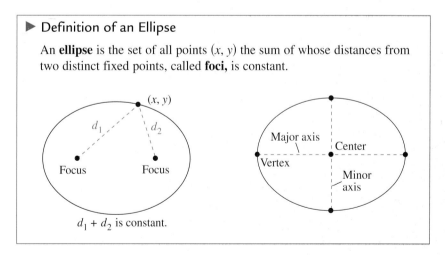

$d_1 + d_2$ is constant.

The line through the foci intersects the ellipse at two points, called the **vertices.** The chord joining the vertices is called the **major axis,** and its midpoint is called the **center** of the ellipse. The chord perpendicular to the major axis at the center is called the **minor axis** of the ellipse.

Figure B.6

You can visualize the definition of an ellipse by imagining two thumbtacks placed at the foci, as shown in Figure B.6. If the ends of a fixed length of string are fastened to the thumbtacks and the string is drawn taut with a pencil, the path traced by the pencil will be an ellipse.

The standard form of the equation of an ellipse takes one of two forms, depending on whether the major axis is horizontal or vertical.

▶ **Standard Equation of an Ellipse (Center at Origin)**

The **standard form of the equation of an ellipse** with the center at the origin and major and minor axes of the lengths $2a$ and $2b$, where $0 < b < a$, is

$$\frac{x^2}{a^2} + \frac{y^2}{b^2} = 1 \quad \text{or} \quad \frac{x^2}{b^2} + \frac{y^2}{a^2} = 1, \quad 0 < b < a.$$

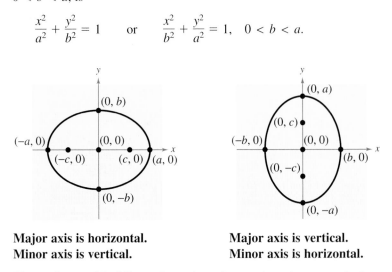

**Major axis is horizontal.**
**Minor axis is vertical.**

**Major axis is vertical.**
**Minor axis is horizontal.**

The vertices and foci lie on the major axis, $a$ and $c$ units, respectively, from the center. Moreover, $a$, $b$, and $c$ are related by the equation $c^2 = a^2 - b^2$.

---

**Example 3** Finding the Standard Equation of an Ellipse

Find the standard form of the equation of the ellipse that has a major axis of length 6 and foci at $(-2, 0)$ and $(2, 0)$, as shown in Figure B.7.

**Solution**

Since the foci occur at $(-2, 0)$ and $(2, 0)$, the center of the ellipse is $(0, 0)$, and the major axis is horizontal. Thus, the ellipse has an equation of the form

$$\frac{x^2}{a^2} + \frac{y^2}{b^2} = 1. \qquad \text{Standard form, horizontal major axis}$$

Since the length of the major axis is 6, we have

$$2a = 6 \qquad \text{Length of major axis}$$

which implies that $a = 3$. Moreover, the distance from the center to either focus is $c = 2$. Finally, we have

$$b^2 = a^2 - c^2 = 3^2 - 2^2 = 9 - 4 = 5$$

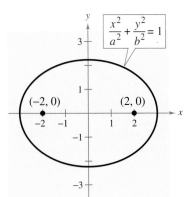

Figure B.7

which yields the equation

$$\frac{x^2}{9} + \frac{y^2}{5} = 1.$$    Standard form

---

**Example 4**   Sketching an Ellipse

Sketch the ellipse given by $4x^2 + y^2 = 36$, and identify the vertices.

**Solution**

We begin by writing the equation in standard form.

$$4x^2 + y^2 = 36$$    Original equation

$$\frac{4x^2}{36} + \frac{y^2}{36} = \frac{36}{36}$$    Divide both sides by 36.

$$\frac{x^2}{3^2} + \frac{y^2}{6^2} = 1$$    Standard form

Since the denominator of the $y^2$-term is larger than the denominator of the $x^2$-term, we conclude that the major axis is vertical. Moreover, since $a = 6$, the vertices are $(0, -6)$ and $(0, 6)$. Finally, since $b = 3$, the endpoints of the minor axis are $(-3, 0)$ and $(3, 0)$, as shown in Figure B.8.

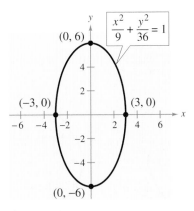

Figure B.8

In Example 4, note that from the standard form of the equation we can sketch the ellipse by locating the endpoints of the two axes. Since $3^2$ is the denominator of the $x^2$-term, we move 3 units to the *right and left* of the center to locate the endpoints of the horizontal axis. Similarly, since $6^2$ is the denominator of the $y^2$-term, we move 6 units *up and down* from the center to locate the endpoints of the vertical axis.

---

## Hyperbolas

The definition of a **hyperbola** is similar to that of an ellipse. The distinction is that, for an ellipse, the *sum* of the distances between the foci and a point on the ellipse is constant, while for a hyperbola, the *difference* of these distances is constant.

▶ Definition of a Hyperbola

A **hyperbola** is the set of all points $(x, y)$, the difference of whose distances from two distinct fixed points, called **foci,** is constant.

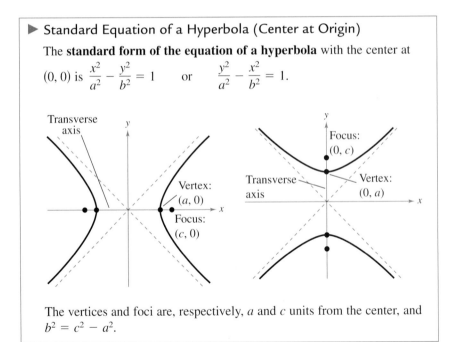

The graph of a hyperbola has two disconnected parts, called **branches.** The line through the two foci intersects the hyperbola at two points, called **vertices.** The line segment connecting the vertices is called the **transverse axis,** and the midpoint of the transverse axis is called the **center** of the hyperbola.

▶ Standard Equation of a Hyperbola (Center at Origin)

The **standard form of the equation of a hyperbola** with the center at $(0, 0)$ is $\dfrac{x^2}{a^2} - \dfrac{y^2}{b^2} = 1$    or    $\dfrac{y^2}{a^2} - \dfrac{x^2}{b^2} = 1$.

The vertices and foci are, respectively, $a$ and $c$ units from the center, and $b^2 = c^2 - a^2$.

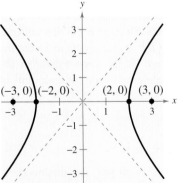

Figure B.9

**Example 5**    Finding the Standard Equation of a Hyperbola

Find the standard form of the equation of the hyperbola with foci at $(-3, 0)$ and $(3, 0)$ and vertices at $(-2, 0)$ and $(2, 0)$, as shown in Figure B.9.

Solution

It can be seen that $c = 3$ because the foci are 3 units from the center. Moreover, $a = 2$ because the vertices are 2 units from the center. So, it follows that

$$b^2 = c^2 - a^2 = 3^2 - 2^2 = 9 - 4 = 5.$$

Since the transverse axis is horizontal, the standard form of the equation is

$$\frac{x^2}{a^2} - \frac{y^2}{b^2} = 1. \qquad \text{Standard form, horizontal transverse axis}$$

Finally, substituting $a^2 = 4$ and $b^2 = 5$, we have

$$\frac{x^2}{4} - \frac{y^2}{5} = 1. \qquad \text{Standard form}$$

An important aid in sketching the graph of a hyperbola is the determination of its **asymptotes,** as shown in Figure B.10. Each hyperbola has two asymptotes that intersect at the center of the hyperbola. Furthermore, the asymptotes pass through the corners of a rectangle of dimensions $2a$ by $2b$. The line segment of length $2b$, joining $(0, b)$ and $(0, -b)$ [or $(-b, 0)$ and $(b, 0)$], is referred to as the **conjugate axis** of the hyperbola.

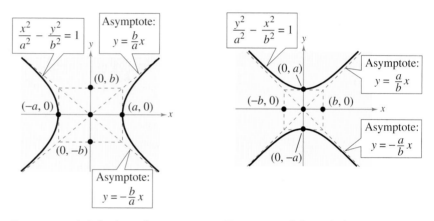

Transverse axis is horizontal

Transverse axis is vertical

Figure B.10

---

▶ **Asymptotes of a Hyperbola (Center at Origin)**

The asymptotes of a hyperbola with center at $(0, 0)$ are

$$y = \frac{b}{a}x \qquad \text{and} \qquad y = -\frac{b}{a}x \qquad \text{Transverse axis is horizontal.}$$

or

$$y = \frac{a}{b}x \qquad \text{and} \qquad y = -\frac{a}{b}x. \qquad \text{Transverse axis is vertical.}$$

---

**Example 6**    Sketching the Graph of a Hyperbola

Sketch the graph of the hyperbola whose equation is $4x^2 - y^2 = 16$.

**Solution**

We begin by rewriting the equation in standard form.

$$4x^2 - y^2 = 16 \qquad \text{Original equation}$$

$$\frac{4x^2}{16} - \frac{y^2}{16} = \frac{16}{16} \qquad \text{Divide both sides by 16.}$$

$$\frac{x^2}{2^2} - \frac{y^2}{4^2} = 1 \qquad \text{Standard form}$$

Because the $x^2$-term is positive, we conclude that the transverse axis is horizontal and the vertices occur at $(-2, 0)$ and $(2, 0)$. Moreover, the endpoints of the conjugate axis occur at $(0, -4)$ and $(0, 4)$, and we are able to sketch the rectangle shown in Figure B.11. Finally, by drawing the asymptotes through the corners of this rectangle, we complete the sketch shown in Figure B.12.

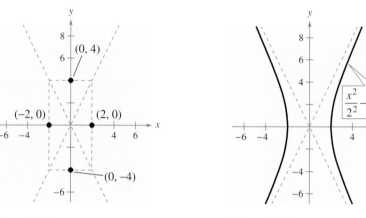

Figure B.11                                            Figure B.12

---

**Example 7**    Finding the Standard Equation of a Hyperbola

Find the standard form of the equation of the hyperbola having vertices at $(0, -3)$ and $(0, 3)$ and with asymptotes $y = -2x$ and $y = 2x$, as shown in Figure B.13.

**Solution**

Since the transverse axis is vertical, we have asymptotes of the form

$$y = \frac{a}{b}x \qquad \text{and} \qquad y = -\frac{a}{b}x. \qquad \text{Transverse axis is vertical.}$$

Using the given equations for the asymptotes, it follows that

$$\frac{a}{b} = 2$$

and since $a = 3$, we can determine that $b = \frac{3}{2}$. Finally, we conclude that the hyperbola has the following equation.

$$\frac{y^2}{3^2} - \frac{x^2}{(3/2)^2} = 1 \qquad \text{Standard form}$$

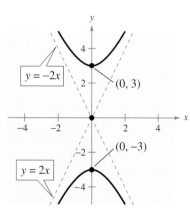

Figure B.13

# Warm Up

The following warm-up exercises involve skills that were covered in earlier sections. You will use these skills in the exercise set for this section.

In Exercises 1–4, rewrite the equations so that they have no fractions.

**1.** $\dfrac{x^2}{16} + \dfrac{y^2}{9} = 1$

**2.** $\dfrac{x^2}{32} + \dfrac{4y^2}{32} = \dfrac{32}{32}$

**3.** $\dfrac{x^2}{1/4} - \dfrac{y^2}{4} = 1$

**4.** $\dfrac{3x^2}{1/9} + \dfrac{4y^2}{9} = 1$

In Exercises 5–8, solve for $c$. (Assume $c > 0$.)

**5.** $c^2 = 3^2 - 1^2$

**6.** $c^2 = 2^2 + 3^2$

**7.** $c^2 + 2^2 = 4^2$

**8.** $c^2 - 1^2 = 2^2$

In Exercises 9 and 10, find the distance between the point and the origin.

**9.** $(0, -4)$

**10.** $(-2, 0)$

## B.1    Exercises

In Exercises 1–8, match the equation with its graph. [The graphs are labeled (a), (b), (c), (d), (e), (f), (g), and (h).]

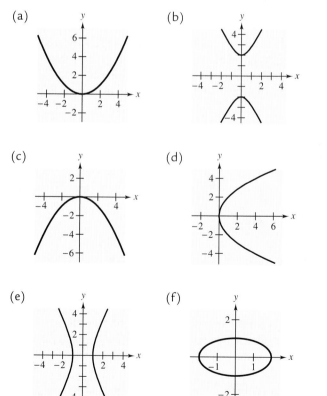

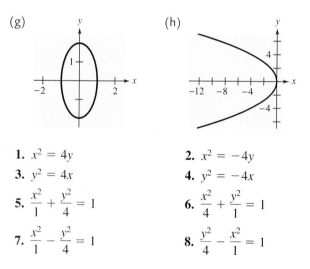

**1.** $x^2 = 4y$

**2.** $x^2 = -4y$

**3.** $y^2 = 4x$

**4.** $y^2 = -4x$

**5.** $\dfrac{x^2}{1} + \dfrac{y^2}{4} = 1$

**6.** $\dfrac{x^2}{4} + \dfrac{y^2}{1} = 1$

**7.** $\dfrac{x^2}{1} - \dfrac{y^2}{4} = 1$

**8.** $\dfrac{y^2}{4} - \dfrac{x^2}{1} = 1$

In Exercises 9–16, find the vertex and focus of the parabola and sketch its graph.

**9.** $y = 4x^2$

**10.** $y = 2x^2$

**11.** $y^2 = -6x$

**12.** $y^2 = 3x$

**13.** $x^2 + 8y = 0$

**14.** $x + y^2 = 0$

**15.** $y^2 - 8x = 0$

**16.** $x^2 + 12y = 0$

In Exercises 17–26, find an equation of the specified parabola with vertex at the origin.

**17.** Focus: $\left(0, -\dfrac{3}{2}\right)$

**18.** Focus: $(2, 0)$

**19.** Focus: $(-2, 0)$

**20.** Focus: $(0, -2)$

**21.** Directrix: $y = -1$

**22.** Directrix: $x = 3$

**23.** Directrix: $y = 2$

**24.** Directrix: $x = -2$

**25.** Horizontal axis; Passes through the point $(4, 6)$

**26.** Vertical axis; Passes through the point $(-2, -2)$

In Exercises 27–34, find the center and vertices of the ellipse and sketch its graph.

**27.** $\dfrac{x^2}{25} + \dfrac{y^2}{16} = 1$

**28.** $\dfrac{x^2}{144} + \dfrac{y^2}{169} = 1$

**29.** $\dfrac{x^2}{16} + \dfrac{y^2}{25} = 1$

**30.** $\dfrac{x^2}{169} + \dfrac{y^2}{144} = 1$

**31.** $\dfrac{x^2}{9} + \dfrac{y^2}{5} = 1$

**32.** $\dfrac{x^2}{28} + \dfrac{y^2}{64} = 1$

**33.** $5x^2 + 3y^2 = 15$

**34.** $x^2 + 4y^2 = 4$

In Exercises 35–42, find an equation of the specified ellipse with center at the origin.

**35.** Vertices: $(0, \pm 2)$; Minor axis of length 2

**36.** Vertices: $(\pm 2, 0)$; Minor axis of length 3

**37.** Vertices: $(\pm 5, 0)$; Foci: $(\pm 2, 0)$

**38.** Vertices: $(0, \pm 8)$; Foci: $(0, \pm 4)$

**39.** Foci: $(\pm 5, 0)$; Major axis of length 12

**40.** Foci: $(\pm 2, 0)$; Major axis of length 8

**41.** Vertices: $(0, \pm 5)$; Passes through the point $(4, 2)$

**42.** Major axis vertical; Passes through the points $(0, 4)$ and $(2, 0)$

In Exercises 43–50, find the center and vertices of the hyperbola and sketch its graph.

**43.** $x^2 - y^2 = 1$

**44.** $\dfrac{x^2}{9} - \dfrac{y^2}{16} = 1$

**45.** $\dfrac{y^2}{1} - \dfrac{x^2}{4} = 1$

**46.** $\dfrac{y^2}{9} - \dfrac{x^2}{1} = 1$

**47.** $\dfrac{y^2}{25} - \dfrac{x^2}{144} = 1$

**48.** $\dfrac{x^2}{36} - \dfrac{y^2}{4} = 1$

**49.** $2x^2 - 3y^2 = 6$

**50.** $3y^2 - 5x^2 = 15$

In Exercises 51–58, find an equation of the specified hyperbola with center at the origin.

**51.** Vertices: $(0, \pm 2)$; Foci: $(0, \pm 4)$

**52.** Vertices: $(\pm 3, 0)$; Foci: $(\pm 5, 0)$

**53.** Vertices: $(\pm 1, 0)$; Asymptotes: $y = \pm 3x$

**54.** Vertices: $(0, \pm 3)$; Asymptotes: $y = \pm 3x$

**55.** Foci: $(0, \pm 8)$; Asymptotes: $y = \pm 4x$

**56.** Foci: $(\pm 10, 0)$; Asymptotes: $y = \pm \frac{3}{4}x$

**57.** Vertices: $(0, \pm 3)$; Passes through the point $(-2, 5)$

**58.** Vertices: $(\pm 2, 0)$; Passes through the point $\left(3, \sqrt{3}\right)$

**59.** *Satellite Antenna*   The receiver in a parabolic television dish antenna is 3 feet from the vertex and is located at the focus (see figure). Find an equation of a cross section of the reflector. (Assume that the dish is directed upward and the vertex is at the origin.)

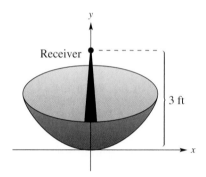

**60.** *Suspension Bridge*   Each cable of a suspension bridge is suspended (in the shape of a parabola) between two towers that are 400 feet apart and 50 feet above the roadway (see figure). The cables touch the roadway midway between the towers. Find an equation for the parabolic shape of each cable.

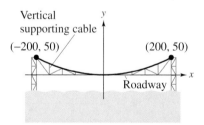

**61.** *Fireplace Arch*   A fireplace arch is to be constructed in the shape of a semiellipse. The opening is to have a height of 2 feet at the center and a width of 5 feet along the base (see figure on next page). The contractor draws the outline of the ellipse by the method shown in Figure B.6. Where should the tacks be placed and what should be the length of the piece of string?

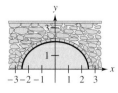

Figure for 61

**62.** *Mountain Tunnel*  A semielliptical arch over a tunnel for a road through a mountain has a major axis of 100 feet, and its height at the center is 30 feet (see figure). Determine the height of the arch 5 feet from the edge of the tunnel.

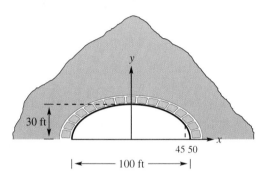

**63.** Sketch a graph of the ellipse that consists of all points $(x, y)$ such that the sum of the distances between $(x, y)$ and two fixed points is 15 units and the foci are located at the centers of the two sets of concentric circles, as shown in the figure.

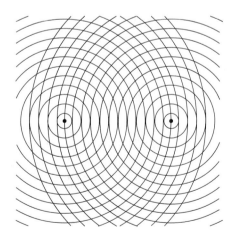

**64.** *Think About It*  A line segment through a focus with endpoints on the ellipse and perpendicular to the major axis is called a **latus rectum** of the ellipse. Therefore, an ellipse has two latera recta. Knowing

the length of the latera recta is helpful in sketching an ellipse because it yields other points on the curve (see figure). Show that the length of each latus rectum is $2b^2/a$.

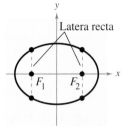

In Exercises 65–68, sketch the graph of the ellipse, making use of the latera recta (see Exercise 64).

**65.** $\dfrac{x^2}{4} + \dfrac{y^2}{1} = 1$

**66.** $\dfrac{x^2}{9} + \dfrac{y^2}{16} = 1$

**67.** $9x^2 + 4y^2 = 36$

**68.** $5x^2 + 3y^2 = 15$

**69.** *Loran*  Long-range navigation for aircraft and ships is accomplished by synchronized pulses transmitted by widely separated transmitting stations. These pulses travel at the speed of light (186,000 miles per second). The difference in the arrival times of these pulses at an aircraft or ship is constant on a hyperbola having the transmitting stations as foci. Assume that two stations 300 miles apart are positioned on the rectangular coordinate system at points with coordinates $(-150, 0)$ and $(150, 0)$ and that a ship is traveling on a path with coordinates $(x, 75)$ (see figure). Find the $x$-coordinate of the position of the ship if the time difference between the pulses from the transmitting stations is 1000 microseconds (0.001 second).

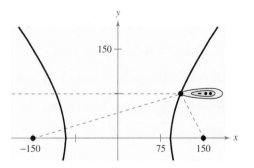

**70.** *Hyperbolic Mirror* A hyperbolic mirror (used in some telescopes) has the property that a light ray directed at one focus will be reflected to the other focus (see figure). The focus of the hyperbolic mirror has coordinates (12, 0). Find the vertex of the mirror if its **mount** has coordinates (12, 12).

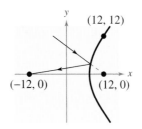

---

## B.2 Conic Sections and Translations

Vertical and Horizontal Shifts of Conics ■ Writing Equations of Conics in Standard Form

## Vertical and Horizontal Shifts of Conics

In Section B.1 we looked at conic sections whose graphs were in *standard position*. In this section we will study the equations of conic sections that have been shifted vertically or horizontally in the plane. The following summary lists the standard forms of the equations of the four basic conics.

> ▶ **Standard Forms of Equations of Conics**
>
> Circle (r = radius)
>
> $(x - h)^2 + (y - k)^2 = r^2$
>
> Ellipse (2a = major axis length,
>         2b = minor axis length)
>
> Center: (h, k)

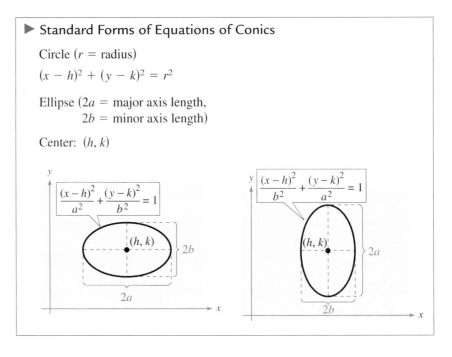

▶ Standard Forms of Equations of Conics   (continued)

Hyperbola ($2a$ = transverse axis length,
  $2b$ = conjugate axis length)
Center: $(h, k)$

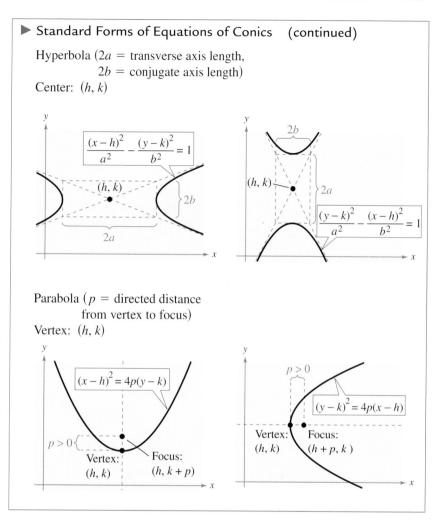

Parabola ($p$ = directed distance
  from vertex to focus)
Vertex: $(h, k)$

**Example 1**   Equations of Conic Sections

**a.** The graph of

$$(x - 1)^2 + (y + 2)^2 = 3^2 \qquad \text{Circle}$$

is a circle whose center is the point $(1, -2)$ and whose radius is 3, as shown in Figure B.14(a) on the following page.

**b.** The graph of

$$\frac{(x - 2)^2}{3^2} + \frac{(y - 1)^2}{2^2} = 1 \qquad \text{Ellipse}$$

is an ellipse whose center is the point $(2, 1)$. The major axis of the ellipse is horizontal with a length of $2(3) = 6$. The minor axis of the ellipse is vertical with a length of $2(2) = 4$, as shown in Figure B.14(b) on the following page.

**c.** The graph of

$$\frac{(x-3)^2}{1^2} - \frac{(y-2)^2}{3^2} = 1 \qquad \text{Hyperbola}$$

is a hyperbola whose center is the point $(3, 2)$. The transverse axis is horizontal with a length of $2(1) = 2$. The conjugate axis is vertical with a length of $2(3) = 6$, as shown in Figure B.14(c).

**d.** The graph of

$$(x-2)^2 = 4(-1)(y-3) \qquad \text{Parabola}$$

is a parabola whose vertex is the point $(2, 3)$. The axis of the parabola is vertical. The focus of the parabola is 1 unit above or below the vertex. Moreover, since $p = -1$, it follows that the focus lies *below* the vertex, as shown in Figure B.14(d).

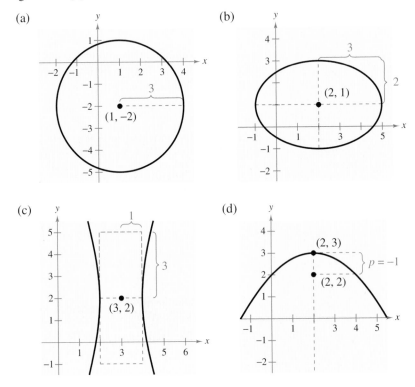

Figure B.14

# Writing Equations of Conics in Standard Form

To write the equation of a conic in standard form, we complete the square, as demonstrated in Examples 2, 3, and 4.

**Example 2**  Finding the Standard Form of a Parabola

Find the vertex and focus of the parabola given by

$$x^2 - 2x + 4y - 3 = 0.$$

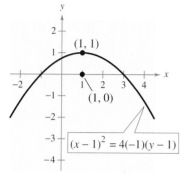

Figure B.15

### Solution

$$x^2 - 2x + 4y - 3 = 0 \qquad \text{Original equation}$$

$$x^2 - 2x = -4y + 3 \qquad \text{Group terms.}$$

$$x^2 - 2x + 1 = -4y + 3 + 1 \qquad \text{Add 1 to both sides.}$$

$$(x - 1)^2 = -4y + 4 \qquad \text{Completed square form}$$

$$(x - 1)^2 = 4(-1)(y - 1) \qquad \text{Standard form}$$

$$(x - h)^2 = 4p(y - k)$$

From this standard form, we see that $h = 1$, $k = 1$, and $p = -1$. Since the axis is vertical and $p$ is negative, the parabola opens downward. The vertex and focus are

*Vertex*:  $(h, k) = (1, 1)$

*Focus*:  $(h, k + p) = (1, 0)$.

The graph of this parabola is shown in Figure B.15.

In Examples 1(d) and 2, $p$ is the *directed distance* from the vertex to the focus. Because the axis of the parabola is vertical and $p = -1$, the focus is 1 unit *below* the vertex, and the parabola opens downward.

### Example 3    Sketching an Ellipse

Sketch the graph of the ellipse whose equation is

$$x^2 + 4y^2 + 6x - 8y + 9 = 0.$$

### Solution

$$x^2 + 4y^2 + 6x - 8y + 9 = 0 \qquad \text{Original equation}$$

$$(x^2 + 6x + \quad) + (4y^2 - 8y + \quad) = -9 \qquad \text{Group terms.}$$

$$(x^2 + 6x + \quad) + 4(y^2 - 2y + \quad) = -9 \qquad \text{Factor 4 out of } y\text{-terms.}$$

$$(x^2 + 6x + 9) + 4(y^2 - 2y + 1) = -9 + 9 + 4(1) \qquad \text{Add 9 and 4 to both sides.}$$

$$(x + 3)^2 + 4(y - 1)^2 = 4 \qquad \text{Completed square form}$$

$$\frac{(x + 3)^2}{4} + \frac{(y - 1)^2}{1} = 1 \qquad \text{Standard form}$$

$$\frac{(x - h)^2}{a^2} + \frac{(y - k)^2}{b^2} = 1$$

From this standard form, we see that the center is $(h, k) = (-3, 1)$. Since the denominator of the $x$-term is $4 = a^2 = 2^2$, the endpoints of the major axis lie 2 units to the right and left of the center. Similarly, since the denominator of the $y$-term is $1 = b^2 = 1^2$, the endpoints of the minor axis lie 1 unit up and down from the center. The graph of this ellipse is shown in Figure B.16.

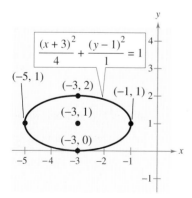

Figure B.16

| Example 4 | Sketching a Hyperbola |

Sketch the graph of the hyperbola given by the equation

$$y^2 - 4x^2 + 4y + 24x - 41 = 0.$$

### Solution

| | |
|---|---|
| $y^2 - 4x^2 + 4y + 24x - 41 = 0$ | Original equation |
| $(y^2 + 4y + \quad) - (4x^2 - 24x + \quad) = 41$ | Group terms. |
| $(y^2 + 4y + \quad) - 4(x^2 - 6x + \quad) = 41$ | Factor 4 out of $x$-terms. |
| $(y^2 + 4y + 4) - 4(x^2 - 6x + 9) = 41 + 4 - 4(9)$ | Add 4, subtract 36. |
| $(y + 2)^2 - 4(x - 3)^2 = 9$ | Completed square form |
| $\dfrac{(y + 2)^2}{9} - \dfrac{4(x - 3)^2}{9} = 1$ | Divide both sides by 9. |
| $\dfrac{(y + 2)^2}{9} - \dfrac{(x - 3)^2}{9/4} = 1$ | Change 4 to $1/(1/4)$. |
| $\dfrac{(y + 2)^2}{3^2} - \dfrac{(x - 3)^2}{(3/2)^2} = 1$ | Standard form |
| $\dfrac{(y - k)^2}{a^2} - \dfrac{(x - h)^2}{b^2} = 1$ | |

From the standard form, we see that the transverse axis is vertical and the center lies at $(h, k) = (3, -2)$. Since the denominator of the $y$-term is $a^2 = 3^2$, we know that the vertices occur 3 units above and below the center.

*Vertices*: $(3, -5)$ \qquad and \qquad $(3, 1)$

To sketch the hyperbola, we draw a rectangle whose top and bottom pass through the vertices. Since the denominator of the $x$-term is $b^2 = \left(\frac{3}{2}\right)^2$, we locate the sides of the rectangle $\frac{3}{2}$ units to the right and left of the center, as shown in Figure B.17. Finally, we sketch the asymptotes by drawing lines through the opposite corners of the rectangle. Using these asymptotes, we complete the graph of the hyperbola, as shown in Figure B.17.

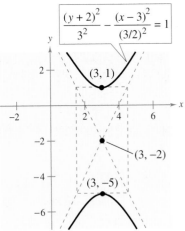

Figure B.17

To find the foci in Example 4, we first find $c$.

$$c^2 = a^2 + b^2 = 9 + \frac{9}{4} = \frac{45}{4} \qquad \Longrightarrow \qquad c = \frac{3\sqrt{5}}{2}$$

Because the transverse axis is vertical, the foci lie $c$ units above and below the center.

*Foci*: $\left(3, -2 + \dfrac{3\sqrt{5}}{2}\right)$ \qquad and \qquad $\left(3, -2 - \dfrac{3\sqrt{5}}{2}\right)$

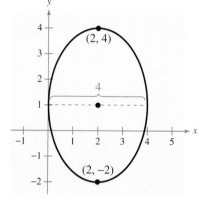

Figure B.18

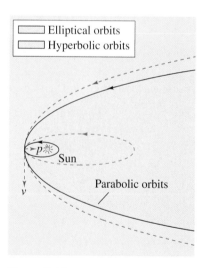

Figure B.19

| Example 5 | Writing the Equation of an Ellipse |

Write the standard form of the equation of an ellipse whose vertices are $(2, -2)$ and $(2, 4)$. The length of the minor axis of the ellipse is 4, as shown in Figure B.18.

### Solution

The center of the ellipse lies at the midpoint of its vertices. Thus, the center is

$$(h, k) = (2, 1). \qquad \text{Center}$$

Since the vertices lie on a vertical line and are 6 units apart, it follows that the major axis is vertical and has a length of $2a = 6$. Thus, $a = 3$. Moreover, since the minor axis has a length of 4, it follows that $2b = 4$, which implies that $b = 2$. Therefore, we can conclude that the standard form of the equation of the ellipse is as follows.

$$\frac{(x - h)^2}{b^2} + \frac{(y - k)^2}{a^2} = 1 \qquad \text{Major axis is vertical.}$$

$$\frac{(x - 2)^2}{2^2} + \frac{(y - 1)^2}{3^2} = 1 \qquad \text{Standard form}$$

An interesting application of conic sections involves the orbits of comets in our solar system. Of the 610 comets identified prior to 1970, 245 have elliptical orbits, 295 have parabolic orbits, and 70 have hyperbolic orbits. For example, Halley's comet has an elliptical orbit, and we can predict the reappearance of this comet every 75 years. The center of the sun is a focus of each of these orbits, and each orbit has a vertex at the point where the comet is closest to the sun, as shown in Figure B.19.

If $p$ is the distance between the vertex and the focus, and $v$ is the speed of the comet at the vertex, then the orbit is:

an *ellipse* if     $v < \sqrt{\dfrac{2GM}{p}}$

a *parabola* if     $v = \sqrt{\dfrac{2GM}{p}}$

a *hyperbola* if   $v > \sqrt{\dfrac{2GM}{p}}$

where $M$ is the mass of the sun and $G \approx 6.67 \times 10^{-8}$ cm$^3$/(gm · sec$^2$).

# Warm Up

The following warm-up exercises involve skills that were covered in earlier sections. You will use these skills in the exercise set for this section.

In Exercises 1–10, identify the conic represented by each equation.

**1.** $\dfrac{x^2}{4} - \dfrac{y^2}{4} = 1$

**2.** $\dfrac{x^2}{9} + \dfrac{y^2}{1} = 1$

**3.** $2x + y^2 = 0$

**4.** $\dfrac{x^2}{9} - \dfrac{y^2}{4} = 1$

**5.** $\dfrac{x^2}{4} + \dfrac{y^2}{16} = 1$

**6.** $4x^2 + 4y^2 = 25$

**7.** $\dfrac{y^2}{4} - \dfrac{x^2}{2} = 1$

**8.** $x^2 - 6y = 0$

**9.** $3x - y^2 = 0$

**10.** $\dfrac{x^2}{9/4} + \dfrac{y^2}{4} = 1$

## B.2    Exercises

In Exercises 1–12, find the vertex, focus, and directrix of the parabola and sketch its graph.

**1.** $(x - 1)^2 + 8(y + 2) = 0$

**2.** $(x + 3) + (y - 2)^2 = 0$

**3.** $\left(y + \tfrac{1}{2}\right)^2 = 2(x - 5)$

**4.** $\left(x + \tfrac{1}{2}\right)^2 = 4(y - 3)$

**5.** $y = \tfrac{1}{4}(x^2 - 2x + 5)$

**6.** $y = -\tfrac{1}{6}(x^2 + 4x - 2)$

**7.** $4x - y^2 - 2y - 33 = 0$

**8.** $y^2 + x + y = 0$

**9.** $y^2 + 6y + 8x + 25 = 0$

**10.** $x^2 - 2x + 8y + 9 = 0$

**11.** $y^2 - 4y - 4x = 0$

**12.** $y^2 - 4x - 4 = 0$

In Exercises 13–20, find an equation of the specified parabola.

**13.** Vertex: $(3, 2)$; Focus: $(1, 2)$

**14.** Vertex: $(-1, 2)$; Focus: $(-1, 0)$

**15.** Vertex: $(0, 4)$; Directrix: $y = 2$

**16.** Vertex: $(-2, 1)$; Directrix: $x = 1$

**17.** Focus: $(2, 2)$; Directrix: $x = -2$

**18.** Focus: $(0, 0)$; Directrix: $y = 4$

**19.**
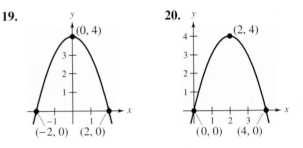

**20.**

In Exercises 21–28, find the center, foci, and vertices of the ellipse and sketch its graph.

**21.** $\dfrac{(x - 1)^2}{9} + \dfrac{(y - 5)^2}{25} = 1$

**22.** $(x + 2)^2 + \dfrac{(y + 4)^2}{1/4} = 1$

**23.** $9x^2 + 4y^2 + 36x - 24y + 36 = 0$

**24.** $9x^2 + 4y^2 - 36x + 8y + 31 = 0$

**25.** $16x^2 + 25y^2 - 32x + 50y + 16 = 0$

**26.** $9x^2 + 25y^2 - 36x - 50y + 61 = 0$

**27.** $12x^2 + 20y^2 - 12x + 40y - 37 = 0$

**28.** $36x^2 + 9y^2 + 48x - 36y + 43 = 0$

In Exercises 29–36, find an equation of the specified ellipse.

**29.** Vertices: $(0, 2), (4, 2)$; Minor axis of length 2

**30.** Foci: $(0, 0), (4, 0)$; Major axis of length 8

**31.** Foci: $(0, 0), (0, 8)$; Major axis of length 16

**32.** Center: $(2, -1)$;  Vertex: $\left(2, \frac{1}{2}\right)$;
Minor axis of length 2

**33.** Vertices: $(3, 1), (3, 9)$;  Minor axis of length 6

**34.** Center: $(3, 2)$;  $a = 3c$;  Foci: $(1, 2), (5, 2)$

**35.** Center: $(0, 4)$;  $a = 2c$;  Vertices $(-4, 4), (4, 4)$

**36.** Vertices: $(5, 0), (5, 12)$;  Endpoints of the minor axis: $(0, 6), (10, 6)$

In Exercises 37–46, find the center, vertices, and foci of the hyperbola and sketch its graph, using asymptotes as an aid.

**37.** $\dfrac{(x - 1)^2}{4} - \dfrac{(y + 2)^2}{1} = 1$

**38.** $\dfrac{(x + 1)^2}{144} - \dfrac{(y - 4)^2}{25} = 1$

**39.** $(y + 6)^2 - (x - 2)^2 = 1$

**40.** $\dfrac{(y - 1)^2}{1/4} - \dfrac{(x + 3)^2}{1/9} = 1$

**41.** $9x^2 - y^2 - 36x - 6y + 18 = 0$

**42.** $x^2 - 9y^2 + 36y - 72 = 0$

**43.** $9y^2 - x^2 + 2x + 54y + 62 = 0$

**44.** $16y^2 - x^2 + 2x + 64y + 63 = 0$

**45.** $x^2 - 9y^2 + 2x - 54y - 107 = 0$

**46.** $9x^2 - y^2 + 54x + 10y + 55 = 0$

In Exercises 47–54, find an equation of the specified hyperbola.

**47.** Vertices: $(2, 0), (6, 0)$;  Foci: $(0, 0), (8, 0)$

**48.** Vertices: $(2, 3), (2, -3)$;  Foci: $(2, 5), (2, -5)$

**49.** Vertices: $(4, 1), (4, 9)$;  Foci: $(4, 0), (4, 10)$

**50.** Vertices: $(-2, 1), (2, 1)$;  Foci: $(-3, 1), (3, 1)$

**51.** Vertices: $(2, 3), (2, -3)$;  Passes through the point $(0, 5)$

**52.** Vertices: $(-2, 1), (2, 1)$;  Passes through the point $(4, 3)$

**53.** Vertices: $(0, 2), (6, 2)$;
Asymptotes: $y = \frac{2}{3}x, y = 4 - \frac{2}{3}x$

**54.** Vertices: $(3, 0), (3, 4)$;
Asymptotes: $y = \frac{2}{3}x, y = 4 - \frac{2}{3}x$

In Exercises 55–62, classify the graph of each equation as a circle, a parabola, an ellipse, or a hyperbola.

**55.** $x^2 + y^2 - 6x + 4y + 9 = 0$

**56.** $x^2 + 4y^2 - 6x + 16y + 21 = 0$

**57.** $4x^2 - y^2 - 4x - 3 = 0$

**58.** $y^2 - 4y - 4x = 0$

**59.** $4x^2 + 3y^2 + 8x - 24y + 51 = 0$

**60.** $4y^2 - 2x^2 - 4y - 8x - 15 = 0$

**61.** $25x^2 - 10x - 200y - 119 = 0$

**62.** $4x^2 + 4y^2 - 16y + 15 = 0$

**63.** *Satellite Orbit*  A satellite in a 100-mile-high circular orbit around the Earth has a velocity of approximately 17,500 miles per hour. If this velocity is multiplied by $\sqrt{2}$, then the satellite will have the minimum velocity necessary to escape the Earth's gravity, and it will follow a parabolic path with the center of the Earth as the focus (see figure).

(a) Find the escape velocity of the satellite.

(b) Find an equation of its path (assume the radius of the Earth is 4000 miles).

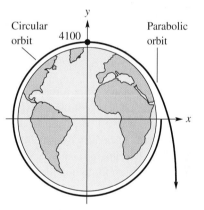

**64.** *Fluid Flow*  Water is flowing from a horizontal pipe 48 feet above the ground. The falling stream of water has the shape of a parabola whose vertex $(0, 48)$ is at the end of the pipe (see figure). The stream of water strikes the ground at the point $\left(10\sqrt{3}, 0\right)$. Find the equation of the path taken by the water.

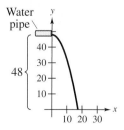

In Exercises 65–71 and 73, $e$ is called the **eccentricity** of the ellipse and is defined by $e = c/a$. It measures the flatness of the ellipse.

**65.** Find an equation of the ellipse with vertices $(\pm 5, 0)$ and eccentricity $e = \frac{3}{5}$.

**66.** Find an equation of the ellipse with vertices $(0, \pm 8)$ and eccentricity $e = \frac{1}{2}$.

**67.** *Orbit of Earth* The Earth moves in an elliptical orbit with the sun at one of the foci (see figure). The length of half of the major axis is $92.957 \times 10^6$ miles and the eccentricity is 0.017. Find the shortest distance (*perihelion*) and the greatest distance (*aphelion*) between the Earth and the sun.

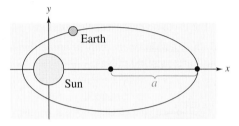

**Figure for 67–69**

**68.** *Orbit of Pluto* The planet Pluto moves in an elliptical orbit with the sun at one of the foci (see figure). The length of half of the major axis is $3.666 \times 10^9$ miles and the eccentricity is 0.248. Find the shortest distance (*perihelion*) and the greatest distance (*aphelion*) between Pluto and the sun.

**69.** *Orbit of Saturn* The planet Saturn moves in an elliptical orbit with the sun at one of the foci (see figure). The shortest distance and the greatest distance between the planet and the sun are $1.3495 \times 10^9$ kilometers and $1.5045 \times 10^9$ kilometers, respectively. Find the eccentricity of the orbit.

**70.** *Satellite Orbit* The first artificial satellite to orbit the Earth was Sputnik I (launched by Russia in 1957). Its highest point above the Earth's surface was 583 miles, and its lowest point was 132 miles (see figure). Assume that the center of the Earth is the focus of the elliptical orbit and the radius of the Earth is 4000 miles. Find the eccentricity of the orbit.

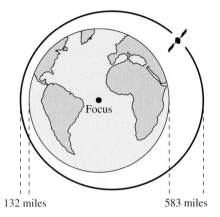

132 miles        583 miles

**Figure for 70**

**71.** Show that the equation of an ellipse can be written as

$$\frac{(x - h)^2}{a^2} + \frac{(y - k)^2}{a^2(1 - e^2)} = 1.$$

Note that as $e$ approaches zero, the ellipse approaches a circle of radius $a$.

**72.** *Australian Football* In Australia, football by *Australian Rules* (or rugby) is played on elliptical fields. The fields can be maximum of 170 yards wide and a maximum of 200 yards long. Let the center of a field of maximum size be represented by the point $(0, 85)$. Find an equation of the ellipse that represents this field. (Source: *Oxford Companion to World Sports and Games*)

**73.** *Comet Orbit* Halley's comet has a elliptical orbit with the sun at one focus. The eccentricity of the orbit is approximately 0.97. The length of the major axis of the orbit is approximately 36.23 astronomical units. (An astronomical unit is about 93 million miles.) Find an equation of the orbit. Place the center of the orbit at the origin and place the major axis on the $x$-axis.

**74.** *Comet Orbit* The comet Encke has a elliptical orbit with the sun at one focus. Encke ranges from 0.34 to 4.08 astronomical units from the sun. Find an equation of the orbit. Place the center of the orbit at the origin and place the major axis on the $x$-axis.

## Appendix C
# Further Concepts in Statistics

## C.1    Representing Data and Linear Modeling

Stem-and-Leaf Plots  ■  Histograms and Frequency Distributions  ■  Scatter Plots  ■
Fitting a Line to Data

### Stem-and-Leaf Plots

Statistics is the branch of mathematics that studies techniques for collecting, organizing, and interpreting data. In this section, you will study several ways to organize and interpret data.

One type of plot that can be used to organize sets of numbers by hand is a **stem-and-leaf plot.** A set of test scores and the corresponding stem-and-leaf plot are shown below.

| *Test Scores* | | *Stems* | *Leaves* |
|---|---|---|---|
| 93, 70, 76, 58, 86, 93, 82, 78, 83, 86, | | 5 | 8 |
| 64, 78, 76, 66, 83, 83, 96, 74, 69, 76, | | 6 | 4 4 6 9 |
| 64, 74, 79, 76, 88, 76, 81, 82, 74, 70 | | 7 | 0 0 4 4 4 6 6 6 6 8 8 9 |
| | | 8 | 1 2 2 3 3 3 6 6 8 |
| | | 9 | 3 3 6 |

Note that the *leaves* represent the units digits of the numbers and the *stems* represent the tens digits. Stem-and-leaf plots can also be used to compare two sets of data, as shown in the following example.

**Example 1**    Comparing Two Sets of Data

Use a stem-and-leaf plot to compare the test scores given above with the following test scores. Which set of test scores is better?

   90, 81, 70, 62, 64, 73, 81, 92, 73, 81, 92, 93, 83, 75, 76,
   83, 94, 96, 86, 77, 77, 86, 96, 86, 77, 86, 87, 87, 79, 88

Solution

Begin by ordering the second set of scores.

   62, 64, 70, 73, 73, 75, 76, 77, 77, 77, 79, 81, 81, 81, 83,
   83, 86, 86, 86, 86, 87, 87, 88, 90, 92, 92, 93, 94, 96, 96

Now that the data have been ordered, you can construct a *double* stem-and-leaf plot by letting the leaves to the right of the stems represent the units digits for the first group of test scores and letting the leaves to the left of the stems represent the units digits for the second group of test scores.

| Leaves (2nd Group) | Stems | Leaves (1st Group) |
|---:|:---:|:---|
| | 5 | 8 |
| 4 2 | 6 | 4 4 6 9 |
| 9 7 7 7 6 5 3 3 0 | 7 | 0 0 4 4 4 6 6 6 6 6 8 8 9 |
| 8 7 7 6 6 6 6 3 3 1 1 1 | 8 | 1 2 2 3 3 3 6 6 8 |
| 6 6 4 3 2 2 0 | 9 | 3 3 6 |

By comparing the two sets of leaves, you can see that the second group of test scores is better than the first group.

---

**Example 2**    Using a Stem-and-Leaf Plot

The table below shows the percent of the population of each state and the District of Columbia that was at least 65 years old in 2000. Use a stem-and-leaf plot to organize the data.    (Source: U.S. Bureau of the Census)

| | | | | | | | | | |
|---|---|---|---|---|---|---|---|---|---|
| AK | 5.7 | AL | 13.0 | AR | 14.0 | AZ | 13.0 | CA | 10.6 |
| CO | 9.7 | CT | 13.8 | DC | 12.2 | DE | 13.0 | FL | 17.6 |
| GA | 9.6 | HI | 13.3 | IA | 14.9 | ID | 11.3 | IL | 12.1 |
| IN | 12.4 | KS | 13.3 | KY | 12.5 | LA | 11.6 | MA | 13.5 |
| MD | 11.3 | ME | 14.4 | MI | 12.3 | MN | 12.1 | MO | 13.5 |
| MS | 12.1 | MT | 13.4 | NC | 12.0 | ND | 14.7 | NE | 13.6 |
| NH | 12.0 | NJ | 13.2 | NM | 11.7 | NV | 11.0 | NY | 12.9 |
| OH | 13.3 | OK | 13.2 | OR | 12.8 | PA | 15.6 | RI | 14.5 |
| SC | 12.1 | SD | 14.3 | TN | 12.4 | TX | 9.9 | UT | 8.5 |
| VA | 11.2 | VT | 12.7 | WA | 11.2 | WI | 13.1 | WV | 15.3 |
| WY | 11.7 | | | | | | | | |

**Solution**

Begin by ordering the numbers, as shown below.

5.7, 8.5, 9.6, 9.7, 9.9, 10.6, 11.0, 11.2, 11.2, 11.3, 11.3,
11.6, 11.7, 11.7, 12.0, 12.0, 12.1, 12.1, 12.1, 12.1, 12.2,
12.3, 12.4, 12.4, 12.5, 12.7, 12.8, 12.9, 13.0, 13.0, 13.0,
13.1, 13.2, 13.2, 13.3, 13.3, 13.3, 13.4, 13.5, 13.5, 13.6,
13.8, 14.0, 14.3, 14.4, 14.5, 14.7, 14.9, 15.3, 15.6, 17.6

Next construct the stem-and-leaf plot using the leaves to represent the digits to the right of the decimal points.

| Stems | Leaves | |
|---|---|---|
| 5. | 7 | Alaska has the lowest percent. |
| 6. | | |
| 7. | | |
| 8. | 5 | |
| 9. | 6 7 9 | |
| 10. | 6 | |
| 11. | 0 2 2 3 3 6 7 7 | |
| 12. | 0 0 1 1 1 1 2 3 4 4 5 7 8 9 | |
| 13. | 0 0 0 1 2 2 3 3 3 4 5 5 6 8 | |
| 14. | 0 3 4 5 7 9 | |
| 15. | 3 6 | |
| 16. | | |
| 17. | 6 | Florida has the highest percent. |

## Histograms and Frequency Distributions

With data such as those given in Example 2, it is useful to group the numbers into intervals and plot the frequency of the data in each interval. For instance, the **frequency distribution** and **histogram** shown in Figure C.1 represent the data given in Example 2.

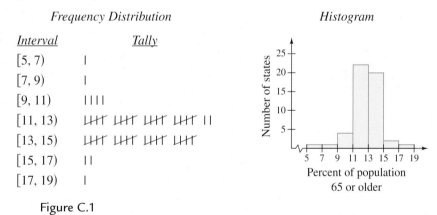

*Frequency Distribution*

| Interval | Tally |
|---|---|
| $[5, 7)$ | I |
| $[7, 9)$ | I |
| $[9, 11)$ | IIII |
| $[11, 13)$ | LHT LHT LHT LHT II |
| $[13, 15)$ | LHT LHT LHT LHT |
| $[15, 17)$ | II |
| $[17, 19)$ | I |

*Histogram*

Figure C.1

A histogram has a portion of a real number line as its horizontal axis. A **bar graph** is similar to a histogram, except that the rectangles (bars) can be either horizontal or vertical and the labels of the bars are not necessarily numbers.

Another difference between a bar graph and a histogram is that the bars in a bar graph are usually separated by spaces, whereas the bars in a histogram are not separated by spaces.

### Example 3    Constructing a Bar Graph

The data below show the average monthly precipitation (in inches) in Houston, Texas. Construct a bar graph for these data. What can you conclude?    (Source: PC USA)

| January | 3.2 | February | 3.3 | March | 2.7 |
|---|---|---|---|---|---|
| April | 4.2 | May | 4.7 | June | 4.1 |
| July | 3.3 | August | 3.7 | September | 4.9 |
| October | 3.7 | November | 3.4 | December | 3.7 |

Solution

To create a bar graph, begin by drawing a vertical axis to represent the precipitation and a horizontal axis to represent the months. The bar graph is shown in Figure C.2. From the graph, you can see that Houston receives a fairly consistent amount of rain throughout the year—the driest month tends to be March and the wettest month tends to be September.

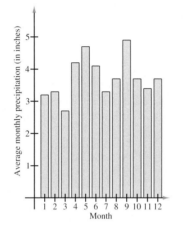

Figure C.2

## Scatter Plots

Many real-life situations involve finding relationships between two variables, such as the year and the amount spent by nonprofit organizations on research and development. In a typical situation, data are collected and written as a set of ordered pairs. The graph of such a set is called a **scatter plot.**

From the scatter plot in Figure C.3, it appears that the points describe a relationship that is nearly linear. (The relationship is not *exactly* linear because the amount spent did not increase by precisely the same amount each year.) A mathematical equation that approximates the relationship between $t$ and $A$ is called a *mathematical model.* When developing a mathematical model, you strive for two (often conflicting) goals—accuracy and simplicity.

Consider a collection of ordered pairs of the form $(x, y)$. If $y$ tends to increase as $x$ increases, the collection is said to have a **positive correlation.** If $y$ tends to decrease as $x$ increases, the collection is said to have a **negative correlation.** Figure C.4, on the next page, shows three examples: one with a positive correlation, one with a negative correlation, and one with no (discernible) correlation.

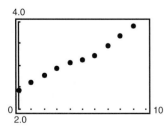

Figure C.3

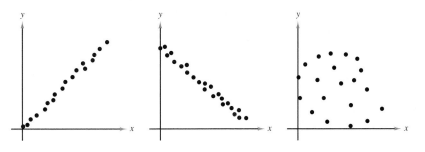

*Positive Correlation*          *Negative Correlation*          *No Correlation*
Figure C.4

## Fitting a Line to Data

Finding a linear model that represents the relationship described by a scatter plot is called **fitting a line to data.** You can do this graphically by simply sketching the line that appears to fit the points, finding two points on the line, and then finding the equation of the line that passes through the two points.

**Example 4**    Fitting a Line to Data

Find a linear model that relates the year with the number of people $P$ (in millions) who were part of the United States labor force from 1989 through 1999. In the table below, $t$ represents the year, with $t = 0$ corresponding to 1989.   (Source: U.S. Bureau of Labor Statistics)

| $t$ | 0 | 1 | 2 | 3 | 4 | 5 | 6 | 7 | 8 | 9 | 10 |
|-----|-----|-----|-----|-----|-----|-----|-----|-----|-----|-----|-----|
| $P$ | 124 | 126 | 126 | 128 | 129 | 131 | 132 | 134 | 136 | 138 | 139 |

Solution

After plotting the data from the table, draw the line that you think best represents the data, as shown in Figure C.5. Two points that lie on this line are $(0, 124)$ and $(10, 139)$. Using the point-slope form, you can find the equation of the line to be

$$P = \frac{15}{10}t + 124 = \frac{3}{2}t + 124.$$    Linear model

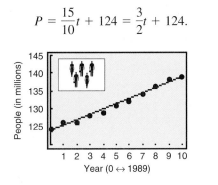

Figure C.5

Once you have found a model, you can measure how well the model fits the data by comparing the actual values with the values given by the model, as shown in the following table.

| t | 0 | 1 | 2 | 3 | 4 | 5 | 6 | 7 | 8 | 9 | 10 |
|---|---|---|---|---|---|---|---|---|---|---|---|
| Actual → P | 124 | 126 | 126 | 128 | 129 | 131 | 132 | 134 | 136 | 138 | 139 |
| Model → P | 124 | 125.5 | 127 | 128.5 | 130 | 131.5 | 133 | 134.5 | 136 | 137.5 | 139 |

The sum of the squares of the differences between the actual values and the model's values is the **sum of the squared differences.** The model that has the least sum is called the **least squares regression line** for the data. For the model in Example 4, the sum of the squared differences is 4.25. The least squares regression line for the data is

$$P = 1.53t + 123.5. \qquad \text{Best-fitting linear model}$$

Its sum of squared differences is approximately 3.07.

▶ **Least Squares Regression Line**

The least squares regression line, $y = ax + b$, for the points $(x_1, y_1)$, $(x_2, y_2)$, $(x_3, y_3)$, . . . , is given by

$$a = \frac{n \sum_{i=1}^{n} x_i y_i - \sum_{i=1}^{n} x_i \sum_{i=1}^{n} y_i}{n \sum_{i=1}^{n} x_i^2 - \left(\sum_{i=1}^{n} x_i\right)^2} \qquad \text{and} \qquad b = \frac{1}{n}\left(\sum_{i=1}^{n} y_i - a \sum_{i=1}^{n} x_i\right).$$

**Example 5** Finding a Least Squares Regression Line

Find the least squares regression line for the points $(-3, 0)$, $(-1, 1)$, $(0, 2)$, and $(2, 3)$.

Solution

Begin by constructing a table like the one shown below.

| $x$ | $y$ | $xy$ | $x^2$ |
|---|---|---|---|
| $-3$ | 0 | 0 | 9 |
| $-1$ | 1 | $-1$ | 1 |
| 0 | 2 | 0 | 0 |
| 2 | 3 | 6 | 4 |
| $\sum_{i=1}^{n} x_i = -2$ | $\sum_{i=1}^{n} y_i = 6$ | $\sum_{i=1}^{n} x_i y_i = 5$ | $\sum_{i=1}^{n} x_i^2 = 14$ |

Applying the formulas for the least squares regression line with $n = 4$ produces

$$a = \frac{n\sum\limits_{i=1}^{n} x_i y_i - \sum\limits_{i=1}^{n} x_i \sum\limits_{i=1}^{n} y_i}{n\sum\limits_{i=1}^{n} x_i^2 - \left(\sum\limits_{i=1}^{n} x_i\right)^2} = \frac{4(5) - (-2)(6)}{4(14) - (-2)^2} = \frac{8}{13}$$

and

$$b = \frac{1}{n}\left(\sum\limits_{i=1}^{n} y_i - a\sum\limits_{i=1}^{n} x_i\right) = \frac{1}{4}\left[6 - \frac{8}{13}(-2)\right] = \frac{47}{26}.$$

Thus, the least squares regression line is $y = \frac{8}{13}x + \frac{47}{26}$. This line is shown in Figure C.6.

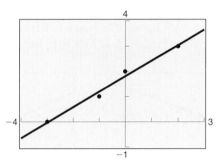

Figure C.6

Many graphing utilities have "built-in" least squares regression programs. If your calculator has such a program, try using it to duplicate the results shown in the following example.

**Example 6**    Finding a Least Squares Regression Line

The following ordered pairs $(w, h)$ represent the shoe sizes $w$ and the heights $h$ (in inches) of 25 men. Use a computer program or a statistical calculator to find the least squares regression line for these data.

(10.0, 70.0), (10.5, 71.0), (9.5, 70.0), (11.0, 72.0), (12.0, 74.0),
(8.5, 66.0), (9.0, 68.5), (13.0, 76.0), (10.5, 71.5), (10.5, 70.5),
(10.0, 72.0), (9.5, 70.0), (10.0, 71.0), (10.5, 69.5), (11.0, 71.5),
(12.0, 73.5), (12.5, 74.0), (11.0, 71.5), (9.0, 67.5), (10.0, 70.0),
(13.0, 73.5), (10.5, 72.5), (10.5, 71.0), (11.0, 73.0), (8.5, 68.0)

Solution

A scatter plot for the data is shown in Figure C.7 on the following page. Note that the plot does not have 25 points because some of the ordered pairs graph as the same point. After entering the data into a statistical calculator, you can obtain

$$a = 1.67 \quad \text{and} \quad b = 53.57.$$

Thus, the least squares regression line for the data is

$$h = 1.67w + 53.57.$$

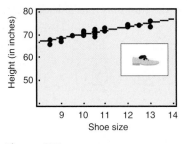

Figure C.7

If you use a statistical calculator or graphing utility to duplicate the results of Example 6, you will notice that the program also outputs a value of $r \approx 0.918$. This number is called the **correlation coefficient** of the data. Correlation coefficients vary between $-1$ and 1. Basically, the closer $|r|$ is to 1, the better the points can be described by a line. Three examples are shown in Figure C.8.

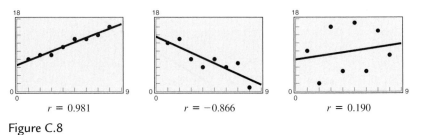

Figure C.8

## C.1    Exercises

*Exam Scores*    In Exercises 1 and 2, use the following scores from a math class with 30 students. The scores are given for two 100-point exams.

*Exam #1:*    77, 100, 77, 70, 83, 89, 87, 85, 81, 84, 81, 78, 89, 78, 88, 85, 90, 92, 75, 81, 85, 100, 98, 81, 78, 75, 85, 89, 82, 75

*Exam #2:*    76, 78, 73, 59, 70, 81, 71, 66, 66, 73, 68, 67, 63, 67, 77, 84, 87, 71, 78, 78, 90, 80, 77, 70, 80, 64, 74, 68, 68, 68

1. Construct a stem-and-leaf plot for exam #1.

2. Construct a double stem-and-leaf plot to compare the scores for exam #1 and exam #2. Which set of test scores is higher?

3. *Insurance Coverage*    The following table shows the total number of persons (in thousands) without health insurance coverage in the 50 states and the District of Columbia in 1998. Use a stem-and-leaf plot to organize the data. (Source: U.S. Bureau of the Census)

| AK | 112 | AL | 714 | AR | 478 | AZ | 1187 | CA | 7373 |
|----|-----|----|-----|----|-----|----|------|----|------|
| CO | 599 | CT | 412 | DC | 87 | DE | 115 | FL | 2564 |
| GA | 1341 | HI | 121 | IA | 265 | ID | 225 | IL | 1842 |
| IN | 839 | KS | 270 | KY | 545 | LA | 817 | MA | 627 |
| MD | 837 | ME | 161 | MI | 1328 | MN | 448 | MO | 570 |
| MS | 554 | MT | 181 | NC | 1111 | ND | 92 | NE | 155 |
| NH | 138 | NJ | 1329 | NM | 386 | NV | 394 | NY | 3177 |
| OH | 1169 | OK | 599 | OR | 481 | PA | 1248 | RI | 96 |
| SC | 594 | SD | 102 | TN | 724 | TX | 4880 | UT | 293 |
| VA | 946 | VT | 58 | WA | 706 | WI | 604 | WV | 302 |
| WY | 82 | | | | | | | | |

Table for 3

4. *Snowfall*    The data below show the seasonal snowfall (in inches) at Lincoln, Nebraska for the years 1968 through 2000 (the amounts are listed in order by year). How would you organize these data? Explain your reasoning. (Source: University of Nebraska-Lincoln)

39.8, 26.2, 49.0, 21.6, 29.2, 33.6, 42.1, 21.1, 21.8, 31.0, 34.4, 23.3, 13.0, 32.3, 38.0, 47.5, 21.5, 18.9, 15.7, 13.0, 19.1, 18.7, 25.8, 23.8, 32.1, 21.3, 21.8, 30.7, 29.0, 44.6, 24.4, 11.9, 37.9

**5.** *Vegetable Crops*   The data below show the cash receipts (in millions of dollars) from vegetable crops of farmers in 1998. Construct a bar graph for these data.   (Source: U.S. Department of Agriculture)

| | | | |
|---|---|---|---|
| Dry beans | 594 | Sweet corn | 686 |
| Potatoes | 2455 | Lettuce | 1577 |
| Broccoli | 554 | Onions | 889 |
| Green peppers | 483 | Tomatoes | 1640 |
| Carrots | 494 | | |

**6.** *Travel to the United States*   The data below give the places of origin and the numbers of travelers (in millions) to the United States in 1998. Construct a bar graph for these data. (Source: U.S. Department of Commerce)

| | | | |
|---|---|---|---|
| Canada | 13.4 | Mexico | 9.3 |
| Europe | 10.7 | Far East | 6.7 |
| Other | 6.3 | | |

*Crop Yield*   In Exercises 7–10, use the data in the table, where $x$ is the number of units of fertilizer applied to sample plots and $y$ is the yield (in bushels) of a crop.

| $x$ | 0 | 1 | 2 | 3 | 4 | 5 | 6 | 7 | 8 |
|---|---|---|---|---|---|---|---|---|---|
| $y$ | 58 | 60 | 59 | 61 | 63 | 66 | 65 | 67 | 70 |

**7.** Sketch a scatter plot of the data.

**8.** Determine whether the points are positively correlated, are negatively correlated, or have no discernible correlation.

**9.** Sketch a linear model that you think best represents the data. Find an equation of the line you sketched. Use the line to predict the yield if 10 units of fertilizer are used.

**10.** Can the model found in Exercise 9 be used to predict yields for arbitrarily large values of $x$? Explain.

*Speed of Sound*   In Exercises 11–14, use the data in the table, where $h$ is altitude in thousands of feet and $v$ is the speed of sound in feet per second.

| $h$ | 0 | 5 | 10 | 15 | 20 | 25 | 30 | 35 |
|---|---|---|---|---|---|---|---|---|
| $v$ | 1116 | 1097 | 1077 | 1057 | 1036 | 1015 | 995 | 973 |

**11.** Sketch a scatter plot of the data.

**12.** Determine whether the points are positively correlated, are negatively correlated, or have no discernible correlation.

**13.** Sketch a linear model that you think best represents the data. Find an equation of the line you sketched. Use the line to predict the speed of sound at an altitude of 27,000 feet.

**14.** The speed of sound at an altitude of 70,000 feet is approximately 971 feet per second. What does this suggest about the validity of using the model in Exercise 13 to extrapolate beyond the data given in the table?

In Exercises 15 and 16, (a) sketch a scatter plot of the points, (b) find an equation of the linear model you think best represents the data and find the sum of the squared differences, and (c) use the formulas in this section to find the least squares regression line for the data and the sum of the squared differences.

**15.** $(-1, 0), (0, 1), (1, 3), (2, 3)$

**16.** $(0, 4), (1, 3), (2, 2), (4, 1)$

In Exercises 17–20, sketch a scatter plot of the points, use the formulas in this section to find the least squares regression line for the data, and sketch the graph of the line.

**17.** $(-2, 0), (-1, 1), (0, 1), (2, 2)$

**18.** $(-3, 1), (-1, 2), (0, 2), (1, 3), (3, 5)$

**19.** $(1, 5), (2, 8), (3, 13), (4, 16), (5, 22), (6, 26)$

**20.** $(1, 10), (2, 8), (3, 8), (4, 6), (5, 5), (6, 3)$

In Exercises 21–24, use a graphing utility to find the least squares regression line for the data. Sketch a scatter plot and the regression line.

**21.** $(0, 23), (1, 20), (2, 19), (3, 17), (4, 15), (5, 11), (6, 10)$

**22.** $(4, 52.8), (5, 54.7), (6, 55.7), (7, 57.8), (8, 60.2), (9, 63.1), (10, 66.5)$

**23.** $(-10, 5.1), (-5, 9.8), (0, 17.5), (2, 25.4), (4, 32.8), (6, 38.7), (8, 44.2), (10, 50.5)$

**24.** $(-10, 213.5), (-5, 174.9), (0, 141.7), (5, 119.7), (8, 102.4), (10, 87.6)$

**25.** *Advertising*  The management of a department store ran an experiment to determine if a relationship existed between sales $S$ (in thousands of dollars) and the amount spent on advertising $x$ (in thousands of dollars). The following data were collected.

| $x$ | 1 | 2 | 3 | 4 | 5 | 6 | 7 | 8 |
|---|---|---|---|---|---|---|---|---|
| $S$ | 405 | 423 | 455 | 466 | 492 | 510 | 525 | 559 |

(a) Use a graphing utility to find the least squares regression line for the data. Use the equation to estimate sales if $4500 is spent on advertising.

(b) Make a scatter plot of the data and sketch the graph of the regression line.

(c) Use a computer or calculator to determine the correlation coefficient.

**26.** *Manufacturers' Shipments*  The table gives the values $y$ (in trillions of dollars) of manufacturers' shipments in the United States from 1995 through 1999, where $t = 0$ corresponds to 1995.  (Source: U.S. Bureau of the Census)

| $t$ | 0 | 1 | 2 | 3 | 4 |
|---|---|---|---|---|---|
| $y$ | 3.6 | 3.7 | 3.9 | 4.1 | 4.3 |

(a) Use a graphing utility to find the least squares regression line for the data. Use the equation to estimate the value of manufacturers' shipments in 2000.

(b) Make a scatter plot of the data and sketch the graph of the regression line.

(c) Use a graphing utility or computer to determine the correlation coefficient.

## C.2    Measures of Central Tendency and Dispersion

Mean, Median, and Mode  ■  Choosing a Measure of Central Tendency  ■  Variance and Standard Deviation

## Mean, Median, and Mode

In many real-life situations, it is helpful to describe data by a single number that is most representative of the entire collection of numbers. Such a number is called a **measure of central tendency.** The most commonly used measures are as follows.

1. The **mean,** or **average,** of $n$ numbers is the sum of the numbers divided by $n$.

2. The **median** of $n$ numbers is the middle number when the numbers are written in order. If $n$ is even, the median is the average of the two middle numbers.

3. The **mode** of $n$ numbers is the number that occurs most frequently. If two numbers tie for most frequent occurrence, the collection has two modes and is called **bimodal.**

---

**Example 1**    Comparing Measures of Central Tendency

You are interviewing for a job. The interviewer tells you that the average income of the company's 25 employees is $60,849. The actual annual incomes of the 25 employees are shown below. What are the mean, median, and mode of the incomes? Was the person telling you the truth?

| | | | | |
|---|---|---|---|---|
| $17,305, | $478,320, | $45,678, | $18,980, | $17,408, |
| $25,676, | $28,906, | $12,500, | $24,540, | $33,450, |
| $12,500, | $33,855, | $37,450, | $20,432, | $28,956, |
| $34,983, | $36,540, | $250,921, | $36,853, | $16,430, |
| $32,654, | $98,213, | $48,980, | $94,024, | $35,671 |

Solution

The mean of the incomes is

$$\text{Mean} = \frac{17{,}305 + 478{,}320 + 45{,}678 + 18{,}980 + \cdots + 35{,}671}{25}$$

$$= \frac{1{,}521{,}225}{25} = \$60{,}849.$$

To find the median, order the incomes as follows.

| | | | | |
|---|---|---|---|---|
| $12,500, | $12,500, | $16,430, | $17,305, | $17,408, |
| $18,980, | $20,432, | $24,540, | $25,676, | $28,906, |
| $28,956, | $32,654, | $33,450, | $33,855, | $34,983, |
| $35,671, | $36,540, | $36,853, | $37,450, | $45,678, |
| $48,980, | $94,024, | $98,213, | $250,921, | $478,320 |

From this list, you can see the median (the middle number) is $33,450. From the same list, you can see that $12,500 is the only income that occurs more than once. Thus, the mode is $12,500. Technically, the person was telling the truth because the average is (generally) defined to be the mean. However, of the three measures of central tendency

*Mean:* $60,849     *Median:* $33,450     *Mode:* $12,500

it seems clear that the median is most representative. The mean is inflated by the two highest salaries.

---

## Choosing a Measure of Central Tendency

Which of the three measures of central tendency is the most representative? The answer is that it depends on the distribution of the data *and* the way in which you plan to use the data.

For instance, in Example 1, the mean salary of $60,849 does not seem very representative to a potential employee. To a city income tax collector who wants to estimate 1% of the total income of the 25 employees, however, the mean is precisely the right measure.

**Example 2**   Choosing a Measure of Central Tendency

Which measure of central tendency is the most representative of the data given in each of the following frequency distributions?

| a. *Number* | *Tally* | b. *Number* | *Tally* | c. *Number* | *Tally* |
|---|---|---|---|---|---|
| 1 | 7 | 1 | 9 | 1 | 6 |
| 2 | 20 | 2 | 8 | 2 | 1 |
| 3 | 15 | 3 | 7 | 3 | 2 |
| 4 | 11 | 4 | 6 | 4 | 3 |
| 5 | 8 | 5 | 5 | 5 | 5 |
| 6 | 3 | 6 | 6 | 6 | 5 |
| 7 | 2 | 7 | 7 | 7 | 4 |
| 8 | 0 | 8 | 8 | 8 | 3 |
| 9 | 15 | 9 | 9 | 9 | 0 |

Solution

a. For these data, the mean is 4.23, the median is 3, and the mode is 2. Of these, the mode is probably the most representative.

b. For these data, the mean and median are each 5 and the modes are 1 and 9 (the distribution is bimodal). Of these, the mean or median is the most representative.

c. For these data, the mean is 4.59, the median is 5, and the mode is 1. Of these, the mean or median is the most representative.

---

# Variance and Standard Deviation

Very different sets of numbers can have the same mean. You will now study two **measures of dispersion,** which give you an idea of how much the numbers in the set differ from the mean of the set. These two measures are called the *variance* of the set and the *standard deviation* of the set.

▶ Definitions of Variance and Standard Deviation

Consider a set of numbers $\{x_1, x_2, \ldots, x_n\}$ with a mean of $\bar{x}$. The **variance** of the set is

$$v = \frac{(x_1 - \bar{x})^2 + (x_2 - \bar{x})^2 + \cdots + (x_n - \bar{x})^2}{n}$$

and the **standard deviation** of the set is

$$\sigma = \sqrt{v}$$

($\sigma$ is the lowercase Greek letter *sigma*).

The standard deviation of a set is a measure of how much a typical number in the set differs from the mean. The greater the standard deviation, the more the numbers in the set *vary* from the mean. For instance, each of the following sets has a mean of 5.

$$\{5, 5, 5, 5\}, \qquad \{4, 4, 6, 6\}, \qquad \text{and} \qquad \{3, 3, 7, 7\}$$

The standard deviations of the sets are 0, 1, and 2.

$$\sigma_1 = \sqrt{\frac{(5 - 5)^2 + (5 - 5)^2 + (5 - 5)^2 + (5 - 5)^2}{4}} = 0$$

$$\sigma_2 = \sqrt{\frac{(4 - 5)^2 + (4 - 5)^2 + (6 - 5)^2 + (6 - 5)^2}{4}} = 1$$

$$\sigma_3 = \sqrt{\frac{(3 - 5)^2 + (3 - 5)^2 + (7 - 5)^2 + (7 - 5)^2}{4}} = 2$$

**Example 3**    Estimations of Standard Deviation

Consider the three sets of data represented by the bar graphs in Figure C.9. Which set has the smallest standard deviation? Which has the largest?

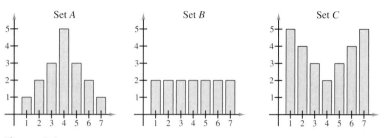

**Figure C.9**

Solution

Of the three sets, the numbers in set $A$ are grouped most closely to the center and the numbers in set $C$ are the most dispersed. Thus, set $A$ has the smallest standard deviation and set $C$ has the largest standard deviation.

**Example 4**    Finding Standard Deviation

Find the standard deviation of each set shown in Example 3.

Solution

Because of the symmetry of each bar graph, you can conclude that each has a mean of $\bar{x} = 4$. The standard deviation of set $A$ is

$$\sigma = \sqrt{\frac{(-3)^2 + 2(-2)^2 + 3(-1)^2 + 5(0)^2 + 3(1)^2 + 2(2)^2 + (3)^2}{17}}$$

$$\approx 1.53.$$

The standard deviation of set $B$ is

$$\sigma = \sqrt{\frac{2(-3)^2 + 2(-2)^2 + 2(-1)^2 + 2(0)^2 + 2(1)^2 + 2(2)^2 + 2(3)^2}{14}}$$

$$= 2.$$

The standard deviation of set $C$ is

$$\sigma = \sqrt{\frac{5(-3)^2 + 4(-2)^2 + 3(-1)^2 + 2(0)^2 + 3(1)^2 + 4(2)^2 + 5(3)^2}{26}}$$

$$\approx 2.22.$$

These values confirm the results of Example 3. That is, set $A$ has the smallest standard deviation and set $C$ has the largest.

The following alternative formula provides a more efficient way to compute the standard deviation.

> ▶ **Alternative Formula for Standard Deviation**
>
> The standard deviation of $\{x_1, x_2, \ldots, x_n\}$ is
>
> $$\sigma = \sqrt{\frac{x_1^2 + x_2^2 + \cdots + x_n^2}{n} - \bar{x}^2}.$$

Because of messy computations, this formula is difficult to verify. Conceptually, however, the process is straightforward. It consists of showing that the expressions

$$\sqrt{\frac{(x_1 - \bar{x})^2 + (x_2 - \bar{x})^2 + \cdots + (x_n - \bar{x})^2}{n}}$$

and

$$\sqrt{\frac{x_1^2 + x_2^2 + \cdots + x_n^2}{n} - \bar{x}^2}$$

are equivalent. Try verifying this equivalence for the set $\{x_1, x_2, x_3\}$ with $\bar{x} = (x_1 + x_2 + x_3)/3$.

**Example 5** Using the Alternative Formula

Use the alternative formula for standard deviation to find the standard deviation of the following set of numbers.

5, 6, 6, 7, 7, 8, 8, 8, 9, 10

Solution

Begin by finding the mean of the set, which is 7.4. Thus, the standard deviation is

$$\sigma = \sqrt{\frac{5^2 + 2(6^2) + 2(7^2) + 3(8^2) + 9^2 + 10^2}{10} - (7.4)^2}$$

$$= \sqrt{\frac{568}{10} - 54.76}$$

$$= \sqrt{2.04}$$

$$\approx 1.43.$$

You can use the statistical features of a graphing utility to check this result.

A well-known theorem in statistics, called *Chebychev's Theorem*, states that at least

$$1 - \frac{1}{k^2}$$

of the numbers in a distribution must lie within $k$ standard deviations of the mean. So, 75% of the numbers in a collection must lie within two standard deviations of the mean, and at least 88.9% of the numbers must lie within three standard deviations of the mean. For most distributions, these percentages are low. For instance, in all three distributions shown in Example 3, 100% of the numbers lie within two standard deviations of the mean.

### Example 6    Describing a Distribution

The following table shows the number of dentists (per 100,000 people) in each state and the District of Columbia. Find the mean and standard deviation of the numbers. What percent of the numbers lie within two standard deviations of the mean?    (Source: American Dental Association)

| AK | 66 | AL | 40 | AR | 39 | AZ | 51 | CA | 62 |
|----|----|----|----|----|----|----|----|----|----|
| CO | 69 | CT | 80 | DC | 94 | DE | 44 | FL | 50 |
| GA | 46 | HI | 80 | IA | 55 | ID | 53 | IL | 61 |
| IN | 47 | KS | 51 | KY | 53 | LA | 45 | MA | 74 |
| MD | 68 | ME | 47 | MI | 62 | MN | 67 | MO | 53 |
| MS | 37 | MT | 62 | NC | 42 | ND | 47 | NE | 63 |
| NH | 59 | NJ | 77 | NM | 45 | NV | 49 | NY | 73 |
| OH | 55 | OK | 47 | OR | 70 | PA | 61 | RI | 56 |
| SC | 41 | SD | 49 | TN | 53 | TX | 47 | UT | 66 |
| VA | 54 | VT | 57 | WA | 68 | WI | 65 | WV | 43 |
| WY | 52 |    |    |    |    |    |    |    |    |

#### Solution

Begin by entering the numbers into a graphing utility that has a standard deviation program. After running the program, you should obtain

$$\bar{x} \approx 56.76 \quad \text{and} \quad \sigma = 12.14.$$

The interval that contains all numbers that lie within two standard deviations of the mean is

$$[56.76 - 2(12.14), 56.76 + 2(12.14)] \quad \text{or} \quad [32.48, 81.04].$$

From the histogram in Figure C.10, you can see that all but one of the numbers (98%) lie in this interval—all but the number that corresponds to the number of dentists (per 100,000 people) in Washington, DC.

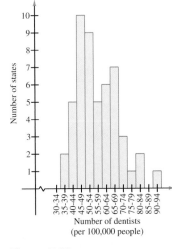

Number of dentists
(per 100,000 people)

Figure C.10

## C.2 Exercises

In Exercises 1–6, find the mean, median, and mode of the set of measurements.

1. 5, 12, 7, 14, 8, 9, 7

2. 30, 37, 32, 39, 33, 34, 32

3. 5, 12, 7, 24, 8, 9, 7

4. 20, 37, 32, 39, 33, 34, 32

5. 5, 12, 7, 14, 9, 7

6. 30, 37, 32, 39, 34, 32

7. *Reasoning* Compare your answers for Exercises 1 and 3 with those for Exercises 2 and 4. Which of the measures of central tendency is sensitive to extreme measurements? Explain your reasoning.

8. *Reasoning*

(a) Add 6 to each measurement in Exercise 1 and calculate the mean, median, and mode of the revised measurements. How are the measures of central tendency changed?

(b) If a constant $k$ is added to each measurement in a set of data, how will the measures of central tendency change?

9. *Electric Bills* A person had the following monthly bills for electricity. What are the mean and median of the collection of bills?

| | | | |
|---|---|---|---|
| January | $67.92 | February | $59.84 |
| March | $52.00 | April | $52.50 |
| May | $57.99 | June | $65.35 |
| July | $81.76 | August | $74.98 |
| September | $87.82 | October | $83.18 |
| November | $65.35 | December | $57.00 |

10. *Car Rental* A car rental company kept the following record of the numbers of miles driven by a car that was rented. What are the mean, median, and mode of these data?

| | | | |
|---|---|---|---|
| Monday | 410 | Tuesday | 260 |
| Wednesday | 320 | Thursday | 320 |
| Friday | 460 | Saturday | 150 |

11. *Six-Child Families* A study was done on families having six children. The table gives the number of families in the study with the indicated number of girls. Determine the mean, median, and mode of this set of data.

| Number of girls | 0 | 1 | 2 | 3 | 4 | 5 | 6 |
|---|---|---|---|---|---|---|---|
| Frequency | 1 | 24 | 45 | 54 | 50 | 19 | 7 |

12. *Baseball* A baseball fan examined the records of a favorite baseball player's performance during his last 50 games. The number of games in which the player had 0, 1, 2, 3, and 4 hits are recorded in the table.

| Number of hits | 0 | 1 | 2 | 3 | 4 |
|---|---|---|---|---|---|
| Frequency | 14 | 26 | 7 | 2 | 1 |

(a) Determine the average number of hits per game.

(b) Determine the player's batting average if he had 200 at bats during the 50-game series.

13. *Think About It* Construct a collection of numbers that has the following properties. If this is not possible, explain why it is not.

Mean = 6, median = 4, mode = 4

14. *Think About It* Construct a collection of numbers that has the following properties. If this is not possible, explain why it is not.

Mean = 6, median = 6, mode = 4

15. *Test Scores* A professor records the following scores for a 100-point exam.

99, 64, 80, 77, 59, 72, 87, 79, 92, 88,
90, 42, 20, 89, 42, 100, 98, 84, 78, 91

Which measure of central tendency best describes these test scores?

16. *Shoe Sales* A salesman sold eight pairs of a certain style of men's shoes. The sizes of the eight pairs were as follows: $10\frac{1}{2}$, 8, 12, $10\frac{1}{2}$, 10, $9\frac{1}{2}$, 11 and $10\frac{1}{2}$. Which measure (or measures) of central tendency best describes the typical shoe size for these data?

In Exercises 17–24, find the mean ($\bar{x}$), variance ($v$), and standard deviation ($\sigma$) of the numbers.

**17.** 4, 10, 8, 2

**18.** 3, 15, 6, 9, 2

**19.** 0, 1, 1, 2, 2, 2, 3, 3, 4

**20.** 2, 2, 2, 2, 2, 2

**21.** 1, 2, 3, 4, 5, 6, 7

**22.** 1, 1, 1, 5, 5, 5

**23.** 49, 62, 40, 29, 32, 70

**24.** 1.5, 0.4, 2.1, 0.7, 0.8

In Exercises 25–30, use the alternative formula to find the standard deviation of the numbers.

**25.** 2, 4, 6, 6, 13, 5

**26.** 10, 25, 50, 26, 15, 33, 29, 4

**27.** 246, 336, 473, 167, 219, 359

**28.** 6.0, 9.1, 4.4, 8.7, 10.4

**29.** 8.1, 6.9, 3.7, 4.2, 6.1

**30.** 9.0, 7.5, 3.3, 7.4, 6.0

**31.** *Reasoning*  Without calculating the standard deviation, explain why the set {4, 4, 20, 20} has a standard deviation of 8.

**32.** *Reasoning*  If the standard deviation of a set of numbers is 0, what does this imply about the set?

**33.** *Test Scores*  An instructor adds five points to each student's exam score. Will this change the mean or standard deviation of the exam scores? Explain.

**34.** *Think About It*  Consider the four sets of data represented by the histograms. Order the sets from the smallest to the largest variance.

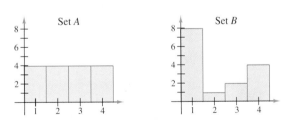

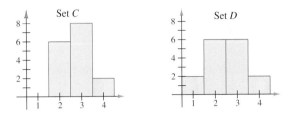

**35.** *Test Scores*  The scores of a mathematics exam given to 600 science and engineering students at a college had a mean and standard deviation of 235 and 28, respectively. Use Chebychev's Theorem to determine the intervals containing at least $\frac{3}{4}$ and at least $\frac{8}{9}$ of the scores. How would the intervals change if the standard deviation were 16?

**36.** *Price of Gold*  The following data represent the average prices of gold (in dollars per fine ounce) for the years 1980 to 1999. Use a computer or calculator to find the mean, variance, and standard deviation of the data. What percent of the data lie within two standard deviations of the mean?  (Source: U.S. Bureau of Mines)

| | | | | |
|---|---|---|---|---|
| 613, | 460, | 376, | 424, | 361, |
| 318, | 368, | 478, | 438, | 383, |
| 385, | 363, | 345, | 361, | 385, |
| 386, | 389, | 332, | 295, | 285 |

**37.** *Price of Copper*  The following data represent the average prices of copper (in cents per pound) for the years 1980 to 1999. Use a computer or calculator to find the mean, variance, and standard deviation of the data. What percent of the data lie within one standard deviation of the mean?  (Source: U.S. Bureau of Mines)

| | | | | |
|---|---|---|---|---|
| 101, | 84, | 73, | 77, | 67, |
| 67, | 66, | 83, | 121, | 131, |
| 123, | 109, | 107, | 92, | 111, |
| 138, | 109, | 107, | 79, | 76 |

# Answers to Warm Ups, Odd-Numbered Exercises, Quizzes, and Tests

## Chapter 1

### Section 1.1   *(page 8)*

---

**Warm Up**   *(page 8)*

1. $-3x - 10$    2. $5x - 12$    3. $x$    4. $x + 26$

5. $\dfrac{8x}{15}$    6. $\dfrac{3x}{4}$    7. $-\dfrac{1}{x(x+1)}$    8. $\dfrac{5}{x}$

9. $\dfrac{7x - 8}{x(x - 2)}$    10. $-\dfrac{2}{x^2 - 1}$

---

1. Identity    3. Conditional equation

5. Conditional equation

7. (a) No    9. (a) Yes    11. (a) Yes    13. (a) No

(b) No    (b) Yes    (b) No    (b) No

(c) Yes    (c) No    (c) No    (c) No

(d) No    (d) No    (d) No    (d) Yes

15. 5    17. $-4$    19. 3    21. 9    23. $-26$

25. $-4$    27. $-\frac{6}{5}$    29. 9    31. No solution

33. 10    35. 4    37. 3    39. 5    41. No solution

43. $\frac{11}{6}$    45. No solution    47. 0

49. All real numbers    51. No solution

53. Extraneous solutions may arise when a fractional expression is multiplied by factors involving the variable.

55. $x \approx 138.889$    57. $x \approx 62.372$    59. $x \approx -2.386$

61. Use the table feature in the ASK mode or evaluate using the scientific calculator part of a graphing utility.

63. (a) $\approx 6.46$    65. (a) $\approx 1.06$    67. (a) $\approx 56.09$

(b) $\approx 6.41$; yes    (b) $\approx 1.06$; no    (b) $\approx 56.13$; yes

69. 2003; $t \approx 12.997$    71. 61.2 inches

73. 1998; $x \approx 7.83$    75. 0.46

### Section 1.2   *(page 19)*

---

**Warm Up**   *(page 19)*

1. 14    2. 4    3. $-3$    4. 4    5. $-2$

6. 1    7. $\frac{2}{5}$    8. $\frac{10}{3}$    9. 6    10. $-\frac{11}{5}$

---

1. $x + (x + 1) = 2x + 1$    3. $50t$    5. $0.2x$    7. $6x$

9. $1200 + 25x$    11. $525 = n + (n + 1)$; 262, 263

13. $5x - x = 148$; 37, 185

15. $n^2 - 5 = n(n + 1)$; $-5, -4$

17. Coworker's check: $300
Your check: $345

19. Coworker's check: $348.65
Your check: $296.35

21. $\approx 96\%$ decrease $(-96.4\%)$    23. 19,300% increase

25. (a) 1995: 7.28 hours
(b) 1999: 7.43 hours

27. Guaranteed investment: 540    29. 0.63
Money fund: 333
Diversified stock: 311
Aggressive stock: 303
Company stock: 296

31. 15 feet $\times$ 22.5 feet    33. $\approx 5.7$ years (5.714 years)

35. 97 or greater    37. $\approx \$22,316.98$    39. $\approx \$1016.77$

41. 38%    43. $361.25    45. 3 hours    47. $\frac{1}{3}$ hour

49. Family 1 (42 miles per hour): 3.8 hours
Family 2 (50 miles per hour): 3.2 hours

51. $\approx 1.29$ seconds    53. 62.5 feet    55. $563,952

57. $4500 at 10.5%    59. 11.43%
$7500 at 13%

61. 8823 units    63. $\approx 48$ feet    65. $\approx 32.1$ gallons

67. A mathematical model should be accurate and reasonable to use. Hence, the major factors are accuracy and ease of use. If the model is very complicated, the users may make errors or choose not to use the model. So sometimes a reasonably accurate model that is easy to use is the best model.

69. $\approx 10.894$ miles per hour    71. $l = \dfrac{P - 2w}{2}$

73. $h = \dfrac{V}{\pi r^2}$    75. $L = \dfrac{S}{1 - R}$    77. $P = \dfrac{A}{\left(1 + \dfrac{r}{n}\right)^{nt}}$

79. $\theta = \dfrac{360A}{\pi r^2}$    81. $r = \dfrac{S - a}{S - L}$

83. $r = \sqrt{\dfrac{A}{4\pi}}$ or $r = \dfrac{\sqrt{A\pi}}{2\pi}$

## Section 1.3 *(page 32)*

**Warm Up** *(page 32)*

**1.** $\dfrac{\sqrt{14}}{10}$ **2.** $4\sqrt{2}$ **3.** $14$ **4.** $\dfrac{\sqrt{10}}{4}$

**5.** $x(3x + 7)$ **6.** $(2x - 5)(2x + 5)$

**7.** $-(x - 7)(x - 15)$ **8.** $(x - 2)(x + 9)$

**9.** $(5x - 1)(2x + 3)$ **10.** $(6x - 1)(x - 12)$

**1.** $2x^2 + 5x - 3 = 0$ **3.** $x^2 - 25x = 0$

**5.** $x^2 - 6x + 7 = 0$ **7.** $2x^2 - 2x + 1 = 0$

**9.** $3x^2 - 60x - 10 = 0$ **11.** $0, -\frac{1}{2}$ **13.** $4, -2$

**15.** $-5$ **17.** $3, -\frac{1}{2}$ **19.** $2, -6$ **21.** $-2, -5$

**23.** $\pm 4$ **25.** $\pm\sqrt{7} \approx \pm 2.65$ **27.** $\pm 2\sqrt{3} \approx \pm 3.46$

**29.** $12 + 3\sqrt{2} \approx 16.24$ **31.** $-2 + 2\sqrt{3} \approx 1.46$

$\quad\;\; 12 - 3\sqrt{2} \approx 7.76$ $\qquad -2 - 2\sqrt{3} \approx -5.46$

**33.** $\pm 5$ **35.** $\pm\dfrac{\sqrt{115}}{5} \approx \pm 2.14$ **37.** $\pm 8$ **39.** $1$

**41.** $\pm\frac{3}{4}$ **43.** $\frac{3}{2}$ **45.** $6, -12$ **47.** $\frac{3}{2}, -\frac{1}{2}$ **49.** $5, -\frac{10}{3}$

**51.** $9, 3$ **53.** $\frac{1}{5}, 1$ **55.** $-1, -5$ **57.** $-\frac{1}{2}$

**59.** Algebra argument:

$\quad (x + 2)^2 = (x + 2)(x + 2)$ Definition of exponent

$\qquad\qquad = x^2 + 2x + 2x + 4$ FOIL

$\qquad\qquad = x^2 + 4x + 4$ Combine like terms.

so $(x + 2)^2 \neq x^2 + 4$.

Graphing utility argument:

(1) Let $y_1 = (x + 2)^2$ and $y_2 = x^2 + 4$. Use the table feature with an example value of $x$ (but not $x = 0$). The table will show that $y_1$ is not the same as $y_2$.

(2) Use the scientific calculator portion to show that if $x = 5$, $(5 + 2)^2 = 49$ and $5^2 + 4 = 29$, so $(x + 2)^2$ is not the same as $x^2 + 4$.

**61.** 34 feet × 48 feet

**63.** Base: $2\sqrt{2}$ feet
Height: $2\sqrt{2}$ feet

**65.** $\dfrac{9\sqrt{13}}{4} \approx 8.11$ seconds **67.** $\approx 1.43$ seconds

**69.** $\approx 3.54$ centimeters **71.** 886 miles

**73.** $\approx 1414$ feet **75.** 50,000 units

**77.** 2005 $(t \approx 15.4)$

**79.** 1987 $(t \approx 18.74)$. The model was a good representation through 2000.

**81.** The model in Exercise 80 is *not* valid for the population in 2050 because it predicts 535,720,250 people (not 404,000,000).

**83.** $\approx 0.4490$

## Section 1.4 *(page 42)*

**Warm Up** *(page 42)*

**1.** $3\sqrt{17}$ **2.** $2\sqrt{3}$ **3.** $4\sqrt{6}$ **4.** $3\sqrt{73}$

**5.** $2, -1$ **6.** $\frac{3}{2}, -3$ **7.** $5, -1$ **8.** $\frac{1}{2}, -7$

**9.** $3, 2$ **10.** $4, -1$

**1.** One real solution **3.** Two real solutions

**5.** No real solutions **7.** Two real solutions

**9.** $\frac{1}{2}, -1$ **11.** $\frac{1}{4}, -\frac{3}{4}$ **13.** $1 \pm \sqrt{3}$

**15.** $-7 \pm \sqrt{5}$ **17.** $-4 \pm 2\sqrt{5}$ **19.** $\dfrac{2}{3} \pm \dfrac{\sqrt{7}}{3}$

**21.** $-\dfrac{1}{3} \pm \dfrac{\sqrt{11}}{6}$ **23.** $-\dfrac{1}{2} \pm \sqrt{2}$ **25.** $\dfrac{2}{7}$

**27.** $2 \pm \dfrac{\sqrt{6}}{2}$ **29.** $6 \pm \sqrt{11}$ **31.** $x \approx 0.976, -0.643$

**33.** $x \approx 0.561, 0.126$ **35.** $x \approx 1.687, -0.488$

**37.** $-11$ **39.** $\pm\sqrt{10}$ **41.** $-\dfrac{3}{2} \pm \dfrac{\sqrt{5}}{2}$ **43.** $-2, 4$

**45.** $\pm 2$ **47.** 50, 50 **49.** 7, 8 or $-8, -7$

**51.** 200 units **53.** 653 units **55.** 9 seats per row

**57.** 14 inches × 14 inches

**59.** Moon: 14.9 seconds
Earth: 2.6 seconds

**61.** Shorter period of time on Earth

**63.** 259 miles, 541 miles

**65.** (a) 1995 $(t \approx 5.14)$

(b) 2005 $(t \approx 14.99)$

**67.** 2000 $(t = 10)$

The model cannot be used to predict the percent of classrooms with Internet access in 2005 because at $t = 15$, the percent exceeds 100%.

**69.** Southbound: $\approx 550$ miles per hour; eastbound: $\approx 600$ miles per hour

**71.** 3761 units or 146,239 units

**73.** In an application, one of the solutions may not be reasonable. In Example 5, the other solution is $t \approx -29.77$. The model is for $t \geq 0$ and not $t < 0$. Additionally, $t \approx -30$ would be about the year 1960, which is not reasonable.

## Mid-Chapter Quiz  *(page 46)*

**1.** $x = 2$    **2.** $x = 6$    **3.** $x = -2$    **4.** No solution

**5.** 180.244    **6.** 431.398

**7.** Use the table feature in ASK mode or the scientific calculator portion of the graphing utility.

**8.** $7.50x + 20,000 = 80,000$; 8000 units

**9.** $250,000 = x(60 - 0.0004x)$; 4290 units ($x = 4289.322$) or 145,711 units ($x = 145,710.678$)

**10.** $\frac{2}{3}, -5$    **11.** $x = \pm\sqrt{5}$; $x \approx \pm2.24$

**12.** $x = -3 \pm \sqrt{17}$; $x \approx -7.12, 1.12$

**13.** $x = \dfrac{-7 \pm \sqrt{73}}{6}$    **14.** $x = -1 \pm \sqrt{6}$

**15.** $x \approx 1.568, -0.068$    **16.** Two real solutions

**17.** No real solutions

**18.** Answers will vary. Use the FOIL method $[(x + 3)^2 = (x + 3)(x + 3) = x^2 + 6x + 9]$; or use the table feature of your graphing utility; or use the scientific calculator portion of your graphing utility to evaluate the solution.

**19.** $\approx 3.95$ seconds    **20.** 6 inches $\times$ 6 inches

## Section 1.5  *(page 55)*

> ### Warm Up  *(page 55)*
>
> **1.** 11    **2.** 20, $-3$    **3.** 5, $-45$    **4.** 0, $-\frac{1}{5}$
>
> **5.** $\frac{2}{3}, -2$    **6.** $\frac{11}{6}, -\frac{5}{2}$    **7.** 1, $-5$    **8.** $\frac{3}{2}, -\frac{5}{2}$
>
> **9.** $\dfrac{3 \pm \sqrt{5}}{2}$    **10.** $2 \pm \sqrt{2}$

**1.** $0, \pm\dfrac{3\sqrt{2}}{2}$    **3.** 3, $-1, 0$    **5.** $\pm3$    **7.** $-3, 0$

**9.** 3, 1, $-1$    **11.** $\pm1$    **13.** $\pm3, \pm1$    **15.** $\pm2$

**17.** $\pm\frac{1}{2}, \pm4$    **19.** 1, $-2$    **21.** 50    **23.** 26

**25.** $-16$    **27.** $\frac{1}{4}$    **29.** 6, 5    **31.** 2, $-5$    **33.** 0

**35.** $-59, 69$    **37.** 1    **39.** $\pm\sqrt{69}$    **41.** $\dfrac{-3 \pm \sqrt{21}}{6}$

**43.** 4, $-5$    **45.** 2, $-\frac{3}{2}$    **47.** $-1$    **49.** 1, $-3$

**51.** 1, $-3$    **53.** 3, $-2$    **55.** $\sqrt{3}, -3$    **57.** 10, $-1$

**59.** The quadratic equation wa before values for $a$, $b$, ... Quadratic Formula. Stan $3x^2 - 7x - 4 = 0$ ($a = $... correct solution is

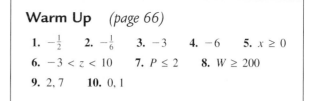

$$x = \frac{-(-7) \pm \sqrt{(-7)^?}}{2(3)}$$

**61.** $x \approx \pm1.038$    **63.** $x \approx 16.756$    **6..**

**67.** 7%    **69.** 19.2%    **71.** 26,250 passengers

**73.** 61 years old    **75.** 2,566,025 units

**77.** $\approx 12.12$ feet    **79.** $t = 13\frac{1}{3}$ minutes

## Section 1.6  *(page 66)*

> ### Warm Up  *(page 66)*
>
> **1.** $-\frac{1}{2}$    **2.** $-\frac{1}{6}$    **3.** $-3$    **4.** $-6$    **5.** $x \geq 0$
>
> **6.** $-3 < z < 10$    **7.** $P \leq 2$    **8.** $W \geq 200$
>
> **9.** 2, 7    **10.** 0, 1

**1.** (a) Yes    (b) No    (c) Yes    (d) No    **3.** c    **4.** h

**5.** f    **6.** e    **7.** g    **8.** a    **9.** b    **10.** d

**11.** $x \geq 3$    **13.** $x < 12$

**15.** $x > -4$    **17.** $x \geq 12$

**19.** $x < -\frac{1}{2}$    **21.** $x \geq \frac{1}{2}$

**23.** $2 \leq x < 4$    **25.** $-1 < x < 3$

**27.** $-\frac{9}{2} < x < \frac{15}{2}$    **29.** $-\frac{3}{4} < x < -\frac{1}{4}$

**31.** $-8 < x < 2$    **33.** $x < -6, x > 6$

**35.** $16 \leq x \leq 24$    **37.** $x < -\frac{1}{2}, x > \frac{11}{2}$

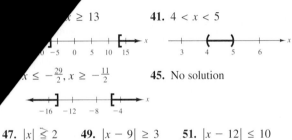

$x \geq 13$      **41.** $4 < x < 5$

$x \leq -\frac{29}{2}, x \geq -\frac{11}{2}$      **45.** No solution

**47.** $|x| \geq 2$      **49.** $|x - 9| \geq 3$      **51.** $|x - 12| \leq 10$

**53.** $|x + 3| > 5$      **55.** $x > 400$ miles      **57.** $r > 12.5\%$

**59.** 24 weeks (or less)

**61.** (a)

| $x$ | 10 | 20 | 30 | 40 | 50 |
|---|---|---|---|---|---|
| $R$ | $1159.50 | $2319.00 | $3478.50 | $4638.00 | $5797.50 |
| $C$ | $1700.00 | $2650.00 | $3600.00 | $4550.00 | $5500.00 |

(b) $x \geq 36$ units

**63.** $m < 24{,}062.5$      **65.** 2003 $(t \geq 13.69)$

**67.** Overcharged about $0.12 or undercharged $0.12

**69.** $[65.8, 71.2]$

---

### Math Matters   *(page 69)*

| Cube | Ratio of $\dfrac{\text{surface area}}{\text{weight}}$ |
|---|---|
| 1 | 6 |
| 2 | 3 |
| 3 | 2 |
| 4 | 1.5 |

---

## Section 1.7   *(page 77)*

### Warm Up   *(page 77)*

**1.** $y < -6$      **2.** $z > -\frac{9}{2}$      **3.** $-3 \leq x < 1$

**4.** $x \leq -5$      **5.** $-3 < x$      **6.** $5 < x < 7$

**7.** $-\frac{7}{2} \leq x \leq \frac{7}{2}$      **8.** $x < 2, x > 4$

**9.** $x < -6, x > -2$      **10.** $-2 \leq x \leq 6$

**1.** $-3 \leq x \leq 3$      **3.** $x < -2, x > 2$

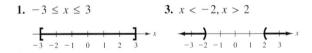

**5.** $-7 < x < 3$      **7.** $x \leq -5, x \geq 1$

**9.** $-3 < x < 2$      **11.** $x < -1, x > 1$

**13.** $-3 < x < 1$      **15.** $x < 0, 0 < x < \frac{3}{2}$

**17.** $-2 \leq x \leq 0, x \geq 2$      **19.** $x < -1, 0 < x < 1$

**21.** $x < -1, x > 4$      **23.** $5 < x < 15$

**25.** $-5 < x < -\frac{3}{2}, x > -1$      **27.** $-\frac{3}{4} < x < 3, x \geq 6$

**29.** $[-2, 2]$      **31.** $(-\infty, 3], [4, \infty)$      **33.** $[-4, 3]$

**35.** All real numbers

**37.** The cube root of any real number is a real number.

**39.** $-3.51 < x < 3.51$      **41.** $-0.13 < x < 25.13$

**43.** $2.26 < x < 2.39$      **45.** $4 < t < 6$ seconds

**47.** 13.8 meters $\leq l \leq$ 36.2 meters

**49.** (a) $90{,}000 \leq x < 100{,}000$      (b) $30 \leq p \leq 32$

**51.** 9.5%      **53.** 2005 $(t > 15.66)$

**55.** 2009 $(t > 19)$

---

## Review Exercises   *(page 82)*

**1.** Conditional equation

**3.** (a) No   (b) Yes   (c) Yes   (d) No

**5.** $-\frac{1}{2}$      **7.** $-10$      **9.** $-\frac{2}{3}$      **11.** $\approx 377.778$      **13.** 12

**15.** May: $121,833.51      **17.** 29.5 feet $\times$ 59 feet
June: $103,558.49

**19.** $12      **21.** $399      **23.** 2 hours      **25.** 2.9 quarts

**27.** $-\frac{1}{2}, \frac{4}{3}$      **29.** 3, 8      **31.** $\pm\sqrt{11}, \pm 3.32$

**33.** $-4 + 3\sqrt{2} \approx 0.24$

$-4 - 3\sqrt{2} \approx -8.24$

**35.** (1) Use the table feature in ASK mode with the variable equal to a solution.

(2) Use the scientific calculator portion to evaluate the quadratic equation at a particular solution.

**37.**

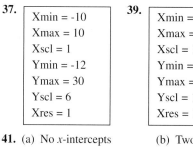

Xmin = -10
Xmax = 10
Xscl = 1
Ymin = -12
Ymax = 30
Yscl = 6
Xres = 1

**39.**

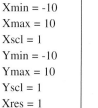

Xmin = -10
Xmax = 10
Xscl = 1
Ymin = -10
Ymax = 10
Yscl = 1
Xres = 1

**59.**

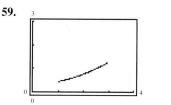

**61.** $2.10 \le x \le 2.11$

**41.** (a) No $x$-intercepts     (b) Two $x$-intercepts

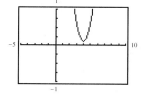   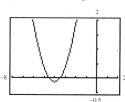

**43.** Three $x$-intercepts

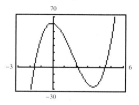

**45.** The graphs of 41(b) and 42(a) cross the $x$-axis and have two $x$-intercepts. The graphs of 41(a) and 42(b) do *not* cross the $x$-axis and have no $x$-intercepts.

**47.**                **49.**

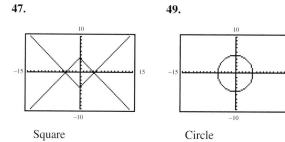

Square                Circle

**51.** (a) $0.587 (in 1982 dollars)

(b) The model is reasonable.

**53.** Answers will vary.

With the table feature: After entering the equation, have the table start at $t = 2$ and let the table step be 1. This will generate a list of ordered pairs.

Without the table feature: Substitute values of $t$ from 2 to 16 in the equation and record the respective ordered pairs.

**55.**                          **57.** 59.49%

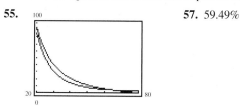

## Mid-Chapter Quiz   *(page 118)*

**1.** (a)

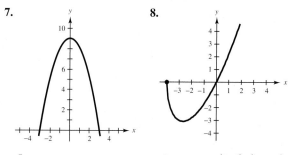

(b) $7\sqrt{2}$

(c) $\left(\frac{1}{2}, -\frac{3}{2}\right)$

**2.** (a)

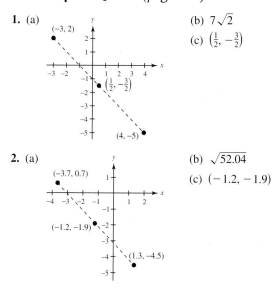

(b) $\sqrt{52.04}$

(c) $(-1.2, -1.9)$

**3.** $1,610,000. Use the Midpoint Formula because 1997 is midway between 1995 and 1999.

**4.** 250 miles     **5.** Right isosceles triangle

**6.** Parallelogram

**7.**                    **8.**

Intercepts:
$(3, 0), (-3, 0), (0, 9)$
Symmetry: $y$-axis

Intercepts: $(0, 0), (-4, 0)$
Symmetry: none

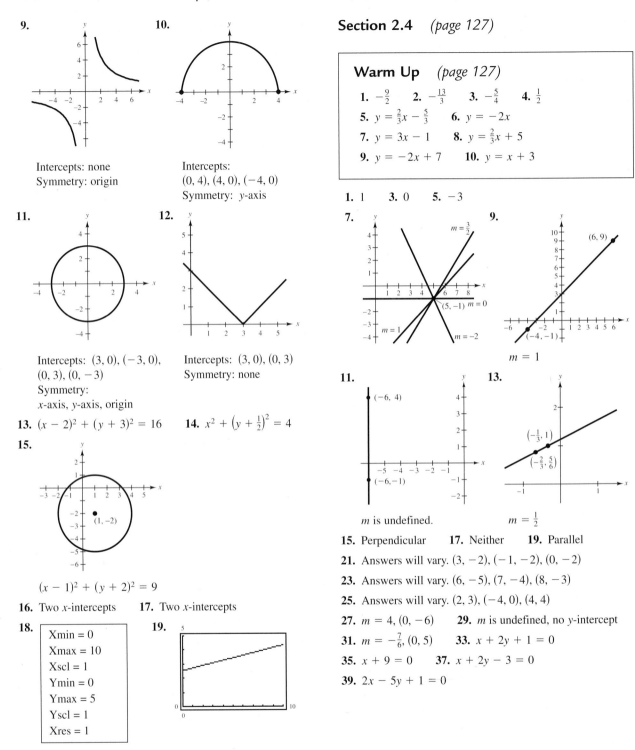

**9.**

Intercepts: none
Symmetry: origin

**10.**

Intercepts:
$(0, 4), (4, 0), (-4, 0)$
Symmetry: $y$-axis

**11.**

Intercepts: $(3, 0), (-3, 0),$
$(0, 3), (0, -3)$
Symmetry:
$x$-axis, $y$-axis, origin

**12.**

Intercepts: $(3, 0), (0, 3)$
Symmetry: none

**13.** $(x - 2)^2 + (y + 3)^2 = 16$   **14.** $x^2 + \left(y + \frac{1}{2}\right)^2 = 4$

**15.**

$(x - 1)^2 + (y + 2)^2 = 9$

**16.** Two $x$-intercepts   **17.** Two $x$-intercepts

**18.**

Xmin = 0
Xmax = 10
Xscl = 1
Ymin = 0
Ymax = 5
Yscl = 1
Xres = 1

**19.**

**20.** \$5.44

## Section 2.4 (page 127)

**Warm Up** (page 127)

**1.** $-\frac{9}{2}$   **2.** $-\frac{13}{3}$   **3.** $-\frac{5}{4}$   **4.** $\frac{1}{2}$

**5.** $y = \frac{2}{3}x - \frac{5}{3}$   **6.** $y = -2x$

**7.** $y = 3x - 1$   **8.** $y = \frac{2}{3}x + 5$

**9.** $y = -2x + 7$   **10.** $y = x + 3$

**1.** 1   **3.** 0   **5.** $-3$

**7.**

**9.**

$m = 1$

**11.**

$m$ is undefined.

**13.**

$m = \frac{1}{2}$

**15.** Perpendicular   **17.** Neither   **19.** Parallel

**21.** Answers will vary. $(3, -2), (-1, -2), (0, -2)$

**23.** Answers will vary. $(6, -5), (7, -4), (8, -3)$

**25.** Answers will vary. $(2, 3), (-4, 0), (4, 4)$

**27.** $m = 4, (0, -6)$   **29.** $m$ is undefined, no $y$-intercept

**31.** $m = -\frac{7}{6}, (0, 5)$   **33.** $x + 2y + 1 = 0$

**35.** $x + 9 = 0$   **37.** $x + 2y - 3 = 0$

**39.** $2x - 5y + 1 = 0$

**41.** $x - y - 7 = 0$

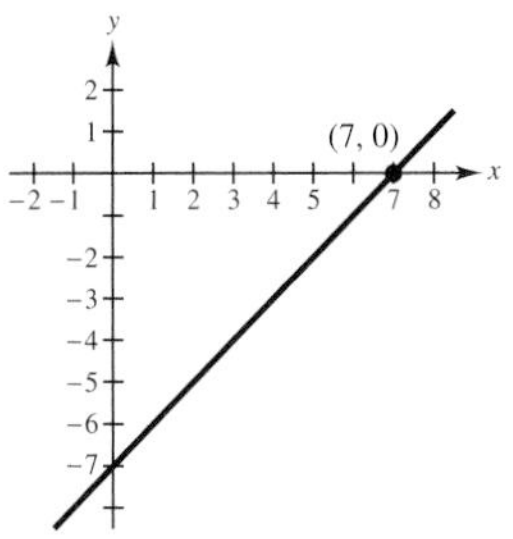

(7, 0)


**43.** $2x + y = 0$

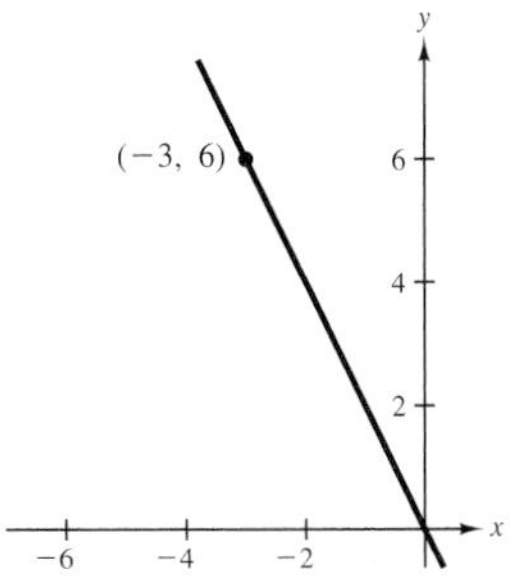

(−3, 6)


**45.** $x + 3y - 4 = 0$

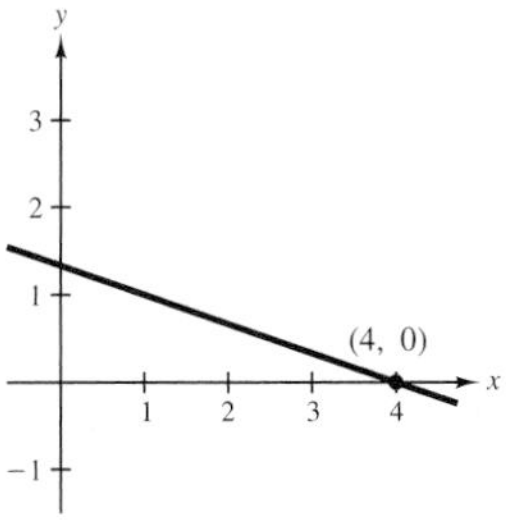

(4, 0)


**47.** $x - 6 = 0$

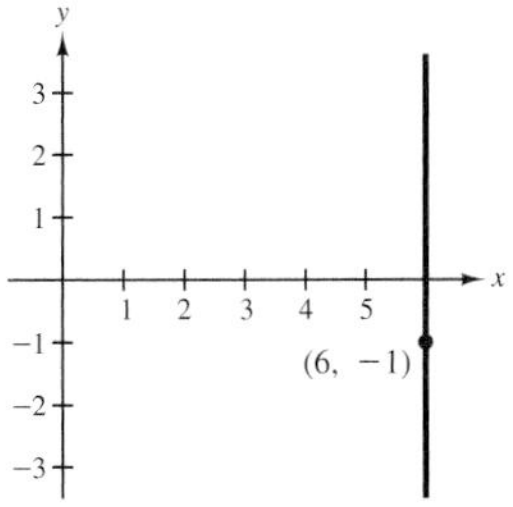

(6, −1)


**49.** $8x - 6y - 17 = 0$

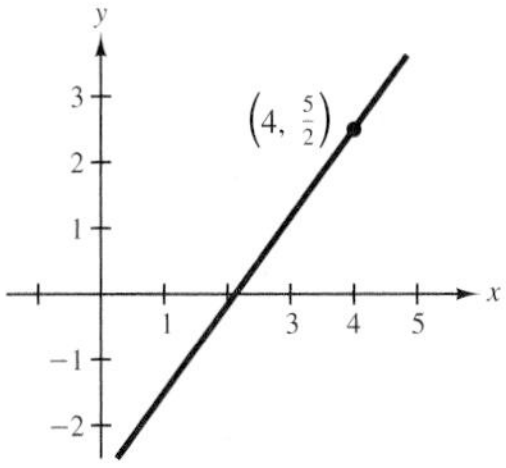

$(4, \frac{5}{2})$


**51.** Answers will vary. You could graph a vertical line and pick two convenient points on the vertical line. Find the slope. Regardless of the vertical line selected and the points selected, the slope will have zero in the denominator. Division by zero is not possible, so the slope does not exist.

**53.** $4x - y - 4 = 0$ **55.** $12x + 3y + 2 = 0$

**57.** (a) $2x - y - 10 = 0$ (b) $x + 2y - 10 = 0$

**59.** (a) $4x - 6y - 5 = 0$ (b) $36x + 24y + 7 = 0$

**61.** (a) $y = 0$ (b) $x + 1 = 0$ **63.** $F = \frac{9}{5}C + 32$

**65.** $A = 4.5t + 750$ **67.** \$166,000 **69.** \$36,200

**71.** \$23,037 million; no **73.** \$29.2 billion; no

**75.** $p = \frac{1}{33}d + 1$; $\frac{1}{33}$ (1 atmosphere for each 33 feet of depth below water level)

## Section 2.5 *(page 138)*

### Warm Up *(page 138)*

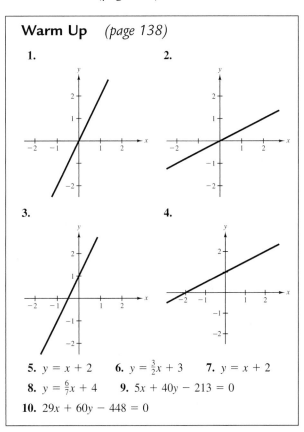

**1.**

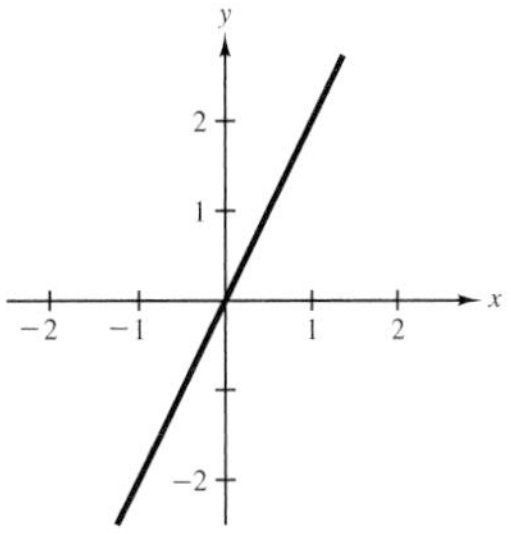

**2.**

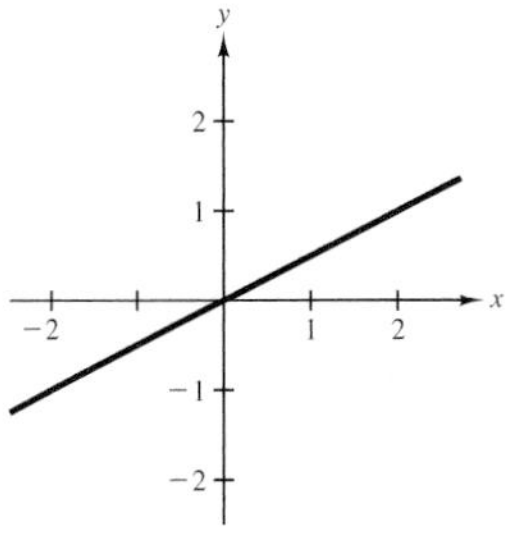

**3.**

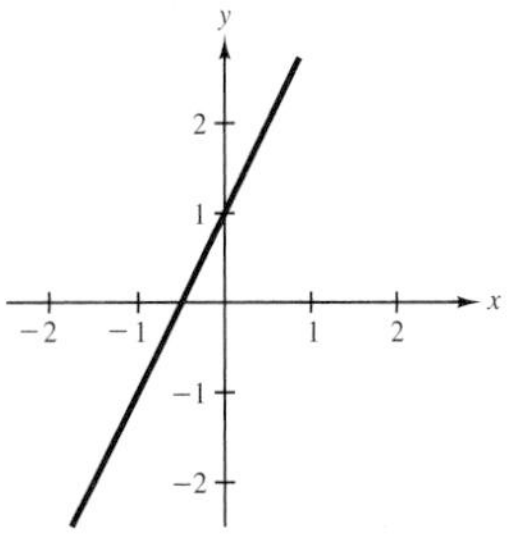

**4.**

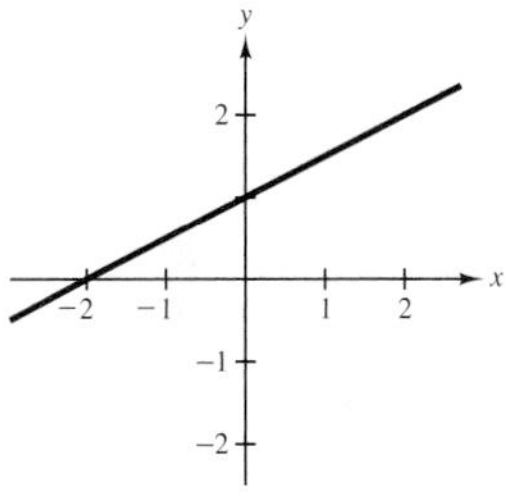

**5.** $y = x + 2$ **6.** $y = \frac{3}{2}x + 3$ **7.** $y = x + 2$

**8.** $y = \frac{6}{7}x + 4$ **9.** $5x + 40y - 213 = 0$

**10.** $29x + 60y - 448 = 0$

**1.**

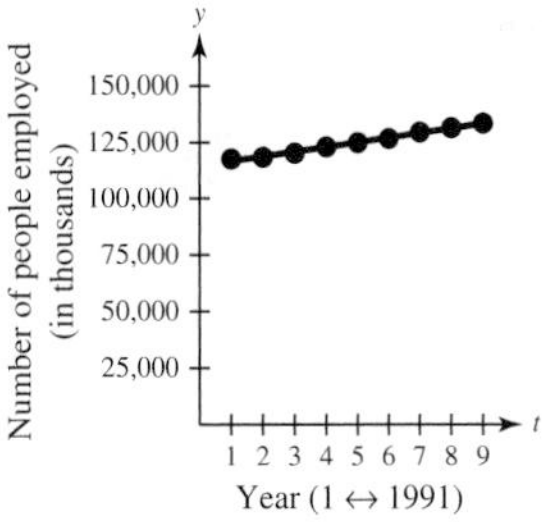

Number of people employed (in thousands)
Year (1 ↔ 1991)


The model is a "good fit" for the actual data.

**3.**

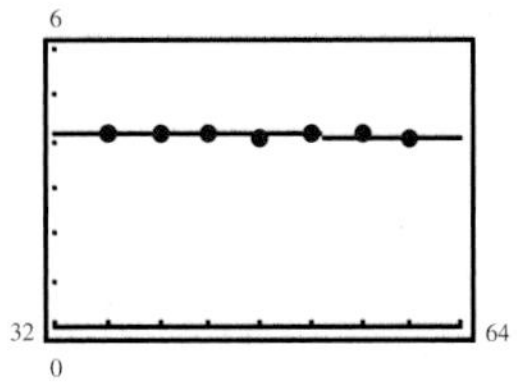

Predicted time: 4.09 minutes. Yes, this is reasonable.

**5.** $y = \dfrac{3}{8}x$ **7.** $y = 20x$ **9.** $y = \dfrac{3.2}{7}x$ or $y = \dfrac{32}{70}x$

**11.** $I = 0.075P$ **13.** (a) $y = 0.0368x$ (b) \$3128

**15.** (a) $C = \frac{33}{13}I$

(b)

| Inches | 5 | 10 | 20 | 25 | 30 |
|---|---|---|---|---|---|
| Centimeters | 12.69 | 25.38 | 50.77 | 63.46 | 76.15 |

**17.** $V = 125t + 2540, \quad 0 \le t \le 5$

**19.** $V = -2000t + 20{,}400, \quad 0 \le t \le 5$

**21.** $V = 12{,}500t + 154{,}000, \quad 0 \le t \le 5$

**23.** $V = 875 - 175t$     **25.** $S = 0.85L$

**27.** $W = 0.75x + 11.50$

**29.** (a) $h = 7000 - 20t$   (b) 2:13:50 P.M.

**31.** b; slope $= -10$; the amount owed *decreases* by $10 per week.

**33.** a; slope $= 0.25$; the amount received *increases* by $0.25 per mile driven.

**35.** (a) $N = 1200 + 50t$

(b) 2200 students

(c)
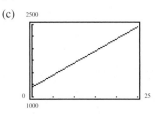

**37.** Yes     **39.** No

**41.** Answers will vary.     **43.** Answers will vary.

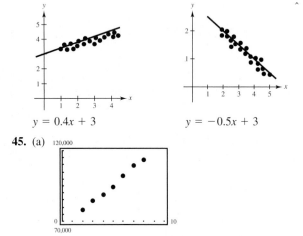

$y = 0.4x + 3$       $y = -0.5x + 3$

**45.** (a)

(b) $y = 6045.04x + 65{,}552.25$

(c) The slope is 6045.04. The slope indicates that the operating revenue for the airline industry should increase by $6,045,040,000 each year.

(d) 1999: $119,957,610,000
2000: $126,002,650,000
Predictions seem reasonable.

**47.** (a)
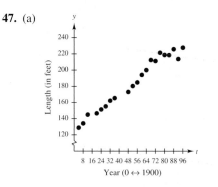

(b) Answers will vary. $y = 1.07t + 124.62$

(c) The slope is 1.07. The slope indicates that the length of the winning discus throw should increase by 4.28 feet every 4 years.

(d) 231.2 feet

(e) The prediction was off by approximately 4 feet.

**49.** (a) $C = 16.75t + 36{,}500$   (b) $R = 27t$

(c) $P = 10.25t - 36{,}500$   (d) 3561 hours

**51.** (a) $x = -\frac{1}{15}p + \frac{226}{3}$

(b) 45 units rented when $p = \$455$

## Review Exercises    *(page 145)*

**1.** 13

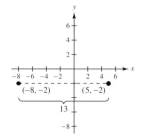

**3.** (a) 4 and 3, hypotenuse $= 5$

(b) 5

**5.** (a)
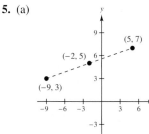

(b) $2\sqrt{53} \approx 14.56$

(c) $(-2, 5)$

**7.** (a)

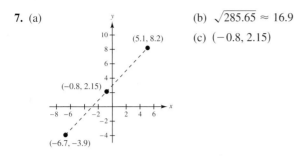

(b) $\sqrt{285.65} \approx 16.9$

(c) $(-0.8, 2.15)$

**9.** Slope of line through $(1, 1)$ and $(8, 2)$ is $\frac{1}{7}$.
Slope of line through $(8, 2)$ and $(9, 5)$ is 3.
Slope of line through $(9, 5)$ and $(2, 4)$ is $\frac{1}{7}$.
Slope of line through $(2, 4)$ and $(1, 1)$ is 3.
Consequently, opposite sides are parallel.

**11.** $-10$ or 30    **13.** 3 or $-7$    **15.** \$725,000

**17.** (a) Yes    (b) Yes

**19.** $x$-intercepts: $(-3, 0), (2, 0)$
$y$-intercept: $(0, -6)$

**21.** $x$-intercepts: $(0, 0), (-3, 0)$
$y$-intercept: $(0, 0)$

**23.** Origin    **25.** None

**27.**

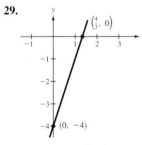

$y$-intercept: $(0, 3)$
Symmetry: $y$-axis

**29.**

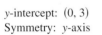

$x$-intercept: $\left(\frac{4}{3}, 0\right)$
$y$-intercept: $(0, -4)$
Symmetry: none

**31.**

$x$-intercept: $(-1, 0)$
$y$-intercept: $(0, 1)$
Symmetry: none

**33.** $(x - 1)^2 + (y - 3)^2 = 25$

**35.** $(x - 1)^2 + \left(y + \frac{1}{2}\right)^2 = \frac{25}{4}$

**37.** $(x - 3)^2 + (y + 2)^2 = 16$

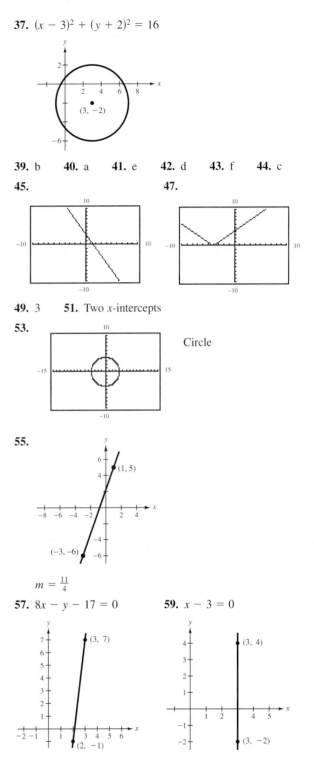

**39.** b    **40.** a    **41.** e    **42.** d    **43.** f    **44.** c

**45.**                                                    **47.**

**49.** 3    **51.** Two $x$-intercepts

**53.**                                                    Circle

**55.**

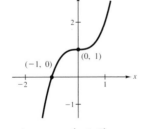

$m = \frac{11}{4}$

**57.** $8x - y - 17 = 0$    **59.** $x - 3 = 0$

**61.** $3x - 2y - 10 = 0$   **63.** $2x + 3y - 6 = 0$

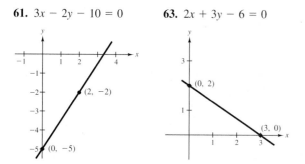

**65.** Slope: $\frac{5}{4}$; $y$-intercept: $\left(0, \frac{11}{4}\right)$

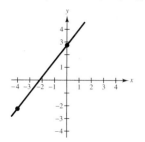

**67.** Parallel   **69.** Neither

**71.** (a) $5x - 4y - 23 = 0$   (b) $4x + 5y - 2 = 0$

**73.** $y = \frac{7}{3}x$   **75.** $y = 348x$   **77.** $I = 0.0626P$

**79.** $y = \frac{85}{3}x$

| Ounces | 6 | 8 | 12 | 16 | 24 |
|--------|---|---|----|----|----|
| Grams | 170 | 226.7 | 340 | 453.3 | 680 |

**81.** $210,000

**83.** $V = 85 + 3.75(t - 2) = 3.75t + 77.50$,
$t = 2$ represents 2002

Value in 2007: $103.75

**85.** $r = 1.25x - 25$; $1725

**87.** (a)

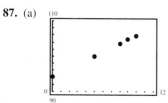

(b) $y = 1.17t + 94.2$, $t = 0$ represents 1990

(c) 2001: 107,700,000
2002: 108,240,000
Predictions seem reasonable.

## Chapter Test   *(page 149)*

**1.** Distance: $4\sqrt{5}$   **2.** Distance: 7.81
Midpoint: $(1, 0)$   Midpoint: $(0.44, 4.34)$

**3.** $x$-intercepts: $(-5, 0), (3, 0)$   **4.** $x$-intercept: $(2, 0)$
$y$-intercept: $(0, -15)$   $y$-intercepts: none

**5.** Symmetric with respect to the origin

**6.** $y - 4 = 0$   **7.** $3x - 4y - 24 = 0$

**8.** $x - 2 = 0$   **9.** $4x - 3y - 12 = 0$

**10.** a   **11.** b   **12.** c   **13.** d

**14.** $(x - 2)^2 + (y - 1)^2 = 9$

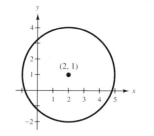

**15.** $220,000   **16.** $W = 0.55x + 8.75$

**17.** $V = 25,000 - 4000x$   **18.** $S = L - 0.30L = 0.70L$

**19.** $S = 141.1x + 1661$

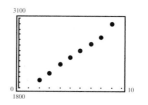

## Cumulative Test: Chapters 1–2   *(page 150)*

**1.** $-4, 4$   **2.** $\frac{12}{7}$   **3.** $\pm 1, \pm 4$   **4.** $-1, 4$

**5.** 1   **6.** $\dfrac{-1 \pm \sqrt{41}}{4}$   **7.** $11,663.24

**8.** $-\frac{7}{2} \le x \le \frac{17}{2}$   **9.** $-6 \le x \le -2$

**10.** $\frac{5}{4} < x < 4$   **11.** $(0, 0), (-4, 0)$

**12.** $y$-axis symmetry

**13.** $(x - 3)^2 + (y + 2)^2 = 16$

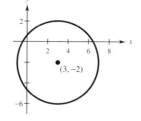

**14.** 11   **15.** $7x + 4y - 13 = 0$   **16.** $2x - 3y - 7 = 0$

**17.** $y + 3 = 0$   **18.** $y = 0.07x + 1000$   **19.** $250,000

**20.** $x = 79,623.23$ or $x = 376.66$. Hence, revenue will equal
cost when either 79,623 units or 377 units are sold.

# Chapter 3

## Section 3.1  *(page 160)*

---

### Warm Up  *(page 160)*

**1.** $-73$    **2.** 13    **3.** $2(x + 2)$    **4.** $-8(x - 2)$

**5.** $y = \frac{7}{5} - \frac{2}{5}x$    **6.** $y = \pm x$    **7.** $x \leq -2, x \geq 2$

**8.** $-3 \leq x \leq 3$    **9.** All real numbers

**10.** $x \leq 1, x \geq 2$

---

**1.** (a) $-6$  (b) 34  (c) $6 - 4t$  (d) $2 - 4c$

**3.** (a) $-1$  (b) $\dfrac{1}{15}$  (c) $\dfrac{1}{t^2 - 2t}$  (d) $\dfrac{1}{t^2 - 1}$

**5.** (a) $-1$  (b) $-9$  (c) $2x - 5$  (d) $-\dfrac{5}{2}$

**7.** (a) 0  (b) 3  (c) $x^2 + 2x$  (d) $-0.75$

**9.** (a) 1  (b) $-7$  (c) $3 - 2|x|$  (d) 2.5

**11.** (a) $\dfrac{1}{7}$  (b) $-\dfrac{1}{9}$  (c) Undefined  (d) $\dfrac{1}{y^2 + 6y}$

**13.** (a) 1  (b) $-1$  (c) 1  (d) $\dfrac{|x - 1|}{x - 1}$

**15.** (a) $-1$  (b) 2  (c) 4  (d) 6

**17.** 5    **19.** $\pm 3$    **21.** $\frac{10}{7}$    **23.** All real numbers

**25.** All real numbers except $t = 0$    **27.** All real numbers

**29.** $-1 \leq x \leq 1$    **31.** All real numbers except $x = 0, -2$

**33.** All real numbers $x \geq -1$ except $x = 2$

**35.** The domain of $f(x) = \sqrt{x - 2}$ is all real numbers $x \geq 2$, since an even root of a negative number is not a real number. The domain of $g(x) = \sqrt[3]{x - 2}$ is all real numbers. $f$ and $g$ have different domains because an odd root of a negative number is a real number but an even root of a negative number is not a real number.

**37.** Not a function    **39.** Function    **41.** Function

**43.** Not a function    **45.** Function

**47.** This is a function from $A$ to $B$, because each element in $A$ is matched with an element in $B$.

**49.** This is a function from $A$ to $B$, because each element in $A$ is matched with an element in $B$.

**51.** Not a function. The relationship assigns two elements of $B$ to the element $c$ in $A$.

**53.** Not a function. The relationship defines a function from $B$ to $A$.

**55.** This is a function from $A$ to $B$, because each element in $A$ is matched with an element in $B$.

**57.** Not a function. The relationship does not
of $A$ with an element of $B$.

**59.** $\{(-2, 4), (-1, 1), (0, 0), (1, 1), (2, 4)\}$

**61.** $\{(-2, 0), (-1, 1), (0, \sqrt{2}), (1, \sqrt{3}), (2, 2)\}$

**63.** (a) $V = 4x(6 - x)^2$  (b) Domain: $0 < x < 6$

**65.** $h = \sqrt{d^2 - 2000^2}$;  Domain: $d \geq 2000$

**67.** 1994: \$1759; 1998: \$1899

**69.** (a)

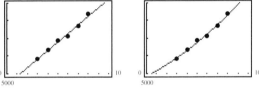

(b) $y = 240.89t + 4695.29$
$y = 8.23t^2 + 150.33t + 4920.31$

(c) Linear                    Quadratic

Answers will vary. The quadratic model is a better fit. The linear model is easier to use but the quadratic model will give more accurate results.

**71.** (a) $C = 35,000 + 1.15x$  (b) $\overline{C} = \dfrac{35,000}{x} + 1.15$

(c)

| $x$ | 100 | 1000 | 10,000 | 100,000 |
|---|---|---|---|---|
| $\overline{C}$ | 351.15 | 36.15 | 4.65 | 1.5 |

(d) Answers will vary. The average cost per unit decreases as $x$ gets larger.

**73.** (a) $r(t) = 0.8t$

(b) $A = \pi(0.8t)^2$

| Time, $t$ | 1 | 2 | 3 | 4 | 5 |
|---|---|---|---|---|---|
| Radius, $r$ | 0.8 ft | 1.6 ft | 2.4 ft | 3.2 ft | 4 ft |
| Area, $A$ | 2.011 ft$^2$ | 8.042 ft$^2$ | 18.096 ft$^2$ | 32.170 ft$^2$ | 50.265 ft$^2$ |

(c) $\dfrac{A(2)}{A(1)} = \dfrac{8.042}{2.011} \approx 3.999$, $\dfrac{A(4)}{A(2)} = \dfrac{32.170}{8.042} \approx 4.000$

Predicted area when $t = 8$: 128.68 square feet
Calculated area when $t = 8$: 128.680 square feet

## ection 3.2 *(page 172)*

**Warm Up** *(page 172)*

**1.** 2    **2.** 0    **3.** $-\dfrac{3}{x}$    **4.** $x^2 + 3$    **5.** $0, \pm 4$

**6.** $\frac{1}{2}, 1$    **7.** All real numbers except $x = 4$

**8.** All real numbers except $x = 4, 5$    **9.** $t \le \frac{5}{3}$

**10.** All real numbers

**1.** Domain: $[1, \infty)$; range: $[0, \infty)$

**3.** Domain: $(-\infty, -2] \cup [2, \infty)$; range: $[0, \infty)$

**5.** Domain: $[-5, 5]$; range: $[0, 5]$

**7.** Domain: $(-\infty, \infty)$; range: $(-\infty, \infty)$

**9.** Function    **11.** Not a function    **13.** Function

**15.** Increasing on $(-\infty, \infty)$; no change

**17.** Increasing on $(-\infty, 0)$ and $(2, \infty)$, decreasing on $(0, 2)$; behavior changes at $(0, 0)$ and at $(2, 4)$

**19.** Increasing on $(-1, 0)$ and $(1, \infty)$, decreasing on $(-\infty, -1)$ and $(0, 1)$; behavior changes at $(-1, -3)$, $(0, 0)$, and $(1, -3)$

**21.** Increasing on $(-2, \infty)$, decreasing on $(-3, -2)$; behavior changes at $(-2, -2)$

**23.** Even    **25.** Odd    **27.** Neither even nor odd

**29.** Even        **31.** Neither even nor odd

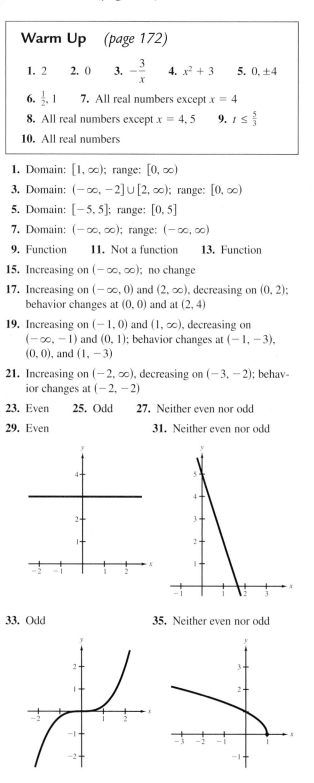

**33.** Odd        **35.** Neither even nor odd

**37.** Neither even nor odd    **39.** Neither even nor odd

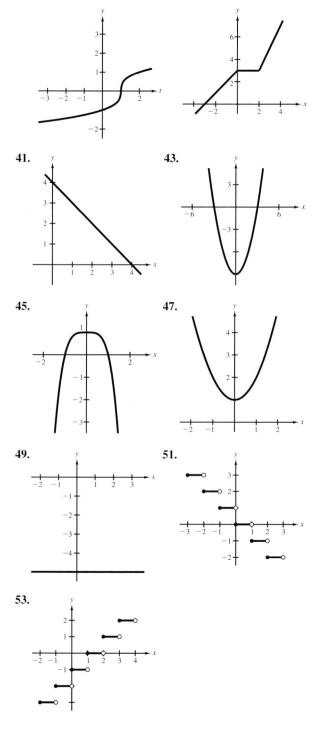

**41.**

**43.**

**45.**

**47.**

**49.**

**51.**

**53.**

**55.** Increasing: $(2, \infty)$
Decreasing: $(-\infty, 2)$

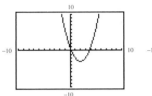

**57.** Increasing: $(-\infty, 0), (2, \infty)$
Decreasing: $(0, 2)$

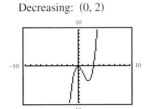

**59.** Increasing: 1990–1995
Decreasing: 1996–1999

**61.**

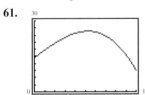

Increasing: 1990–1995
Decreasing: 1996–2000
Maximum revenue: $2,558,200
Minimum revenue: $891,000

**63.** Maximum or minimum values may occur at endpoints.

**65.** Approximately 350,000 units

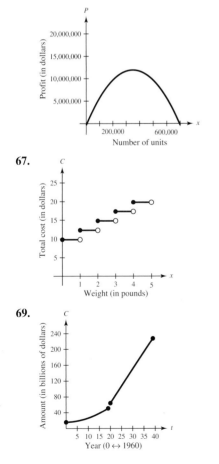

**67.**

**69.**

**71.** (a) $y = 27.583t^3 - 335.37t^2 + 1292.2t + 9618$

(b) $y = -25.5t^2 + 286.1t + 10{,}189$

(c)

$$y = \begin{cases} 27.583t^3 - 335.37t^2 + 1292.2t + 9618, & 0 \le t \le 6 \\ -25.5t^2 + 286.1t + 10{,}189, & 7 \le t \le 10 \end{cases}$$

## Section 3.3   *(page 182)*

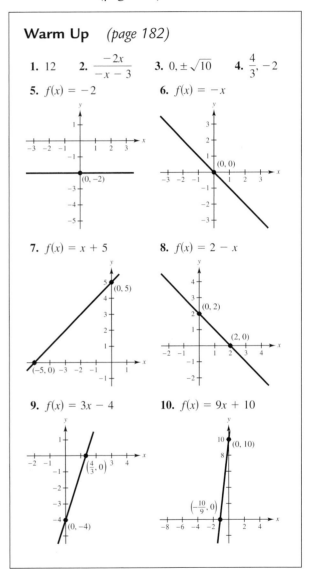

**Warm Up**   *(page 182)*

**1.** 12    **2.** $\dfrac{-2x}{-x-3}$    **3.** $0, \pm\sqrt{10}$    **4.** $\dfrac{4}{3}, -2$

**5.** $f(x) = -2$

**6.** $f(x) = -x$

**7.** $f(x) = x + 5$

**8.** $f(x) = 2 - x$

**9.** $f(x) = 3x - 4$

**10.** $f(x) = 9x + 10$

**1.**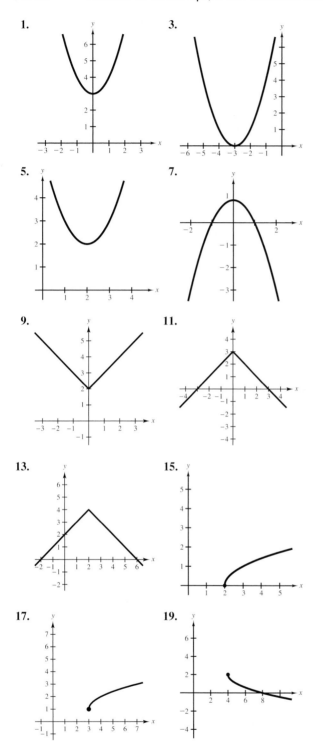

**3.**

**5.**

**7.**

**9.**

**11.**

**13.**

**15.**

**17.**

**19.**

**21.**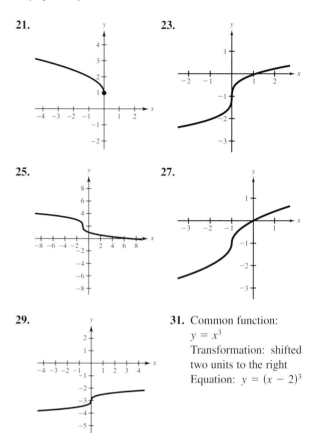

**23.**

**25.**

**27.**

**29.**

**31.** Common function:
$y = x^3$
Transformation: shifted
two units to the right
Equation: $y = (x - 2)^3$

**33.** Common function: $y = x^2$
Transformation: reflection about the $x$-axis
Equation: $y = -x^2$

**35.** Common function: $y = \sqrt{x}$
Transformation: reflection about the $x$-axis and shifted one
unit up
Equation: $y = -\sqrt{x} + 1$

**37.** (a) Vertical shift of two units

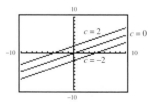

(b) Horizontal shift of two units

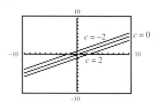

(c)  Slope of the function changes

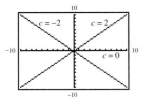

**39.** (a) $g(x) = (x - 1)^2 + 1$

  (b) $g(x) = -(x + 1)^2$

**41.** (a) $y = f(x) + 2$          (b) $y = -f(x)$

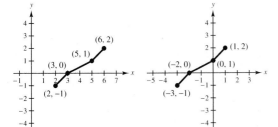

  (c) $y = f(x - 2)$          (d) $y = f(x + 3)$

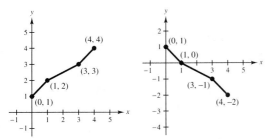

  (e) $y = 2f(x)$          (f) $y = f(-x)$

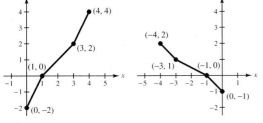

**43.** $g(x) = |x - 2| + 1$     **45.** $g(x) = -3|x|$

**47.** $h(x) = \sqrt{x + 3} + 2$     **49.** $h(x) = -4\sqrt{x}$

**51.** $g(x) = -x^3 + 3x^2 + 1$

**53.** Shifted one unit to the right and two units down

  $g(x) = (x - 1)^2 - 2$

**55.** (a)

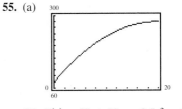

  (b) $P(x) = 80 + 20x - 0.5x^2 - 25$

  $P(x) = 55 + 20x - 0.5x^2, \quad 0 \le x \le 20$

  Shifted 25 units down

  (c) $P\left(\dfrac{x}{100}\right) = 80 + \dfrac{x}{5} - 0.00005x^2$

  Vertical shrink

**57.** (a), (c), and (e) are odd functions; (b), (d), and (f) are even. Also, (a), (c), and (e) are increasing for all real numbers; (b), (d), and (f) are decreasing for all $x < 0$ and increasing for all $x > 0$.

**59.**

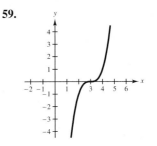

## Mid-Chapter Quiz   *(page 185)*

**1.** 16     **2.** 4     **3.** 1     **4.** $2a^2 - a + 1$

**5.** $-2, 0, 2$     **6.** $\frac{4}{3}$     **7.** $P = 5.49x - 85,000$

**8.** All real numbers except 1 and 0        **9.** $x \le 5$

**10.** $x \ge 1$

**11.** Increasing: $x < 2.5$

  Decreasing: $x > 2.5$

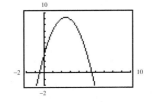

**12.** Increasing: $x < 0, x > \frac{5}{3}$

Decreasing: $0 < x < \frac{5}{3}$

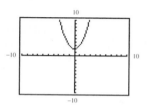

**13.** Increasing: $x > 0$

Decreasing: $x < 0$

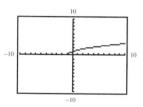

**14.** Increasing: $x > -1$

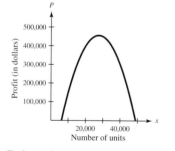

**15.** $g(x)$ is shifted two units to the right.

**16.** $g(x)$ is reflected about the $x$-axis and shifted one unit to the left.

**17.** $g(x) = (x - 2)^3 - 1$

**18.** $y = (x - 3)^2 - 2$        **19.** $y = -\sqrt{x + 2}$

**20.** $P = x(80 - 0.001x) - (300,000 + 25x)$

$P = -0.001x^2 + 55x - 300,000$

Estimated number of units to produce a maximum profit: $\approx 27,500$

**Section 3.4**    *(page 192)*

**Warm Up**    *(page 192)*

**1.** $\dfrac{1}{x(1 - x)}$    **2.** $-\dfrac{12}{(x + 3)(x - 3)}$    **3.** $\dfrac{3x - 2}{x(x - 2)}$

**4.** $\dfrac{4x - 5}{3(x - 5)}$    **5.** $\sqrt{\dfrac{x - 1}{x + 1}}$    **6.** $\dfrac{x + 1}{x(x + 2)}$

**7.** $5(x - 2)$    **8.** $\dfrac{x + 1}{(x - 2)(x + 3)}$    **9.** $\dfrac{1 + 5x}{3x - 1}$

**10.** $\dfrac{x + 4}{4x}$

**1.**

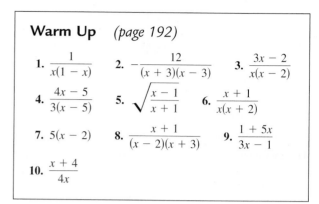

**3.**

**5.** (a) $2x$    (b) $2$    (c) $x^2 - 1$

(d) $\dfrac{x + 1}{x - 1}$; domain: $(-\infty, 1) \cup (1, \infty)$

**7.** (a) $x^2 - x + 1$    (b) $x^2 + x - 1$    (c) $x^2 - x^3$

(d) $\dfrac{x^2}{1 - x}$; domain: $(-\infty, 1) \cup (1, \infty)$

**9.** (a) $x^2 + \sqrt{1 - x} + 5$    (b) $x^2 - \sqrt{1 - x} + 5$

(c) $x^2\sqrt{1 - x} + 5\sqrt{1 - x}$

(d) $\dfrac{x^2 + 5}{\sqrt{1 - x}}$; domain: $(-\infty, 1)$

**11.** (a) $\dfrac{x + 1}{x^2}$    (b) $\dfrac{x - 1}{x^2}$    (c) $\dfrac{1}{x^3}$

(d) $x, x \neq 0$; domain: $(-\infty, 0) \cup (0, \infty)$

**13.** $9$    **15.** $4t^2 - 2t + 5$    **17.** $0$    **19.** $26$

**21.** $5$    **23.** $\frac{3}{5}$    **25.** (a) $x^2 - 2x + 1$    (b) $x^4$

**27.** (a) $20 - 3x$    (b) $9x + 20$

**29.** (a) $\sqrt{x^2 + 4}$    (b) $x + 4, x \geq -4$

**31.** (a) $x - \frac{8}{3}$    (b) $x - 8$    **33.** (a) $\sqrt[4]{x}$    (b) $\sqrt[4]{x}$

**35.** (a) $|x + 6|$    (b) $|x| + 6$    **37.** (a) $3$    (b) $0$

**39.** (a) $0$    (b) $4$

**41.** (a) $x \geq 0$, or $[0, \infty)$

(b) All real numbers, or $(-\infty, \infty)$

(c) All real numbers, or $(-\infty, \infty)$

**43.** (a) All real numbers except $x = \pm 1$, or
$(-\infty, -1) \cup (-1, 1) \cup (1, \infty)$

(b) All real numbers, or $(-\infty, \infty)$

(c) All real numbers except $x = -2$ and $x = 0$, or
$(-\infty, -2) \cup (-2, 0) \cup (0, \infty)$

**45.** Answers will vary. $f(x) = x^2$, $g(x) = 2x + 1$

**47.** Answers will vary. $f(x) = \sqrt[3]{x}$, $g(x) = x^2 - 4$

**49.** Answers will vary. $f(x) = \dfrac{1}{x}$, $g(x) = x + 2$

**51.** Answers will vary. $f(x) = x^2 + 2x$, $g(x) = x + 4$

**53.** $T = \dfrac{3}{4}x + \dfrac{1}{15}x^2$

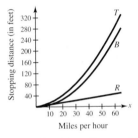

**55.** Total sales: $R_1 + R_2 = -0.8t^2 + 0.78t + 753.9$,
$t = 7, 8, 9, 10, 11, 12$
Total sales have been decreasing.

**57.** (a)

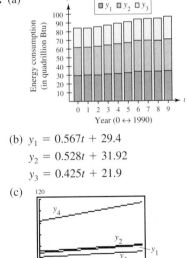

(b) $y_1 = 0.567t + 29.4$
$y_2 = 0.528t + 31.92$
$y_3 = 0.425t + 21.9$

(c)

$y_4 = 1.52t + 83.22$
Estimated total energy consumption in 2005:
106 quadrillion Btu
Estimated total energy consumption in 2009:
112.1 quadrillion Btu

**59.** (a) $N(T(t)) = 100t^2 + 275$

(b) Approximately 21.8 hours

**61.**

| Year | $P$ | $E$ | $P/E$ |
|------|-----|-----|-------|
| 1992 | $11.05 | $0.65 | 17 |
| 1993 | $13.07 | $0.73 | 17.9 |
| 1994 | $14.36 | $0.84 | 17.1 |
| 1995 | $18.81 | $0.99 | 19 |
| 1996 | $23.87 | $1.11 | 21.5 |
| 1997 | $24.15 | $1.15 | 21 |
| 1998 | $31.12 | $1.26 | 24.7 |
| 1999 | $42.26 | $1.39 | 30.4 |
| 2000 | $33.43 | $1.46 | 22.9 |

**63.** $(C \circ x)(t) = 3000t + 750$
$C \circ x$ represents the cost of producing $x$ units in $t$ hours.

**65.** Domain of $(f/g)(x)$: all real numbers $0 \leq x < 3$, or $[0, 3)$
Domain of $(g/f)(x)$: all real numbers $0 < x \leq 3$, or $(0, 3]$
The two domains differ because if $x = 3$, $(f/g)(x)$ is undefined (division by zero), and if $x = 0$, $(g/f)(x)$ is undefined (division by zero).

## Section 3.5    *(page 203)*

**Warm Up**    *(page 203)*

**1.** All real numbers    **2.** $[-1, \infty)$

**3.** All real numbers except $x = 0, 2$

**4.** All real numbers except $x = -\dfrac{5}{3}$    **5.** $x$

**6.** $x$    **7.** $x$    **8.** $x$    **9.** $x = \dfrac{3}{2}y + 3$

**10.** $x = \dfrac{y^3}{2} + 2$

**1.**

**3.**

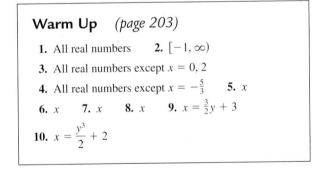

**5.**

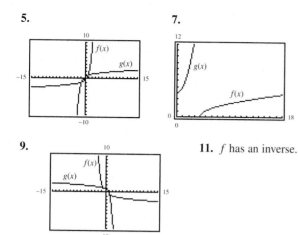

**7.**

**9.**

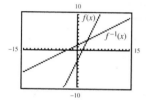

**11.** $f$ has an inverse.

**13.** $f$ doesn't have an inverse.     **15.** $g$ has an inverse.

**17.** $h$ doesn't have an inverse.

**19.** $f$ doesn't have an inverse.

**21.** Error: $f^{-1}$ does not mean to take the reciprocal of $f(x)$.

**23.** $f^{-1}(x) = \dfrac{x+3}{2}$     **25.** $f^{-1}(x) = \sqrt[5]{x}$

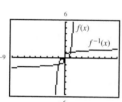

**27.** $f^{-1}(x) = x^2,\ x \geq 0$     **29.** $f^{-1}(x) = \sqrt{4 - x^2}$,
$0 \leq x \leq 2$

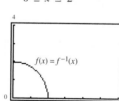

**31.** $f^{-1}(x) = x^3 + 1$     **33.** $f$ doesn't have an inverse.

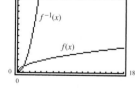

**35.** $g^{-1}(x) = 8x$     **37.** $p$ doesn't have an inverse.

**39.** $f^{-1}(x) = \sqrt{x} - 3,\ x \geq 0$     **41.** $h^{-1}(x) = \dfrac{1}{x}$

**43.** $f^{-1}(x) = \dfrac{x^2 - 3}{2},\ x \geq 0$

**45.** $g$ doesn't have an inverse.

**47.** $f^{-1}(x) = -\sqrt{25 - x},\ x \leq 25$

**49.**

| $x$ | 0 | 1 | 2 | 3 | 4 |
|---|---|---|---|---|---|
| $f^{-1}(x)$ | $-2$ | 0 | 1 | 2 | 4 |

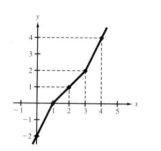

**51.** 32     **53.** 600

**55.** $(g^{-1} \circ f^{-1})(x) = \dfrac{x+1}{2}$     **57.** $(f \circ g)^{-1}(x) = \dfrac{x+1}{2}$

**59.** (a)

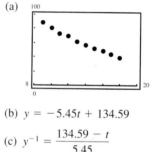

(b) $y = -5.45t + 134.59$

(c) $y^{-1} = \dfrac{134.59 - t}{5.45}$

(d) 2000

**61.** After graphing $f(x) = x^2,\ x \geq 0$, and $f^{-1}(x) = \sqrt{x}$, it is observed that $f(x)$ and $f^{-1}(x)$ are reflections of each other about the line $y = x$. Because of this reflection, interchanging the roles of $x$ and $y$ seems reasonable.

**63.** $f^{-1}(x) = \dfrac{x^2 + 0.019}{0.047}$

The earnings per share in 1996 were approximately $0.66.

## Review Exercises     *(page 208)*

**1.** $y$ is a function of $x$.     **3.** $y$ is a function of $x$.

**5.** Function. Every element in $A$ is assigned to an element in $B$.

**7.** (a) $-2$   (b) $-11$   (c) $3m - 5$   (d) $3x - 2$

**9.** (a) 5   (b) 8   (c) $5\frac{1}{2}$   (d) $x^2 + 5$

**11.** $-\frac{7}{2}$     **13.** All real numbers     **15.** $[-9, \infty)$

**17.** $[3, 5) \cup (5, \infty)$

**19.** The domain of $h(x)$ is all real numbers except $x = 0$, since division by zero is undefined. The domain of $k(x)$ is all real numbers except $x = -2$ and $x = 2$, since if $x = 2$ or $x = -2$, then $x^2 - 4$ equals zero, and division by zero is undefined. Using a graphing utility and the table feature, $h(0)$ results in an error and $k(-2)$ or $k(2)$ also results in an error.

**21.** (a) $V = x(20 - 2x)(20 - 2x) = 4x(10 - x)^2$

(b) Domain: $(0, 10)$

(c)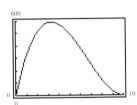

**23.** (a) 16 feet per second   (b) 1.5 seconds

(c) $-16$ feet per second

**25.** (a) Domain: all real numbers
Range: $[1, \infty)$

(b) Decreasing: $(-\infty, 0)$
Increasing: $(0, \infty)$

(c) Even

**27.** (a) Domain: all real numbers
Range: all real numbers

(b) Decreasing: $\left(0, \frac{8}{3}\right)$
Increasing: $(-\infty, 0) \cup \left(\frac{8}{3}, \infty\right)$

(c) Neither

**29.** **31.** **33.** **35.**

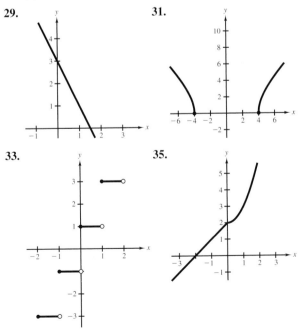

**37.** **39.**

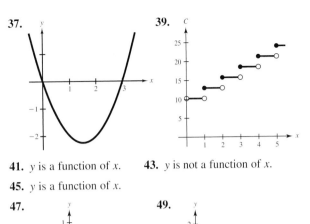

**41.** $y$ is a function of $x$.   **43.** $y$ is not a function of $x$.

**45.** $y$ is a function of $x$.

**47.** **49.**

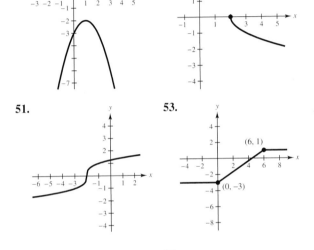

**51.** **53.** **55.** **57.**

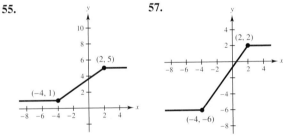

**59.** Common function: $y = x$
Transformation: shifted three units to the left
Equation: $y = x + 3$

**61.** Common function: $y = x^2$
Transformation: shifted down one unit
Equation: $y = x^2 - 1$

**63.** $(f + g)(x) = 2x - 1$

$(f - g)(x) = 5$

$(fg)(x) = x^2 - x - 6$

$(f/g)(x) = \dfrac{x + 2}{x - 3}$

Domain of $f/g$: all real numbers except $x = 3$, or $(-\infty, 3) \cup (3, \infty)$

**65.** $(f + g)(x) = \sqrt{x^2 - 4} + 3$

$(f - g)(x) = \sqrt{x^2 - 4} - 3$

$(fg)(x) = 3\sqrt{x^2 - 4}$

$(f/g)(x) = \dfrac{\sqrt{x^2 - 4}}{3}$

Domain of $f/g$: $x \le -2, x \ge 2$, or $(-\infty, -2] \cup [2, \infty)$

**67.** 7    **69.** 40

**71.** (a) $x^2 + 6x + 9$   (b) $x^2 + 3$

**73.** (a) $|x - 5| + 2$   (b) $|x| - 3$

**75.** (a) $x^2 - 2x + 1$   (b) $x^2 - 1$

**77.** Answers will vary. $f(x) = \dfrac{1}{x^2}$,   $g(x) = x - 2$

**79.** Answers will vary. $f(x) = x^2 - 6x + 9$,   $g(x) = \sqrt{x}$

**81.** $R = R_1 + R_2$

$R = 600.52 + 0.22t - 0.3t^2$

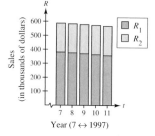

Total sales are decreasing.

**83.** (a) $r = \dfrac{x}{2}$   (b) $A = \pi r^2$   (c) $A \circ r(x) = \dfrac{\pi x^2}{4}$

$A \circ r(x)$ is the area of the base of the tank in terms of $x$, where $x$ is the side of the square base.

**85.** $(C \circ x)(t) = 5000t + 900$

$(C \circ x)(t)$ is the cost of producing $x$ units in $t$ hours.

**87.** $f(g(x)) = x = g(f)(x)$,

$f(g(x)) = 3\left(\dfrac{x - 5}{3}\right) + 5 = x$

$g(f(x)) = \dfrac{3x + 5 - 5}{3} = x$

so $f$ and $g$ are inverse functions of each other.

**89.** $f(x)$ does not have an inverse.

**91.** $f^{-1}(x) = \dfrac{1}{x}$        **93.** $f^{-1}(x) = \dfrac{3x + 5}{2}$

($f$ is its own inverse.)

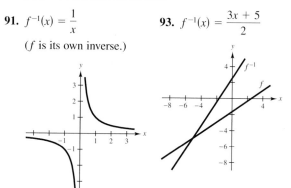

**95.** (a) $f^{-1}(x) = 2(x + 3) = 2x + 6$

(b)

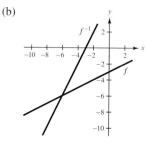

(c) $f \circ f^{-1}(x) = \frac{1}{2}(2x + 6) - 3$

$= x$

$f^{-1} \circ f(x) = 2\left(\frac{1}{2}x - 3\right) + 6$

$= x$

**97.** (a) $f^{-1}(x) = x^2 - 1, x \ge 0$

(b)

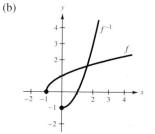

(c) $f \circ f^{-1}(x) = \sqrt{(x^2 - 1) + 1}$

$= x, x \ge 0$

$f^{-1} \circ f(x) = \left(\sqrt{x + 1}\right)^2 - 1$

$= x, x \ge -1$

**99.** (a) $f^{-1}(x) = \sqrt{x}, x \geq 0$

(b)

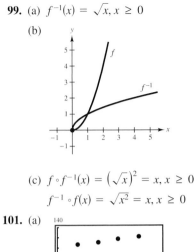

(c) $f \circ f^{-1}(x) = \left(\sqrt{x}\right)^2 = x, x \geq 0$

$f^{-1} \circ f(x) = \sqrt{x^2} = x, x \geq 0$

**101.** (a)

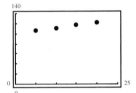

(b) $y = 1.049t + 100.42$

(c) $y^{-1} = \dfrac{t - 100.42}{1.049}$

The energy consumption will be 114 quadrillion Btu in 2013.

## Chapter Test    *(page 212)*

**1.** $(-\infty, \infty)$    **2.** $[7, \infty)$    **3.** $[2, 7) \cup (7, \infty)$

**4.** $(-\infty, -2) \cup (-2, 2) \cup (2, \infty)$

**5.** True. To each value of $x$ there corresponds exactly one value of $y$.

**6.** False. The $\pm$ indicates that to a given value of $x$ there correspond two values of $y$.

**7.** False. The element $-9$ is not included in set $B$.

**8.** (a) Domain: $(-\infty, \infty)$    (b) Increasing: $(0, \infty)$
     Range: $[2, \infty)$       Decreasing: $(-\infty, 0)$

(c) Even

**9.** (a) Domain: $(-\infty, -2] \cup [2, \infty)$
     Range: $[0, \infty)$

(b) Decreasing: $(-\infty, 2)$    (c) Even
    Increasing: $(2, \infty)$

**10.** (a) 0    (b) 2    (c) 11

**11.**

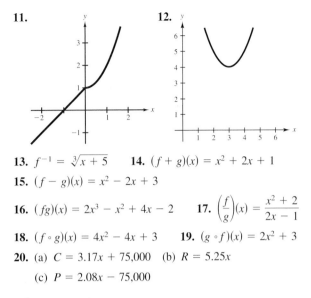

**12.**

**13.** $f^{-1} = \sqrt[3]{x + 5}$    **14.** $(f + g)(x) = x^2 + 2x + 1$

**15.** $(f - g)(x) = x^2 - 2x + 3$

**16.** $(fg)(x) = 2x^3 - x^2 + 4x - 2$    **17.** $\left(\dfrac{f}{g}\right)(x) = \dfrac{x^2 + 2}{2x - 1}$

**18.** $(f \circ g)(x) = 4x^2 - 4x + 3$    **19.** $(g \circ f)(x) = 2x^2 + 3$

**20.** (a) $C = 3.17x + 75{,}000$    (b) $R = 5.25x$

(c) $P = 2.08x - 75{,}000$

# Chapter 4

## Section 4.1    *(page 222)*

**Warm Up**    *(page 222)*

**1.** $\frac{1}{2}, -6$    **2.** $-\frac{3}{5}, 3$    **3.** $\frac{3}{2}, -1$    **4.** $-10$

**5.** $3 \pm \sqrt{5}$    **6.** $-2 \pm \sqrt{3}$    **7.** $4 \pm \dfrac{\sqrt{14}}{2}$

**8.** $-5 \pm \dfrac{\sqrt{3}}{3}$    **9.** $-\dfrac{3}{2} \pm \dfrac{\sqrt{5}}{2}$    **10.** $-\dfrac{3}{2} \pm \dfrac{\sqrt{21}}{2}$

**1.** g    **2.** e    **3.** c    **4.** f    **5.** b    **6.** a

**7.** h    **8.** d    **9.** $y = -(x + 2)^2$

**11.** $y = (x - 3)^2 - 9$    **13.** $y = -2(x + 3)^2 + 3$

**15.** Intercept: $(0, 2)$    **17.** Intercepts: $(\pm 4, 0), (0, 16)$
     Vertex: $(0, 2)$       Vertex: $(0, 16)$

**19.** Intercepts:
$\left(-5 \pm \sqrt{6}, 0\right), (0, 19)$
Vertex: $(-5, -6)$

**21.** Intercepts:
$\left(2 \pm \sqrt{2}, 0\right), (0, 2)$
Vertex: $(2, -2)$

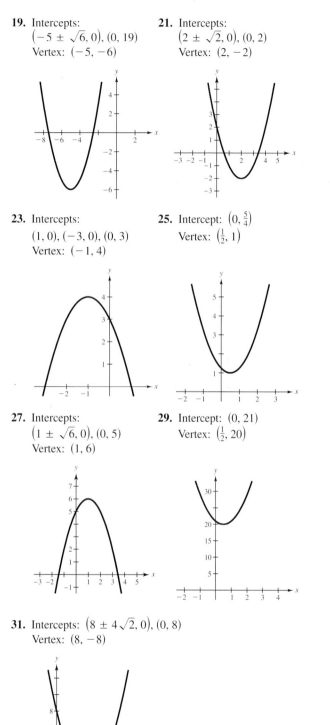

**23.** Intercepts:
$(1, 0), (-3, 0), (0, 3)$
Vertex: $(-1, 4)$

**25.** Intercept: $\left(0, \frac{5}{4}\right)$
Vertex: $\left(\frac{1}{2}, 1\right)$

**27.** Intercepts:
$\left(1 \pm \sqrt{6}, 0\right), (0, 5)$
Vertex: $(1, 6)$

**29.** Intercept: $(0, 21)$
Vertex: $\left(\frac{1}{2}, 20\right)$

**31.** Intercepts: $\left(8 \pm 4\sqrt{2}, 0\right), (0, 8)$
Vertex: $(8, -8)$

**33.** $y = -\frac{1}{2}(x - 2)^2 - 1$   **35.** $y = \frac{3}{4}(x - 5)^2 + 12$

**37.** Answers will vary.        **39.** Answers will vary.
$f(x) = x^2 - x - 2$            $f(x) = x^2 - 10x$
$g(x) = -x^2 + x + 2$           $g(x) = -x^2 + 10x$

**41.** Answers will vary.
$f(x) = 2x^2 + 7x + 3$
$g(x) = -2x^2 - 7x - 3$

**43.** $A = 50x - x^2$; 25 feet × 25 feet

**45.** $x = 25$ feet, $y = 33\frac{1}{3}$ feet; 50 feet × $33\frac{1}{3}$ feet

**47.** 45,000 units       **49.** 20 fixtures       **51.** 14 feet

**53.** (a)

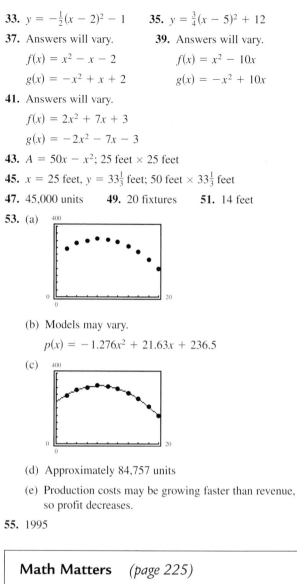

(b) Models may vary.
$$p(x) = -1.276x^2 + 21.63x + 236.5$$

(c)

(d) Approximately 84,757 units

(e) Production costs may be growing faster than revenue, so profit decreases.

**55.** 1995

---

**Math Matters**   *(page 225)*

Both regions have the same area, $9\pi r^2$.

---

**Section 4.2**   *(page 233)*

---

**Warm Up**   *(page 233)*

**1.** $(3x - 2)(4x + 5)$       **2.** $x(5x - 6)^2$

**3.** $z^2(12z + 5)(z + 1)$       **4.** $(y + 5)(y^2 - 5y + 25)$

**5.** $(x + 3)(x + 2)(x - 2)$       **6.** $(x + 2)(x^2 + 3)$

**7.** No real solution       **8.** $3 \pm \sqrt{5}$

**9.** $-\frac{1}{2} \pm \sqrt{3}$       **10.** $\pm 3$

**1.** e    **2.** c    **3.** g    **4.** d    **5.** f    **6.** h

**7.** a    **8.** b    **9.** Rises to the left
                          Falls to the right

**11.** Rises to the left    **13.** Rises to the left
     Falls to the right         Rises to the right

**15.** Rises to the left    **17.** Falls to the left
     Rises to the right         Falls to the right

**19.** $\pm 4$    **21.** $-4$    **23.** $1, -2$    **25.** No real zeros

**27.** $2, 0$    **29.** $\pm 1$    **31.** $\pm\sqrt{5}$    **33.** $3$

**35.**    **37.**

**39.**    **41.**

**43.**    **45.**

**47.**    **49.**

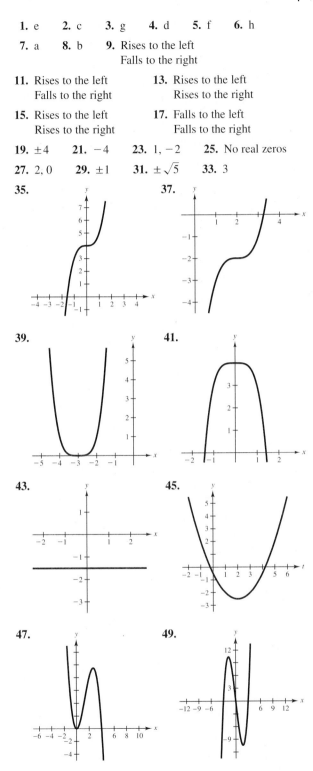

**51.**    **53.**

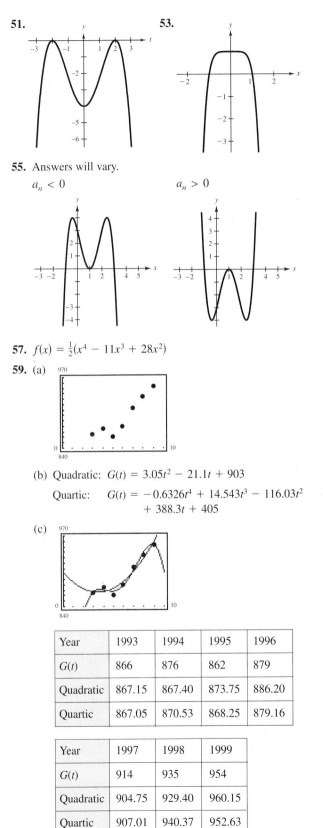

**55.** Answers will vary.
    $a_n < 0$                      $a_n > 0$

**57.** $f(x) = \frac{1}{2}(x^4 - 11x^3 + 28x^2)$

**59.** (a)

(b) Quadratic: $G(t) = 3.05t^2 - 21.1t + 903$

    Quartic:   $G(t) = -0.6326t^4 + 14.543t^3 - 116.03t^2$
                             $+ 388.3t + 405$

(c)

| Year | 1993 | 1994 | 1995 | 1996 |
|---|---|---|---|---|
| $G(t)$ | 866 | 876 | 862 | 879 |
| Quadratic | 867.15 | 867.40 | 873.75 | 886.20 |
| Quartic | 867.05 | 870.53 | 868.25 | 879.16 |

| Year | 1997 | 1998 | 1999 |
|---|---|---|---|
| $G(t)$ | 914 | 935 | 954 |
| Quadratic | 904.75 | 929.40 | 960.15 |
| Quartic | 907.01 | 940.37 | 952.63 |

(d) The quartic model represents the data more accurately because the values are closer to the actual values, $G(t)$.

**61.** $(200, 320)$

**63.** The functions have a common shape. The graphs are not identical.

## Section 4.3 *(page 243)*

> ### Warm Up *(page 243)*
>
> **1.** $x^3 - x^2 + 2x + 3$  **2.** $2x^3 + 4x^2 - 6x - 4$
>
> **3.** $x^4 - 2x^3 + 4x^2 - 2x - 7$
>
> **4.** $2x^4 + 12x^3 - 3x^2 - 18x - 5$
>
> **5.** $(x - 3)(x - 1)$  **6.** $2x(2x - 3)(x - 1)$
>
> **7.** $x^3 - 7x^2 + 12x$  **8.** $x^2 + 5x - 6$
>
> **9.** $x^3 + x^2 - 7x - 3$
>
> **10.** $x^4 - 3x^3 - 5x^2 + 9x - 2$

**1.** $3x - 4$  **3.** $x + 3$  **5.** $x^3 + 3x^2 - 1$

**7.** $7 - \dfrac{11}{x + 2}$  **9.** $3x + 5 - \dfrac{2x - 3}{2x^2 + 1}$

**11.** $x^2 - 3x + 2$

**13.** $2x^3 + 4x^2 - 2x - 8 - \dfrac{10x - 7}{x^2 - 2x + 1}$

**15.** $2x^2 - 3x + 5$  **17.** $4x^2 - 9$

**19.** $-x^2 + 10x - 25$  **21.** $5x^2 + 14x + 56 + \dfrac{232}{x - 4}$

**23.** $10x^3 + 10x^2 + 60x + 360 + \dfrac{1360}{x - 6}$

**25.** $2x^4 + 8x^3 + 2x^2 + 8x - 5 - \dfrac{7}{x - 4}$

**27.** $-3x^3 - 6x^2 - 12x - 24 - \dfrac{48}{x - 2}$

**29.** $-x^2 + 3x - 6 + \dfrac{11}{x + 1}$  **31.** $4x^2 + 14x - 30$

**33.** $(x + 2)(x + 2)(x - 4)$ or $(x + 2)^2(x - 4)$

**35.** $(2x - 3)(x + 4)(x - 1)$

**37.** $(x + 2)(x + \sqrt{3})(x - \sqrt{3})$

**39.** $f(x) = (2x - 1)(x^2 - 4x + 3) - 6$, $f\left(\frac{1}{2}\right) = -6$

**41.** $f(x) = (x - \sqrt{2})[x^2 + (3 + \sqrt{2})x + 3\sqrt{2}] - 8$, $f(\sqrt{2}) = -8$

**43.** (a) $-69$  (b) $-6306$  (c) $-6$  (d) $446$

**45.** (a) $1$  (b) $-267$  (c) $-\frac{11}{3}$  (d) $-3.8$

**47.** (a) $72$  (b) $0$  (c) $37.648$  (d) $30$

**49.** b; $3, \dfrac{-1 \pm \sqrt{17}}{2}$  **50.** d; $-2, \dfrac{3 \pm \sqrt{5}}{2}$

**51.** a; $-1, -2 \pm \sqrt{2}$  **52.** c; $3, 1 \pm \sqrt{5}$

**53.** $f(x) = 3x^3 - 13x^2 + 4x + 20$

$f(x) = -3x^3 + 13x^2 - 4x - 20$

Infinitely many polynomial functions

**55.** $x^2 + x - 2$  **57.** $3x^2 + 5x + 7$  **59.** $x^2 + 3x$

**61.** $x^2 + 10x + 24$ square feet  **63.** (a) \$192,116

## Section 4.4 *(page 255)*

> ### Warm Up *(page 255)*
>
> **1.** $f(x) = 3x^3 - 8x^2 - 5x + 6$
>
> **2.** $f(x) = 4x^4 - 3x^3 - 16x^2 + 12x$
>
> **3.** $x^4 - 3x^3 + 5 + \dfrac{3}{x + 3}$
>
> **4.** $3x^3 + 15x^2 - 9 - \dfrac{2}{x + (2/3)}$
>
> **5.** $\frac{1}{2}, -3 \pm \sqrt{5}$  **6.** $10, -\frac{2}{3}, -\frac{3}{2}$  **7.** $-\frac{3}{4}, 2 \pm \sqrt{2}$
>
> **8.** $\frac{2}{5}, -\frac{7}{2}, -2$  **9.** $\pm\sqrt{2}, \pm 1$  **10.** $\pm 2, \pm\sqrt{3}$

**1.** Possible: $\pm 1, \pm 2, \pm 4$  **3.** Possible: $\pm\frac{1}{4}, \pm\frac{1}{2}, \pm 1, \pm 2, \pm 4$

Actual: $-1, \pm 2$  Actual: $\pm\frac{1}{2}, \pm 2$

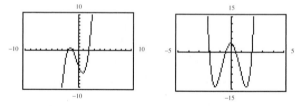

**5.** $1, 2, 3$  **7.** $-1, -10$  **9.** $\frac{1}{2}, -1$  **11.** $\pm 1, \pm\sqrt{2}$

**13.** $-1, 2$  **15.** $-6, \frac{1}{2}, 1$  **17.** $-2, 0, 1$

**19.** (a) $\pm 1, \pm 3, \pm\frac{1}{2}, \pm\frac{3}{2}, \pm\frac{1}{4}, \pm\frac{3}{4}, \pm\frac{1}{8}, \pm\frac{3}{8}, \pm\frac{1}{16}, \pm\frac{3}{16}, \pm\frac{1}{32}, \pm\frac{3}{32}$

(b)  (c) $1, \frac{3}{4}, -\frac{1}{8}$

**21.** $\pm 2, \pm \frac{3}{2}$    **23.** Real zero $\approx 0.7$    **25.** Real zero $\approx 3.3$

**27.** e; $-1.769$    **28.** c; 0.755    **29.** d; 0.206

**30.** a; 0.266, 1.175, 2.559    **31.** f; 2.769

**32.** b; $-1.675, -0.539, 2.21$

**33.** $-1.164, 1.453$    **35.** 0.529    **37.** $-1.453, 1.164$

**39.** $-2.177, 1.563$

**41.** (a) $V = x(12 - 2x)(10 - 2x)$

Domain: $0 \le x \le 5$

(b)

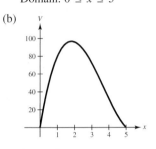

Approximate measurements:
2 inches $\times$ 8 inches $\times$ 6 inches or
1.81 inches $\times$ 8.38 inches $\times$ 6.38 inches

(c) $x \approx 1.628, 2, \approx 7.372$

$x \approx 7.372$ inches is physically impossible because $10 - 2(7.372)$ and $12 - 2(7.372)$ would yield a negative length and width.

**43.** 18 inches $\times$ 18 inches $\times$ 36 inches

**45.** 4.5 hours

**47.** (a)

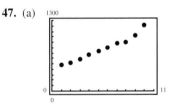

(b) Models may vary.

Linear: $I = 74.78t + 371.6$

Quadratic: $I = 4.183t^2 + 28.77t + 463.6$

Cubic: $I = 0.9923t^3 - 12.190t^2 + 104.28t + 378.5$

Quartic: $I = 0.49254t^4 - 9.8436t^3 + 67.109t^2$
$- 112.43t + 547.5$

(c) Linear          Quadratic

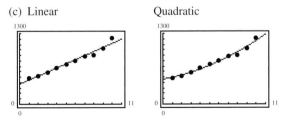

Cubic          Quartic

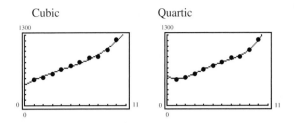

(d) The quartic model fits the data best because it follows the scatter plot most accurately. According to the model, imports reached \$1500 billion in 2001.

**49.** (a)

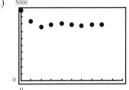

(b) Models may vary.

Cubic: $y = -11.602t^3 + 173.22t^2 - 766.4t$
$+ 4885$

Quartic: $y = 3.8995t^4 - 73.993t^3 + 483.51t^2$
$- 1252.2t + 4978$

(c) Cubic          Quartic

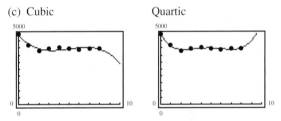

Answers will vary. The quartic model fits the data well. The cubic model does not fit the data as well as the quartic model.

(d) 2000: Cubic: 2941 tons

Quartic: 5809 tons

2005: Cubic: $-6793$ tons

Quartic: 42,671 tons

Answers will vary. The estimates for each model are not reasonable for years after 1998. The cubic model quickly turns in the negative direction and the quartic model quickly turns in the positive direction. Judging from the points in the scatter plot, future emissions would not likely change so drastically, without major regulation changes.

**51.** 1.5 inches $\times$ 5 inches $\times$ 7 inches or
1.445 inches $\times$ 5.11 inches $\times$ 7.11 inches

## Mid-Chapter Quiz    *(page 259)*

**1.** Vertex: $(-1, -2)$

Intercepts: $\left(-1 \pm \sqrt{2}, 0\right), (0, -1)$

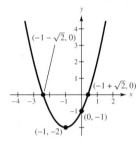

**2.** Vertex: $(0, 16)$

Intercepts: $(\pm 4, 0), (0, 16)$

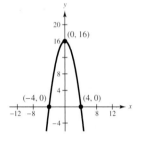

**3.** Falls to the left      **4.** Falls to the left
   Rises to the right            Falls to the right

**5.** $(x + 2)(x - 3)(2x + 1)$

**6.** 13

**7.** $f(x) = (x - 1)(x^3 + x^2 - 4x - 4) + 0; f(1) = 0$

**8.** $f(x) = (x + 3)(x^2 + 2x - 8) + 0; f(-3) = 0$

**9.** $2x^2 + 5x - 12$      **10.** $P = \$2,534,375,\ x \approx 33.76$

**11.** $\pm\sqrt{5}, -\frac{7}{2}$      **12.** $\pm 3, \pm\frac{1}{2}$      **13.** $1, -\frac{4}{3}$      **14.** $\frac{3}{2}$

**15.** (a)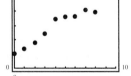

(b) Models may vary.

Quadratic: $L = -18.45t^2 + 354.9t + 404$

Cubic: $L = -5.784t^3 + 50.95t^2 + 145.6t + 501$

Quartic: $L = 1.7963t^4 - 34.525t^3 + 193.89t^2$
$- 78.2t + 544$

(c) Quadratic          Cubic

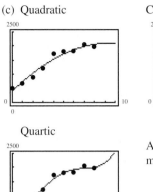

Quartic

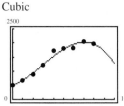

Answers will vary. Each
model fits the data well.

(d) 2000: Quadratic: 2108 leases

Cubic: 1268 leases

Quartic: 2589 leases

2001: Quadratic: 2075 leases

Cubic: 569 leases

Quartic: 3491 leases

Answers will vary. The predictions given by both the
quadratic and the quartic models could be accurate,
depending on the rate at which the number of consumer new passenger car leases increased. The cubic
model does not follow the path indicated by the scatter
plot after 1998, so its predictions are not reasonable.

## Section 4.5    *(page 268)*

### Warm Up    *(page 268)*

**1.** $2\sqrt{3}$    **2.** $10\sqrt{5}$    **3.** $\sqrt{5}$    **4.** $-6\sqrt{3}$

**5.** 12    **6.** 48    **7.** $\dfrac{\sqrt{3}}{3}$    **8.** $\sqrt{2}$

**9.** $-\dfrac{1}{2} \pm \dfrac{\sqrt{5}}{2}$    **10.** $-1 \pm \sqrt{2}$

**1.** $i, -1, -i, 1, i, -1, -i, 1,\ i, -1, -i, 1, i, -1, -i, 1$

**3.** $a = 3, b = 12$    **5.** $a = 4, b = -3$

**7.** $9 + 4i, 9 - 4i$    **9.** $-3 - 2\sqrt{3}i, -3 + 2\sqrt{3}i$

**11.** $-21, -21$    **13.** $-1 - 6i, -1 + 6i$    **15.** $-5i, 5i$

**17.** $-3, -3$    **19.** $2 + i$    **21.** $5 + 6i$

**23.** $3 - 3\sqrt{2}i$    **25.** $\frac{1}{6} + \frac{7}{6}i$    **27.** $-2\sqrt{6}$    **29.** $-10$

**31.** $5 + i$    **33.** 41    **35.** $21 + 42i$    **37.** 8

**39.** $\left(16 + 4\sqrt{3}\right) + \left(-16\sqrt{2} + 2\sqrt{6}\right)i$    **41.** $-9 + 40i$

**43.** $\frac{3}{5} + \frac{4}{5}i$    **45.** $1 + \frac{1}{2}i$    **47.** $-7 - 6i$    **49.** $\frac{1}{8}i$

**51.** $-\frac{5}{4} - \frac{5}{4}i$    **53.** $\frac{35}{29} + \frac{595}{29}i$    **55.** $-10$

**57.** Error: $(3 - 2i)(3 + 2i) = 9 - 4i^2 = 9 + 4 = 13$
(not $9 - 4 = 5$)

**59.** $1 \pm i$    **61.** $-2 \pm \frac{1}{2}i$    **63.** $-\frac{3}{2}, -\frac{5}{2}$

**65.** $\frac{1}{8} \pm \frac{\sqrt{11}}{8}i$    **67.**

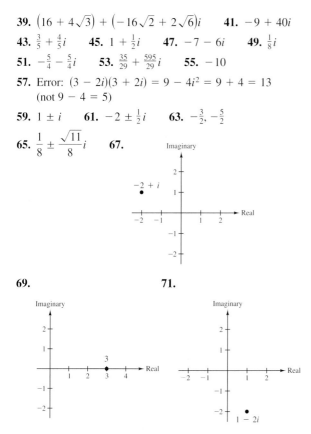

**69.**    **71.**

**73.** The complex number 0 is in the Mandelbrot Set because, for $c = 0$, the corresponding Mandelbrot sequence is 0, 0, 0, 0, 0, 0, which is bounded.

**75.** The complex number $\frac{1}{2}i$ is in the Mandelbrot Set because, for $c = \frac{1}{2}i$, the corresponding Mandelbrot sequence is $\frac{1}{2}i$, $-\frac{1}{4} + \frac{1}{2}i$, $-\frac{3}{16} + \frac{1}{4}i$, $-\frac{7}{256} + \frac{13}{32}i$, $-\frac{10,767}{65,636} + \frac{1957}{4096}i$, $-\frac{864,513,055}{4,294,967,296} + \frac{46,037,845}{134,217,728}i$, which is bounded.

**77.** The complex number 1 is not in the Mandelbrot Set because, for $c = 1$, the corresponding Mandelbrot sequence is 1; 2; 5; 26; 677; 458,330; . . . ; which is unbounded.

## Section 4.6   (page 276)

### Warm Up   (page 276)

**1.** $4 - \sqrt{29}i, 4 + \sqrt{29}i$    **2.** $-5 - 12i, -5 + 12i$

**3.** $-1 + 4\sqrt{2}i, -1 - 4\sqrt{2}i$    **4.** $6 + \frac{1}{2}i, 6 - \frac{1}{2}i$

**5.** $-13 + 9i$    **6.** $12 + 16i$    **7.** $26 + 22i$

**8.** 29    **9.** $i$    **10.** $-9 + 46i$

**1.** $\pm4i, (x + 4i)(x - 4i)$

**3.** $\frac{5 \pm \sqrt{5}}{2}, \left(x - \frac{5 + \sqrt{5}}{2}\right)\left(x - \frac{5 - \sqrt{5}}{2}\right)$

**5.** $\pm4, \pm4i, (x - 4)(x + 4)(x - 4i)(x + 4i)$

**7.** $0, \pm\sqrt{5}i, x\left(x - \sqrt{5}i\right)\left(x + \sqrt{5}i\right)$

**9.** $2, 2 \pm i, (x - 2)(x - 2 + i)(x - 2 - i)$

**11.** $-5, 4 \pm 3i, (t + 5)(t - 4 + 3i)(t - 4 - 3i)$

**13.** $-10, -7 \pm 5i, (x + 10)(x + 7 - 5i)(x + 7 + 5i)$

**15.** $-\frac{3}{4}, 1 \pm \frac{1}{2}i, (4x + 3)(2x - 2 + i)(2x - 2 - i)$

**17.** $-2, 1 \pm \sqrt{2}i, (x + 2)\left(x - 1 + \sqrt{2}i\right)\left(x - 1 - \sqrt{2}i\right)$

**19.** $-\frac{1}{5}, 1 \pm \sqrt{5}i, (5x + 1)\left(x - 1 + \sqrt{5}i\right)\left(x - 1 - \sqrt{5}i\right)$

**21.** $2, \pm2i, (x - 2)^2(x + 2i)(x - 2i)$

**23.** $\pm i, \pm3i, (x + i)(x - i)(x + 3i)(x - 3i)$

**25.** $-4, 3, \pm i, (t + 4)^2(t - 3)(t + i)(t - i)$

**27.** $x^3 + 3x^2 + 36x + 108$    **29.** $x^3 - 5x^2 + 9x - 5$

**31.** $x^5 + 4x^4 + 13x^3 + 52x^2 + 36x + 144$

**33.** $x^4 + 8x^3 + 9x^2 - 10x + 100$

**35.** $16x^4 + 36x^3 + 16x^2 + x - 30$

**37.** (a) $(x^2 - 8)(x^2 + 1)$

(b) $\left(x - 2\sqrt{2}\right)\left(x + 2\sqrt{2}\right)(x^2 + 1)$

(c) $\left(x - 2\sqrt{2}\right)\left(x + 2\sqrt{2}\right)(x - i)(x + i)$

**39.** (a) $(x^2 - 2x - 2)(x^2 - 2x + 3)$

(b) $\left(x - 1 + \sqrt{3}\right)\left(x - 1 - \sqrt{3}\right)(x^2 - 2x + 3)$

(c) $\left(x - 1 + \sqrt{3}\right)\left(x - 1 - \sqrt{3}\right) \cdot$
$\left(x - 1 + \sqrt{2}i\right)\left(x - 1 - \sqrt{2}i\right)$

**41.** $\pm4i, \frac{5}{3}$    **43.** $\pm6i, 1$    **45.** $-3 \pm i, \frac{1}{4}$

**47.** $1, 2, -3 \pm \sqrt{2}i$    **49.** $\frac{3}{4}, \frac{1}{2} \pm \frac{\sqrt{5}}{2}i$

**51.** Answers will vary.

$f(x) = x^4 - 7x^3 + 17x^2 - 17x + 6$
four real zeros: 1, 1, 2, 3

$g(x) = x^4 - 3x^3 + 3x^2 - 3x + 2$
two real zeros: 1, 2; two complex zeros: $\pm i$

$h(x) = x^4 + 5x^2 + 4$
four complex zeros: $\pm i, \pm4i$

Similarities: $f$, $g$, and $h$ all rise to the left and rise to the right, and are similar in shape.

Differences: differences in the number of intersections with the $x$-axis

**53.** Solving the equation for $P = 9,000,000$ and graphing shows no real solutions.

**55.** No. The conjugate pair statement specifies polynomials with *real* coefficients. $f(x)$ has imaginary coefficients.

**57.** No. $\left(x + \sqrt{3}\right)$ and $\left(x - \sqrt{3}\right)$ are not rational factors.

## Section 4.7    *(page 286)*

### Warm Up    *(page 286)*

**1.** $x(x - 4)$     **2.** $2x(x^2 - 3)$     **3.** $(x - 5)(x + 2)$

**4.** $(x - 5)(x - 2)$     **5.** $x(x + 1)(x + 3)$

**6.** $(x^2 - 2)(x - 4)$

**7.**                        **8.**

**9.**                        **10.**

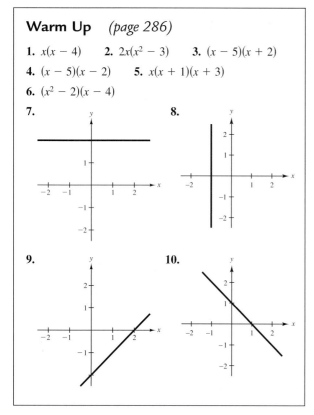

**1.** Domain: all $x \neq -2$    **3.** Domain: all $x \neq 3$

**5.** Domain: all reals    **7.** Domain: all reals    **9.** f

**10.** e    **11.** a    **12.** b    **13.** c    **14.** d

**15.** $g(x)$ shifts down 2 units.    **17.** $g(x)$ is a reflection about
the $x$-axis.

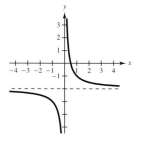    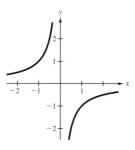

**19.** $g(x)$ shifts up 3 units.    **21.** $g(x)$ is a reflection about
the $x$-axis.

**23.** $g(x)$ shifts up 5 units.    **25.** $g(x)$ is a reflection about
the $x$-axis.

**27.**    **29.**

**31.**    **33.**

**35.**    **37.**

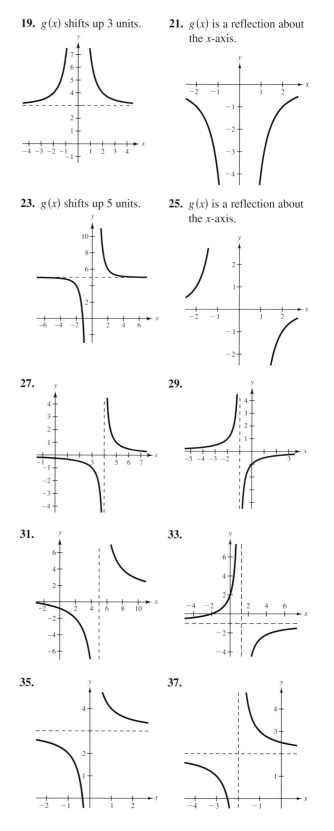

**39.**
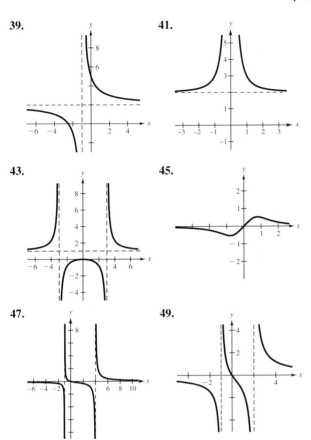

**41.**

**43.**

**45.**

**47.**

**49.**

**51.** Asymptotes: none

Graph is equivalent to the graph of $y = x - 4$, $x \ne 3$.

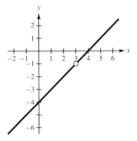

**53.** Asymptotes: vertical, $x = -1$; slant, $y = x - 1$

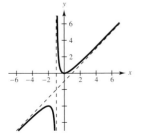

**55.** Answers will vary.

$$f(x) = \frac{1}{x^2 + 1}$$

**57.** Answers will vary.

$$f(x) = \frac{x}{x^2 - 2x - 3}$$

**59.** (a) $176 million
(b) $528 million
(c) $1584 million
(d) No. The model would generate a zero in the denominator.

**61.** (a) 167, 250, 400    (b) 750

**63.** (a)
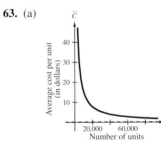

(b)

| $x$ | 1000 | 10,000 | 100,000 |
|---|---|---|---|
| $\overline{C}$ | $150.25 | $15.25 | $1.75 |

Eventually, the average cost per unit will approach the horizontal asymptote of $0.25.

**65.** (a)

| $n$ | 1 | 2 | 3 | 4 | 5 |
|---|---|---|---|---|---|
| $P$ | 0.50 | 0.74 | 0.82 | 0.86 | 0.89 |

| $n$ | 6 | 7 | 8 | 9 | 10 |
|---|---|---|---|---|---|
| $P$ | 0.91 | 0.92 | 0.93 | 0.94 | 0.95 |

(b) 100%

**67.** (a)
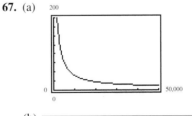

(b)

| $x$ | 1000 | 10,000 | 100,000 |
|---|---|---|---|
| $\overline{C}$ | $355 | $40 | $8.5 |

Eventually, the average recycling cost per pound will approach the horizontal asymptote of $5.

**69.** The sales of long-playing albums dropped after cassette tapes and compact discs were introduced.

**71.** (a) $\approx 1.56$ hours

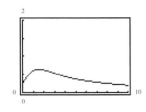

(b) Yes, the function has a horizontal asymptote of $y = 0$, which means that eventually the concentration of the medication will be untraceable.

## Review Exercises   *(page 292)*

**1.**

**3.**

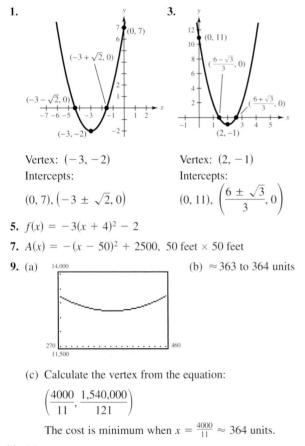

Vertex: $(-3, -2)$

Vertex: $(2, -1)$

Intercepts:

Intercepts:

$(0, 7), \left(-3 \pm \sqrt{2}, 0\right)$

$(0, 11), \left(\dfrac{6 \pm \sqrt{3}}{3}, 0\right)$

**5.** $f(x) = -3(x + 4)^2 - 2$

**7.** $A(x) = -(x - 50)^2 + 2500,\ 50\ \text{feet} \times 50\ \text{feet}$

**9.** (a)

(b) $\approx 363$ to $364$ units

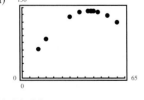

(c) Calculate the vertex from the equation:

$$\left(\frac{4000}{11}, \frac{1,540,000}{121}\right)$$

The cost is minimum when $x = \frac{4000}{11} \approx 364$ units.

**11.** (a)

(b) Models may vary.

$d(x) = -0.077x^2 + 6.59x + 2.4$

(c)

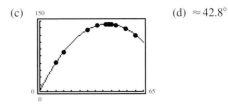

(d) $\approx 42.8°$

**13.** Falls to the left    **15.** Rises to the left

Rises to the right          Falls to the right

**17.** $\pm 5$     **19.** $0, 2, 5$     **21.** $x^2 - 3x + 1 - \dfrac{1}{2x + 1}$

**23.** $x^2 + 11x + 24$     **25.** $x^2 - 3x + 3$

**27.** $(x - 5)(x - 2)(x + 3)$     **29.** (a) $-10$   (b) $11$

**31.** $x^2 + 11x + 28$ square feet

**33.** $\pm 1, \pm 3, \pm 5, \pm 15, \pm\frac{1}{2}, \pm\frac{3}{2}, \pm\frac{5}{2}, \pm\frac{15}{2}, \pm\frac{1}{4}, \pm\frac{3}{4}, \pm\frac{5}{4}, \pm\frac{15}{4}$

From the graph: $x \approx 2.357$

**35.** $x \approx -2.3$     **37.** $-3, -1, 2$     **39.** $\pm 2, \pm\sqrt{5}$

**41.** $-2, 1, \pm\dfrac{\sqrt{3}}{3}$     **43.** $-1.321, -0.283, 1.604$

**45.** \$212,000 or \$509,000

**47.** $4\sqrt{2}i, -4\sqrt{2}i$     **49.** $-3 + 4i, -3 - 4i$

**51.** $5 + i$     **53.** $11 + 9\sqrt{3}i$     **55.** $89$     **57.** $-10 - 8i$

**59.** $-7 + 24i$     **61.** $3 - 2i$     **63.** $-3 - 4i$     **65.** $10$

**67.** $\dfrac{1 \pm \sqrt{23}i}{4}$     **69.** $\dfrac{-11 \pm \sqrt{73}}{8}$

**71.**

**73.** $(x - 3)(x + 3)(x - 3i)(x + 3i)$

**75.** $(t + 5)\left(t - \sqrt{3}i\right)\left(t + \sqrt{3}i\right)$

**77.** $(x - 2)(2x - 3i)(2x + 3i)$     **79.** $x^3 - 3x^2 + 16x - 48$

**81.** (a) $(x^2 + 8)(x^2 - 3)$

(b) $(x^2 + 8)\left(x - \sqrt{3}\right)\left(x + \sqrt{3}\right)$

(c) $\left(x + 2\sqrt{2}i\right)\left(x - 2\sqrt{2}i\right)\left(x - \sqrt{3}\right)\left(x + \sqrt{3}\right)$

**83.** $\pm 4i, \frac{1}{4}$     **85.** $-1 \pm 3i, -1, -4$

**87.** Domain: all $x \neq 6$

Vertical asymptote: $x = 6$

Horizontal asymptote: $y = 0$

**89.** Domain: all $x \neq \pm 2$
Vertical asymptotes: $x = -2, x = 2$
Horizontal asymptote: $y = 1$

**91.**

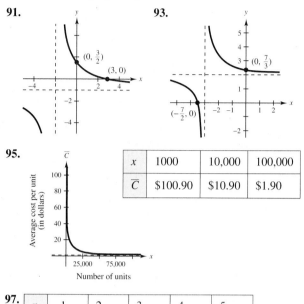

(0, $\frac{3}{2}$)
(3, 0)

**93.**

(0, $\frac{7}{3}$)
$\left(-\frac{7}{2}, 0\right)$

**95.**

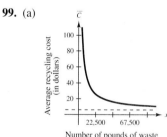

Average cost per unit (in dollars)

Number of units

| $x$ | 1000 | 10,000 | 100,000 |
|---|---|---|---|
| $\overline{C}$ | \$100.90 | \$10.90 | \$1.90 |

**97.**

| $n$ | 1 | 2 | 3 | 4 | 5 |
|---|---|---|---|---|---|
| $P$ | 0.5 | 0.737 | 0.821 | 0.865 | 0.891 |

| $n$ | 6 | 7 | 8 | 9 | 10 |
|---|---|---|---|---|---|
| $P$ | 0.909 | 0.922 | 0.932 | 0.939 | 0.945 |

**99.** (a)

$\overline{C}$

Average recycling cost (in dollars)

100
80
60
40
20

22,500  67,500

Number of pounds of waste

(b)
| $x$ | 1000 | 10,000 | 100,000 |
|---|---|---|---|
| $\overline{C}$ | \$456 | \$51 | \$10.50 |

Eventually, the average recycling cost will approach the horizontal asymptote of \$6.

**Chapter Test** *(page 296)*

**1.** Vertex: $(-1, 2)$
Intercepts: $\left(-1 \pm \sqrt{2}, 0\right), (0, 1)$
Line of symmetry: $x = -1$

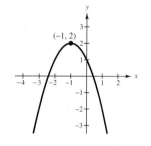

(−1, 2)

**2.** (a) Falls to the left
Rises to the right

(b) Rises to the left
Rises to the right

**3.** $\pm 1, \pm 5, \pm\frac{1}{2}, \pm\frac{5}{2}$
$f\left(\frac{5}{2}\right) = 0$
$f(x) = (2x - 5)(x - i)(x + i)$

**4.** $x^2 + 7x + 12$   **5.** $16 - 3i$   **6.** $7 - 9i$

**7.** $27 - 4\sqrt{3}i$   **8.** $23 - 14i$   **9.** $-5 - 12i$

**10.** $21 + 20i$   **11.** $i$   **12.** $-2 - 5i$

**13.** $x = \dfrac{-5 \pm \sqrt{3}i}{2}$   **14.** $x = \dfrac{5 \pm 3\sqrt{7}i}{4}$

**15.** $\pm\sqrt{5}i, -2$   **16.** $x^4 - 7x^3 + 19x^2 - 63x + 90$

**17.**

Intercept: $(0, 0)$
Vertical asymptote:
  $x = -1$
Horizontal asymptote:
  $y = 2$
Domain: all $x \neq -1$

**18.** $x = 2500$ units

**19.**

| $x$ | 10,000 | 100,000 | 1,000,000 |
|---|---|---|---|
| $\overline{C}$ | \$50 | \$9.50 | \$5.45 |

Conclusion: High volume yields lower average cost of recycling.

**20.** \$200,000 $(x = 20)$ and \$777,250 $(x = 77.725)$ both result in profits of \$222,000. The company should spend \$200,000, which is the smaller amount.

# Chapter 5

## Section 5.1    (page 306)

---

**Warm Up**    (page 306)

**1.** $5^x$    **2.** $3^{2x}$    **3.** $4^{3x}$    **4.** $10^x$    **5.** $4^{2x}$

**6.** $4^{10x}$    **7.** $\left(\frac{3}{2}\right)^x$    **8.** $4^{3x}$    **9.** $2^{-x}$    **10.** $16^{x/4}$

---

**1.** 3.463    **3.** 95.946    **5.** 0.079    **7.** 54.598

**9.** 1.948    **11.** g    **12.** e    **13.** b    **14.** h    **15.** d

**16.** a    **17.** f    **18.** c

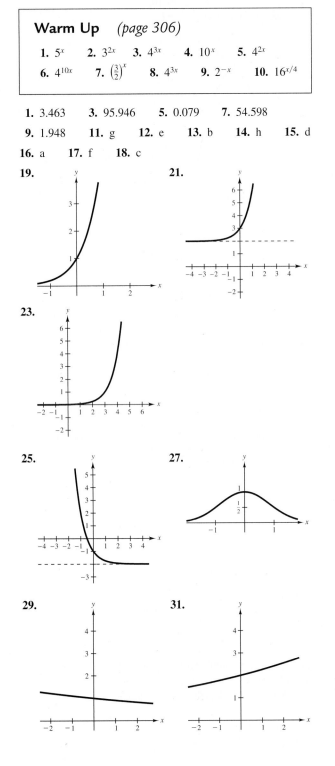

**19.**    **21.**

**23.**

**25.**    **27.**

**29.**    **31.**

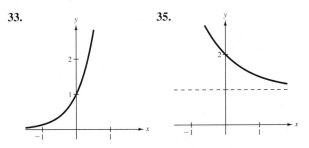

**33.**    **35.**

**37.** The graphs have similar shapes and are increasing as $x$ gets larger. The larger the base, the more quickly the $y$-values increase. The graphs have similar shapes and are increasing as $x$ gets larger, but the larger the base, the smaller the $y$-values and the closer the graph is to the $x$-axis.

**39.**

| $n$ | 1 | 2 | 4 |
|---|---|---|---|
| $A$ | $7346.64 | $7401.22 | $7429.74 |

| $n$ | 12 | 365 | Continuous compounding |
|---|---|---|---|
| $A$ | $7449.23 | $7458.80 | $7459.12 |

**41.**

| $n$ | 1 | 2 | 4 |
|---|---|---|---|
| $A$ | $24,115.73 | $25,714.29 | $26,602.23 |

| $n$ | 12 | 365 | Continuous compounding |
|---|---|---|---|
| $A$ | $27,231.38 | $27,547.07 | $27,557.94 |

**43.**

| $t$ | 1 | 10 | 20 |
|---|---|---|---|
| $P$ | $91,393.12 | $40,656.97 | $16,529.89 |

| $t$ | 30 | 40 | 50 |
|---|---|---|---|
| $P$ | $6720.55 | $2732.37 | $1110.90 |

**45.**

| $t$ | 1 | 10 | 20 |
|---|---|---|---|
| $P$ | $90,521.24 | $36,940.70 | $13,646.15 |

| $t$ | 30 | 40 | 50 |
|---|---|---|---|
| $P$ | $5040.98 | $1862.17 | $687.90 |

**43.** One possible solution:

$C = -\dfrac{a}{2} + 250{,}000$

$M = \dfrac{a}{2} + 125{,}000$

$B = -a + 125{,}000$

$G = a$

**45.** $y = 0.079x^2 + 0.63x + 2.9$

**47.** $y = -0.207x^2 - 0.89x + 5.1$

**49.** Models may vary.

$y = 0.0136x^2 + 1.201x + 7.94;$

17.0134 million personal computers in 2005

**51.** (a)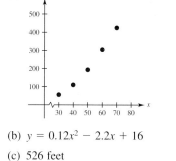

(b) $y = 0.12x^2 - 2.2x + 16$

(c) 526 feet

**53.** (a)

(b) $y = -0.93x^2 + 25.6x + 192$

(c) 1999: \$347,070,000,000

2000: \$355,000,000,000

**55.** $s = -16t^2 - 20t + 220$

$(g = -32, v_0 = -20, s_0 = 220)$

## Mid-Chapter Quiz  *(page 395)*

**1.** Answers will vary.

$\begin{aligned} 3x + y &= 11 \\ x - 2y &= -1 \end{aligned}$

**2.** Answers will vary.

$\begin{aligned} x + 2y + z &= 4 \\ 2x - y - z &= 9 \\ 3y + 2z &= 0 \end{aligned}$

**3.** $(1, 3)$    **4.** $(0.927, 2.854), (-1.727, -2.454)$

**5.** 1342 units   **6.** 500,000 units   **7.** $(2, -1)$

**8.** $\left(1, \dfrac{3}{2}\right)$   **9.** $x = 18{,}333$ units

$p = \$26.67$

**10.** $x = 71{,}429$ units   **11.** $(1, -2, 3)$

$p = \$80.71$

**12.** Answers will vary. $(a + 6, a + 6, a)$, $a$ is any real number

**13.** Inconsistent

**14.** Answers will vary.        **15.** Answers will vary.

$a = 1: (1, -1, 3)$           $a = 0: (0, 5, 0)$

$a = 0: (0, -2, 0)$           $a = 2: (4, 7, 2)$

$a = 3: (3, 1, 9)$            $a = 1: (2, 6, 1)$

**16.** $s = -16t^2 + 10t + 150$        **17.** Infinitely many

$(g = -32, v_0 = 10, s_0 = 150)$

**18.** $3\tfrac{1}{3}$ gallons of 25% and $6\tfrac{2}{3}$ gallons of 40%

**19.** Yes    **20.** No

## Section 6.4   *(page 403)*

### Warm Up   *(page 403)*

**1.** Line     **2.** Line     **3.** Parabola     **4.** Parabola

**5.** Circle    **6.** Ellipse    **7.** $(1, 1)$    **8.** $(2, 0)$

**9.** $(2, 1), \left(-\dfrac{5}{2}, -\dfrac{5}{4}\right)$    **10.** $(2, 3), (3, 2)$

**1.** d    **2.** b    **3.** a    **4.** c    **5.** f    **6.** e

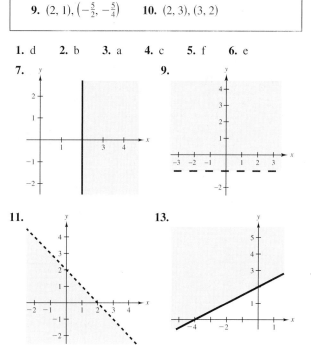

**15.**
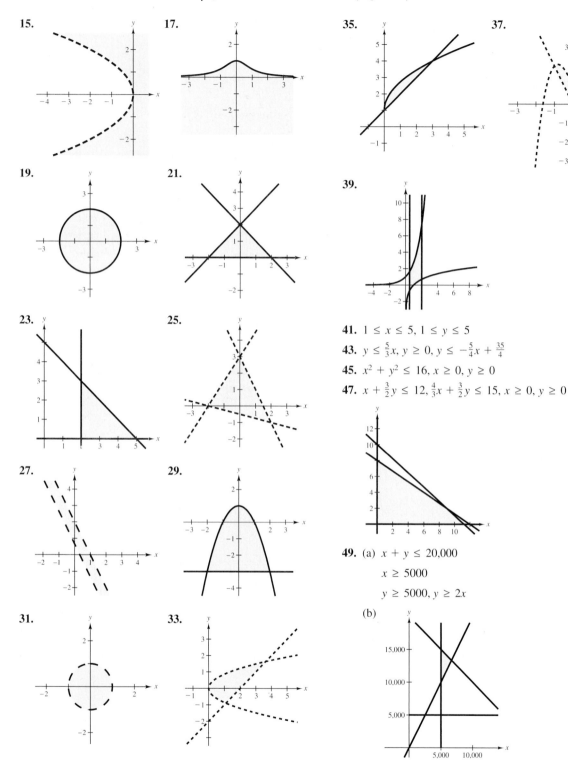

**17.**

**19.**

**21.**

**23.**

**25.**

**27.**

**29.**

**31.**

**33.**

**35.**

**37.**

**39.**

**41.** $1 \le x \le 5, 1 \le y \le 5$

**43.** $y \le \frac{5}{3}x, y \ge 0, y \le -\frac{5}{4}x + \frac{35}{4}$

**45.** $x^2 + y^2 \le 16, x \ge 0, y \ge 0$

**47.** $x + \frac{3}{2}y \le 12, \frac{4}{3}x + \frac{3}{2}y \le 15, x \ge 0, y \ge 0$

**49.** (a) $x + y \le 20{,}000$

$x \ge 5000$

$y \ge 5000, y \ge 2x$

(b)

**61.** (a)

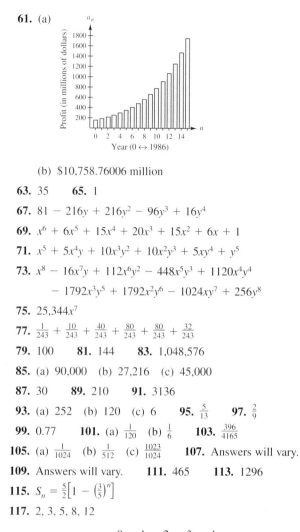

(b) \$10,758.76006 million

**63.** 35     **65.** 1

**67.** $81 - 216y + 216y^2 - 96y^3 + 16y^4$

**69.** $x^6 + 6x^5 + 15x^4 + 20x^3 + 15x^2 + 6x + 1$

**71.** $x^5 + 5x^4y + 10x^3y^2 + 10x^2y^3 + 5xy^4 + y^5$

**73.** $x^8 - 16x^7y + 112x^6y^2 - 448x^5y^3 + 1120x^4y^4$
$\quad\quad - 1792x^3y^5 + 1792x^2y^6 - 1024xy^7 + 256y^8$

**75.** $25,344x^7$

**77.** $\frac{1}{243} + \frac{10}{243} + \frac{40}{243} + \frac{80}{243} + \frac{80}{243} + \frac{32}{243}$

**79.** 100     **81.** 144     **83.** 1,048,576

**85.** (a) 90,000   (b) 27,216   (c) 45,000

**87.** 30     **89.** 210     **91.** 3136

**93.** (a) 252   (b) 120   (c) 6     **95.** $\frac{5}{13}$     **97.** $\frac{2}{9}$

**99.** 0.77     **101.** (a) $\frac{1}{120}$   (b) $\frac{1}{6}$     **103.** $\frac{396}{4165}$

**105.** (a) $\frac{1}{1024}$   (b) $\frac{1}{512}$   (c) $\frac{1023}{1024}$     **107.** Answers will vary.

**109.** Answers will vary.     **111.** 465     **113.** 1296

**115.** $S_n = \frac{5}{2}\left[1 - \left(\frac{3}{5}\right)^n\right]$

**117.** 2, 3, 5, 8, 12

$$n: \quad 0 \quad 1 \quad 2 \quad 3 \quad 4$$
$$a_n: \quad 2 \quad 3 \quad 5 \quad 8 \quad 12$$

*1st differences:*    1   2   3   4

*2nd differences:*    1   1   1

Quadratic model

## Chapter Test   *(page 570)*

**1.** 3, 5, 7, 9, 11     **2.** $-1, 4, -9, 16, -25$

**3.** 2, 6, 24, 120, 720     **4.** $\frac{1}{2}, \frac{1}{4}, \frac{1}{8}, \frac{1}{16}, \frac{1}{32}$

**5.** Arithmetic, $d = 5$     **6.** Neither

**7.** Geometric, $r = 2$     **8.** Neither     **9.** 2600

**10.** $\approx 29,918.311$     **11.** $\frac{1}{2}$     **12.** \$22,196.40

**13.** \$41,174.69

**14.** $x^5 + 10x^4 + 40x^3 + 80x^2 + 80x + 32$

**15.** $-192,456$     **16.** 120     **17.** $2^{10} = 1024$

**18.** 45,000     **19.** $\frac{1}{24}$     **20.** $\frac{1}{32}$

## Cumulative Test: Chapters 6–8
*(page 571)*

**1.** $x = 2, y = -3, z = 5$     **2.** $x = -2, y = 4, z = -5$

**3.** $\begin{bmatrix} 1 & 4 & 8 \\ -1 & 4 & 3 \\ 0 & 1 & 5 \end{bmatrix}$     **4.** $\begin{bmatrix} 16 & 9 \\ 8 & 7 \\ 7 & 2 \end{bmatrix}$     **5.** $\begin{bmatrix} 2 & 12 \\ -5 & -3 \\ 1 & 9 \end{bmatrix}$

**6.** $\begin{bmatrix} \frac{1}{2} & -\frac{1}{2} & 0 \\ \frac{1}{4} & \frac{1}{4} & -1 \\ -\frac{1}{2} & \frac{1}{2} & 1 \end{bmatrix}$     **7.** $-39$     **8.** $-7$

**9.** $x = 18,000$     **10.** \$20,000 at 7.5%
    $p = 66$           \$25,000 at 8.5%

**11.** $BA = \begin{bmatrix} 170,000 & 230,000 & 237,000 \end{bmatrix}$
\$170,000 is the value of the inventory in warehouse 1.
\$230,000 is the value of the inventory in warehouse 2.
\$237,000 is the value of the inventory in warehouse 3.

**12.** 5, 8, 11, 14, 17     **13.** 3, 1, $\frac{1}{3}, \frac{1}{9}, \frac{1}{27}$

**14.** $y^7 - 28y^6 + 336y^5 - 2240y^4 + 8960y^3$
$\quad\quad - 21,504y^2 + 28,672y - 16,384$

**15.** 720     **16.** 18,000     **17.** $\approx 0.978$     **18.** $\frac{1}{4}$

**19.** 2, 4, 7, 11, 16

$$n: \quad 1 \quad 2 \quad 3 \quad 4 \quad 5$$
$$a_n: \quad 2 \quad 4 \quad 7 \quad 11 \quad 16$$

*1st differences:*    2   3   4   5

*2nd differences:*    1   1   1

Quadratic model

# Appendix A

## Section A.1   *(page A7)*

**1.** (a) Natural: $\{5\}$
   (b) Integer: $\{-9, 5\}$
   (c) Rational: $\left\{-9, -\frac{7}{2}, 5, \frac{2}{3}, 0.1\right\}$
   (d) Irrational: $\left\{\sqrt{2}\right\}$

**3.** (a) Natural: $\left\{3, \frac{6}{3}\right\}$ *(Note:* $\frac{6}{3} = 2$*)*
   (b) Integer: $\left\{3, -1, \frac{6}{3}\right\}$
   (c) Rational: $\left\{3, -1, \frac{1}{3}, \frac{6}{3}, -7.5\right\}$
   (d) Irrational: $\left\{-\frac{1}{2}\sqrt{2}\right\}$

**5.** $\frac{3}{2} < 7$

**7.** $\frac{5}{6} > \frac{2}{3}$

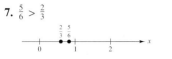

**9.** $x \le 5$ denotes all real numbers less than or equal to 5.

**11.** $x > 3$ denotes all real numbers greater than 3.

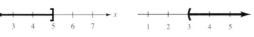

**13.** $-2 < x < 2$ denotes all real numbers greater than $-2$ and less than 2.

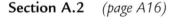

**15.** $x < 0$     **17.** $3.5\% \le r \le 6\%$     **19.** 10     **21.** $-1$

**23.** $-9$     **25.** $|-4| = |4|$     **27.** $-|-6| < |-6|$

**29.** 4     **31.** $\frac{7}{2}$     **33.** 51     **35.** $\left|z - \frac{3}{2}\right| > 1$

**37.** $|y - 0| \ge 6 \Rightarrow |y| \ge 6$     **39.** $\frac{127}{90}, \frac{584}{413}, \frac{7071}{5000}, \sqrt{2}, \frac{47}{33}$

**41.** 0.625

| $|a - b|$ | $0.05b$ | Passes Budget Variance Test |
|-----------|---------|------------------------------|
| **43.** $127.88 | $1250 | Yes |
| **45.** $572.59 | $470 | No |
| **47.** $671.75 | $1882 | No |

| | Median Income, $y$ | Household Income, $x$ | $|y - x|$ |
|---|---|---|---|
| **49.** | $52,725 | $65,400 | $12,675 |
| **51.** | $52,984 | $37,300 | $15,684 |
| **53.** | $36,538 | $21,300 | $15,238 |

## Section A.2     (page A16)

> **Warm Up**     (page A16)
>
> **1.** $-4 < -2$     **2.** $0 > -3$     **3.** $\sqrt{3} > 1.73$
> **4.** $-\pi < -3$     **5.** $|6 - 4| = 2$     **6.** $|2 - (-2)| = 4$
> **7.** $|0 - (-5)| = 5$     **8.** $|3 - (-1)| = 4$
> **9.** $|-7| + |7| = 7 + 7 = 14$
> **10.** $-|8 - 10| = -|-2| = -2$

**1.** $7x, 4$     **3.** $x^2, -4x, 8$     **5.** (a) $-10$  (b) $-6$

**7.** (a) 0  (b) 0     **9.** Commutative (addition)

**11.** Inverse (addition)     **13.** Inverse (multiplication)

**15.** Inverse (multiplication)     **17.** Identity (multiplication)

**19.** Identity (multiplication)     **21.** Associative (addition)

**23.** $\frac{1}{7}(7 \cdot 12) = \left(\frac{1}{7} \cdot 7\right)12$ Associative (multiplication)

$= 1 \cdot 12$     Inverse (multiplication)

$= 12$     Identity (multiplication)

**25.** $2 \cdot 3 \cdot 5$     **27.** $-9$     **29.** 40     **31.** $-2$     **33.** $\frac{59}{66}$

**35.** $\frac{5}{16}$     **37.** $-\frac{14}{5}$     **39.** $-13.67$     **41.** $-0.45$

**43.** 41.14     **45.** 17.5%

**47.** Social Security spent 22.7% of the total budget.
National Defense: 290 billion dollars
Social Security: 406 billion dollars
Income Security: 251 billion dollars
Medicare: 202 billion dollars
Health: 154 billion dollars
Education: 63 billion dollars
Veterans Benefits: 47 billion dollars
Transportation: 47 billion dollars
Net Interest: 220 billion dollars
Other: 111 billion dollars

**49.** $2(-4 + 2)$

**51.** Scientific:  6 $\boxed{x^2}$ $\boxed{+/-}$ $\boxed{-}$ $\boxed{(}$ 7 $\boxed{+}$ $\boxed{(}$ 2 $\boxed{+/-}$ $\boxed{)}$ $\boxed{x^y}$
3 $\boxed{)}$ $\boxed{=}$

Graphing:  $\boxed{(-)}$ 6 $\boxed{x^2}$ $\boxed{-}$ $\boxed{(}$ 7 $\boxed{+}$ $\boxed{(}$ $\boxed{-}$ 2 $\boxed{)}$ $\boxed{\wedge}$ 3
$\boxed{)}$ $\boxed{\text{ENTER}}$

**53.** $\approx 5237$ students

## Section A.3     (page A24)

> **Warm Up**     (page A24)
>
> **1.** 1     **2.** 5     **3.** 4     **4.** $\frac{1}{4}$     **5.** 0     **6.** 1
> **7.** $\frac{3}{7}$     **8.** 4     **9.** $-\frac{1}{8}$     **10.** 1

**1.** 64     **3.** 729     **5.** $-81$     **7.** 8     **9.** $\frac{9}{16}$

**11.** 5184     **13.** 36     **15.** 1     **17.** 18     **19.** $\frac{7}{16}$

**21.** $-125z^3$     **23.** $5x^6$     **25.** $-3z^7$     **27.** $3x^2$     **29.** $\frac{7}{x}$

**31.** $1, x \ne -5$     **33.** $-2x^3$     **35.** $\frac{10}{x}$     **37.** $3^{3n}$

**39.** $5.75 \times 10^7$ square miles     **41.** $9.461 \times 10^{15}$ kilometers

**43.** $1.3837 \times 10^{-2}$ inch     **45.** $13,000,000\ ^\circ\text{C}$

**47.** 0.00000000048 electrostatic unit

**49.** (a) 7697.125   (b) 954.448

**51.** (a) $19,055.59   (b) $19,121.84

**53.** Insert parentheses around the exponent.

$$A = 2000\left(1 + \frac{0.06}{12}\right)^{(12 \cdot 5)}$$

**55.** About 6.79%     **57.** $(5.1 - 3.6)^5$

**25.** Center: $(1, -1)$

Foci: $\left(\frac{1}{4}, -1\right), \left(\frac{7}{4}, -1\right)$

Vertices: $\left(-\frac{1}{4}, -1\right), \left(\frac{9}{4}, -1\right)$

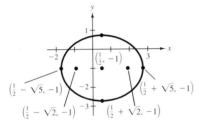

**27.** Center: $\left(\frac{1}{2}, -1\right)$

Foci: $\left(\frac{1}{2} - \sqrt{2}, -1\right), \left(\frac{1}{2} + \sqrt{2}, -1\right)$

Vertices: $\left(\frac{1}{2} - \sqrt{5}, -1\right), \left(\frac{1}{2} + \sqrt{5}, -1\right)$

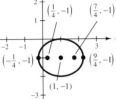

**29.** $\dfrac{(x-2)^2}{4} + \dfrac{(y-2)^2}{1} = 1$   **31.** $\dfrac{x^2}{48} + \dfrac{(y-4)^2}{64} = 1$

**33.** $\dfrac{(x-3)^2}{9} + \dfrac{(y-5)^2}{16} = 1$   **35.** $\dfrac{x^2}{16} + \dfrac{(y-4)^2}{12} = 1$

**37.** Center: $(1, -2)$

Vertices: $(-1, -2), (3, -2)$

Foci: $\left(1 - \sqrt{5}, -2\right), \left(1 + \sqrt{5}, -2\right)$

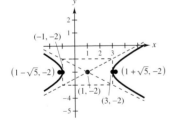

**39.** Center: $(2, -6)$

Vertices: $(2, -7), (2, -5)$

Foci: $\left(2, -6 - \sqrt{2}\right), \left(2, -6 + \sqrt{2}\right)$

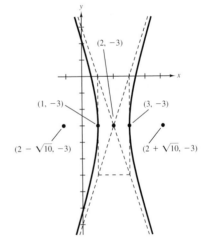

**41.** Center: $(2, -3)$

Vertices: $(3, -3), (1, -3)$

Foci: $\left(2 - \sqrt{10}, -3\right), \left(2 + \sqrt{10}, -3\right)$

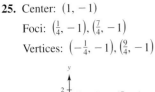

**43.** Center: $(1, -3)$

Vertices: $\left(1, -3 - \sqrt{2}\right), \left(1, -3 + \sqrt{2}\right)$

Foci: $\left(1, -3 - 2\sqrt{5}\right), \left(1, -3 + 2\sqrt{5}\right)$

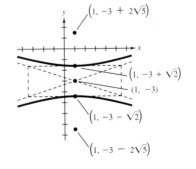

**45.** Center: $(-1, -3)$

Vertices: $(-1 - 3\sqrt{3}, -3), (-1 + 3\sqrt{3}, -3)$

Foci: $(-1 - \sqrt{30}, -3), (-1 + \sqrt{30}, -3)$

**47.** $\dfrac{(x-4)^2}{4} = \dfrac{y^2}{12} = 1$    **49.** $\dfrac{(y-5)^2}{16} - \dfrac{(x-4)^2}{9} = 1$

**51.** $\dfrac{y^2}{9} - \dfrac{(x-2)^2}{9/4} = 1$    **53.** $\dfrac{(x-3)^2}{9} - \dfrac{(y-2)^2}{4} = 1$

**55.** Circle    **57.** Hyperbola    **59.** Ellipse

**61.** Parabola    **63.** (a) 24,748.74 miles per hour

(b) $x^2 = -16,400(y - 4100)$

**65.** $\dfrac{x^2}{25} + \dfrac{y^2}{16} = 1$

**67.** Perihelion: $9.138 \times 10^7$ miles    **69.** $e \approx 5.431 \times 10^{-2}$

Aphelion: $9.454 \times 10^7$ miles

**71.** $\dfrac{(x-h)^2}{a^2} + \dfrac{(y-k)^2}{b^2} = 1$

$\dfrac{(x-h)^2}{a^2} + \dfrac{(y-k)^2}{a^2 - c^2} = 1$

$\dfrac{(x-h)^2}{a^2} + \dfrac{(y-k)^2}{a^2/a^2(a^2 - c^2)} = 1$

$\dfrac{(x-h)^2}{a^2} + \dfrac{(y-k)^2}{a^2(1 - c^2/a^2)} = 1$

$\dfrac{(x-h)^2}{a^2} + \dfrac{(y-k)^2}{a^2[1 - (c/a)^2]} = 1$

$\dfrac{(x-h)^2}{a^2} + \dfrac{(y-k)^2}{a^2(1 - e^2)} = 1$

**73.** $\dfrac{x^2}{328.15} + \dfrac{y^2}{19.39} = 1$

# Appendix C

## Section C.1    *(page A84)*

**1.**

| Stems | Leaves |
|---|---|
| 7 | 0 5 5 5 7 7 8 8 8 |
| 8 | 1 1 1 1 2 3 4 5 5 5 5 7 8 9 9 9 |
| 9 | 0 2 8 |
| 10 | 0 0 |

**3.**

| Stems | Leaves |
|---|---|
| 0 | 82 58 87 92 96 |
| 1 | 12 38 21 61 81 02 15 55 |
| 2 | 70 65 25 93 |
| 3 | 86 94 02 |
| 4 | 12 78 81 48 |
| 5 | 99 54 94 99 45 70 |
| 6 | 04 27 |
| 7 | 14 24 06 |
| 8 | 39 37 17 |
| 9 | 46 |
| 10 | |
| 11 | 69 11 87 |
| 12 | 48 |
| 13 | 41 29 28 |
| 18 | 42 |
| 25 | 64 |
| 31 | 77 |
| 48 | 80 |
| 73 | 73 |

**5.**

Vegetable crops

**7.**

Units of fertilizer

**9.** $y = 57 + 1.4x$; 71

**11.**

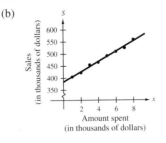

**25.** (a) $S = 384 + 21.2x$; $479,400

(b)

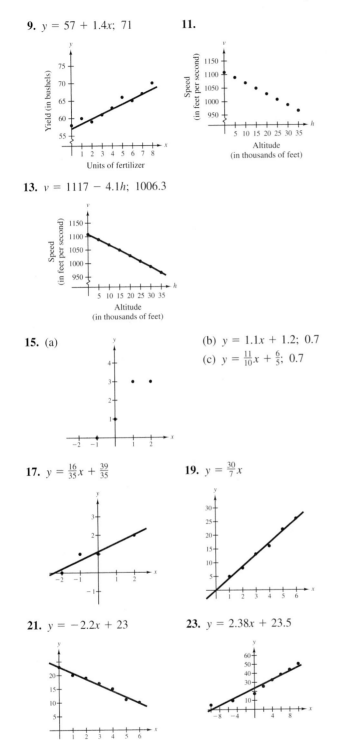

**13.** $v = 1117 - 4.1h$; 1006.3

(c) $r \approx 0.996$

## Section C.2   *(page A92)*

**1.** Mean: 8.86; median: 8; mode: 7

**3.** Mean: 10.29; median: 8; mode: 7

**5.** Mean: 9; median: 8; mode: 7

**7.** The mean is sensitive to extreme values.

**9.** Mean: $67.14; median: $65.35

**11.** Mean: 3.07; median: 3; mode: 3

**13.** One possibility: $\{4, 4, 10\}$

**15.** (a)

(b) $y = 1.1x + 1.2$; 0.7

(c) $y = \frac{11}{10}x + \frac{6}{5}$; 0.7

**15.** The median gives the most representative description.

**17.** $\bar{x} = 6$, $v = 10$, $\sigma = 3.16$

**19.** $\bar{x} = 2$, $v = \frac{4}{3}$, $\sigma = 1.15$     **21.** $\bar{x} = 4$, $v = 4$, $\sigma = 2$

**23.** $\bar{x} = 47$, $v = 226$, $\sigma = 15.03$     **25.** 3.42

**27.** 101.55     **29.** 1.65

**31.** $\bar{x} = 12$ and $|x_i - 12| = 8$ for all $x_i$.

**33.** It will increase the mean by 5, but the standard deviation will not change.

**17.** $y = \frac{16}{35}x + \frac{39}{35}$

**19.** $y = \frac{30}{7}x$

**35.** With $\bar{x} = 235$ and $\sigma = 28$:

At least $\frac{3}{4}$ of the scores in $[179, 291]$.
At least $\frac{8}{9}$ of the scores in $[151, 319]$.
With $\bar{x} = 235$ and $\sigma = 16$:
At least $\frac{3}{4}$ of the scores in $[203, 267]$.
At least $\frac{8}{9}$ of the scores in $[187, 283]$.

**37.** $\bar{x} = 96.05$, $v = 508.58$, $\sigma \approx 22.55$
60% lie within one standard deviation.

**21.** $y = -2.2x + 23$

**23.** $y = 2.38x + 23.5$

# Index of Applications

## BIOLOGY AND LIFE SCIENCE APPLICATIONS

Average life expectancy, 97
Bacteria count, 191, 194, 195, 307, 328, 347, 355, 356, 357
Comparing protein intake of pigs and shoulder weight, 292
Endangered species, 348
Estimating the time of death, 350
Forest yield, 337
Galloping speed of animals, 326
Human height, 10, 69
Human memory model, 288, 295, 316, 318, 337, 354, 357
Learning curve, 347, 356
Life expectancy, 56, 97, 117
Life expectancy of a child, 108
Native prairie grasses, 338
Number of air sacs in lungs, A24
Number of insect species, A24
Oxygen level in a pond, A55
Population of deer, 288
Population of elk, 288
Population of fish, 295
Skill retention model, 319
Spread of a virus, 344
Stocking a lake with fish, 348
Suburban wildlife, 338
Tree height, 22
Ways of enjoying nature, 21
Width of human hair, A24
Wildlife management, 356

## BUSINESS APPLICATIONS

Acquisition of raw materials, 463
Advertising expenditures, 369
Advertising expenses, 131, 235, 294
Advertising and sales, A86
Annual operating cost, 68
Annual payroll of car dealerships, 502
Annual profit, 567
Annual profit for Microsoft Corporation, 520
Annual profit for Procter & Gamble Company, 519
Annual revenue for American Express, 129
Annual revenue for McDonald's Corp., 511
Annual revenue for SBC Communications, Inc., 511
Annual sales, 567
Annual sales for Home Depot, Inc., 511
Annual sales for La-Z-Boy, Inc., 520
Annual sales for Newell Rubbermaid, Inc., 520
Annual sales for Wal-Mart Stores, Inc., 510
Assembly line production, 538
Average production costs, 35, 284, 288, 295
Bicycle sales, 158
Borrowing money, 438, 439, 486
Break-even analysis, 68, 84, 365, 368
Break-even point, 368, 395, 419
Budget variance, A6, A7
Business headaches, 21
Cash receipts from vegetable crops, A85
Cigarette production, 225
Compact disc sales, 419
Company profits, 78, 85, 86
Company revenue, 174
Company sales, 118, 145
Comparing product sales, 419
Comparing profits, 193
Comparing sales, 194, 210
Comparing units produced and profit, 224
Computer inventory, 404
Computer storage devices, 163
Concert ticket sales, 405, 421
Consumer and producer surplus, 401, 405, 422
Contracting purchase, 142, 148
Cookie sales, A17
Corporate retirement plans, 20
Cost of a new product, 164
Cost of producing a new game, 164
Cost of renting, 148
Cost, revenue, and profit, 211, 212
Daily doughnut sales, 68
Declining balances depreciation, A33
Defective product, A55
Defective units, 538, 547, 551, 568, 569, 571
Demand function, 307, 328, 337, 355
Demand for a pager model, 257
Discount rate, 21, 22, 82
Dividends, 502
Dollar value of a product, 148
Earnings and dividends for Procter & Gamble Company, 117
Earnings-dividend ratio for Wal-Mart Stores, Inc., 205
Earnings per share
    for Microsoft Corporation, 108
    for Minnesota Mining and Manufacturing Company, 118
    for Wal-Mart Stores, Inc., 108
Equipment purchase, 149
Exports, 380
Factory production, 452, 487, 490, 571
Fourth quarter sales, 129, 148
Furniture production, 404
Grades of paper, 392
Hotel pricing, 452
Imports to the U.S., 258, 380
Increasing profit, 75
Internet commerce, 10
Inventory of computers, 452
Inventory levels, 487
Labor/wage requirements, 453, 464, 488
List price of a compact disc player, 21
List price of a microwave oven, 82
List price of a swimming pool, 21
Making a sale, 551
Market research, 54, 57, 84
Market share, 425, 484
Maximum profit, 175, 185, 224, 292, 296, 411, 414, 415, 416, 422, 423, 424
Maximum revenue, 223, 292, 415, 423
Media selection, 415
Minimum cost, 223, 292, 415, 423
Monthly profit, 20, 82
Multiplier effect, 491, 564
New York Stock Exchange, 308, 326
Number of years of service, 545
Office rent, 140, 148
Operating revenue of the airline industry, 141
Payout ratio, 117
Percent of a benefit package, 13
Price-earnings ratio
    for McDonald's Corporation, 195
    for Walt Disney Corporation, 195
Price of gold, 98, A93
Product types, 392, 420
Production costs, 195, 211

Production limit, 23, 46
Profit, 184, 185, 209, 277, 293, 358
Profit and advertising expenses, 254, 259, 296
Profit in producing calculators, 245
Profit in producing compact discs, 245
Projected expenses, 23
Projected revenue, 23, 83
Quarterly sales, 149, 150
Real estate purchase, 142
Recycling costs, 288, 289, 295, 296
Research and development costs, 175
Research and development expenditures, 462
Restaurant sales, 381
Retail sales, 95
Retail trade sales, 149
Revenue, 209, 277
Revenue and cost, 86, 150
Sale of long-playing albums, 289
Sale price, 12
Sale price and list price, 139, 149
Sales, 347
Sales and advertising, 349, 356
Sales of college textbooks, 347
Sales commission, 140
Sales of in-line skates, 347
Sales of morning and evening newspapers, 421
Sales prediction, 122
Sales of recreational vehicles, 348
Sales of sound recordings, 356
Sales of sports clothes, 348
Shipments of computers, 335
Shoe sales, 380, A92
Straight-line depreciation, 135, 139, 148
Supply and demand, 380, 395, 420, 424, 571
Supply and demand of a calculator, 377
Television advertising, 9
Ticket sales, 380
Total investment capital for Northwest Natural Gas Company, 235
Total profit, 511
Total revenue, 34, 45, 46, 83
Total revenue for WorldCom, Inc., 502
Total sales, 508, 510, 511
Total sales for Microsoft Corporation, 142
Transportation costs, 257
U.S. manufacturing sales, 394
U.S. manufacturing shipments, A86
Value of sound recordings, 295
Videocassette recorder inventory, 421
Watching the ads, 552

Worker's productivity, 350
Worldwide e-commerce, 326
Years of service, 545

## CHEMISTRY AND PHYSICS APPLICATIONS

Acid mixture, 379, 392, 395, 419
Air quality standards, 151, 206
Automobile aerodynamics, 184
Bacteria growth, 307, 328, 347, 355, 356, 357
Bouncing ball, 213, 290
Carbon dating, 342, 357
Chemical concentration in the bloodstream, 257, 289
Comparing the angle of a baseball and distance, 224, 292
Cost-benefit model for smokestack emission, 283, 287, 295
Crop spraying, 392
Depth of a marine salvage ship, 34
Depth of a submarine, 34
Depth of an underwater cable, 83
Earthquake magnitude, 345, 349, 356
Electron charge, A24
Falling object, 29, 33, 46, 511
Fluid flow, A75
Fuel efficiency, 98
Fuel mixture, 379
Gasoline and oil mixture, 24
Grand Canyon, 83
Height of a flare, 85
Height of a parachutist, 140
Height of a projectile, 77, 85
Hot air balloon, 44
Humidity control, 69
Hyberbolic mirror, A68
Intensity of sound, 350
Lead emissions, 258
Light year, A24
Loran long-range navigation, A67
Maximum height of a baseball, 219
Maximum height of a diver, 224
Maximum height of a shot-put, 224
Mixture, 23, 83
Newton's Law of Cooling, 297, 351
Noise level, 350
Olympic diver, 33
Orbit of Earth, A76
Orbit of Pluto, A76
Orbit of Saturn, A76
Orbits of comets in solar system, A73, A76
Owl and mouse, 33

Path of a baseball, 157
Period of a pendulum, A34
Position equation, 394, 395
Radio waves, 22
Radioactive decay, 305, 308, 346, 347, 355, 357
Relative density of hydrogen, A24
Ripples in a pond, 164, 195
Royal Gorge Bridge, 33
Satellite orbit, A75, A76
Scuba diving pressure and depth, 129
Temperature conversion, 128, 129
Thawing a package of steaks, 350, 356
Throwing an object on the moon, 41, 43, 84
Throwing a rock on the surface of the Earth, 44
Time for refrigerated food to cool, A55
Vertical motion, 208
Water depth in a trough, 23
Work, 317

## CONSTRUCTION APPLICATIONS

Brick pattern, 511
Cost of running power lines, 57
Fireplace arch, A66
Height of the Aon Center Building, 15
Height of a building, 22

## CONSUMER APPLICATIONS

Annual food budget, 21
Annual salary, 82, 129, 148
Average cost of a new car, 79
Average cost of a new imported car, 31
Average income, A86
Average monthly cellular phone bills, 205
Bonus, 211
Borrowing money, 56
Cash advance on a credit card, 56, 84
Cash or charge?, 552
Charitable contributions, 221, 231
Charter bus fares, 164
College tuition, A25
Comparative shopping, 64, 67
Consumer credit outstanding, 359, 417
Cost of
    dental care, 258
    overnight delivery, 175, 209
    a telephone call, 169
Delivery service, 380
Endowment income, 129
Federal income tax liability, 242
Flexible work hours, 552

Health care costs, 163
Hospital costs, 502
Hourly wages, 139, 148, 149
Housing costs, 164
Insurance coverage, A84
Job applicants, 537, 539, 568
Job offer, 510, 522, 567
Life insurance, 529
Loan payments as a percent of annual
    income, 21
Median family income, 205
Median income in relation to sporting
    goods purchases, A8
Median prices of new homes, 423
Minimum cost, 412
Minimum wage, A25
Monthly electric bills, A92
Monthly payment
    for a car loan, A55
    of a home mortgage, 319
Monthly salary, 150
Mountain tunnel, A67
Paycheck amount, 11
Percent of a raise, 13
Personal income, 85
Price comparison from 1917 to 1994, 20
Price
    of gold, 98
    of mobile homes in the U.S., 393
    of one-family houses, 168
    of a product, 85
Property tax, 139, 147
Purchase of sound recordings, 369
Purchasing power of the dollar, 108, 116,
    258
Salary increase, 12, 68
Satellite antenna, A66
Ski club trip, 52
Social Security benefits, 10
Spectator amusements, 405
State income tax, 132
State sales tax, 139
Sunday newspapers, 142
Suspension bridge, A66
Vacation packages, 452
Weekly earnings, 7
Weekly pay, 20
Weekly salary, 22
Would you take this job?, 520

### GEOMETRY APPLICATIONS

Area, 520
Area
    of a box, 43
    of a concrete foundation, 211
    of a region, 482
    of a shaded region, A41

Centimeters and inches, 139
Circles and pi, 225
Collinear points, testing for, 476, 489,
    490
Cutting across a lawn, 30
Diagonals of a polygon, 539
Grams and ounces, 148
Height of a balloon, 163
Length of a tank, 23
Liters and gallons, 139
Maximum area
    of a corral, 223
    of a rectangle, 223, 292
Measurements
    of a billboard, 33, 83
    of a box, 256, 257
    of a building, 33
    of a building lot, 33
    of a container, 157
    of a corral, 43
    of a cube, A33
    of a house, 245
    of a package, 257
    of a picture frame, 21
    of a room, 14, 21, 28, 82, 245, 293,
        A46
    of a square, A33
    of a square base, 46
    of a triangle, 33, 34
    of a triangular sign, 33
    of a volleyball court, 82
Ratio of volume to surface area, A20
Rectangular
    playing field, 78
    room, 78
Sailboat stay, 57
Similar triangles, 15
Snowflake curve, 521
Volume
    of a box, 43, 84, 162, 208, A39, A41
    of a cylinder, 18
    of a package, 162

### INTEREST RATE APPLICATIONS

Balance in an account, 208, A23, A25
Bond investment, 463
Comparing investment returns, 23
Compound interest, 53, 56, 78, 85, 86,
    150, 304, 307, 309, 328, 337,
    346, 357, 501, 517, 519, 522,
    566, 567, 570, A40, A41
Doubling an investment, 334
Investment mixture, 23, 362, 392, 405,
    416, 420, 424
Investment portfolio, 369, 379, 388, 392,
    419, 420, 489

Investment time, 318, 354
Rate of inflation, A23
Savings plan, A38, A40
Simple interest, 16, 21, 67, 82, 129, 139,
    147, 571
Trust fund, 307, 353

### TIME AND DISTANCE APPLICATIONS

Airplane speed, 376, 379
Average speed
    of a marathon runner, 24
    of a truck, 22
Distance
    between airplanes, 118
    from a boat to a dock, 44
    to a star, 22
    from the sun, 323, A76
    traveled by a car, 140
Driving distance, 97
Flight time
    between New York and San Francisco,
        14
    between Orlando and Denver, 22
Flying distance, 44
Flying distance between Atlanta and
    Buffalo, 34
Flying speed, 45
Football passes, 93, 97
Miles per gallon, 175
Miles per hour related to kilometers per
    hour, 133, 148
Olympic swimming times, 138
Speed
    of light, A21
    of sound, A85
Stopping distance, 193, 393
Travel time, 22
Travel time for two cars, 83
Winning Olympic discus throw, 141
Winning times for the 1500-meter run,
    289

### U.S. DEMOGRAPHICS APPLICATIONS

American males and females who have
    ever married, 117
Annual cost of physicians' services in
    the U.S., 45
Automated teller machines, 205
Average age
    of American man at first marriage, 309
    of American woman at first marriage,
        309
    of groom for a given age of bride, 294

Average height
  of American females, 338
  of American males, 337
Average salary
  of professional baseball players, 1, 80
  of professors, 439
  of teachers in the U.S., 68
Average size of U.S. households, 87, 143
Average time spent watching TV, 20
Births in District of Columbia, 405
Cable television
  subscribers in the U.S., 68
  systems, 175
Cellular subscribers, 136
Colleges and universities, 543
Comparing populations, 369
Computer use in schools, 393
Employed civilians in the U.S., 138
Energy consumption, 194, 211
Ever been married?, 117
Federal deficit, 57
Federal government expenses, A17
Female labor force, 194
Football game attendance, 98
Gallons of regular ice cream produced in
  the U.S., 235
Golf winnings in the U.S., 34
Grade level salaries, 175
Grocery stores in the U.S., 56
Gross domestic product, 369
Height
  of American men, 348
  of American women, 348
High school graduates, 349
Internet access in the classroom, 45
Median age of U.S. population, 319, 354
Monthly precipitation in Houston, Texas,
  A80
New house sales, 394
Number of colleges and universities in
  the U.S., 543
Number of dentists in the U.S., A91
Number of doctorates in mathematics in
  the U.S., 90
Number of mobile homes manufactured,
  163
Number of new passenger car leases,
  259
Number of people who were part of the
  labor force, A81
Number of service industry employees,
  462
Number of shopping centers, 129

Occupied housing units, 148
Per capita land area, 284
Percent of college graduates, 79
Percent of the population that was 65
  years old or older, A78
Political makeup of the U.S. Senate, 190
Population
  of Omaha, Nebraska, 135
  of Pennsylvania, 174
  of Rhode Island, 174
  of the U.S., 35, 498
  of the U.S. 30 years old or older, 546
Professional football salaries, 20
Radio stations with a country format, 174
Ratio of men to women, 502
Recreational spending in the U.S., 44
SAT scores, 343, 381
Saving, 85
School expenditures in the U.S., 40
Seasonal snowfall for Lincoln, Nebraska,
  A84
Students per computer, 380
Temperature in Baltimore, Maryland,
  338
Travel to the U.S., A85
U.S. exports and general imports, 380
Vehicle registration, 79

## MISCELLANEOUS APPLICATIONS

Accuracy of measurement, 65, 69, 84
Airline flight, 35
Alumni association, 550
Apartment rental, 129
Australian football, A76
Average waiting time between incoming
  calls, 328
Backup system, 551
Backup vehicle, 551
Baling hay, 510
Baseball player getting a hit, 551, 552
Batting average, 57
Boy or girl?, 551, 569
Choice of two jobs, 369, 419
Choosing officers, 538
Coin toss, 540, 541, 549, 551, 570
College bound, 550
College enrollment, 129, A17
Combination lock, 538
Comparing speed to mileage of an
  automobile, 220
Computer models, 488
Computer-related expenditures, 129

Computer systems, 537, 568
Concert seats, 538
Contract bonuses, 453
Counting card hands, 536
Counting horse race finishes, 533
Course grade, 21
Course schedule, 537
Diet supplement, 405
Drawing cards from a deck, 541, 544,
  549, 550, 568, A52, A55
Drawing marbles, 549
Exercise and diet program, 64, 67
Favorite baseball player's performance,
  A92
Finding a pattern, 529
Forming a committee, 538, 539
Forming an experimental group, 538
Full-length CDs, 349
Gambler's ruin, 520
Game show, 550, 569, 570, 571
Guess the number, 474
Height of a mountain climber, 134
High school enrollment, 140
Hydroelectric power, 225
Land area of Earth, A24
Letter mix-up, 550
License plate numbers, 537
Lottery choices, 538
Man and mouse, 69
Meal menus, 536
Miles driven by a rented car, A92
Misplaced test, A55
Nail length, 326
News programs, 421
Note on a musical scale, A34
Number of airline passengers, 56
Number combinations, 537, 538, 568,
  570, 571
Number of logs, 511
Number of possible telephone numbers,
  531
Number of subsets, 538
Number of telephones, 552
Nutrition, 402
Ocean area of Earth, A24
Order of college theater performances,
  539
Pairs of numbers, 530
Payroll mix-up, 551, 569
Poker hand, 539, 551, 569
Population
  of a city, 347, 350, 355, 357, 358
  growth, 308, 317, 353
  increase, 340

Posing for a photograph, 538
Preparing for a test, 550, 569
Prize money at the Indianapolis 500, 137
Probability, 10
Probability
    of an event, 541, A52
    of a parachutist landing in the center,
        A53
    of winning a lottery, 543, A55
Random number generator, 546, 550
Riding in a car, 538
Roller coaster ride, 568
School lunch snacks, 20

Shoe sizes and heights of men, A83
Six-child families, A92
Seating capacity, 507, 510, 567
Seating capacity of a classroom, 43
Seizure of illegal drugs, 287
Sharing the cost, 56, 84
Short deck, 35
Single file, 538
Softball team expenses, 449
Sound systems, 568, 570
Starting lineup, 568
Study habits, 82
Temperature of the sun, A24

Test questions, 538, 568
Test scores, 356, A77, A84, A92, A93
Toboggan ride, 537
Tossing a die, 542, 549, 569
True-false exam, 537, 568, 570
Voting preference, 453
Winning an election, 550
Work rate, 57
Worker's productivity, 350
World coal production, 366
World population, 79, 340

# Index

## A

Absolute value, A4
  definition of, A4
  equations, 50
  inequality, 63
    solution of, 63
  properties of, A5
  solving equations involving, 51
Addition, 12, A9
  of complex numbers, 261
  of a constant, 60
  of fractions, A12
  of functions, 186
  of inequalities, 60
  of matrices, 441
  of polynomials, A36
  of rational expressions, A49
Additive identity
  for a complex number, 261
  for a matrix, 444
Additive Identity Property, A10
Additive inverse for a complex number,
    261
Additive Inverse Property, A10
Adjoining matrices, 456
Air speed, 376
Algebra
  basic rules of, A10
  Fundamental Theorem of, 270
Algebraic expression, A8
  domain of, A47
  equivalent, A47
  evaluate, A9
  terms of, A8
Algebraic function, 298
Alternative formula for standard
    deviation, A90
Annual percentage rate, 53
Apollonius (262–190 B.C.), A56
Approximately equal to, A2
Archimedes and the Gold Crown, 327
Area
  formulas for, 17
  of a triangle, 475
Arithmetic
  combinations of functions, 186
  Fundamental Theorem of, A13
  sequence, 503
    common difference of, 503
    nth term of, 504
  series, 506

## B

Back-substitution, 361, 431
Balance in an account, A22
Bar graph, A79
Base, A18
  of an exponential function, 298
Basic rules of algebra, A10
Bell-shaped curve, 343
Bimodal, A86
Binomial, 523, A35
  coefficients, 523
  cube of, A37
  expansion, 526
  square of, A37
  Theorem, 523
Bounded interval, 58
Bounded intervals on the real number
    line, 58
Bounded sequence, 265
Branch of a hyperbola, A62
Break-even point, 365

## C

Carbon dating, 342
Cartesian plane, 88
Cayley, Arthur (1821–1895), 440
Center
  of a circle, 104

Associative Property of Addition, A10
  for complex numbers, 262
  for matrices, 443
Associative Property of Multiplication,
    A10
  for complex numbers, 262
  for matrices, 447
Associative Property of Scalar
    Multiplication for matrices, 443
Asymptote
  horizontal and vertical, 279
  of a hyperbola, A63
  slant, 285
Augmented matrix, 427
Average, A86
Axis
  coordinate, 178
  imaginary, 265
  of a parabola, 215, A57
  real, 265
  of symmetry, 215

  of an ellipse, A59
  of a hyperbola, A62
Certain event, 541, A52
Change-of-base formula, 320
Chebychev's Theorem, A91
Checking a solution, 4
Chui-chang suan-shu (250 B.C.), 382
Circle
  center of, 104
  general form of the equation of, 104
  radius of, 104
  standard form of the equation of, 103,
    104
Closed interval, 58
Coded row matrices, 478
Coefficient, A9, A34
  binomial, 523
  leading, A35
Coefficient matrix, 427
Cofactor
  expansion by, 468
  of a matrix, 467
Collinear points, test for, 476
Column matrix, 426
Combination of $n$ elements taken $r$ at a
    time, 535, 536
Common difference, 503
Common formulas, 17
Common graphs of functions, 171
Common logarithmic function, 311
Common ratio, 512
Commutative Property of Addition, A10
  for complex numbers, 262
  for matrices, 443
Commutative Property of Multiplication,
    A10
  for complex numbers, 262
Complement, probability of, 547
Complement of an event $A$, 547
Complete the square, 36, 104, 217
Completely factored, A42
Complex
  plane, 265
  zeros in conjugate pairs, 273
Complex conjugate, 263, 273
Complex fraction, A50
Complex number
  addition of, 261
  definition of, 260
  division of, 263
  equality of, 260

multiplication of, 262
operations of, 261
standard form, $a + bi$, 260
subtraction of, 261
Composite, A13
Composition of two functions, 188
Compound interest, 303
equation involving, 53
formula for, 53, 304
Conditional equation, 2
Conic, A56
degenerate, A56
section, A56
standard forms of equations of, A68
Conjugate, A28
complex, 263
pairs, 273
Conjugate axis of a hyperbola, A63
Consecutive, 12
Consistent system, 373
Constant
addition of, 60
of an algebraic expression, A8
multiplication of, 60
of proportionality, 132
of variation, 132
Constant function, 167, 214
Constant term, A9, A35
Constraints, 406
Consumer surplus, 401
Continuous compounding, 303
Continuous function, 226
Coordinate, A3
Coordinate axes, reflection in, 178
Coordinate of a point, 88
Correlation
coefficient, A84
negative, A80
positive, A80
Critical number, 70, 74
Cross-multiplication, 6
Cryptogram, 478
Cryptography, 478
Cube(s)
of a binomial, A37
sum or difference of two, A43
Cube root, A26
Curve
bell-shaped, 343
logistics, 344
sigmoidal, 344

**D**

d'Alembert, Jean Le Rond (1717–1783),
270
Decoded message, 478

Decreasing function, 167
Definition
of absolute value, A4
of an arithmetic sequence, 503
of a complex number, 260
of the composition of two functions,
188
of the determinant of a $2 \times 2$ matrix,
465
of an ellipse, A59
of an exponential function, 298
of factorial, 494
of a function, 152
of a geometric sequence, 512
of horizontal asymptote, 279
of a hyperbola, A62
of intercepts, 101
of the inverse of a function, 197
of the inverse of a square matrix, 454
of a linear equation, 3
of a logarithmic function, 310
of a matrix, 426
of matrix addition, 441
of matrix multiplication, 445
of the $n$th root of a number, A26
of order on the real number line, A3
of a parabola, A57
of permutation, 532
of a polynomial function, 214
of a polynomial in $x$, A35
of a quadratic equation, 25
of a quadratic function, 214
of rational exponents, A29
of scalar multiplication, 441
of a sequence, 492
of the slope of a line, 119
of standard deviation, A88
of summation notation, 496
of symmetry, 102
of variance, A88
of vertical asymptote, 279
Degenerate conic, A56
Degree of a polynomial, A35
Degree of a polynomial term, A34
Denominator, A10
least common, A13
rationalizing, A28
Dependent system, 373
Dependent variable, 153, 159
Descartes, René (1596–1650), 88
Determinant, 465
of a square matrix, 468
of a $3 \times 3$ matrix, 470
of a triangular matrix, 471
of a $2 \times 2$ matrix, 458
of a $2 \times 2$ matrix, definition of, 465

Diagonal matrix, 471
Difference
of complex numbers, 261
of functions, 186
of two cubes, A43
of two squares, A43
Differences
first, 560
second, 560
Direct variation, 132
Directed distance, A69
Directly proportional, 132
Directrix of a parabola, A57
Discriminant, 37
Distance
between two numbers, 91, A5
Formula, 92
formula for, 17
Distinguishable permutations, 534
Distributive Property, 443, A10
left, 447
of Multiplication over Addition for
complex numbers, 262
right, 447
Dividend, 237
Divides evenly, 237
Division, 12, A9
Algorithm, 237
of complex numbers, 263
of fractions, A12
of functions, 186
of polynomials, long, 236
synthetic, 239
results of, 241
Divisor(s), 237, A13
Domain
of an algebraic expression, A47
of a function, 152, 159, 165
implied, 156, 159
of a logarithmic function, 315
of a rational function, 278
Double inequality, 62, A4
Double solution, 26

**E**

$e$, natural base, 302, 314
Eccentricity, A76
Elementary row operations, 427
Elimination
Gauss-Jordan, 433
Gaussian, 383
with back-substitution, 431, 432
method of, 370, 372
Ellipse
center of, A59
definition of, A59

Producer surplus, 401
Product
  of functions, 186
  of matrices, 445
  of two polynomials, A36
Proper fraction, 237
Proper rational expression, 237
Properties
  of absolute value, A5
  of equality, A14
  of exponents, A19, A30
  of fractions, A12
  of inequalities, 60
  of logarithms, 321
  of matrix addition and scalar
    multiplication, 443
  of matrix multiplication, 447
  of natural logarithms, 314
  of negation, A11
  of the probability of an event, 541,
    A52
  of radicals, A27
  of sums, 497
  of zero, A12
Proportion, 15
Proportionality, constant of, 132
Pure imaginary number, 260
Pythagorean Theorem, 30, 130
Pythagorean triple, 31

### Q

Quadrants, 88
Quadratic equation, 25
  definition of, 25
  function
    definition of, 214
    graph of, 216
    standard form, 217
  solving
    by extracting square roots, 27
    by factoring, 26
  type, 48
Quadratic Formula, 36
Quotient, 237
Quotient of functions, 186

### R

Radical
  equations involving, 49
  index of, A26
  properties of, A27
  symbol, A26
Radicand, A26
Radius of a circle, 104
Range, 152, 159
  of a function, 165
Rate of change, 134

Ratio and proportion, 15
Rational
  exponent
    definition of, A29
    an equation involving, 49
  expression, A48
    adding, A49
    combining, A50
    dividing, A12
    improper, 237
    multiplying, A12
    proper, 237
    reducing, A48
  function, 278
    domain of, 278
    graph of, 281
    horizontal asymptote of, 280
  inequality, 74
    solution of, 74
  numbers, A2
  Zero Test, 246
Rationalizing the denominator, A28
Real axis, 265
Real number line, A3
  bounded intervals on, 58
  unbounded intervals on, 59
Real numbers, set of, A1
Reciprocal, A10
Reciprocal power, 49
Rectangular coordinate plane, 88
Recursion formula, 505
Red herring, 18
Reduced row-echelon form, 430
Reflection
  in the coordinate axes, 178
  in the $x$-axis, 178
  in the $y$-axis, 178
Regression analysis, 136
Regression line, least squares, A82
Regular polygons and regular polyhedra,
  394
Remainder, 237
Remainder Theorem, 240
Repeated
  solution, 26
  zero, 231
    multiplicity of, 231
Right Distributive Property, 447
Rigid transformation, 180
Root key, 252
Roots
  cube, A26
  even, A27
  $n$th root of $a$, A26
  odd, A27
  principal $n$th, A26
  square, A26

Rounded, A2
Rounding, A15
Row and column of a matrix, 426
Row-echelon form, 382, 430
Row-equivalent, 427
Row matrices, coded and uncoded, 478
Row matrix, 426
Row operations, 383

### S

Sample space, 540, A51
Satisfy an inequality, 58
Scalar, 441
Scalar Identity, 443
Scalar multiple, 441
Scalar multiplication
  definition of, 441
  of matrices, 441
Scalar multiplication and matrix
    addition, properties of, 443
Scatter plot, 136, A80
Scientific notation, A21
Second-degree polynomial equation
    in $x$, 25
Second differences, 560
Sequence
  arithmetic, 503
    $n$th term of, 504
  definition of, 492
  finite, 492
  geometric, 512
    common ratio of, 512
  infinite, 492
  terms of, 492
Series
  arithmetic, 506
  finite, 496
  geometric, 515
  infinite, 496
Set of real numbers, A1
Setting the viewing window of your
    graphing utility, 112
Shift, 176
  downward, 176
  horizontal, 176
  left, 176
  right, 176
  upward, 176
  vertical, 176
Sigma notation, 496
Sigmoidal curve, 344
Simple interest, formula for, 17
Simplest form of a radical expression,
    A27
Single fraction, 6

Singular matrix, 455
Sketching the graph of an inequality in
     two variables, 396
Sketching graphs of lines, 123
Slant asymptote, 285
Slope-intercept form of the equation of a
     line, 123
Slope of a line, 119
Snowflake curve, 521
Solution
     of an absolute value inequality, 63
     checking of, 4
     of a double inequality, 62
     of an equation, 2
     of equations involving radicals, 49
     of an inequality in $x$ and $y$, 396
     of a linear inequality, 61
     of a polynomial equation, 47
     of a polynomial inequality, 72
     of a quadratic equation, 37, 48
     of a rational inequality, 74
     of a system of equations, 360
     of a system of inequalities, 398
Solution point, 99
Solution set of an inequality, 58
Special products, A37
Square
     of a binomial, A37
     completing the, 36, 104, 217
     root, A26
     root of a negative number, 264
     roots, extracting, 27
     setting for viewing window, 113
     system, 387
Square matrix, 426
     determinant of, 468
     inverse of, 454
     minors and cofactors of, 467
Standard deviation, A88
     alternative formula for, A90
Standard form, A35
     $a + bi$, of a complex number, 260
     of the equation of a circle, 103, 104
     of the equation of an ellipse, A60
     of the equation of a hyperbola, A62
     of the equation of a parabola, A57
     of equations of conics, A68
     of a polynomial, A35
     of a quadratic function, 217
Standard normal distribution, 343
Standard viewing window, 110
Statistics
     average, A86
     bar graph, A79
     bimodal, A86

Chebychev's Theorem, A91
     correlation
         coefficient, A84
         negative, A80
         positive, A80
     fitting a line to data, A81
     frequency distribution, A79
     histogram, A79
     mean, A86
     measure of central tendency, A86
     measures of dispersion, A88
     median, A86
     mode, A86
     normally distributed, 343
     regression line, least squares, A82
     scatter plot, 136, A80
     standard deviation, A88
     standard normal distribution, 343
     stem-and-leaf plot, A77
     sum of the squared differences, A82
     variance, A88
Stem-and-leaf plot, A77
Step function, 169
Strategy for solving word problems, 18
Subset, A1
Substitution, method of, 360
Substitution Principle, A14
Subtraction, 12, A9
     of complex numbers, 261
     of fractions, A12
     of functions, 186
     of polynomials, A36
Sum
     of complex numbers, 261
     of a finite arithmetic series, 506
     of a finite geometric series, 515
     of functions, 186
     of an infinite geometric series, 516
     of matrices, 441
     of powers of integers, 557
     of the squared differences, A82
     of two cubes, A43
Sum and difference of two terms, A37
Summation
     index of, 496
     lower limit of, 496
     notation, 496
     upper limit of, 496
Sums, properties of, 497
Surplus
     consumer, 401
     producer, 401
Symmetry
     definition of, 102
     with respect to the origin, 102

     tests for, 103
     with respect to the $x$-axis, 102
     with respect to the $y$-axis, 102
Synthetic division, 239
     results of, 241
System of equations, 360
     consistent, 373
     dependent, 373
     equivalent, 371, 383
     inconsistent, 373
     independent, 373
     nonsquare, 387
     row-echelon form, 382
     row operations on, 383
     solution of, 360
     solving, 360
     square, 387
System of inequalities, 398
     solution of, 398
System of linear equations, 459

**T**

Temperature, formula for, 17
Terms of an algebraic expression, A8
Terms of a sequence, 492
Test
     for collinear points, 476
     for even and odd functions, 170
     intervals for a polynomial, 70
     for symmetry, 103
Theorem, Pythagorean, 30
Transcendental function, 298
Transformations, 176
     downward, 176
     horizontal, 176
     left, 176
     nonrigid, 180
     reflection, 178
     right, 176
     rigid, 180
     upward, 176
     vertical, 176
     vertical shrink, 180
     vertical stretch, 180
Transitive property, 60
Translating key words and phrases, 12
Transverse axis of a hyperbola, A62
Triangle, area of, 475
Triangular matrix, 471
Trinomial, A35
     perfect square, A43
True equation, 2
Two-point form of the equation of a line,
     121, 477

## U

Unbounded
  interval, 59
  region, 410
  sequence, 265
Unbounded intervals on the real number
  line, 59
Uncoded row matrices, 478
Union symbol, 63
Upper limit of summation, 496
Upper triangular matrix, 471

## V

Value of $f$ at $x$, 154, 159
Variable, A8
  dependent, 153, 159
  expression, 5
  independent, 153, 159
  term, A9
Variance, A88
Variation
  constant of, 132
  direct, 132
Verbal model, 11

## Vertex

Vertex
  of an ellipse, A59
  of a hyperbola, A62
  of a parabola, 215, A57
Vertical
  asymptote, 279
  Line Test for functions, 166
  shifts, 176
  shrink, 180
  stretch, 180
Volume, formulas for, 17

## W

Window feature of a graphing utility,
  110
With replacement, 530
Without replacement, 530
Word problems, strategy for solving, 18

## X

$x$-axis, 88
  reflection in, 178
$x$-coordinate, 88
$x$-intercepts, 101

## Y

$y$-axis, 88
  reflection in, 178
$y$-coordinate, 88
$y$-intercepts, 101

## Z

Zero, A1
  of a function, 165
  key, 252
  matrix, 444
  polynomial, A35
  of a polynomial function, 230, 250,
    251, 271
  properties of, A12
  real, 230
  repeated, 231
Zero-Factor Property, A12
Zoom-and-trace technique, 250

## Creating a Viewing Rectangle

Each of the viewing rectangles shows portions of the graph of $y = 0.1x^4 - x^3 + 2x^2$.
The first viewing rectangle shows the most complete graph of the equation.

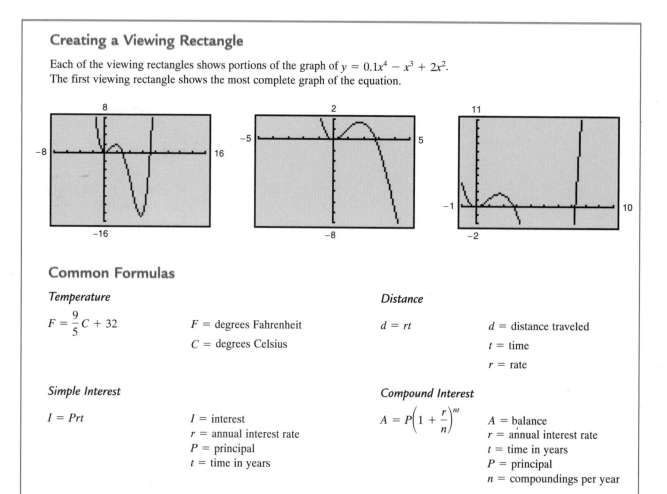

## Common Formulas

### Temperature

$$F = \frac{9}{5}C + 32$$

$F$ = degrees Fahrenheit

$C$ = degrees Celsius

### Distance

$$d = rt$$

$d$ = distance traveled

$t$ = time

$r$ = rate

### Simple Interest

$$I = Prt$$

$I$ = interest

$r$ = annual interest rate

$P$ = principal

$t$ = time in years

### Compound Interest

$$A = P\left(1 + \frac{r}{n}\right)^{nt}$$

$A$ = balance

$r$ = annual interest rate

$t$ = time in years

$P$ = principal

$n$ = compoundings per year

### Coordinate Plane: Distance Formula

$$d = \sqrt{(x_2 - x_1)^2 + (y_2 - y_1)^2}$$

$d$ = distance between points $(x_1, y_1)$ and $(x_2, y_2)$

### Coordinate Plane: Midpoint Formula

$$\left(\frac{x_1 + x_2}{2}, \frac{y_1 + y_2}{2}\right)$$

midpoint of line segment joining $(x_1, y_1)$ and $(x_2, y_2)$

## Conversions

1 foot = 12 inches
1 mile = 5280 feet
1 kilometer = 1000 meters
1 kilometer ≈ 0.6214 miles
1 meter ≈ 39.370 inches
1 foot ≈ 30.480 centimeters
1 liter ≈ 1.057 quarts
1 ton = 2000 pounds
1 kilogram ≈ 2.205 pounds

1 yard = 3 feet
1 mile = 1760 yards
1 meter = 100 centimeters
1 mile ≈ 1.609 kilometers
1 centimeter ≈ 0.3937 inch
1 inch ≈ 2.540 centimeters
1 gallon ≈ 3.785 liters
1 pound = 16 ounces
1 pound ≈ 0.454 kilogram

1 acre = 4840 square yards
1 square mile = 640 acres
1 meter = 1000 millimeters
1 meter ≈ 3.282 feet
1 foot ≈ 0.305 meter
1 gallon = 4 quarts
1 quart ≈ 0.946 liter
1 kilogram = 1000 grams
1 gram ≈ 0.035 ounce